621.38 0625   F V
OPTICAL-FIBER TRANSMISSION
                69.95

St. Louis Community
College

Library

5801 Wilson Avenue
St. Louis, Missouri 63110

Optical-Fiber Transmission

Other Reference Books Available from Sams

To order these and other Sams books, call toll-free 1-800-428-SAMS.

Optical-Fiber Transmission

COMPILED AND EDITED BY

E. E. Bert Basch

Howard W. Sams & Co.
A Division of Macmillan, Inc.
4300 West 62nd Street, Indianapolis, IN 46268 USA

International Standard Book Number: 0-672-22301-5
Library of Congress Catalog Card Number: 86-62436

Acquired by: *Charlie Dresser*
Edited by: *J. L. Davis*
Designed and illustrated by: *T. R. Emrick*
Cover art by: *Stephanie Ray*
Compositor: *Shepard Poorman Communications Corp.,
Indianapolis*

Printed in the United States of America

CONTENTS

CONTRIBUTORS

Peter Bark was born and educated in Germany. In 1963 he received a bachelor of science degree (Diplom-Ingenieur) from the Technical University at Munich. In 1968 he received his doctorate in engineering from the same university. From 1968 to 1971 he was involved in research in molecular beam physics in Peymenade in France.

In 1971 he joined Siemens AG in Munich and became involved in the development of telecommunications cable and related hardware. Since 1973 he has been working exclusively in optical-fiber technology. In 1980 he became Vice President and General Manager for Optical Cable Manufacturing. His key task was to build a new factory in Hickory, North Carolina. In 1985 he became Vice President and Director of Research, Development, and Engineering at Siecor Corporation.

He has authored and coauthored numerous papers and made a number of key contributions in optical communications.

E. E. Bert Basch received the Degree of Engineer in electrical engineering from the Technical University of Delft in 1967. In 1966 he joined the Communication System Laboratories of Sylvania Electronic Systems, where he worked on error-correcting codes. From 1968 to 1970 he led investigations in

digital signal processing at the Armed Forces Laboratory for Electronic Development, The Netherlands.

In 1970 he joined GTE Laboratories, Waltham, Massachusetts, where he was an individual contributor doing research on broadband communication, digital transmission techniques, and fiber-optic transmission systems. In 1976 he became the manager for exploratory research on optical communication and the development of experimental fiber-optic transmission systems. He has directed all of GTE's major fiber-optics field trials. In 1980 he received the Leslie H. Warner Technical Achievement Award, GTE's major technical award, for his outstanding contributions to the development of fiber-optic transmission systems.

He has published numerous original papers in fiber optics, as well as review articles, and has contributed two chapters to a book. He is a member of the OSA, Sigma Xi, the Royal Institute of Engineers of The Netherlands, and is a senior member of the IEEE.

Thomas G. Brown received the baccalaureate degree in physics in 1979 from Gordon College, Wenham, Massachusetts, and served as a member of the technical staff at GTE Laboratories, Waltham, Massachusetts, until August, 1981.

He is currently completing PhD research at the Institute of Optics, University of Rochester, where he is conducting experimental research in silicon-based integrated optics.

Mr. Brown has authored papers on fiber-optic system performance and coherent optical communication systems, as well as papers related to his thesis work.

While at Rochester, he has acted as consultant in the areas of fiber-optic systems modeling and fiber-optic sensor design.

Howard A. Carnes is currently the director of Systems Engineering for Telco Systems Fiber-Optic Corporation. In this capacity he is responsible for establishing technical requirements and qualification of all new products manufactured by Telco Systems FOC.

Formerly he was a senior member of the Technical Staff at GTE Laboratories, Inc. During his association with GTE Laboratories he was involved with R&D fiber-optic systems research, catv, and optical vision systems for robots. Mr. Carnes was presented a Leslie H. Warner Award for early fiber-optic systems research.

He was educated at Northeastern University in Boston, Massachusetts, where he received a bachelor of science degree in electrical engineering in 1968. He has publication credits for about 20 papers.

Mark L. Dakss received the BEE degree from The Cooper Union, New York, New York, in 1960. He received the AM and PhD degrees in physics from Columbia University in 1962 and 1966, respectively.

He was a Research Staff member at IBM Thomas J. Watson Research Center, Yorktown Heights, New York, from 1966 to 1971, where he did applied research on integrated optics, acoustic surface waves, optical information processing and scanning lasers. From 1971 to the present he has been with GTE Laboratories, Waltham, Massachusetts, where he is now a senior member of the Technical Staff. At GTE Laboratories he has been involved in

applied research on optical-fiber measurements, diode laser amplifiers, fiber Raman amplifiers, coupling of light sources to fibers, and fiber splices and connectors.

Dr. Dakss has served on the Electronics Industries Association P-6.6.5 task group on single-mode fibers, which is involved with setting standards on single-mode fiber characteristics and measurements. He is a member of the Optical Society of America, the IEEE Laser and Electro-Optics Society, and Sigma Xi.

James C. Daly is a professor of electrical engineering at the University of Rhode Island, Kingston, Rhode Island. Dr. Daly received his BS degree in electrical engineering from the University of Connecticut in 1960 and his MEE and PhD degrees, both in electrical engineering, from Rensselaer Polytechnic Institute in 1962 and 1967, respectively. Before coming to the University of Rhode Island he was a member of the technical staff of Bell Telephone Laboratories.

Dr. Daly is a member of the Institute of Electrical and Electronic Engineers and the honor societies, Tau Beta Pi, Eta Kappa Nu, and Sigma Xi.

His current research interest involves integrated-circuit design for fiber optics. He has published over 20 papers in optical communications and electronic systems. He holds a patent on an optical waveguide.

David Davidson received the BS degree from Trinity College, the MS degree from University of Michigan, and the MA and PhD degrees (applied physics) from Harvard University. For over 40 years he has been involved in electromagnetic wave propagation research and application, for wavelengths ranging in size from 6 Mm down to a few micrometers. He was one of the principal developers of the Loran navigation system, and has been variously involved with radio-communication and navigation system design and application.

Since 1969 he has been with GTE Laboratories in Waltham, Massachusetts, where he is a senior scientist in telecommunications. He has been working on propagation characterization of single-mode optical fibers, and on communications-satellite link-performance problems, especially concerning rain attenuation effects at frequencies above 10 GHz.

Dr. Davidson is a life member of the IEEE, and a member of the IEEE Wave Propagation Committee, the American Geophysical Union, the Bioelectromagnetics Society, and the C95.4 Subcommittee of the American National Standards Institute (Radiofrequency Protection Standards). He is a registered professional engineer in Massachusetts.

 **Niloy K. Dutta** received the BSc degree in physics in 1972, and the MSc degree in physics in 1974, from St. Stephen's College, New Delhi, India. He received the PhD degree in physics in 1978 from Cornell University.

Dr. Dutta was a postdoctoral associate at Cornell University for one and a half years. During his doctoral and postdoctoral work at Cornell, he worked on spin-flip Raman lasers, infrared spectroscopy, soft X-ray lasers, and nonlinear optics with nonmonochromatic waves.

Since 1979 he has been a member of the technical staff at AT&T Bell Laboratories, where he has contributed extensively to the research and development of InGaAsP semiconductor lasers.

He has authored or coauthored more than 80 publications on his work and one book on long-wavelength semiconductor lasers, which is to be published. He is a senior member of the IEEE and a member of the American Physical Society and the Optical Society of America.

 **Charles Kuen Kao,** who was born in 1933 in Shanghai, received a BSc degree and a PhD degree in electrical engineering in 1957 and 1965, respectively, from the University of London.

In 1957 he joined ITT. Dr. Kao pioneered optical-fiber communications and has worked in this field since 1963. His experience includes theoretical studies and basic research in optical communication systems, fiber-optic waveguide communications, circuits and systems design, and quasioptical techniques applicable to microwave systems.

Since 1982 Dr. Kao has been ITT Executive Scientist. He is a fellow of the IEEE and has been elected a fellow of the IEE. He has won many prestigious awards and honors for his work in optical communications and holds 29 patents to date. He has published numerous papers and the book *Optical Fiber Systems—Technology, Design, and Applications.*

 **Brian Kawasaki** received the BSc degree from University of Toronto in 1964 and the PhD degree in physics from McMaster University, in Hamilton, Ontario, in 1972. From 1972 to 1982 he was a research scientist at the Communication Research Centre, Canadian Department of Communications, Ottawa, Canada. The areas of research included nonlinear effects in fibers, fiber coupling structures, network configurations, and modal noise.

In 1983 he joined Northern Telecom Canada and is at present the manager of advanced optical product development in the Optical Devices Group. His areas of activity have included development of measuring techniques and development of advanced transmitter and detector devices for high–bit-rate transmission.

He has contributed numerous original papers in fiber optics and holds more than 15 patents in the field. He has served on international committees on fiber-optics technology and is now involved in standards activities with the International Electrotechnical Commission.

 **Robert F. Kearns** received the BS degree in electrical engineering from the University of Massachusetts and the MS degree in electrical engineering from Northeastern University. In 1958 he joined the Ap-

plied Research Laboratories of Sylvania Electronic Systems, where he worked on a variety of projects. These included electronic countermeasures systems, a novel sonically scanned magnetic memory element, and an infrared detection study to characterize the electronic signature of blackbody radiators.

In 1970 he joined the Space Surveillance Department of Honeywell's Electro-Optics Laboratory, where he worked on a 27-channel infrared detection system for a suborbital atmospheric probe.

In 1973 he joined GTE Laboratories, where he worked on developing a new-generation robot, and since 1978 he has conducted research in fiber-optic transmission. He is currently a senior member of technical staff in the telecommunications systems department conducting theoretical modeling of both analog and digital transmission systems. He has published numerous technical articles and is a member of IEEE and Tau Beta Pi.

John W. Ketchum was born in Boston, Massachusetts, in 1951. He received the BS and MS degrees in electrical engineering from Northeastern University in 1979, and the PhD degree in electrical engineering, also from Northeastern University, in 1982.

In 1982 he joined the faculty of Northeastern University and served for three years in the position of assistant professor. During this time he taught courses in systems theory, detection and estimation theory, and data communication networks, as well as doing sponsored research in the area of communication theory and signal processing. In addition, during this time he acted as a consultant to GTE Laboratories, Waltham, Massachusetts, in the areas of communication theory and signal processing. In 1985, he joined GTE Laboratories full time as a member of the technical staff in the Network and Communication Theory Department with responsibilities for research in the area of communication theory. He is also currently a member of the part-time faculty at Northeastern University. His current research interests include

bandwidth efficient modulation, trellis coding, optical communication, and adaptive signal processing.

Dr. Ketchum is a member of IEEE and Sigma Xi.

S. Lakshmanasamy was born in Tamilnadu, India, on July 23, 1960. He received the BE degree in electronics and communication engineering from Madurai University, Madurai, India, in 1981 and the M.Tech degree in microwave and optical communications engineering from the Indian Institute of Technology, Kharagpur, India, in 1983. He is currently pursuing the PhD degree in electrical engineering at the University of Rhode Island, Kingston, Rhode Island, specializing in fiber-optic communication. His fields of interest are nonlinear phenomena in optical fibers, soliton propagation, optical communication theory, planar microwave circuits, computer-aided design of microwave circuits, and applied mathematics.

He is a student member of the IEEE, Optical Society of America, and the Photoinstrumentation Engineers. His extracurricular interests include Tamil literature and long-distance running.

Derek Lawrence was born and educated in England. He received a degree in electrical engineering from The City University. He began working in the cable industry in 1970 and was first involved in optical-fiber technology in 1973.

Since 1975 he has worked exclusively in this technology, first in Canada and since 1978 in the United States. He joined Siecor Corporation in 1979, and for the last six years has been engineering manager at one of the world's largest optical cable facilities. His responsibilities include product and manufacturing engineering.

He has authored and coauthored numerous papers on optical-fiber cable technology and is a member of IEEE.

M. Farooque Mesiya received a BE degree from the University of Karachi, Pakistan, in 1968 and the MSc and PhD degrees in electrical engineering from Queen's University, Kingston, Ontario, Canada, in 1971 and 1975, respectively.

During 1976 and 1977 he was a postdoctoral research fellow in the Department of Electrical Engineering, Queen's University. He joined Canada Wire & Cable Company, Limited, Toronto, in 1977, where he was involved in digital fiber-optic trunking and data communication applications and products. From 1979 to 1980 he worked at GTE Laboratories, where he investigated long-wavelength receiver design and applications of fiber-optic technology for interframe cabling in a digital end office.

Since 1981 he has been with Times Fiber Communication, Inc., Wallingford, Connecticut, where he is now Director of Electronics & Systems Engineering, Communications Systems Division.

Tran Van Muoi was born in Phan Thiet, Vietnam, in 1951. He received the BE and PhD degrees in electrical engineering from the University of Western Australia in 1974 and 1977, respectively.

From 1978 to 1981 he was with Bell Laboratories, where he worked on measurement techniques and test instruments for optical-fiber communication systems. He was also involved in fiber-optic digital transmission system analysis. In 1981 he joined TRW Electro-Optics Research Center, where he worked on high-speed optical transmitter and receiver design and specialized optical-fiber system development for military applications. In 1984 he joined PlessCor Optronics, Chatsworth, California, where he is Director of Lightwave Circuit Development.

Dr. Muoi is a member of IEEE and Optical Society of America (OSA). He has served as program committee member for the Conference on Optical Fiber Communication (OFC) in 1985, 1986, and OFC/IOOC (Integrated Optics and Optical Communications) in 1987. His invited paper entitled "High-Speed and High-Sensitivity Receiver Design for Optical-Fiber Communication Systems," won the Best Paper Award for Fiber systems at the Conference on Optical Fiber Communications (OFC) in 1983.

Arnab Sarkar is currently Vice President, Research & Development, and Member of the Board of Lightwave Technologies, Inc., Van Nuys, California, a privately owned company involved in manufacturing single-mode fibers. In the past Dr. Sarkar has held technical management positions in Corning Glass Works and General Cable Corporation in the field of fiber optics. He obtained his PhD from Catholic University of America, Washington, D.C., in Materials Science and his B.Tech and M.Tech degrees in chemical engineering from I.I.T. Kharagpur.

Dr. Sarkar is one of the few scientists in the world with extensive R&D experience in MCVD, OVD and VAD processes. He has 10 U.S. patents issued in his name and has 17 technical publications in glass science and fiber optics. He has also coauthored a chapter on the OVD process in the book *Optical Fiber Communications,* Volume 1, edited by Tinge Li, and published by Academic Press (1985).

Gregory E. Stillman received the BSEE degree with distinction from the University of Nebraska, Lincoln, in 1958. He received the MS and PhD degrees in electrical engineering in 1965 and 1967, respectively, from the University of Illinois.

He then joined the Applied Physics Group of MIT Lincoln Laboratory, where he worked on characterization of high-purity gallium arsenide, far-infrared photoconductivity, and compound-semiconductor avalanche photodiodes. In 1975 he returned as Professor of Electrical Engineering to the University of Illinois, where he is currently involved in research on characterization and crystal growth of compound semiconductors and semiconductor alloys, transport measurements on these materials, and near-infrared avalanche photodiodes and related photodetectors.

Dr. Stillman was elected to the grade of Fellow in the IEEE in 1977 and he is a past president of the Electron Devices Society. In 1985 he was elected to membership in the National Academy of Engineering. He is treasurer of the U.S. Gallium Arsenide Symposium Committee, a member of the program committee of the Device Research Conference, and was technical program chairman of the 1982 International Symposium on GaAs and Related Compounds.

In 1956, **Samuel M. Stone** joined the technical staff of GTE Laboratories, where he currently is engaged in research and development of coherent optical-fiber transmission systems. His initial work in optical communications started in 1961, when he did research and development on HeNe lasers and electro-optic modulators, which resulted in an operational experimental system in which a laser beam was modulated at 3 GHz. He initiated work on optical-fiber systems in 1971 which led to the introduction of optical-fiber transmission by the GTE Telephone operating companies.

In addition to optical communications, Mr. Stone's work has included high-density electron beams for traveling-wave amplifiers, 18- to 41-GHz backward-wave oscillators, submillimeter-wave generation, high pulsed magnetic fields, argon and krypton ion lasers, low-loss dielectric coated mirrors, laser color tv projection systems, which received an IR 100 award, and flat-panel plasma display devices.

Mr. Stone received his BSEE degree from the Polytechnic Institute of Brooklyn (now Polytechnic University), where he did graduate studies with a fellowship award. He is a member of Tau Beta Pi, Eta Kappa Nu, and Sigma Xi, and has been a member of IEEE and SID. He has authored and coauthored over 15 technical papers and has been awarded 7 patents.

Joseph Straus received the BSc degree in 1968 and the PhD degree in 1976 in physics from University of Alberta in Edmonton, Alberta, Canada. From 1974 to 1982 he was employed at Bell Northern Research, Ltd., in Ottawa, Ontario, Canada, engaged in various aspects of fiber optics related research activities including development of light sources, passive optical components, and system applications.

In 1982 he was transferred to Northern Telecom Canada, Ltd., where he is the director of device technology responsible for fabrication of laser diodes, as well as development of advanced optical components required for high–bit-rate transmission technology.

He has served on several international panels and conference committees involved with fiber-optic activities in research and industrial environments.

He has contributed numerous papers in fiber optics and holds more than ten patents in the field.

Herbert G. Winful is a Principal Member of the Technical Staff at GTE Laboratories, Waltham, Massachusetts, where he conducts research in nonlinear optics. He has made contributions to the theory of optical bistability, degenerate four-wave mixing, nonlinear propagation in optical fibers, instabilities and chaos in semiconductor lasers, and wave propagation in nonlinear periodic structures. He has published over 30 journal articles, holds one patent, and has presented numerous invited and contributed papers at international conferences.

Dr. Winful was born in London, England, on December 3, 1952. He grew up in Ghana, West Africa, then came to the United States in 1971. He earned a BS in electrical engineering from the Massachusetts Institute of Technology, and an MSEE and PhD from the University of Southern California, Los Angeles. He joined GTE Laboratories in 1980.

PREFACE

Optical-fiber transmission has emerged as a major innovation in telecommunications and is in the process of revolutionizing the telecommunication industry. Since its inception, optical-fiber communication technology has shown significant and rapid advances in performance. The exponential growth in the state of the art has been driven by the development of a firm theoretical and experimental understanding of the dependence of system performance and its constituents, combined with significant advances in underlying device and material technologies. The future for optical-fiber transmission is bright, and society will benefit since it is an essential facilitator to usher in the new information era.

The material presented in this book is intended to provide the reader with fundamental and practical engineering principles for all parts of a state-of-the-art optical-fiber communication system. There are, of course, discussions on the relevant characteristics of fiber, sources, detectors, and passive components, as well as measurement techniques, and analog and digital system design. In addition, cable design, nonlinear optical phenomena, communication theory as applied to optical systems, and the electromagnetic-field theory of single-mode fibers are reviewed. Finally, in the last chapter an introduction is provided to coherent optical communication, which is the subject of a growing research interest. Each chapter has been written by one or more researchers in his own specialty. As editor I gratefully acknowledge the quality of their contributions and their wholehearted cooperation in the preparation of this book.

E. E. Bert Basch

Introduction to Fiber Optics

C. KAO

ITT Corp. Advanced Technology Center

E. BASCH

GTE Laboratories, Inc.

1. A View of the Future

Optical fiber is, and will remain, a transmission medium par excellence. Its primary advantages of low loss, wide bandwidth, small size, strength, flexibility, and low cost, coupled with its special advantages of immunity to electromagnetic noise and to electronic eavesdropping, are unique and meet the demanding requirements for transmission and other applications, such as sensors and signal processing.

As originally envisaged for guiding coherent electromagnetic waves at optical wavelengths, fiber has taken only 20 years of development to reach its place in the field. And, at this stage, practical optical sources and detectors for the generation of coherent electromagnetic waves (stimulated emission of photons) and detection can be controllably fabricated, so that the full potential of optical fiber can now be realized.

The future of optical fiber is bright. In this postindustrial age we are becoming an information-intensive society for which optical fiber is an essential facilitator. Just as interstate highways led to the modern transportation system, optical fiber will form our arterial information highway, providing transport for our information traffic. In sensing and optical signal processing, fiber extends its usefulness to transducers and processors, being compatible with the transmission medium. Thus, as fiber performance continues to improve and the infrastructure of related technology continues to build up, the application range of optical fiber will continue to widen.

The history of fiber transmission loss over the past and its projected realization are illustrated in Fig. 1.

Transmission loss in modern silica fiber is close to its asymptotic theoretical value of around 0.14

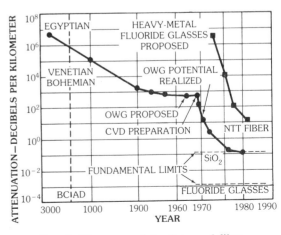

FIG. 1. The past and the future of fiber transmission loss.

dB/km at 1.55-μm wavelength. With this achievement a practical single-mode fiber with low dispersion over a very wide spectral region can be manufactured at low cost. Basic work on exact electromagnetic solutions of light waveguides with frequency-dependent refractive-index profiles is leading to optimum single-mode fiber designs. Basic research on glass material formation mechanisms and their relationship to scattering and radiation hardness will substantially increase our ability to tailor fibers to specific applications. The continuing work on new fiber fabrication processes will result in further cost reductions for optical fibers so that the cost of single-mode fiber will one day be similar to that of a pair of copper wires.

Both the sol-gel process, in which the glass constituent material starts in liquid form, and the mechanically shaped preform technique, which starts with soot material in powder form, are volume production techniques requiring low capital investment, with efficient use of raw materials. Also, improvement of traditional fabrication techniques alone indicates cost reductions by a factor of 2 to 4 through upscaling and increases in fabrication speed and yield.

Silica, with its excellent mechanical characteristics and its abundance as a raw material, will remain as the key medium for transmission fibers which will be single mode. Exceptions may appear for "repeaterless" applications, where repeater spacing is to be as long as possible, as in intercontinental trunking, transoceanic, or interisland links with spans greater than 200 km, for which fluoride type fibers may be developed. For transoceanic and island-hopping applications an ultralow-loss fiber with a repeater spacing times bandwidth product of 1000 GHz·km is needed. This is indeed a challenge, even if current material work indicates that 10^{-3} dB/km is in principle possible by operation at 4- to 10-μm wavelength. Indeed, fluoride glasses, chalcogenide glasses and infrared transmitting crystals should realize low losses as predicted if the material structure can be controlled and impurities differentiated and removed. Once that is accomplished the low-loss operating wavelength must be constrained to coincide with the zero-dispersion point and a nominally single-frequency source

developed. The physical properties of the fiber must also be addressed. The manufacturing difficulties are numerous, since many of the materials have undesirable properties. They may have toxic constituents or be unstable, hygroscopic, or susceptible to environmental attack.

Special single-mode fibers are needed for sensor applications where polarization maintenance is crucial. This requirement is also called for when fiber is used as a signal-processing element, although for a signal delay-line, polarization is not an issue. Multimode fibers are now relegated to component roles, serving as short interconnection wires and as components where multimode operation facilitates the realization of dispersive effects. Moreover, the special properties of the multimode fibers can be used to make the fiber more versatile. For example, large-diameter, graded-index fiber can be used as an excellent lens.

Other special single-mode fibers, needed for component applications, may be fabricated from a variety of materials. A germania (GeO_2) fiber can be used as a Raman amplifier. A low-temperature glass fiber can be the basis for making a range of microwave-equivalent components, such as couplers, attenuators, and phase shifters. The fiber material is selected to enable easier fabrication of components, rather than ensuring, perhaps, low-loss characteristics. In component applications, 3 dB/km translates to less than 10^{-3} dB/10 cm, a negligible loss over an unusually large distance encountered in a component.

For use in hostile environments and some sensor applications, silica fiber requires a coating of non-glassy material. Silicon nitride (SiN_2) and metals have been successfully used to coat the fiber to form a hermetic shield. Depending on the coating method, however, the intrinsic strength of the fiber can then be lowered, presumably due to surface crack formation. Lack of coating uniformity can also cause excessive temperature-dependent bending losses. Research in this area includes controlling deposition of material on glass surface through controlling, both or separately, the homogeneous and heterogeneous chemical reactions involved. This should lead to improved deposition uniformity, adhesion, as well as broadening the range of materials useful for fiber coating. Acoustically insensitive materials, materials with opposite thermal expansion to glass, magnetic materials, and electrostrictive materials can allow different types of fiber sensors to be more readily realized and with increased sensitivity.

Key milestones for fiber performance projection, expressed as a product of bandwidth times repeater spacing, together with those for developing of key components and techniques in associated technologies, are given in Fig. 2.

It is appropriate to note that the bandwidth-distance product is capable of handling 1000 GHz over 1 km. This suggests that the signal-processing speeds at the terminals should be very substantially increased. Currently fiber systems have been designed to handle signals up to 2.24 Gb/s. The highest–bit-rate signal transmission demonstrated is below 10 Gb/s. Two pertinent questions are: (1) is 10^{12} b/s (terabit rates) possible for future optoelectronic components; and, (2) if 10^{12} b/s can indeed be attained, how should the design of signal-handling systems be changed to take advantage of such high signal-processing speeds? These questions are particularly timely since the transmission capabilities of the fiber will surely cater to these rates. Moreover, several incipient applications with their involved information capacity, transmission and processing rates can make use of such high signaling rates. For example, if video phones at 70 Mb/s per channel are as densely installed as are telephones at 64 kb/s per channel today, the transmission requirement would be 1000 times the current trunk transmission rates of 560 Mb/s. Hence, a 560-Gb/s pathway would be needed. Certainly, it would be profitable to revise the basis of signal-processing deployment, hitherto one of conserving bandwidth. With bandwidth available in abundance, signal-processing schemes which are wasteful in bandwidth but simplify the overall system design and are more cost-effective, should be entertained.

Not surprisingly, optical fiber in an information society is stimulating research into high-speed optoelectronic components. We can speculate that the combination of electrons and photons used appropriately can result in effective signal-processing rates well beyond what can be achieved in pure electronics for both silicon and gallium-arsenide technologies as shown in Fig. 3.

A revolution in optoelectronics is taking place. Advances in electronic material fabrication techniques such as MBE and MOCVD are leading to

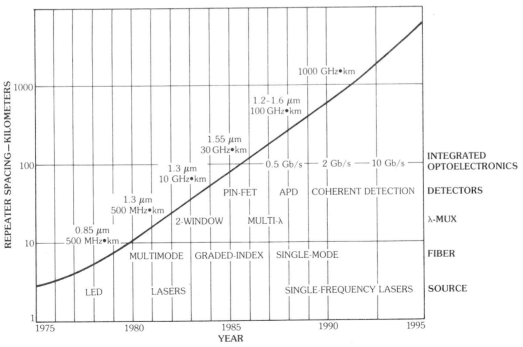

FIG. 2. Past and projected milestones of fiber performance.

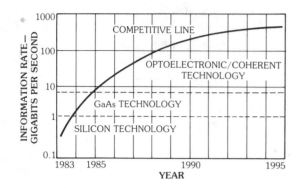

FIG. 3. Projected high-bit-rate technologies.

semiconductors with tailored properties. This is sometimes referred to as "band-gap" engineering. The study of electron-transport mechanisms in semiconductor material and availability of ultrafast optical spectroscopy diagnostic techniques are significant steps towards realizing truly integrated optoelectronic system building blocks for ultrafast systems.

In the meantime a substantial research interest has developed in applying optical-fiber technology to the "local loop," to determine if a broadband integrated services digital network can be provided in a cost-effective way to meet the needs of the private subscriber. Field trials by many national entities are providing opportunities to test not only technical feasibility but also customer reaction to various experimental services. These activities will increase significantly in the near future, when the benefits of such systems begin to be perceived not in terms of the extension of conventional communication services such as telephones to video phones, data to ISDN, etc., but in terms of new services based on fulfilling real information needs.

2. Summary of Chapters

In the following paragraphs the reader is provided with an expanded outline to the book. We hope that these brief chapter summaries will facilitate browsing by the reader and help him or her find the information he or she needs.

Chapter 2, "Properties of Optical Fibers," provides a general discussion of fiber properties.

Bandwidth, dispersion mechanisms, and attenuation properties limiting information transmission capacity of fibers are covered. The classification of fibers as multimode and single mode, together with propagation effects in different types of fibers, is discussed. This includes polarization-preserving fibers. Particular attention is paid to the properties of single-mode fibers. This includes the use of dispersion-shifted and dispersion-flattened index profiles to maximize the bandwidth at low-loss wavelengths. Noise sources, such as modal noise and mode-partition noise, due to propagation effects, are reviewed. Attenuation properties of multimode and single-mode fibers, such as material absorption, linear and nonlinear scattering, waveguide attenuation, leaky modes, and the effects of ionizing radiation, are covered.

Chapter 3, "Single-Mode Wave Propagation in Cylindrical Optical Fibers," examines how light propagates in cylindrical, low-loss, glass fibers whose refractive-index cross sections are so specified to confine the light to a narrow region at the fiber core, with maximum intensity at the center. The propagation attributes of these fibers are determined by solving the fundamental wave equation of Maxwell's theory for radially symmetric cases and also for fibers of elliptical cross section. The boundary-value problem defined by the fiber's profile is analyzed and used to determine "eigenvalues," those characteristic light frequencies for which only the lowest electromagnetic mode propagates, as well as the cutoff frequency of the adjacent, higher, unwanted mode. With the electromagnetic wave functions (eigenfunctions) that correspond, the distribution of light fields in any part of the fiber cross section can be determined. Of primary interest for single-mode fibers is the choice of a refractive-index profile that controls the group or pulse transmission delay as a function of wavelength. Finally, a number of powerful approximation methods to determine fiber attributes from the fiber profile are described.

Chapter 4, "Fiber Design Considerations for Optical Telecommunications," is meant for readers with prior knowledge in the physics of signal propagation in glass fibers and the basic characteristics of high-silica materials. The first section of the chapter reviews the chemical vapor deposition processes used to manufacture high-quality optical fibers and briefly compares them. It outlines the evolution of fiber designs from the onset of the technology in the early seventies. The second section reviews the materials and structural considerations for glass fibers, including a section on polymeric coatings used to preserve the optical and mechanical characteristics of the fiber. The third section reviews design considerations of multimode and single-mode fibers currently being used in telecommunication systems worldwide. The final section briefly looks at the emerging fiber designs that are expected to be used in the industry in the future.

Chapter 5, "Fiber-Optic Cables," provides a complete review of optical-cable technology. Basic analyses of fiber stress considerations and cable design principles are presented to form a detailed understanding of cable types for specific applications. The telephony, computer, utility, and government and military applications are reviewed with descriptions of the optical-cable technology optimized for each. Optical-cable testing techniques are described and cable placement techniques are illustrated. Splicing and connection technology are discussed and some transmission subsystem performance information is reviewed. Finally, requirements for future applications in the feeder and subscriber loop are addressed.

Chapter 6, "Optical-Fiber Measurements," reviews recently developed techniques for optical measurements on fibers. Since the principal parameters and measurement techniques for single-mode and multimode fibers can differ, these fiber types are treated separately.

For multimode fibers, after a discussion on modal distributions and steady-state effects, techniques for measuring fiber near-field and far-field patterns, and from the latter, numerical aperture, are described. Then fiber attenuation measurements, of the cutback, substitution, and optical time-domain reflectometry types, are reviewed. Methods for measuring losses of joints (splices and connectors) are also given. Finally, bandwidth measurement techniques are discussed.

For single-mode fibers, techniques for measuring fiber attenuation and joint losses are covered. In addition, techniques for measuring cutoff wavelength and mode-field diameter are reviewed. Finally, fiber dispersion measurements of the time domain, phase shift, interference and spectral

mode
mul
pro

s

linea
in the design
systems.

Among the effects consideri Raman and Brillouin scattering. The ba and frequency-shifted Brillouin light has deleterioul effects on a transmission system. The Brillouin gain, however, may also be used to amplify the carrier and improve receiver sensitivity in coherent optical communication. Stimulated Raman scattering leads to crosstalk in wavelength-division--multiplexed systems but can also be used in high-gain amplifiers and for generating new desirable wavelengths.

A class of phenomena that arise from the dependence of the fiber's refractive index on the light intensity, including self-phase modulation, soliton propagation and intensity-dependent polarization changes, and parametric processes, such as four-photon mixing, are also treated. Lastly, a prognosis for the future role of nonlinear effects in single-mode–fiber communications is given.

Chapter 8, "Passive Optical Components," is a review of the development of passive fiber-optic components used for processing of optical signals in fiber-optic links and networks. Components such as multiport couplers, wdm devices, and attenuators can be used to distribute signals to several receivers, multiplex/demultiplex, and optimize signal levels, respectively.

In general, the development of these components has followed the trends in fiber development. When the dominant fiber type evolved from multimode to single-mode, passive components evolved to follow in terms of both mode containment and the wavelength ranges of operation. Similarly, the perform-

5

ms of factors such as
of coupling to the fiber
em performance demands

pment in this area shows no
t. More recent areas being
e of single-mode components as
sed devices and in components
structed of specially processed opti-

r 9, "Optical Sources for Lightwave Sys-
plications," describes the principles of oper-
, fabrication and performance characteristics
ifferent types of semiconductor lasers. Empha-
s is on lasers fabricated using the InGaAsP material system and operating in the wavelength range of
1.3 μm and 1.55 μm since these are the regions of
interest for modern fiber-optic transmission systems. The concept of stimulated emission, the calculation of optical gain and the calculation of
radiative and nonradiative recombination rates and
their relation to the operational characteristics, such
as threshold current and efficiency of semiconductor lasers, are discussed. The various real index-
guided laser structures that have been fabricated for
high-performance applications are described. The
fabrication and performance characteristics of single-frequency lasers utilizing the coupled cavity or
the distributed-feedback mechanism for frequency
selection are described. The potential advantages of
quantum-well structures with regard to the possibilities of lower threshold current, higher quantum efficiency, etc., are discussed in relation to the
modification of the band structure in the quantum
well. The degradation mechanisms of semiconductor lasers and the development of a suitable reliability assurance strategy for applications such as the
submarine optical-fiber cable are discussed.

Chapter 10, "Modulation of Optical Sources,"
describes various methods of impressing information on optical sources for use in communication
systems. Factors which limit the information
bandwidth in directly modulated semiconductor
lasers and light-emitting diodes are discussed, and
the results of research and development to optimize their performance are given. Also discussed
are the various types of modulation which can be
achieved by the direct modulation of these optical
sources.

Indirect modulation of lasers with devices
which utilize various material phenomena are
described. The factors which limit their
bandwidths are discussed and examples of devices
which have been developed to optimize their characteristics are presented.

Chapter 11, "Detectors for Optical-Waveguide
Communications," provides the basis for understanding the operation and performance limitations
of photoconductive and reverse-biased pn-junction
semiconductor detectors for optical-waveguide communications. After a review of the basic physics
important for the operation of these devices, the
performance requirements of detectors in optical-
waveguide communication systems are expressed in
terms of bit error rate and detector signal-to-noise
ratio. The minimum received optical power required
to obtain a given signal-to-noise ratio is calculated
for both photovoltaic and avalanche photodiode
depletion-mode devices. Separate absorption and
multiplication (SAM) avalanche photodiodes for
long-wavelength optical-waveguide communication
systems are described, and, based on recent accurate
measurements of the impact ionization coefficients,
the improvement in system performance obtainable
with these devices over that obtained with pin detectors is predicted. The near-unity ratio of the electron and hole impact ionization coefficients in most
compound semiconductors limits the improvement
that can be obtained with SAM devices. New pro-
posals for artificially structured materials which
enhance the ratio of electron and hole impact ionization coefficients are reviewed, and the current status of development of these devices is described.

Chapter 12, "Receiver Design of Optical-Fiber
Systems," discusses the analysis and design of
optical receiver for optical-fiber communication
systems. In particular, the design of the front-end
low-noise receiver amplifier is considered in detail.
Signal and noise analysis of two popular receiver
amplifier designs, namely high-impedance and transimpedance amplifiers, are covered using one single
theory. Receiver design using both field-effect transistors (FETs) and bipolar junction transistors
(BJTs) are analyzed. The calculation of receiver sensitivity and its dependence on various system
parameters are discussed in detail. The impact of
other receiver requirements, such as dynamic range,
bit rate transparency, and acquisition time on its

design, are also considered. Throughout most of the chapter the discussion is based on digital communication applications. However, the theory developed is also applicable to analog communication systems.

Chapter 13, "Communication Theory for Fiber-Optic Transmission Systems," is intended as a tutorial introduction to the fundamental concepts of both analog and digital communication theory, presented in the context of the more specific problem of optical communication over glass fibers. The discussion is based on classical communication theory, which deals with the problem of reliable communication over channels whose primary impairment is white Gaussian noise. The limitations of this model for the optical-fiber medium are discussed where appropriate, and the material from classical communication theory is supplemented by discussion of problems which are specific to the optical fiber medium. The review of analog modulation covers amplitude modulation, double-sideband/supressed-carrier modulation, single-sideband modulation, frequency modulation, and intensity modulation. The spectral occupancy and performance in white noise of these techniques are presented and compared. The discussion of digital modulation methods begins by presenting a general model for digital signals, then proceeds with a brief examination of pulse shaping and the concept of inter-symbol interference. In addition, the phenomena of linear and quadratic delay distortion due to the properties of single-mode fibers are discussed. A derivation of the bit error probability performance for various digital modulation methods is also provided.

The chapter concludes with a review of fundamental limitations on communication over optical fibers. The concept of channel capacity is presented, and the channel capacities associated with coherent modulation with homodyne and heterodyne detection, and direct detection, are given.

Chapter 14, "Digital Optical System Design," examines how individual components in a system interact and how their characteristics determine the design and performance of an optical-fiber

communication system. The chapter starts with a brief overview of the evolution of optical communication systems through five generations of technology, from a first generation operating at a wavelength of 850 nm over multimode fiber to a fifth generation using single-mode fiber, operating at 1550 nm and employing heterodyne receiving techniques. To provide a background for design issues, the main features of each generation are noted, along with the reasons for its development. Next, basic system engineering considerations, such as power budget, receiver sensitivity, and fiber dispersion, are introduced and their influence on system performance examined. Finally, a number of system impairments, such as modal noise, mode-partition noise, and chirping, are discussed.

Chapter 15, "Design of Multichannel Analog Fiber-Optic Transmission systems," reviews techniques for transmitting a number of analog signals over an optical fiber. Both am and fm subcarriers are considered. The degradation of system performance due to intrinsic noise, linearity, and bandwidth characteristics of the system components are analyzed. The effects of noise and distortion generated by source-fiber interaction in ILD-based systems are also considered. A comparison of the performance limits of various system configurations is presented.

Chapter 16, "Introduction to Coherent Fiber-Optic Communication," presents, in introductory form, the principles of coherent carrier transmission applied to fiber-optic communication systems. The advantages and trade-offs for such systems are presented, as well as an overview of the major approaches to modulation/detection. Emphasis is placed on the critical role of the source line width in these systems and the effects of the associated phase noise on receiver sensitivity. General requirements for the major system components, as well as practical problems of implementation, are outlined. The chapter concludes with a section on quantum statistics in which the concept of "squeezed states" is introduced.

Properties of Optical Fibers

J. C. DALY & S. LAKSHMANASAMY

University of Rhode Island

1. Introduction

Optical fibers are dielectric waveguiding structures used to confine and guide light. They consist of an inner dielectric material called a *core* which is surrounded by another dielectric (called a *cladding*) with a smaller refractive index. A plastic and lossy *jacket* is commonly applied to the outside of the fiber to prevent crosstalk with other guides and to keep the fiber strong by preventing chemical and abrasive attack on its surface. Fig. 1 gives the cross-sectional view of a fiber [1].

Generally, optical waveguides have to meet the following requirements at the wavelength of interest [2]:

1. Low transmission loss
2. High bandwidth and data rate
3. High mechanical stability
4. Easy and reproducible fabrication
5. Low optical and mechanical degradation under all anticipated operational conditions
6. Low-loss coupling to system components such as sources and detectors

Optical fibers have properties that make them attractive replacements for copper wire and coax-

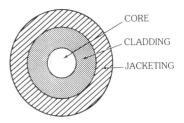

FIG. 1. Cross-sectional view of an optical fiber.

ial communication systems. They have extremely wide bandwidth. Their small diameter and high tensile strength result in smaller, lighter-weight cables and connectors. They do not radiate electromagnetic energy. They are usually made of glass, a stable material that resists corrosion and is tolerant of temperature extremes. This results in high reliability and easy maintenance. Since glass is an electrical insulator, optical fibers are immune to electromagnetic interference and are not subject to ground loop problems. The raw material used to fabricate glass fibers is sand, an abundant resource.

The wide-bandwidth property of optical fibers has made them attractive for systems requiring high-capacity transmission, such as telephone connections between central offices or between computers and high-capacity remote data-entry stations. Its wide bandwidth allows a single fiber to transmit multiple channels of information and replace many parallel copper wires or coaxial cables.

High-capacity, low-loss properties of fibers have been utilized for telephone applications. General Telephone of California installed the first link carrying regular telephone service in Long Beach, California, on April 22, 1977 [3]. Data was transmitted at 1.544 Mb/s over a graded-index fiber with a 6.2-dB/km mean loss. Light-emitting-diode sources and avalanche-photodiode detectors were used. Since then fiber-optic communication links have become commonplace. In 1984, systems operating at speeds of 400 Mb/s and repeater spacings of 25 km were installed by a number of long-distance telephone carriers. The rate of 400 Mb/s represents 6000 simultaneous digitized voice circuits. The current trend is toward single-mode fibers operating at wavelengths of 1300 nm (for minimum dispersion) and 1550 nm (for minimum attenuation) [4].

Optical fibers are used in computer systems [5]. Multiple-conductor parallel interconnects can be replaced by a single fiber carrying serial data. Cable and connector bulk and weight are significantly reduced. Reliability is improved. Fiber-optic high data rates, noise immunity, and low loss make it possible to extend high-data-rate channels beyond the confines of the computer room. The use of smart terminals for applications such as high-resolution graphics increases the need for high-bandwidth local networks. Smart terminals process and store information. Relatively large amounts of information are transmitted to the host in bursts. A wide bandwidth is required if excessive response times are to be avoided. Military applications of fibers range from communication links to rotation and sonar sensors. The dramatic weight and bulk reduction as well as the high reliability and large bandwidth of fibers are important for military applications. A 64-km fiber-optic field link used by the Army transmits 2.3 Mb/s, requires seven repeaters, and can be transported on one 2½-ton (2268-kg) truck. The equivalent coaxial link requires 39 repeaters and four 2½-ton trucks for transportation [6]. Fiber-optic links have been developed for submarines [7]. Information from external sensors is transmitted through the submarines hull to onboard signal processors. These links reduce the size of submarine hull penetrators in addition to improving the system performance.

This chapter discusses general fiber properties. Bandwidth and dispersion are covered in Section 2. Modal classification and properties of multimode, single-mode, dispersion-shifted, and single-polarization fibers are discussed in Section 3. Section 4 contains a description of numerical aperture (NA). Attenuation mechanisms are discussed in Section 5. Modal noise and mode-partition noise are discussed in Section 6.

2. Bandwidth Properties

Bandwidth determines information transmission capacity. High-capacity systems use signals coded into binary ones and zeros. A pulse of light represents a one. The absence of light represents a zero. The number of pulses (bits) per second possible is inversely proportional to the pulse duration (pulse width). A Fourier analysis of pulses shows that the bandwidth occupied by a pulse is also inversely proportional to its duration. The bandwidth properties of fibers can best be understood in terms of the minimum pulse width that can be used. If a sufficient time is not allowed between pulses, pulses will interfere. This results in intersymbol interference and errors in the transmission.

Dispersion

The width (duration) of a pulse propagating in an optical fiber increases with distance of propagation. The pulse of light is composed of photons. The propagation velocity is not the same for all photons. This phenomenon is called *dispersion*.* The photons in a given propagation mode have the same velocity. Consider a narrow (short-duration) pulse injected into a fiber at the sending end. The photons will arrive at the receiving end at different times due to their different velocities. The arrival time difference between the faster and the slower photons increases with distance traveled. That is, the duration of the pulse at the receiver increases with distance traveled. In situations where there is a random coupling of energy

---

*Photons of different frequencies or photons following different paths have different velocities.

between high-velocity and low-velocity propagation modes, the pulse width increases as the square root of the distance of the transmission. The duration that must be allotted between pulses to avoid intersymbol interference is proportional to the length of the fiber.

There are three classes of dispersion: (a) intermodal, (b) material and (c) waveguide dispersion. A *mode* is a transverse pattern of energy propagating at a specific velocity.

Modal properties can be understood by considering ray propagation. Different modes can be represented by different sets of rays. In step-index fibers, the transmission time for a ray depends on the angle it makes with the fiber axis. Rays propagating along the axis have the shortest transmission time, while rays zigzagging across the fiber and experiencing multiple reflections travel farther and therefore take longer to reach the receiving end.

Intermodal dispersion, the dominant source of dispersion in multimode fibers, is due to different velocities of different modes. Single-mode fibers are not subject to intermodal dispersion. Graded-index fibers greatly reduce intermodal dispersion. Rays that deviate from the axis travel farther, but, also, when they move away from the guide axis they move into a region of lower index of refraction where the velocity is greater. This situation is illustrated in Fig. 2. A suitably chosen index profile will result in nearly the same average velocity for all rays [8]. The optimum index profile is nearly parabolic, but small deviations from this optimum cause significant pulse broadening. Precise control of the index profile is difficult to achieve and measure in production. Profiles are optimized in practice by adjusting process parameters and measuring dispersion. The set of process parameters that result in minimum dispersion is used for production. A precise determination of the resulting index profile is difficult.

Material dispersion is due to the variation of velocity with wavelength. It causes an optical pulse to spread, even when all the light follows the same path. This dispersion is greater when the optical source is distributed in wavelength. Material dispersion has a null close to 1.3 μm.

Waveguide dispersion results from the guiding structure and is important in single-mode fibers.

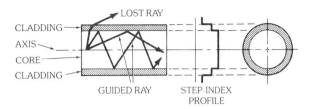

(a) Step-index fiber.

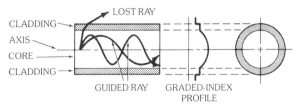

(b) Graded-index fiber.

FIG. 2. Propagation aspects of fibers with different structures.

The guided energy is divided between the core and the cladding. The propagation velocity in the cladding is greater than in the core. Each mode, however, has a distinct velocity that is determined by the distribution of its energy between the core and cladding. This distribution and therefore the velocity of the mode is a function of wavelength and of the index of refraction structure used to guide the light.

Since waveguide and material dispersion depend on the source's spectral width as well as the length of the fiber, they are measured in picoseconds per nanometer per kilometer (pulse spreading divided by source spectral width divided by fiber length).

3. Classification of Fibers

Fibers are classified according to the core refractive-index profile. As shown in Fig. 3 we can distinguish between *step-index* fibers, in which the refractive index is homogeneous in the core and changes abruptly from core to cladding, and *graded-index* fibers, in which the refractive index decreases continuously from core center to a region within the cladding. A further distinction can be made depending on whether only a single mode or double mode or several modes can propagate [9] in a given fiber. Such fibers are called *single-mode*, *double-mode*, and *multimode* fibers, respectively.

Multimode Fibers

This type of fiber supports a number of modes in addition to the dominant hybrid mode HE_{11} or the linearly polarized LP_{01} mode [10]. Multimode fibers can be either step-index or graded-index [11, 12]. The step-index multimode fiber consists of a homogeneous core of refractive index n_1 surrounded by a cladding of a slightly lower refractive index n_2. The relative index difference

$$\Delta = (n_1 - n_2)/n_1 \qquad (1)$$

is usually of the order of 1% or less for low-loss fibers based on fused silica [9]. Different partial rays propagate at different angles and therefore exhibit different transit times, resulting in a limited bandwidth of the fiber.

In the case of graded-index multimode fibers a reduction of the difference of transit times between propagating modes is accomplished by a gradation of the refractive index from the center to the outside. This causes the light to be guided not by total reflection, but by distributed refraction.

All propagating modes can have approximately the same average velocity, if the index profile is suitably chosen [8]. The optimum index profile is near-parabolic, but small deviations from this optimum cause significant pulse broadening. Sufficiently accurate control of the index profile is required, but is difficult to achieve in production. Figs. 3a and 3b illustrate the refractive-index profile of step-index and graded-index multimode fibers.

The propagation properties of a fiber can be described by the so-called V value sometimes referred to as the *normalized frequency*. The V value is an important fundamental fiber property, because it contains the most important fiber parameters [13]. It is defined by

$$V = 2\pi(a_1/\lambda)(n_1^2 - n_2^2)^{1/2} = 2\pi(a_1/\lambda) \times NA . \quad (2)$$

Here a_1 is the radius of the core, n_1 and n_2 are respectively the refractive indexes of core and cladding, λ is the wavelength of light propagating, and NA is the numerical aperture, a measure of the light-gathering capability of a fiber (which is defined in Section 4).

Among the other factors the V value determines the number of modes that a fiber can support. The concept of modes is important because it determines the bandwidth that is achievable over a certain fiber length. A multimode fiber with a circular cross section can support a finite number of guided modes and an infinite number of unguided or radiative modes. The guided modes can be divided into a family of $H_{0n} E_{0n}$ modes as well as a family of hybrid HE_{mn} modes. In addition, there are TE_{mn} and TM_{mn} modes. Theoretically a waveguide may be capable of carrying thousands of different modes [13].

Normally, it is desirable to reduce the number of modes carried to a minimum. In a step-index fiber the number of modes N_m is related to the V value by

$$N_m = V^2/2 , \qquad (3)$$

which shows that the number of possible modes increases with the square of the fiber diameter. So, in order to minimize the number of modes it is

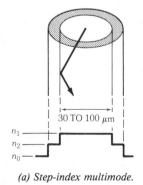

(a) Step-index multimode.

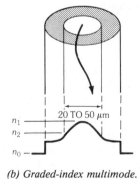

(b) Graded-index multimode.

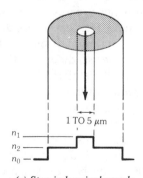

(c) Step-index single-mode.

FIG. 3. Refractive-index profiles and classification of fibers.

necessary to reduce the fiber diameter. This, however, has important economic implications because of fabrication difficulties and reduction in source-fiber coupling efficiency. Additionally, an increase in wavelength tends to decrease the number of possible modes [14].

Multimode fibers offer the following advantages [13]. They can be coupled to multimode lasers and incoherent sources as well as single-mode lasers with reasonable efficiency. Much wider tolerance for the alignment of these sources with the multimode fiber may be allowed than for a single-mode laser with a single-mode fiber. To connect and splice multimode fibers also requires less precision than for single-mode fibers.

Typical properties of multimode graded-index fibers include the following [15]:

Attenuation	Scattering limit, increases with NA, 0.5–2.0 dB/km
Bandwidth	Greater than 1 GHz·km for a profile optimized for maximum bandwidth. Limited by profile control and profile change with wavelength. Bandwidth decreases with an increase in NA
Core	50 to 60 μm diameter with an NA of 0.2 to 0.3
Cladding	10λ cladding thickness and 2:1 ratio of outer diameter to core is adequate for low microbending loss (λ is the wavelength of the source)
Coupling	To Lambertian source of size equal to core, the coupling efficiency is $(NA)^2/2$. To laser source with coupling lens the coupling efficiency is greater than 90%. Fiber-to-fiber coupling requires alignment accurate to 10% for coupling loss to be within 1 dB. Loss is dependent on mode distribution within the fiber at the coupling point

The following are typical properties of multimode step-index fibers [15]:

Attenuation	Close to the theoretical Rayleigh scattering limit
Bandwidth	Approximately 20 MHz·km, limited by modal dispersion
Core	Greater than 80-μm diameter with NA of 0.3 to 0.6
Cladding	2:1 ratio of cladding outer diameter to core diameter is adequate for low microbending loss
Coupling	To Lambertian source of size equal to core the coupling efficiency is equal to $(NA)^2/2$. Fiber-to-fiber coupling is most sensitive to lateral misalignment (1-dB loss with 10% offset). Loss is dependent on mode distribution within the fiber at the coupling point

Single-Mode Fibers

A single-mode fiber has the ultimate wide-bandwidth capability and has a completely defined propagation characteristic. It is ideally suited for long-haul and high-capacity applications. The structure for the common step-index, single-mode fiber is shown in Fig. 3.

The single-mode fibers carry only the dominant HE_{11} mode, but are doubly dengerate in polarization. This degeneracy results from the fact that in circular waveguides all orientations are equivalent, thus permitting two orthogonal polarization modes to exist with the same wave number. Fig. 4 presents a plot of effective modal index β/k_0 for step-index profile fibers as a function of the normalized frequency, V. Here k_0 is the wave number of free space and β is the propagation constant of a mode. Below $V = 2.405$ a single mode exists, designated HE_{11}, while for $V > 2.405$ other modes are possible [13].

Single-mode propagation is realized by designing core sizes to be a few wavelengths in cross-sectional dimensions and by having small index differences between the core and cladding. Since

$$V = (2^{3/2} \pi a_1/\lambda) \Delta^{1/2},$$

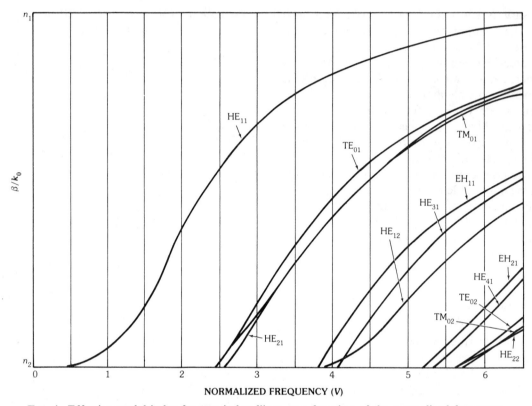

FIG. 4. Effective modal index for step-index fibers as a function of the normalized frequency.

one sees that the physical core size a_1 and core-cladding index difference Δ may be varied over a considerable range. Fig. 5 indicates the relationship between the normalized frequency V and the physical extent ($1/e$ points) of the guided HE_{11} mode. By making either Δ or a_1 small it is possible to have the guided fields extend significantly into the cladding and increase the physical extent of the mode [16]. This feature is of interest in fiber splicing since it affords a means of relaxing the translational dimensional tolerances. However, as V becomes small the wave in the fiber becomes loosely guided and thus susceptible to bending losses.

Another feature of single-mode fibers is the dependence of propagation loss on wavelength [17]. This loss is a strong function of wavelength through the dependence of V on λ. On the long-wavelength side, scattering into the cladding dominates, whereas when the wavelength decreases, the fiber becomes multimoded. An attractive feature of single-mode fibers is their relative insensitivity to microbending losses: losses induced by local lateral microdisplacements of the fiber from a mean axis. These losses are directly related to the fiber's statistical lateral displace-

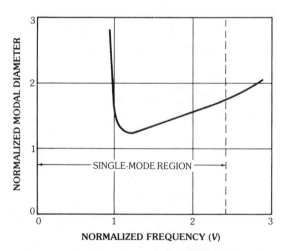

FIG. 5. Normalized beam size as a function of normalized frequency. (*After Giallorenzi [16], © 1978 IEEE*)

ment power spectrum, $F(\beta - \beta_0)$, which is defined [18] as

$$F(\beta - \beta_0) = C_0 / (\beta - \beta_0)^{4+2p} . \qquad (4)$$

Here, β_0 is the propagation constant of the HE_{11} mode, C_0 is a measure of the magnitude of the displacement spectrum, and p is the power spectrum exponent. In order to minimize microbending losses, $F(\beta - \beta_0)$ must be minimized. This is possible to the greatest extent in single-mode fibers for modal propagation in the vicinity of $V=2.405$.

So far we have discussed the step-index, single-mode fibers. We have found in the case of step-index fibers that the fiber parameter value near cutoff of the next higher-order mode at $V=2.405$ yields a favorable fundamental-mode characteristic in the single-mode range. This point of single-mode operation below the cutoff for the next higher-order mode will, in general, and for the same reasons, also be favorable in graded-index fibers. In the truncated parabolic profile with

$$n(r) = \begin{cases} n_1[1 - 2\,\Delta\,(r/a_1)^2] & 0 \le r \le a_1 , \\ n_2 & r > a_1 . \end{cases} \qquad (5)$$

the fiber parameter at cutoff of the LP_{11} mode has the value $V=3.53$. Compared with the limiting value $V=2.405$ for the step-index fiber, this represents an increase by a factor of 1.47 for the single-mode limit of this particular index profile. However, this higher single-mode limit does not necessarily mean that the fundamental-mode fields extend over a wider fiber cross section. Due to the graded index the fields actually confine themselves more to the center of the profile and, although the relative core radius is larger in the truncated parabolic profile at the single-mode limit, the mode radius remains nearly the same as in the step-index fiber.

Single-mode fibers do not suffer from intermodal dispersion. In the case of elliptical deformation of the nominally circular fiber cross section, the two orthogonal polarization states would travel at slightly different velocities, causing an intermodal delay problem. Aside from this difficulty, dispersion in single-mode fibers is due to the dispersive properties of the fiber material (material dispersion) and, to a lesser degree, to chromatic dispersion inherent to waveguiding processes (waveguide dispersion). A

zero-dispersion wavelength can be obtained at any desired wavelength between 1300 and 1700 nm by canceling out material dispersion with waveguide dispersion [19]. The signal distortion due to polarization mode dispersion can in principle be removed by using only one polarization mode for transmission [20].

Dispersion-Shifted and Dispersion-Flattened Fibers—Single-mode fibers have been developed that shift the zero-dispersion wavelength to the minimum-attenuation wavelength of 1550 nm or broaden the dispersion minimum so that a low-dispersion characteristic is available over a range of frequencies. Losses of 0.16 dB/km at 1550 nm have been reported [21]. This is close to the theoretical minimum of 0.15 dB/km. With losses approaching the theoretical minimum of 0.15 dB/km, further system improvement can only be obtained by improving the dispersion characteristic. Dispersion produces signal distortion that increases with distance of propagation. Reducing this distortion in addition to reducing attenuation increases the distance a signal can propagate before a repeater is required.

Step-index single-mode fiber dispersion passes through a null near 1300 nm. Unfortunately, minimum attenuation is at 1550 nm, where dispersion is higher. The best situation is to have a dispersion null at the wavelength of minimum attenuation. Dispersion-shifted fibers achieve this. In these fibers the wavelength of minimum dispersion is shifted to 1550 nm, the minimum-attenuation wavelength. This is accomplished by adjusting the index profile. Zero dispersion occurs when waveguide dispersion cancels material dispersion. At 1550 nm material dispersion is larger than at 1300 nm. Therefore waveguide structures (index profiles) that achieve the larger waveguide dispersion required to cancel material dispersion are used to produce zero dispersion at 1550 nm.

Decreasing the core diameter and increasing the refractive-index difference between the core and the cladding increases waveguide dispersion but also increases attenuation and reduces the cutoff wavelength for the second-order mode. Fibers with low-cutoff wavelength are subject to increased microbending losses. Better fiber characteristics are achieved by adjusting index profiles to achieve the desired dispersion characteristics. Index profiles, such as those shown in Fig. 6, produce the desired dispersion at 1550 nm.

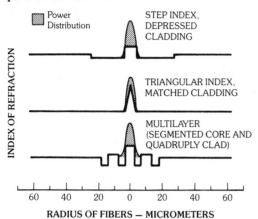

FIG. 6. Low-dispersion refractive-index profiles for single-mode fibers at 1550 nm. (*Adapted from Kapron [23], © 1985 IEEE*)

Dispersion-flattened fibers have index profiles designed to achieve low dispersion over wavelengths from 1300 nm to 1550 nm. Over this range waveguide dispersion is tailored to compensate material dispersion. There are a number of interacting design objectives for both dispersion-shifted and dispersion-flattened fibers. Designers adjust the index profile with the objective of reducing the total dispersion, achieving a reasonable spot size to reduce splicing losses, minimizing microbending losses, and reducing losses by minimizing the doping required to achieve the refractive differences between the core and the cladding.

Multimode Effects in Single-Mode Fibers—There are two mechanisms for multimode effects in single-mode fibers. The first is polarization, and the second is the existence of higher-order modes. Since two independent polarizations are possible, the fundamental mode of a single-mode fiber is really two modes. Fiber dispersion and attenuation will be different for each polarization. Imperfections cause energy to transfer between the two polarizations. Modal dispersion results when different polarizations have different velocities. A modal noise results when polarization is changing and polarization-dependent loss mechanisms are present.

When fibers operate close to the cutoff of the higher-order mode, lossy propagation of the higher-order mode results. This results in modal noise when short lengths of fibers, such as in pigtails, are used. Mode-conversion mechanisms, such as bends and splices, cause energy to be transferred from the fundamental to the higher-order mode. Modal noise occurs when the energy in the modes varies with time in the presence of mode-dependent losses [22]. Mode-dependent losses are speckle-dependent losses and can occur at splices and microbends. Energy in the higher-order mode can also propagate for some distance before being lost or transferring back to the fundamental mode. This can cause some pulse spreading (modal dispersion) of the signal propagating in the fundamental mode.

Typical single-mode fiber properties include the following [15, 21, 23]:

Attenuation	Scattering limit 0.15 dB/km at 1550 nm
Bandwidth	Theoretically 1000 GHz·km or more, but limited by the spectral width of the source and second-order material and waveguide dispersion effects
Core	One to several wavelengths in diameter. Core size can be increased by lowering NA, but the radiative loss will increase at a bend and/or with dimensional variations
Cladding	Thickness must be at least ten times the core radius. Increased thickness eases handling and reduced susceptibility to microbending. The influence of lossy outer jacket on propagation will also be reduced
Coupling	Requires light-source emission spot size to be equal to or smaller than the fiber core area for maximum efficiency

Single-Polarization Fibers

Fibers generally exhibit some ellipticity of the core and/or some anisotropy in the refractive-index distribution to anisotropic stresses. This results in two different propagation constants for the x-polarized and y-polarized HE_{11} or LP_{01} modes, leading to perturbations of the state of polarization of the light transmitted by the fiber. The origin of birefringence when anisotropic stresses are present in fiber is that, under such conditions, the dielectric permittivity is no longer a scalar, but a tensor. This leads to different propagation constants for TE and TM modes, even with a circularly symmetric waveguide [23].

Three methods have been proposed and studied so far to increase the birefringence between two eigenpolarization modes in a single-mode fiber [24]. The first method is the use of asymmetric index profiles, such as an elliptical core [25] and an index side pit [24], as shown in Figs. 7a and 7b. The birefringence introduced by the asymmetric index profiles is generally insufficient for the above application. Loss reduction seems to be difficult in these structures, since a large and/or complex refractive-index discontinuity is located at a core-cladding boundary.

The second method is the use of uniaxial internal stress-induced birefringence [26] as shown in Figs. 7c and 7d. This scheme is more advantageous than the simple anisotropic index profile scheme with respect to both birefringence enhancement and loss reduction. A beat length of less than 1 mm was reported in a fiber with a B_2O_3-doped elliptic jacket. A loss as low as 0.53 dB/km was reported in a fiber with B_2O_3-doped side pit. Loss reduction and birefringence enhancement was involved in the trade-off relation of these fibers. The distributed-mode crosstalk between two linear polarization modes in these fibers is caused mainly by unintentionally introduced bends and twists.

The third method is the use of a twisted single-mode fiber as shown in Fig. 7e. This scheme is different from the above two methods in that the two eigenpolarization modes are not linear polarizations but circular polarizations. The difficulties in developing a splicing technique with matched anisotropic axes and in matching the polarization axis between fiber output signal waves and local-oscillator waves in a heterodyne receiver are removed in this method [23].

A polarization control device is an alternative approach to overcoming the polarization-state

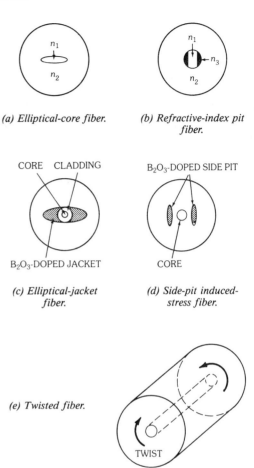

(a) Elliptical-core fiber.

(b) Refractive-index pit fiber.

(c) Elliptical-jacket fiber.

(d) Side-pit induced-stress fiber.

(e) Twisted fiber.

FIG. 7. Single-polarization fibers. (*After Kimura et al. [24]*)

fluctuation of fiber output signals. Electromagnetic-squeezer–induced birefringence, the use of anistropic fiber, twist- or bending-induced birefringence, and an electro-optic polarization transformer were all studied for this application [25].

4. Numerical Aperture

Numerical aperture is a measure of a fiber's light-gathering ability. When coupling sources to fibers, fibers to fibers, and fibers to detectors, the coupling efficiency depends on the angular distribution of energy as well as on alignment. Numerical aperture (NA) describes the angular distribution of light accepted by a fiber.

For efficient coupling to a fiber, two requirements must be satisfied. The light must be incident on the core, and it must also be injected into the fiber at an angle relative to the fiber axis that is less than the acceptance angle. Only light that is within the acceptance cone, formed by rotating a ray at the acceptance angle ϕ'_c about the axis as shown in Fig. 8, will propagate in the fiber.

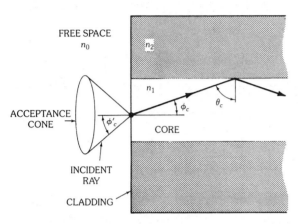

FIG. 8. Source-fiber coupling in an optical fiber.

In step-index fibers the guidance mechanism is total internal reflection at the core-cladding interface. Rays injected into the core at angles greater than the acceptance angle strike the cladding at angles less than the critical angle for total internal reflection. These rays pass out of the core and into the cladding, where they are subject to high loss.

In graded-index fibers the guidance mechanism is the gradual decrease of index of refraction as a function of distance from the axis. Rays follow trajectories that oscillate about the fiber axis as a function of distance along the axis. If a ray trajectory strikes the cladding, light may pass into the cladding and be lost. The amplitude of oscillation, and therefore the extent to which energy deviates from the fiber axis, is a function of the input injection angle. The oscillation amplitude also depends on the point at which light is injected into the core, that is, the displacement of the ray from the fiber axis at the begining of the fiber. The acceptance angle for graded-index fibers is a function of where on the core face the light enters the fiber. Just as with step-index fibers, however, there is an overall limit on the angles at which light will be accepted by the fiber.

An expression for numerical aperture in step-index fibers follows from Fig. 8. The light incident at the core-cladding interface must have an angle of incidence greater than the critical angle θ_c, where

$$\sin \theta_c = n_2/n_1 \qquad (6)$$

is the sine of the critical angle. Angles ϕ'_c, ϕ_c, and θ_c in Fig. 8 are related by

$$n_0 \sin \phi'_c = n_1 \sin \phi_c = n_1 (1 - \sin^2 \theta_c)^{1/2}, \qquad (7)$$

where n_0 is the refractive index of free space. The term on the left side of the equation is the numerical aperture or NA of the fiber and can be written

$$NA = (n_1^2 - n_2^2)^{1/2}. \qquad (8)$$

For the usual case of small differences between n_1 and n_2,

$$NA = n_1 (2\Delta)^{1/2}, \qquad (9)$$

where $\Delta = (n_1 - n_2)/n_1$.

Light exiting from a fiber is confined to the same acceptance angle as rays entering the fiber. Numerical-aperture measurements are usually made on light exiting a fiber excited by light completely filling the acceptance cone at the sending end of a fiber.

5. Attenuation

Two properties of primary importance are signal attenuation and dispersion. Attenuation (or loss) determines the distance over which a signal can be transmitted. Dispersion determines the number of bits of information that can be transmitted over a given fiber in a specified period. While attenuation and dispersion are separate physical phenomena, they both cause signal distortion. In a communication system where bit error rate is a measure of system performance, the effects of attenuation can be traded off against the effects of dispersion. Attenuation, α, is expressed in terms of decibels per unit length, and is given by

$$\alpha = 10 \ln (P_{in}/P_{out})/L , \qquad (10)$$

where P_{out} is the output power, P_{in} the input power, and L the length of fiber [2]. If α refers to a standard length L, it is usually expressed in decibels per kilometer (dB/km).

Low material absorption has been achieved by reducing impurity absorption due to transition metal ions, such as iron, chromium, cobalt and copper. Absorption from OH^- ions due to water impurity is also an important factor. Parts per billion purity of iron and chromium is required if their loss contributions are to be kept below 1 dB/km. An important scattering loss is due to density fluctuations in the material (Rayleigh scattering). These fluctuations are frozen into the glass at the time it was formed. Raleigh scattering places a fundamental lower bound on attenuation. This type of scattering loss varies inversely as the fourth power of the wavelength. This is why designers look to long-wavelength systems for lowest attenuation. However, if the wavelength becomes too long, coupling to molecular resonances of the glass occurs. This limits low-loss transmission to wavelengths less than 1800 nm in silica fibers. Additional scattering can occur due to large inhomogenities, such as bubbles and crystallites. This loss can be eliminated with good fabrication process control. Waveguide losses can be caused by irregularities at the core cladding interface. These can also be significantly reduced by proper process control. Microbending losses are a type of waveguide loss.

The dependence of attenuation on wavelength is shown in Fig. 9. There are three transmission windows. The first window is in the 810- to 850-nm region. Early systems using AlGaAs sources and silicon photodiodes operated in this window. In early 1980, optical sources and detectors using InGaAs and InGaP operating in the 1200- to 1600-nm region became available. This opened up the second and third transmission windows. Losses and dispersion are lower in these long-wavelength systems.

There are several mechanisms that lead to transmission losses in fibers. These are: (a) material absorption, (b) linear scattering, (c) nonlinear scattering, (d) waveguide attenuation, (e) leaky modes and fiber design, and (f) exposure to ioniz-

ing radiation. The first five (a–e) listed above are termed *intrinsic losses*, and the last one is due to factors external to the fiber.

Material Absorption

Material absorption is a loss mechanism by which part of the transmitted power is dissipated as heat in the guide. The natural mechanical resonances of the silica (SiO_2) crystalline structure and resonances due to impurities lie outside the wavelengths used for transmission. However, the tails of these resonant absorption peaks and their harmonics affect transmission losses. The absorption loss in high-silica glasses is composed mainly of ultraviolet and infrared absorption tails of pure silica, absorption due to impurities, and infrared absorption tails due to the presence of dopants [25].

The infrared absorption tail of silica has been studied [27] and it is shown that the vibrations of the basic silica tetrahedron are responsible for strong resonances around 9000 to 13 000 nm. This results in losses of about 10^{10} dB/km at these wavelengths. Overtones and combinations of these fundamental vibrations lead to various absorption peaks at shorter wavelengths, among which the two limiting ones are around 3000 nm (5×10^4 dB/km) and 3800 nm (6×10^5 dB/km). The tail of these various absorption peaks leads to typical values of 0.02 dB/km at 1550 nm, 0.1 dB/km at 1630 nm, and 1 dB/km at 1770 nm. This is responsible for the long-wavelength cutoff of the transmission in high-silica optical fibers around 1800 nm. The ultraviolet absorption leads to an absorption tail in the usual wavelengths of interest. Ultraviolet absorption decreases with increasing wavelength to typical values of 1 dB/km at 620 nm and 0.02 dB/km at 1240 nm. Fig. 9 shows the observed-loss spectrum in a germano-silicate single-mode fiber.

The usual impurities that lead to spurious absorption effects in the wavelength range of interest are the transition metal ions and water in the form of OH^- ions [25]. The concentration of transition metal ions has been reduced to a negligible amount during the past decade, but the presence of water been decreased significantly only recently. The presence of OH^- ions from water in silica leads to an absorption peak centered at 2730 nm. This produces overtones and combinations with silicon resonances to produce absorption

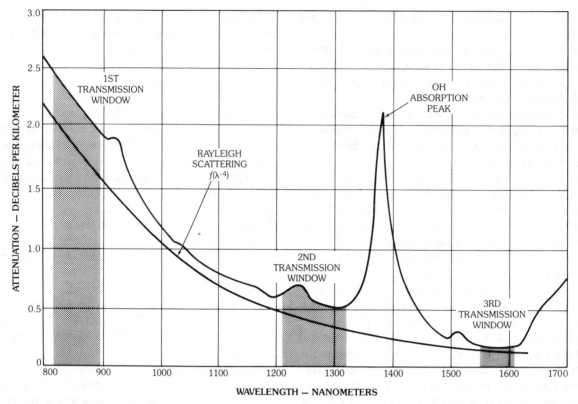

FIG. 9. Optical-fiber attenuation as a function of wavelength. (*After Basch and Carnes [34]*)

peaks at 950, 1240, and 1390 nm. For a concentration of 1 ppm the corresponding attenuations are about 1 dB/km (950 nm), 3 dB/km (1240 nm) and 40 dB/km (1390 nm), and it thus appears necessary to keep the OH^- concentration at levels below 0.1 ppm if ultralow losses are desired in the range of 1200 to 1600 nm. Water thus remains practically the only realistic source of parasitic absorption, and it still requires careful attention during the manufacturing process.

Dopants such as GeO_2, P_2O_5, F, and B_2O_3 are incorporated into the deposited silica to decrease the temperature of manufacturing, especially in the inside vapor deposition (IVD) process and to modify its refractive index (and thus design the waveguide structure). The presence of P_2O_5 brings an absorption peak at 3800 nm with an intensity of 10^6 dB/km for a 7-mol·% concentration, whereas B_2O_5 brings absorption peaks at 3200 and 3700 nm with intensities of 10^5 and 4×10^6 dB/km, respectively, for a 5-mol·% concentration [28]. Ultralow losses at 1300 nm require that B_2O_3 not

be present at radial distances smaller than five times the core radius, and P_2O_5 should be kept out of the core.

Linear Scattering

Linear scattering occurs when part of the power carried by one mode of the fiber is transferred linearly (proportional to that power) into another mode [28]. As in all linear processes no change in frequency is involved. This coupling among modes occurs because the guide is not a mathematically perfect cylindrical structure. Imagine, for example, that a parameter of the fiber varies sinusoidally with period Λ along the fiber. This parameter can be, for example, the departure of the axis from a straight line, or, in a straight fiber, the departure of the index from constant along a line parallel to the axis, etc. Then a guided mode with period λ_g couples strongly to another mode of period λ_{gc} along the axis, only if

$$\Lambda = |1/\lambda_g - 1/\lambda_{gc}|^{-1}, \qquad (11)$$

that is, if Λ is equal to the beat wavelength of the two modes. The relation between Λ and the direction in which light is coupled or scattered can be deduced from the above formula.

If the scattering angle is greater than $(2\Delta)^{1/2}$, where Δ is the refractive-index difference between core and cladding, the coupled mode is unguided and the fiber exhibits scattering losses.

Index fluctuations of a random nature occurring on a small scale compared with the wavelength λ/n_1, where n_1 is the refractive index of the core, scatter light almost omnidirectionally. This is Rayleigh scattering and produces an attenuation proportional to λ^{-4}. Therefore Rayleigh scattering is strongly reduced by operating at the longest possible wavelength. Fig. 9 gives the Rayleigh scattering loss variation with the wavelength [29].

These losses can be reduced by following design criteria that essentially tend to (a) increase guidance by increasing Δ, (b) stiffen the fiber by increasing the fiber's cross section, and (c) filter out the high-frequency components of the mechanical spectrum by housing each fiber in soft plastic and surrounding it with stiffening members, such as rigid plastics, graphite, metal, etc.

Nonlinear Scattering

If the field intensity in a fiber is very high, nonlinear phenomena occur, and power from a mode can be transferred to the same or other modes traveling either in the forward or backward direction, shifted in frequency [28]. Unlike linear scattering, where the coupling coefficients are independent of the optical powers in the modes involved, the coupling coefficients in nonlinear scattering are functions of those powers.

Stimulated Brillouin and Raman scattering are nonlinear, and as in any stimulated phenomena they have a threshold. The minimum threshold, and consequently the most critical for optical communication, occurs in a single-mode, high-silica fiber transmitting monochromatic light, where the threshold power in watts for Brillouin scattering is

$$P_T = 8 \times 10^{-5} \alpha / W^2 . \qquad (12)$$

In this expression α, in decibels per kilometer, is the linear attenuation coefficient of the fiber and

W is the beam's full width at the half-power density in micrometers. For a realistic single-mode fiber having $\alpha = 2$ dB/km and $W = 10$ μm the threshold power P_T is 16 mW.

Raman scattering occurs in the forward direction. For the same fiber the Raman threshold is about three orders of magnitude higher than the Brillouin threshold. For a broad-spectrum source and a multimode fiber the Brillouin and Raman thresholds are even higher. Losses introduced by nonlinear scattering can be avoided by the judicious choice of core diameter and signal level.

See Chapter 7 for more information on nonlinear effects.

Waveguide Attenuation

The losses due to waveguide structure arise from bending, microbending of the fiber's axis and defects at joints between axes.

The simplest qualitative description of bending losses in a fiber can be obtained by assuming that in the bent fiber the field is not significantly changed compared with the field in the straight fiber. The plane wavefronts associated with the guided mode are pivoted at the center of the curvature of the bent fiber, and their longitudinal velocity along the local fiber axis increases with the distance from the center of curvature. As the phase velocity in the core is slightly smaller than that of a plane wave in the cladding, there must be a critical distance from the center of curvature above which the phase velocity would exceed that of a plane wave in the cladding. The electromagnetic field resists this phenomenon by radiating power away from the guide, causing radiation losses. The bending losses increase when the radius of curvature decreases. Also, a mode close to cutoff is affected more than a mode far from cutoff. A high-index difference will decrease bending losses.

Microbending losses correspond to a fiber randomly oscillating around its nominal position with small deviations. Despite the small deviation the typical periods of the oscillations may be small and therefore the fiber may have sharp local bending. There are thus two loss sources, one arising from the permanent coupling between the LP_{01} mode and leaky and radiation modes, and another arising from the pure bending-loss effect of the LP_{01} mode, which becomes leaky at some radial distance in a curved fiber.

Alignment problems, geometry and refractive-index differences between fibers can lead to losses when fibers are joined or spliced [30]. At present multimode-fiber splices with average losses in the vicinity of 0.2 dB are obtained routinely in the field. Splicing single-mode fibers requires more care; nevertheless, in a laboratory environment an average loss of less than 0.1 dB has been achieved using the fusion technique.

Leaky Modes

In a multimode fiber, modes can be excited which are neither refracted nor completely guided. These are referred to as *leaky modes*. These modes radiate according to a phenomenon similar to tunneling through a potential barrier in quantum-mechanical systems. The majority of these modes have high attenuation coefficients, but still can propagate over kilometer lengths [28, 30, 31]. Leaky modes are especially excited by incoherent sources and must be taken into account in certain fiber measurements that involve short lengths of fiber.

An additional loss is produced by the fiber design. The power in a fiber is not totally confined within the core but a certain fraction travels in the cladding. This fraction is attenuated according to the cladding loss rather than the core loss. This means that if a small amount of power is propagated in the cladding, the cladding material should have losses comparable to those of the core.

Radiation Effects

Optical-waveguide attenuation may be strongly affected by exposure to nuclear radiation. Ionizing effects cause light-absorbing defect centers in the material. There is a correlation between the level of radiation, induced loss, and the intrinsic material loss. Fibers with higher intrinsic loss have a higher radiation sensitivity, which is probably due to a higher impurity concentration [32]. Pure silica-core fibers show substantially less damage than doped silica fibers.

Another point of interest is the wavelength dependence of radiation damage. For a germanium-doped silica fiber, codoped with boron and phosphorous, it turns out that there is a minimum of induced loss at the 1050-nm wavelength [1].

The damage at the short wavelength is dominated by intense absorption in the ultraviolet with a tail extending into the infrared. The increase in absorption for wavelengths greater than 1050 nm suggests a broadband absorption above 1700 nm, possibly caused by a change in the glass structure which results in an induced vibrational spectrum. In addition, fibers with low OH^- content are much more sensitive to radiation than fibers with higher OH^- content [1]. When fibers with pure silica cores and borosilicate cladding are considered, this behavior reverses at longer wavelengths so that the damage in dry silica fibers is less than in wet fibers near 1300 nm.

In cases where a fiber-optic system is exposed to radiation from a nuclear explosion, gamma rays and neutrons cause damage. Gamma rays cause the most serious damage in balanced hardened cases. (Balanced hardening occurs when all components, including people, are protected so that they have approximately equal probability of surviving.) A pulse of gamma rays causes the fiber's glass core to luminesce with blue Cerenkov radiation. This light can be received by the detector and mistaken for a signal. It is usually of secondary importance since the detector itself will probably also be subject to the gamma pulse and will have photocurrents generated in it. A more important effect of a gamma pulse is the generation of defect centers in the fiber. These centers can cause an increase in attenuation sufficient to cause the communication system to fail. Usually defects generated by gamma radiation anneal with time to levels somewhat worse than preradiation values. There is, however, a permanent degradation in attenuation. Higher temperatures and higher signal levels shorten recovery time. Germanium and germanium-boron–doped glasses generally show shorter recovery times than phosphorous-doped glass. Gamma-ray damage is more pronounced at shorter wave-lengths. Systems operating at 1300 nm suffer less damage from gamma rays than shorter-wavelength systems.

6. Fiber Noise

Two types of noise, *modal* (or *speckle*) *noise* and *mode-partition noise*, are a result of fiber propagation properties. Modal noise occurs when the distribution of energy among the fiber propagating modes changes in the presence of mode-dependent losses. Mode-partition noise is due to a change in the distribution of energy in the laser longitudinal modes rather than in the fiber propagating modes. A change in longitudinal laser modes together with fiber chromatic dispersion results in fluctuations in the received power even though the total power injected into the fiber from the laser is constant.

Modal noise is the result of speckle patterns produced by interfering propagating modes. Speckle patterns occur when the time delay between fiber modes is less than the coherence time of the source. Narrow-band, high-coherence sources, such as lasers, result in more modal noise than broadband sources. Incoherent sources do not produce modal noise. Speckle effects are less pronounced after long distances when relative delay times exceed the source coherence time [33]. For modal noise to occur, losses dependent on speckle must be present. Splices, connectors, microbends, and detectors with nonuniform responsivity have speckle-dependent losses. At a splice, for example, a small transverse misalignment or a core-size mismatch produces a speckle-dependent loss. Propagating-mode changes produce speckle-pattern changes. This results in a change in the energy transmitted through the splice. Changes in the relative strengths of propagating modes can be caused by changes in the spatial distribution of the laser output or by bending, and by pressure or temperature changes in the fiber. Wavelength changes in the source, due to changes in the dielectric constant in the presence of laser drive-current electrons, are a prime cause of modal changes in the fiber.

Mode-partition noise differs from modal noise in that it is caused by changes in the longitudinal modes of the source rather than changes in fiber propagation modes [34]. Mode-partition noise is the dominant noise in single-mode fibers. Laser output at discrete frequencies is determined by the longitudinal resonances of the laser cavity. Light propagating at these discrete frequencies is subject to relative delays determined by the chromatic dispersion of the fiber. At the laser the light is a clean pulse in time, but after propagation in the fiber, interference causes distortion. Changes in laser spectral distribution result in a change in the interference pattern at the detector. This causes a fluctuation in the detected signal even though the total laser output power is constant. Mode-partition noise is proportional to the signal power. Therefore the signal to mode-partition noise ratio cannot be improved by increasing signal power.

7. Conclusion

Optical fibers are attractive for the transmission of information. They are rapidly replacing copper wires and coaxial cables in a variety of applications. In addition to providing increased information capacity, they weigh less and occupy less space. Optical fibers have proven to be highly reliable.

There are trade-offs between ease of coupling and bandwidth that should be considered when picking a fiber for a particular application. Fibers with a large core diameter are easy to couple to but are subject to more modal dispersion than fibers with smaller cores. These large-core fibers are used for short-distance transmission at data rates in the megabits-per-second range, where ease of coupling is an important factor. Small-core fibers are more difficult to couple to but have a greater bandwidth and therefore a higher capacity. These are the fibers used for the transmission of large amounts of information between cities.

The bandwidth of fibers is limited because all photons do not have the same velocity (dispersion). All the light that enters the fiber at the same instant will not arrive at the receiving end at the same time. This pulse spreading limits the number of pulses per second that can be transmitted. In single-mode fibers, dispersion is practically eliminated. Data rates are limited by the laser sources and the detection methods used at the receiver.

Fiber losses have steadily declined since the first introduction of low-loss optical fibers. Current low-loss fiber-optic systems are based on silica fiber with a minimum attenuation of 0.15 dB/km at 1550 nm. Future systems may use different materials with significantly lower loss. New materials are under investigation that offer theoretical attenuations of

0.0003 dB/km for wavelengths in the region from 2000 to 12 000 nm [25].

Three groups of materials are being considered: heavy-metal oxides, heavy-metal halides, and chalcogens. Heavy-metal oxide fibers are formed from mixtures of the oxides of such metals as aluminum, barium, calcium, germanium, potassium, lanthamum, lead, tin, tantalum, tungsten, and zinc. These fibers have theoretical minimum attenuations below 0.1 dB/km. Heavy-metal halides have the extremely low theoretical losses of 0.0003 dB/km [23]. Chalcogenide glass fibers contain arsenic, germanium, phosphorus, sulphur, selenium, and telluride. Theoretical attenuations of 0.01 dB/km are predicted for these fibers at wavelengths of 3000 to 5000 nm. By far the most work is being done on heavy-metal fluoride glasses [35, 36].

Light sources and detectors for these long wavelengths exist. At wavelengths greater than 2000 nm directly modulated diode sources composed of mercury cadmium telluride and lead salts are available. Lasers that operate continuously at 147 K have been constructed. Indium-antimonide detectors operate at wavelengths up to 5000 nm. Longer-wavelength detectors are fabricated using mercury-cadmium-telluride (HgCdTe) photoconductors.

8. References

[1] Technical staff of CSELT, *Optical Fiber Communication* New York: McGraw-Hill Book Co., 1981.

[2] D. MARCUSE, "Optical Fibers for Communications," *Radio Electron., Eng.*, Vol. 43, No. 11, p. 655, 1973.

[3] E. E. BASCH, R. A. BEAUDETTE, and H. A. CARNES, "Optical Transmission for Interoffice Trunks," *IEEE Trans. Commun.*, Vol. COM-26, No. 7, pp. 1007–1014, July 1978.

[4] T. E. BELL, "Communications," *IEEE Spectrum*, Vol. 22, No. 1, pp. 56–59, January 1985.

[5] J. D. CROW and M. W. SACHS, "Optical Fibers for Computer Systems," *Proc. IEEE*, Vol. 68, No. 10, pp. 1275–1280, 1980.

[6] M. K. BARNOWSKI, "Fiber Systems for the Military Environment," *Proc. IEEE*, Vol. 68, No. 10, pp. 1315–1320, 1980.

[7] F. C. ALLARD, L. D. OLIN, and E. F. MANES, "A Fiber-Optic Sonar Link: Fiber-Optic Components

Design Considerations and Development Status," *Proc. EO/Laser 1980 Conf.*, Boston, November 19–21, 1980.

[8] K. OKAMOTO and T. OKOSHI, "Computer-Aided Synthesis of the Optimum Refractive-Index Profile for a Multimode Fiber," *IEEE Trans. Microwave Theory Tech.*, Vol. MTT-24, pp. 213–221, 1977.

[9] H. F. WOLF, *Handbook of Fiber Optics: Theory and Applications*, New York: Garland STPM Press, 1979.

[10] D. GLOGE, "Optical Power Flow in Multimode Fibers," *Bell Syst. Tech. J.*, Vol. 51, p. 1767–1783, 1972.

[11] D. GLOGE and E. A. J. MARCATILE, "Multimode Theory of Graded-Core Fibers," *Bell Syst. Tech. J.*, Vol. 52, p. 1563–1578, 1973.

[12] P. J. B. CLARRICOATES et al., "Propagation Behavior of Cylindrical-Dielectric-Rod Waveguide," *Proc. IEE*, Pt. H, Vol. 120, No. 11, p. 1371, 1973.

[13] H. G. UNGER, *Planar Optical Waveguides and Fibers*, Oxford: Clarendon Press, 1977.

[14] S. CHOUDHARY and L. B. FELSEN, "Guided Modes in Graded-Index Optical Fibers," *J. Opt. Soc. Am.*, Vol. 67, pp. 1192–1196, 1977.

[15] M. J. HOWES and D. V. MORGAN, *Optical Fiber Communications*, New York: John Wiley & Sons, 1980.

[16] T. G. GIALLORENZI, "Optical Communications Research and Technology," *Proc. IEEE*, Vol. 66, pp. 744–780, 1978.

[17] S. MORSLOWSKI, "High-Capacity Communication Using Monomode Fibers," *Proc. 2nd Eur. Conf. Optical-Fiber Transmission*, Paper XII, p. 373, Paris, France, September 27–30, 1976.

[18] R. OLSHANSKY, "Microbending Loss in Single-Mode Fibers," *Proc. 2nd Eur. Conf. Optical-Fiber Transmission*, Paper VI, pp. 3, 101, Paris, France, September 27–30, 1976.

[19] C. T. CHANG, "Minimum Dispersion at 1.55 µm for Single-Mode Step-Index Fibers," *Electron Lett.*, Vol. 15, pp. 765–767, 1979.

[20] T. MIYA et al. "Fabrication of Low-dispersion Single-Mode Fiber Over a Wide Spectral Range," *IEEE J. Quantum Electron.*, Vol. QE-17, 1981.

[21] D. B. KECK, "Fundamentals of Optical Waveguide Fibers," *IEEE Commun. Mag.*, Vol. 23, No. 5, pp. 17–22, May 1985.

[22] N. K. CHANG et al., "Observation of Modal Noise

in Single-Mode Fiber Transmission Systems," *Electron. Lett.*, Vol. 21-1, pp. 5–6, 1985.

[23] F. P. KAPRON, "Fiber-Optic System Trade-offs," *IEEE Spectrum*, pp. 68–75, March 1985.

[24] T. KIMURA et al., "Review: Progress of Coherent Optical Fiber Communications Systems," *Optical and Quantum Electron.*, Vol. 15, pp. 1–39, London: Chapman and Hall, 1983.

[25] L. B. JEUNHOMME, *Single-Mode Fiber Optics*, New York: Marcel Dekker, 1983.

[26] V. RAMASAMY et al., "Polarization Characteristics of Noncircular Core Single-Mode Fibers," *Appl. Opt.*, Vol. 17, p. 3014–3017, 1978.

[27] T. IZAWA, N. SHIBATA, and A. TAKEDA, "Optical Attenuation in Pure and Doped Fused Silica in the IR Wavelength Region," *Appl. Phys. Lett.*, Vol. 31, No. 1, 1977.

[28] S. MILLER and A. G. CHYNOWETH, *Optical Fiber Telecommunication*, New York: Academic Press, 1979.

[29] T. MIYA, Y. TERUNMA, T. HOSAKA, and T. MIYASHITA, "Ultimate Low-Loss Single-Mode Fiber at 1.55 μm," *Electron Lett.*, Vol. 15, pp. 106–108, February 1979.

[30] TINGYE LI, "Structure, Parameters and Transmission Properties of Optical Fibers," *Proc. IEEE*, Vol. 68, No. 10, pp. 1175–86, October 1980.

[31] J. D. LOVE and C. PASK, "Universal Curves for Power Attenuation in Ideal Multimode Fibers," *Electron. Lett.*, Vol. 12, pp. 254–255, May 1976.

[32] B. BENDOW and S. S. MITRA, *Fiber Optics*, New York: Plenum Press, 1978.

[33] KEN-ICHI SATO and KOICHI ASATANI, "Speckle Noise Reduction in Fiber-Optic Analog Video Transmission Using Semiconductor Laser Diodes," *IEEE Trans. Commun.*, Vol. COM-29, No. 7, pp. 1017–1024, July 1981.

[34] E. E. BASCH and H. A. CARNES, "Digital Optical Communication Systems," in *Fiber Optics*, ed. by J. C. Daly, Boca Raton, Florida: CRC Press, p. 163, 1984.

[35] M. G. DREXHAGE, "Heavy-Metal Fluoride Glasses," *Treatise on Materials Science and Technology*, Vol. 26, ed. by M. Tomozawa and R. H. Doremus, New York: Academic Press, pp. 151–243, 1985.

[36] R. N. BROWN and J. J. MUTTA, "Material Dispersion in High-Quality Heavy-Metal Fluoride Glasses," *Appl. Opt.*, Vol. 24, No. 24, pp. 4500–4503, December 15, 1985.

3

Single-Mode Wave Propagation in Cylindrical Optical Fibers

D. DAVIDSON

GTE Laboratories, Inc.

1. Introduction

This chapter examines how light is propagated in a cylindrical fiber and those fiber attributes leading to light confinement within the central core with minimum dispersion in the transmission time. The theory involves modal solutions to the fundamental wave equation for electric and magnetic fields. The basic approach taken is similar to that used in analyzing radio-frequency propagation in coaxial and hollow waveguides, and there are also similarities with the field treatment of radio-wave propagation in the earth-ionosphere waveguide. The special aspect of fiber, though, is that the medium is a pure dielectric, or layers of dielectric, comprising carefully deposited glasses of well-defined purity.

Much about the path of light in optical waveguides can also be deduced by a geometric-optical approach employing Snell's law and tracing ray refraction and reflection, except that some of the resulting relations are approximate and the theory requires considerable extension to account for diffraction, leaky modes, and the division of power within the fiber cross section. Some authors, like Marcuse [1] and Adams [2], have employed a dual approach, starting with rays in parallel-plane waveguides, then transferring to a full field theory,

especially when treating the cylindrical optical waveguide. The more subtle features of waveguide propagation, such as mode conversion, mode coupling, and the precise nature of the fields and cutoff frequencies are, at the least, difficult to explain with a pure geometric-optical treatment in the cylindrical fiber. Our aim here is to provide a concise picture of modal-field behavior of the important single propagation mode that takes full advantage of the extremely low intrinsic losses found in today's best fibers. Particularly emphasized are the several methods that have been employed to find solutions for fibers with rather arbitrary refractive-index profiles, as might be encountered in practical fibers.

In Section 1 the basic electromagnetic-wave equation is examined for adjoining dielectric regions of a cylindrically symmetric guide and the character of the field quantities in each region is readily identified by applying the electromagnetic boundary conditions. For the circularly homogeneous fiber, the modes are linearly polarized, and by assuming "weakly guiding" approximations, unique single-mode fields and modal eigenvalues are found, leading in Section 2 to detailed solution for the ideal step-index (SI) fiber, including its dispersion properties. In Section 3 we consider fibers with cores having arbitrary index profiles, including a central dip in the core index, an artifact of fiber fabrication. Sections 4 and 5 treat the W-fiber, one having an inner cladding that improves light confinement and allows a greater latitude of compensation of dispersion, and we consider also the effect of a W-fiber with central dip. Mode spot size and the concept and utility of using an equivalent-step-index (ESI) fiber are covered in Section 6. Section 7 contains a brief review of several powerful methods for obtaining dispersion and propagation properties of fibers with arbitrary profiles, including "resonance" methods using a transmission-line approach, variational methods, and the Numerov method. The last two are applied by converting the wave equation for the cylindrical fiber to the form of the one-dimensional Schroedinger equation of quantum physics. In several of the methods discussed, having an approximate idea of the fiber field existing in an "equivalent" step-index fiber is shown to be extremely useful in converging on the actual fiber's field and the essential, desired, design parameters. Section 8 is devoted to a brief summary of polarization effects in elliptical and near-circular symmetric optical fibers, with illustrations of the differences in the dispersion characteristics of the two lowest orthogonally polarized modes.

2. Electromagnetic-Wave Propagation

Electromagnetic waves (light waves or radio waves) are represented by electric and magnetic fields that are mathematical descriptions of the two significant measurement parameters that enable us to detect the wave's existence: voltage and current, or possibly combinations of these such as power or power density.

An electromagnetic-field element ψ propagating in an unbounded homogeneous medium obeys a scalar wave equation, derivable from the set of Maxwell's equations.

With the time dependence exp $(j\omega t)$ assumed, and with ∇ the Laplacian operator, the wave equation has the form*

---

*Derivation of the wave equation from Maxwell's equations will be found in the annex at the end of Section 2.

$$\nabla^2\psi + k^2\psi = 0. \qquad (1)$$

Here k is the wave number, and in free space, $k = k_0 = 2\pi/\lambda$, with λ being the wavelength. Thus k indicates the periodicity of phase in the direction of propagation. In free space far from a source the solutions for the electric field ($\psi=E$) or magnetic field ($\psi=H$) are such that $\mathbf{E}$ and $\mathbf{H}$ are orthogonal and are perpendicular to the direction of propagation.

For propagation in a most common form of optical fiber, we deal with a cylindrical bounded medium or, more exactly, bounded media immersed in an unbounded medium (air). The fiber's cross section has a refractive-index profile that may consist of only two zones, or multiple zones or annuluses, or zones with some complicated radial variation. In each local region (1) will hold but now k is described by

$$k = k_0 n(r), \qquad (2)$$

where $n(r)$ is the refractive-index variation with r, the fiber radius in the region, assuming that the fiber's cross section is azimuthally uniform.

Light is assumed to have been launched centrally into the cylindrical fiber well behind the particular cross section being studied, so that we are to deal with a far-field problem. We are interested in the distribution of the light fields over the cross section and, more importantly, what happens to these fields as they propagate forward along the fiber, in particular, how the light fields and light power are distributed over the fiber's cross section.

To solve the wave equation, we specialize to a natural system of orthogonal cylindrical coordinates, (r,θ,z) with z the axial propagation direction. The fiber is assumed perfectly cylindrical. The $\mathbf{E}$ and $\mathbf{H}$ fields of interest are those lying in cross sectional (r,θ) plane transverse to z, the direction of propagation, along with the orthogonal components that arise from symmetry and the basic theory. From symmetry, possible solutions could involve TE modes, with field elements E_r, E_θ, H_r, H_θ, and H_z, while with TM modes $H_z=0$ and E_z is present.

Solutions now take the (scalar) form

$$\psi(r,\theta,z)=R(r)\,\Theta(\theta)\,Z(z)\,, \qquad (3)$$

achieving "separation" of variables. The function ψ may also represent electromagnetic-field quantities other than **E**- and **H**-field components.

Because each of ψ's constituent wave functions is a function only of its own coordinate, we must have

$$d^2Z/dz^2=-\beta^2\,, \qquad (4)$$

with β a constant, so that physically the $Z(z)$ solution for forward propagating waves has the form $\exp(-j\beta z)$.

If the fiber is lossy in the z direction, then β is a complex constant whose imaginary part represents the loss per unit distance, and the real part the phase progression per unit distance. For the study of fiber modes and distribution of fields and light intensity in the cross section, the loss is such a small quantity that within dimensions of light wavelengths it can be neglected.

With β so taken, the radial and angular wave functions $R(r)$ and $\Theta(\theta)$ are linked by the reduced wave equation

$$R_{rr}+R_r/r+(1/r^2)\,\Theta_{\theta\theta}/\Theta+(k^2-\beta^2)\,R=0, \quad (5)$$

where the subscripts indicate partial differentiation with respect to the indicated coordinates. As with $Z(z)$ we set

$$d^2\Theta/d\theta^2=-m^2\,. \qquad (6)$$

Solutions for the angular wave function will be of form $\sin m\theta$ or $\cos m\theta$, or combinations of these, because of the cross-section's circular symmetry. Clearly, m can only have integer values $m=0, 1, 2 \ldots$ since Θ must be single-valued in the fiber cross section.

If the fiber comprises a core and one or more concentric regions, each with constant refractive index n_i, then in each region the remaining differential equation for $R(r)$ is the most general form of Bessel's equation,

$$R_{rr}+R_r/r+(\gamma_i^2-m^2/r^2)\,R=0\,, \qquad (7)$$

applying to each region if we take $\gamma_i^2=(k_0 n_i)^2-\beta^2$.

The Form of $R(r)$ in Various Regions

The solution for the radial wave function in a region of constant index involves combinations of Bessel functions. In a propagating region the appropriate pair of Bessel functions consists of

$$A\,J_m(\gamma r)+B\,Y_m(\gamma r)\,,$$

while in an outer or cladding region, modified Bessel functions (often called *Hankel functions* [3]) apply:

$$C\,I_m(j\gamma r)+D\,K_m(j\gamma r)\,.$$

In each region, γ and coefficients A,B,C,D are to be determined from the associated boundary conditions.

For a fiber with multiple concentric zones, intervening zones may have either of these indicated Bessel function combinations, depending on whether the zone functions to propagate light or to confine it.

For each value of m we have a principal propagating mode, but because of the periodicity there will be azimuthal nulls or minima in $\Theta(\theta)$ corresponding to higher-order modes when m is not identically zero. For long-haul communication, higher-order modes are less efficient and are a source of unwanted multipath—if one imagines rays in the fiber—that restrict the bandwidth. True monomode transmission will occur when $m=0$, giving azimuthal independence to the field distribution in the fiber cross section.

Looking at $J_m(\gamma r)$, the choice $m=0$ reveals an important aspect of the fiber field. For m not equal zero, all Bessel functions $J_m(0)$ are zero, whereas $J_0(0) = 1$. The lowest-order mode ensures an axial central maximum in the field as amplitude coefficient B must be zero in the core because $Y_0(0)$ is not finite.

In the outermost, or cladding region of the fiber, which is assumed to have indefinite extent, only Bessel function $K_0(\cdot)$ is appropriate, since it is a monotonically decaying (or evanescent) function for large values of argument, an essential requirement for light confinement.

Boundary Conditions for Dielectrics

These intuitive choices can be rigorously established as we define and apply the boundary conditions of the problem:

(1) The field must be finite at the origin (the axis). It may be convenient to let the axial field have unit amplitude. We have seen that $J_0(\cdot)$ meets this requirement. (With certain fiber configurations to be discussed later, the axial field might also take the form $I_0(\cdot)$ since $I_0(0)=1$, but in that case $I_0(0)$ is a local minimum, which does not match core concentration of the light from a driving source with light lobe maximum at the center.)

(2) At indefinitely great radial distances from the fiber axis all fields must vanish (Sommerfeld's radiation condition). As we have seen, this limits the outer radial wave function to $K_0(\cdot)$ in a light-guiding fiber. Otherwise, a radiative form of the field would apply, such as $J_0(\cdot)$, or $Y_0(\cdot)$, or both.

(3) According to electromagnetic theory [4], at the boundary between dissimilar dielectrics the normal component of the **E** field is discontinuous, while all other field components are continuous. Rather than avoid accounting for this discontinuity at this stage by describing the field in terms of the electric displacement, the **D** field, we note that for the type of fibers we are considering the difference in adjacent dielectric constants or refractive indexes is extremely small, so that we can assume, conveniently, that *all* fields are continuous. Also, although a complete, accurate, solution indicates the existence of longitudinal **E**- and **H**-field components (i.e., in the z direction), these are extremely small as they are proportional to the index difference and so generally may be neglected in dealing with the boundary conditions. This leads to much simplified expressions for $R(r)$.

As a result of the relations between **E** and **H** fields in Maxwell's equations, continuity is required of a field and its derivative at dielectric boundaries, and this is compactly and conveniently expressed through the continuity of logarithmic derivatives. Physically this is equivalent to requiring matched wave impedances at the dielectric boundaries to obtain a unique solution.

Loss Spectrum in a Typical Fiber

While the loss per unit distance is negligible for the study of waveguide modes and determining modal eigenvalues, the loss spectrum is very important in determining the total transmission loss of a fiber span, and for locating the least lossy part of the spectrum. It is within such a "transmission window" that the designer wishes to transmit the widest information-bearing spectrum with the least distortion. (In Fig. 1 very-wide-bandwidth transmission is possible as a result of a profile design that invokes very low chromatic dispersion in window B. The absorption peak is an OH-radical contamination phenomenon of silica fibers.) The solution of this problem is the real subject of the rest of this chapter.

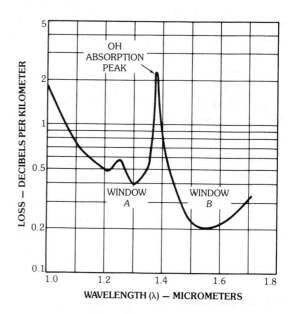

FIG. 1. Transmission loss spectrum of a high-quality optical fiber with a tailored refractive-index profile, produced by the MCVD process. *(After Shang et al. [5]. Acknowledgment is made of the prior publication of this material by the IEE)*

Annex: Derivation of the Wave Equation from Maxwell's Equations—In vector differential form Maxwell's equations relate the fundamental measurable vector electromagnetic quantities **E**, the electric-field strength, and **H**, the magnetic-field strength, and their respective flux densities **D**, the electric displacement, and **B**, the magnetic flux density, as follows. (In the SI system of units the units of |**E**| are volts per meter; |**H**|, amperes per meter.)

$$\nabla \mathbf{D}=0, \qquad \nabla \mathbf{B}=0, \qquad (8a, 8b)$$

$$\nabla \times \mathbf{E}=-d\mathbf{B}/dt, \quad \nabla \times \mathbf{H}=d\mathbf{D}/dt. \qquad (9a, 9b)$$

The operator ∇ is compact vector notation for the spatial gradient

$$\nabla = \hat{\mathbf{x}}\,\partial/\partial x + \hat{\mathbf{y}}\,\partial/\partial y + \hat{\mathbf{z}}\,\partial/\partial z$$

in Cartesian coordinates, or

$$\nabla = \hat{\mathbf{r}}\,\partial/\partial r + (\hat{\theta}/r)\,\partial/\partial \theta + \hat{\mathbf{z}}\,\partial/\partial z$$

for cylindrical coordinates (r, θ, z).

The relations between $\mathbf{D}$ and $\mathbf{E}$, and between $\mathbf{B}$ and $\mathbf{H}$, are known as the *constitutive relations*, because they involve the properties of the physical medium in which the field quantities are defined:

$$\mathbf{B}=\mu \mathbf{H} \quad \text{and} \quad \mathbf{D}=\epsilon \mathbf{E}. \qquad (10a, 10b)$$

Here ϵ is the permittivity of the medium (in farads per meter) and μ its magnetic permeability (in henrys per meter). In nonmagnetic media, $\mu=\mu_0$. In nonpolarizing media $\epsilon=\epsilon_0$. The following relations exist in free space:

The square root of the ratio of μ_0 to ϵ_0 is

$$\eta_0=(\mu_0/\epsilon_0)^{1/2}=120\,\pi \simeq 377 \text{ ohms}, \qquad (11)$$

which is known as the *impedance of free space*.

The reciprocal of the product of the square roots of μ_0 and ϵ_0 is

$$c=(\mu_0\epsilon_0)^{-1/2} \simeq 3\times10^8 \text{ m/s}, \qquad (12)$$

which is the speed of light.

Operating on both sides of (9a) and (9b) with $\nabla \times$, using the vector identity

$$\nabla \times \nabla \times \mathbf{E}=\nabla(\nabla \cdot \mathbf{E})-\nabla^2\mathbf{E},$$

and remembering that $\nabla \cdot \mathbf{E}=0$, we obtain the wave equation in the form

$$\nabla^2\mathbf{E}=(-d/dt)(\nabla \times \mu_0 \mathbf{H}).$$

With the time-dependent term $\exp(j\omega t)$ as a common factor the derivative d/dt can be replaced by the simple factor $j\omega$, whence the wave equation becomes

$$(\nabla^2+\mu_0\epsilon\,\omega^2)\mathbf{E}=0. \qquad (13)$$

An identical wave equation holds for $\mathbf{H}$ if we start with (10a) and (10b).

Note that ∇^2 is a scalar operator and therefore operates on all components of the indicated field vector.

If we take $k_0=\omega/c=2\pi/\lambda$, the free-space wave number, and $n^2=\epsilon_r=\epsilon/\epsilon_0$, with n the refractive index and the medium's relative dielectric constant ϵ_r, we have, finally, (1), where ψ represents any field component (of $\mathbf{E}$ or $\mathbf{H}$) expressible as shown in (3) above. In the special geometries adopted to obtain tractable solutions and under appropriate field excitation, one or more of the orthogonal components of $\mathbf{E}$ and $\mathbf{H}$ may be zero.

3. The Step-Index Fiber

To fix ideas consider the step-index (SI) fiber in Fig. 2, an idealized model extremely useful in understanding propagation in fibers, including those with arbitrary index profiles. The step-index fiber comprises two regions: a *core* with refractive index n_1 constant for $0 \leq r \leq a$, and an outer, or *cladding*, region, where $r > a$, extending out indefinitely with n_2=constant refractive index. To confine the field as much as possible to the core, $n_1 > n_2$.

Ultimately we are interested in the functional dependence of β on the fiber dimensions, the light wavelength, and the refractive-index profile $n(r)$ with reference to index n_1 or n_2. How β varies with wavelength determines substantially the dispersive character of transmission within the defined fiber, and this property will be very useful in studying real fibers.

For any set of (input) fiber parameters, after application of the boundary conditions, there is a unique value, the *eigenvalue β*, that represents the solution at a given frequency or wavelength.

Normalized Variables

To carry on the analysis in a generalized form it is convenient to introduce reduced or dimensionless

31

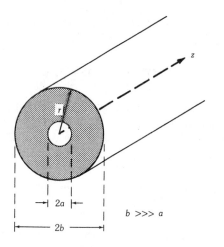

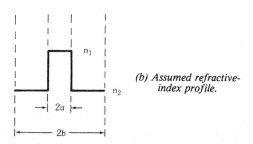

(a) Cross section of fiber.

(b) Assumed refractive-
index profile.

FIG. 2. The step-index fiber and its refractive-
index profile.

parameters [6, 7], readily defined for the ideal step-index fiber, and these are profitably utilized in the analysis of more complicated fibers. (Gloge's notation is now almost universally used in the literature of fiber wave propagation.)

If we let

$$V^2 = (k_0 a)^2 (n_1^2 - n_2^2) , \qquad (14)$$

then V is a *normalized frequency*. The coefficient $k_0 a$ expresses the core's "electrical" radius in radians.

Let

$$b = (\beta^2/k_0^2 - n_2^2)/(n_1^2 - n_2^2) . \qquad (15)$$

This is the *normalized eigenvalue* we seek, and its behavior with frequency will characterize the dispersion.

We define a *phase refractive index* $n_e = \beta/k_0$; thus

$$n_e^2 = b (n_1^2 - n_2^2) + n_2^2 , \qquad (16)$$

which will be useful in subsequent discussions on dispersion.

Guided propagation occurs when β lies between $k_0 n_1$ and $k_0 n_2$, so that we have $0 \leq b \leq 1$. Core propagation cutoff (for mode $m=0$) occurs when $b=0$ or $\beta = k_0 n_2$; just beyond this limit light propagates essentially in the cladding.

Subsidiary or related radial eigenparameters can then be defined:

In the core, instead of γ we choose a dimensionless parameter, u:

$$u^2 = a^2 (k_0^2 n_1^2 - \beta^2) . \qquad (17)$$

In the cladding,

$$w^2 = a^2 (\beta^2 - k_0^2 n_2^2) . \qquad (18)$$

with the result that

$$w^2 + u^2 = V^2 \qquad (19)$$

and

$$b = (w/V)^2 = 1 - (u/V)^2 . \qquad (20)$$

Thus at the $b=0$ cutoff, $u=V$ and $w=0$. At the other extreme, when $\beta = k_0 n_1$, $b=1$.

A determination of $b(V)$ sets the arguments of all the wave functions and defines the "waveguide" dispersion properties.

With these dimensionless parameters the argument of the Bessel functions J_0 and K_0 become $u\rho$ and $w\rho$ respectively, where $\rho = r/a$ is the radial distance from the center, normalized with respect to the core radius.

Gloge originated the description "weakly guiding fiber," one where the difference $n_2 - n_1$ is very small, leading to the advantageous property discussed for the third boundary condition. Symbolically,

$$\Delta \equiv (n_1^2 - n_2^2)/2 n_2^2 \lll 1$$

$$\simeq (n_1 - n_2)/n_2 = \Delta n/n_2 . \qquad (21)$$

This leads to often useful approximations for V and b:

$$V \simeq 2^{1/2} k_0 \, a \, n_2 \, \Delta \qquad (22)$$

and

$$b \simeq (\beta/k_0 - n_2)/n_2 \, \Delta \, . \qquad (23)$$

A typical value for Δ in a weakly guiding fiber is 0.003, and n_2 has a nominal value of about 1.5 for pure silica fiber substrate.

Note that V combines all the "input" quantities, wavelength, core radius, cladding index and index difference, into one fundamental parameter. If the wavelength is chosen to fall in a low-attenuation region or "window," then the choice of V determines substantially the core radius for given fiber composition.

Eigenexpression for the Step-Index Fiber

At the boundary between core and cladding, logarithmic derivatives of $R(r)$ are equated. Formally

$$\lim_{\epsilon \to 0} [R'(u \, \rho)/R(u \, \rho)]_{\rho=1-\epsilon} =$$

$$[R'(w \, \rho)/R(w \, \rho)]_{\rho=1+\epsilon} \, , \qquad (24)$$

leading to

$$u \, J_{m+1}(u)/J_m(u) = w \, K_{m+1}(w)/K_m(w) \, . \qquad (25a)$$

When $m=0$ this becomes

$$u \, J_1(u)/J_0(u) = w \, K_1(w)/K_0(w) \, . \qquad (25b)$$

Solution of (25a, 25b) yields a pair of eigenvalues u, w or u, V to be used in computing b. Solving this equation and the more complicated ones arising when there are more than two regions requires an efficient root-searching method that proceeds, for example, by choosing a value for b, using two guesses for V that give opposite sign for the residual in (25a) or (25b), and then adjusting the choices for V to converge on a single value (eigenvalue) that makes the residual as small as desired, with few iterations. An excellent method is Muller's [8], which utilizes a second-degree polynomial to fit three points near a root; from the quadratic's proper zero an improved estimate of the root is obtained.

Since the fields have to be continuous at $\rho=1$ we have

$$A \, J_0(u_s) = B \, K_0(w_s) \, . \qquad (26)$$

If we specify unit amplitude of the field on the axis, then $A=1$ and then

$$B = J_0(u_s)/K_0(w_s) \, , \qquad (27)$$

where the subscripts to u and w indicate that particular roots were found.

Power Distribution in the Step-Index Fiber

The power flow is found by integrating the cross product ("Poynting vector") of the transverse electric and magnetic fields over the fiber cross section; this yields an expression involving the sum $J_1^2(u) + J_0^2(u)$ in the core, and the difference $K_1^2(w) - K_0^2(w)$ in the cladding. Using the relation between A and B in the eigenexpression and considering the roles of u, V, and w, the relative fractions of light power in core and cladding may be calculated. Thus, with $p_1 = P_{core}/P_{total}$ for the core and $p_2 = 1 - p_1$ for the cladding, it can be shown that for the $m=0$ case,

$$p_1 = [1 + J_0^2(u)/J_1^2(u)] \, (w/V)^2 \, , \qquad (28)$$

$$p_2 = [1 - K_0^2(w)/K_1^2(w)] \, (u/V)^2 \, . \qquad (28b)$$

Note that throughout the discussion we have assumed refractive indexes that are constant, independent of wavelength, a property only of an idealized fiber.

Gambling and colleagues [9] show that the propagation factor b is directly equal to the ratio of the field $\psi(r)$ itself integrated over the core to the field $\psi(r)$ integrated over the entire fiber cross section. In the later discussion on dispersion (Section 3) we shall see that p_1 is also to be directly associated with the fiber's dispersive properties through the contribution of profile dispersion.

Examples for the Idealized Step-Index Fiber

The basic dispersion function $b(V)$ has the form shown in Fig. 3. Shown in Fig. 4 is the related function $u(V)$. Low-frequency cutoff occurs when $V=0$, and the slope near that cutoff is very gradual.

At high frequencies the growth of b tapers off, with the curvature reaching a maximum for a range of V. (This property, crucial to the determination of dispersion and dispersion compensation, will be discussed in detail in Section 3 under "Dispersive Properties of the Optical Fiber.") Note that while in the definition (15) b can reach unity when $\beta = kn_1$, in a given fiber b may not ever reach unity for finite V, because the appropriate u,V or w,V values are governed by the eigenexpression. Certain aspects of this behavior will be discussed presently.

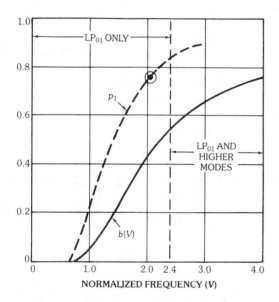

FIG. 3. Dispersion function $b(V)$ and the fraction of power p_1 carried by the core in a step-index fiber.

A corresponding curve for p_1 is also shown in Fig. 3. If a useful operating region is defined as one for which 75% of the power flows in the core, then the dot on the curve indicates the V value below which the core carries most of the light power.

Cutoffs: Modes, Frequencies, and Related Eigenparameters

Applying small-argument Bessel-function approximations to (25b) for near low-frequency cutoff ($w \to 0$) we obtain

$$\ln w = -J_0(u)/u\,J_1(u)$$

or

$$w = \exp\,[-J_0(u)/u\,J_1(u)] \simeq \exp\,[-J_0(V)/V\,J_1(V)]\,.$$

For example, when $V=0.2$, $w \simeq 10^{-22}$. Thus $b(V)$ will be extremely close to zero for a range of values of V somewhat above zero. (*Cf.* Marcuse [1])

As V becomes very large (high frequency), w asymptotically overtakes V, with b tending to unity. From the eigenexpression (25) u progresses much more slowly, tending toward a value u_{max} determined in the eigenexpression by $J_0(u_{max})=0$. The smallest or first of these zeros is $u_{max}=2.405$, so that below u_{max} the behavior of u can be gauged by introducing asymptotic values into (25) with the result

$$u = u_{max} \exp\,(-1/w) \qquad (u \le u_{max})$$

$$= 2.405\,\exp\,(-1/V)\,. \tag{29}$$

For $w \to 0$ and $V \to 0$, as we saw, the field in the fiber is no longer confined to the core, spreading well into the cladding. Then $K_0(w\rho)$ in the cladding behaves as

$$K_0(w\rho) \simeq (1/w\rho)\exp\,(-w\rho) \tag{30}$$

for a large argument. For a location where $w\rho=1$ but $\rho>1$ we find appreciable field even at distances far beyond the core. This agrees with what we found for p_2, the fraction of power carried in the cladding as $w \to 0$.

Had we taken $m \ne 0$, the next higher mode would be $m=1$ governed by the boundary condition at $\rho=1$ through eigenexpression (25a) as

$$u\,J_2(u)/J_1(u) = w\,K_2(w)/K_1(w)\,. \tag{31}$$

For $w \to 0$, the right-hand side approaches the value 2, and the left-hand side will equal 2 when $u=(u_1)_{co}=V=2.405$. This is the cutoff value of the radial eigenparameter u for the *next higher-order mode*, $m=1$. Thus guaranteed single-mode propagation with the lowest mode $m=0$ in the step-index fiber lies in the frequency range $0 \le V < 2.405$. Just beyond $V=2.405$, both modes $m=0$ and $m=1$ are present. For the $m=0$ mode itself, when $V=2.405$, $u=u_0=1.647$, and $b \simeq 0.53$. See Fig. 4.

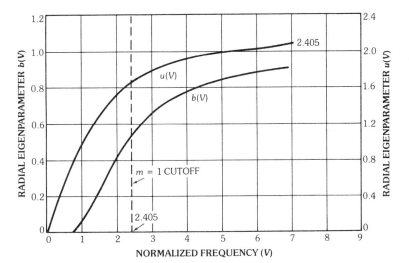

FIG. 4. Behavior of the radial eigenparameter $u(V)$ for a step-index fiber.

The parameter u does not reach 2.405 until $b \to 1$. At that extreme, $J_0(2.405)=0$ and the core field just reaches zero at the core-cladding interface. Thus, with pure single-mode propagation the core field is never zero at the core/cladding interface. This property would be difficult to establish on the basis of geometrical optics alone.

It is important not to confuse the conditions where the special value 2.405 appears (a) in V, the normalized frequency, (b) in u_0 for the wanted lowest-order mode, and (c) in u_1 for the unwanted higher-order mode.

For the step-index fiber the useful range $2 < V < 2.405$ is found to correspond, approximately, to $1.532 < u_0 < 1.647$. The fraction of power carried by the core shifts within this range, from about 26% at $V=2$ to about 16% at $V=2.405$. Operation very close to $V=2.405$, while achieving good core light confinement, would risk introducing the $m=1$ mode if fiber characteristics turn out not to be exactly uniform throughout a span.

The limited practical frequency range for the ideal step-index fiber as indicated by just these considerations has led investigators to look for a relatively simple algebraic expression for determining $u(V)$ or $w(V)$ rather than resorting to solving the eigenexpression. Rudolph and Neumann [10] discovered a linear approximation, asserted to fit within 0.1% for $w(V)$:

$$w = 1.1428\,V - 0.9960$$
$$(1.5 \le V \le 2.5). \qquad (32)$$

This approximation is widely employed since it covers the region of good light confinement and where, for the step-index fiber, the fiber dispersion is lowest.

Mode Nomenclature

Gloge [6, 7] introduced the designation LP_{01} for the $m=0$ mode to indicate that it is the lowest-order *linearly polarized* mode for the weakly guiding fiber, where field pairs E_x, H_y or E_y, H_x exist in the cross-sectional plane. Their normalized values in the core lie between 0 and 1 since they are uniquely described by $J_0(x)$ with $x \le 2.405$. When $m=1$ the modes arising are designated LP_{11}, LP_{12}, etc., depending on the Bessel functions involved. In the prior notation, often used interchangeably, LP_{01} is called HE_{11}, LP_{11} is called HE_{21}, and LP_{12} is called HE_{22}. As noted, the eigenexpression (25) is a simplified version of the exact eigenexpression, for which the approximation $n_1 \simeq n_2$ in certain of the sums and ratios is not taken. When $m>0$ in the exact case, there are two solutions for each value of m, giving slightly different eigenvalues. Detailed discussions on mode nomenclature can be found in Gloge [6], Marcuse [1], and Adams [2].

The basis for Gloge's nomenclature is summarized in Chart 1.

CHART 1. Summary of Gloge's Nomenclature
for Linearly Polarized Modes

Mode Nomenclature in Optical Fibers

Mode regions can be defined as a result of the interleaving of the zeros of Bessel functions $J_0(u)=0$ and $J_1(u)=0$. Let the zeros of $J_0(u)$ be called u_{01}, u_{02}, u_{03}, and so on, while those of $J_1(u)$ be called u_{11}, u_{12}, etc.

For $J_0(u)$ the zeros are at $u=2.405$, 5.520, 8.65, ...; and for $J_1(u)$, the values are $u=0$, 3.813, 7.016, ... These eigenvalues form the basis of Gloge's nomenclature LP_{mp} for linearly polarized modes.

With $m=0$ there is no zero in the core light field; with $m=1$ a zero would occur at some radius within the core. The ρ subscripts result from defining possible field configurations in the fiber, because two linear polarizations are possible, and there are also the possible zeros in the azimuthal functions $\cos\phi$ and $\sin\phi$.

Mode Designation	Sign of $J_0(u)$	Sign of $J_1(u)$	Range of u	Other Nomenclature
LP_{01}	+	+	$u_{11} \longleftrightarrow u_{01}$	HE_{11}
LP_{11}	−	+	$u_{01} \longleftrightarrow u_{12}$	$HE_{21}(TM_{01}, TE_{01})$
LP_{02}	−	−	$u_{12} \longleftrightarrow u_{02}$	HE_{12}
LP_{12}	+	−	$u_{02} \longleftrightarrow u_{13}$	$HE_{22}(TM_{02}, TE_{02})$
LP_{03}	+	+	$u_{13} \longleftrightarrow u_{03}$	HE_{13}

Dispersion Properties of the Optical Fiber

Three apparent dispersion sources in the mono-mode fiber can be nominally identified:

Waveguide dispersion, evident in the broadening of short pulses as they travel long distances in the fiber, is contained within the function $b(V)$, the normalized longitudinal propagation factor.

Material dispersion occurs because the refractive indexes in each concentric section of a real fiber depend on the wavelength. If the index is measured over a wide range of wavelengths, several "resonances" may occur.

Profile or composite dispersion, usually much smaller than the others, appears because the index differences Δ_{ij} between adjacent sections i and j are not constant with wavelength.

If a pulse or a group of adjacent frequencies, is transmitted over a fiber of length L, the *group transmission time* per unit fiber length (sometimes called the *group delay*) is given by

$$\tau = /v_g^{-1} = N_g/c, \tag{33}$$

where c is the in-vacuo velocity of light, v_g is the group velocity ($v_g < c$) and N_g is an effective group refractive index. Group time τ can be directly measured in a coiled fiber; it can be calculated by relating it to the variation of β with wavelength λ, or with radian frequency ω, or with wave number $k = 2\pi/\lambda$. See Stratton [4], for example. Taking $k = k_0$, then

$$\tau = d\beta/d\omega = (1/c)\, d\beta/dk$$

$$= (1/c)\, d(k\, n_e)/dk, \tag{34}$$

since by definition $\beta = k\, n_e$. Thus $N_g = d\beta/dk$.

The three sources of dispersion for the step-index fiber are determined from the function $b(V)$, obtained from the solution of the eigenexpression, and knowledge of the variation of the refractive indexes with wavelength, usually in the form of the Sellmeier three-term expansion, with constants A_j and λ_j, for the indexes [2]:

$$n_2 = 1 + \sum_j A_j \lambda^2/(\lambda^2 - \lambda_j^2). \tag{35}$$

The Sellmeier constants result from careful measurements of the refractive indexes of the constituent glasses. Representative values for the term coefficients A_j and the "resonant wavelengths" λ_j may be found in the literature [2, 11]. A few values are given in Table 1. For some doped glasses these are known with considerable precision. For other glasses effective values may be found by interpolation methods.

The full determination of the dispersion D requires that the derivative of τ be computed according to

$$D = d\tau/d\lambda = (1/c)\, dN_g/d\lambda \tag{36a}$$

$$= -(\lambda/c)\, d^2 n_e/d\lambda^2. \tag{36b}$$

A complete derivation of D can be found, for example, in South [12] and elsewhere. Here only the outline will be given in a compact form originated by Francois [13] which will be useful later when we discuss more complicated fibers.

Let

$$D = D_{wg} + D_m + D_{pro}. \tag{37}$$

The waveguide dispersion is

$$D_{wg} = -(V n_2 \Delta / \lambda c) d^2 (V b) / dV^2 . \qquad (38)$$

The material dispersion is

$$D_m = (\lambda / c) [n_2'' + (n_1'' - n_2'') \Gamma] , \qquad (39)$$

where the primes denote differentiation with respect to λ. (The term Γ will be discussed presently.)

The profile dispersion is

$$D_{pro} = [(\Delta n)' / 2 c] [V d^2 (V b) / dV^2$$

$$+ d(V b) / dV - b + 2 V d\Gamma / dV] . \qquad (40)$$

The quantity Γ differs from zero in a fiber whose index is not constant within the core. It is a mean value of the index ratio $n_1(r)/n_1(0)$ obtained with respect to the power density distribution in the fiber. For wave function $\psi(r)$

$$\Gamma \equiv \int_0^a f(r) \psi^2 r \, dr \Big/ \int_0^a \psi^2 r \, dr \qquad (41)$$

with

$$n^2(r) = n_2^2 + (n_1^2 - n_2^2) f(r) ,$$

and $f(r)$ is a shape factor, with $0 \leq f(r) \leq 1$ for $r \geq a$, and $f(r) = 0$ elsewhere.

In an ideal step-index fiber $f(r) = 1$ for $r \leq a$, $\Gamma = 1$, and

$$D_m = (\lambda / c) n_1'' , \qquad (42)$$

$$D_{pro} = [(\Delta n)' / 2 c]$$

$$\times [V d^2 (V b) / dV^2 - d(V b) / dV - b] . \qquad (43)$$

Thus, to determine the complete dispersion we require $b(V)$, its slope, and the rate of change of the slope. The computation is not as straightforward as it might seem. The function $b(V)$ is obtained by solving the governing eigenexpression for a number of frequencies V. Each solution bears with it accumulated errors arising from the need to evaluate (four) Bessel functions. An attempted numerical differentiation has to be done with care, because inherently the process propagates and possibly increases the errors. Representing $b(V)$ by a best-fit polynomial expression offers no relief, for such a polynomial is allowed to have rather arbitrary excursions in between the chosen values of the independent variable V as long as the fit fulfills a minimum deviation requirement. The resulting second derivative can have wide variations or swings in the interesting range of V.

As Hildebrand points out [14]: "...if we visualize a curve, representing an approximating function and oscillating about the curve representing the function to be approximated, we may anticipate the fact that, even though the deviation ... be small throughout an interval, still the *slopes* of the two curves representing them may differ quite appreciably. Furthermore, it is seen that round-off errors (or errors of observation) of alternating sign in consecutive ordinates could affect the calculation of the derivatives quite strongly if those ordinates were fairly closely spaced." Later he says, "In particular, numerical differentiation should be avoided whenever possible, particularly when the data are empirical and subject to errors of observation. When such a calculation must be made, it is desirable to *smooth* the data to a certain extent." We can regard the errors developed from the Bessel function calculations as essentially errors of observation. The problem in calculating D is compounded because two successive numerical differentiations are needed.

Compensation to Attain Minimum Dispersion

Since the waveguide dispersion term has a sign opposite to the others, under appropriate choices of materials and dimensions (core diameter and wavelength), waveguide dispersion can compensate most of the other dispersion, leading to a region of zero or minimum dispersion. The second derivative of Vb not only has a rather sharp maximum; it becomes negative for values of V around 2.5. The first derivative has an S-shaped behavior reaching its maximum for $2 < V < 2.5$, approximately. Material dispersion is a steadily rising function passing through zero at some wavelength which is offset from the resulting wavelength of minimum total dispersion. The object of fiber design is to place the wavelength of minimum total dispersion as close as possible to the wavelength region where fiber attenuation is a minimum, far from wavelengths of hydroxyl (OH) or other absorption peaks.

Fig. 5 shows the behavior of the first and second derivatives of Vb for step-index fiber. Fig. 6 shows typical material dispersions for core, cladding and guide [12]; the smallness of profile dispersion is evident. The $(V, \lambda)_{min}$ curves of Fig. 7 show how dopant content affects the dispersion minimum. The locus on the extreme left has the widest minimum-dispersion region.

Material dispersion for pure silica runs from -40 ps/nm·km at around $\lambda = 1.0$ μm, to $+20$ ps/nm·km at 1.6 μm, while for the step-index fiber the peak value of $V d^2(V b)/dV^2$ is [15] about 1.3 at $V \simeq 1.35$.

Because of the discussed difficulties with numerical differentiation a number of investigators have attempted to retain analytical methods as much as possible in calculating dispersion and the wavelength of minimum dispersion. Thus, instead of numerically differentiating a $b(V)$ curve, obtained after invoking the weakly guiding approximation, analytical expressions are sought for $b(V)$ and its derivatives, usually over the practical ranges of V. In the process numerical methods are limited to the less sensitive parts of the calculation. Pires and colleagues [16] used this approach and compared their results to those

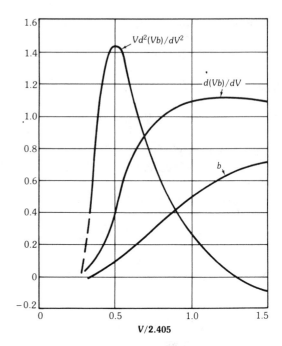

FIG. 5. First and second derivatives of $b(V)$ for step-index fiber, needed to calculate waveguide disperson.

obtained by Marcuse [17], Chang [11], and South [12]. They show excellent agreement with Marcuse's composite approach to calculation of the dispersion constituents and wavelength of minimum dispersion. Fig. 8 shows their dispersion comparisons with those of Marcuse and South, for a fiber with a core of 13.5 m/o germania and

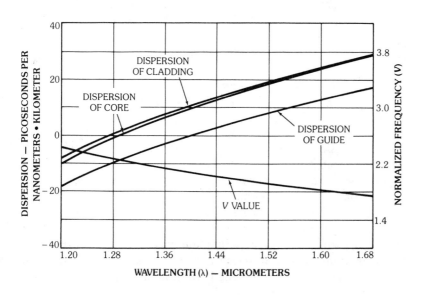

FIG. 6. Typical material dispersions for core and cladding. (*After South [12]. Acknowledgment is made of the prior publication of this material by the IEE*)

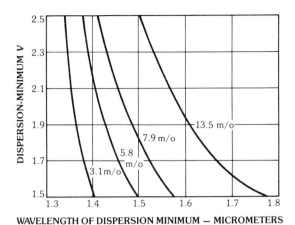

FIG. 7. Minimum dispersion loci for step-index fiber, for cores of differing dopant concentrations. (*After South [12]. Acknowledgment is made of the prior publication of this material by the IEE*)

86.5 m/o silica, and cladding of pure silica. Both Marcuse and South included the variation of the index difference with wavelength. Marcuse relied on an approximation for $u=u(V)$ and computation of derivatives from polynomial fits. South utilized approximations for $w(V)$ and then of $b(V)$ and its derivatives.

Finally, Table 1 shows Sellmeier constants for a variety of glasses including those considered by Pires and colleagues along with the index-difference range for wavelengths between 0.8 and 1.6 μm for the latter. This table of Sellmeier constants is taken from Adams [2]. (In his book, Adams' table has misplaced decimal points for λ_1 for the first two doped glasses shown. Table 1 below has corrected these.)

TABLE 1. Sellmeier Constants for Pure Silica and a Doped Glass. (*Reprinted and corrected from Adams [2], ©1981, by permission of John Wiley & Sons, Inc.*)

	A_1	A_2	A_3	λ_1	λ_2	λ_3
Pure Silica (annealed):	0.696 166 3	0.407 942 6	0.897 479 4	0.068 043	0.116 241 4	9.896 161
13.5 m/o GeO$_2$+86.5 m/o SiO$_2$	0.734 543 95	0.427 108 28	0.821 033 99	0.086 976 93	0.111 951 91	10.846 54
7.0 m/o GeO$_2$+93.0 m/o SiO$_2$	0.686 982 9	0.444 795 05	0.790 735 12	0.078 087 582	0.115 518 40	10.436 628
4.1 m/o GeO$_2$, 95.9 m/o SiO$_2$	0.686 717 749	0.434 815 05	0.896 565 82	0.072 675 189	0.115 143 51	10.002 398
9.1 m/o GeO$_2$, 7.7 m/o B$_2$O$_3$, 83.2 m/o SiO$_2$	0.723 938 84	0.411 295 41	0.792 920 34	0.085 826 532	0.107 052 60	9.377 295 9
4.03 m/o GeO$_2$, 9.7 m/o B$_2$O$_3$, 86.27 m/o SiO$_2$	0.704 204 20	0.412 894 13	0.952 382 53	0.067 974 973	0.121 477 38	9.643 621 9
0.1 m/o GeO$_2$, 5.4 m/o B$_2$O$_3$, 94.5 m/o SiO$_2$	0.696 813 88	0.408 651 77	0.893 740 39	0.070 555 513	0.117 656 60	9.875 480 1
13.5 m/o B$_2$O$_3$, 86.5 m/o SiO$_2$	0.707 246 22	0.394 126 16	0.633 019 29	0.080 478 054	0.109 257 92	7.890 806 3
13.5 m/o B$_2$O$_3$, 86.5 m/o SiO$_2$ (chilled)	0.676 268 34	0.422 131 13	0.583 397 70	0.076 053 015	0.113 296 18	7.848 609 4
3.1 m/o GeO$_2$, 96.9 m/o SiO$_2$	0.702 855 4	0.414 630 7	0.897 454 0	0.072 772 3	0.114 308 5	9.896 161
3.5 m/o GeO$_2$, 96.5 m/o SiO$_2$	0.704 203 8	0.416 003 2	0.907 404 9	0.051 441 5	0.129 160 0	9.896 156
5.8 m/o GeO$_2$, 94.2 m/o SiO$_2$	0.708 887 6	0.420 680 3	0.895 655 1	0.060 905 3	0.125 451 4	9.896 162
7.9 m/o GeO$_2$, 92.1 m/o SiO$_2$	0.713 682 4	0.425 480 7	0.896 422 6	0.061 716 7	0.127 081 4	9.896 161
3.0 m/o B$_2$O$_2$, 97.0 m/o SiO$_2$	0.693 540 8	0.405 297 7	0.911 143 2	0.071 702 1	0.125 639 6	9.896 154
3.5 m/o B$_2$O$_2$, 96.5 m/o SiO$_2$	0.692 964 2	0.404 746 8	0.915 406 4	0.060 484 3	0.123 960 9	9.896 152
3.3 m/o GeO$_2$, 9.2 m/o B$_2$O$_3$, 87.5 m/o SiO$_2$	0.695 880 7	0.407 658 8	0.940 109 3	0.066 565 4	0.121 142 2	9.896 140
2.2 m/o GeO$_2$, 3.3 m/o B$_2$O$_3$, 94.5 m/o SiO$_2$	0.699 339 0	0.411 126 9	0.903 527 5	0.061 748 2	0.124 240 4	9.896 158
Quenched SiO$_2$	0.696 750	0.408 218	0.890 815	0.069 066	0.115 662	9.900 559
13.5 m/o GeO$_2$, 86.5 m/o SiO$_2$	0.711 040	0.451 885	0.704 048	0.064 270	0.129 408	9.425 478
9.1 m/o P$_2$O$_5$, 90.9 m/o SiO$_2$	0.695 790	0.452 497	0.712 513	0.061 568	0.119 921	8.656 641
13.3 m/o B$_2$O$_3$, 86.7 m/o SiO$_2$	0.690 618	0.401 996	0.898 817	0.061 900	0.123 662	9.098 960
1.0 m/o F, 99.0 m/o SiO$_2$	0.691 116	0.399 166	0.890 423	0.068 227	0.116 460	9.993 707
16.9 m/o Na$_2$O, 32.5 m/o B$_2$O$_3$, 50.6 m/o SiO$_2$	0.796 468	0.497 614	0.358 924	0.094 359	0.093 386	5.999 652

Index fractional difference, Δ, for step-index fiber with silica substrate, for $0.8 \leq \lambda \leq 1.6$ μm:

Core with 13.5/86.5 mixture: 0.0147 5–0.0155   m/o = mole percent
Core with 7.0/93.0 mixture: 0.0073–0.0080

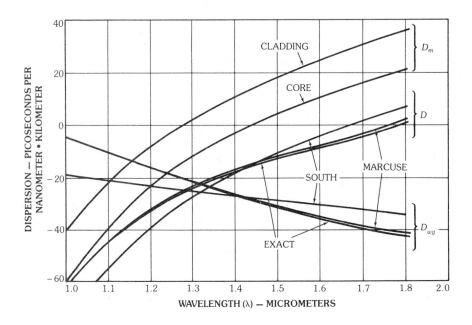

FIG. 8. Comparison of approximate and exact dispersion calculations. (*After Pires et al. [16],* © *1982 IEEE*)

Transmission Bandwidth for Fiber Driven by Wide-Spectrum Source

When the spectral spread of the driving light source (like an LED) is much larger than that of the impressed signaling, the resulting transmission bandwidth over a fiber length is determined by the baseband transfer functions, whose kernel is a representation of the source power spectral distribution. The essentials have been given by Cohen, Mammel, and Lumish [18]. The light source power distribution is generally well described by a Gaussian distribution about the operating wavelength λ_1:

$$P(x)=P_0 \exp\left(-\tfrac{1}{2}x^2\right) \quad \text{with } x=(\lambda-\lambda_1)/\delta\lambda. \quad (44)$$

where $\delta\lambda$ is the source rms spectral bandwidth.

In the frequency domain the baseband transfer function is

$$H_c(\nu)=\int P(\omega) \exp\left\{jz[\beta(\omega-\nu)-\beta(\omega)]\right\} d\omega, \quad (45)$$

where $\omega=2\pi f=2\pi c/\lambda$ is the radian light source frequency, ν is a radian baseband frequency, and z is the fiber length.

By expanding $\beta(\omega)$ in a Taylor series and retaining only terms through third order, the resulting integration yields

$$H_c(\nu,\lambda_1)=(1+j\eta)^{-1/2}\exp\left[-\zeta^2/2(1+j\eta)\right]. \quad (46)$$

Here $\zeta=z\nu\beta/T$, $\eta=\zeta/T$, and the constant $T=\lambda_1^2/(\pi c\,\delta\lambda)$.

The derivatives of β have been encountered earlier: The first derivative is the unit group delay τ, the second is proportional to the dispersion D of (36), and the third describes the curvature.

In the analysis λ_1 is taken to be positioned at the group delay minimum, and the fiber profile and core dimension have been selected to have D go through zero at the same location. For $\lambda_1=1.36\ \mu$m, Fig. 9 shows transfer function curves (absolute values) from [18] for source widths $\delta=4$, 20, and 100 nm. These curves show that while the bandwidth-distance product may be large right around λ_1, it can fall to much, much lower values either side of λ_1, thus limiting the useful bandwidth of an impressed signal for certain fiber lengths.

The single peak in Fig. 9 is characteristic of a step-index fiber and owes its origin to the single extremum (minimum) in the group delay behavior with wavelength. In fibers with an interior, second cladding, known as *depressed-index fibers* or *W-fibers*, to be discussed in the next section, there is a nonzero LP_{01} cutoff of the normalized frequency V, and, as a result, the group delay for

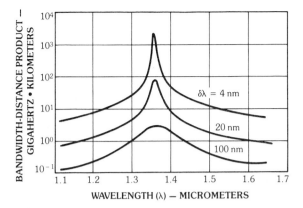

FIG. 9. Transmission bandwidth-distance product for fiber driven by light sources with spectral widths $\delta\lambda$ = 4, 20, 100 nm, centered at λ_1 = 1.36 nm. (*After Cohen, Mammel, and Lumish [18],* © *1981 IEEE*)

optimal designs [19] has two extrema: a minimum, usually around $\lambda \simeq 1.3$ μm, and a local maximum at around 1.55 μm, just before LP_{01} cutoff occurs. The bandwidth-distance product then has two peaks (around 1.3 and 1.55 μm) with a flat saddle in between, covering a very wide spectrum. This feature has made W-fibers very attractive for extremely wide band single-mode transmission.

4. Beyond the Step-Index Fiber: Radially Inhomogeneous Profiles

In this section we study the more complex situation of fibers having radially inhomogeneous refractive-index profiles.

Graded-Index Core Profile (Power Law)

Since real fibers though carefully produced do not have step refractive-index profiles, numerous investigators have used graded-index representations for the fiber core. Furthermore, real fibers usually have a central dip in the core's refractive-index profile as a result of dopant depletion during the various stages of fiber manufacture. These two features of fiber realism cause profound changes in fiber eigenvalues, in modal characteristics, and in selection of the operating wavelength for minimum dispersion.

An early representation of the core profile

[20, 21] employed a power law for function $\epsilon(r)=n^2(r)$ with an exponent α, such that $\alpha \to \infty$ describes the step-index fiber, while $\alpha=1$ represents a linear profile in the core. Of considerable interest was the parabolic profile, $\alpha=2$.

$$\epsilon(\rho)/\epsilon_1(\rho)=\begin{cases}1-2\,\Delta\,(\rho^{\alpha}), & 0\le\rho\le1, \\ 1-2\Delta, & \rho>1.\end{cases} \quad (47)$$

The definitions (14) through (18) for normalized eigenparameters b, u, V, w of the step-index fiber are retained, as they involve only the two end-point indexes n_1 and n_2, and Δ. This permits us to see how eigenvalues behave under shifts in the profile.

The wave equation for the radial wave function $R=R(\rho)$ in the cladding is the same as before, leading to $R(\rho)\simeq K_m(w\rho)$ while in the core the power law shows up in the coefficient of V^2:

$$R''+R'/\rho+[u^2-V^2\rho^{\alpha}-(m/\rho)^2]\,R=0. \quad (48)$$

With the boundary conditions the eigenexpression is very similar to (24) except that known functions cannot be employed for $R(\rho)$, for $\rho<1$. For the most general case

$$w\,K_{m+1}(w)/K_m(w)=m-[R'(\rho)/R(\rho)]_{\rho=1}. \quad (49)$$

The right-hand side cannot be expressed in terms of known functions for arbitrary values of α, except for the step-index fiber, $\alpha=\infty$. Solution in series for $R(\rho)$ in the core yields a set of recursion relations in u, V, and α in which the coefficients in the series expansion change character in the qth term, when $q=1+\alpha$. When $\alpha\to\infty$, the series expansion is just that for the ordinary Bessel function $J_0(u\rho)$ when $m=0$.

The results are depicted in Fig. 10, which is taken from Gambling and Matsumura [20]. The boundary labeled V_{co} is the locus of cutoff of the next higher-order mode, so the region to its right is a multimode region with poor dispersion properties. The $\alpha=\infty$ locus is the same as the step-index $u(V)$ curve shown in Fig. 4. The locus $V=V_{sm}$, nearly parallel to V_{co}, is that which leads to a minimum spot size, to be discussed in Section 6. The parabolic profile $\alpha=2$ has a mode $m=1$

cutoff $V=3.518$, to be compared with $V=2.405$ for the step-index fiber.

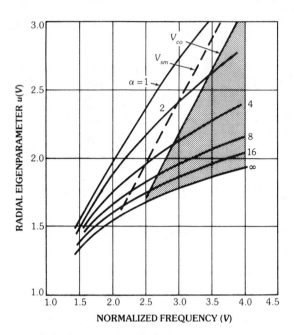

FIG. 10. Cutoff loci for fiber with "power-law" core. (*After Gambling and Matsumura [20]*)

Studies of the power confinement in the core with power-law profile by Paek, Peterson, and Carnevale [22] show that confinement is maximized for $\alpha=2$, the parabolic case, with around 40% improvement over the step-index case if the core radius is optimized for the selected value of α to make the fiber "dispersion-free."

For the power-law profiled core Krumbholz, Brinkmeyer, and Neuman [23] show that (28) can be generalized to show the connection between the core's fraction p_1 of the light power and the dispersion:

$$p_1=(1/2)\,[(1+2/\alpha)\,d(Vb)/dV+b\,(1-2/\alpha)]\,.\quad(50)$$

Setting $\alpha=2$ shows that the role of the derivative term is enhanced by a factor of two compared with the step-index fiber. For the step-index fiber ($\alpha\to\infty$), this expression for p_1 can be derived on the basis of conservation laws without requiring the weakly guiding approximation. (See [23].) In the special case for $\alpha=2$ the relation is simply

$$p_1=d(Vb)/dV\,,\quad(51)$$

which shows that the normalized group delay for any mode of this fiber equals the fraction of power flowing in the $\alpha=2$ core.

The light field from an exciting laser source is usually represented by a Gaussian field function and therefore the efficiency with which such a field couples to or is launched into the fiber is of considerable interest (Section 6.) Paek and colleagues, investigating the extent to which a Gaussian description fitted the actual field in the fiber for arbitrary α, found that the Gaussian approximation is a fair description of the core field for the step-index fiber, but a poor representation for the $\alpha=2$ fiber except close to the core-cladding interface. But for all values of α, the Gaussian description fails for the cladding region; this, of course, is evident from the exponentially decaying or evanescent character of the modified Bessel function $K_0(w\,r/a)$ in the cladding.

T-Fibers

There has been considerable interest in fibers with $\alpha=1$, sometimes called T or *triangular fibers*. These fibers have much higher second-order mode cutoff, given approximately as $V_{co}/2.405=(1+2/\alpha)^{1/2}$, or by scaling an estimate from Fig. 10. Furthermore, when fabricated, these fibers tend to have lower attenuation per unit length than do step-index fibers, possibly in part because of lower stress at the core-cladding interface. A zero-dispersion design may be achieved at $V/V_{co}=0.5$ to 0.6, which leads to a larger core diameter than with the step-index fiber, easing splicing in the field. These fibers also appear to show lower sensitivity of the dispersion compensation to slight variations in the design parameters [24]. With these T-fibers the resulting zero dispersion can be positioned at $\lambda=1550$ nm, exactly where the fiber attenuation is very low [25]. In one case [26] the specific attenuation was a minimum, 0.23 dB/km, essentially due to Rayleigh scattering.* Profile control requirements during fabrication are eased because of the graded core-

---

*In early 1985, median losses of 0.21 dB/km were reportedly achieved with long spans, and some fibers were produced at the theoretical minimum of 0.17 dB/km [27].

cladding boundary. And, with this type of fiber, bandwidth requirements on the driving laser can be less stringent.

Power-Law Core Profile With Central Dip

To achieve more realism and see how a dopant-depleted core center affects the propagation features of a two-zone fiber, Gambling, Matsumura, and Ragdale [21] employed an "inverse" power-law profile containing an extra coefficient δ to depict the magnitude of the central dip with respect to the core's nominal index maximum n_1:

$$\epsilon(\rho)/\epsilon_1 = \begin{cases} 1-2\delta\,(1-\rho)^\alpha, & 0\le\rho\le1 \\ \epsilon_1 = 1{=}2\varDelta, & \rho\ge1. \end{cases} \qquad (52)$$

This time, too, $\alpha\to\infty$ leads to a step-index fiber, while $\alpha=1$ produces a linear ramp in the core that starts at a depth $n_1(1-\delta)$ and rises to n_1 at the boundary $\rho=1$. Thus for all α there is the same jump from n_1 to n_2 at the core-cladding interface.

Again a power series expansion for the core wave function is used to solve the wave equation and the eigenvalues are obtained for the single-mode case, $m=0$, from our expression (24). The geometric parameter of interest now is $\gamma=\delta/\varDelta$, the fractional depth of dip in the core. A (u,V) plot (Fig. 11) exhibits the movement of $m=1$ mode cutoff as α decreases, and by picking a critical value of the normalized frequency and letting γ vary, it can be seen (Fig. 12) that u is made to increase, markedly so for low values of α.

For these radially inhomogeneous cases simple approximations are sought for the cutoff frequency V_{co} above which the $m=1$ mode will appear. Gambling and colleagues introduced the notion of weighting the refractive-index variation over the core to arrive at a "guidance factor," constituting an approximate correction to $(V_{co})_{SI}=2.405$.

Setting

$$G=\int_0^1 [(\epsilon(\rho)-\epsilon_2)/(\epsilon_1-\epsilon_2)]\,\rho\,d\rho\,, \qquad (53)$$

Then, because $G=1/2$ for the step-index fiber,

$$V_{co}=2.405\,(2\,G)^{-1/2}\,. \qquad (54)$$

The result is

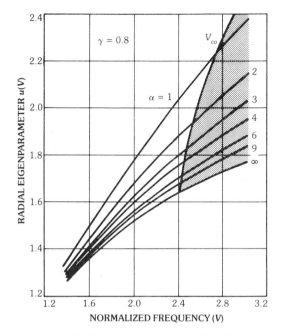

FIG. 11. Radial eigenparameter $u(V)$ behavior for fiber with inhomogeneous core having a central dip. (*After Gambling, Matsumura, and Ragdale [21]*)

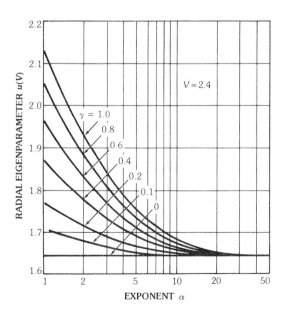

FIG. 12. Effect of central dip depth on the behavior of the radial eigenparameter $u(V)$ in a fiber with inhomogeneous core. (*After Gambling, Matsumura, and Ragdale [21]*)

$$(V_{co}/2.405)^2=(\alpha+1)(\alpha+2)/[\alpha^2+3\alpha+2(1-\gamma)]. \quad (55)$$

This approximation method for estimating $m=1$ cutoff is quite general. The error was shown to be within 5% for the cases investigated, covering the whole range of α. Studies of the field distribution within the core showed that when $\alpha>20$ the fiber acts as a step-index fiber.

Note that the integrand in the integral definition (53) is just the shape factor $f(r)$ utilized in the dispersion discussion, (41).

Generalization in the Theory of Moments

An effective normalized frequency can be defined for the fiber using G:

$$\bar{V}^2=2k^2(\epsilon_1-\epsilon_2)G. \quad (56)$$

This leads to $\bar{V}=2.405$ if the profile is that of a step-index fiber.

Hussey and Pask [28] have shown that this formulation of G is a special case in a theory of profile moments, where instead of G the moments Ω_j are defined through

$$\Omega_j=\int_0^1 s(\rho)\rho^{1+j}d\rho, \quad (57)$$

where $s(\rho)$, a reduced shape factor, is equivalent to the integrand in (53) above; it differs from zero only in the core.

Only the first three even moments ($j=0, 2, 4$) are needed for computation of all the essential properties of a fiber with arbitrary core profile: cutoffs of the modes above LP_{01}, dispersion properties, spot size (see Section 6), and an interpretation in terms of an equivalent step-index fiber. By inspection of the effects of rapid profile changes (index jumps, slopes, proximity of the core-cladding interface, etc.) on the moments, the likely influence on fiber properties can be assessed. The shape influence is concisely described through the profile-type indicator x:

$$x=1-2\bar{\Omega}_2/3(\bar{\Omega}_4)^{1/2}. \quad (58)$$

The overbars indicate normalization with respect to Ω_0. For step-index fibers $x=0$, while when "shoulders" occur in the core, $x>0$. Dip profiles produce $x<0$. The equivalences to step-index work out thus:

For $x\leq0$, the cladding eigenparameter w depends on $\bar{\Omega}_2$,

$$(w/w_{SI})^2\simeq1/2\bar{\Omega}_2, \quad (59)$$

where w_{SI} is taken via $w_{SI}=w_{SI}(V)$

When $x>0$ the right-hand side of (57) becomes

$$[1+xu_{SI}^2/4J_1^2(u_{SI})]/(3)^{1/2}\bar{\Omega}_4, \quad (60)$$

where $u_{SI}=u_{SI}(V)$ is the step-index fiber eigenparameter in the core. These relations are used to determine modal properties as well as the behavior of $b(V)$ and its derivatives.

Interestingly, applying the Hankel transform [29] to the shape function $s(\rho)$ leads to

$$S(\kappa)=\int J_0(\kappa\rho)s(\rho)\rho\,d\rho \quad (61)$$

$$=\sum_m[(-1)^m/(m!)^2][\Omega_{2m}(\kappa/2)^{2m}],$$

which is a directly measurable quantity. Here κ is a normalized spatial "frequency" parameter related to the minima in the diffraction pattern obtained by illuminating the fiber at normal incidence with a laser beam. (See [30].) This expansion shows why only the even moments play the important role in fiber characterization.

The theory of moments clearly shows that single-mode fibers have smoothly shaped core fields, much like Gaussian in form, that do not "see" the finer details of the profile but only its integrated properties. Fig. 13 (from Hussey and Pask) illustrates the effect that several different core-profile artifacts can have on the normalized dispersion, here defined through the second derivative of $b(V)/2\Omega_0$, in the range corresponding to pure LP_{01} mode, where the equivalent normalized frequency $\bar{V}<2.405$. In this figure the reference moment is $\bar{\Omega}_0=5/12$. Note that the largest effects occur around $\bar{V}=1.1$ to 1.2, while at higher frequencies the differences are much less pronounced. Higher-order moments are enhanced by abruptness near the core-cladding interface. Central dips principally affect the lower-order moments.

Whenever more than one mode is present, as in

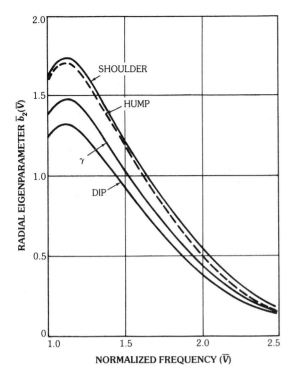

FIG. 13. Effect of refractive-index profile's shape on (normalized) dispersion of the LP$_{01}$ mode, as determined from the theory of moments by Hussey and Pask [28]. (*Acknowledgment is made of the prior publication of this material by the IEE*)

multimode fibers or in single-mode fibers operated just above LP$_{02}$ cutoff, the composite-core field is no longer smoothly shaped and details of the core profile will profoundly affect propagation properties, especially dispersion.

5. Beyond the Step-Index Fiber: The W-Fiber

In the design of step-index and graded-indexd fibers there is a rather narrow range of control over dispersion. The search for a refractive-index profile that provides more flexibility in compensating dispersion and in setting the minimum-dispersion wavelength reached a remarkable stage with the development of the depressed-index-cladding (DIC) or W-fiber. An idealization of a depressed-index-cladding profile is shown in Fig. 14; the similarity to the shape of a W is evident.

The inner cladding functions qualitatively as a "shield," to decrease the amount of light power in the outer cladding, thus improving guidance in the core. The appearance of two additional design parameters, the index contrast ratio between core and inner cladding, and the diameter ratio of the inner zones, provide the needed flexibility.

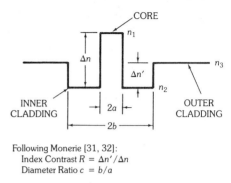

Following Monerie [31, 32]:
Index Contrast $R = \Delta n' / \Delta n$
Diameter Ratio $c = b/a$

FIG. 14. Refractive-index profile schematic for a W-fiber, showing inner cladding with depressed index.

The earliest treatment of the W-fiber with stepped indexes was given by Kwakami and Nishida [33] in 1974, but only relatively recently was this advance fully appreciated and were comprehensive measurements made on model fibers. The development here follows Monerie [31, 32], and will be extended to investigate the effect of a central dip on the W-fiber.

Fundamental W-Fiber Normalized Parameters

The fundamental normalized parameters are the same as with the step-index fiber except that we have to redefine subscripts because now there are three refractive index zones, n_1, n_2, and n_3, starting from the core. Thus

$$V^2 = (k\,a)^2\,(n_1^2 - n_3^2) \qquad (62)$$

and

$$u^2 = a^2\,(k^2\,n_1^2 - \beta^2) \qquad (63)$$

for the core region ($\rho \le 1$).

The inner cladding region ($1 < \rho \le b/a$) has its own radial eigenparameter s, via

$$s^2 = b^2 (\beta^2 - k^2 n_2^2) . \tag{64}$$

And

$$w^2 = b^2 (\beta^2 - k^2 n_3^2) \tag{65}$$

for the outer cladding, where $\rho > b/a$. Here the unsubscripted b is used to indicate the radius of the boundary between inner and outer cladding.

The normalized propagation factor is defined as before but renamed b_{13} to indicate interest in propagation between the index limits of core and *outer* cladding. Thus the normalized propagation factor in (15) becomes

$$b_{13} = (\beta^2/k^2 - n_3^2)/(n_1^2 - n_3^2) . \tag{66}$$

The relations between parameters are as follows:

$$V^2 = u^2 + w^2 c^2 \tag{67}$$

and

$$b_{13} = (w/c V)^2 = 1 - (u/V)^2 . \tag{68}$$

Let $\Delta n = n_1 - n_3$, $\Delta n' = n_2 - n_3$, with $\Delta n < 0$. The contrast ratio (used by Monerie) is defined as

$$R = \Delta n'/\Delta n , \tag{69}$$

which is negative ($R < 0$) for depressed-index-cladding fibers.

[Some workers use the notation $\sigma = -\Delta n'/(\Delta n + \Delta n')$, a positive number instead of R. The relationship is $R = -\sigma/(1+\sigma)$ so that the frequent choice $R = -0.5$ corresponds to $\sigma = 1.0$, where the index depression in inner cladding just equals the index elevation of the core with respect to the outer cladding.]

Then

$$s \simeq (V/c) [b_{13} - |R|/(1+R)^{1/2}] . \tag{70}$$

Here $c = 2a/2b$ is the diameter ratio.

The forms that $R(r)$ must take in the several zones are

$$A\, J_m(u\, r/a) , \qquad\qquad 0 \le r \le a , \tag{71}$$

$$B\, I_m(s\, r/b) + C\, K_m(s\, r/b) , \qquad a \le r \le b, \tag{72}$$

$$D\, K_m(w\, r/b) , \qquad\qquad b \le r . \tag{73}$$

Applying boundary conditions at $r = a$ and $r = b$ leads to a 4-by-4 determinant, set equal to zero for a solution, to yield the wanted eigenvalues and the field coefficients A, B, C, D.

The resulting eigenexpression may be compactly written by introducing "compressed" Bessel functions in the form

$$\tilde{Z}_m(x) = Z_m(x)/x\, Z_{m+1}(x) \tag{74}$$

for any Bessel function $Z_m(x)$.

The eigenexpression may then be cast in the following computationally useful form:

$$1/J_m(u) = (1 - Q)/[Q\, \tilde{I}_m(s\, c) + \tilde{K}_m(s\, c)] , \tag{75}$$

where the ratio $Q = q_1/q_2$, with

$$q_1 = K_{m+1}(s)\, I_{m+1}(s\, c)\, [\tilde{K}_m(V) - \tilde{K}_m(s)]$$

and

$$q_2 = K_{m+1}(s\, c)\, I_{m+1}(s)\, [\tilde{K}_m(V) + \tilde{I}_m(s)] .$$

The form (75) has the core specification all on the left-hand side. The right-hand side is seen to depend strongly on the specifications of the width of the inner cladding and its radial eigenparameter s, and on the normalized frequency V. The right-hand side is the replacement for the right-hand side of (25a) for the step-index fiber.

Fig. 15 shows the $b_{13}(V)$ curve for particular choices of R and c. Notice that the W-fiber has a nonzero cutoff for the fundamental $m = 0$ mode. This cutoff $V = V_{co}$ is a quasi-linear function of R for a large range in R, as shown in Fig. 16. Monerie found the dependence of this LP_{01} cutoff frequency to be approximately linear for a wide range of the contrast ratio R. In particular, for a large-diameter ratio ($c^{-1} \to \infty$),

$$V_{co} = 1.075\,(1 - R) . \tag{76}$$

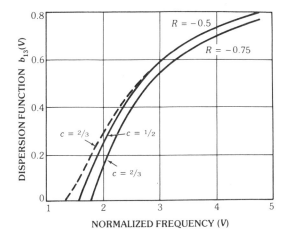

FIG. 15. Behavior of the dispersion function $b_{13}(V)$ of the LP$_{01}$ fundamental mode for a W-fiber with $R = -0.5$ and $R = -0.75$. (*After Monerie [31], © 1982 IEEE*)

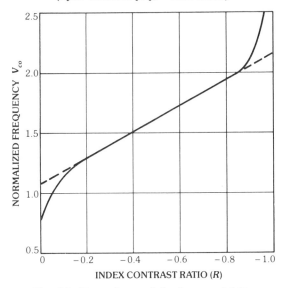

FIG. 16. Dependence of the (nonzero) LP$_{01}$ cutoff in a W-fiber on the index contrast ratio R for a very large inner-cladding width. (*After Monerie [31], © 1982 IEEE*)

Monerie examined the relation between the normalized propagation parameter b_{13} and the one defined with respect to the inner cladding alone, b_{12}, and by using Rudolf and Neumann's approximation [(32) above], arrived at a more general expression for the LP$_{01}$ cutoff:

$$V_{co} = V_{co/\infty}\{1 - 2.008\,(1+R)\,\exp\,[h(c,R)]\} \quad (77)$$

where

$$h(c,R) = [-1.992\,(1/c-1)|R|^{1/2}]/(1.1428 - |R|^{1/2}).$$

Both these formulas are limited to the linear domain represented by (76), $-0.85 < R < -0.20$. This more than covers the range of practically useful W-fibers.

Fig. 17 shows the cutoffs for the LP$_{01}$ and LP$_{11}$ modes for several index contrasts as a function of b/a. When $b/a=1$, the W-fiber is reduced to a step-index fiber, so $V_{co}=2.405$ for LP$_{11}$; this value does not change markedly until R assumes rather large (negative) values and b/a is around 1.5.

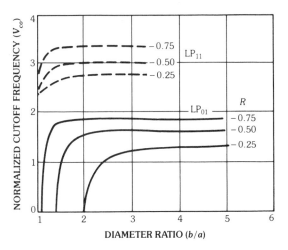

FIG. 17. Comparison of LP$_{01}$ and LP$_{11}$ mode cutoffs for the W-fiber. (*After Monerie [31], © 1982 IEEE*)

Fig. 18 shows the behavior of $V\,d^2(V\,b_{13})/dV^2$ for $b/a=1.5$ as a function of V. Note that when $R=-0.75$ the region of large second derivative is extremely narrow, while for $R=-0.50$ the peak is lower but the region is broader, a more practical choice for delay compensation that is less sensitive to small changes in parameters. The curves in Fig. 18 were obtained from the slopes in Fig. 15. Figs. 19 and 20 illustrate two cases of optimized W-fibers, showing how two zero crossings occur in the dispersion curve, leading to an optimum choice of fiber diameter at the mid-wavelength in the low-loss spectral window. The predicted curve

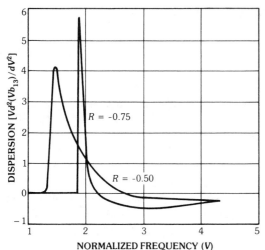

FIG. 18. The principal contribution $V d^2(V b_{13})/dV^2$ for a W-fiber. (*After Monerie [31], © 1982 IEEE*)

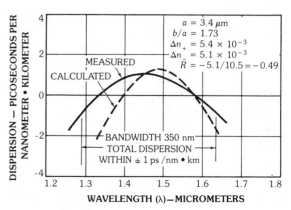

FIG. 19. Total dispersion design-optimized in a W-fiber to occur in the fiber's low-loss spectral window. (*After Bernard et al. [34]*)

in Fig. 19 is based on measurements of the fiber's preform [34, 35].

The W-fiber will actually function below the nonzero mathematical cutoff, that is, when $V < V_{co}$ for the LP_{01} mode. In that case $\beta < k n_3$, and this leads [see (65)] to an imaginary value for w, the eigenparameter in the outer cladding. This "converts" $K_0(w r/b)$ to $(\pi/2j)[J_0(w r/b) - j Y_0 (w r/b)]$, a quasi-oscillatory radiative-type field,

whose envelope for very large argument behaves as $(w r/b)^{-1/2}$ rather than evanescently as $[\exp(-w r/b)](w r/b)^{-1/2}$. This radiative field is largest just below the LP_{01} cutoff. An approximate boundary between the radiative or "leaky" mode (when $\beta < k n_3$) and the guiding mode ($\beta > k n_3$) can be determined by letting $V \to 0$ in the eigenexpression itself. The result is defined by a relation between b/a and R:

$$b/a = |R|^{-1/2}, \quad \text{(W-fiber: } R < 0). \qquad (78)$$

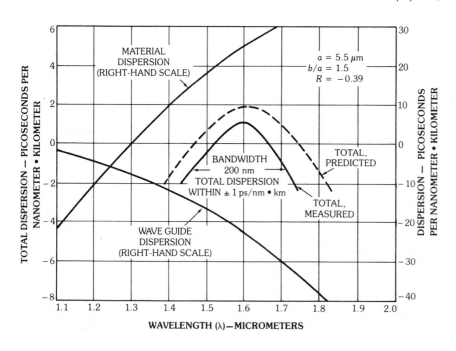

FIG. 20. Material, waveguide, and total dispersion in an experimental W-fiber (*After Jang et al. [35]*)

Thus, to ensure a guided LP_{01} mode, a necessary condition would be that

$$c > |R|^{1/2}$$

as shown in Fig. 21. Kwakami and Nishida [33] give a more exact statement of the condition for ensured guided mode (needed when c is larger than about 0.3) as

$$|R| < c^2 (1-c^2). \qquad (79)$$

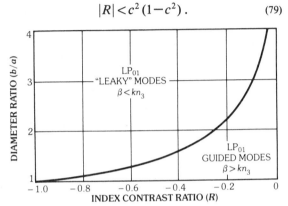

FIG. 21. Delineation of the possible leaky- and guided-mode regimes in W-fibers. (*After Monerie [31, 32],* © *1982 IEEE and acknowledgment is made of the prior publication of this material by the IEE*)

As (77) indicates, V_{co} moves away more and more from zero as $|R|$ increases and as the width of the inner cladding is made larger. Consequently, the conditions $0 \leq V < V_{co}$ and (79) together are necessary and sufficient to establish a leaky fundamental mode; this mode could propagate and still encounter low losses along the fiber *if* the width of the inner-cladding region is sufficiently large.

Limits of Monomode Operation in the W-Fiber

With the step-index fiber the monomode range of operation is readily determined because, since the LP_{01} mode has a zero-frequency cutoff, the upper limit $V = 2.405$ is set by the eigenexpression (25a) when $m = 1$. As we have seen above, the W-fiber has a nonzero cutoff that depends on R and $c = a/b$.

To determine the onset of the LP_{11} mode, the eigenexpression (75) should be solved for $m = 1$,

but because of the nonzero LP_{01} cutoff the upper value of V will, in contrast to the step-index fiber case, vary upwards as the LP_{01} cutoff increases. Hussey and deFornel [36] have studied this variation by invoking the theory of moments described above in Section 3 under "Power-Law Core Profile with Control Dip." They used the zero moment Ω_0 [see (57)] derived from integrating the shape factor $s(\rho)$ over the fiber cross section; Hussey and deFornel find that Ω_0 consolidates the R and c dependence:

$$\Omega_0 = c^2 (1 + R/c^2)/2. \qquad (80)$$

The W-fiber parameter limits to be considered are then

$$0 < c < 1 \qquad (81)$$

and

$$-1 < R < 0. \qquad (82)$$

The resulting limits for Ω_0 are

$$0.5 \geq \Omega_0 \geq -0.5. \qquad (83)$$

By examining the limiting cases Hussey and deFornel show that for all practical ranges of R and c (given by $\Omega_0 < -0.07$) not easily treated by using an equivalent–step-index approach (see Section 6), a "good working expression" for the intermodal ($m = 0$ to $m = 1$) cutoff difference is

$$\delta V_{co} = (V_{11} - V_{01})_{co} \approx 1.427. \qquad (84)$$

This difference is useful only if either one of the two cutoffs is known. The second-order mode's cutoff may be easier to determine experimentally, perhaps by attenuation measurements as a function of V, so that (84) would aid in determining the approximate location of the LP_{01} cutoff. Note, though, that in practical cases of fibers laid with bends the *observed* LP_{11} cutoff is sensitive to bend radius and fiber length, the effect of which is to permit monomode working at shorter wavelengths for the longer lengths of fiber [37].

Fraction of Power in the Core for W-Fibers

Calculations for a particular W-fiber with $R=-0.67$, $c=0.71$, and core radius $a=5$ μm in [38] showed the sensitivity of the fraction of core power p_1 to the choice of frequency V:

V	p_1
1.61	0.09
1.62	0.13
1.73	0.75
1.88	0.9813

Leakage Loss in W-Fibers Due to Bending

Bending a cylindrical fiber introduces distortion of the refractive-index profile with respect to a local radial direction. In such a guide the effective refractive index n' is deformed from the initial profile in a dependence on the radius of curvature R_b according to

$$n'=n(r)\,[1+(r\cos\phi)/R_b]\,, \qquad (85)$$

where $n(r)$ is the original profile, and ϕ is an azimuthal reference direction such that $\phi=0$ is in the plane of the bend. The condition $n'=\beta/k$ determines when light will escape from the guiding part of the fiber into the outer substrate. This will occur at the inner-outer cladding boundary $(r=b)$ in the W-fiber whenever

$$R_b/b=n_0/(\beta/k-n_0)\,, \qquad (86)$$

where n_0 is the substrate (silica) index. If the right-hand side reaches about 1000, as it could when β/k is close to n_0, the guided light will escape if the radius of curvature is 1000 times the radius at the boundary between the claddings. Leakage is more likely in W-fibers with a wide region $(a\le r\le b,\ b>5a)$ of depressed index where bends of 3 to 10 cm would be critical. Estimates of likely leakage losses in these types of W-fibers have been given by Cohen, Marcuse, and Mammel [38] for the interesting case of a fiber with contrast ratio $R=-0.5$, $a=0.375$ μm, and $c=b/a$ in the range from 5 to 9. The losses can reach 1 to 2 dB/km at wavelengths of around 1.5 μm.

W-Fiber With Shaped Profile in the Core

In the W-fiber we may replace the core of constant refractive index with one having a power-law variation of index of the form of (47). A solution in series is involved along with a nesting of Bessel functions that governs behavior of the fields in the inner and outer cladding. Of principal interest for this case is how the nonzero cutoff frequency of the fundamental (LP$_{01}$) mode is shifted as a function of the exponent α in the core, Fig. 22 shows V_{co} as a function of α computed for a W-fiber with index contrast ratio $R=-0.5$ (referred to the peak index in the core) and a cladding-core diameter ratio of 3:1. The figure shows that very high values of α (greater than 15 to 20) result in behavior essentially the same as for a step-index W-fiber. On the other hand, low values of α, particularly the parabolic profile ($\alpha=2$) move the LP$_{01}$ cutoff to a much higher frequency (lower wavelength).

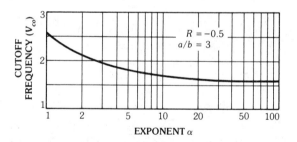

FIG. 22. The LP$_{01}$ cutoff behavior in W-fibers having a "power-law" core, as a function of the exponent α.

With $\alpha=1$ the core has a triangular profile. The combination of triangular profile and depressed-index inner cladding has been used to great advantage [5] to produce a monomode fiber with a loss of about 54 dB in a 227-km length, with LP$_{11}$ cutoff at 1.1 μm, with average zero chromatic dispersion at 1.544 μm, and highly resistant to possible bending and cabling losses that would occur in a simple T-fiber (Section 3 under "T-Fibers") for which LP$_{11}$ cutoff occurs at around 0.85 μm.

6. Extensions of the W-Fiber

This section covers two profile variations of the W-fiber: the W-fiber with a central index dip and multiply clad fibers.

W-Fiber With Central Dip

The foregoing application of Bessel functions can be extended to estimate the effect of a central index dip in the W-fiber. The core is partitioned to have a dip with index difference δn (with same sign as Δn) from the rest of the core. (See Fig. 23.) Core, inner cladding, and outer cladding bear indexes n_1, n_2, and n_3, respectively, as before. The dip's index is designated n_d. The normalized frequency and propagation parameters retain their definitions. The dip occupies a fraction μ of the core, and the fractional contrast ratio applying to the dip is

$$\gamma = \delta n / \Delta n .$$

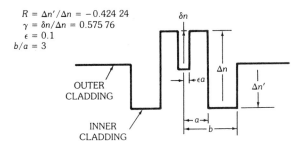

$R = \Delta n' / \Delta n = -0.424\ 24$
$\gamma = \delta n / \Delta n = 0.575\ 76$
$\epsilon = 0.1$
$b/a = 3$

FIG. 23. Cross section of step-type W-fiber with rectangular dip, showing the defined dimensionless parameters with specific values for calculation.

Successively, from dip outwards, the wave functions are

$$A\, J_m(\sigma\,\rho) , \qquad (87a)$$

$$B\, J_m(u\,\rho) + C\, Y_m(u\,\rho) , \qquad (87b)$$

$$D\, I_m(s\,\rho) + E\, K_m(s\,\rho) , \qquad (87c)$$

$$F\, K_m(w\,\rho) . \qquad (87d)$$

The new eigenparameter σ for the dip region is defined via

$$\sigma^2 = u^2 - \sigma\, V^2/(1+R) . \qquad (88)$$

There are now three eigenexpressions corresponding to the three interfaces, leading to a 6-by-6 determinantal equation to be solved. By using compressed Bessel functions as defined in (74) and rearranging to place only functions with dip-related arguments on the left-hand side, of type $J_m(\sigma\,\mu)$, root-searching is simplified.

As with the "pure" W-fiber, there is a nonzero LP_{01} cutoff. With certain choices of the μ, γ, R, and c, the dip eigenparameter σ becomes imaginary for V beyond some critical value, and $J_0(\sigma p)$ goes over to $I_0(\sigma p)$. The on-axis field ψ is actually no different since $I_0(0) = J_0(0) = 1$, but its coefficient A will be changed in relation to the other coefficients because of the boundary condition at $\rho = \mu$. The field in the dip will grow with ρ as the dip-core interface is approached, instead of diminishing; the dip region is acting like a cladding.

Fig. 24 compares the $b_{13}(V)$ behavior for a W-fiber with and without a dip. The dip shifts the curve almost (but not quite) uniformly from left to right, to lower wavelengths. This indicates that if design intends to place the minimum-dispersion region at longer wavelengths, the unexpected appearance of a dip during fabrication would shift the dispersion's minimum and possibly increase it.

Francois examined this effect in the W-fiber by identifying sensitivity factors on which to base tolerance requirements [39]. These factors are just the partial derivatives of the dispersion D with respect to the important design parameters of the fiber: the two index differences, the core radius a, the inner cladding's radius b, and the wavelength. As an example, the sensitivity to wavelength is near zero for index contrasts R between -0.14 and -0.38 but takes on large negative values for $R < -0.4$.

Both Francois and Kwakami and Nishida [33] caution that W-fiber designs with high-index-contrast ratios have a high fundamental-mode (nonzero) cutoff, as we see from (77), and this heightens fiber sensitivity to any change in design parameters. See Section 7, Fig. 29.

The central dip does have a pronounced effect on the LP_{01} cutoff for the higher values of R. Fig.

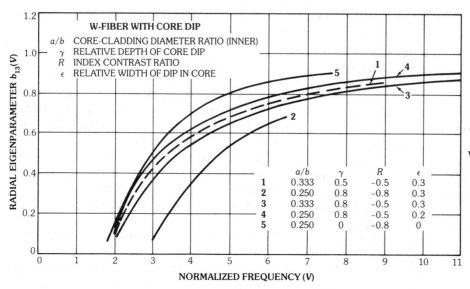

FIG. 24. Fundamental-mode dispersion function $b = b_{13}(V)$ for W-fibers, showing effect of rectangular central dips.

25 shows a plot of LP_{01} cutoff versus R for a fiber with a 50% central dip and a occupying 20% of the core, and it is compared with Monerie's approximation, (77) above.

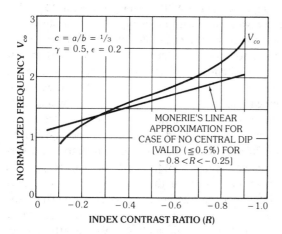

FIG. 25. The LP_{01} cutoff frequency as a function of index contrast ratio R for W-fiber with central dip, compared with the same fiber with no dip.

Multiply Clad Fibers

Success with the W-fiber in confining light within the core and in "tuning" the dispersion compensation to the region of lowest intrinsic fiber loss has been followed by studies of fibers with multiple claddings. One of these fibers, an extension of the W-fiber, has a region of relatively high index,

but not as high as the core index, interposed between the W-fiber's inner and outer claddings. The field in the outer cladding is much reduced, compared with the W-fiber, reducing the sensitivity to fiber curvature. Waveguide dispersion $V d^2(V b)/dV^2$ for a typical multiply clad (MC) fiber will be a maximum at a much higher normalized frequency (lower wavelength) than for the W-fiber on which it is built. The advantageous result is that dispersion compensation can be achieved over a very wide wavelength range.

The most striking feature, however, is the very small separation between the multiply clad fiber's LP_{01} and LP_{02} mode $b(V)$ characteristics, and, further, the cutoff of the LP_{02} mode occurs at a low value of V, close to the fundamental-mode cutoff of the constituent W-fiber. Because the resulting region for fundamental-mode operation is narrow (Fig. 26), and because in mass-produced fiber nonuniformities (affecting the frequency range where dispersion compensation would be optimized) can lead to mode coupling or mode conversion, deployment of multiply clad fibers may not be very favorable [13].

7. Spot Size and Equivalent Step-Index Fiber

An important consideration in selecting fiber parameters (e.g., core radius, wavelength) is how

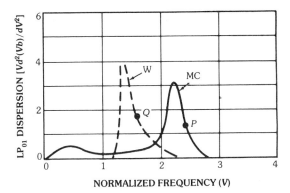

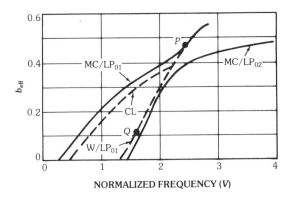

(a) Dispersion properties of MC fiber compared to its W-fiber components. The MC fiber supports two modes in the region just beyond the disperson peak (point P). Compare with point Q appropriate to the W-fiber. CL applies to the cladding.

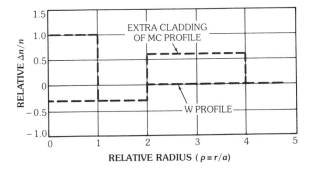

(b) Profile structures.

FIG. 26. Dispersion curves pertaining to a multiply clad (MC) fiber, showing proximity of branches of the LP_{01} and LP_{02} modes. (*After Francois [13]. Acknowledgment is made of prior publication of this material by the IEE*)

these influence the launching efficiency of the fields of a laser light source into the fiber. The spatial distributions from laser resonator sources have been studied and described by Kogelnik and Li [40]. For driving monomode fibers one deals with TEM_{00} laser resonators.

The nominal half-width of the input excitation field is known as the *spot size*, designated ω; its definition derives from the intensity distribution transverse to the forward propagation axis. As Kogelnik and Li point out, a Gaussian intensity profile, though not the only allowed mode, nevertheless characterizes the fundamental mode of the laser resonator. For optimum coupling, the spot size radius should approximate the core radius of the fiber.

If $E(r)$ is the fiber's cross-sectional far field, then the *launching efficiency* η of a source field may be defined in more than one way. A commonly used definition due to Marcuse [41] is, if P_0 is the reference integrated intensity,

$$\eta = (1/P_0) \int_0^a (A/\omega) g(r/\omega) E(r) r \, dr , \qquad (89)$$

with $g(r/\omega) = \exp\left[-\tfrac{1}{2}(r/\omega)^2\right]$ the Gaussian excitation, and ω is defined as the spot size.

For a maximum, $d\eta/d\omega = 0$, giving

$$\omega^2 \int_0^a E(r) g(r/\omega) r \, dr = \int_0^a E(r) g(r/\omega) r^3 \, dr. \quad (90)$$

If (90) can be solved, the spot size ω can be determined. Clearly if $E(r)$ itself is Gaussian, then a direct solution results. In that case $\eta \simeq 0.99$ [42].

Matsumura and Suganama [43] estimated spot size as a function of normalized frequency V for three types of fiber: (*a*) a graded-index fiber, (*b*) a step-index fiber with a central dip, and (*c*) a W-fiber with $R = -0.5$. In all cases normalized spot size ω/a decreases with increasing normalized frequency V. This is shown in Fig. 27 for case (*b*). This common behavior indicated that any fiber can be represented by an equivalent-step-index (ESI) fiber through an appropriate transformation of basic parameters.

The transformation takes the form (where the subscript s is for ESI fiber)

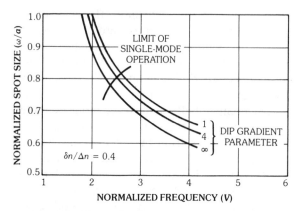

FIG. 27. Normalized spot size for step-index fiber with central dip of relative depth $\delta n/\Delta n$ = 0.4 and several dip gradients. (*After Matsumura and Suganama [43]*)

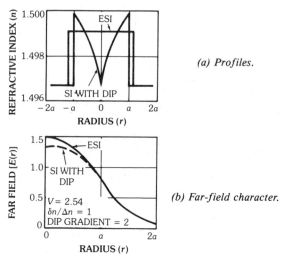

(a) Profiles.

(b) Far-field character.

FIG. 28. Equivalent-step-index profile obtained for a step-index fiber with deep central dip, using the spot-size equivalence method, and the relative light-field distribution $E(r)$. (*After Matsumura and Suganama [43]*)

$$
\left.\begin{aligned}
a_s &= X\,a,\\[6pt]
V_s &= Y\,V,\\[6pt]
\mathrm{NA}_s &= (Y/X)\,\mathrm{NA},
\end{aligned}\right\} \tag{91}
$$

since NA, the numerical aperture, is

$$
\mathrm{NA} = V/k\,a = (n_1^2 - n_2^2)^{1/2} . \tag{92}
$$

These equivalent step-index fiber representations are valid only at a given value of V; there is a different X, Y pair for each wavelength. The transformation can be determined either on the basis of compared radii or index differences. Fig. 28 shows the case of the equivalent step-index for the step-index fiber with central dip. The two fields are also shown, and they are seen to differ slightly only right on the axis. Streckert and Brinkmeyer [42] and others prefer to normalize ω with respect to an optimum effective core radius a_{eff}. The latter quantity is derived from the first minimum in the diffraction pattern measured under transverse illumination of a fiber immersed in an index-matching fluid:

$$
a_{\mathrm{eff}} = 3.832/(k \sin\theta_{\min}) . \tag{93}
$$

To complete comparison with step-index

fiber the results are expressed in terms of V_{eff} relating the cutoff frequencies of the two fibers:

$$
V_{\mathrm{eff}} = 2.405\,(V/V_{co}) = 2.405\,(\lambda_{co}/\lambda) . \tag{94}
$$

The numerical quantities 2.405 and 3.832 will be recognized as the first zeros of $J_0(u\,r)$ and $J_1(u\,r)$, respectively. With this method the dependence of ω/a_{eff} on V_{eff} is almost identical for a wide class of profiles spreading out in the range of interest for minimizing dispersion, $1.5 < V_{\mathrm{eff}} < 2.4$.

There are practical difficulties with using the various definitions of spot size to determine an equivalent step-index. For example, in transverse illumination to obtain $\theta_{\min}$ no real minimum may be encountered because the diffraction pattern may be bell-shaped. Fox [44] shows that a least-squares fit to one of the $\omega(V)$ approximations for the step-index fiber, that found by Marcuse [45], yields "best" values for the equivalent step-index diameter and index difference. This approximation is

$$
\omega/a = 0.65 + 1.619/V^{1.5} + 2.879/V^6 . \tag{95}
$$

The sense of "best" must suit the application; the

least-squares method is most useful in seeking an estimation of losses due to joining of fiber lengths.

A definition for ω by Petermann [46] has also been used:

$$\omega^2 = \int_0^a r^3 E_y^2 \, dr \bigg/ \int_0^a r E_y^2 \, dr. \qquad (96)$$

Here E_y is the linearly polarized field in the fiber. For a range of 97% to 99% launching efficiency in the step-index fiber Gambling and Matsumura [20] show that optimum normalized spot size (ω/a) changes from 0.8 to 1.2.

8. Other Methods for Fibers of Arbitrary Profiles

This section deals with three ways of obtaining solutions for fibers of arbitrary profiles: the variational, "resonance," and Numerov methods.

Variational Techniques

The wave equation for a fiber's radial wave function $\psi(r)$ can be rearranged into the form of the one-dimensional Schroedinger equation of quantum mechanics, in which the refractive-index profile $n^2(r)$ is analogous to the potential function and the square of the eigenvalue β is the wanted "energy level." The same methods of solution apply for determining the essential properties of the fundamental mode LP_{01}. A variational principle [47, 48] is constructed from the wave equation by multiplying throughout by $\psi(r)$ and integrating the result over the entire fiber cross section, with the sought-for eigenvalue placed on the left-hand side. With application of Green's theorem and Sommerfeld's radiation condition for the behavior of $\psi(r)$ for $r \to \infty$ the formal result is

$$\beta^2 = \int_0^\infty [\psi \nabla^2 \psi + \psi^2 k^2 n^2(r)] \, r \, dr \bigg/ \int_0^\infty \psi^2 r \, dr, \qquad (97)$$

where for the radial dimension involved, $\nabla^2 \to d^2\psi/dr^2$.

Even though the refractive index is allowed to have arbitrary variation in the core, and perhaps a fixed value in other regions of the cross section, a solution would require that we know $\psi(r)$. If we did, β^2 would be a maximum. The method works by employing a trial wave function that is as close to $\psi(r)$ as can be contrived. For example, as we have seen in the previous sections, Bessel functions from a step-index fiber with a choice of fiber parameters that makes β^2 a maximum. That step-index fiber is then the equivalent step-index fiber for our case. Since the equivalent step-index fiber and the fiber under study each obey the wave equation, the extremum takes the form

$$\beta^2 = \bar{\beta}^2 + \int_0^\infty k^2 (n^2 - \bar{n}^2) \, \bar{\psi}^2 \, r \, dr \bigg/ \int_0^\infty \bar{\psi}^2 r \, dr. \qquad (98)$$

The terms with overbar are those of the step-index fiber, and the n, $\bar{n}$ are the profiles of the studies and equivalent step-index fiber, respectively. Of all possible $\bar{\psi}^2$, the one with value maximizing the left-hand side is closest to the wanted field of the studied fiber.

To obtain refined estimates of the transverse fields in the core a Rayleigh-Ritz variational procedure could be employed, in which the trial field is expressed as a linear sum of an initial field and perturbation fields, ideally in the form of orthonormal functions [48].

Examples for the two-region fiber with radially inhomogeneous core have been worked out by Okamoto, Okoshi, Hotate, Snyder, and Sammut [49–52]. In the development of the theory of moments discussed above in Section 3, Hussey and Pask also employed a variational approach to obtain eigenvalues.

Francois and colleagues [53] went further by treating the case of the W-fiber with arbitrary core. For trial function they used that of the step-type W-fiber, like those in Section 4, and fixed all parameters for that W-fiber except its core radius, which became the allowed degree of freedom. The eigenvalue parameter whose extremum was sought was u^2, the radial eigenparameter (squared) in the core. It was found to be expressible in terms of two parameters of the reference W-fiber: (a) the ratio of radii for the two fibers, and (b) two integrals that are the first moment of the core profile-shape functions involved with respect to the power density [or $\psi^2(r)$] in the fibers. In this way (Fig. 29) the LP_{01} cutoff-frequency behavior was obtained as a function of the index contrast ratio ($R = \Delta n'/\Delta n$) and of the diameter ratio in the equivalent step-index W-fiber.

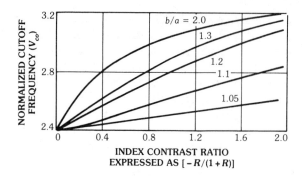

FIG. 29. The LP_{11} cutoff-frequency sensitivity to the index contrast ratio R and the diameter ratio for a step-index W-fiber. (*After Francois et al. [53]*).

The results are exceptionally complete so that without many further assumptions Francois and colleagues were able to find a "reference" equivalent step-index fiber by letting the contrast ratio $R \to 0$ for the inhomogeneous case. In the process they showed Gambling's "guidance factor" [see (53)] to be a natural consequence of the theory.

Fig. 30 is from Okoshi and Okamoto [49], and shows a set of assumed core profiles, the core fields for them as determined by a Rayleigh-Ritz variational procedure, and the eigenparameter

$u(V)$ for these fibers. The field in the cladding behaves as $K_0(w)$ in each case, as discussed in Section 2. In Fig. 30a, for the shown P, Q values,

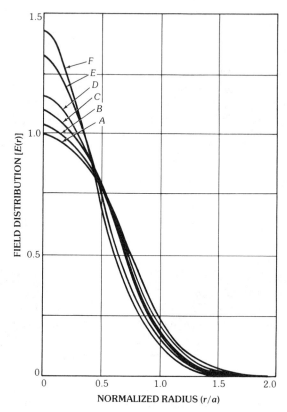

(b) Core fields determined for the fibers of (a).

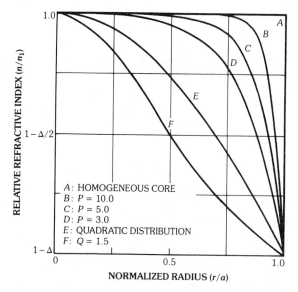

(a) Index profiles for fibers whose properties are to be determined by a variational method in a Rayleigh-Ritz procedure.

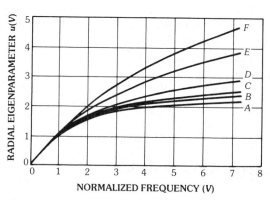

(c) Radial eigenparameters $u(V)$ for the fibers of (a).

FIG. 30. Assumed core profiles, their core fields, and their eigenparameters. (*After Okoshi and Okamoto [50], © 1976 IEEE*)

profiles B,C,D follow

$$n/n_1 = 1 - \Delta\,[\sinh^2(P\,r/a)\,\sinh^2 P],$$

profile E follows $n/n_1 = 1 - \Delta\,(r/a)^2$,

profile F follows

$$n/n_1 = 1 - \Delta\,[(\tanh^2 Q\,r\,a)/\tanh^2 Q]\ .$$

"Resonance" Methods

In setting up the eigenexpression for given fibers the logarithmic derivative was made continuous at the boundaries. For E- (or H-) fields the logarithmic derivative is proportional to the wave impedance (or admittance) at the boundary; the proportionality constant involves the "free-space impedance" ($120\,\pi$ ohms) and the (average) refractive index at the boundary. It is feasible therefore to convert the wave equation, a second-degree differential equation, into a first-order equation in the impedance (admittance). Solutions are found by considering limiting cases (value in the core, value far out in the cladding, etc.), leading to a fairly simplified characteristic equation. Shibanuma and colleagues [54] used the (normalized) wave susceptance B looking down the fiber. For mode number m it is defined as

$$B = -a\,[(1/R)\,dR/d\rho + m/p]\ . \tag{99}$$

This method appears to work for any index profile and enables computation of mode cutoffs to high accuracy, especially for low-order modes.

If the fiber cross section is represented by a very large number of homogeneous rings, then the impedance seen at the core center looking outwards and that at the outer boundary looking inwards can be connected by an iterative expression resembling that for a sectioned transmission line. If the rings are sufficiently small in width (compared with the local wavelength) Maxwell's equations can be reduced to a pair of uncoupled transmission-line equations. Similar techniques [55] have been used in the analysis of radio and telluric waves propagating into a layered earth. In this way it is not necessary to resort to using Bessel functions in each of the rings, and the computation effort is greatly reduced. Boucouvalas and Papageorgiou [56] employed this form of the resonance technique for the computation of cutoff frequencies for fibers of arbitrary profile. Each homogeneous ring is represented by

an equivalent-filter T-section. Resonance is sought for the condition of short-circuit termination at the core center, and open circuit at $r\to\infty$. An efficient root-searching method is required. This technique requires around 400 to 500 layers to reproduce cutoffs to within 0.1%.

The Numerov Method

The variational method is not the only one involving application of the one-dimensional Schroedinger equation of quantum physics to light transmission in the optical fiber. Another method, adapted for modern computational capabilities, is the Numerov method [57] as explained originally by Chow [58]. Davies, Davidson, and Singh [59] have developed this method to obtain waveguide dispersion for fibers with arbitrary profiles, particularly W-fibers with central dip. The Numerov method utilizes numerical integration applied to a second-order differential equation which has no first-derivative term. The advantage of this method is that the neglected terms in a Taylor series expansion are of order h^6, where h is the step size during integration. The method requires continuity of logarithmic derivatives at any interface and so is appropriate within the weakly guiding approximation. A normalized wave equation results on introduction of the substitutions

$$\xi = k\,r \quad \text{and} \quad \phi(\xi) = \psi(\xi)\,\xi^{1/2} \tag{100}$$

and with replacement of β/k by $\overset{*}{\beta}$, giving

$$f(\xi) = \phi''/\phi = \overset{*}{\beta}^2 - n^2(\xi) + (m^2 - \tfrac{1}{4})/\xi^2 , \tag{101}$$

where the primed quantities represent derivatives with respect to ξ.

For $m=0$ we require that as $\xi\to 0$, $\psi(\xi)\to 1$, so that $\theta(\xi)\to \xi^{1/2}\to 0$.

The Numerov method uses a three-point recursive relation,

$$\phi/(\xi+h)\,[1 - F(\xi+h)] + \phi(\xi-h)\,[1 - F(\xi-h)] =$$

$$\phi(\xi)\,[2 + 10\,F(\xi)]\ , \tag{102}$$

in which

$$F(\xi) = (h^2/12)\,f(\xi) , \tag{103}$$

where $h=\Delta\xi$ is an appropriately small step size chosen in the numerical integration procedure. (A suitably small step size would be one much smaller than the local wavelength.)

For bound modes trial eigenvalues must be employed to initialize the procedure. Equation 102 is applied in two directions: (a) outward, by assuming $\phi(\xi)$ and $\phi(\xi-h)$, one obtains $\phi(\xi+h)$, (b) inward, by obtaining $\phi(\xi-h)$, given $\phi(\xi)$ and $\phi(\xi+h)$. At a suitable location in the profile the results of the integration in the two directions are made to agree by imposing continuity of the logarithmic derivative at the matching point. This requires the second Numerov approximation, involving the first derivative ϕ' of ϕ:

$$\phi'(\xi)=\tfrac{1}{2}h\left\{[1-2F(\xi+h)]\,\phi(\xi+h)\right.$$

$$\left.-[1-2F(\xi-h)]\,\phi(\xi-h)\right\}.\quad(104)$$

The result for the trial eigenvalue $\overset{*}{\beta}{}^2$ can be improved, as in variational techniques, by seeking

$$\overset{*}{\beta}{}_1^2=\overset{*}{\beta}{}_0^2+\Delta(\overset{*}{\beta}{}^2)\,,\qquad(105)$$

where the last term is found iteratively seeking a minimum value of

$$DD=[\phi'_i/\phi_i]_{\xi=\xi_0}-[\phi'_o/\phi_o]_{\xi=\xi_0}$$

and then finding

$$\Delta(\overset{*}{\beta}{}^2)=DD\left/\left\{\left[\int_0^{\xi_0}\phi_o^2\,d\xi/\phi_o^2(\xi_0)\right]\right.\right.$$

$$\left.+\left[\int_{\xi_0}^{\infty}\phi_i^2\,d\xi/\phi_i^2(\xi_0)\right]\right\}.\quad(106)$$

Here ϕ_i, ϕ_o, are inward, outward versions of $\phi(\xi)$, respectively, as the matching point $\xi=\xi_0$ is approached.

For $m\geq0$, starting values for $\phi(\xi\to0)$ can be obtained from the solution of (101) for small ξ, as $A\xi^{m+1/2}$. To lowest order, this is consistent with solutions $J_m(\cdot)$ or $I_m(\cdot)$ for $\psi(\xi)$. In practice, it is best not to start the Numerov process too close to $\xi=0$, but rather handle this region with a power-series expansion for $\phi(\xi)$.

For large ξ the asymptotic form used to start the inwardly integrated solution of $\phi(\xi)$ is $\exp[-\xi(\overset{*}{\beta}{}^2-n_0^2)^{1/2}]$, which is consistent with the

solution $K_m[\xi(\overset{*}{\beta}{}^2-n_0^2)^{1/2}]$ for $\psi(\xi)$, as with the step-index fiber.

Figs. 31a and 31b compare results of applying the Numerov method to a step-type W-fiber with central dip with exact methods using nested Bessel functions. Fig. 31a plots the behavior $b(V)$, while Fig. 31b shows the resulting dispersion, expressed as $V\,d^2(Vb)/dV^2$. The characteristics of the fiber studied (shown in Fig. 23) were $R=-0.424\,24$, $\gamma=0.575\,76$, $c=1/3$, and $\mu=0.1$. The rms deviation between the two curves for $b(V)$ was found to be $0.007\,52$, over the range $1.575\leq V\leq2.4$.

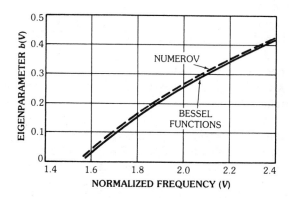

(a) Comparison of the reduced propagation constant b(V) obtained in the Numerov and Bessel function calculations.

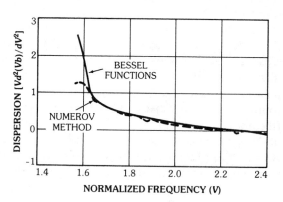

(b) Comparison of the reduced second derivatives V d²(bV)/dV² from the Numerov and Bessel function calculations.

FIG. 31. Applying the Numerov method to a step-type W-fiber with exact methods. (After Davies, Davidson, and Singh [59], © 1985 IEEE)

9. Polarization in Near-Cylindrical Optical Fibers

The fibers considered thus far have been assumed to be perfectly circular in cross section, with radial index profile independent of azimuth and independent of distance along the fiber. Real fibers will not be as ideal as this. A long-haul fiber comprises many joined spans, each departing from circularity to greater or lesser extent, and each having certain differences in fabrication. And during installation of the fiber, mechanical stress may cause distortion of the fiber cross section.

The Source of Polarization in a Fiber

In Section 2 we noted that in the lowest-order mode either or both of two polarizations could propagate down the fiber, depending on the light polarization at launching. Once launched, the light would seemingly be sustained in its polarization. In a real fiber, if the original fields in the cross section were E_x and H_y, then orthogonal fields E_y and H_x will appear if there is asymmetry of anisotropy. The fiber polarization quality may be characterized by the discrimination, the ratio of unwanted-to-wanted polarized fields at points along the fiber. One can imagine a fiber with asymmetry regular all along its length. For such a fiber two separated propagation factors apply, β_x and β_y, the subscripts indicating the direction of the **E** field that is aligned with one of the fiber's cross-sectional principal axes. Since $\beta_x \neq \beta_y$, there is a "fast" and a "slow" mode. The cross-polarized component could have come from the source itself or from the fiber. Assessment of fiber polarization performance by measurement thus requires that source purity be determined. Because deformations and anisotropy may vary from one fiber section to the next, the coupling from one mode to the other, and hence the discrimination, may have a random (spatial) component along the length of the fiber.

Since, at any wavelength, the propagation factor characterizes the phase shift per unit fiber length, the two independently propagating "signals" will periodically be in phase, with spacing $S = 2\pi/(\beta_x - \beta_y)$, known as the *beat length*. The *modal birefringence*, $B = \lambda/S$, is just the differ-

ence in the effective phase refractive indexes for the two polarizations. If the light excitation is plane polarized and launched at some angle with respect to a principal axis, then after a length z of the fiber the light will be elliptically polarized, with the state of elliptic polarization depending on the beat phase $2\pi z/S$. These definitions implicitly assume, however, that both polarization constituents are coherent as generated.

As Adams and colleagues [2, 60] have pointed out there are applications where a long beat length S is desired to prevent significant variations in the sensed polarization along a given length. In other applications a short beat length is required, which can be assured by deliberately fabricating a fiber with asymmetrical core or inducing asymmetrical stress in a fiber section. Considerable efforts are being devoted to producing fibers with high "polarization-holding power" h_p over long lengths. (See [61, 62].) With certain of these fibers information capacity might be doubled by transmitting two wholly independent signals (with opposite polarization) without significant cross coupling. Kaminow [63] has shown that h_p is proportional to the spatial power spectrum $|\Gamma(S)|^2$ of the autocorrelation function of the random polarization contributions, and S is the characteristic beat length for the two modes arising from coupling.

Light sources have a spectral width $\Delta\lambda$, the spacing between source spectral components with "zero" autocorrelation, and the coherence time is approximately $\Delta\lambda/c$, where c is the speed of light. The maximum fiber length over which mutual coherence of two polarizations can be maintained is [63]

$$L_c \simeq \lambda^2 / B \, \Delta\lambda$$
$$\simeq c / B \, \Delta f \,. \tag{107}$$

This characteristic is important when an information signal and a coherently related "local-oscillator" signal to be used for heterodyne detection are to be transmitted on opposite polarizations.

In experimental fibers B may be anywhere between 10^{-9} and 10^{-3}. With $\Delta f \simeq 1$ MHz for coherent excitation of a monomode fiber, and with $B < 10^{-3}$, $L_c > 300$ km.

Cross-Polarization Discrimination

Calculations [64] have shown that high-grade fibers for telecommunication have discriminations of about 70 dB, although measured values are around 60 dB because of imperfection in the measuring polarizers. Fibers with large B (around 3×10^{-4}) show [63] discriminations between 38 and 45 dB. (Note: In the literature of optics the term *extinction ratio* is often used to describe cross-polar discrimination.)

Monomode fibers will have low birefringence, with $B \simeq 10^{-7}$ to 10^{-6}. The transmission delay (per unit length of fiber) between orthogonal modes is

$$d\tau/dL = (1/c)(B - \lambda \, dB/d\lambda) . \qquad (108)$$

Far from cutoff in the normal operating region the second term in (108) is much smaller than the first. Thus, pulse spreading would be between 0.3 ps/km to 3 ps/km, with the corresponding bandwidth-distance limit 1500 GHz·km to 150 GHz·km due to polarization properties alone. For long fiber spans, say about 100 to 200 km, modal birefringence will have to be held to the lower end of the range just cited for rates of 0.5 to 1 Gb/s to be transmitted without significant introduction of polarization modal noise.

Full-Mode Theory Solutions for Fibers With an Elliptically Shaped Core

A formal solution for the propagating modes in a dielectric guide with elliptical cross section was obtained by Yeh [65]. With confocal elliptical (orthogonal) coordinates (ξ, η, z) (see [4], or [47]), the wave equation is solvable in terms of Mathieu functions [3]. Parametrically, in the xy plane (fiber cross section),

$$x = h \cosh \xi \cos \eta, \quad y = h \sinh \xi \sin \eta ,$$

with $2h$ the interfocal distance. A locus $\xi = \xi_0$ is the elliptically shaped core-cladding interface, and $\eta = $ constant identifies a particular pair of hyperbolas along which the field of one fundamental mode is oriented.

Normalized propagation factors can be defined as before, with the normalized frequency referred to the major axis as

$$V = k \, n_1 \, (h \cosh \xi_0) \, 2^{1/2} \, \Delta , \qquad (109)$$

where n_1 is the core index, $h \cosh \xi_0$ the half-width of the core along the major axis, and Δ the index difference, and $k = 2\pi/\lambda$.

This fiber may be described as an "elliptical step-index fiber."

Core eigenparameter u, similar to that of the circular step-index fiber, is also referred to the half-width $h \cosh \xi_0$:

$$u^2 = (k^2 \, n_1^2 - \beta^2)(h \cosh \xi_0)^2 , \qquad (110)$$

and

$$b = (u/V)^2$$

is the normalized axial propagation factor, the sought-for eigenvalue.

The wave equation for fields in the fiber takes the form

$$\nabla^2 \psi + f^2 (\cosh 2\xi - \cos 2\eta) \, \psi = 0 \qquad (111)$$

with $f^2 = u^2/(2 h^2 \cosh^2 \xi_0)$ and the Laplacian operator $\nabla^2 = \partial^2/\partial\xi^2 + \partial^2/\partial\eta^2$.

Mathieu functions are expandable in terms of an infinite series of Bessel functions, and the full determinantal eigenexpression is of infinite extent. For most practical cases, however, it is possible to achieve adequate convergence of the eigenexpression solution with a relatively small number of elements in the determinant. For cases of small ellipticity the eigenexpression converges to one similar to that of the circular step-index fiber (25b) except that additional perturbation terms appear, involving Bessel functions as well as coefficients that depend on Δ and the ellipticity e, which is just $e = 1 - \tanh \xi_0$. Perturbation terms with coefficients up to the second power in e generally suffice [66].

Physically, the two degenerate or indistinguishable HE_{11} modes of the circular step-index fiber are now resolved into the two orthogonal modes, eH_{11} the "even" mode, and oHE_{11} the "odd" mode, denoted according to the type of Mathieu function characterizing the axial magnetic (H_z) field. The respective **E**-field configurations in the xy cross sections are depicted in Fig. 32. Since the fiber is a source-free region without metallic con-

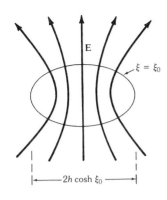

(a) The eHE$_{11}$ field.

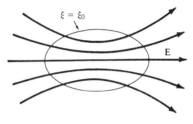

(b) The oHE$_{11}$ field.

Fig. 32. Elliptically shaped core nomenclature and a sketch of the **E** fields of the two polarizations. (*After Yeh [65]*)

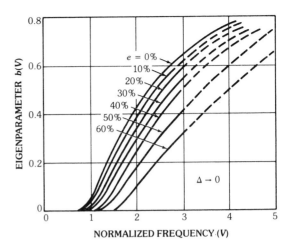

Fig. 33. The $b(V)$ behavior in a weakly guiding fiber with elliptically shaped core, as a function of ellipticity. The notation 3×3, etc., indicates the size of the determinant needed to establish accurately the eigenfunctions. (*After Fuji and Sano [67]*)

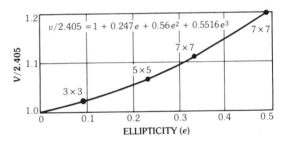

Fig. 34. Next-higher-mode cutoff frequency's dependence on the ellipticity. (*Adapted from Rengaragan and Lewis [68]. Acknowledgement is made of the prior publication of this material by the IEEE*)

fines, the **E** and **H** fields in the cross section are portions of continuous loops returning along the outer cladding of the fiber.

Fig. 33 from [67] shows the shift in $b(V)$ for one of the two modes in a weakly guiding fiber as ellipticity increases. Like the circular step-index fiber, these fibers have zero-cutoff HE$_{11}$modes, but their next higher-order modes appear at $V > 2.405$, with approximately a cubic dependence on ellipticity e, for $0 \le e \le \frac{1}{2}$, corresponding to aspect ratios between 1:1 and 2:1. Details of this increase in V_{co} are shown in Fig. 34, adapted from Rengarajan and Lewis [68]. The numerals adjacent to the plotted points indicate the size of the determinant needed to achieve convergence in the third decimal place. (Note that [68] computes normalized frequency V' referred to the *semiminor axis* rather than the semimajor axis. Since $V' = V(1-e)$, it will decrease with increasing ellipticity.)

More interesting is Fig. 35, also from [67], showing the differential normalized propagation factor $\delta b = (\beta_o - \beta_e)/k\, n_1\, \Delta = B/(n_1 - n_2)$ as a function of ellipticity. Alternatively, the scale in Fig. 34 could have been rendered in terms of the modal birefringence B; for the top curve at its maximum, B is about 4.5×10^{-7}. Note that for each curve the largest differential phase occurs in the region around $V = 2.405$, the cutoff frequency for the next higher mode in step-index monomode fibers, and including the monomode operating region usually chosen to minimize dispersion.

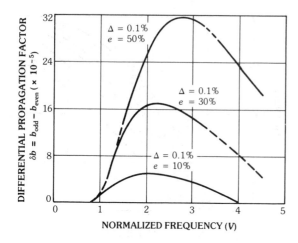

FIG. 35. Differential normalized propagation factor as a function of ellipticity for weakly guiding fiber with elliptically shaped core. (*After Fuji and Sano [67]*)

An approximate, simplified method for computing the differential phase $\beta_o - \beta_e$ has been developed by Kumar, Varshney, and Thyagarajan [69] by finding the eigenvalues of an equivalent rectangular guide, one with a rectangular core of area equal to that of the elliptical core and having the same aspect ratio. Only simple transcendental equations are involved. The fit to the more accurate solution is good only in the region of the peak ($1 \le V \le 2$) and then only for aspect ratios of less than 2:1.

10. References

[1] D. MARCUSE, *Theory of Dielectric Optical Waveguides*, New York: Academic Press, 1974.

[2] M. J. ADAMS, *An Introduction to Optical Waveguides,* Chichester: John Wiley & Sons, Ltd., 1981.

[3] M. ABRAMOWITZ and I. A. STEGUN, *Handbook of Mathematical Functions*, Washington: National Bureau of Standards, 1964.

[4] J. A. STRATTON, *Electromagnetic Theory*, New York: McGraw-Hill Book Co., 1941.

[5] H.-T. SHANG et al., "Design and Fabrication of Dispersion-Shifted Depressed-Clad Triangular-Profile (DDT) Single-Mode Fiber," *Electron. Lett.*, Vol. 21, pp. 201–203, February 28, 1985.

[6] D. GLOGE, "Weakly Guiding Fibers," *Appl. Opt.*, Vol. 10, pp. 2252–2258, 1971.

[7] D. GLOGE, "Dispersion in Weakly Guiding Fibers," *Appl. Opt.*, Vol. 10, pp. 2242–2445, 1971.

[8] C. F. GERALD, *Applied Numerical Analysis*, Reading, MA: Addison-Wesley, p. 34, 1980.

[9] W. A. GAMBLING, H. MATSUMURA, and C. M. RAGDALE, "Zero Total Dispersion in Graded-Index Single-Mode Fibers," *Electron. Lett.*, Vol. 15, pp. 474–476, 1979.

[10] H. RUDOLPH and E. NEUMANN, "Approximations for the Eigenvalues of the Fundamental Mode of a Step-Index Glass-Fiber Waveguide," *Nachrichtentechn. Z.*, Vol. 29, pp. 328–329, 1976.

[11] C. T. CHANG, "Minimum Dispersion in a Single-Mode Step-Index Optical Fiber," *Appl. Opt.*, Vol. 18, pp. 2516–2522, 1979.

[12] C. R. SOUTH, "Total Dispersion in Step-Index Monomode Fibers," *Electron. Lett.*, Vol. 15, No. 13, pp. 394–395, June 21, 1979.

[13] P. FRANCOIS, "Propagation Mechanisms in Quadruple-Clad Fibers: Mode Coupling Dispersion and Pure Bend Losses," *Electron. Lett.*, Vol. 19, pp. 885–886, October 13, 1983.

[14] F. B. HILDEBRAND, *Introduction to Numerical Analysis*, New York: McGraw-Hill Book Co., 1956.

[15] L. JEUNHOMME, "Dispersion Minimization in Single-Mode Fibers Between 1.3 μm and 1.7 μm," *Electron. Lett.*, Vol. 15, No. 15, pp. 478–479, July 19, 1979.

[16] P. PIRES et al. "Prediction of Laser Wavelength for Minimum Total Dispersion in Single-Mode Step-Index Fibers," *IEEE Trans. Microwave Theory Tech.*, Vol. MTT-30, pp. 131–139, February 1982.

[17] D. MARCUSE, "Interdependence of Waveguide and Material Dispersion," *Appl. Opt.*, Vol. 18, pp. 2930–2932, September 1979.

[18] L. G. COHEN, W. L. MAMMEL, and S. LUMISH, "Dispersion and Bandwidth Spectra in Single-Mode Fibers," *IEEE J. Quantum Electron.*, Vol. QE-18(1), pp. 49–53, 1981.

[19] L. G. COHEN, W. L. MAMMEL, and S. LUMISH, "Tailoring the Shapes of Dispersion Spectra to Control Bandwidths in Single-Mode Fibers," *Opt. Lett.*, Vol. 7(4), pp. 183–185, 1982.

[20] W. A. GAMBLING and H. MATSUMURA, "Propagation in Radially Inhomogeneous Single-Mode Fiber," *Optical and Quantum Electron.*, Vol. 10, pp. 31–40, London: Chapman and Hall, 1978.

[21] W. A. GAMBLING, H. MATSUMURA, and C. M. RAGDALE, "Wave Propagation in a Single-Mode

Fiber With Dip in the Refractive Index," *Optical and Quantum Electron.*, Vol. 10, pp. 301–309, London: Chapman and Hall, 1978.

[22] U. G. PAEK, G. E. PETERSON, and A. CARNEVALE, "Electromagnetic Fields, Field Confinement, and Energy Flow in Dispersionless Single-Mode Lightguides With Graded-Index Profiles," *Bell Syst. Tech. J.*, Vol. 60, pp. 1727–1743, 1981.

[23] D. KRUMBHOLZ, E. BRINKMEYER, and E. G. NEUMANN, "Core/Cladding Power Distribution, Propagation Constant, and Group Delay: Simple Relation for Power-Law Graded-Index Fibers," *J. Opt. Soc. Am.*, Vol. 70, p. 179, 1980.

[24] M. A. SAIFI et al., "Triangular-Profile Single-Mode Fiber," *Opt. Lett.*, Vol. 7, pp. 43–45, 1982.

[25] K. I. WHITE, "Design Parameters for Dispersion—Shifted Triangular-Profile Single-Mode Fibers," *Electron. Lett.*, Vol. 18, pp. 725–726, August 19, 1982.

[26] B. J. AINSLIE et al., "Monomode Fiber With Ultralow Loss and Minimum Dispersion at 1.55 μm," *Electron. Lett.*, Vol. 18, pp. 843–844, 1982.

[27] T. D. CROFT, J. E. RITTER, and A. BHAGAVATULA, "Low-Loss Dispersion-Shifted Single-Mode Fiber Manufactured by the OVD Process," *Proc. OFC '85,* Paper WD2, 1985.

[28] C. D. HUSSEY and C. PASK, "Theory of the Profile-Moments Description of Single-Mode Fibers," *IEEE Proc.*, Vol. 129, Pt. II, pp. 123–134, 1982.

[29] A. PAPOULIS, *Systems and Transforms With Applications in Optics*, New York: McGraw-Hill Book Co., 1968.

[30] E. BRINKMEYER, "Spot size of Graded-Index Single-Mode Fibers: Profile-Independent Representation and New Determination Method," *Appl. Opt.*, Vol. 18, pp. 932–937, 1979.

[31] M. MONERIE, "Propagation in Doubly-Clad Single-Mode Fibers," *IEEE Trans. Quantum Electron.*, Vol. QE-18, No. 4, pp. 535–542, 1982.

[32] M. MONERIE, "Fundamental-Mode Cutoff in Depressed Inner Cladding Fibers," *Electron. Lett.*, Vol. 18, pp. 642–644, 1982.

[33] A. KWAKAMI and S. NISHIDA, "Characteristics of a Doubly-Clad Optical Fiber With a Low-Index Cladding," *IEEE Trans. Quantum Electron.*, Vol. QE-10, No. 12, pp. 879–887, 1974.

[34] J. J. BERNARD et al., "Etude des Proprietes des Fibers Unimodales à Gaine Interne Deprimée,"

Ann. Telecomm., Vol. 38, pp. 47–52, January–February, 1983.

[35] S. I. JANG et al., "Experimental Verification of Ultrawide-Bandwidth Spectra in Double-Clad Single-Mode Fiber," *Bell Syst. Tech. J.*, Vol. 61, pp. 385–390, 1982.

[36] C. D. HUSSEY and F. deFORNEL, "Effective Refractive Index and Range of Monomode Operation for W-Fibers," *Electron. Lett.*, Vol. 20, pp. 346–347, February 16, 1984.

[37] Y. KITAYAMA and S. TANAKA, "Length Dependence of Cutoff Wavelength for Single-Mode Fiber," *Electronics and Communications in Japan*, Vol. 68, Pt. I, pp. 104–113, 1985.

[38] L. G. COHEN, D. MARCUSE, and W. MAMMEL, "Radiating Leaky-Mode Losses in Single-Mode Lightguides With Depressed-Index Claddings," *IEEE Trans. Microwave Theory Tech.*, Vol. MTT-30, pp. 1455–1460, 1982.

[39] P. FRANCOIS, "Tolerance Requirements for Dispersion-Free Single-Mode Fiber Design: Influence of Geometrical Parameters, Dopant Diffusion, and Axial Dip," *IEEE Trans. Microwave Theory Tech.*, Vol. MTT-30, No. 10, pp. 1478–1487, October 1982.

[40] H. KOGELNIK and T. LI, "Laser Beams and Resonators," *Appl. Opt.*, Vol. 5, pp. 1550–1567, 1966.

[41] D. MARCUSE, "Gaussian Approximation of the Fundamental Modes of Graded Fibers," *J. Opt. Soc. Am.*, Vol. 68, pp. 103–109, 1978.

[42] J. E. STRECKERT and E. BRINKMEYER, "Characteristic Parameters of Monomode Fibers," *Appl. Opt.*, Vol. 21, pp. 1910–1915, 1982.

[43] H. MATSUMURA and T. SUGANAMA, "Normalization of Single-Mode Fibers Having an Arbitrary Index Profile," *Appl. Opt.*, Vol. 19, pp. 3151–3158, 1980.

[44] M. FOX, "Calculation of Equivalent Step-Index Parameters for Single-Mode Fibers," *Optical and Quantum Electron.*, Vol. 15, pp. 451–455, 1983.

[45] D. MARCUSE, "Loss Analysis of Single-Mode Fiber Splice," *Bell Syst. Tech. J.*, Vol. 56, pp. 703–718, 1977.

[46] K. PETERMANN, "Microbending Loss in Monomode Fibers," *Electron. Lett.*, Vol. 12, pp. 107–109, 1976.

[47] P. MORSE and H. FESHBACH, *Methods of Theoretical Physics*, New York: McGraw-Hill Book Co., 1953.

[48] R. F. HARRINGTON, *Time-Harmonic Electromag-*

netic Fields, New York: McGraw-Hill Book Co., 1961.

[49] T. OKOSHI and K. OKAMOTO, "Analysis of Wave Propagation in Optical Fibers Using a Variational Method," *IEEE Trans. Microwave Theory Tech.*, Vol. MTT-22, pp. 938–945, 1974.

[50] K. OKAMOTO and T. OKOSHI, "Analysis of Wave Propagation in Optical Fibers Having Core With α-Power Refractive-Index Profile and Uniform Cladding," *IEEE Trans. Microwave Theory Tech.*, Vol. MTT-24, pp. 416–421, 1976.

[51] K. HOTATE and T. OKOSHI, "Formula Giving Single-Mode Limit of Optical Fiber Having Arbitrary Refractive-Index Profile," *Electron. Lett.*, Vol. 14(8), pp. 246–248, April 13, 1978.

[52] A. W. SNYDER and R. A. SAMMUT, "Fundamental (HE_{11}) Modes of Graded Optical Fibers," *J. Opt. Soc. Am.*, Vol. 69, pp. 1663–1671, 1979.

[53] P. FRANCOIS et al., "Equivalent Step-Index Profile for Graded W-Fibers: Application to TE_{01} Mode Cutoff," *Optical and Quantum Electron.*, Vol. 14, pp. 483–499, London: Chapman and Hall, 1982.

[54] N. SHIBANUMA et al., "Analysis of Graded-Index Fibers by Means of the Transverse Resonance Method," *J. Opt. Soc. Am.*, Vol. 72, pp. 1502–1505, 1982.

[55] J. R. WAIT, *Electromagnetic Waves in Stratified Media*, Chapter 1, New York: MacMillan Co., 1962.

[56] A. C. BOUCOUVALAS and C. D. PAPAGEORGIOU, "Cutoff Frequencies in Optical Fibers of Arbitrary Refractive-Index Profile Using the Resonance Technique," *IEEE Trans. Quantum Electron.*, Vol. QE-18, pp. 2027–2031, 1982.

[57] B. NUMEROV, *Publ. Obs. Central Astrophys.* (USSR), Vol. 2, p. 188, 1933.

[58] P. C. CHOW, "Computer Solutions to the Schroedinger Problem," *Am. J. Phys.*, Vol. 40, pp. 730–734, 1972.

[59] R. W. DAVIES, D. DAVIDSON, and M. P. SINGH, "Single-Mode Optical Fiber With Arbitrary Refractive-Index Profile: Propagation Solution by the Numerov Method," *IEEE J. Lightwave Tech.*, Vol. LT-3(3), pp. 619–627, 1985.

[60] M. J. ADAMS, D. N. PAYNE, and C. M. RAGDALE, "Birefringence in Optical Fibers With Elliptical Cross Section," *Electron. Lett.*, Vol. 15, pp. 298–299, 1979.

[61] N. SHIBATA, K. OKAMOTO, and Y. SASAKI, "Structure Design for Polarization-Maintaining and Absorption-Reducing Optical Fibers," *Rev. Electron. Commun. Lab.* (Japan), Vol. 31, pp. 393–399, 1983.

[62] S. C. RASHLEIGH and M. J. MARRONE, "Polarization-Holding in a High-Birefringence Fibre," *Electron Lett.*, Vol. 18, pp. 326–327, April 15, 1982.

[63] I. P. KAMINOW, "Polarization in Optical Fibers," *IEEE Trans. Quantum Electron.*, Vol. QE-17, pp 15–22, January 1981.

[64] M. P. VARNHAM, D. N. PAYNE, and J. D. LOVE, "Fundamental Limits to the Transmission of Linearly Polarised Light by Birefringent Optical Fibres," *Electron. Lett.*, Vol. 20, pp. 55–56, January 5, 1984.

[65] C. YEH, "Elliptical Dielectric Waveguides," *Appl. Phys.*, Vol. 33, pp. 3235–3243, 1962.

[66] J. D. LOVE, R. A. SAMMUT, and A. W. SNYDER, "Birefringence in Elliptically Deformed Optical Fibers," *Electron. Lett.*, Vol. 15, pp. 615–616, September 27, 1979.

[67] Y. FUJII and K. SANO, "Polarization Transmission Characteristics of Optical Fibers With Elliptical Cross Section," *Electron. Commun. Japan*, Vol. 63, p. 87, 1980.

[68] S. R. RENGARAJAN and J. E. LEWIS, "First Higher-Mode Cutoff in Two-Layer Elliptical Fibre Waveguides," *Electron, Lett.*, Vol. 16, pp. 263–264, March 27, 1980.

[69] A. KUMAR, R. K. VARSHNEY, and K. THYAGARAJAN, "Birefringence Calculations in Elliptical-core Optical Fibers," *Electron. Lett.*, Vol. 20, pp. 112–113, February 2, 1984.

4

Fiber Design Considerations for Optical Telecommunications

ARNAB SARKAR

Lightwave Technologies, Inc.

1. Introduction

The use of glass fibers as a medium for the transmission of light signals became an exciting possibility with the invention of semiconductor lasers. The high losses of glass fibers, however, initially seemed to prevent their use as a practical transmission medium in competition with established means of electromagnetic-wave propagation over copper conductors. In 1966 Charles Kao [1] first established that the high loss was due to transition metal-ion impurities in the glass and that this was not intrinsic to all glass. Soon thereafter the first high-silica, low-loss fiber, fabricated from glass made by a vapor-deposition technique, was realized in the laboratories of Corning Glass Works [2]. In the decade of the seventies dramatic improvements in fiber performance, along with in-depth understanding of fiber design considerations, have made optical fibers the well-accepted and practical transmission media they are today.

In order to appreciate the constraints of fiber design it is necessary to identify their sources. The most critical constraints occur due to transmission system needs, basic properties of the materials used, and the processes used to manufacture optical fibers. Before we proceed to discuss these in detail the reader is reminded that this discussion presupposes prior knowledge of fiber structure and the theory of propagation of optical signals along glass fibers, and a familiarity of the optical transmission system components that have been discussed in prior chapters.

Fiber Manufacturing Processes

The preeminence of vapor-deposition technology in fiber manufacturing is due to its ability to produce high-purity glass and its ability to control refractive-index profiles and dimensions to form precise waveguide structures. The four processes that have matured into fully automated industrialized processes are known as the modified chemical vapor-deposition (MCVD) process [3], the outside vapor-deposition (OVD) process [4], the vapor-axial-deposition (VAD) process [5], and the plasma chemical vapor-deposition (PCVD) process [6]. One other version of the vapor-deposition process that is still in developmental stages, but which has shown potential as a high-rate process, is the plasma-enhanced modified chemical vapor-deposition (PMCVD) process [7]. Although fiber-performance trends from all these processes are converging because of the common technology base, each of them imposes some restrictions on fiber design. Therefore some products or specific fiber designs are manufactured more conveniently by a specific process. This has caused some functionally equivalent products from a system point of view to have different fiber structures.

These processes have been qualitatively compared in terms of design flexibility, product performance, and cost effectiveness [8]. A brief summary of the analysis, updated from the time of that review is presented below.

Design flexibility constitutes ability to incorporate diverse dopants to different extents, along with precision of the refractive-index profile and dimensional control. All of the established processes have successfully produced all the fiber designs used or contemplated for use in telecommunications. However, the degree of difficulty of fabricating fibers of different designs is different for different processes. For graded-index multimode fibers, precise control of the refractive-index profile is necessary for wide bandwidth. This is best accomplished by the PCVD process, which deposits many hundreds of layers of glass with near 100% collection efficiency. Thus it is only necessary to control the reactant composition accurately. In MCVD one must control deposition temperature as well, because it affects the relative collection efficiency ratio of the glass components. Furthermore, in MCVD the number of layers is typically around 50 to 70. Thus striae due to intralayer-composition variation are also significant. In the OVD process the number of layers is higher than in MCVD, but one is required to control many burner-flow parameters and the exhaust conditions as well. On top of this, profile distortion in the sintering process step due to volatilization and redeposition of germania has to be taken into account and controlled. Also, closing of the central hole after mandrel removal imposes additional constraints. In the VAD process the refractive-index profile is a strong function of the temperature profile of the deposition surface and the composition profile of the soot stream. Although controlling these requires control of many parameters, some of the widest-bandwidth fibers have been made by this process.

Fabrication of nominally step-index single-mode fibers does not require accurate control of the profile. The degree of dimensional control required for single-mode fiber manufacturing does not significantly tax the capability of MCVD, PCVD, or the OVD process. The VAD process, however, has enough variation in the core-cladding ratio that it is difficult to control the cutoff wavelength of the fiber very closely. This is reflected in a rather loose cutoff-wavelength specification in Japan.

Fabrication of dispersion-shifted or dispersion-flattened fibers is more easily accomplished in the PCVD and MCVD processes. The dispersion-shifted fibers are easily fabricated by OVD and VAD processes as well. But fabrication of the dispersion-flattened fibers requires a very complex manufacturing procedure using VAD or OVD processes.

The performance distributions of fibers of all designs made by the various processes are one of convergence. Basically they all have the common vapor-deposition technology and use the same quality of raw materials. Thus, improved control of defect density as the processes have matured has ensured almost identical loss distributions. Control of other parameters, such as bandwidth or dispersion, fiber geometry, or strength, has been developed to more or less the same level, so that there is little product differentiation possible based on use of one process or the other.

The cost effectivenesses of each of these processes have been established by the ability of all of them to be used by various manufacturers competitively in the market place. In the absence of detailed information on fabrication rates, process selects and efficiencies, and capital and labor intensity, the most important parameters to observe may be deposition rate, preform size, and the number of process steps. Deposition rate and the number of process steps affect the cost effectiveness of preform fabrication. The preform size has a significant effect on the economics of the downstream process steps of fiber drawing and measurements.

The deposition rate in these processes depends on the rate of soot generation, deposition surface area per unit time, and the thermophoretic forces. Based on these considerations OVD has the highest deposition-rate potential, primarily because of surface-area considerations. Thermophoretic forces do not favor the VAD process as the deposition surface is constantly being heated. MCVD and PCVD processes face constraints of the substrate tube. The number of process steps favor MCVD, as preform fabrication is done in one step. This makes it easier to control process yields and minimizes labor content. In PCVD the tube is collapsed into a solid preform in a separate step. The OVD process steps consist of deposition, mandrel removal, and sintering at the least. In VAD process the steps are deposition, sintering, elongation, and sleeving. On the other hand, OVD and VAD processes have made the largest preforms. In MCVD and PCVD processes the preform size is restricted by tube parameters.

It is probable that all the processes will continue to be used for the next few years. However, a very large increase in deposition rate (greater than 20 g/min) and very large preforms (longer than 200 km) can give a process a cost advantage over the others. If this happens, it will be in the category of soot processes—probably one of the existing processes or hybrids of them.

The Evolution of Fiber Designs

The evolution of optical telecommunication systems has been extremely rapid since their inception, primarily due to improvements in almost all of their components. The initial and the predominant application to date has been in telephone trunking. This is because optical fibers lend themselves much more conveniently to point-to-point communication systems, compared with distribution networks. The initial systems used graded-index multimode fibers and operated at 0.82 to 0.85 μm in wavelength, with bit rates between 1.5 and 45 Mb/s [9, 10]. With the advent of low-loss, low-water, single-mode fibers, single-mode lasers operating around 1.3 μm and a suitable splice technique, today's systems use single-mode fibers at operating wavelengths of 1.3 μm with bit rates of 135 to 565 Mb/s [11, 12]. However, fiber designs for trunking applications are at this point far from being static. More on this topic will be discussed under "Technology Trends."

As the use of optical fibers is increasing in telecommunication trunking applications, costs of all the system components have been dramatically reduced. This and the experience in optical telecommunications system design are making it extremely attractive to use fibers in the distribution systems as well. Thus there is an increasing use of fibers in the feeder portion of the distribution systems. These systems are still point-to-point communication systems. Further incorporation of fibers into the distribution systems will require a further reduction of device and termination costs, as well as more ele-

gant solutions in the problems of switching, splitting, and other distribution functions.

Optical Telecommunications Systems Needs

For a transmission medium to be practical it has to have the following features:

1. It has to be effectively coupled to the electro-optic transmitters and receivers
2. It has to be able to transmit signals over long distances
3. It has to be able to transmit large numbers of signals simultaneously
4. It has to be packaged into cables that can withstand normal operating conditions of such transmission media
5. It has to be spliced efficiently for installation in the field
6. It has to be stable over the designed life of the system

The above system requirements impose certain restrictions on fiber design which are of a conflicting nature and require significant trade-off considerations. It is the author's intent to emphasize these trade-offs, as only then does one appreciate the fact that fibers cannot be designed effectively without due considerations to system performance requirements and selection of other system components.

2. Basic Design Considerations

The basic structure of fibers and the rationale behind them have been introduced in Chapter 2 of this book. The glass fiber comprising the core and the cladding is typically 125 μm in diameter. The fiber is coated with two layers of polymeric coatings between 230 to 900 μm in diameter. The polymeric coatings serve the dual purpose of preserving the optical transmission characteristics and the mechanical integrity of the fibers. Coatings also enhance handleability of glass fibers for ease of cabling and installation. Thus the designing of glass fibers includes the design of the glass fiber as well as its coating. These will be treated separately in this section under "Design of Fiber Coatings."

One aspect of fiber design that is common for almost all telecommunication fibers is the criteria for selecting materials. Since the optical properties of glasses are wavelength-dependent, selection of material is strongly influenced by the selection of operating wavelength. Today's fibers of oxide glasses are operated at the wavelength ranges of 0.85 and 1.3 μm. In the near future such fibers may be operated in the 1.55-μm wavelength range. At this time intense research is going on in developing fluoride glasses that have lower potential losses between 2- to 10-μm wavelength range [13]. The discussion in this chapter will be restricted to oxide glass fibers that are considered practical today.

The other aspects of glass-fiber design are the design of refractive-index profile and dimensions. As mentioned previously, these aspects of fiber design are extremely dynamic. Therefore all the different products that are already obsolete, such as the step-index fiber [14] and the double-window, graded-index, multimode fiber [15], will not be discussed in this section. Neither will we discuss design considerations for single-polarization, single-mode fibers that are being considered for very long distance, coherent-communication systems [16]. The focus of this chapter will be on design of multimode and single-mode fibers being used in today's trunking applications. This area will be covered in detail in Section 3.

Characteristics of Fiber Materials

The factors to be considered for materials selection are all systems-need related, i.e., the loss characteristics of the material, dispersion properties, and mechanical or chemical stability. Because high-silica glasses compatible with the vapor-deposition processes have the best mechanical and chemical properties of all glasses, these two considerations have never been very critical to material selection. In reality it is the loss considerations that have dominated fiber-material selection.

Loss Considerations—Fig. 1 shows the intrinsic loss mechanisms of oxide glasses. These consist of two intrinsic absorption mechanisms. A fundamental absorption band occurs at the ultraviolet wavelengths, due to stimulation of electron transitions. This phenomenon has been considered empirically by Urbach [18] and more analytically

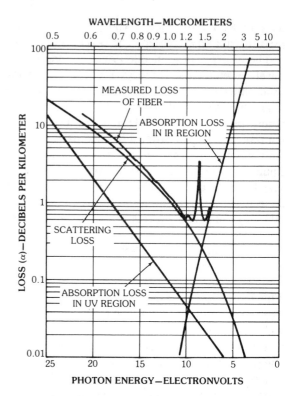

FIG. 1. Spectral-attenuation curve of germania–silica–core optical fiber separated into intrinsic loss components. (*After Osanai et al. [17],* © *Institution of Electrical Engineers*)

by Dow and Redfield [19]. The proposed model for the ultraviolet edge can be expressed in terms of an absorption loss α_{uv}, in decibels per kilometer per percent weight of germanium as

$$\alpha_{uv} = \alpha_{uv_0} \exp{(E/E_0)}, \qquad (1)$$

where E and α_{uv} are material-dependent and E is the photon energy. The ultraviolet edge shown in Fig. 1 is an extrapolation based on measurements by Schultz [20] in the range of 0.15 to 0.25 μm on germania-silica (GeO$_2 \cdot$SiO$_2$) glass. Direct absorption measurements at around 1-μm wavelength are extremely difficult due to a sharp drop of intensity of the band. However, attempted measurements of single-mode germania-silica fibers have shown values five to seven times greater than that obtained from extrapolation of data from Schultz. The degree of difficulty in resolving this discrepancy and the relatively small impact of the absorption band in the operating wavelengths have caused this area to remain largely unexplored.

The infrared absorption edge, at longer wavelengths, is due to cation-oxygen vibration bands. In pure silica the three fundamental vibrations associated with its tetrahedral structure show absorption peaks at wavelengths of 9.1, 12.5, and 21 μm [21]. The edges of the band extends into the 1.3-μm wavelength range. Miya and colleagues [22] have empirically modeled the infrared edge for germania-silica fibers. The absorption loss α_{ir} is expressed in decibels per kilometer as

$$\alpha_{ir} = 7.81 \times 10^{11} \exp\left(-48.48/\lambda\right). \qquad (2)$$

The infrared edge influences selection of the dopant material used in a fused silica-base glass very strongly. Fig. 2 shows the impact of the dopant selection on long-wavelength fiber performance, due to infrared-band edges caused by individual dopants.

The other intrinsic mechanism of fiber loss is Rayleigh scattering. In glass, light scattering is associated with frozen-in thermally induced local fluctuations in its density and composition about its equilibrium value. A review of the theory of Rayleigh scattering as applied to fiber materials has been presented by Olshansky [24]. The Rayleigh scattering loss α_{RS} for a single-component glass can be expressed as

$$\alpha_{RS} = (8\pi/3\lambda^4)(n^2-1)^2 \beta_c k T_F, \qquad (3)$$

where β_c is the isothermal compressibility at the fictive temperature T_F and k is Boltzmann's constant. The Rayleigh scattering coefficient in glass fibers is a strong function of the dopant used and increases with the proportion of the dopant. Fig. 3 shows measurements of the Rayleigh scattering coefficient of fibers with different dopant combinations. In this figure $\Delta = (n_1 - n_2)/n_1$ in percent.

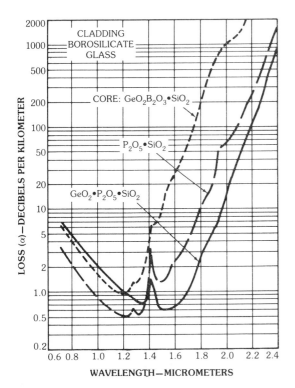

FIG. 2. Impact of choice of dopants on spectral attenuation. (*After Horiguchi and Osanai [23],* © *Institution of Electrical Engineers*)

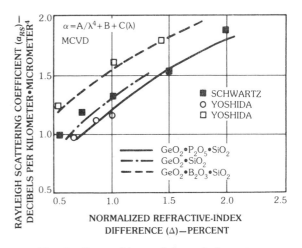

FIG. 3. Composition and numerical-aperture dependence of the Rayleigh scattering coefficient. (*After Yoshida et al. [25],* © *Institution of Electrical Engineers*)

In designing fibers one also needs to keep in mind the extrinsic loss mechanisms that are material-related and affect signal propagation. The biggest hurdle that had to be overcome in producing low-loss glass fibers was absorption bands in the wavelength range of interest, due to transition metal-ion impurities. Fig. 4 shows relative absorption loss due to trace contamination of certain tran-

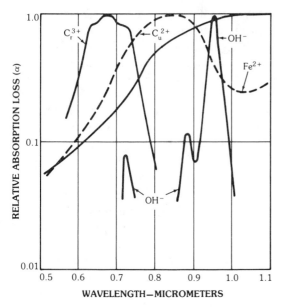

FIG. 4. Relative absorption loss of impurity ions. (*After Miller, Marcatili, and Li [26], © 1973 IEEE*)

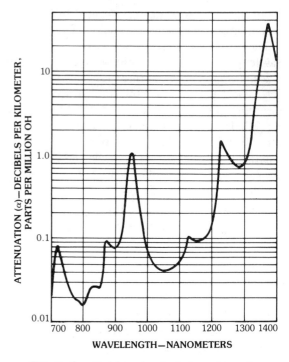

FIG. 5. Spectral loss due to hydroxyl ions in fused silica. (*After Barnoski and Personick [27], © 1973 IEEE*)

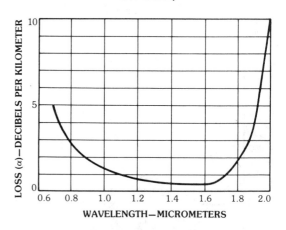

FIG. 6. Loss spectrum of VAD fiber with hydroxyl-ion content of less than 0.8 ppb. (*After Hanawa et al. [28], © Institution of Electrical Engineers*)

sition metal ions. It also includes absorption bands due to presence of hydroxyl ions in silica glass. A more detailed version of the absorption bands due to hydroxyl ions is shown in Fig. 5. It is critical that these absorption bands be minimized for optimum signal propagation. In today's fibers these absorption bands are well under control. Except for hydroxyl-ion absorption bands their presence can no longer be detected. An example of a fiber free from all extrinsic absorption-loss mechanisms is shown in Fig. 6.

Stability of transmission-loss characteristics of optical fibers was not considered to be an issue in this industry until recently. The first observation of fiber degradation in an installed experimental system was in Japan. In 1983 Uchida and colleagues first reported this observation [29]. Beales and colleagues [30] reported similar results in fibers by laboratory experiments. Both groups concluded that loss increase was an aging phenomenon, due to migration of hydrogen into the fiber. Intensive research in this field since then traced the sources of hydrogen in the cable structure to silicone coating of a specific type and the use of aluminum tape [31]. It has also been demonstrated that the severity of this phenomenon is related to the fiber composition

[32], with the fibers codoped with germania and phosphorus being most affected. Results of accelerated aging tests in a hydrogen atmosphere are shown in Fig. 7. Several mathematical models have been proposed to fit the aging characteristics. With

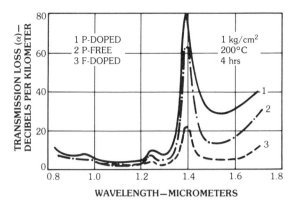

(a) With small amounts of additional dopants, after accelerated testing in hydrogen atmosphere.

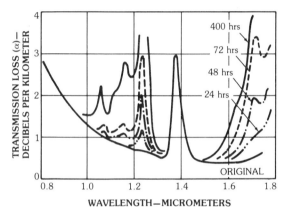

(b) In hydrogen atmosphere at 20°C as a function of time.

FIG. 7. Transmission loss of germania-doped silica core. (*After Ogai et al. [32]*)

the use of these models and with accelerated-test data several laboratories have concluded that, with proper safeguards in fiber design, fibers are quite safe for field use in telephony [31, 32]. A summary of these safeguards would be to avoid sources of hydrogen generation in the fiber coating and cable structure and to avoid a fiber core, codoped with germania and phosphorous.

Dispersion Considerations—The impact of material selection on dispersion is due to the fact that in practice the transmitted signal has a finite spectral width within which the refractive index of the glass changes. This results in temporal broadening of the transmitted signal. This is conventionally called *material dispersion* and can be expressed as

$$d\tau/d\lambda = -c^{-1}\, d^2n/d\lambda^2 . \qquad (4)$$

The computed curves for material dispersion for silica and three different doped silica compositions are shown in Fig. 8.

Characteristics of Fiber Structure

The fiber structure and its key parameters influence all of the propagation characteristics. The design parameters are the dimensions of the core and the cladding, the normalized refractive-index difference, and the index profiles of the core and cladding. The influence of the choice of these parameters on transmission loss, dispersion, coupling loss, and splicing characters will be discussed. The discussion in this subsection will be generalized for all types of fibers and will be elaborated on for specific designs in subsequent subsections. The reader will also note that trade-offs in structural considerations are not necessarily independent of considerations of materials.

Loss Considerations—These are the considerations of bending loss, microbending loss, and structural imperfections.

Bending Loss—In practice the bending of fiber is unavoidable. Bending fiber causes loss of transmitted power by radiation. In the framework of geometrical optics this can be expressed in terms of the refraction of light: that due to the curvature of the fiber the transmitted light is not under the condition of total internal reflection. In the use of fiber in the cable structure and in installations one has to make sure the bending loss is negligible. The design parameter that controls bending loss in fiber is the refractive-index difference between the core and the cladding. The higher the refractive index difference, the lower is the bending loss.

Microbending Loss—As opposed to continuous bends of a certain curvature microbending is a phenomenon where, with the fiber in contact with any surface and under small compressive forces, the fiber axis is randomly distorted due to defects on the surface. This phenomenon was observed at the very early stages of low-loss fiber development [34, 35] and is caused by the lack of rigidity of the fiber due to its very small diameter. In order to minimize this effect the refractive-index difference needs to be raised, the cladding diameter needs to

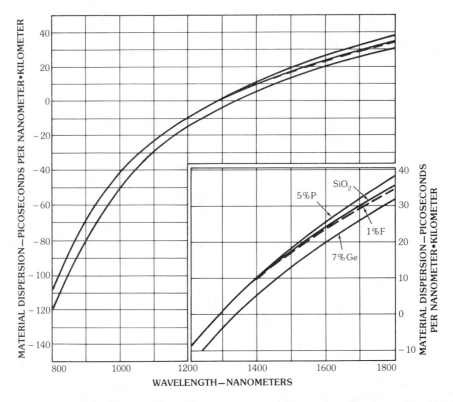

FIG. 8. Effect of doping on material dispersion of fused silica. (*After Garrett and Todd [33]*)

be increased, and the cladding-to-core diameter ratio needs to be increased. The sensitivity of fibers to microbending has been effectively controlled by suitable coating designs in conjunction with appropriate cable structures.

Structural Imperfections—Any structural imperfection can cause scattering loss in optical fibers. These imperfections can be variations in fiber diameter, cladding-to-core diameter ratio, and refractive-index profiles. They can be bubbles or devitrified particles as well. In spite of the dramatic improvements in control of these parameters the effect of structural loss in today's fibers is significant in terms of the total intrinsic loss [36]. Furthermore, structural imperfections are the primary sources of variation of loss of fibers of a specific design.

The Design of Fiber Coatings

Polymeric coatings were first applied on fibers to facilitate winding, by preventing fiber breakage due to abrasion of the fiber's surface. Today it is recognized that fiber coating serves many other functional needs as well. It has to densensitize the fiber from increase in loss as a result of microbending in the cable structure. Also it has to facilitate handling characteristics of the fiber in cable manufacturing. All this has to be done without adversely affecting the performance of the fiber over its operating temperature range. The complexity of this task can be appreciated by observing the magnitudes of the differences of the physical characteristics of glass and organic polymers.

Therefore coating design considerations to optimize fiber performance are complex. In real life one needs to superimpose on these considerations the constraints imposed by different cable designs and also those of an economical fiber-drawing process. For simplicity we will subdivide this discussion into structural, optical, and material considerations.

Structural Considerations—The first fiber coatings were of single materials. Some of these, such as lacquer, were applied as very thin coatings and others, such as ethylene vinyl acetate (eva), were applied as thicker coatings. These coatings were far from being satisfactory with respect fo the functional considerations. Gloge [35] first proposed that to minimize microbending loss one had to use a composite coating of at least two concentric layers. He proposed that the composite could be a hard layer over a soft inner layer, or vice versa. Today a hard layer over soft layer is the preferred combination. In such coatings the inner coating is referred to as the *primary coating,* and the outer as the *secondary coating.* One reason for the preference of this design is that the harder materials are less sticky and are easier to handle. Also they are easier to extrude over as they have a higher use temperature.

Optical Considerations—The most discussed optical transmission characteristic related to fiber coating is the low-temperature attenuation of the fiber. At low temperatures the coating shrinks and imposes a compressive force on the fiber, since normally the coating materials have a much higher expansion coefficient than glass. This causes microbending and results in increased attenuation. The low-temperature moduli of the coating materials, particularly that of the primary coating and its thickness, are critical for low-temperature attenuation performance.

Resistance to excess loss due to external microbending forces is the next most important consideration. Increases in the coated-fiber diameter and the secondary-coating modulus are most critical in controlling this excess loss. The upper limits of these parameters are restricted, due to coating-induced microbending and economic considerations.

The other optical characteristic of coating that is important is the refractive index of the material. If the refractive index is lower than that of the cladding glass, the cladding acts as a waveguide as well. This causes significant problems in accurate measurements of fiber properties. Therefore all primary-coating materials used today have refractive indexes higher than that of the glass.

Material Considerations—Polymeric coating materials used today can be divided into three categories: thermally cured materials, ultraviolet-radiation–cured materials, and materials that are applied as hot melts. From the point of view of curing or cooling rate the ultraviolet-cured materials are most advantageous, followed by thermally cured materials and hot melts, respectively.

The considerations for optical performance are more involved. For primary coatings the most important parameter is the modulus of the material in the operating temperature range of $-40°$ C to $60°$ C. This is important for low-temperature attenuation performance. From this point of view silicone rubber has the best performance, as its modulus is around 100 psi and stays relatively low at low temperatures. Commercially available ultraviolet-cured urethane acrylates used for optical fibers have had a relatively high modulus of around 500 psi, but at low temperatures the modulus of these materials increases by several orders of magnitude. Recently, polybutadiene-acrylate compositions that have an improved low-temperature modulus have shown excellent low-temperature performance [37]. At the same time urethane acrylates have been formulated that have improved low-temperature moduli, improved stability and have shown excellent low-temperature performance [38]. The temperature dependences of the moduli of primary coatings of urethane and polybutadiene acrylates are compared in Fig. 9. Low-temperature attenuation performances of fibers with these primary coatings and urethane-acrylate secondary coatings are shown in Fig. 10. Hot melts are used by Western Electric as primary coating material [39]. Their use, however, has not spread in the industry, because of the limited potential of hot melts for use in conjunction with high-speed drawing.

The other parameter that is important for primary coatings is the force of adhesion of the coating to the glass. This is important for high-strength–fiber manufacturing.

In the case of secondary coating the handling considerations are more critical. Can the coated fibers be extruded over for cabling without degradation? Does the coating provide adequate abrasion resistance to prevent strength degradation? Does the coating material get brittle at low temperatures? Is it stable in its characteristics over the life of the product? The answers to these questions require a thorough understanding of organic materials of several different families. For our purposes it suffices to say that materials of all the

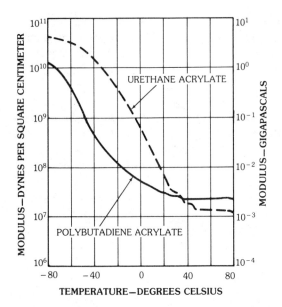

Fig. 9. Temperature dependence of modulus of ultraviolet-cured acrylate primary coatings. (*After Kimura and Yamakawa [37], © Institution of Electrical Engineers*)

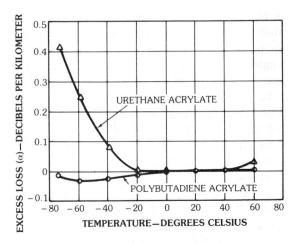

Fig. 10. Loss increase of acrylate-coated fibers at low temperature. (*After Kimura and Yamakawa [37], © Institution of Electrical Engineers*)

families have been satisfactorily used in fibers. Table 1 shows the materials commonly used for primary and secondary coatings in the industry today.

Today's Coatings and Technology Trends—For telecommunication-grade optical fibers there are two kinds of coatings design. In the United States fibers with 230- to 250-μm–diameter composite coatings are used. They are used in conjunction with loose-tube [40], slotted-core [41] and ribbon-cable structures [42]. These coatings are applied online at the fiber draw tower. The secondary coating is normally ultraviolet-cured acrylate material, but the primary coating can be rubber or acrylate. In Japan silicone-nylon coating of 900-μm diameter is used, primarily in conjunction with tight buffer-cable structure [43]. Since this structure imposes significant microbending forces

an extra microbending resistance of this fiber-coating design becomes necessary. Normally the nylon is applied in a separate extrusion operation, subsequent to fiber draw.

It is this writer's belief that as the ultraviolet-cured materials improve their low-temperature behavior and as their performance over the lifetime of the cables is satisfactorily demonstrated, their use will increase. This is primarily due to their fast curing speeds. It is also possible that as fibers are designed with improved microbending sensitivity and the coating materials become more stable in modulus with respect to temperature, the economics of fiber drawing could drive the lowest-cost fibers to be of single coatings.

3. Fiber Designs for Trunk Applications

From their inception to about 1983 fiber-optic trunks were almost all graded-index multimode fibers. The bit rates were significantly higher than

TABLE 1. Fiber-Coating Materials

Coating	Ultraviolet Cure	Thermal Cure	Hot Melt
Primary	Acrylates	Silicones	Thermoplastic rubber
Secondary	Acrylates		Nylon

in the existing copper systems and increased to about 140 Mb/s. Repeater spans increased to about 15 to 20 km. However, with the advent of single-mode laser diodes at 1.3 μm and the development of effective splicing techniques, single-mode fibers with potential bit rates of 2 Gb/s and repeater spans of up to 200 km have dominated the market. In the following material on fiber designs for trunk applications both multimode and single-mode fibers will be discussed.

Multimode Fiber Design Considerations

Graded-index multimode fibers were first used in the wavelength range of 0.85 μm. In the United States and western Europe fiber design then developed an upgradable double-window fiber. With such fibers systems could be installed at 0.85 μm and then, as laser diodes operating at 1.3 μm became available, the systems could be upgraded to take advantage of the lower loss in this wavelength range. In Japan, however, the transition was directly to fiber designs optimized for operation at the 1.3-μm wavelength range, without the use of special designs for double-window operation. Today graded-index fiber systems are predominantly used in conjunction with laser diodes operating in the 1.3-μm wavelength range.

The two alternate designs that are functionally equivalent are shown in Fig. 11, made by different processes. In this section we will review the basic design considerations with a focus on the trade-offs between functional parameters and fiber design parameters. These design parameters are the normalized refractive-index difference Δ between the core and the cladding, core diameter $2a$, and fiber diameter $2b$. An excellent review paper on optimization of fiber design [46] presented a quantitative analysis of optimization of fiber design parameters based on system design considerations.

Loss Considerations—These are considerations of transmission loss, microbending loss, fiber loss, splice loss, and coupling loss.

Transmission Loss—The transmission loss in decibels of any repeater section of length L can be expressed as [47]

$$\alpha_L = \alpha_f L + \alpha_m L + \alpha_s (L/\ell - 1) + \alpha_c , \qquad (5)$$

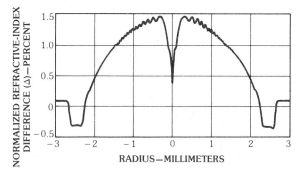

(a) Of depressed-index barrier layer, graded-index, multimode fiber fabricated by the MCVD process, measured by York Technology preform analyzer. (After [44])

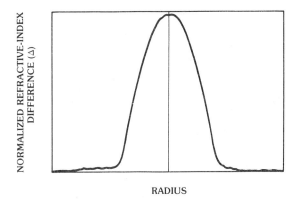

(b) Of graded-index multimode fibers made by OVD process. (After A. J. Morrow et al. [45])

Fig. 11. Typical refractive-index profiles.

where

α_f = fiber loss in decibels per kilometer,
L = repeater span length in kilometers,
α_m = microbending loss in decibels per kilometer,
α_s = unit splice loss in decibels,
ℓ = distance between splices in kilometers,
α_c = transmitter-to-fiber coupling loss in decibels

To be able to analyze the pertinent trade-offs one needs to understand the functional behavior of these loss mechanisms with respect to the fiber design parameters. This is presented below:

Fiber Loss—The generalized form for fiber loss α_f is

$$\alpha_f = a_{RS}/\lambda^4 + b + c(\lambda), \qquad (6)$$

where

a_{RS} = Rayleigh scattering coefficient,
λ = wavelength,
b = waveguide structural loss,
$c(\lambda)$ = wavelength-dependent absorption loss, principally due to OH-absorption bands.

By statistical analysis of large number of fibers Swartz and Buckler [47] found

$$\alpha_f = 0.05 + 0.4\,\Delta^{0.7} \qquad (7)$$

in decibels per kilometer, where Δ is the normalized refractive-index difference $(n_1 - n_2)/n_1$.

Microbending Loss—This loss, due to distortion of fiber axis by external forces, can be expressed in terms of the fiber design parameters [34] as shown below:

$$\alpha_m \propto (1/\Delta^3)[(2\,a)^4/(2\,b)^6], \qquad (8)$$

where $2\,a$ is the core diameter and $2\,b$ is the fiber diameter. It is important to note that microbending loss in multimode fibers is a wavelength-independent function.

Splice Loss—The predominant slice loss is caused by axial mismatch of the fiber core. It has been found that in practice one can express splice loss α_s as

$$\alpha_s = k_s/a, \qquad (9)$$

where k_s is a constant and a is the core radius.

Coupling Loss—This loss between fiber and electro-optic terminal devices is mainly the loss between the transmitter and the fiber. For small-area emitters, such as the laser diodes, the coupling loss does not depend on the core radius. But for large-area emitters, such as light-emitting diodes, the core radius (a) is a key parameter. The coupling losses can be expressed as

$$\alpha_c = 10 \log (k_{c_1}/\Delta) \qquad (10)$$

for laser diodes, and

$$\alpha_c = 10 \log (k_{c_2}/\Delta a^2) \qquad (11)$$

for light-emitting diodes, where Δ is the normalized refractive-index difference.

To summarize the loss considerations let us review the response of the different loss mechanisms to changes in the design parameters. If Δ is increased, the scattering loss increases, but the microbending loss decreases along with the coupling loss from an LED. If the core diameter is increased, the microbending loss increases, but the splice loss and the coupling loss from an LED decrease. If the fiber diameter is increased, the microbending loss decreases, but the fiber cost rises very quickly.

Bandwidth Considerations—The principal fiber parameter that controls bandwidth is the precision of the refractive-index profile, which is given by

$$n(r) = n_0 [1 - 2\,\Delta\,(r/a)^\alpha]^{1/2}. \qquad (12)$$

The optimum index-profile exponent α_0 depends on the core composition and Δ. Bandwidth versus profile exponent α for fibers with different Δ are shown in Fig. 12. In fiber manufacturing one has to control α and localized perturbations from the optimum profile. The effect of perturbations on bandwidth response in shown in Fig. 13. From these generalized effects based on statistical data Swartz and Buckler [47] have shown that

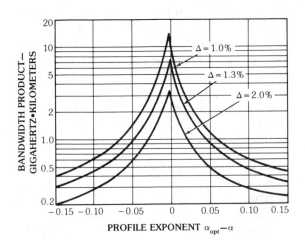

FIG. 12. Bandwidth versus profile exponent as a function of refractive-index difference (*After Nagel [48]*)

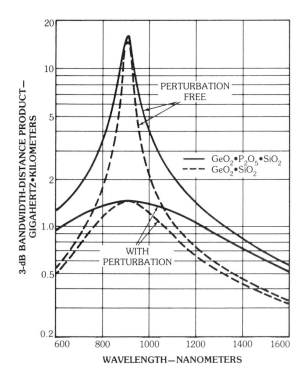

FIG. 13. Effect of perturbations on bandwidth spectra of fibers with two core compositions. (*After Blankenship et al. [15]*)

$$bw \simeq k_\alpha \Delta^{-z}, \qquad (13)$$

where k_α and z are constant for a given α. Therefore it is more difficult to make wide-bandwidth fibers with high Δ.

Although the theory of minimizing modal dispersion to maximize bandwidth of graded-index fibers is well understood, it is very difficult to predict the link bandwidth when fibers are concatenated together. It is this uncertainty that has caused the rapid obsolescence of this type of fiber in high–bit-rate trunk applications. We will look at the concatenation behavior of multimode fiber in detail. In addition to individual fiber bandwidths the parameters that affect link bandwidth are excited-mode distribution, differential-modal attenuation, mode conversion, and modal equalization.

Modal Excitation—In fibers that are not free from perturbations in refractive-index profile or those that cannot be described by a single profile exponent, measured bandwidth is a function of modal excitation. In other words, if a fiber is measured with a mode scrambler [49], there is a high probability that the same fiber will have a significantly different measured bandwidth if measured with a pigtailed laser diode. Since fiber and cable bandwidths in factories are measured in manufacturing facilities with mode scramblers for improved measurement reproducibility, this effect makes concatenated-link bandwidth less predictable. It should be emphasized that the less the localized perturbations are in the fiber, the less is the impact of this effect.

Differential Modal Attenuation—This causes modal power distribution to change along the fiber length [50]. This then changes the impact that existing perturbations would have on measured fiber bandwidth. For example, if a group of fibers have large perturbations at the core-cladding interface, their measured bandwidth, using a mode scrambler and with the fibers on measurement spools free from microbending forces, would be relatively low. If the same fibers are cabled rather poorly, causing relatively high microbending forces, such that little power continues to propagate in the higher-order modes, the measured fiber bandwidths would increase significantly. The converse is true if the perturbation is close to the fiber axis. As some degree of microbending loss occurs in almost all cabling operation and almost all fibers have some degree of index perturbation at the core-cladding interface, this phenomenon further complicates bandwidth predictability.

Mode Conversion—It has been demonstrated that if the modes are mixed at splice points the link bandwidth is enhanced [51]. If this is done using a short length of step-index core fiber, an increase of link bandwidth of up to 20% can be achieved. This, however, causes a loss penalty. Thus this approach is no substitute for producing high-performance fiber, but it can be used effectively as a restoration technique for systems that are bandwidth-limited but with significant attenuation margins.

Mode Equalization—By mixing fibers with profile exponents higher and lower than the optimum value, the link bandwidth can be maximized [52]. However, this approach requires a significant amount of fiber selection and results in increased operational complexity.

In summary, the inability to consistently produce fibers with no index perturbations and profile exponents with not more than 1% error from their optimum values has caused system engineers to look a lot deeper at production test data to ensure some degree of predictability. The most effective tool is a spectral-bandwidth measurement using a Raman fiber laser system [53]. Others have attempted to use two bandwidth measurements, one at 0.85 μm and the other at 1.3 μm [54]. In this case the bandwdith-versus-wavelength model along with the two data points are used to compute spectral bandwidth response. It is this information with which bandwidth concatenation has been best predicted. One can define a *link bandwidth* (bw) as shown below:

$$bw \simeq L^{-\gamma}, \qquad (14)$$

where γ is called the length exponent and L is the link length.

The length exponent γ theoretically should have values between 0.5 and 1.0, the two extremes being the limits of all perturbations being random or systematic, respectively. In practice numbers beyond this range can be obtained due to differences of launch conditions between individual fibers and link test or differential modal attenuation. What we would like to explore is how the length exponent γ behaves in today's fibers and what needs to be done to improve this performance.

Let us first examine the effect of profile exponent error. If the profile exponent is not optimum for the operating wavelength, the fiber bandwidth peaks at a wavelength different from the operating wavelength. Therefore, if fibers are selected, concatenated, and the length exponent is measured as a function of wavelength, one can estimate the effect of profile exponent error on length exponent of bandwidth. This is shown in Fig. 14. This observation of $\gamma=0.5$ for the optimum exponent is observed only when fibers do not have any systematic perturbations. For example, fibers with an optimum index profile but with large axial index depressions exhibit $\gamma=1.0$. With fibers properly selected for an optimum exponent

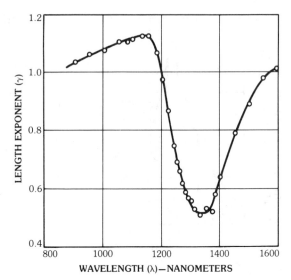

FIG. 14. Length exponent versus wavelength for a 13.3-km link of OVD fibers. (*After Love [54]*)

and by avoiding fibers with systematic index perturbations, a 20-km link with a peak bandwidth product of 4 GHz·km has been demonstrated [54]. The result of such a performance is shown in Fig. 15.

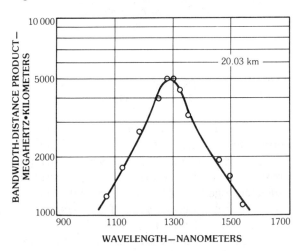

FIG. 15. Bandwidth sepctrum for a 20-km OVD fiber link. (*After Love [54]*)

In practice, however, one can obtain length exponents of 0.6 to 0.7 on a regular basis, with some exceptions. The search for a method to identify fibers that would concatenate unfavorably has been ongoing [55] and should continue until predictability is established. The pressure on this work has dropped off very significantly, with the growing use of single-mode fibers for all high-capacity links.

Mechanical-Strength Considerations—The fiber parameter that most strongly influences the mechanical properties is the fiber diameter (*2b*). This is because fiber in use is expected to be bent. The tensile stress σ_f in the presence of a small axial force and bends is expressed as

$$\sigma_f = W/\pi b^2 + E_b/R, \qquad (15)$$

where

W = tensile force,
R = bend radius,
E_b = modulus.

Using practical values of W and R as 0.2 kg and 20 to 200 mm, Murata [46] has concluded that a diameter greater than 110 μm is safe. In the early stages of fiber development this led to the use of 110-, 125- and 150-μm–diameter fibers. After the two additional considerations of microbending and cost were taken into account, the fiber diameter was standardized at 125 μm.

Summary—Based on all these considerations telecommunication-grade, multimode, graded-index fibers have the following nominal parameters: diameter of 125 μm, core diameter of 50 μm, and refractive-index difference of 1.0 to 1.3%.

Single-Mode Fiber Design Considerations

Single-mode fibers for trunk applications are designed for operating wavelength range of 1.3 μm. Such fibers have low loss and low dispersion at this wavelength range, but also have low loss at the 1.55-μm wavelength range. The three alternate designs used for this application are called *matched-cladding, depressed-cladding,* and *depressed refractive index* and are shown in Figs. 16a, 16b, and 16c, respectively. The matched-cladding design is produced by OVD and VAD processes. Some MCVD fibers are manufactured in this design as well. Both the depressed-cladding and the depressed–refractive-index designs are manufactured by the MCVD process. All three designs are used for similar applications, but with different cable and system configurations. Single-mode fiber designs for tele-

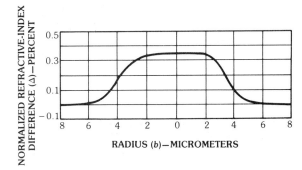

(a) An OVD matched-cladding, single-mode fiber. (After Berkey [56])

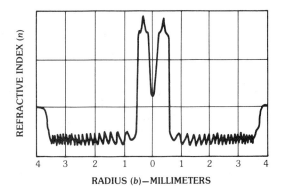

(b) A depressed-cladding, single-mode fiber fabricated by MCVD process. (After Anderson et al. [57])

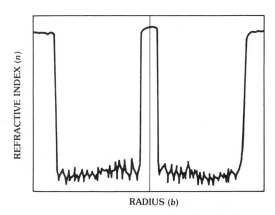

(c) A depressed–refractive-index single-mode fiber, fabricated by the MCVD process. (After Csencsitis et al. [58])

FIG. 16. Typical refractive-index profiles of single-mode fiber of different fabrications.

communications have been extensively reviewed by Sarkar [59].

In single-mode fiber design one has to ensure

that the signal propagated is indeed single-mode. The condition of single-mode propagation is shown below:

$$\lambda_{co} = 2\pi n a \Delta^{1/2}/V_{co}, \qquad (16)$$

where

n = refractive index of the core,
a = core radius,
Δ = normalized refractive-index difference,

and where λ_{co} is the cutoff wavelength (defined as the wavelength above which only the fundamental mode propagates) and V_{co} is the normalized frequency V at the cutoff wavelength. The value of V_{co} depends on the refractive-index profile of the core and is 2.404 for a step-index profile.

From this it is apparent that the control parameters for single-mode fibers are the normalized refractive-index difference Δ, the core diameter, the fiber diameter, and the refractive-index profile.

The functional parameters for single-mode fibers are essentially the same as for multimode fibers, but they are represented a little differently. The functional parameters are fiber loss, dispersion, and mode-field radius. Although these fibers were designed to have a step-index core, process constraints caused the actual profiles to have significant deviations from a step profile, and so they can at best be called nominally step-index fibers. These deviations from step index have a significant impact on functional parameters and will be pointed out from time to time.

Because functional properties such as dispersion and mode-field radius are affected by fiber index profiles, it was important to find a convenient way to represent the arbitrary refractive-index profiles of fibers and express them in terms of step-index profile parameters. This is most elegantly done by converting the fiber parameters in terms of equivalent–step-index (ESI) parameters [60]. Thus (16) can be changed to

$$\lambda_{co} = 2\pi n a_{ESI}(\Delta_{ESI})^{1/2}/V_{co}. \qquad (17)$$

Accuracy of prediction of functional parameters from the ESI parameters depends on computation technique used. Refinement of the computation

technique for ESI parameters to improve predictability of functional parameters of fibers has continued. A recent work that includes the fourth moment of the refractive-index profiles in the computation of the equivalent step-index parameters [61] has improved the predictability of functional parameters a great deal.

Effective Cutoff Wavelength—To be a single-mode fiber system the cutoff wavelength of the fiber must be shorter than the operating wavelength all among the transmission line. If not, second-order modes will propagate, causing modal noise in the system [62]. As techniques for the measurement of the cutoff wavelength were being developed, it was found that the measured cutoff wavelength was dependent on sample length, as shown in Fig. 17, as well as its radius of curvature. Furthermore, this measured cutoff wavelength was found to be significantly shorter than the theoretical cutoff wavelength calculated from (17). Cutoff wavelengths measured on 2-m relatively straight fibers are normally between 160 and 180 nm shorter than the theoretical cutoff wavelength [64]. Thus the measured cutoff wavelength was called the effective cutoff wavelength of the fiber. Initially it was proposed that the effective cutoff wavelength of the entire fiber link had to be lower than the operating wavelength of the system to control modal noise. Others felt it is sufficient to ensure that the effective cutoff wavelength of the link is less than the operating wavelength, even if the effective cutoff wavelengths of individual fibers in the link have cutoff wavelengths longer than the operating wavelengths. After measuring system modal noise with fibers having high effective cutoff

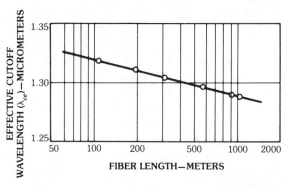

FIG. 17. Length dependence of the cutoff wavelength of single-mode fibers. (*After Kitayama and Tanaka [63]*)

wavelengths, Cheung and colleagues [65] concluded that fibers with effective cutoff wavelengths as high as 1.35 μm can be used in practical systems operating in the 1.3-μm wavelength range. Use of higher effective cutoff wavelength values is particularly attractive, because it improves the microbending loss and dispersion characteristics of the fiber. Recently, NTT in Japan has adopted the use of a cutoff wavelength of 1.34 μm as well.

Mode-Field Radius—Like the effective cutoff wavelength the mode-field radius is a critical functional parameter for single-mode fibers. Conceptually this parameter simply defines the power distribution of the propagating mode. Although this can be precisely calculated for step-index fiber, Marcuse [66] used a definition for mode-field radius (ω_0) as the width of a Gaussian beam which will optimally excite the single-mode fiber. He approximated ω_0 by the following expression:

$$\omega_0/a = 0.65 + 1.619\,V^{-1.5} + 2.879\,V^{-6}. \qquad (18)$$

Since the normalized frequency V is inversely proportional to the wavelength, the mode-field radius ω_0 is also a strong function of wavelength. This functional behavior is shown in Fig. 18.

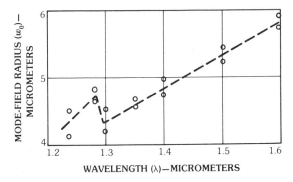

FIG. 18. Wavelength dependence of the mode-field radius of single-mode fibers. (*After Pocholle et al. [67]*)

The transmitted-signal field distribution of single-mode fibers, which is characterized by the mode-field radius, is significantly different from that of multimode fibers. In multimode fibers the field of the propagating signal is almost entirely limited to the core. In single-mode fibers a significant amount of the signal propagates in the clad-

ding. The fractional power propagating in the cladding is also a function of wavelength. This is shown in Fig. 19. A consequence of this is that all functional properties of single-mode fibers are wavelength dependent.

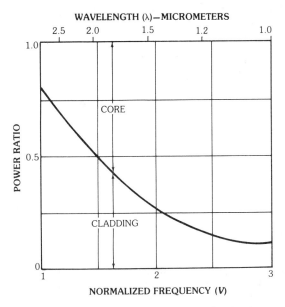

FIG. 19. Fractional power propagating in the core of a single-mode fiber as a function of its V value. (*After Kawachi [68], © Institution of Electrical Engineers*)

Loss Considerations—The system-loss equation for single-mode fibers is the same as that of multimode-fiber systems. All the loss mechanisms, however, have different functional behaviors.

Scattering Loss—The scattering coefficient a_{sm} of single-mode fibers has been expressed as [69]

$$a_{sm}(\lambda) = \{\,a_{core}\,P_{core}(\lambda) + a_{clad}\,[1 - P_{core}(\lambda)]\,\}\,, \qquad (19)$$

where

a_{core} = scattering coefficient of core composition,

a_{clad} = scattering coefficient of the cladding composition,

P_{core} = fractional power in core.

From this conceptual framework the scattering coefficient of OVD single-mode fibers with germania-silica core and silica cladding has been accurately estimated and verified experimentally. The wavelength dependence of the scattering coefficient of OVD fibers is shown in Fig. 20.

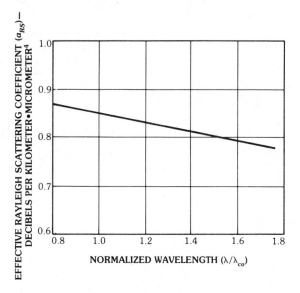

FIG. 20. Wavelength dependence of effective Rayleigh scattering coefficient of OVD single-mode fibers (*After Bhagavatula et al. [69]*)

Bending and Microbending Losses—As in multimode fibers, cables are conveniently designed and manufactured without bending the fibers to a point that bending losses become significant. Theoretical analyses of bending loss characteristics are abundant in the literature. It suffices to say that for the existing single-mode fiber designs the bending loss is not going to be significant if the fiber radius of curvature is kept greater than 5 cm.

Microbending of single-mode fibers, on the other hand, is a very complex phenomenon and must be understood well. Furuya and Suematsu [70] have modeled microbending of single-mode fibers and have shown good agreement with cabling losses observed in real life. The contrast of microbending losses in single- and multimode fibers is dramatic. In multimode fibers one observes a wavelength-independent excess loss of a small value. In single-mode fibers a sharp increase in loss is observed at a certain wavelength, as shown in Fig. 21. Increased microbending causes the low-loss operating wavelength

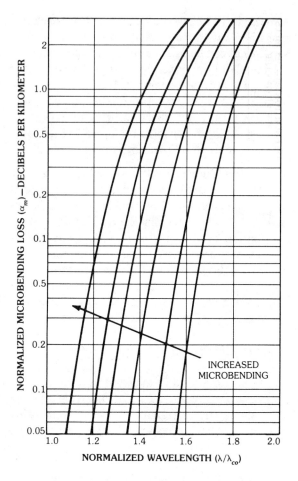

FIG. 21. Wavelength dependence of microbending-loss edge for Gaussian perturbation spectra. (*After Furuya and Suematsu [70]*)

range to decrease. This is observed in spectral attenuation measurement of the low-V value region of single-mode fibers as a function of temperature. The effect of increased microbending at low temperature is shown in Fig. 22. A very important observation is that cabling-induced microbending loss in single-mode fibers is negligible at wavelengths close to the cutoff wavelength. According to the Furuya and Suematsu model, the microbending loss α_m in the case of an assumed Gaussian power spectrum can be expressed as

$$\alpha_m = X N \overline{(1/R_1)^2} \, \bar{L}_c^2 \, \Delta^{-1} \left[-Y (\bar{L}_c n_1 / \lambda_c)^2 \, \Delta^2 \right], \quad (20)$$

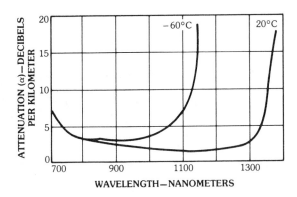

FIG. 22. Change in position of microbending-loss edge as a function of the extent of microbending, caused by exposure to low temperature. (*After Sarkar et al. [38]*)

where

N = average number of bends per meter,

$\overline{(1/R_1)^2}$ = mean square of the curvature of the fiber axis,

$\bar{L}_c$ = correlation length,

λ_{co} = cutoff wavelength,

n_1 = refractive index of core,

Δ = normalized refractive-index difference,

X, Y = factors shown in Fig. 23.

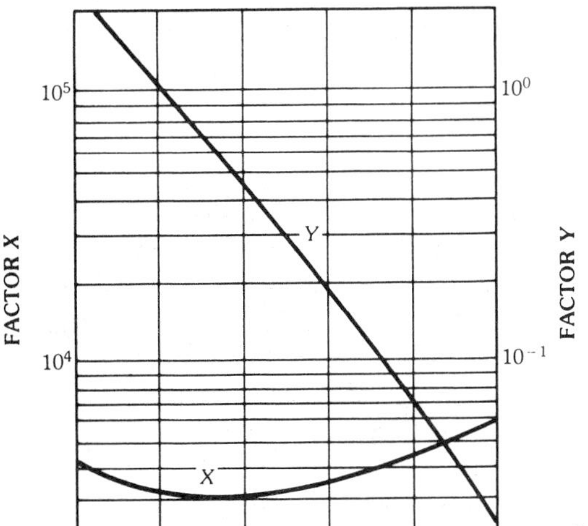

FIG. 23. Factors X and Y of the microbending-loss model. (*After Furuya and Suematsu [70]*)

Coupling Loss—In single-mode fiber systems mostly laser diode sources are used. The emitting areas of these lasers are small. Therefore one does not have to consider cross-sectional area mismatch between laser and fiber. Also, since lasers have a narrow angle of emission, the coupling losses from laser to fiber are relatively small. The source-to-fiber coupling loss can be expressed as

$$\alpha_c = 10 \log \left[(\omega_1/\omega_0 + \omega_0/\omega_1)/2 \right], \qquad (21)$$

where ω_0 is the mode-field radius of fiber, and ω_1 is the mode-field radius of source.

Splice Loss—The splice loss of single-mode fibers is governed by essentially three factors: axial offset loss, α_0, angular offset loss, α_θ, and loss due to mismatch of mode-field radius, α_ω. Therefore the splice loss α_s is expressed as [37]

$$\alpha_s = \alpha_a + \alpha_\theta + \alpha_\omega, \qquad (22)$$

where

$\alpha_a = 4.3 \, (a/\omega_0)^2$,

$\alpha_\theta = 0.02 \, \theta^2$,

$\alpha_\omega = 10 \log \left[(\omega_2/\omega_1 + \omega_1/\omega_2)/2 \right]$.

where a indicates axial offset and θ angular offset. The response of splice loss due to mode-field–radius mismatch is shown in Fig. 24. Splice-loss distributions of fibers with a mode diameter tolerance of

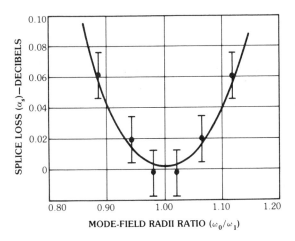

FIG. 24. Effect of mode-field diameter mismatch on splice loss. (*After Keck [71]*)

±0.5 μm are shown in Fig. 25 and are considered to be fully satisfactory from the standpoint of system considerations.

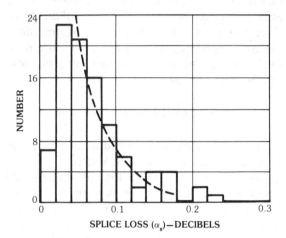

FIG. 25. Splice-loss histogram of single-mode fibers. (*After Tateda et al. [36]*)

Dispersion Considerations — The dispersive properties of single-mode fibers are a consequence of the wavelength dependence of refractive index of the core and the cladding glasses of the fiber structure, and the interaction of the fiber structure with the spectral width of the light source. This causes temporal broadening of the launched pulse. Accurate estimation of this broadening in a repeater section and its compatibility with the system information-carrying capacity are necessary for control of the bit-error rate of the system.

Total dispersion of a single-mode fiber is conventionally expressed as temporal broadening per unit length of the fiber, per unit width of the light source used. This total dispersion caused by material and structural properties of the fiber is in fact totally coupled. However, Gambling and colleagues [72], using the approximation of weak guidance, partially separated these terms. In practice this separation is widely used and the total dispersion D is expressed as

$$D = D_m + D_{\omega g}, \quad (23)$$

where D_m is the material dispersion and is expressed as

$$D_m = -(\lambda/c) \{ (P_{\text{core}}/P) (d^2 n_{\text{core}}/d\lambda^2)$$
$$+ [1 - (P_{\text{core}}/P)] (d^2 n_{\text{clad}}/d\lambda^2) \}, \quad (24)$$

and $D_{\omega g}$, the waveguide dispersion, as approximated by White and Nelson [73], can be expressed as

$$D_{\omega g} \simeq (4 \pi^2 n_{\text{clad}} c)^{-1} (\lambda/a^2). \quad (25)$$

The wavelength dependence of these two terms and the consequent total dispersion of a germania-silica–core, silica-cladding, nominally step-index, single-mode fiber is shown in Fig. 26.

Summary of Single-Mode Fiber Design Considerations—Fig. 27 shows a plot of Δ versus core radius. This diagram is the conceptual framework for single-mode fiber design. But because the core diameter cannot be accurately measured, fiber manufacturers use Δ and the cutoff wavelength λ_{co} as the control parameters. The profile is essentially dictated by the choice of process and attempts are made to keep it closely controlled.

The range of Δ and λ_{co} that can be used to produce high-quality, single-mode fibers are limited by the four limits shown in the figure. Attenuation consideration limits the upper limit of Δ for matched-cladding designs, as the Rayleigh scattering coefficient of the core increases with increasing dopant content of the core. Dispersion considerations limit the high-Δ low-cutoff corner of the specification box. In fact, dispersion limits the upper limit of Δ that can be used. This is particularly true for fibers made by the OVD and VAD processes, where the core-cladding interface is considerably diffused due to migration of germania during sintering. The lower limit of Δ as well as the lower limit of λ_{co} is due to microbending considerations, where the normalized spot size ω_0/a has the highest value. The upper limit of λ_{co} is of course limited by the modal-noise considerations mentioned previously.

In depressed-cladding fiber designs one has to also define two additional parameters. First, the thickness of deposited cladding, which is primarily decided by attenuation considerations. Maintaining a cladding thickness to core radius ratio (t/a) of 7 ensures minimum loss. A second consideration is the magnitude of index depression that

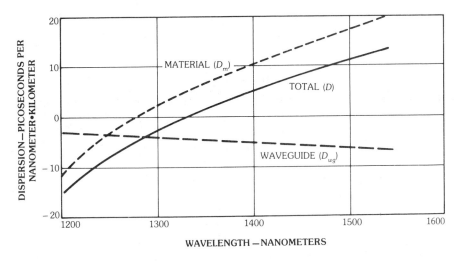

FIG. 26. Components of dispersion of single-mode fibers. (*After Keck [71]*)

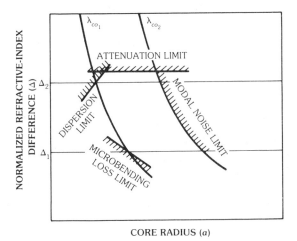

FIG. 27. Schematic diagram of single-mode fiber design constraints. (*After Sarkar [59]*)

could be permitted without having fundamental-mode cutoff problems causing high attenuation in the longer wavelengths. The extent of index depression permitted without affecting long-wavelength attenuation is a function of the deposited cladding thickness used. For t/a of 7 an index depression of more than half the fiber Δ is not recommended.

Present Status of Nominal Step-Index Single-Mode Fiber Designs—In the absence of any world-wide standardization of single-mode fibers, different users choose different ranges of functional parameters, and also specify these parame-

ters differently. Table 2 is an attempt to list the functional parameters of single-mode fibers that emerged at three different laboratories in the world, using a common set of terminologies.

TABLE 2. Single-Mode Fiber Functional Parameters

Parameter (μm)	NTT	Corning	AT&T
ω_0	4.5–5.5	4.5–5.5	4.0–4.7
λ_0	1.305–1.337	1.305–1.325	1.30–1.32

In order to understand the reasons for these differences it is necessary to first look at the corresponding differences in the control parameters used as well as the manufacturing process used. See Table 3.

TABLE 3. Single-Mode Fiber Control Parameters

Parameter	NTT	Corning	AT&T
λ_{co} (μm)	1.1–1.34	1.13–1.27	1.17–1.33
Δ (%)	0.30±0.04	0.30±0.04	0.37 nom

Both NTT and Corning use the matched-cladding design. The cutoff wavelength specification of NTT reflects a recent change in upper limit from 1.28 μm to 1.34 μm, based on experiments on modal noise. The total range of cutoff wavelength reflects the inherent variability of the λ_{co} of fibers made by the VAD process. The NTT lower limit of cutoff wavelength of 1.1 μm is due to microbending considerations. This wide cutoff wavelength range along with a rather diffused core-cladding interface of VAD single-mode fibers causes the dispersion distribution to be adversely affected. In order to use such fibers in NTT's 400-Mb/s trunk line, laser selection is used in conjunction with span lengths to ensure system performance.

The Corning specification is driven by dispersion considerations, which causes the lower limit of the cutoff wavelength to be 1.13 μm. The upper limit of the cutoff wavelength is yet to be changed to reflect recent improvement in modal-noise considerations. This rather stringent cutoff-wavelength specification is not an issue in the OVD process used by Corning, as control of refractive-index profile and dimensional tolerances is comparatively easier.

The AT&T design with its comparatively small spot size is based on AT&T's desire to have very high microbending protection for ease of economic cabling. However, because of use of depressed-cladding design in conjunction with the MCVD process, AT&T does not have to pay any attenuation penalty. This is because reduction of spot size is accomplished by reducing the refractive index of the cladding, which does not increase the scattering loss of the fiber. Furthermore, AT&T has demonstrated excellent splicing performance with fiber of this design. Their range of zero dispersion was selected based on the distribution of operating wavelength of their lasers. The longer cutoff wavelength range was helpful in controlling the zero-dispersion wavelength within the desired range.

In summary, all the designs are functional. In the future one expects further convergence in the specifications of the functional parameters. Acceptance of 1.34 μm as the upper limit of the effective cutoff wavelength is expected to spread, as it improves the dispersion characteristics of the fiber and at the same time reduces sensitivity to microbending forces.

For completeness it should be mentioned that the diameter of single-mode fibers is 125 μm, just as in the case of multimode fibers. Although this is primarily due to standardization considerations, one has to remember that the mechanical considerations are identical.

4. Technological Trends in Fiber Designs

In this section we will look at trends in two contemporary applications of fiber optics.

Trunk Applications

Since most trunk applications today are single-mode systems, in this section we will discuss trends in only single-mode fiber designs. Although today's step-index fiber is only three years old, many alternative design concepts are being vigorously pursued in research and development. Because the dominant fiber design of the future depends on system needs and advances in other system components, the different design concepts being pursued today will be listed and their key features will be pointed out.

Silica-Core Fluorosilicate-Clad Fiber—This fiber design is potentially the lowest-loss oxide-glass fiber. The two alternative refractive-index profiles for this design are shown in Figs. 16c and 28. The best-performance fibers of this design were made by the MCVD process, but such fibers have also been made by the OVD and VAD processes as well.

This design has the lowest scattering coefficient of all oxide-glass fibers, and its low-loss potential has been realized, with demonstrated median loss at 1.57 μm of 0.19 dB/km [56]. The design can be made more microbending insensitive by raising its refractive-index difference, without paying any penalties in attenuation or dispersion. In this design the zero-dispersion wavelength is in the 1.3-μm range. If it is assumed that all systems will eventually move to the 1.55-μm wavelength range to take advantage of the lowest-loss window of oxide fibers, then large-scale use of this design hinges on successful development and availability of narrow–line-width laser diodes at 1.55 μm. If the operating wavelength continues to be 1.3 μm, it is safe to say that this design will dominate as soon as it provides today's fiber performance level in production quantities.

Dispersion-Shifted Fibers—The design objective

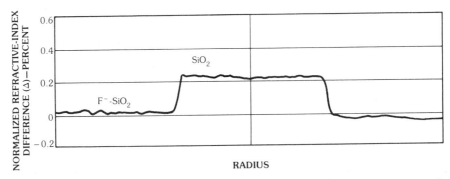

FIG. 28. Refractive-index profile of silica-core, fluorosilicate-cladding, single-mode fiber made by the OVD process. (*After Berkey [74]*)

of such fibers is to have low dispersion and low loss in the 1.55-μm wavelength range. The alternative design approaches that permit obtaining these characteristics are many. The ones that have demonstrated the best chance of commercial use are shown in Fig. 29.

Corning commercialized this type of fiber with a design shown in Fig. 29b in 1985. With the availability of single-mode lasers at 1.55 μm and 0.22-dB/km average loss with excellent control of zero dispersion, this system concept is very attractive for repeater spans of greater than 45 km at high bit rates. The specific design that Corning has commercialized is relatively complex and consequently the fibers are more expensive to manufacture than the standard design. The rationale behind the raised–refractive-index annular ring is that it desensitizes spot size and dispersion to variations of cutoff wavelength and that it makes the fiber more insensitive to microbending. Recently it has been reported [77] that it is not necessary to have the raised–refractive-index annular ring for microbending considerations, as fibers without them are more insensitive to microbending forces than today's single-mode fibers. If it is possible to control dispersion of fibers of design shown in Fig. 29a, it will be a more economical fiber for this system design.

The lowest losses in these designs to date have been higher than those obtained in the silica-core fluorosilicate-cladding fibers mentioned above. These designs are also significantly more complex than today's step-index fibers, where the need for control of refractive-index profile is much less. Therefore, only if the development of narrow–linewidth lasers runs into unexpected difficulties will

these designs have a long product life in the broadest applications.

Dispersion-Flattened Fibers—The design objectives of this fiber type are to have low loss and low dispersion in both the 1.3- and 1.55-μm wavelength ranges. The refractive index-profile of such fibers is shown in Fig. 30. This design is also very complex in structure. Moreover, the design tolerances for dispersion control are extremely stringent. In addition, the design is constrained in the size of mode-field diameter it can have, and this could affect splicing performance. Whether or not high-performance fiber of this design can be produced in quantity is also yet to be demonstrated.

Successful completion of the development of dispersion-flattened fibers will make them useful for systems where large-scale wavelength division multiplexing will be required to meet the system's capacity requirements. At this point it is not clear how many systems will have such capacity requirements.

Summary of Trunk Applications—Starting in 1983 most trunk applications began converting from multimode to step-index single-mode fibers operating at 1.3 μm. Today few trunk applications are being planned for use with multimode fibers. Simultaneously vigorous research and development is ongoing on the different single-mode fiber designs mentioned previously. Also the trend in glass composition research is in fluoride glasses for operation in the 2- to 10-μm wavelength range. A trend in system research is in coherent communications, for which single-polarization, single-mode fibers are being developed. At this point in time it is difficult to assess which of these fibers will replace today's single-mode fibers.

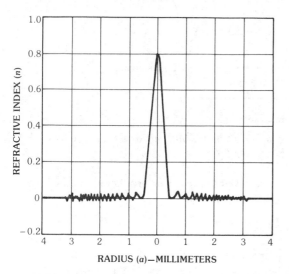

(a) Of triangular-core fiber. (After Saifi et al. [75])

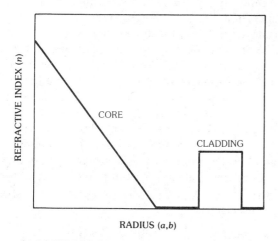

(b) Of SEGCOR fiber. (After Bhagavatula et al. [76])

FIG. 29. Refractive-index profiles of dispersion-shifted, single-mode fibers.

Distribution Systems

In the introduction section of this chapter it was mentioned that the feeder portion of the distribution system has started to use single-mode fibers. Depending on the span length these systems can use laser diodes or LEDs. Recently, edge-emitting LEDs have been developed that have been coupled to single-mode fibers [79, 80] with enough power

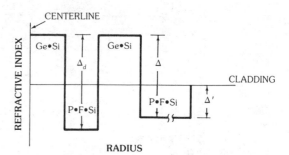

FIG. 30. Refractive-index profile of SEGCOR dispersion-flattened fiber. (*After Bhagavatula et al. [76]*)

to transmit 140 Mb/s over 50 km [81] and 560 Mb/s over 25 km [82]. The emergence of single-mode fiber systems with LEDs is accelerating the utilization of single-mode fibers in the feeder portions of the distribution systems. The fibers used in the feeder systems will be essentially the same as those in the trunk systems.

The primary question, regarding portions of distribution systems downstream of feeders, is whether they will use single-mode or multimode fibers. Electro-optic terminal costs and costs of connectorization considerations favor multimode systems. The costs of fiber and system expandability considerations favor single-mode fibers. To date most experimental systems in distribution networks downstream of the feeder systems in telecommunications have used multimode fibers. These decisions were in part based on economics and more so from considerations of availability of components. In the future many of these choices will be dictated by whether one uses distribution systems for telecommunication or whether integrated network systems are used. It is unlikely that these potential distribution systems, whatever they may be, will require a new fiber design.

5. References

[1] K. C. KAO and G. A. HOCKMAN, "Dielectric Fibre Surface Waveguides for Optical Frequencies," *Proc. IEEE,* Vol. 113, pp. 1151–1158, 1966.

[2] F. P. KAPRON and D. B. KECK, "Radiation Losses in Glass Optical Waveguides," *Proc. Trunk Communications by Guided Waves,* London, 1970.

[3] J. B. MacChesney, P. B. O'Connor, F. V. DiMarcello, J. R. Simpson and P. D. Lazay, "Preparation of Low-Loss Optical Fibers Using Simultaneous Deposition and Sintering," *Proc. Tenth Intl. Cong. on Glass,* Kyoto, Japan, 1974.

[4] D. B. Keck, P. C. Schultz and F. W. Zimar, U.S. Patent 3,737,393.

[5] T. Izawa et al., "Progress in Continuous Fabrication Process of High-Silica Fiber Preform," *Proc. Fourth ECOC* (1978), Genoa.

[6] P. Geittner, D. Kuppers, and H. Lydtin, "Low-Loss Optical Fibres prepared by Plasma-Activated Chemical Vapor Deposition (CVD)," *Appl. Phys. Lett.,* Vol. 28, pp. 645–646, 1976.

[7] J. W. Fleming, J. B. MacChesney and P. B. O'Connor, U.S. Patent 4,331,462 (1982).

[8] A. Sarkar, "Fabrication Techniques of High-Quality Optical Fibers," *Fiber and Integrated Optics,* Vol. 5, No. 2, pp. 135–149, 1985.

[9] E. E. Basch et al., "The GTE Fiber-Optic System," *Proc. NTC '77* (1977), Los Angeles.

[10] I. Jacobs and J. R. McRory, "Atlanta System Experiments Overview," *Proc. Conf. Optical Transmission* (1977), Williamsburg.

[11] T. Ito et al., "Results and Experience of the Field Trial for the First Fully Engineered Single-Mode Fiber Cable Transmission System at 400 Mb/s," *Proc. IOOC* (1983), Tokyo.

[12] W. Bambach et al., "565-Mb/s Single-Mode Transmission System With Monolithic Integrated Circuits," *Proc. ECOC* (1984), Stuttgart.

[13] S. Yoshida, "Review of New Materials for Infrared Fibers," *Proc. IOOC* (1983), Tokyo.

[14] J. A. Olszewski et al., "Optical Fiber Table for T1 Carrier Systems," *Proc. Intl. Wire and Cable Symp.,* Atlantic City, 1976.

[15] M. G. Blankenship et al., "High-Phosphorus Containing P205-Ge02-Si02 Optical Waveguide," *Proc. OFC* (1979), Washington D.C.

[16] T. Okoshi, "Review of Polarization-Maintaining Single-Mode Fiber," *Proc. IOOC* (1983), Tokyo.

[17] H. Osanai et al., "Effect of Dopants on Transmission Loss of Low-OH-Content Optical Fibres," *Electron. Lett.,* Vol. 12, No. 21, pp. 549–550, 1976.

[18] F. Urbach, "The Long-Wavelength Edge of Photographic Sensitivity and the Electronic Absorption of Solids," *Phys. Rev.,* Vol. 92, p. 1324, 1953.

[19] J. D. Dow and D. Redfield, "Theory of Exponential Absorbtion Edges in Ionic and Covalent Solids," *Phys. Rev. Lett.,* Vol. 26, No. 13, pp. 762–764, March 29, 1971.

[20] P. C. Schultz, "UV Extinction Coefficients for Titanium- and Germanium-Doped Fused Silica," *Proc. Eleventh Intl. Glass Cong.,* Vol. 3, pp. 155–163, 1977.

[21] T. Izawa, N. Shibata, and A. Takeda, "Optical Attenuation in Pure and Doped Fused Silicon in the IR Wavelength Region," *Appl. Phys. Lett.,* Vol. 31, No. 1, pp. 33–35, July 1, 1977.

[22] T. Miya et al., "Fabrication of Single-Mode Fiber for 1.5-μm Wavelength Region," *Trans. IECE* (Japan), Vol. E63, pp. 514–519, 1980.

[23] M. Horiguchi and H. Osanai, "Spectral Losses of Low-OH–Content Optical Fibers," *Electron. Lett.,* Vol. 12, pp. 310–312, 1976.

[24] R. Olshansky, "Propagation in Glass Optical Waveguides," *Rev. Mod. Phys.,* Vol. 51, pp. 341–367, 1979.

[25] K. Yoshida et al., "Low-Loss Fiber Prepared Under High Deposition Rate by Modified CVD Techniques," *Electron. Lett.,* Vol. 13, pp. 608–610, 1977.

[26] S. E. Miller, E. A. J. Marcatili, and T. Li, "Research Toward Optical-Fiber Transmission Systems," *Proc. IEEE,* Vol. 61, No. 12, pp. 1703–1751, 1973.

[27] M. K. Barnoski and S. D. Personick, "Measurements in Fiber Optics," *Proc. IEEE,* Vol. 66, No. 4, pp. 429–441, 1978.

[28] F. Hanawa et al., "Fabrication of Completely OH-Free VAD Fibre," *Electron. Lett.,* Vol. 16, pp. 699–700, 1980.

[29] N. Uchida et al., "Infrared Loss Increase in Silica Optical Fiber due to Chemical Reaction of Hydrogen," *Proc. ECOC* (1983), Genoa.

[30] K. J. Beales, D. M. Cooper, and J. D. Rush, "Increased Attenuation of Optical Fibers Caused by Diffusion of Hydrogen," *Electron. Lett.,* Vol. 19, pp. 917–919, 1983.

[31] E. Miles et al., "Hydrogen Susceptibility Studies Pertaining to Optical Fiber Cables," *Proc. OFC* (1984), New Orleans.

[32] M. Ogai et al., "Infrared Loss Increase of Silica Fiber in Hydrogen Atmosphere," *Proc. OFC* (1984), New Orleans.

[33] I. Garrett and T. J. Todd, "Review: Components and Systems for Long-Wavelength Monomode Fibre Transmission," *Optical and Quantum Electron.,*

Vol. 14, pp. 95–143, London: Chapman and Hall, 1982.

[34] R. OLSHANSKY, "Distortion Losses in Cabled Optical Fibres," *Appl. Opt.*, Vol. 14, No. 1, pp. 20–21, 1975.

[35] D. GLOGE, "Optical-Fiber Packaging and Its Influence on Fiber Straightness and Loss," *Bell Syst. Tech. J.*, Vol. 54, No. 2, pp. 245–262, 1975.

[36] M. TATEDA et al., "Design Consideration on Single-Mode Optical-Fiber Parameters," *J. Electron. Commun. Soc.* (Japan), Vol. J65, No. 3, pp. 324–331, 1982.

[37] T. KIMURA and S. YAMAKAWA, "New UV-Curable Primary Coating Material for Optical Fibre," *Electron. Lett.*, Vol. 20, pp. 201–202, 1984.

[38] A. SARKAR et al., "High Performance UV-Cured Optical-Fiber Primary Coating," to be published, in *Fiber and Integrated Optics.*.

[39] A. C. LEVY, U.S. Patent 4,432,607 (1984).

[40] U. OESTREICH et al., "Optical Waveguide Cable," *Proc. ECOC* (1977), Munich.

[41] G. LeNOANE, "Optical-Fibre Cable and Splicing Techniques," *Proc. ECOC* (1976), Paris.

[42] I. JACOBS, "Lightwave Communications Passes Its First Test," *Bell Lab. Rec.*, pp. 291–296, 1976.

[43] S. TANAKA et al., "Properties of Cable Low-Loss Silicone-Clad Optical Fiber," *Proc. ECOC* (1976), Paris.

[44] *Catalog,* York Technology, Ltd.

[45] A. J. MORROW et al., "Outside Vapor Deposition," Chapter 2, *Optical Fiber Communications,* Vol. 1, Fiber Fabrication, New York: Academic Press, 1985.

[46] H. MURATA, manuscript of speech, "Optical Fiber Design," Twentieth General Assembly of USRI, Commission C, Optical Communications (1981), Washington, D.C.

[47] M. I. SWARTZ and M. J. BUCKLER, "The Choice of Refractive-Index Difference for Multimode Fibers Operated at 1.3 micron," *ICC '81,* pp. 27.1.1–27.1.4, 1981.

[48] S. R. NAGEL, minitutorial on fiber fabrication, OFC (1985), San Diego.

[49] W. A. LOVE, U.S. Patent 4,229,070.

[50] R. OLSHANSKY, S. M. OAKS, and D. B. KECK, "Measurement of Differential Mode Attenuation in Graded-Index Fiber Optical Waveguide," *Proc. Conf. Optical Fiber Transmission* (1977), Williamsburg.

[51] S. SENTSUI and T. SHIBA, "Broadened Bandwidth of Concatanated Fiber," *Proc. OFC* (1985), San Diego.

[52] M. EVE et al., "Wavelength Dependence of Light Propagation in Long Fiber Links," *Proc. ECOC* (1970), Genoa.

[53] L. G. COHEN and C. LIN, "Pulse Delay Measurements in the Zero Material Dispersion Wavelength Region for Optical Fibers," *Appl. Opt.*, Vol. 16, No. 12, pp. 3136–3139, 1977.

[54] W. A. LOVE, "Bandwidth Spectrum Model for Multimode Fibers," *Proc. OFC* (1983), New Orleans.

[55] W. A. LOVE and D. A. NOLAN, "Wavelength Dependence of the Bandwidth Concatenation Exponent in Multimode Optical Fibers," *Proc. ECOC* (1984), Stuttgart.

[56] G. E. BERKEY, "Single-Mode Fibers by the OVD Process," *Proc. OFC* (1982), Phoenix.

[57] W. T. ANDERSON et al., "Thermally Induced Refractive-Index Changes in a Single-Mode Optical-Fiber Preform," *Proc. OFC* (1984), New Orleans.

[58] R. C. CSENCSITS et al., "Fabrication of Low-Loss Single-Mode Fibers," *Proc. OFC* (1984), New Orleans.

[59] A. SARKAR, "Design of Single-Mode Fibers for Telecommunications," to be published in *JIETE* (India).

[60] C. A. MILLAR, "Direct Method of Determining Equivalent–Step-Index Profiles for Monomode Fibers," *Electron. Lett.*, Vol. 17, pp. 458–460, 1981.

[61] F. MARTINEZ and C. D. HUSSEY, "Enhanced ESI for Prediction of Waveguide Dispersion in Single-Mode Fibres," *Electron. Lett.*, Vol. 20, 1984.

[62] S. HECKMANN, "Modal Noise in Single-Mode Fibers Operated Slightly Above Cutoff," *Electron. Lett.*, Vol. 17, pp. 499–500, 1981.

[63] Y. KITAYAMA and S. TANAKA, "Effective Cutoff Wavelength of Single-Mode Fiber," *Proc. OFC* (1984), New Orleans.

[64] Y. KATO et al., "Effective Cutoff Wavelength of the LP_{11} Mode in Single-Mode–Fiber Cable," *IEEE J. Quantum Electron.*, Vol. 17, pp. 35–39, 1981.

[65] N. K. CHEUNG et al., "Observation of Modal Noise in Single-Mode Fibre Transmission Systems," *Electron. Lett.*, Vol. 21, pp. 5–6, 1985.

[66] D. MARCUSE, "Loss Analysis of Single-Mode–Fiber Splices," *Bell Syst. Tech. J.*, Vol. 56, pp. 703–718, 1977.

[67] J. P. POCHOLLE et al., "A New Method to Determine Fundamental Parameters of Single-Mode Fibers," *Proc. IOOC* (1983), Tokyo.

[68] M. KAWACHI et al., "Low Loss Single Mode Fibre at Material-Dispersion-Free Wavelength of 1.27 μm," *Electron. Lett.,* Vol. 13, pp. 442–443.

[69] V. A. BHAGAVATULA et al., "Scattering Loss in Single-Mode Fibers by the Outside Process," *Proc. OFC* (1983), New Orleans.

[70] K. FURUYA and Y. SUEMATSU, "Random-Bend Loss in Single-Mode and Parabolic-Index Multimode Optical-Fiber Cables," *Appl. Opt.,* Vol. 19, No. 9, pp. 1493–1500, 1980.

[71] D. B. KECK, minitutorial "Fiber Theory Overview," OFC (1985), San Diego.

[72] W. A. GAMBLING et al., "Mode Dispersion, Material Dispersion, and Profile Dispersion in Graded-Index Single-Mode Fibers," *Microwave Optics and Acoustics,* Vol. 3, pp. 239–246, 1979.

[73] K. I. WHITE and B. P. NELSON, "Zero Total Dispersion in Step-Index Monomode Fibers at 1.3 and 1.55 Microns," *Electron. Lett.,* Vol. 15, pp. 396–397, 1979.

[74] G. E. BERKEY, "Fluorine-Doped Fibers by the Outside Vapor Deposition Process," *Proc. OFC* (1984), New Orleans.

[75] M. A. SAIFI et al., "Triangular-Profile Single-Mode Fiber," *Opt. Lett.,* Vol. 7, p. 43, 1982.

[76] V. A. BHAGAVATULA et al., "Bend-Optimized Dispersion-Shifted Segmented Core Designs for Specialized 1550-nm Operation," *Proc. OFC* (1985), San Diego.

[77] R. YAMAUCHI et al., "Effect of an Additional Ring Profile on Transmission Characteristics of 1.55-Micron Dispersion-Shifted Fibers," *Proc. OFC* (1986), Atlanta.

[78] V. A. BHAGAVATULA, "Single-Mode Fiber With Segmented Core," *Proc. OFC* (1983), New Orleans.

[79] O. KRUMPHOLM, "Subscriber Links Using Single-Mode Fibres and LEDs," *Proc. IOOC-ECOC* (1985), Venice.

[80] G. ARNOLD et al., "1.3-μm Edge-Emitting Diodes Launching 250 μW Into a Single-Mode Fiber at 100 mA," *Electron. Lett.,* Vol. 21, No. 21, pp. 993–994, October 1985.

[81] J. L. GEMLETT et al., "Transmission Experiments at 560 Mb/s and 140 Mb/s Using Single-Mode Fibre and 1300-nm LEDs," *Electron. Lett.,* Vol. 21, No. 25/26, pp. 1198–1200, December 1985.

[82] G. MESLENER and E. E. BASCH, unpublished work.

5

Fiber-Optic Cables: Design, Performance Characteristics, and Field Experience

P. R. BARK & D. O. LAWRENCE

Siecor Corporation

1. Introduction

In the more than 10 years since optical waveguides became a reality for practical applications, there have been tremendous strides in the development of cabling. The goal of cabling is to enable the multitude of advantages of optical waveguides to be fully realized.

The benefits of optical cables include such attributes as light weight, small diameter, and excellent transmission characteristics. The enormous information-carrying capacity over long distances offers a capability unmatched by any other guided-communication medium. However, the transmission characteristics of optical waveguides are sensitive to mechanical and environmental influences and their conservation is a major challenge in cabling.

The brittle nature of the material used for most optical waveguides is in direct contrast to the ductile behavior of the conductors in almost all electrical cables. The stress resistance of waveguides under tension and bending has improved over recent years due to advances in the waveguide and coating materials and in manufacturing techniques, but an optical conductor cannot be cabled without due consideration of its sensitivity. Nevertheless, an optical cable is expected to withstand as severe or more severe handling than equivalent copper cables.

Historically, the approach to cabling of optical waveguides tended to be one of designing the cable structure based on the ability of that structure to be applied to the fiber without impairing the inherent transmission capability. The degree and area of application for such a cable design was then evaluated. More recently the understanding of cabling technology, materials, and manufacturing methods has improved to the point where the application and environment can be considered first. This progression has led to the development of an increasing variety of different cable designs, each with its own set of capabilities optimized for a given application.

In the early stages of optical-waveguide technology development, economic justifications were rare since new technology almost always cost more than a more conventional approach. The first applications used characteristics of optical waveguide cables that could not be matched by conventional technology at any price. Such characteristics include immunity from electromagnetic and radio-frequency interference (emi and rfi), small size and low weight. Process-control and very specialized scientific, analytical, and military applications were the forerunners for widespread utilization of the technology.

As the cost of cabled fiber came down, the applications where the cost-effectiveness of the technology is significant have broadened noticeably. Trunk and toll-grade telephony applications have dominated the recent volume explosion and the cable costs have been driven down. This has led to a recent emergence of the computer and communications field as a potential large-volume application.

2. Cable Design Considerations

The specific requirements for an optical-fiber cable are diverse and vary greatly depending on the exact application. There are, however, some fundamental objectives that are applicable to every cable design. They are as follows:

1. The cable can be handled in a straightforward practical manner as with most other communications cables (wire, coax, and so on)

2. The cable has the mechanical, optical, and environmental characteristics compatible with the specific use and application

3. The cable's optical properties are as close as possible to, or identical with, the properties of the input fiber

4. The cable can be spliced and/or connectorized in the field or application with minimal difficulty and time

The key factors that must be known when designing an optical fiber cable are the fundamental properties and behavior of the particular fiber type to be cabled and the exact requirements and conditions of the specific applications. Anything less than a thorough knowledge of both of these can lead to a cable design that is less than optimal.

Fiber Stress

There are three major sources of fiber stress that must be analyzed as part of any optical-cable design. Tensile stress occurs when the fiber is axially strained. This can occur during the cable-making process and at any time when a tensile load is applied to the cable. Torsional stress occurs if the fiber is twisted. Special stranding techniques are frequently employed to ensure that the cable-making process does not contribute a torsional component. Two principal stranding techniques are commonly being used: unidirectional stranding, achieved by a planetary strander, and stranding with alternating lay, achieved by a neutralizing or an SZ (oscillating) strander. Bending or flexural stress occurs when the fiber is subjected to a bend radius. Such stress is often unavoidable in cables where stranding is used but must be recognized and included in the analysis together with cable bending.

Tensile Stress—The tensile stress that a fiber is subjected to is a function of many parameters. In the simple case when a bare fiber alone is subjected to a tensile load L (in newtons) the stress σ_t is given in newtons per square millimeter by

$$\sigma_t = L/A , \qquad (1)$$

where A is the cross-sectional area of the fiber. The tensile strain ε_t is related to the stress by

$$\varepsilon_t = \sigma_t/E , \qquad (2)$$

where E is the Young's modulus of the fiber material. For example, a typical silica-glass fiber has a Young's modulus of $E \simeq 70\ 000$ N/mm².

For practical handling and strength preservation, fibers are invariably protected by application of a coating. The coating may be comprised of more than a single component. The tensile load is shared between the fiber and the coating in the ratio of the EA (Young's modulus times cross section) product. Since the modulus of the coating in most cases is an order of magnitude or more lower than the glass fiber, its contribution can essentially be ignored.

Despite the relatively high modulus of the fiber material the load bearing capabilities of the fiber in absolute terms are quite small. For example, a fiber with the common outside diameter of 125 μm can only bear a load of 8.6 N at a strain as large as 1%. Thus in almost all cases the cable's strain behavior is a function of the other cable materials.

During manufacturing, the fiber is subjected to a screen or proof test. This is a localized, continuous, tensile force that is applied throughout the entire length of fiber after it has been drawn and coated. The purpose of the test is to strain the fiber to a certain known level which will break the fiber at any location that does not have the equivalent minimum strength. The time and magnitude of the screen test are important. A stress of 0.35 GPa for approximately 1 s is frequently used for fibers coated to 250 μm. A more meaningful way of practically stating the screen test is in terms of fiber strain. The 0.35-GPa stress on a fiber having a Young's modulus of 70 000 N/mm^2 and a diameter of 0.125 mm (or 125 μm) equates using (2) to a strain of

$$\varepsilon_t = L/A\,E = 350/70\ 000 = 0.5\% \ . \qquad (3)$$

To relate subsequent failure probability to the variables of time, length, and stress, the Weibull statistical analysis is frequently employed. This analysis requires the measurement of failure frequency under known sets of time, length, and stress conditions. From this data the failure probability is predicted for a given set of application conditions. Measured failure frequency for fiber lengths in meters (1 to 20 m), times in seconds and minutes, and stress levels of up to several times the screen test are used to predict for system lengths (30 to 100 km) and lifetimes (20+ yr) the allowable stress or strain level relative to the screen test. These predictions are the foundation for cable-stress analysis.

By reviewing the cable application the economic expectations for lifetime and reliability are first established. This allows a requirement for maximum fiber stress to be determined. Typically this will be in the range of $\frac{1}{10}$ to $\frac{2}{3}$ the screen level. With this data the design analysis may proceed.

In practical applications the tensile load on the cable may well be hundreds, thousands, or even tens of thousands of newtons. To achieve this capability while staying within the established strain limitations, tensile reinforcing materials must be included in the cable structure. The materials utilized for this purpose are selected for their high Young's modulus E, which relates to low elongation at high stress. Commonly used materials include tempered spring steel

(E=170 000 to 190 000 N/mm^2) and aramid yarns (E=110 000 to 130 000 N/mm^2).

In a cable containing more than one type of reinforcing material, the tensile behavior may be expressed

$$L = \varepsilon_t\,(E_1 A_1 + E_2 A_2 + \cdots + E_n A_n)\,, \qquad (4)$$

where
L = the total load in newtons,
ε_t = the total elongation (tensile),
E_n = the Young's modulus of the nth material,
A_n = the cross-sectional area of the nth material.

Bending Stress—In most cable designs a great deal of attention is paid to the tensile conditions and performance. Fiber stress due to bending can be a crucial factor in some cable designs and applications, but it does not always get the attention it deserves.

The stress induced in a fiber due to bending is given by

$$\sigma_b = E\,d_f/2\,R\,, \qquad (5)$$

where
σ_b = the bending stress in newtons per square millimeter,
d_f = the fiber diameter in millimeters,
R = the bend radius in millimeters,
E = the Young's modulus of the fiber in newtons per square millimeter.

In most cables containing more than one fiber the conductors are stranded together to render a flexible cable. A central member is often used around which the fibers are stranded. In this case the fibers are subjected to a bending stress by virtue of this stranding.

The radius of bend induced by stranding may be expressed

$$R = (D/2)\,[1 + (P/\pi D)^2]\,, \qquad (6)$$

where D is the pitch circle diameter, and P is the stranding pitch.

Substituting the effective bend radius R induced by stranding into the bend stress equation allows the stress σ_b to be calculated.

Torsional Stress—In most cabling techniques the fiber is rendered essentially torsion-free by using planetary, neutralizing, or SZ stranding. However, in the case where the fiber is subjected to torsion, the stress induced σ'_t is given by

$$\sigma'_t = G(\theta/\ell)\,d_f/2\,, \qquad (7)$$

where

G = the shear modulus,
θ = the torsion angle,
ℓ = the length of fiber.

The shear modulus G may be expressed in terms of the Young's modulus E and the Poisson's ratio γ as follows:

$$G = E/(2+2\gamma)\,. \qquad (8)$$

Fiber Reliability

The strength of optical fibers is mainly determined by randomly distributed surface cracks, which can grow under permanent strain and in the presence of moisture, causing delayed, spontaneous fiber fracture (static fatigue). The fracture probability F depends on the fiber stress σ and on the loading time t and is usually represented as a Weibull distribution [1]:

$$F = 1 - \exp\left[(-L/L_0)\,(\sigma/\sigma_0)^a\,(t/t_0)^b\right]\,. \qquad (9)$$

The parameters of this distribution have to be obtained by experimentation, i.e., by loading a statistically significant amount of randomly selected samples and comparing (9) with the measured time- and load-dependent cumulative frequency distribution of the fiber breakage. Because (9) is used to extrapolate from short lengths and short time frames to practically significant long time frames (i.e., $t=20+$ yr) and to long link lengths (greater than 30 km) it is very important to match (9) very carefully to the low-frequency tails of the measured distribution, not taking into account the high-strength modes around the mean value of the mea-

sured distribution. In fiber fabrication plants with a normal quality assurance program, all fibers are passing a 100% proof test in which they are screened at a test level σ_T for a short test time t_T. By rejecting all fiber pieces which do not pass this proof test level, the strength distribution is truncated, resulting in a reduced fracture probability F_t.

$$F_t = [F(\sigma,t) - F(\sigma_T,t_T)]/[1 - F(\sigma_T,t_T)]\,. \qquad (10)$$

This truncation of the original distribution has to be taken into account when computing the Weibull parameters by matching (9) and (10) to the measured results (Fig. 1). When doing this a summary of published data shows

$\sigma_0 = 2000 \pm 800$ N/mm^2,
$a = 3 \pm 1$,
$b = 0.2 \pm 0.05$.

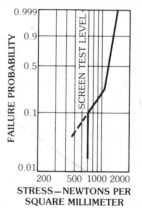

FIG. 1. Typical fiber-strength distribution with high inherent strength, weak low-frequency tail, and a truncation by the screen test.

In this case $L_0 = 20$ m, $t_0 = 10$ s for telephone-type fibers (high-grade–silica, abrasion-resistant coating). The uncertainties in the exponents a and b have a dramatic influence on the long-term behavior of a fiber link and are mainly caused by the fact that the measurements are made with a too small sample size and with restricted variations in test parameters (gauge length, load time, stress).

Optical Requirements

At the point of manufacture an optical waveguide fiber has a certain set of optical parameters associated with it. Commonly these will include attenuation (loss) and bandwidth or dispersion. In most cases the cost of the fiber is strongly influenced by the values of these parameters. It is of utmost importance for the cable designer to ensure that these parameters are not compromised during cable manufacturing or subsequent use in application.

During the fiber-manufacturing process a buffer coating is applied. The purpose of the coating is to protect the fiber surface and preserve the intrinsic high strength. In addition, it must not negatively impact the fundamental transmission properties built into the fiber. In order to achieve this over a wide range of environmental influences, the coating must be very carefully selected.

Under certain conditions microscopic distortion of the guiding part of the fiber can lead to induced attenuation increase. This microbending loss can result from a number of factors, including improperly selected coating, improperly applied coating, variation of coating modulus or dimension with temperature or humidity, coating asymmetry, inclusions, voids, and bubbles or outside distortions transferred to the coating. This last factor is the manner in which the cable and cabling process can affect the transmission parameters. The necessity to eliminate or minimize the effect of the coating on the fiber itself imposes some stringent requirements on the selection of coating materials and in many cases may lead to the use of composite (dual-layer or multilayer) coatings. Often a soft cushioning inner layer is combined with a harder protection layer to achieve optimum results. Some of the desirable properties of coatings are (a) good abrasion resistance, (b) ease of mechanical stripping, (c) chemical resistance, (d) ease of processing, and (e) environmental stability.

Two of the material systems that are most commonly selected are acrylates (heat and ultraviolet cured) and silicones or silicone/nylon combinations.

The optical fibers are generally of two types:

1. Multimode fibers having either step-index or graded-index profiles. Both fibers have a cylindrical glass cladding with a constant refractive index over the cladding radius. The step-index fiber has a constant refractive index over the core radius while the refractive index of the graded-index fiber varies with the core radius in an approximately parabolic function, having a maximum at the center of the core. Multimode step-index fibers are mainly used in nontelephone systems because of smaller bandwidth requirements

2. Single-mode fibers today are step-index fibers with a very small core diameter so that only one mode is able to propagate. The vast majority of cables which are being manufactured for telecommunication applications contain fibers of this type

3. Cable Design Principles

Every set of performance, application, and economic requirements will dictate a somewhat unique approach to designing a cable optimized for the particular combination. Cable design approaches however, can be broadly categorized to permit a basic analysis of techniques.

Without exception some kind of protective coating is applied during the fiber-manufacturing process to preserve the inherent strength of the glass fiber. In some cases the protective coating is sufficient protection to allow the fiber to be directly employed in the cable structure without further processing. In other cases an additional protection is utilized, commonly referred to as *buffering*.

Fiber Buffering

When additional protection is required, an additional layer or layers of material are applied to enhance the mechanical and environmental resistance of the fiber. The buffering is generally one of two types (Fig. 2):

1. Tight buffering, usually a harder plastic, which is in intimate contact with the coated fiber. Normally an extruded annular layer which results in a diameter of 0.5 to 1.0 mm. Materials commonly used are nylons and polyester. A special case of "tight buffering" can take the form of a laminated tape in which multiple fibers are sandwiched between two layers of tape which are in intimate contact with the fibers to form a ribbon structure. This design will be specifically addressed later in the chapter

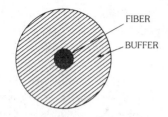

(a) Tight buffering.

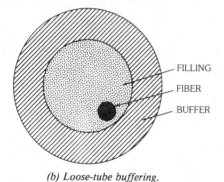

(b) Loose-tube buffering.

FIG. 2. Buffering techniques.

2. Loose buffering, which produces an oversized cavity to house the fiber and decouple it from external forces of cable strain and temperature contraction. The loose-buffering materials are selected to be hard, smooth, and flexible; and the buffering dimensions are usually in the 1.0- to 3.0-mm range. The buffer-tube size will vary, depending on a number of factors, including whether one or several fibers are to be housed and what mechanical and environmental characteristics are to be achieved. Commonly this oversized cavity is in the form of an extruded tube and is filled with special moisture-resistant compounds to exclude the presence of water, which would cause microbending on freezing. The filling compound must be soft, self-healing, and stable over a wide temperature range. Specially blended petroleum-based and silicone compounds are commonly used. The desirable properties of buffer filling compounds include [2]:

viscosity stability from −40°C to +70°C, or −55°C to +85°C, depending on application,

nondripping at +60°C to +80°C,

minimum shrinkage,

compatible with other cable components,

ease of removal,

dermatologically safe,

ease of processing,

low costs.

A secondary direct benefit of filled loose buffers is the lubrication effect of the compound. Cables whose performance is highly stable over a wide range of temperatures require that the fibers move around within the cable structure to compensate for cable expansion and contraction. The filling compound facilitates easy movement of the fiber compared with an unfilled cable, where the coefficient of friction between the fiber and the cable structure is higher.

Alternative "loose buffer" designs include folded or formed cavity tapes and extruded cavity profiles (slotted) and these will be addressed later.

Structural Member

Both of the above-mentioned buffering techniques render the fiber more resistant to mechanical influences from the application. Neither, however, do much to provide practical tensile strength. Furthermore, the high thermal coefficient of expansion of typical buffering materials compared with that of glass can often upset the temperature/attenuation performance of the fiber. Due to these facts one or more structural components may be included in a cable. Such members may serve one or both of the following purposes: (*a*) to serve as a core foundation, such as a central cylindrical component around which the buffered fibers are stranded or a slotted component into which they are laid (Fig. 3), and (*b*) to enhance the axial properties by counteracting tension due to loading applied to the cable and compression due to temperature contraction. Thus the structural members serve as an "antibuckling" element.

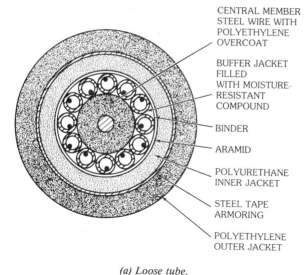

CENTRAL MEMBER
STEEL WIRE WITH
POLYETHYLENE
OVERCOAT

BUFFER JACKET
FILLED
WITH MOISTURE-
RESISTANT
COMPOUND

BINDER

ARAMID

POLYURETHANE
INNER JACKET

STEEL TAPE
ARMORING

POLYETHYLENE
OUTER JACKET

(a) Loose tube.

EXTRUDED
PLASTIC

OPTICAL FIBER

TENSILE MEMBER

HEAT BARRIER

COMPOSITE
JACKET

(b) Slotted core.

FIG. 3. Structural-member application.

The structural element may be metallic (e.g., steel wire, solid or stranded) nonmetallic (e.g., plastic aramid, fiberglass) or both (plastic overcoated metal). If more than one element is used, they should be symmetrically arranged about the cable axis. The dimensions of the structural member can vary considerably, depending on the number of fibers in the cable and the specification requirements for the cable's tensile rating and operating temperature range.

Cable Core

When cables are bent, all components of the cable not situated on the neutral axis are subjected to either compressive or tensile forces. It is standard practice in almost all types of cable to strand conductors together with a sufficiently short lay in comparison to expected bend radii such that the average position of all components approximates the neutral axis and thus the forces resulting from bending of the cable are minimized.

A fiber-optic cable core typically comprises fibers buffered using one of the previously described techniques stranded around a central component.

Practical installation conditions dictate that fiber-optic cables must be able to withstand lifetime bending radii in the region of 10 to 20 times the cable diameter. This, in turn, dictates a stranding lay length for the fibers in these cables. Such stranding bends the fibers, resulting in a strain level which must be taken into account for the allowable long-term fiber stress previously discussed.

In the case of "tight buffer" designs an attempt is sometimes made to loosely strand the buffered fibers around the central component. This is done in an attempt to create a "window" of allowable cable expansion (elongation under load) before strain is transferred to the fiber. The window in this case is typically 0.05 to 0.15% and provides only minor benefit.

In the case of loose buffer designs a significant benefit is gained from the stranding process. The fact that the fiber is free to move relative to the buffer itself allows the fiber to be essentially decoupled from the cable structure [3, 4, 5].

Elongation/contraction windows which allow fibers to remain stress free are found in stranded cables. These windows are determined by the fiber clearance inside the buffer tube W, by stranding pitch P, and by pitch circle diameter D. See Fig. 4, where η_s is the allowed elongation/contraction window as determined by W and D. This stranding-induced fiber excess length is considerably bigger than in unstranded loose-tube cable designs. When a stranded cable is strained, the fiber moves radially toward the center of the cable core. The fiber remains unstressed until it contacts the inner wall of the buffer jacket next to the central member. Typical values in stranded cables range from 0.3 to 0.8%. Cables with a broad range of allowable strain values can be

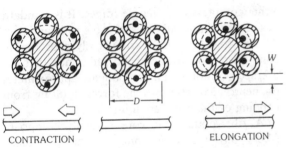

(a) Geometry of stranded loose buffer tubes.

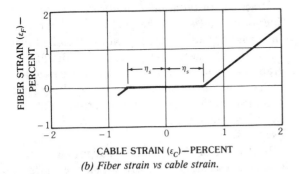

(b) Fiber strain vs cable strain.

FIG. 4. Elongation/contraction window in stranded cable.

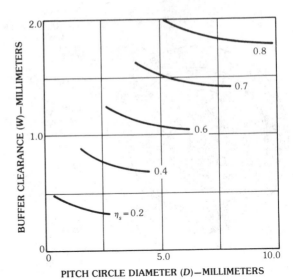

FIG. 5. Elongation/contraction window η_s dependence on pitch circle diameter and fiber clearance for constant fiber curvature (bend radius of 80 mm).

designed by selecting appropriate cabling parameters. Fig. 5 demonstrates the dependence of the elongation/contraction window on pitch circle diameter D and clearance W for constant fiber curvature (constant fiber-bending stress).

Similarly, the fiber has the ability to move radially away from the center of the cable. It remains unstressed until it contacts the outer wall of the buffer jacket. This enhances the behavior of the cabled fiber at low temperatures when all of the cable plastics tend to contract. The amount of cable contraction with temperature in a stranded cable is primarily determined by the central element around which the buffered fibers are stranded. This central element, usually consisting of steel or fiberglass, acts as a stiffening member and shows an effective coefficient of thermal expansion similar to that of silica fiber.

By analyzing the mechanics of this stranding excess length it is possible to develop the appropriate equations to allow the degree of elongation and contraction prior to fiber strain or compression to be calculated (Fig. 6).

By solving

$$P^2(\varepsilon_{c,t}{}^2 - 2\,\varepsilon_{c,t}) + \pi^2\,W(2\,D \pm W) = 0 \quad (11)$$

the following approximations can be derived:

$$\varepsilon_c = (\pi^2\,W/P^2)(D + W/2) \quad (12)$$

and

$$\varepsilon_t = (\pi^2\,W/P^2)(D - W/2), \quad (13)$$

where

P = stranding pitch,
D = pitch circle diameter,
W = fiber clearance inside buffer tube,
ε_c = contraction strain,
ε_t = tensile strain (elongation).

It can be seen clearly that the operating window can be enlarged by:

increasing W,
reducing P,
increasing D.

However, increasing W or D makes the cable larger,

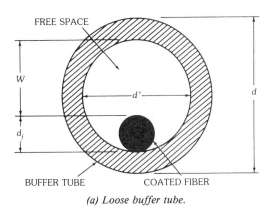

(a) Loose buffer tube.

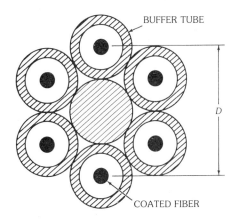

(b) Stranded-core configuration.

FIG. 6. Loose buffer-tube and stranded-core geometries.

and P can only be reduced consistent with the fiber bend-radius limitations from stress considerations.

The elongation and contraction window expressions can be related practically to cable performance. In a real cable the tensile performance is a function of the allowable fiber strain, the construction and creep stretch, and the tensile window:

$$\varepsilon = \varepsilon_f + \varepsilon_t , \qquad (14)$$

where

ε = total cable elongation,
ε_f = allowable fiber strain,
ε_t = tensile window.

Also,

$$\varepsilon_t = \varepsilon_s + \varepsilon_r , \qquad (15)$$

where

ε_s = construction stretch and creep,
ε_r = reversible tensile strain,
ε_t = tensile window.

The temperature performance is a function of the coefficients of thermal expansion and Young's moduli for the component materials. A simplified expression describing the temperature range of the cable is

$$\Delta T = \varepsilon_c / [\alpha_1 - (\alpha_2 - \alpha_1)/(E_1 A_1 / E_2 A_2 - 1)] , \qquad (16)$$

where ΔT is the temperature range compared to room temperature and α is the temperature coefficient of the material (e.g., α_1 could be the central member, and α_2 plastic). In almost all cases α and E are not constant with temperature and require approximation or modeling.

Strength Members

From the mechanical properties of the fiber it can be clearly seen that specific load-bearing components must be incorporated in a cable if the allowable stress is not to be readily exceeded in any practical application. It has been shown that cable elongations must be limited to the 0.3 to 0.8% range for reliable long-term operation. Thus the cable-reinforcing materials of most interest in optical cables are those having an elastic region up to at least 1% with as high as possible a Young's modulus. At the same time, size, weight, and cost have a great bearing on the material selection.

The most common materials used in fiber-optic cable are steels, aramid yarns, and, to a lesser extent, glass yarns. Glass and aramid yarns may appear in their raw yarn form or held in epoxy/plastic composites, which also addresses the needs of a temperature-stable central element. The moduli ranges for the common tensile materials are

steel: 170 to 190 kN/mm²,
aramid: 110 to 130 kN/mm²,
glass: 60 to 80 kN/mm².

Steel wire may be utilized either as a central single

element or, in smaller diameters, in an annular ring. The wires may bridge or may be separated, depending on the requirements.

Aramid and glass yarns are often used in an annular ring around the cable core.

Filling Compound

To retard the ingress and axial migration of water, moisture-resistant filling compounds may be used to fill voids in the cable's cross section. The filling compound must not inhibit identification or handling of the individual optical fibers and should be readily removable. Specifically formulated petroleum-based compounds or silicone compounds are most frequently used.

In nonpressurized optical cable, if metallic armoring or strength members are used which are subject to corrosion in the environment for which the cable is intended, those metallic elements should be protected by flooding with a suitable compound or by a bonded plastic coating. To prevent degradation the flooding compound must be compatible with all other components of the cable.

Filling and flooding compounds have to be nontoxic, noncorrosive, and dermatologically safe to exposed skin in normal cable handling. In addition they must be temperature stable and not flow excessively from the cable under the elevated temperatures that the cable could experience.

Cable Sheath

The function of the sheath is to protect the cable core, which contains the transmission medium. The sheath can be optimized to meet the mechanical and environmental conditions that the cable will experience. In outside plant duct installations the tension imparted to the cable is a function of cable weight and the effective coefficient of friction between the cable and the duct material. Thus, a single jacket of low-friction, abrasion-resistant material, such as polyethylene, is a good choice. With no metal, weight is minimized and the resulting small size allows the use of subducts, which is becoming a standard practice. Aerial cables see only limited tension during installation but are subjected to high ice and wind loading. These loadings are dependent on the size of the cable, which should be minimized. Again, a

sheath without metal minimizes weight and size and a single sheath has proven more than adequate in numerous installations. Direct-burial cables can be subjected to high crushing forces during installation and their subsequent lifetimes, and also to attack by gnawing rodents. Thus steel-tape armoring over a cushioning inner jacket with a durable outer jacket provides significant cut-through and crush resistance.

Cable sheaths for other nontelephony applications vary greatly. Many cables for computer and indoor applications need to meet flammability requirements, which dictate the use of special materials. In most cases a single sheath of some type is utilized. The flammability and chemical resistance properties of some common materials [6] are shown in Table 1.

4. Cable Types by Application

This section deals with cables used in telephony, computers, utility, and military applications.

Telephony

The telephony environment is the one which to date has seen the most widespread application of optical cables. In almost all cases the utilization has been in a point-to-point toll or trunk application although optical cables are now starting to appear in the feeder routes and to a lesser extent the numerous subscriber routes.

The telephony environment may be described as outside plant with one of three common installation methods. In metropolitan areas, cables are mostly installed in underground ducts either directly or within plastic pipes called "subducts." As the routes become more rural, the utilization of direct burial by plow or trench and aerial by hanging from a messenger increases significantly.

The factors which typify almost all telephony applications are

20- to 40-yr lifetime expectancy,

duct, burial, or aerial,

high installation cost,

fiber counts varying from very high to very low,

TABLE 1. Flammability and Chemical Resistance Properties of Some Common Materials*
(*After Belden Corp. [6]*)

Characteristics	PVC	Low-Density Polyethylene	High-Density Polyethylene	Polyurethane	Fluoropolymer
Oxidation resistance	E	E	E	E	O
Heat resistance	G-E	G	E	G	O
Oil resistance	E	G-E	G-E	E	O
Low temperature flexibility	P-G	G-E	E	G	O
Weather, Sun resistance	G-E	E	E	F-G	O
Ozone resistance	E	E	E	E	E
Abrasion resistance	F-G	F-G	E	O	G-E
Electrical properties	F-G	E	E	P-F	E
Flame resistance	E	P	P	P	O
Nuclear radiation resistance	P-F	G	G	G	P-F
Water resistance	E	E	E	P	E
Acid resistance	G-E	G-E	G-E	F	E
Alkali resistance	G-E	G-E	G-E	F	E
Gasoline, kerosene, etc., (aliphatic hydrocarbons) resistance	G-E	P-F	P-F	F	E
Benzol, tuluol, etc., (aromatic hydrocarbons) resistance	P-F	P	P	P	E
Degreaser solvents (halogenated hydrocarbons) resistance	P-F	P	P	P	E
Alcohol resistance	G-E	E	E	P	E

*In this table P=poor, F=fair, G=good, E=excellent, O=outstanding. These ratings are based on average performance of general-purpose compounds.

long lengths,

permanent splicing vs connectors,

high cost component of system.

Several basic designs have now emerged as standard with the telephony environment. Typical specifications are shown in Chart 1.

Loose-tube cables have seen widespread application. Their excellent isolation of the fiber from environmental stresses yields extremely stable fiber performance. The development of a high-fiber-density, miniunit variation with multiple fibers per buffer tube (2 to 12) allows large fiber counts within a relatively small size.

The central member is usually a steel wire overcoated as required with polyethylene. For applications in high-lightning areas a fiber-glass/epoxy composite rod is often substituted.

The cable core can readily allow the use of a twisted pair of copper conductors, if required for communication while splicing. Fully filled cable is almost always employed for its obvious cost and maintenance advantages over pressurized cable.

The steel central member is often not sufficient to meet the 2500- to 5000-N tensile force requirements, which are typical for this application. An annular layer of aramid yarn may be used to increase the tensile capability and provide cushioning and protection to the core.

CHART 1. Typical Telephony Specifications

Multimode	Attenuation	3.5–4.5	dB/km @ 850 nm
		1.0–3.5	dB/km @ 1300 nm
	Bandwidth	600–1000	MHz·km
Single-Mode	Attenuation	0.35–0.5	dB/km @ 1300 nm
		0.25–0.4	dB/km @ 1550 nm
	Dispersion (max)	3.5–6.0	ps/nm·km @ 1300 nm
		15–20	ps/nm·km @ 1550 nm

CHART 1 cont. Typical Telephony Specifications

Fiber count	30– 40	40– 50	70– 80	140–150
Weight (kg/km)	120–122	150–275	250–375	300–450
Diameter (mm)	12– 18	15– 19	17– 21	18– 23

Temperature (°C)			
Operating	−40 to +70	Aerial	
	−30 to +60	Underground	
Storage	−40 to +70		
Tensile force (N)			
Installation	1300–2700		
Installed	300– 600		
Bend radius			
Free	10×diameter		
Installation	15×diameter		
Crush resistance* (N/cm)	600/750		
Impact resistance*	50×3 Nm/200×5 Nm		
Flex resistance*	1000/100		

*Single-jacket/armored.

For duct and aerial applications a single polyethylene jacket is used. The more stringent requirements for crush and cut-through resistance usually dictate the use of an armored cable for direct burial. An inner polyethylene jacket is used to support a longitudinally applied, overlapped, corrugated, steel-tape armor with an additional outer polyethylene jacket (Fig. 7).

More recently, a modification of the loose-tube cable concept has been developed. Several smaller stranded tubes have been replaced by one large central circular cavity surrounded by the outside sheath comprised of layers of steel wires and polyethylene jackets. Fig. 8 depicts a cross-sectional diagram of this loose fiber-bundle cable.

A variation of the loose-structure cable for telephony applications is the slotted core (Fig. 9). In this design a "castellated" or "starfished" profile is helically extruded (continuous or oscillating) around a central steel wire or strand. The coated fibers are laid unbuffered into the slots. Filling is again employed and, after core wraps or binders, the various jacketing options previously mentioned may be utilized.

The most commonly utilized tight structure for outside plant telephony is the stacked ribbon (Fig. 10). In this design, ribbons of fibers are stacked to form a core matrix. The ribbons are laminated sandwiches of two tapes and the appropriate number of fibers. The stacked matrix is coaxial with the center of the cable, and the tensile reinforcing is applied as layers of stranded steel wires embedded in polyethylene jackets. For direct burial additional tape armoring may be employed.

Computer Applications

In computer applications the characteristics of significance differ from those of telephone applications. The transmission rates are typically no higher than several hundred kilobits per second and very often the system length does not exceed a few kilometers. The optical sources are commonly LEDs and the transmission medium is predominantly multimode fiber cable. Larger fiber cores and higher numerical apertures than the telephony counterparts facilitate power coupling and minimize connector loss.

Other specific requirements for indoor computer cables include small size, flexibility, and ease of installation in trays and in small enclosures requiring small bending radii. One-fiber jumper cables and other low-fiber-count cables are being used to connect transmitters and receivers to patch panels. These cables are preconnectorized and commonly use tight-buffered fibers. Because of the relatively short distances involved, the slightly higher attenuation levels that result from microbending in a tight structure are acceptable. Considerably improved microbending resistance can be achieved by a composite buffer-

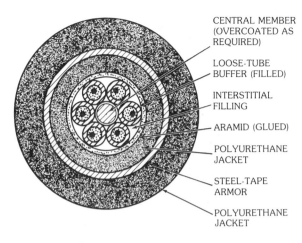

CENTRAL MEMBER (OVERCOATED AS REQUIRED)

LOOSE-TUBE BUFFER (FILLED)

INTERSTITIAL FILLING

ARAMID (GLUED)

POLYURETHANE JACKET

STEEL-TAPE ARMOR

POLYURETHANE JACKET

(a) Siecor® standard loose-tube cable: 6-fiber, double-jacket, steel-tape armored.

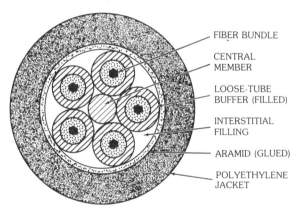

FIBER BUNDLE

CENTRAL MEMBER

LOOSE-TUBE BUFFER (FILLED)

INTERSTITIAL FILLING

ARAMID (GLUED)

POLYETHYLENE JACKET

(b) Siecor® Minibundle® (loose-tube) cable: up to 30 fibers, single jacket.

FIG. 7. Siecor® loose-tube cables. (*Courtesy Siecor Corp.*)

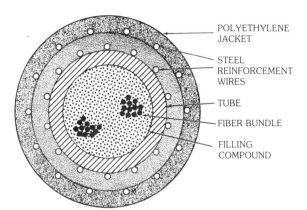

POLYETHYLENE JACKET

STEEL REINFORCEMENT WIRES

TUBE

FIBER BUNDLE

FILLING COMPOUND

FIG. 8. Loose–fiber bundle cable (AT&T Lightpack™ cable).

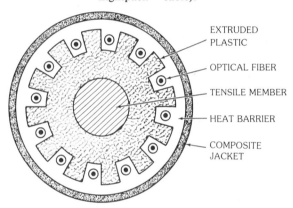

EXTRUDED PLASTIC

OPTICAL FIBER

TENSILE MEMBER

HEAT BARRIER

COMPOSITE JACKET

FIG. 9. Cavity cable (NTL slotted core).

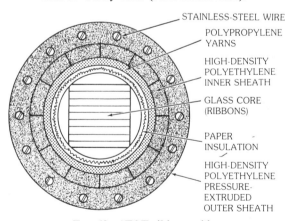

STAINLESS-STEEL WIRE

POLYPROPYLENE YARNS

HIGH-DENSITY POLYETHYLENE INNER SHEATH

GLASS CORE (RIBBONS)

PAPER INSULATION

HIGH-DENSITY POLYETHYLENE PRESSURE-EXTRUDED OUTER SHEATH

FIG. 10. AT&T ribbon cable.

ing [7]. In this case the fiber is protected by a three-layer buffering, consisting of an acrylate coating surrounded by a low-viscosity or softly crosslinked compound that is enclosed within a hard, double-layered, plastic shell. Fig. 11 depicts the cross section of a composite fiber buffering.

Representative two-fiber cables are shown in Fig. 12. Fig. 12a is a duplex intrabuilding (DIB®) cable containing either two tightly or composite buffered fibers stranded together with aramid reinforcing yarns in a round configuration. For connectorization either a duplex connector or a special reinforced fanout tubing is required. Performance characteristics for this cable are sum-

marized in Table 2. Another approach is shown in Fig. 12b. This fanout cable consists of self-contained one-fiber subunits readily available for connectorization. The diameter is larger than for the DIB cable.

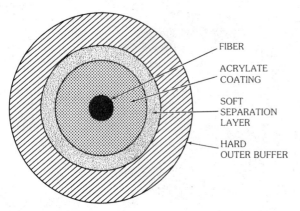

FIG. 11. Composite buffer, with a 1-mm outside diameter.

For installations in the plenum above ceilings, cables which develop no toxic smoke in case of a fire are of great importance. Fluoropolymer materials have been used so far for these plenum applications, but they do not meet the requirements for low toxicity and low smoke. New materials and cable designs are under development in the industry at this time.

Outdoor interbuilding connections are mainly based on multifiber telephony-type cable designs as described previously. In some applications hybrid designs are being used containing both fibers and twisted copper pairs. These cables are installed today for existing computer systems with an option for future expansion based on fiber optics.

Utility

The application of optical-fiber cables in the power utility environment is beginning to increase. The cables may be self-contained or composite in the ground wire or phase conductor.

The environment for the utility cable is similar to the outside-plant telephony aerial application.

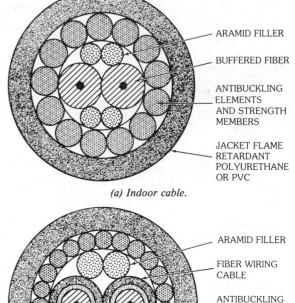

(a) Indoor cable.

(b) Ruggedized fan-out cable.

FIG. 12. Representative two-fiber cables.

It, however, is typically more severe since span lengths between supports may be 500 m rather than 50 to 100 m and more severe wind, ice, and temperature conditions often exist. Fig. 13 shows different fiber-optic utility cable design options.

The overhead power ground wire (OPGW) containing optical fibers is attractive for new installations. In this case a fiber-optic core is housed at the center of a modified ground wire. The tensile

TABLE 2. Performance Characteristics of a Two-Fiber Intrabuilding Cable

Characteristics	Performance*	
Tensile strength @ 0.6% cable strain	1000 N	$\Delta\alpha = 0.5$ dB/km, reversible
Temperature performance	(a) $-20°C$ to $+70°C$	$\Delta\alpha \leq 0.8$ dB/km
	(b) $-40°C$ to $+70°C$	$\Delta\alpha \leq 0.8$ dB/km
Crush	(a) ≤ 100 N/cm	$\Delta\alpha = 0$
	(b) ≤ 1000 N/cm	$\Delta\alpha = 1.4$ dB/km, reversible
Impact	50×2.2 Nm	No failure

*Fiber 50/125 μm, NA=0.21.

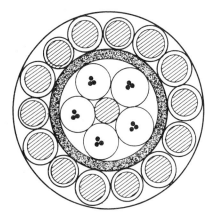

(a) OPGW.

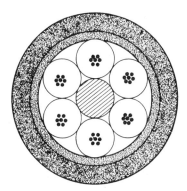

(b) All-dielectric self-supporting.

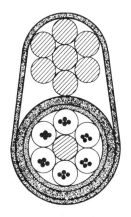

(c) Clipped.

FIG. 13. Fiber-optic cable designs for utility applications.

properties of the two must be carefully matched (Fig. 13a).

Self-supporting all-dielectric cables are also suitable for retrofit where ground wires already exist. In this case the tensile design of the cable must include the full range of installation, wind, ice, and creep conditions. Large aramid-yarn cross sections are employed (Fig. 13b).

A unique concept for utility cables is the use of an all-dielectric fiber-optic cable which is discreetly attached (clipped) to the ground wire (Fig. 13c). This technique takes advantage of the fact that the major part of the ground-wire elongation happens prior to attachment of the fiber cable. Thus the requirements are less stringent than for a composite cable where the full installation strain is applied. The discreet attachment avoids any problems that would be associated with broken lashing wires (see also Fig. 27).

In both of the above designs special jacketing materials must be employed to survive the extreme corona effects which are inevitable.

Military Applications

Fiber optics offers many advantages over the conventional metallic transmission medium that are especially applicable for military applications. These advantages include size, weight, and rfi/emp immunity. These characteristics have stimulated the design and use of fiber-optic cables for different applications where especially high reliability and more security are required. Highly sophisticated cable designs and materials have to be selected in order to meet the extreme environmental characteristics of applications under severe low and/or high temperatures, extreme tensile loads, in the upper atmosphere, or undersea at great depths.

Because of the broad range of applications there are a multitude of unique fiber types and cable design possibilities. The entire range of available multimode and single-mode fibers is being used. For most applications high proof strengths greater than 0.7 GPa (100 kpsi) and radiation "hard" performance are additional requirements.

In the following the requirements of some of the more elaborate military cables are summarized. These designs are being used today [8].

Ground-Launched Cruise Missile—The ground-launched cruise missile (GLCM) was one of the first weapon systems to be interconnected using fiber optics. Low weight, small size, and nonmetalic are the key requirements for the cable design. The cable must survive the nuclear battlefield environmental conditions, including temperature extremes of −46°C to +71°C, wind, snow, rain, hail, ice accumulation, sand/dust, salt, fog, solar radiation and altitudes of 12 000 m. In addition, the GLCM cable has to go through a series of tests which range from temperature life at 105°C to compression loading and abrasion (Table 3). The special features of the design are a flame-retardant polyurethane inner jacket, an annular layer of impregnated antibuckling elements together with aramid yarns to provide radial and axial protection, and an outer fluoropolymer jacket to withstand the thermal blast.

Underground Nuclear Tests—For cables used in nuclear experiments that are conducted in deep underground holes, special requirements have to be met. In particular, a stringent requirement for gas blocking along the cable exists (Table 4). This

TABLE 3. Ground-Launched Cruise Missile Cable Test Parameters

Test Description	Cycles	Time	Temperature	Mandrel	Mass/Load
Fiber proof test	1	na	23°C		
Temperature life	1	96 hr	105°C		
Tensile loading & elongation	1	na	23°C		1800N
Low temperature flexibility (cold bend)	1	20 hr	−46°C	40-mm diameter	12.5 kg
Impact	25	1 min	23°C	12.5-mm radius	3.0 kg
Compressive loading	1	5–10 s	23°C		700 N
Fluid immersion					
Oil	1	18 hr	121°C		
Others	1	46 hr	23°C		
Abrasion					
Before immersion	4	na	23°C		
After immersion	4	na	23°C		
Cyclic flexing	500	17 min	23°C	20-mm radius	12.5 kg
Twist bending	2000	67 min	23°C	40-mm diameter	12.5 kg
Corner bend	1	1 min	23°C		400 N
Knot test	1	1 min	23°C		400 N

TABLE 4. Typical Requirements for Underground Nuclear Test Fiber-Optic Cable

Characteristic	Requirement	Performance
Tension	100-m sample, 1100-N pull force for 1 hr, reduce to 450 N for 1 hr	Attenuation change ≤ 0.2 dB
Impact	Impact level of 1.4 N·m 200 impacts at +20°C 100 impacts at +71°C 100 impacts at −20°C	Light transmission must be maintained
Reverse bend	50 reverse bends on mandrel with radius 10 × cable diameter	Mechanical and optical integrity must be maintained
High-temperature bend	Specimen coiled on a maximum radius of 10 × cable diameter at 74°C for 6 hr	Mechanical and optical integrity must be maintained
Low-temperature bend	Same as high-temperature bend but at −29°C	Mechanical and optical integrity must be maintained
Gas block	3-m specimens temperature cycled: +74°C for 4 hr and ambient temperature for 3 hr, 6 cycles. −20°C for 4 hrs and ambient temperature for 3 hr, 1 cycle. Gas pressure of 0.862 MPa is then applied in the axial direction for 24 hr	Zero gas leakage

CHART 2. Performance Parameters for Two-Fiber Tactical Cable Utilizing Composite Buffer Techniques

Transmission Specification	
Number of fibers, multimode*	2
Lowest attenuation available at 850/1300 nm	2.8/1.0 dB/km
Highest bandwidth available	1.0 GHz·km
Performance and Use	
Storage temperature range	−70°C to +85°C
Operating temperature range, change in attenuation ≤ 0.2 dB/km	−55°C to +85°C
Maximum tensile load for deployment	1800 N
Maximum tensile load, long term, change in attenuation=0	500 N
Mechanical Properties	
Cable outside diameter, nominal	6.0 mm
Cable weight, nominal	30 kg/km
Compressive strength, change in attenuation ≤ 0.3 dB	2000 N/cm
Cold bend, change in attenuation=0	−46°C/30 mm
Twist bend, 2000 cycles	12.7 kg/30 mm
Impact, 100 cycles, change in attenuation=0	3 N·m
Fiber Data—Multimode*	
Core diameter, nominal	50 μm
Cladding diameter, nominal	125 μm
Numerical aperture	0.20
Coating diameter	500 μm
Buffer diameter	1000 μm

* Single-mode, radiation-hard, and large-core fiber also possible in this cable configuration.

gas blocking feature is required to guarantee that no radiation leaks from the experiment to the surface via the cable structure. Every cable must pass a test in which a gas pressure of 125 psi (0.862 MPa) is applied along the axis of a 3-m specimen for 24 hr without evidence of gas leakage.

To guarantee pulse delay times in multimode fibers not exceeding 10 ns, loose-tube cable designs providing a stress-free window and no measurable mode mixing are mandatory for the high cable tensile forces. In order to meet the gas blocking requirements, a special tube- and core-filling compound had to be developed. Eight-fiber cables have been developed by several U.S. manufacturers and are successfully being used in the field.

Ground Tactical Communications—One of the most promising military applications for fiber optics in the United States is the replacement of conventional metallic cables used in battlefield communications. Redeployable two-fiber tactical cable offers higher reliability at a much lower price, a weight reduction of approximately 80% over its twin-coaxial counterpart, and a dramatic reduction in the number of repeaters required in a long-haul link due to the lower attenuation levels that are possible.

Fig. 14 shows a drawing of one design option for a two-fiber ruggedized field cable. Two composite buffered fibers are stranded together with aramid fillers and are jacketed with a thin layer of polyurethane. Alternating aramid yarns and impregnated fiberglass yarns are stranded around them to provide tensile strength and temperature stability. An outer polyurethane jacket completes the cable. This very compact cable design combines excellent mechanical and environmental performance with reliability as shown in Chart 2. Most of the performance data given is based on DOD-STD-1678 or EIA RS-455 standard test procedures. Of particular significance is the excellent temperature performance of this cable. The maximum change of attenuation is 0.2 dB/km at 850 nm over 5 cycles at either end of the temperature extremes from −55°C to +85°C.

Cable designs using the tight buffer approach are also being developed.

Sonar-Buoy Undersea Cables—For sonar-buoy undersea applications very small, flexible, and lightweight single-fiber cables containing single-mode fibers are under development. These cables

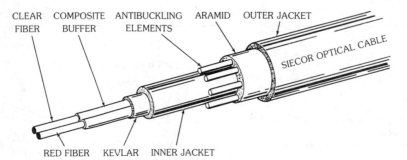

FIG. 14. Two-fiber ruggedized field cable.

are designed to meet low-loss performance requirements at both 1300 nm and 1550 nm. The application in the deep-sea environment requires continuous cable lengths ranging from 10 km to 50 km. Fig. 15 depicts a cross-sectional diagram

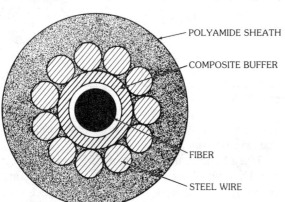

FIG. 15. Sonar-buoy undersea cable.

of a steel-wire–armored single-fiber cable. Design features include the very microbending-insensitive composite buffering and thin steel wires for armoring. The cable has a very small diameter of 2.3 mm or less and its specific gravity is 1.8 or less.

There are many more cables for different military applications available or under development. For some applications new materials and processes will need to be invented.

5. Cable Testing

In order to determine standards of performance for fiber-optic cables, test procedures have to be established simulating the environmental conditions to be encountered by the particular cable

type. Two categories of tests are of importance for the final functioning of the product.

1. *Materials Tests:* Standardized test methods determining the properties of each cable component/material in respect to its function. A number of materials being used in fiber-optic cables have already been proved in conventional copper cables. Established standards for material tests are available, e.g., ASTM, REA, DIN specifications for cable jackets and core filling and flooding compounds (see Chart 3). For the unfamiliar cable components/materials, existing standards have to be modified or new ones introduced

2. *Performance Tests:* Standardized test methods describing the properties of each cable type in respect to its use. The two categories of importance are optical and mechanical cable performance. The most comprehensive compilation of optical and mechanical test procedures currently available for fiber optics is published by the Electronic Industries Association (EIA). Working Group FO-6.4 of EIA Committee FO-6 on Fiber Optics is primarily responsible for establishing standardized test methods and instrumentation applicable for fiber-optic components (fibers, cables, connectors, etc.) for a variety of environmental uses. Chart 4 lists a number of optical and mechanical cable performance parameters which have to be determined by standard test methods. A comprehensive review of optical measurement techniques is provided in the next chapter

CHART 3. Materials Tests for Fiber-Optic Telecommunications Cables

Cable Jacket	Shrinkage
	Stress cracking
	Tensile strength/elongation
	Elastic modulus
	Tear propagation resistance
	Abrasion resistance
	Resistance against oils, acetones, gasolines, benzenes
	Ultraviolet resistance
	Fungus tests
Core Filling/Flooding	Compound drip test
	Compatibility
	Toxicity
Optical Fiber	Strength proof test
	Coating compatibility
Strength Member	Stress-strain characteristic
	Young's modulus

CHART 4. Optical and Mechanical Performance Tests for Fiber-Optic Telecommunication Cables

Optical Performance
Attenuation, bandwidth, dispersion, numerical aperture, refractive-index profile, refractive-index difference, mode-field diameter, cutoff wavelength

Mechanical Performance

Physical dimensions	Fiber
	Cable
Tensile strength	Structural
	Short-term
	Long-term
Bending	Short-term
	Long-term
Twist bend	
Cold bend	
Flexing	
Temperature (shipping, installation, operating)	High-temperature
	Low-temperature
	Cycling
Crush	
Ice crush	
Impact	
Humidity	
Wicking	
Flammability	

The most relevant cable performance test methods will be described and their results discussed.

The pass/fail criteria for mechanical performance tests are stability/change of optical transmission parameters, and/or fiber break, and the degree of visible (jacket) damage or deformation. The only optical-transmission parameter in the context of these tests is the cable attenuation. During and/or after the test the attenuation of each optical fiber in the cable should be as required in the pertinent specification. The method of measurement of attenuation or change of attenuation as proposed by EIA is based on an insertion-loss measurement using a reference length. In addition, the following conditions have to be specified: the wavelength(s) at which the attenuation is to be measured, the type of mode stripping/mode mixing, and the launching condition to be used. The accuracy of these measurements is dependent on the quality of the fiber-end preparation and the cable length.

Combined Cable Tensile and Bend Resistance

For determining the structural cable strength, a property which is mainly controlled by cross section and Young's modulus of the reinforcement elements, a short-length tensile test may be used. This test, however, characterizes the tensile force at fiber/cable break, a value which is only of secondary importance for the cable user. To ensure reliability and to simulate actual installation and handling conditions, a long-length tensile test is considerably more meaningful. The cable will be tested to determine the allowable tensile forces which, even in combination with bending, result in no significant increase in attenuation (for use) and no permanent irreversible increase in attenuation (for installation).

The test apparatus is shown schematically in Fig. 16a, and its functioning is described elsewhere. By connecting the cable to an optical power transmission and a pulse response measuring device, not only attenuation change but also fiber elongation, as a function of tensile force and cable elongation, can be measured. The results of a typical stranded multifiber underground or aerial cable are shown in Fig. 16b. Within a certain region, fiber elongation and change in attenuation are zero and independent of cable loads. Point A represents a typical value for an allowable permanent cable load for long-term use and takes into consideration a safety margin for

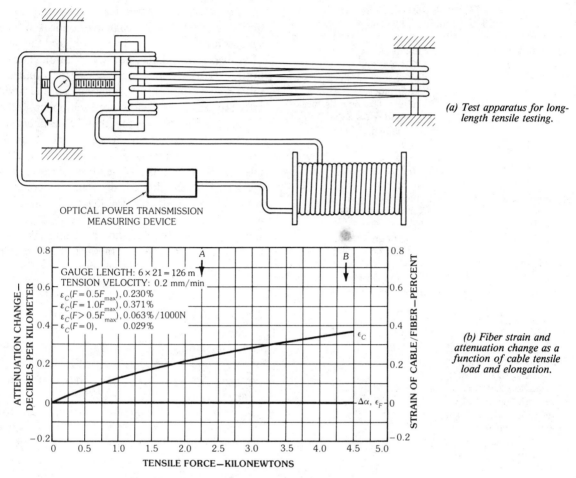

(a) Test apparatus for long-length tensile testing.

OPTICAL POWER TRANSMISSION
MEASURING DEVICE

GAUGE LENGTH: $6 \times 21 = 126$ m
TENSION VELOCITY: 0.2 mm/min
$\varepsilon_C(F = 0.5F_{max})$, 0.230%
$\varepsilon_C(F = 1.0F_{max})$, 0.371%
$\varepsilon_C(F > 0.5F_{max})$, 0.063%/1000N
$\varepsilon_C(F = 0)$, 0.029%

ATTENUATION CHANGE—DECIBELS PER KILOMETER

STRAIN OF CABLE/FIBER—PERCENT

TENSILE FORCE—KILONEWTONS

(b) Fiber strain and attenuation change as a function of cable tensile load and elongation.

Fig. 16. Measurement of fiber attenuation and elongation.

manufacturing variations and creep. Beyond a certain cable load, fiber strain and attenuation increase with increasing cable strain. These changes are reversible for short-term cable loads as applied during installation. Point *B* represents a typical value for a maximum installation load.

Impact Resistance

The cable should be tested for its ability to maintain optical transmission within specified limits when subjected to repetitive localized impacts. Fig. 17a schematically shows a test apparatus. The drop head consists of a round steel hammer with a 12.5-mm radius. The required mass is proportional to the cable cross section to be tested, and the excursion of the drop hammer from the top of its travel to the top of the test sample is constant

(150 mm). In Fig. 17b the relationship between specified impact energy in newton·meters and cable geometry is shown. Because the impact resistance of a cable is somewhat proportional to its volume, the impact energy bears a linear relationship to cable cross section for different cable sizes of a similar construction. More rugged cables (e.g., buried cables) must be designed to withstand higher impact energies. With increasing cable ruggedness, both the impact energy *E* and the number of impacts *n* increases. Actually, the total impact energy, a product of *E* and *n*, describes the impact resistance of a cable design.

Compressive Load (Crush) Resistance

The cable should be tested for its ability to maintain optical transmission within specified limits

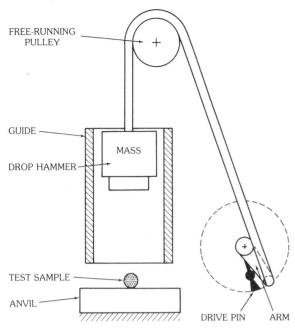

(a) Test apparatus.

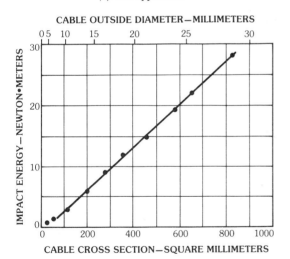

(b) Relationship between specified impact energy and cable geometry.

FIG. 17. Measurement of cable impact resistance.

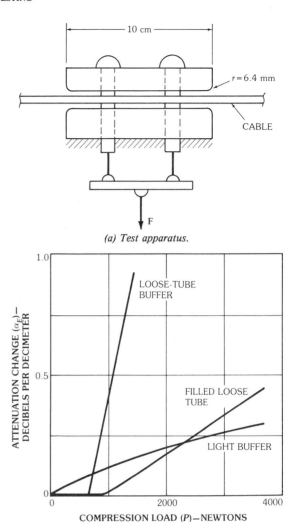

(a) Test apparatus.

(b) Attenuation change as a function of compressive load for differently buffered fibers.

FIG. 18. Measurement of crush resistance.

when subjected to a compressive or crushing load. Fig. 18a schematically shows the test setup. The cable to be tested is placed between two steel plates each 100 mm long. A constant load, again a function of cable geometry as in the impact test procedure, is applied over a certain length of time. Typical results for various types of buffered fibers are shown in Fig. 18b. While a tightly buffered fiber shows a continuous increase in attenuation with compressive load, a structure consisting of a filled buffer tube provides a certain "window," where no change of optical transmission occurs. In summary, the crush-resistance tests characterize the cable ruggedness and the fiber protection against microbending.

Temperature Performance Tests

The temperature performance of fiber-optic cables, as previously referred to, is mainly determined by microbending because of contraction forces at lower temperatures and irreversible material shrinkage at elevated temperatures. Three categories of temperature performance requirements are of importance during cable life.

Shipping—During shipping the cable is subjected to high and low temperature extremes. Attenuation change due to material contraction and shrinkage is minimized when cables are shipped/stored tightly wound on the original shipping reel. Shipping reels consisting of wood, cardboard, etc., exhibit lower expansion and contraction with temperature than the plastic material of the cable. Attenuation changes at extreme temperatures during shipping/storage are acceptable as long as they are reversible.

Installation—The temperature range for cable installation is determined by the cable material properties. The lower temperature point for cable installation is given by the jacket flexibility (jacket cracking) and the higher temperature point by the viscosity (compound dripout) of core-filling/flooding compounds.

Operating—The cable may be tested for its ability to maintain optical transmission within specified limits at temperature extremes and during cycling between them. As the cable contracts, the fiber eventually sees buckling forces and attenuation increases. The degree of sensitivity of temperature dependence is strongly a function of both the fiber type and the cable structures. The loose constructions which mechanically decouple the fiber from the cable structure allow some of the most stable performances.

The sensitivity of multimode fibers decreases with increasing cladding-to-core diameter ratio and numerical aperture. In the case of single-mode fibers the sensitivity increases as the wavelength of operation becomes much greater than the cutoff wavelength and comes closer to the bending edge, and with decreasing refractive-index difference. Fig. 19 shows a typical temperature/attenuation performance of single-mode-fiber Minibundle® cables at 1300 nm and 1550 nm. This attenuation stability is excellent and variations are less than 0.05 dB/km.

6. Cable Placement

In the following sections field experiences with optical cable and related hardware will be described. The main emphasis is put on large-scale toll and trunk applications in the United States, where since 1983–1984 most of the requirements have been met by single-mode fiber cable technology.

Fiber-optic cable does not require special handling techniques other than compliance with the tensile and minimum–bend-radius specifications. Cable has been installed and is successfully operating under widespread climatic conditions. Duct, aerial, direct buried, trenched, and vertical installations have all been successfully implemented. Fiber counts from 12 to 156 have been installed, spliced, and put into operation. The handling of 2- to 12-km continuous lengths is routine. The benefits of long cable lengths are elimination of splice points, time and cost savings, and increased reliability [9–16].

Placement of Duct Cable

Optical cable is pulled into underground ducts or plastic subducts using modified conventional cable-pulling methods. A sturdy pulling eye is attached to the strength elements of the optical cable. A pulling rope or winch line is then connected to this pulling eye with a swivel. To minimize friction of the cable with the conduit walls a pulling lubricant can be used. Because of the small size, light weight, and flexibility of optical cable, the necessary pulling line force is considerably less than that used with conventional copper cables. During the pulling-in operation the cable's pulling line tension should not be allowed to exceed the "installation rating" given in the specifications. Typical underground cable ratings are in the 1300-to 4500-N (300- to 1000-lbf) range. Tension monitoring may be done simply and accurately with commercial dynamometers or load-cell instruments. Fig. 20 shows the strain of a duct cable that is pulled with the "installation rating" of 2700-N pulling force into a straight duct. During installation the cable strain increases along the run and is maximum at the pulling head. After installation the cable ends are released, but in the center of the pull friction may cause up to half of the strain to be permanently present.

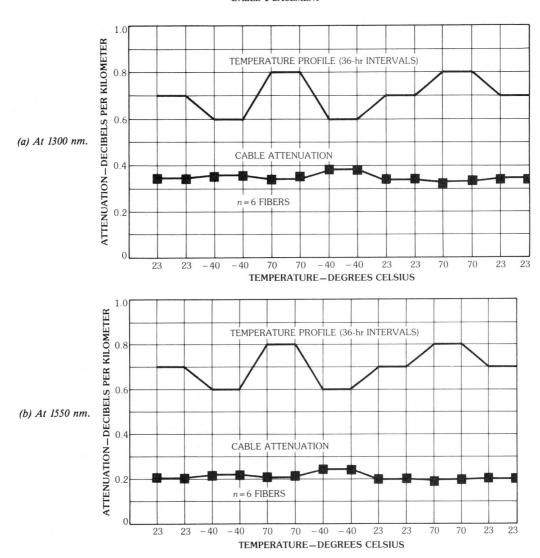

FIG. 19. Temperature/attenuation performance of single-mode–fiber Minibundle® cables.

The optical cable must not be forced around sharp corners of a duct terminus at either feed or pulling end. Pulling sheaves and cable guides are commonly used to maintain appropriate bending radii, typically 20 cm. Because of the small size of optical cables valuable duct space is commonly used more economically by pulling optical cables in 1-in (2.54-cm) inside-diameter innerducts. The preferred maximum degree of duct fill is 50% by area, which allows for a 0.7-in or 18-mm maximum cable outside diameter.

Center pulling of cable is recommended for long, continuous runs. At the span midpoint half of the cable length is pulled from the shipping reel into the conduit in the usual fashion. When this portion is complete the remaining half of the cable must be removed from the reel and the other end made available for pulling in the opposite direction. This is done by hand-pulling the cable from the reel and laying it, as shown in Fig. 21, in large figure-8 loops. The purpose of the figure-8 pattern is to avoid cable twisting and kinking. The advantages of this method are reduced crew size and reduction in the amount of equipment necessary.

In some instances small cable capstan drives on pulling machines are used for intermediate pulling

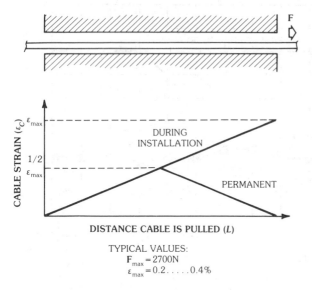

FIG. 20. Strain of duct cables, with typical values of $F_{max} = 2700$ N and $\varepsilon_{max} = 0.2$ to 0.4%.

of long cable lengths (see Fig. 22). When, during placement, the cable reaches its recommended "installation rating," an intermediate pulling machine is installed in the last manhole. The next section of the route can then be installed using the standard pulling technique until it becomes necessary again to install the next intermediate pulling machine, etc. Depending on the nature of the route, the route pulling tension diagram [17] may have the characteristics shown in Fig. 23.

Placement of Buried Cable

To provide protection against rodents and gophers, buried cables contain an additional metallic or all-dielectric armoring (see Section 3, under "Cable Sheath."). These cable types can be placed by means of a conventional cable-laying plow. As with conventional cables the reel is placed on the plow. The optical cable is then guided over a deflector or a wheel that is large enough to ensure that the minimum bend radius is not exceeded. The cable is then

FIG. 21. Laying out the cable in a figure-8 for pulling into the cable duct system.

Fig. 22. Intermediate takeup unit, cable guide, and drive (direction of operation reversible) in a manhole.

which cable plows are typically designed. Thus several optical and/or power cables can be plowed in in a single pass. Generally, optical cables have sufficiently high tensile strength and crush resistance to withstand the plowing-in forces. In very cold parts of North America, burying up to 1.5 m in depth can be done without any problems or change of performance characteristics. A warning tape is usually buried along with the cable, but at a lesser depth.

More recently, trials have been undertaken to plow unarmored optical cables that have been put into an innerduct. Thus, the outside diameter of the composite cable-innerduct is larger and reduces the risk of rodent attack because of the size and gnawing angle of the rodent's teeth. More field experiences with this method have to be accumulated in order to decide whether this is a reliable alternative for burying cables.

Instead of burying cables by plowing, the whole installation can be carried out by trenching. In case of limited distances or in rough terrains, trenching may provide the more economical or the preferred method.

fed into the plow blade and through the entrance conduit entering buildings or repeater locations.

The overall diameter of optical cable is much smaller than that of conventional copper cable for

Placement of Aerial Cable

Aerial installation of optical cables has been shown by actual field experience to be a very fast

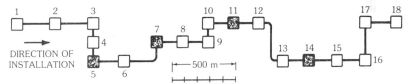

(a) Cable duct with 18 manholes.

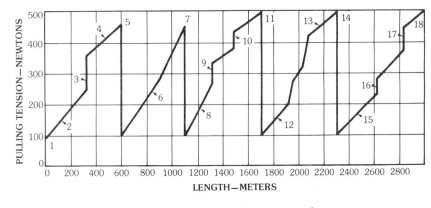

(b) Pulling tension-length diagram.

Fig. 23. Route pulling tension diagram for a 3-km placement section.

and effective method. The standard method is lashing of the optical cable to a messenger wire. A steel messenger wire is strung along the pole spans, and the optical cable is subsequently lashed to it. The messenger wire need not be handled with any great care, and the spools of optical cable are much smaller and easier to handle than in the case of self-supporting cable. Furthermore, both the optical cable and the messenger wire can be separately specified for the particular requirements of the installation.

Appropriate messenger wire is standard high-strength steel wire, such as grade 180, in the range of ¼-inch (6.35-mm) diameter. This is installed with standard eyebolts and clamps between the poles with the tension calculated for the span length and anticipated climatic conditions. High cable strain will occur in aerial installations where additional wind and ice loads are present (see Fig. 24). In many parts of North America ice loads in

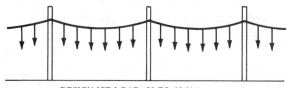

DESIGN ICE LOAD: 30 TO 60 N/m
MAXIMUM CABLE STRAIN: 0.4 TO 0.8%

FIG. 24. Strain of aerial cables.

the range from 2.2 to 4.4 lbs/ft (32 to 64 N/m) can be expected, depending on the geographical area and cable size. Under these conditions and for normal pole distances of approximately 200 ft (61 m), the strain of optical cables can reach values in the order of 0.5%. The corresponding cable sag is approximately 5 ft (1.5 m).

Depending on the terrain, aerial installation can be performed in either a "pull-in" or "drive-off" mode. In a pull-in installation, rollers are hung periodically from the previously installed messenger wire. A craftsperson in an aerial bucket moves along the route with the end of the optical cable and hangs it into the rollers as a second craftsperson pays it off from the *stationary* spool. When the end of the route is reached and enough excess cable is pulled through, the craftspersons retrace the route to clamp the cable to the poles and lash it to the messenger wire with a lashing machine. A drive-off

installation tends to go more quickly. The first end of the cable is clamped to a pole and to the messenger wire and then pulled along. The optical cable is fed from a spool on a *moving* truck which precedes the zone of lashing. In Fig. 25 a typical installation is shown.

FIG. 25. Installation of an aerial fiber-optic cable which is first pulled along the messenger on rollers and then lashed.

Other aerial cable designs are figure-8 cable, self-supporting steel-wire armored, or self-supporting cable. Some of these cables are designed for larger pole distances and for utility applications. For these cables the installation methods require more equipment and the attachment hardware is more rugged and complicated to handle the larger stresses and withstand the high electrical fields.

A unique way of adding high–information-transmission capacity to existing power lines is clipping an all-dielectric fiber-optic cable to the overhead ground wire or even to one of the electrical conductors. A specially designed "clipping" machine lashes the optical cable to the overhead ground wire or conductor using individual metallic clips. As shown in Fig. 26, the preformed clips are bent around both the optical and the metallic cable as the clipping machine travels along the pole span. The advantages of this method are that a standard optical cable can be used without adding an expensive amount of dielectric strength members to the cable and that no short circuit can occur because of a broken lashing wire connecting two high-voltage conductors.

7. Fiber Splicing

In long-haul communication systems where high–fiber-count cables and high traffic density are typical, reliable splice loss and relatively short splicing times are very important. Several methods of joining optical cables have been established and optimized over the past 5 to 10 years. The most common are thermal (fusion) and mechanical splicing. During the last two years single-mode–fiber performance has improved to the point where attenuations at 1300 nm and 1550 nm of 0.4 dB/ km and 0.25 dB/km, respectively, are now commonplace. That means that splice-loss targets of 0.1 dB to 0.2 dB are necessary. In order to achieve these performance levels very accurate fiber geometry, tight tolerances, and perpendicular fiber-end cuts/surfaces are required. As a result fibers, splice parts, and splicing tools have to meet and handle tolerances in the micrometer or even submicrometer range.

Fusion Splicing

For optimized thermal splicing of optical fibers, fusion splicing devices have been developed worldwide [18]. Fig. 27 shows a typical fusion splicer. Such machines utilize an electric arc for thermally

FIG. 26. Clipping machine lashes optical cable to overhead ground wire. (*Courtesy Siemens AG*)

FIG. 27. Typical single-mode fusion splicer.
(*Courtesy Siecor Corp.*)

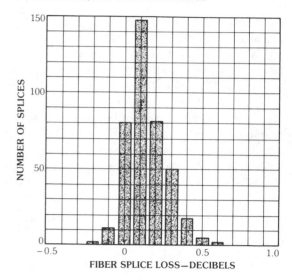

FIG. 28. Typical single-mode fusion splice-loss distribution (OTDR).

fusing both fiber ends permanently together into a mechanically stable connection. Integrated cutting tools yield fiber-end angles of 1° or less, values which are required to achieve low single-mode–fiber splice losses. Most of the fusion splicers are semiautomated or fully automated to achieve reliable operator-independent splices under field conditions.

Single-Fiber Fusion Splicing—A typical splice-loss histogram of single-mode fusion splices is depicted in Fig. 28. This distribution represents three splice points of a 132-fiber cable, i.e., a total of 396 fiber splices. The average splice loss is 0.09 dB; the standard deviation is 0.12 dB. About 94% of all splices were below 0.3 dB. The measurements were taken with an optical time-domain reflectometer (OTDR). Therefore, individual numbers may be somewhat in error; but the average for all fiber splice losses is accurate. During the course of splicing some splices were remade. The average number of splice attempts was approximately 1.4. Splicing times, including setup of the splice location and equipment, cable-sheath removal, fiber stripping, fiber-end preparation and performance testing, splice closure preparation, and sealing and closing the manhole, are in the range from 10 to 12 min per splice. In total, a 132-fiber cable can be spliced by three splice crews in approximately 24 hr. The actual fiber splicing time, i.e., fiber stripping, and preparation and fusion is of the order of 4 to 6 min.

After splicing, all fibers are stored in individual splice trays (see Fig. 29). Within the splice tray

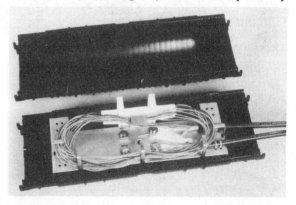

FIG. 29. Spliced single-mode fiber in splice tray.

there is typically enough fiber slack available to remake a splice several times, if necessary. All splice trays are stored inside a universal plastic closure suitable for direct buried, underground, and aerial nonpressurized applications.

In order to increase productivity during field splicing, several local splice-monitoring systems have been developed. "Local injection detection" or LID systems for single-mode–fiber field splicing have been developed by several manufacturers and successfully proved in the field. These modules inject light into the fiber core on one side of

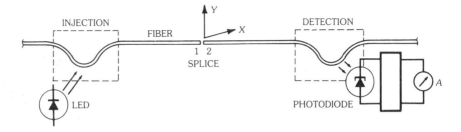

FIG. 30. Principal elements of an LID system.

the splice and detect the light in the fiber core on the other side of the splice. Light injection and extraction is facilitated by bending the fiber around a very small mandrel (see Fig. 30). Use of the "LID" system results in both significantly higher productivity through fewer remakes and an improved splice performance.

Multifiber Fusion Splicing—For splicing high-fiber-count cables, such as in subscriber distribution networks, reduced splicing time is most desirable. Several companies have developed multifiber fusion machines. As most of the processing steps are being carried out simultaneously, highly reliable methods have to be used for coating removal, fiber cutting, fusion, and fiber protection. Critical are the reliable functioning of the 12-fiber cutting tool and equal heat distribution for all fibers in the electrical arc. By means of specially designed fiber-holding fixtures these requirements can be met. The actual fusion process includes the same basic elements of prefusion and automatic feed as in the single-fiber fusion process. Typical laboratory results with this process for multimode fibers are as follows:

$$\bar{\alpha}=0.16 \text{ dB}, \qquad \sigma=0.11 \text{ dB}.$$

Mechanical Splicing

There are two major categories of mechanical splicing: multifiber array splicing and mechanical single-fiber splicing.

Multifiber Array Splicing—For less demanding applications, i.e., in most subscriber loop systems where splice-loss requirements are not critical, splicing time in the field can be reduced by factory-installed silicon array connectors. For both cable types, single-mode minibundle and ribbon cables, array connectors have been used (see Fig. 31). Six- and twelve-fiber arrays for minibundle

cables and twelve-fiber arrays for ribbon cables have been field proved. After preconnectorization, including end polishing in the factory, the cables are pulled through ducts using specially designed pulling eyes which protect the connectors. Fig. 32 is a histogram for the splice loss for

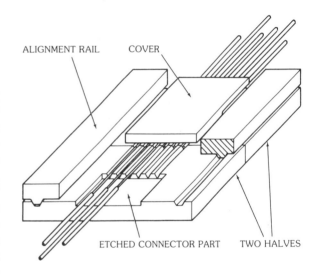

FIG. 31. Silicon array splice for six single-mode fibers.

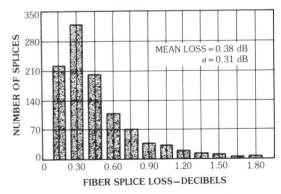

FIG. 32. Performance histogram of silicon array chip splicing of single-mode ribbon cable. (*After Gartside and Baden [19]*)

1030 fibers using twelve-fiber array chips [19]. The mean loss and standard deviation are 0.38 and 0.31 dB, respectively. Not counting the time for connectorization in the factory, the setup of splice location, splice closure preparation, and performance testing a 144-fiber ribbon cable with twelve-chips for twelve single-mode fibers each can be spliced in less than 3 hr by craft personnel in the field.

Mechanical Single-Fiber Splices—For lower–fiber-count cables, several single-fiber connectors/splices have been developed and used in the field. The following summarizes the basic principles and performance characteristics of the most commonly used elastomeric, bonded, and rotary connectors.

Elastomeric Splice—The GTE elastomeric splice is designed to be used in a field environment by unskilled operators [20]. The primary function of the elastomeric splice is to align two fibers to a common centerline. This self-alignment capability of the splice is due to the geometry of the three component parts. These are injection molded using an elastomeric polyester. Fig. 33 includes an end view of an assembled splice. There are two center inserts and an outer sleeve. One of the inserts has a precision 60° groove molded along its length. The opposing insert is molded with a flat surface along its length. When these two parts are mated they form a triangular cross-sectional hole into which the fibers are inserted. The external geometry of these two inserts together is a hexagon which is inserted into the elastic sleeve. The dimensions of this hexagon are such that its vertices deform slightly when inserted into the sleeve and the resulting forces hold the entire assembly together.

The splice can be used by itself or in conjunction with a plastic housing (closure) for additional protection. Aside from cable preparation, a splice can be completed in approximately 5 min. Typical splice losses in the temperature range from −30°C to +85°C are 0.25 dB.

Bonded Splice—The bonded splice, designed at AT&T Bell Laboratories for low splice loss, provides optimum fiber alignment and bonds the

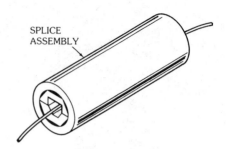

(a) Three quarter view.

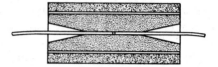

(b) Cross section.

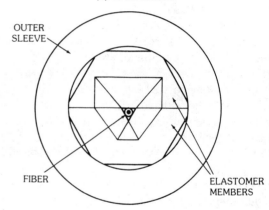

(c) End view.

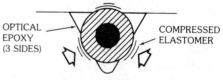

(d) Compression detail.

Fig. 33. Optical-fiber elastometric splice. *(After Knecht, Carlson, and Melman [20])*

fiber with an ultraviolet-curable adhesive [21]. The basic component of the single-mode bonded splice is a cylindrical glass plug, called the *glass terminus*, with a precise capillary hole to accept an uncoated fiber. The terminus, shown in

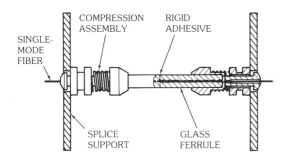

FIG. 34. Single-mode bonded splice. (*After Reynolds and Cagen [21], ©1984 IEEE*)

Fig. 34, is packaged in a plastic housing with a compression spring and the complete product is called a *compression assembly.*

Rotary Splice—The rotary splice is based on the principle of rotation of low-eccentricity splice parts [22]. This allows for precise alignment to within 0.1 μm without elaborate alignment equipment and without having tolerances tighter than 0.1 μm. Alignment of the rotary splice relies on precision glass ferrules installed on the fibers and a triangular alignment sleeve. Built-in eccentricities of the bore within the glass ferrules and an offset built into the alignment sleeve allow relatively large rotational movements to provide small movements of the fiber cores relative to each other. A splice-loss histogram for 226 splices is shown in Fig. 35. Typical splicing times are 15 to 20 min.

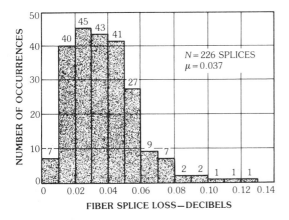

FIG. 35. Splice-loss histogram of single-mode rotary splices.(*After Miller and DeVeau [22]*)

8. Performance of Optical Passive Transmission Subsystems

Worldwide there has been a large number of fiber-optic cable system installations and the transmission characteristics of some of them have been monitored for up to ten years. In the following pages the field testing, performance, maintenance/restoration, and reliability of passive transmission subsystems (PTSSs) will be discussed.

Field Testing

Field testing of fiber-optic cables is usually performed in order to monitor the fiber-splicing process, to ensure the end-to-end transmission specifications (attenuation, bandwidth, and dispersion) are met, and to provide baseline data for cable plant maintenance and reliability [9].

The most common field tests before, during, and after cable installation are summarized in Table 5. For each test one or more methods are given. Whether one elects to do all or any of the tests depends on the specifications and the margins involved.

Multimode and single-mode optical time-domain reflectometers (OTDRs) are commercially available for 850-nm, 1300-nm, and 1550-nm wavelength ranges, respectively. They are suitable for cable testing, splice testing, system performance evaluation, and fault locating.

Short- and long-wavelength attenuation test sets for both multimode and single-mode fibers are commercially available. Test sets with dynamic ranges from 30 to 60 dB are typical for measuring insertion loss. The dynamic ranges of these test sets are determined by the type of diodes used in the transmitters and receivers. While LEDs are commonly used for multimode systems, laser transmitters are required for testing long single-mode–fiber systems.

For verifying end-to-end bandwidth performance of an installed multimode fiber cable system, portable test sets consisting of transmitter and receiver units are commercially available for 850 nm, 1300 nm, and 1550 nm. With the help of a laser transmitter at either wavelength the amplitude response in the frequency domain is measured. The accuracy of the measurement is in the order of ±10 MHz.

TABLE 5. Summary of Fiber-Optic Field Tests

Time	Test	Method
Prior to installation	Attenuation	OTDR
Splicing	Splice attenuation	1. OTDR 2. Local monitor
	Connector/end splice attenuation	1. OTDR 2. Insertion loss 3. Local monitor
After installation	End-to-end attenuation	Insertion loss
	End-to-end dispersion	Phase measurement

The end-to-end dispersion of single-mode fiber systems can be measured using various techniques. So far, only a few test devices are commercially available. All methods use three to five lasers operating at different wavelengths centered around the operating system wavelength. The techniques are based on (a) time-domain pulse-delay measurement, (b) single-frequency phase measurement, and (c) swept-frequency measurement. The only field units available, however, are based on phase measurements of sinusoidally modulated lasers [23].

System Performance

In order to be able to evaluate passive transmission subsystem performance one has to know the performance of individual cables as manufactured. Fig. 36 shows a typical attenuation distribution of single-mode–fiber Minibundle® for two wavelengths. The mean for thousands of fiber kilometers of cable is 0.358 dB/km at 1300 nm and 0.209 dB/km at 1550 nm with very small σ values. Very similar values have been reported for loose–fiber-bundle cable (AT&T Lightpack™) [24].

In Table 6 performance data for a few single-mode passive transmission subsystems are summarized [25]. The given end-to-end attenuation figures are averages for all fibers in each system and represent the kilometric loss of the installed and spliced cable for the given span length. The table shows span lengths, cable type, and manufacturer, number of splices in the system, splicing method being used, and performance at 1300 nm and 1550 nm, where available. These examples demonstrate the excellent performance possible for installed passive transmission subsystems at both wavelengths, independent of cable type and splicing method.

To prove the end-to-end system bandwidth, dispersion has to be measured. A typical laboratory test setup for bandwidth measurements of single-mode fibers is based on the use of a Raman laser. The expected field performance based on 40 randomly selected, matched-clad, single-mode fibers measured with the Raman laser and 10 000 computer simulations resulted in a mean system dispersion of 2.0 ps/nm·km at 1285 nm and 1.7 ps/nm·km at 1330 nm [26]. The probability of a 30-km span exceeding the dispersion of 3.5 ps/nm·km within the wavelength range from 1285 nm to 1330 nm is 10^{-10}. Based on these calculations the system dispersion performance can be expected to be excellent.

Recent measurements of passive transmission subsystem dispersion confirmed these expectations. A Siecor SM Dispersion Test Set M75 was used to measure a 36.6-km span and a 73.2-km span. The specified dispersion performance in the wavelength range from 1285 nm to 1330 nm is

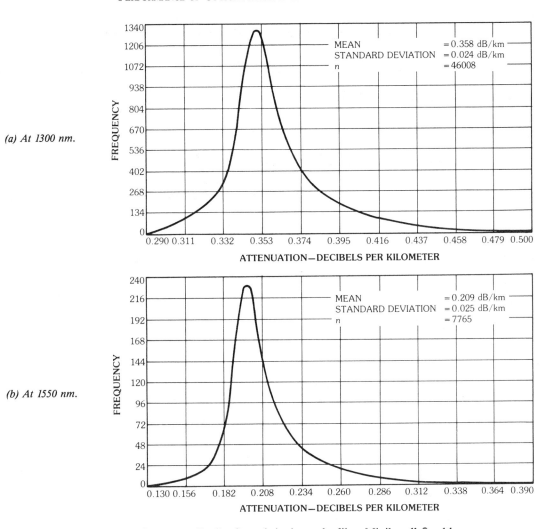

(a) At 1300 nm.

(b) At 1550 nm.

FIG. 36. Performance distribution of single-mode–fiber Minibundle® cable.

±3.5 ps/nm·km. The actual measured values for several fibers of both spans were −2.6 to −2.8 ps/nm·km at 1280 nm and +1.3 to +1.6 ps/nm·km at 1330 nm. The repeatability of the measurements for all fibers was excellent and the dispersion characteristics were better than the manufacturer's specifications.

System Reliability

In general, installed fiber-optic cable systems have an enviable record for reliability although some failures have been reported [27, 28]. As far as the user is concerned, the reliability of the entire passive transmission subsystem is of importance. In this context the PTSS is defined as the fiber-optic transmission medium between fiber distribution frames or between transmitters and receivers.

Cable system reliability is influenced by the reliability of the system components, namely the input/output connectors/pigtails, splices, and the cables themselves. In the following the reliability of installed cables and splices is summarized. Connector reliability is not addressed.

The largest number of fiber failures have been in the "uncontrollable extrinsic" category. Extrinsic failures include uncontrollable breaks, such as those due to digups, breaking poles, collapsing ducts, etc. All types of cable, fiber and copper, are subject to such failures. Most of the fiber failures have been due to complete cable cuts. The second largest

TABLE 6. Typical Single-Mode–Fiber System Performance Data

System	Manufacturer	Cable Type	Length (km)	No. of Splices	Type of Splices	Average End-to-End 1300 nm	Performance (dB/km) 1550 nm
A	AT&T	Ribbon	14.4	8	Bonded	0.43	0.31
B	AT&T	Stranded	27.0	14	Rotary	0.51	0.38
C	AT&T	Stranded	30.0	38	Bonded	0.57	0.41
D	Siecor	Minibundle	31.4	9	Fusion	0.45	0.33
E	Siecor	Minibundle	34.0	8	Fusion	0.44	0.31
F	Siecor	Minibundle	37.0	10	Fusion	0.43	Not measured
G	Siecor	Minibundle	43.5	8	Fusion	0.41	Not measured
H	Siecor	Minibundle	22.3	14	Fusion	0.41	0.29

source of extrinsic failures has been gunshot damage to aerial cable. In one instance, for example, two out of eight buffer tubes in a 46-fiber single-mode cable were damaged. To a certain degree, steel-tape armoring can reduce this failure mode.

The field splice reliability data shown in Table 7 relates mostly to single-mode fusion splicing. There were a total of two failures out of 2889 fusion splices representing over 37 million splicehours. Both of these were avoidable. This extremely low failure rate is remarkable especially because only six of the systems were spliced by experienced fiber splicers. The remainder, on the average, were spliced by personnel who received three days of training and field supervision.

Restoration

To maintain systems in service it is clear that troubleshooting and repair procedures have to be established.

Once it is determined that the trouble is in the cable the problem fibers are accessed with an OTDR. Cable faults, increased attenuation, and failed splices or connectors are readily identifiable. If failures impair the transmission performance of a small fraction of the fibers or splices in a minibundle cable, for example, the repair, in most cases, can be handled without service interruption, as long as the system is protected or there are spare fibers in the cable. Repairs of defective fibers or splices can be carried out without disturbing the good fibers.

Most of the cable failures are catastrophic and uncontrollable, being caused by backhoes and other extrinsic events. Emergency restoration procedures are necessary for such events. The requirements are

(*a*) rapid temporary restoration and (*b*) the ability to transition to a permanent repair without disturbing any of the fibers except the one being worked on. In the case of fusion splicing, procedures and emergency restoration kits have been developed to meet these requirements. In applications where preconnectorized cables are being used, the entire defective cable section can easily be replaced by a new spare preconnectorized cable.

9. Future Applications and Requirements

The use of fiber-optic technology for the transmission of digital voice, data, and video communications is commonplace throughout the telecommunications industry. A major trend towards the utilization of single-mode fiber-optic cables for toll and long- and short-haul trunk applications has been observed. Decreasing cost trends of both cable and terminal electronics and the elimination of large numbers of digital repeaters are the main reasons for the success of fiber technology.

The next levels of integration for fiber-optic technology are subscriber feeder and distribution networks. In the following the requirements are defined and cable design considerations are discussed.

Subscriber Feeder

Fiber-optic subscriber feeder network topologies are significantly different from those of trunk applications. Unlike trunk applications, where circuit demands are point-to-point with both end

TABLE 7. Summary of Reliability Data for Some Siecor Optical Cable Systems

System	Total Fiber km·hr	Fiber Failures After Certification			Splice Failures/ Total Splices M—Mechanical F—Fushion	Comments
		Extrinsic		Intrinsic		
		Uncontrollable	Avoidable			
1	1 003 968				2/50M	Mechanical splice failure due to improper handling
2	1 331 712				0/72F	
3	134 784				0/3F	
4	767 232		1		0/12F	Cable kinked during re-reeling, failure occurred 7 months after installation
5	929 664			1	1/100M	Factory fusion splice failed after installation. One mechanical splice failed
6	2 437 344	11			0/114 M&F	Cable damaged by rodent Birdshot damage to aerial cable
7	2 124 864	6				Buried cable severed by backhoe
8	536 544				0/54M	
9	2 742 336	6			0/78F	Duct severed
10	628 704		4		0/56F	Cable twisted—swivel inoperative
11	525 312		8			Cable kinked—inadequate protection in central office
12	990 720				0/30F	
13	1 683 360				0/70F	
14	990 720				0/20F	
15	1 050 624	6			0/60F	Backhoe
16	807 864				0/50F	
17	673 680				0/42F	
18	4 071 408				2/304F	Improper cable anchoring in closure
19	1 765 536				0/136F	
20	1 135 800				0/64F	
21	772 416	8			0/80F	Backhoe
22	95 808				0/18F	
23	586 080				0/60F	
24	753 312				0/64F	
25	195 840					
26	486 072				0/60F	
27	988 992	14			0/90F	Backhoe (twice). Shotgun
28	412 800				0/72F	
29	1 209 600				0/78F	
30	677 280				0/90F	
31	230 376				0/24F	
32	4 193 280				0/560F	
33	329 640				0/80F	
34	128 352				0/30F	
35	218 880				0/60F	
36	155 664				0/12F	
37	409 248				0/130F	
38	1 749 840				0/270F	
Total	39 925 656	51	13	1	5/3093	The 3093 splices represent 42 447 966 splicehours

locations well defined, subscriber feeder routes have one end point located in the central office and several remote points located throughout the exchange area. Fig. 37 shows a typical feeder topology [29].

The initial step in developing a new design technique for the subscriber feeder environment was introduced as a result of the large scale use of digital subscriber carrier equipment. The carrier serving area (CSA) concept was introduced and adopted by many operating companies to address the different feeder topology and subscriber pair gain requirements.

Feeder cables will typically range between 4 km and 20 km in length. Distribution beyond the remote equipment location typically employs voice-frequency copper pairs but can also use 1.544-Mb/s lines.

As mentioned previously, one of the most obvious differences between feeder and trunk applications relates to the end points of the cables. As with trunk systems one end of the feeder terminates in a central office (CO). The other end of the conductors, whether copper or fiber, may terminate in one of a number of different remote terminal locations. Given that the forecasted growth requirements and the location of these requirements are well defined, the feeder design is only slightly more complex than trunk applications. However, not only are access line requirements difficult to forecast, it is also hard to predict with any amount of certainty where along

the route they will occur. Therefore, the feeder design and equipment must be flexible to allow reconfiguration to satisfy changing requirements.

The requirements for single-mode fiber-optic cable are as follows: (a) a modular concept with varying fiber counts from four to six fibers per unit for highest flexibility, (b) unique unit identification distinguishing working and protection channels, and (c) easy access to units for rearrangement purposes without disrupting service. Designs with stranded buffer tubes or minibundles can meet the requirements listed above. Because there usually will be many remote equipment locations along a feeder, the cables used will typically taper to smaller fiber counts as they progress from the central office toward the exchange boundary. Fig. 38 shows a typical cable assignment scheme with tapering to smaller–fiber-count cables. For each of the remote locations A, C and E, four working and four protection fibers are provided. As a result, for example, fibers in bundle 6 in the cable leaving the central office may be spliced to fibers in bundle 4 of the next cable at the first taper point. Typical layered-cable design options are shown in Fig. 39. These designs can provide the flexibility required for the feeder loop application.

Distribution Network (Local Loop)
In the distribution network high–fiber-count cables are required to run through the main servicing areas and one- to four-fiber drop cables are needed for

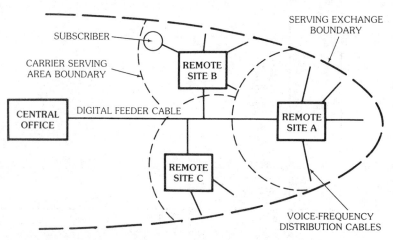

FIG. 37. Typical feeder topology.

DIGITAL SUBSCRIBER CARRIER TERMINALS ARE LOCATED AT CENTRAL OFFICE AND REMOTE SITES.

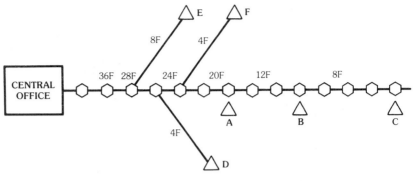

FIG. 38. Feeder-cable assignment scheme.

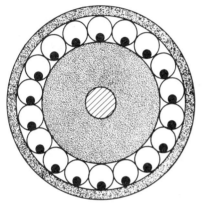

(a) Single-layer design.

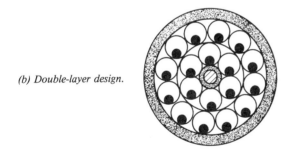

(b) Double-layer design.

FIG. 39. Layered feeder-cable designs.

connection to the subscribers. Although the economics do not justify using fiber-optic technology in the local network today, several field trials in the United States, Europe, and Japan have been under way for several years.

Again, proved loose-tube, slotted-core, and ribbon-cable designs are conceivable for the high–fiber-count cables required for this application. A multifiber subunit can form the basic module in fiber-optic cables for the local network. This subunit may consist of up to 10 or 12 fibers housed in a loose tube, inside a cavity of a slotted core, or in the form of a tape. Several of these subunits are stranded around a central member and form a 50- to 120-fiber subunit. Several of these subunits will be stranded, again, around a central member—if necessary in several layers—to build the final cable core. Core filling, tensile strength reinforcement, and outer polyethylene jacket complete the cable. Fig. 40 shows a cross-sectional diagram of this unit cable concept. Fig. 41 shows a photograph of a unit-based cable with a total of 800 fibers [30]. Cables of this type can be made either all-dielectric or metal-containing. Multifiber splicing will be prerequisite for a successful application in the next decade.

Acknowledgments

The authors wish to thank many coworkers at Siecor Corporation and at Siemens AG for providing valuable inputs, contributions, and suggestions for this chapter.

10. References

[1] P. R. BARK, D. O. LAWRENCE, O. I. SZENTESI, and G. H. ZEIDLER, "Reliability of Fiber-Optic Cable Systems," *ICC 1980/IEEE Intl. Conf. Commun.,* pp. 10.3.1–10.3.4, Seattle, June 1980.

[2] P. R. BARK and D. O. LAWRENCE, "Design and Performance of Filled Fiber-Optic Cables," *Globecom 1982,* pp. 275–277, Miami, 1982.

[3] P. R. BARK, U. OESTREICH and G. ZEIDLER, "Stress-Strain Behavior of Optical-Fiber Cable," *28th Intl. Wire & Cable Symp.,* pp. 385–389, Cherry Hill, November 1979.

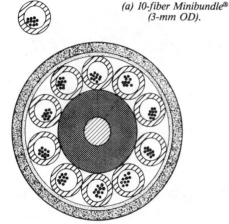

(a) 10-fiber Minibundle® (3-mm OD).

(b) 100-fiber subunit cable (15-mm OD).

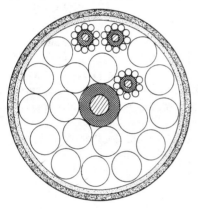

(c) 2000-fiber unit cable (55-mm OD).

FIG. 40. Unit cable concept for distribution network applications.

FIG. 41. Fiber-optic cable with 800 fibers grouped in units, (*After Mayr, Schinko, and Schoeber [30]*)

[4] P. R. BARK, U. OESTREICH and G. ZEIDLER, "Fiber-Optic Cable Design, Testing and Installation Experiences," *27th Intl. Wire & Cable Symp.,* pp. 379–384, Cherry Hill, November 1978.

[5] P. R. BARK and D. O. LAWRENCE, "Emerging Standards in Fiber-Optic Telecommunications Cable," *Proc. SPIE Tech. Symp. East*, 1980.

[6] Belden Corporation, *Design Guide for Electronic Wire & Cable.*

[7] P. R. BARK, D. O. LAWRENCE, U. OESTREICH and E. MAYR, "Composite Buffering for High-Performance Fiber Cables, *33rd Intl. Wire & Cable Symp.,* pp. 98–101, Reno, 1984.

[8] M. G. MASSARANI, "Fiber Optics—Communica-
tions of the Future," *Mil. Tech.,* Vol. VIII, Issue 6, pp. 31–40, 1984.

[9] O. I. SZENTESI, "Singlemode Cable Technology, Installation, Splicing, Testing and Maintenance," *Tech. Bull. Siecor Corp.,* May 1984.

[10] R. SOYKA, "Optical Fiber Cable Placing Techniques —Long Section Lengths," *Proc. 26th Intl. Wire & Cable Symp.,* pp. 281–286, Cherry Hill, November 15–17, 1977.

[11] F. KRAHN, W. MEININGHAUS, D. RITICH, K. SERAPINS, and J. GLADENBECK, "Installation and Field Measurement Equipment for Optical Communication Cables," *Proc. 28th Intl. Wire & Cable Symp.,* pp. 357–363, Cherry Hill, November 13–15, 1979.

[12] A. L. HALE, D. L. POPE, and D. R. RUTLEDGE, "Lightguide Cable Installation in Underground Plant," *Proc. Fiber Opt. Commun. Expo.,* pp. 227–231, San Francisco, September 16–18, 1980.

[13] D. L. POPE and N. E. FLENNIKEN, "Installation of

Lightguide Cable," *West. Elect. Eng.,* Vol. XXIV, pp. 103–108, Winter 1980.

[14] J. B. MASTERSON, "Innovative Methods for Installing Fiber-Optic Cables," *Proc. 29th Intl. Wire & Cable Symp.,* pp. 299–305, Cherry Hill, November 18–20, 1980.

[15] O. I. SZENTESI, "Installation and Testing of Fiber-Optic Cable Systems," *Proc. Fiber Opt. Commun. Exp.,* pp. 18–21, Los Angeles, September 15–17, 1982.

[16] E. Y. LOYTTY and R. K. LUX, "A Planner's Guide to Fiber Optics—Installation Considerations and Case Histories," *Telephony,* Vol. 203, pp. 42–55, September 20, 1982.

[17] H. GOLDMANN, "Installation of Fiber-Optic Cables," *Telcom Rep.,* Siemens, Vol. 6, pp. 46–49, October 1983, special issue.

[18] H. LIERTZ, "Fusion Splicing Machines for Optical Fibers," *Electron. Lett.,* pp. 426–427, Vol. 16, May 22, 1980.

[19] C. H. GARTSIDE III and J. L. BADEN, "Singlemode Ribbon Cable and Array Splicing," *OFC 1985,* pp. 106–107, San Diego, 1985.

[20] D. M. KNECHT, W. J. CARLSON, and P. MELMAN, "Fiber-Optic Field Splice," *Proc. SPIE Intl. Soc. Opt. Eng.,* Vol. 326, pp. 57–60, Los Angeles, 1982.

[21] M. R. REYNOLDS and P. F. CAGEN, "Field Splicing of Singlemode Lightguide Cable," *IEEE-ICC,* pp. 1071–1074, Vol. 3, Amsterdam, 1984.

[22] C. M. MILLER and G. F. DEVEAU, "Simple High-Performance Mechanical Splice for Singlemode Fibers," *OFC 1985,* pp. 26–27, San Diego, 1985.

[23] P. R. BARK and D. O. LAWRENCE, "How Singlemode Cable Performs Its Task," *Telephone Engineer & Management,* Vol. 89, No. 12, pp. 66–73, June 15, 1985.

[24] C. H. GARTSIDE and C. F. COTTINGHAM, "Production and Field Experience With AT&T Lightpack Cable," *OFC 1986,* pp. 112–113, Atlanta.

[25] A. G. VEDEJS, "Design and Performance of Optical Fiber Cables," presented at OFC 1985, San Diego, 1985.

[26] Corning Glass Works, "Dispersion in Singlemode Fibers," *Product Information Sheet III,* Issue 1/84.

[27] O. I. SZENTESI, "Field Experience with Fiber-Optic Cable Installation, Splicing, Reliability and Maintenance," *IEEE J. Selected Areas in Commun.,* Vol. SAC-1, pp. 541–546, April 1983.

[28] O. I. SZENTESI, "Reliability of Optical Fibers, Cables, and Splices," to be published in the *IEEE J. Lightwave Tech.*

[29] M. R. AMICONE, "Requirements for Subscriber Feeder Applications," presented at USTA '85, Las Vegas, May 21–23, 1985.

[30] E. MAYR, M. SCHINKO and G. SCHOEBER, "Fiber-Optic Unit-Based Cables in the Local Network," *Telcom Rep.,* Siemens, Vol. 6, pp. 37–40, October 1983, special issue.

Optical-Fiber Measurements

M. L. DAKSS

GTE Laboratories, Inc.

1. Introduction

This chapter will describe recently developed techniques for the characterization of the optical properties of fibers. Since the techniques for measuring multimode and single-mode fibers can differ and, in fact, the key parameters required to describe the fiber differ for the two types, multimode and single-mode fiber characterization techniques are treated in separate sections. Where overlaps occur, they are indicated.

The emphasis is to describe techniques tutorially so that those without much experience in the field can learn their implications and what is involved in their implementation. An attempt to describe the most recent developments in measurements is made. Descriptions of earlier techniques are found in the review papers, which are referenced. Standardization aspects are also mentioned.

Certain subjects are not included, however. One is measurements of geometrical properties of fibers, preforms, and protective coatings or buffers on fibers; some aspects of these are treated in [1, 2]. Another is refractive index profile measurements, which is covered for preforms in [1, 3, 4] and for fibers in these references and also in [3, 5]. Also not covered are mechanical properties or microbending and macrobending properties (for

the latter two in single-mode fiber, see [4, 5, 6]). Among single-mode fiber measurements, those of birefringence and polarization properties are not covered. The reader is referred to [7, 8].

2. Multimode-Fiber Measurements

Since the distribution of modes excited in and propagating down a fiber can have profound effects on the light transmission properties in the fiber, it is worthwhile to review some relevant modal effects, particularly differential modal attenuation and mode coupling. The appropriate background will be given in the following subsection. The subsequent three sections treat near- and far-field pattern, attenuation (including the loss of joints, i.e., splices and connectors) and bandwidth measurements.

General multimode fiber reviews appear in [1, 9, 10, 11, 12]. Those relating to standards work are [13, 14, 15, 16, 17, 18].

Modal Properties and Steady State

To understand the evolution of the modal distribution as light propagates down the fiber, we look first at differential modal attenuation (DMA). Fig. 1 shows the attenuation of one type of graded-index fiber as a function of m/M, where m is the principal mode number and M is the maximum principal mode number for the fiber [19]. The higher-order (larger m/M) modes show a higher loss. This is because they have a larger percentage of light concentrated in the outer part of the core and into the cladding. These "outer" modes have a greater sensitivity to microbending and to geometrical variations of the fiber than lower ones, and some see the absorption in the fiber coating. (However, note that more recent fibers operated at 1.3 μm or beyond show quite a bit less DMA than that of Fig. 1.) The DMA tends to "clip off" the outer modes after a sufficient length.

the light propagates down the fiber, mode coupling produces a spread in the spot size [21] because, although mode coupling occurs to modes of smaller as well as larger mode indices, there are more modes excited at small indices to begin with so there is a net flow outward. The spot continues to grow until it reaches a size at which DMA clips off its outer modes. A "steady-state" spot size results, after which no change in the distribution occurs with further distance, only a lessening of mode power. The steady-state distribution has also been called the equilibrium mode distribution, and the distance at which it occurs, the steady state length.

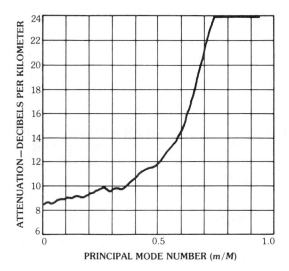

FIG. 1. Total losses at 632.8 nm versus mode number for graded-index fiber. (*After Olshansky [19]*)

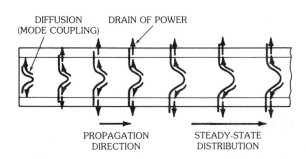

FIG. 2. Approach to steady-state distribution.

The other phenomenon, mode coupling, couples power between different modes, usually most strongly among modes of neighboring modal index. This can be a result of microbending or macrobending in the fiber, perturbations in the fiber such as random diameter variations, or scattering, e.g., Rayleigh scattering, in the fiber.

The effect of mode coupling and DMA is shown schematically for a graded-index fiber in Fig. 2 (see [20, 21]). The fiber is shown with an initially "underexcited" modal distribution, corresponding to predominantly low-order (inner) modes. For a graded-index fiber this would be produced by an excitation with a suitably small, low–cone-angle axial beam, such as that obtained from a laser. As

The approach to steady state is also indicated in the power versus distance curve of Fig. 3. For an overexcited distribution the loss is larger because the outer modes are lossier via DMA. These get clipped off and eventually steady state is reached. For underexcitation the loss is less than the steady-state value because the lossier outer modes included in the steady state are not present. Once the distribution spreads to steady state the attenuation goes up and remains constant at the steady-state value.

The positive or negative loss added to the steady-state one for the approach to steady state has been called the transient loss [10, 22]. These correspond to the extra or smaller power lost in the approach to steady state. The steady-state length for overexcited conditions differs from that for underexcited conditions because of the different mechanisms involved. In many recent fibers designed for long-distance transmission the mode coupling is so low at the popularly used near-infrared wavelengths that the lengths required to approach steady state for negative transient losses

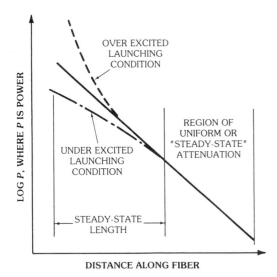

FIG. 3. Power versus distance along a multimode fiber for different input-launching conditions. (*Adapted from Cherin [11], ©1983, with permission of McGraw-Hill Book Co.*)

can be in the tens of kilometers and can reach 100 km [23, 24]. The distances over which clipping occurs, however, are much shorter (typically a few kilometers) and a "quasi-equilibrium" distribution very close to the steady state (because of small DMA) is reached [10, 23]. It is noted that the steady-state is the unique (source-independent) fiber attenuation. The transients, which occur after sources and other discontinuities like joints (the biggest transients occurring with LED sources), should be taken into account in the source or joint specifications [10, 25]. For the joint, transient losses give rise to a difference between measurements with long and short fiber lengths. It will be discussed below that the steady-state attenuation is the one which is additive between concatenated fibers.

It is noted that not all communication fibers have long distances to reach steady-state conditions. For fibers designed for shorter-distance applications (e.g., in local-area networks), much shorter steady-state lengths can be involved, since fibers with more scattering could be used.

Near- and Far-Field Patterns— Numerical Aperture

The near-field distribution is the light-intensity distribution at the exit face of the fiber waveguide.

At increasing distances past the exit face the intensity pattern changes from the Fresnel to the Fraunhofer regime and approaches the "far-field" pattern, which remains constant with respect to angle as the distance increases farther [26]. Generally, the far-field distribution is reached at a distance $R \gg x^2/\lambda$, where x is the transverse dimension of the near-field pattern and λ is wavelength [27]. For a multimode fiber with a 50-μm core diameter x and $\lambda = 1$ μm, we get $R \gg 2.5$ mm. For a single-mode fiber, $R \gg 0.1$ mm. In both cases the far field is reached in a convenient distance for making measurements.

Both the near- and far-fields are direct indicators of the fiber modal power distribution [2, 28, 29, 30] and these patterns are utilized in measurements of several fiber parameters [27].

Near-Field Pattern—A straightforward procedure for measuring the near-field distribution, the transmitted near-field technique [27, 31, 32], is shown schematically in Fig. 4. The test fiber, typically 2 m long, is overfilled with light, i.e., light having a spot size greater than the core diameter, and a cone angle greater than the acceptance angle of the fiber (see next section) is centered on the fiber core. The light beam should be uniform, so that it will uniformly excite all modes in the fiber. (More details on how this launching can be achieved are given in the section on attenuation.) The source can be an LED or a filtered white-light source. Then a lens is used to form a magnified image of the near field. An apertured detector is scanned in the image plane to obtain the near-field pattern. A chopper/lock-in amplifier combination can be used for noise reduction, which is useful since only a small part of the light pattern is detected. See [27] for a typical pattern.

A cladding-mode stripper should be included to prevent the cladding light from contributing to the pattern. To do the stripping the coating or buffer on the fiber is removed and the fiber is immersed in a groove some 15 cm long and slightly deeper than the fiber. The groove is filled with a liquid of refractive index slightly higher than the cladding index and covered. Glycerine is a good, inexpensive choice but this liquid should be changed frequently since it absorbs water vapor, which lowers its index. An alternative to the trough is a black

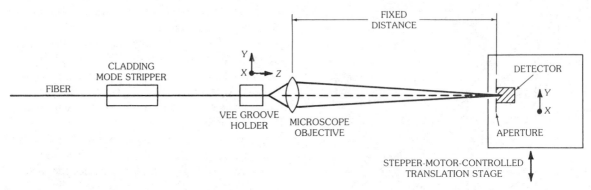

FIG. 4. Near-field measurement system based on a radial scan of a magnified near-field image, where the launching optics are not shown. (*After Kim and Franzen [27]*)

velvet pad which is soaked with the fluid, the fiber being held down with a small weight. Care has to be taken to minimize stress on the fiber, which can change the modal distribution being measured, for example, by producing microbends or macrobends.

In many fibers the coating or buffer acts as an excellent mode stripper, eliminating the need for an external one. This should be investigated for the particular fiber type.

The near-field measurement has several potential areas of concern. These include the need for fiber-end stability, a high-quality cleave and a high-quality lens giving a distortion-free, flat-field image. Locating the image plane accurately is difficult. Also, a proper calibration of lens magnification is needed. The scanning detector could be replaced by a vidicon tube but problems of detection linearity and geometric distortions can occur. Several other considerations on the near-field technique, such as resolution and dynamic range, are described in [27].

Several other types of near-field measurements have also been developed (e.g., the refracted near-field measurement [3]). The transmitted near-field measurement has been standardized by the Electronics Industries Association (EIA) in their Fiber-Optic Test Procedure (FOTP) RS-455-43 [33]. This measurement has been utilized to determine the index profile of graded-index fibers [1, 3, 31] and also the core diameter from this profile [2, 32]. The EIA has standardized using the transmitted near-field (and refracted near-field) techniques for measuring core diameter, while the CCITT chooses the refracted near field as the Reference Test Method (RTM) and the transmitted near field as an Alternative Test Method (ATM) [17, 18].

Far-Field Pattern: Numerical Aperture—The most popular technique for measuring the far-field pattern is the rotated-detector technique [27], which is shown in Fig. 5a. Fiber excitation (not shown) is the same as for the near-field measurement. The need for proper mode stripping and gentle clamping is also the same. An apertured detector is located on a turntable with the fiber positioned so that its end face is on the axis of rotation. The detector is scanned and the detected power (actually, power per unit area, or intensity) versus the angle is the far-field pattern. The angular resolution is D_a/d_f, where D_a is the aperture diameter and d_f is the distance to the fiber end (small-angle approximation), and should be smaller than the smallest important detail in the pattern to get a meaningful measurement. Other areas such as resolution, precision, accuracy and dynamic range are discussed in [27]. A typical far-field pattern for a graded-index multimode fiber is shown in Fig. 5b.

This far-field technique provides a determination of numerical aperture (NA) and has been standardized as test procedure RS-455-47 by the EIA and as an Alternative Test Method by CCITT. In both, a short (2-m) length is measured [34]. For the EIA method the angles $\theta_{0.05a}$ and $\theta_{0.05b}$, at which the pattern has 5% of its maximum value (see Fig.

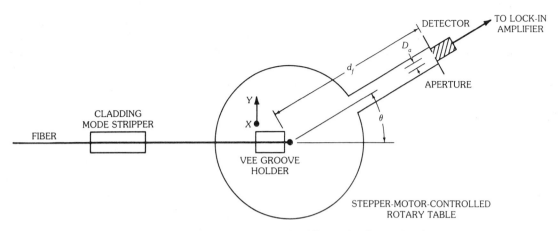

(a) Far-field measurement system using a fixed fiber end and a rotating detector.

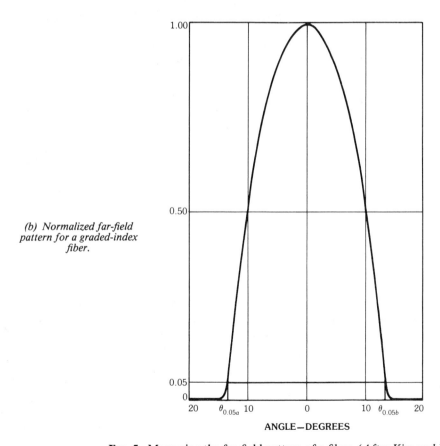

(b) Normalized far-field pattern for a graded-index fiber.

FIG. 5. Measuring the far-field pattern of a fiber. (*After Kim and Franzen [27]*)

5b), are determined, and their average value $\theta_{0.05}$ calculated. Then

$$NA = \sin \theta_{0.05}. \qquad (1)$$

The advantage of using the angles at the 5% value, rather than those for smaller intensity, is that the latter would be harder to judge because of the curve's flatness and because of signal-to-noise–ratio limitations. This truncation, however, produces a difference between this NA and the theoretical NA that would be determined from the index profile (which would correspond to the $\theta_{0.00}$ value for very short lengths [35]); other factors also contribute to the difference [17]. This difference was measured in [36] and is about 10% for several types of fibers, but is not critical as long as all fibers are measured in the same way and are compared with other fibers of the same type.

It is noted that the angular scanning method of Fig. 5a, as opposed to a linear scan perpendicular to the fiber axis, eliminates the need to correct for an angular illumination factor and for the dependence of detection efficiency on illumination, angle. Other considerations are given in [37].

Attenuation Measurements

In this section, after a discussion on techniques for launching light into fibers, three classes of attenuation measuring techniques are discussed— cutback, substitution, and optical time-domain reflectometry. Then techniques for measuring the loss in joints (splices and connectors) are discussed. Measurements of the individual scattering and absorption components of the fiber attenuation, which can be of interest to fiber designers, are not covered here. They are treated in [30].

Introduction and Launching Techniques—We have seen in the section on modal properties and steady state that the only attenuation unique to a multimode fiber, i.e., independent of the source or of fiber discontinuities, is the steady-state attenuation (or, very close to it for many fibers, the quasi-equilibrium attenuation). Losses based on this attenuation are also additive in concatenated links [23, 38], i.e., if fibers A and B are spliced together and fiber A is excited with the steady-state distribution, then

$$\ell = \alpha_A L_A + \ell_{AB} + \alpha_B L_B, \qquad (2)$$

where ℓ is the overall loss in decibels, α_A and α_B are the attenuations in decibels per kilometer of fibers A and B (assumed uniform), respectively, L_A and L_B are the fiber lengths, and ℓ_{AB} is the splice loss, which could include a transient loss or gain. Implications of this additivity is that an attenuation measured for a nominal fiber length (e.g., 1 km) is valid for a much longer fiber, i.e., is length scalable, and, if measurements are made on several fibers, the loss of a fiber concatenated from these lengths is predictable, which is important for sound system design. Source and joint transients, however, must be handled properly.

A steady-state launching condition can be achieved in several ways. The two which are now standardized by the EIA in its RS-455-50 are the limited–phase-space (LPS) method and the mode-filter method [35]. In the LPS method [39, 40], also called the restricted launch method, light is launched into the fiber with a restricted spatial extent and angular distribution chosen to match as closely as possible to the steady-state mode distribution. The most popular LPS approach is the "beam-optics" one [39] shown in Fig. 6 in a setup to measure spectral attenuation (attenuation versus wavelength). Generally, a broad-wavelength–based source, such as a tungsten-halogen lamp, is used and followed by a spectral filter F, such as a monochromator or set of interference filters, to select particular wavelengths for measurement. The source is imaged by lens L_1 to the plane of the aperture A_1, preferably with the source image larger than the aperture and uniformly filling it [41]. The uniformity is difficult to achieve with a coiled-filament lamp so ribbon-filament lamps (with the disadvantage of a lower brightness) must be used. An alternative is to use a uniformizing mixer fiber [42] or a ground-end mixer rod [43], or to tilt the lamp so the coils act to fill the gaps in the pattern [44]. With the collimating lens L_2 and the focusing lens L_3 the illuminated aperture A_1 is demagnified to give an appropriate launch spot diameter at the entrance face of the fiber. (The focal lengths of L_2 and L_3 are f_2 and f_3, respectively.) Control of the cone angle of the launched light (launch NA) is achieved with the aperture A_2 in the collimated beam. The technique provides essentially independent control of launch spot diameter and launch NA. The EIA, in its RS-455-50, specifies that for the graded-index fiber of the standard 50-μm core diameter, the launch spot diameter should be $70 \pm 5\%$ of the core diameter and the launch NA should be $70 \pm 5\%$ of the fiber NA [16]. This gives the length-scaling additivity discussed above.

The beam-optics method has the advantages of an ability to vary launch conditions to adapt to different types of fiber. Its disadvantages are the need for bulky equipment that must be precisely aligned on a stable platform. The cleave on the input face of the test fiber has to be of high quality; a tilted end, for example, will skew the mode pattern. A simple and useful test for fiber-end quality is to observe the reflection of a visible laser beam from the end [45]; major faults on the end and its centering on the optical beam can also be viewed by a microscope (Fig. 6). Another concern is that chromatic aberrations in the lenses can produce significant changes in the launch spot and NA with wavelength. It has been found that with a proper choice of lenses, this can be made insignificant over a large range [42], and a mixer rod can also help [43]. One approach to eliminating the lens setup, with its bulk, cost, and stability problems, is to use a specially constructed "70% coupling fiber," which has a core diameter 70% of that of the test fiber and an NA that is 70% of the test fiber's NA. This has been shown to give good agreement with the beam-optics method in the cases looked at [46].

The other launching technique standardized by the EIA is the mode-filter technique [40, 47]. In this technique, which is shown schematically in Fig. 7, the test fiber is overfilled with light and is subjected to a mode filter which effectively attenuates and thus removes higher-order modes of the fiber. In one popular type (Fig. 7a) the fiber is wrapped around a mandrel [47]. A useful criterion [16, 47] to test the effectiveness of a particular mode filter involves the measurement of the far-field pattern of a short (1- or 2-m) sample of the test fiber which is overfilled and subjected to the mode filter. If the output far-field angle $\theta_{0.05}$, at the 5% points, agrees with that, $\theta'_{0.05}$, measured on the long fiber without a filter, to within an amount

$$\delta = (\theta_{0.05} - \theta'_{0.05})/\theta'_{0.05} = -3 \pm 3\% \,,$$

the mode filter is deemed adequate. Attenuation measurements made with the mandrel-wrap mode filter have been found to give results very consistent with length scalability described at the beginning of this section [47, 48]. Note that the overfilled condition cannot be easily achieved for

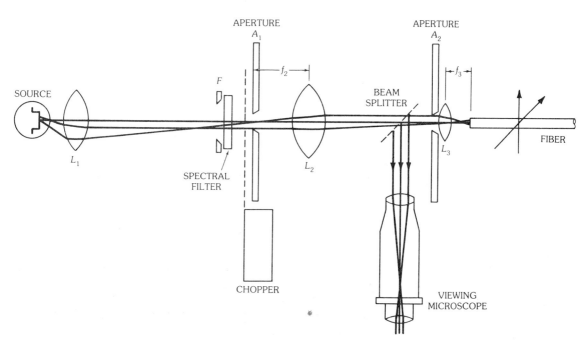

FIG. 6. Light launching with the beam-optics method. (*Adapted from Reitz [39], ©1981, with permission from Optical Publishing Co., Inc.*)

certain sources, namely laser sources, which tend to underfill fibers. A mode scrambler (such as that made by splicing 1-m step-index, graded-index, and step-index fiber sections consecutively together [49], to be discussed below) can be used before the mode filter to help achieve the overfilled condition. Very good agreement between attenuations measured for beam-optics and mandrel-wrap light launching have been reported in round-robin inter-laboratory comparisons [40, 50], with the agreement best at $\delta = -3\%$.

type. Also, the good performance of the mandrel-wrap technique depends on a careful prescription of the mandrel diameter and of the fiber-winding pitch, tension, and number of turns. This was indicated in an AT&T interlocation round-robin study [13]. The mandrel-wrap prescription would depend on the fiber type and the fiber coating or buffer, and its effectiveness should be tested periodically using the criterion mentioned above.

Another mode filter is a "dummy" fiber (Fig. 7b) of the same type as the test fiber, and long

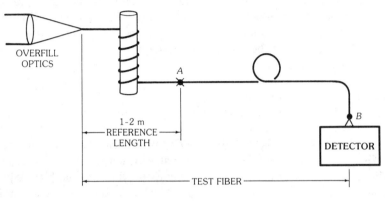

(a) Mandrel-wrap type.

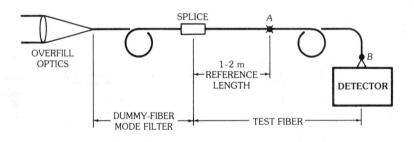

(b) Dummy-fiber type.

FIG. 7. Schematic diagram of mode filters. (*Adapted from Cherin [11], ©1983, with permission of McGraw-Hill Book Co.*)

The mandrel-wrap mode-filter technique avoids the bulkiness and stability requirements of the beam-optics technique and is therefore better suited to the field. The latter technique, however, is advantageous for research or factory testing, where a variety of fiber types must be tested and the apertures can be set appropriately for each

enough to approximate the steady-state distribution [51]. This is overfilled and spliced to the test fiber. This technique has the advantage of simplicity and does not require breaking into the cable to get at a length of fiber for a mandrel-wrapped mode filter, giving it usefulness in the field [14]. However, the fiber must be well matched to the test fiber, a requirement that may not be able to be met consistently because of manufacturing variations, and the splice must not cause any significant mode mixing. The loss of the dummy fiber can also limit the dynamic range of the measurement set. Both the dummy-fiber and the mandrel-wrap mode filters are included in the EIA RS-455-50.

Other mode filters include a splice with one fiber end roughened by etching [52], the wrapping of a fiber alternately around rollers [40, 53], or the placement of a fiber in a ball-bearing bed [54], all of which give mode scrambling as well as mode mixing. Most of these methods, as well as the beam-optics and mandrel wrap, are compared in [55], where it is shown that only the biconical taper, the ball-bearing filter and (under certain circumstances) the roller mode filter bring the fibers truly close to the steady-state distribution.

surements could be made with the actual field source, or a very similar one (since transient losses are quite source-dependent) together with appropriate fiber lengths to determine these transient losses.

Transmission Measurements—This subsection discusses transmission methods for measuring fiber attenuation. It concentrates on measuring spectral attenuation, rather than attenuation at single wavelengths, since this gives measurements for a range of light-source wavelengths. This is particularly useful for LEDs, where the fiber can act as a spectral filter [10], and for diagnosing the causes of attenuation.

The two major methods for measuring spectral attenuation by transmission are the cutback and substitution techniques. The cutback, or two-point technique [40], is a CCITT RTM and the EIA RS-

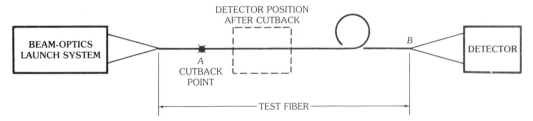

FIG. 8. Attenuation measurement by cutback.

It is pointed out, however, that with recent fibers having low differential modal attenuation, all of these launching techniques give attenuation values very close to the steady-state one. The important thing for attenuation measurements is to avoid exciting the lossier highest-order modes [17].

The preceding techniques help bring the modal distribution to the steady-state distribution for attenuation measurements. In applying the attenuation results to determine power budgets for field systems, however, the transient losses must also be taken into account. Those could be included in the proper characterization of the light source (through its fiber coupling loss [25]) and of joints (discussed below). Transient effects would, of course, be dependent on the length of fiber used. Ideally, source and joint manufacturers should give such figures (or at least ranges) for different types of fiber [10, 25]. If this is not possible, mea-

455-47. It is illustrated schematically in Fig. 8. Light from a spectrally broad source followed by a wavelength filter is launched into the test fiber by the methods of the previous subsection. The power out of the fiber is measured at the detector at end B (call this P_B). The fiber is then cleaved at point A, to produce a reference of about 1 to 2 m in length, without changing the launching conditions. The detector is then used to measure the power P_A emitted at A (dotted detector position). The attenuation of the fiber in decibels per kilometer is then calculated from the relation

$$\alpha = -(10/L) \log_{10} (P_B/P_A), \qquad (3)$$

where L is the length from A to B in kilometers. Fig. 8 shows the cutback technique schematically. For the special case of a mode-filter launch, the points A and B are shown in Fig. 7.

The purpose of cutting back is that it eliminates needing to know how much of the light power is coupled from the source into the fiber. The use of the short reference length also eliminates some of the leakiest modes, although this is not of importance if steady-state launching is used. The technique is also simple. The disadvantage is that the cable must be broken into and fiber removed. Sometimes the fiber can simply not be spared.

where L is the length of the test fiber minus that of the reference length. Since this technique does not involve cutback, it is more amenable for use in the field [14], especially if it is used with connectors. It is, however, less accurate and precise than the cutback technique since the reference fiber may not always be accurately the same type as the test fiber. This, together with varying misalignment at the connectors, may not give connector losses that are the same for the reference and

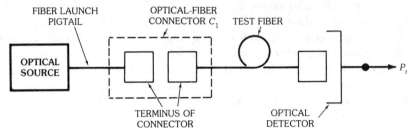

(a) Output measurement.

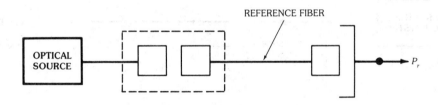

(b) Reference measurement.

FIG. 9. Measurement of fiber attenuation by substitution. (*Adapted from Cherin [11], ©1983, with permission of McGraw-Hill Book Co.*)

An alternative is the substitution technique [11], also called the *insertion-loss technique,* given in the EIA RS-455-53 and illustrated in Fig. 9. In the usual version of this method the source optics and a pigtail fiber with any mode filter terminate on a connector C_1 (a splice could also be used, preferably a remakable splice like the elastomeric splice [56]). The power P_t out of a connectorized test fiber, connected with C_1, is measured with a detector at the other fiber end (Fig. 9a). A reference fiber of the usual reference length and ideally of the same type as the test fiber is then used and its output power P_r measured (Fig. 9b). The attenuation is then determined by the relation

$$\alpha = -(10/L)\log_{10}(P_t/P_r),\qquad(4)$$

test fiber. This produces errors in the apparent test fiber loss. Also, the connection C_1 may interfere with obtaining the proper steady-state modal distribution in the test fiber. However, in some field measurement situations, a less precise measured attenuation may be acceptable.

Another nondestructive method, which is based on side illumination, has been recently reported [57]. In this, light is launched at the side of the fiber via bends of radius 2 mm at various points along the fiber and detected at the end. Knowing the position of the injection points permits calculation of loss. A lot of fluctuation occurs due to the nature of the launching but these can be averaged out with a large number of launch points. This technique is unwieldy, is not possible for cabled or even heavily buffered fiber, and is made less accurate by the inability to launch the proper distribution.

A third nondestructive technique, optical time-domain reflectometry, is described in the next subsection.

Several other considerations influence the accuracy and precision of the transmission measurements. Since they involve more than one measurement at any one wavelength, it is important that the system stays sufficiently stable between the measurements. One instability problem is drift of the source power. This can be handled by the use of highly stable source power supplies or power monitoring (e.g., via the beam splitter shown in Fig. 6). References 58 and 59 give other suggestions for improving stability. Another concern is that the detector be linear, sufficiently spatially uniform, and intercept all the fiber output power, the latter even in spite of unavoidable variations of end-cleave quality and fiber positioning. Large-area detectors can help in this regard, as can integrating spheres, but the latter cause a major reduction in dynamic range (one type has 20-dB reduction), which may make this method not useful for fiber measurements but acceptable for joint-loss measurements (see below).

Concern must also be given to extra losses and mode conversion produced by macrobending and microbending at improperly designed fiber holders or by tightly spooling the fiber. A proper cladding-mode stripper, just as in near- and far-field measurements, has to be used if the fiber coating does not mode-strip adequately. Finally, chopper/lock-in amplifier combinations and/or cooled detectors can be used for improvement of signal-to-noise ratios.

A study of the accuracy of attenuation measurements is given in [1]. Many automated setups have been made, e.g., see [12, 60].

Optical Time-Domain Reflectometry—The optical time-domain reflectometry (OTDR) technique, reviewed in [1, 61] and standardized in the EIA RS-455-59, is a technique that permits measuring the length-dependent loss of fibers, locating faults, and measuring and optimizing the loss of joints. The technique, also called the *backscattering technique,* shown schematically in Fig. 10, involves

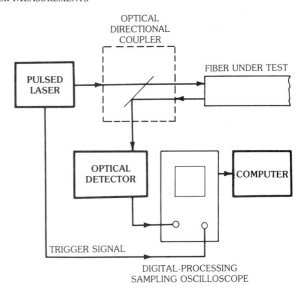

Fig. 10. Schematic diagram of optical time-domain reflectometer. (*Adapted from Cherin [11], ©1983, with permission of McGraw-Hill Book Co.*)

sending a light pulse into a fiber via a directional coupler (e.g., a beam splitter) and detecting the return signal from the fiber [62]. This signal, viewed as a function of time, is shown in Fig. 11a and is also displayed in a log format in Fig. 11b. Unlike pulse-echo techniques for electrical cables, which detect discontinuities only, this technique permits the examining of points successively further along the fiber by using Rayleigh-scattered light. A certain fraction, the "capture fraction," of this scattered light is scattered backwards within the acceptance NA of the fiber and propagates down the fiber in the reverse direction. In Fig. 11a the pulse at t_0 is a result of reflection from the front facet of the fiber and the pulse at t_e is that from the far end. A point a distance z within the fiber will give rise to a signal at time $t_0 + 2z/v_g$ where v_g is the group velocity and the factor of 2 comes from the fact than a round trip of light is needed for viewing. The general relation for the back-scattered light power P_S at time t is given by [63]

$$P_S = P_0 S \alpha_s w v_g \exp\left[-2\int_0^z \xi(z')\,dz'\right], \quad (5)$$

where the pulse launched into the fiber has power P_0 and width w. It is assumed that w is sufficiently narrow so that insignificant attenuation occurs over the pulse width. The term α_s represents the fiber scattering loss and $\xi(z)$ is the fiber attenuation (nepers per kilometer) at a distance $z = t/2\,v_g$. The capture fraction S is given by [64]

$$S = \tfrac{3}{8}\,[(NA)^2/n_0^2]\,\alpha/(\alpha+1)\,, \qquad (6)$$

where a graded-index fiber of grading constant $\alpha = 2$ is assumed and where n_0 is the index at the fiber center. This expression assumes isotropic scattering and a uniform excitation.

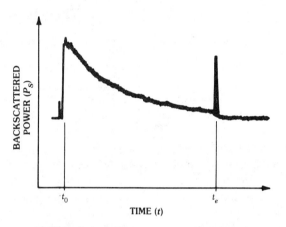

(a) Viewed on oscilloscope. (Adapted from Cherin [11], ©1983, with permission of McGraw-Hill Book Co.)

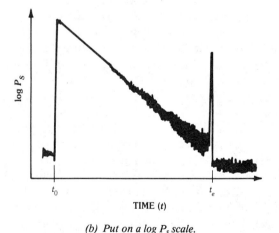

(b) Put on a log P_s scale.

FIG. 11. Backscatter signal.

The ratio of the backscattered signal power P_S to a Fresnel-reflected signal P_R coming from the same point on the fiber (i.e., assuming the fiber

ends there with a clean, perpendicular break, yielding a -14-dB return) is given by [61, 64]

$$P_S/P_R = \alpha_s\,v_g\,w\,S/2\,R\,, \qquad (7)$$

where R is the reflection coefficient (0.04). Typical values for pulses of 100-ns width are 30 to 40s dB and are of course fiber and wavelength dependent [64].

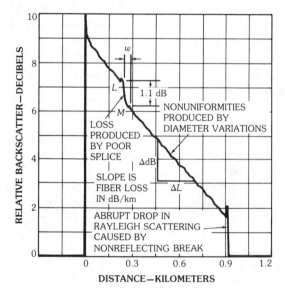

FIG. 12. Log plot versus distance (backscatter signature) of a fiber showing a poor splice. *(After Gentile [61])*

The curve in Fig. 12 displays the data (backscatter "signature") for a fiber against a scale of true distance. The ordinate is that from a plot vs. time like that of Fig. 11b divided by 2. It is seen that it is possible to extract the different loss features of the fiber. The loss of the fiber in decibels per kilometer in any region is found by taking the slope of the curve, and the loss in decibels at the discontinuities (joints, faults, etc.) are found from the jumps in the curve.

A major advantage of the OTDR technique for measuring loss is that it permits looking at the length dependence of losses in the fiber, e.g., nonuniformities in losses, loss components due to imperfections (bubbles, etc.) or faults, or the losses introduced at splices or connectors in a link of jointed fibers. The position of the individual joints or faults is also determined as is the fiber length. Knowing how the losses are distributed is of great value for the fiber manufacturer and for the user. For the latter the technique permits the determination of faults for repair as well as the evaluation and, perhaps, optimization of joints. Other advantages of the OTDR technique include the fact that only one end of the fiber need be accessed, cutting down the number of personnel needed in field installations, and the fact that it is nondestructive.

Knowing the position of any joint or fault accurately depends on an accurate knowledge of the fiber group velocity. While certain fiber manufacturers provide accurate values for this, others have not done so for proprietary reasons [65]. The distance resolution also depends on the width w of the probe pulse. We have seen in (5) that the backscattered signal increases with w; this can be understood if one considers that each time segment dw of the pulse produces a separate backscattered signal, appropriately delayed, and all are added up (convolved) in the fiber. This convolution spreads out the discontinuity from any joint over a width w (e.g., LM in Fig. 12). However, the pulse rise time, related to the OTDR system bandwidth, can be the most important factor in accurately locating the start of the discontinuity and this gives the defect position since the discontinuity will start when the first part of the return pulse arrives [65]. If there are two discontinuities within the rise time, they will not be resolvable.

A major area of concern is the range of the OTDR. Since, as is seen in Figs. 11a and b, the signal falls with distance because of attenuation, there is a maximum distance above which the signal-to-noise ratio is too small for proper measurement. The system range is usually stated in terms of the one-way or round-trip losses that are permissible to perform a desired function. Actually one or more of three distinct range types are usually stated in published papers, or even in the specifications of OTDR equipment. One is the range of detecting a reflecting fault. Usually the Fresnel reflection from a clean, perpendicular fiber end (-14-dB reflectivity) is what is specified. This is, of course, a reflection phenomenon and gives a much stronger signal than backscatter. The second range is that for a nonreflecting break, such as that for a fiber end

that is surrounded in oil or broken badly, and is characterized by a change in noise as in signal. This change of noise is shown in Fig. 11b which, however, also shows an end-reflection pulse. Ranges corresponding to a certain signal-to-noise ratio, or, less carefully, to a "minimum discernible end," are usually given. The one-way range for this can be 15 to 20 dB smaller than for the reflecting fault, as expected from our discussion of (7). The third range is that for accurate measurement of localized fiber losses or joint losses, and is stated in terms of the loss resolution possible, e.g., 0.1 dB. This is equivalent to a signal-to-noise ratio (snr) of 2:1 if the noise corresponds to 0.05 dB. It is noted that on a log scale the noise increases with distance (Fig. 11b). As of this writing a typical set of ranges for commercial equipment [65, 66, 67] operating at 0.8-μm wavelength and using recent Corning graded-index (G1) fiber is 40-dB one-way range for the reflection type, 20 dB for the nonreflecting end detection (2:1 snr) and 13 dB for 0.2-dB resolution splice-loss detection. It is noted that the latter two ranges are backscattering ranges, and since the backscatter signal depends on the fiber's scattering coefficient and capture fraction the ranges are fiber-dependent. Thus, in comparing the results of two papers, or the specifications of two commercial machines, the fiber used must be taken into account. We also may note that for backscatter measurement the name OTDR is somewhat incorrect since "reflectometer" implies measuring reflections. Although backscatter is being measured with an instrument that is a reflectometer, it is not a reflectometry measurement.

Going to a 1.3-μm wavelength implies a decrease in range as compared to the 0.8-μm wavelength. This comes from the smaller amount of backscattering. If all the scattering were Rayleigh scattering, then from the λ^4 dependence one expects about a 10-dB reduction in α_s and this is typically what is seen. Also, there is usually smaller source pulse power and smaller detector sensitivity. A decrease in range of 10 dB is possible [65]. Because of the lower loss at 1.3 μm, however, this can still correspond to a greater length of fiber. To go even greater distances, techniques at 1.5 μm can be used, and they are discussed below.

Techniques for Improving Range—There are several ways the range can be improved. One is through wider pulses and lower system bandwidths. Thus, there is a trade-off between range and resolution (see the comments on resolution above). Another, obvious, way to increase range is to use larger laser powers and more-sensitive and lower-noise detectors. Another way is to use averaging to improve the signal-to-noise ratio. With N averages the improvement is by a factor of $N^{0.5}$. For example if 2^{16} averages are taken, the improvement is 24 dB. Taking many averages can substantially increase the measurement time, and, among averaging techniques, digital averaging requires much less time than analog averaging with a boxcar integrator [68]. Increasing the pulse-repetition rate reduces the measurement time but the maximum rate is limited to $2v_g/L$, where L is the length desired to be measured. For 10 km this is about a 10-kHz rate. If the OTDR is used for optimizing splice losses, the measurement has to be in real time so that averaging is limited. However, optimization can be done by observing the end reflection. This gives the largest and most easily observed signal and may not require much averaging. Another way of minimizing measurement time is to average only those areas of the signal where enhancement is needed, e.g., in the farthest-out region [65] (see Fig. 11b) or right at a splice or other region of interest. Another point is that the decibel range need only be half of the fiber (or fiber link) loss if measurements can be made from both ends.

We now describe some other, specific techniques for improving OTDR range to above the typical value of 20-dB one-way loss at 0.8-μm wavelength and 15 dB at 1.3 μm [1, 69]. (From here on, a range that is otherwise unspecified is taken to mean the one-way range for a nonreflecting break). One technique is *two-point processing* [70, 71]. In this, attenuation is derived from the ratio of two samples, corresponding to two distances z into the fiber, taken for the same laser pulse. Stability and noise problems, including those due to pulse-to-pulse variations in the source and receiver, are much reduced. In one theoretical and experimental study at 0.9 μm [71], localized losses could be measured to 0.1 dB/km accuracy at a one-way range of 40 dB, which agreed with theoretical results, including a detailed signal-to-noise–ratio study [71, 72]. The corresponding nonreflecting break range is 47 dB, where we note that signal averaging (100 000 averages) was used. Another method for improving range is the *photon-counting technique* [73] in which an APD is used in a ''Geiger-tube breakdown mode,'' which gave a 40-dB range. For 1.3 μm a pulse-compression coding technique, with a 24-bit code, was used in conjunction with photon counting, giving a 27-dB range [74].

The preceding techniques utilized diode lasers as sources. Another class of techniques increase range by using powerful lasers and, in some cases, nonlinear optical effects in fibers. In one experiment, 1-kW light pulses from an Nd:YAG laser at 1.3 μm were launched into a fiber and a series of longer-wavelength spectral lines (1.40, 1.49 and 1.59 μm) were generated in the fiber by the phenomenon of stimulated Raman scattering [75]. This was used to probe a test fiber at 1.49 μm, taking advantage of the lower loss there. In another experiment a similar laser generated a quasi-continuous spectrum resulting from a superposition of stimulated Raman scattering and stimulated four-photon mixing [76]. This permitted measuring fiber loss over the range 1.32 to 1.70 μm in fibers as long as 13 km and for which measurements in some parts of the region were not possible with the cutback method because of its limited signal-to-noise ratio. Yet another technique [76] made use of nonlinear generation in the fiber being tested. In one use of this Raman OTDR technique, 1.6-μm returning light was used to find nonreflecting ends at 80-km length (projected maximum length was 90 km).

While these high-power laser techniques are capable of going large distances in fibers, they are not amenable to field use because of the bulk and the power and water-cooling needs of the lasers, as well as safety considerations. Their use is thereby restricted to that of the research laboratory and for certain applications in quality control.

Choice of couplers—The choice of the directional coupler (shown for the beam splitter case in Fig. 10) is very important. It is necessary to have maximum isolation between forward and backscattered light beams so as to keep any forward power from getting into the receiver and swamping it. The insertion loss of the coupler should also be low to permit maximum fiber range. Also, if possible, means should be provided to suppress the reflection from the input face of the fiber, as this is about 30 to 40 dB bigger [see (7)] than the back-scatter signal from the nearest point within the fiber, and produces an undesirable saturation of the receiver, which distorts the backscatter trace until the receiver has recovered. Among the techniques used [1] to minimize the reflection are an index-match fluid cell, a gated photomultiplier [63], switchable directional couplers such as an acousto-optic deflector, and a polarizing beam splitter which permits viewing only the polarization perpendicular to that of the reflection from the front face [68].

Fluctuations in Fiber Parameters—The back-scatter signatures in OTDR are affected not only by the loss composition of the fiber and its joints but also by the fluctuations in other parameters along the fiber, namely, NA, core diameter, and scattering coefficients. The effects on backscatter, if not taken into account, can lead to erroneous interpretation of fiber and joint losses. Proper treatment of the data, however, can lead to a determination of both the losses and (in certain cases) the fluctuations. This can be seen by the model of [77], which uses a ray approach in a fiber in which uniform modal excitation is assumed. A backscatter power expression similar to that in (5) is obtained. It can be seen from (5) and (6) that the backscattered power is a function of the fiber NA and index profile (through the capture fraction S and, via the dopants, through α_s). The core diameter also has an effect, via the doping, on the average scattering coefficient (α_s varies with radius). As one example, Fig. 13, taken from [77] for a case with unusually large fluctuations, shows how NA variations can affect the signature. While some of the fluctuations can be averaged out, those slowly varying with distance z will give apparent changes in loss, sometimes with negative apparent losses occurring. It is suggested [77] that, to remove the ambiguities, backscattering signatures be taken from both ends of the fiber. If the forward and backward powers, labeled $P_f(z)$ and $P_b(z)$, are plotted against the distance from the same input face (that for the forward face), then a separation between the actual decay of optical power due to losses and the parameter fluctuation effects can be obtained by calculating the following quantities [77]:

$$F(z) = [P_f(z)/P_b(z)]^{1/2} = \text{constant}$$
$$\times \exp\left[-2\int_0^z \xi(z')\,dz'\right], \quad (8)$$

and

$$I(z) = [P_f(z)\,P_b(z)]^{1/2} = \text{constant}$$
$$\times S(z)\,\alpha_s(z). \quad (9)$$

The term $F(z)$ yields the power decay while $I(z)$ contains the parameter-variation effect (S and α_s are written as functions of z here). This separation is shown in Fig. 13. From (9) it can be seen that if more than one parameter varies with z, it is not possible to separate the individual variations out.

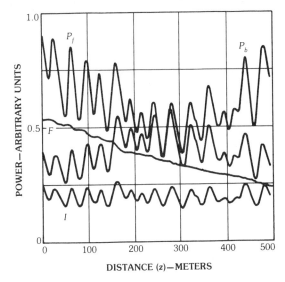

FIG. 13. Backscattered optical power (P_f and P_b) from each end face of a fiber mainly affected by fluctuations of numerical aperture, together with geometrical mean I and ratio F. (*After DiVita and Rossi [77], ©Institution of Electrical Engineers*)

While the above separation into power-decay and parameter-variation contributions works well in some cases, modal effects, which are not taken into account in the above "uniform excitation" theory, can be significant [78]. For example, when differential modal attenuation is taken into account, the above separation would not be possible under certain modal excitations; direction-dependent effects, such as transient losses at splices, confuse the situation. Also, the extraction of parameter information from backscatter traces would be inaccurate. Reference 78 worked out a theoretical treatment which included modal effects. It was found that if a proper "mode-filtered" excitation of the fiber, in which the launched beam has a smaller diameter than the smallest-fiber core diameter, is used, and this is done for both the backward and forward signatures, the separation of (8) and (9) is correct. In addition, a more accurate expression to determine parameter-variation effects than in (9) can be derived, e.g., with core-diameter variations included. This mode-filtered condition is maintained in the usual mode-filtered excitation described above for the cutback technique. Very good agreement with experiment has been obtained [77, 78]. In another study [79] a standard dummy fiber was used as mode filter to launch light in a test fiber containing variations in core diameter and index profile. A separation was obtained and the resulting measured fiber loss was in very good agreement with that by the cutback technique, and the determined parameter-variation effects in (9) were as predicted via a somewhat different theory than that of [78]. If more than one parameter was varying, however, it was not usually possible to separate the effects and thereby determine all types of variations from the backscatter signature. None of the above techniques, however, include mode-coupling effects. Such effects can produce very significant errors in measured losses, especially those for splices [1].

Mode-filtered excitation is useful for launching light in any OTDR fiber-loss measurement, for the same reasons as in the cutback technique. However, the backscattered beam generally has a larger higher-order-mode content [80, 81], giving rise to some loss inaccuracy. Also, the use of a mode filter in the proper way would require the use of a mode scrambler preceding it since the light sources used in OTDR underexcite rather than overexcite the fiber. The combination of scrambler and filter leads to losses which diminish the range of the OTDR, and the need to match filter to fiber diminishes the simplicity and field usability usually associated with OTDR use.

The need for a forward and backward measurement where major parameter fluctuations exist cause interpretation difficulties and also has major

disadvantages for field use. It increases the measurement and processing times and requires that the OTDR range cover the whole fiber length rather than half the length.

In a recent study with fibers up to 2 km long from several manufacturers, it was shown that with appropriate mode filtering the mean difference between OTDR-measured losses and those measured by cutback was 0.01 dB/km, with a standard deviation of 0.05 dB/km [82]. This, however, may be a reflection of the low differential modal attenuation and mode coupling in recent fibers. Also, fluctuations have been much reduced in many recent types of fibers.

Joint Losses: Transmission Measurements—The simplest measurement of joint (splice or connector) loss is as shown in Fig. 14a. Here, a fiber is broken at a certain point, the new ends cleaved and jointed. The loss is then determined from the ratio of the power transmitted through the jointed fiber to that before the fiber was broken. However, it will be found that the loss actually measured will depend on the type of source and the lengths L_1 and L_2 [83]. This is a result of modal effects. For example, for an LED source and a small length L_1 the proportion of light in higher-order modes will be larger than that for a large L_1 (for example, large enough to reach steady state), and the apparent loss of a misaligned joint will be correspondingly larger, since the outer modes that exist in the shorter length but not the longer one are more easily shorn off at a joint. Also, lossy outer modes excited at such a joint would not make it down a long fiber but could down a short fiber, making the apparent joint loss more. Thus there is a difference between short-short, short-long, and other length combinations. The difference is less with a laser source, unless it is poorly aligned to the fiber. Some of the effects are more correctly regarded as transient losses, some associated with light sources coupled to the fiber and some with the joints.

A more typical joint loss measurement is as shown in Fig. 14b. Here, fibers *A* and *B*, whose attenuations are presumably known, are jointed together, and the output power is measured. Perhaps the beginning of fiber *A* is cut back to form a reference, or, alternatively, the source has a pigtail which is spliced to *A* with a splice loss pre-

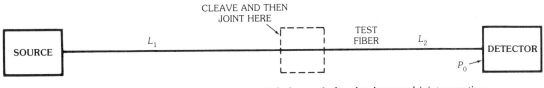

(a) Joint-loss measurement—ratio of powers P_0 before and after the cleave-and-joint operation.

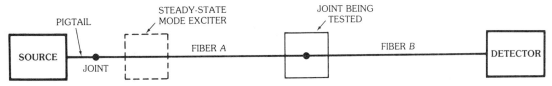

(b) Measurement with fibers A *and* B *of known losses.*

FIG. 14. Transmission measurements of joint loss.

viously measured. From the output power the insertion loss ℓ is measured. The joint loss in decibels is then given by

$$\ell_J = \ell - \alpha_A L_A - \alpha_B L_B, \qquad (10)$$

where α_A, L_A and α_B, L_B are the attenuations and lengths of fibers A and B, respectively. The most meaningful measurements would be for large L_A and large L_B since in that case the transient losses in the fiber are taken into account. This is the "long-long" condition, which is being standardized by the EIA. If L_A and L_B are short (meters long), a steady-state mode exciter or exciters (dotted line in Fig. 14b) could be used to simulate the transient effects in long fiber lengths. A mode scrambler followed by a mode filter can constitute such an exciter. It has been found by many connector manufacturers that the loss for the long-long condition is very well approximated by that for the short-short condition [10]. For measurements of maximum accuracy and precision the considerations (stability, etc.) described at the end of the subsection on measuring fiber attenuation by transmission measurements also hold here.

Because of modal effects the presence of joints can affect other joints further downstream. Also, there are times when joint losses for other than steady-state conditions in the fiber are of interest, e.g., when we wish to make a measurement of the loss of a joint right after a transmitter or right before a receiver. In that case, a direct measurement should be made.

Joint Losses: OTDR Measurements—It has been mentioned in the previous discussion on OTDR that this technique can provide a measurement of joint losses. Minimization of the joint loss is also possible if communication with the OTDR operator at either fiber end is possible. The easiest way to optimize the joint loss via OTDR is to maximize the reflection from the far fiber end.

The extraction of the joint loss from discontinuities in the signature is shown in Fig. 12, where the discontinuity, like other features in the signature, must be divided by two to get the one-way loss. Although there is an advantage in measuring loss by OTDR in that the measurement is made from one end of the fiber and the fiber losses do not have to be known, in contrast to transmission measurements, the OTDR-measured loss is subject to a number of inaccuracies, which can be seen with reference to Fig. 15. This shows schematically that the difference in returning power comes from the extra two-way trip through the joint for path B. First, the scattering coefficient α_s can differ in the two fibers being joined. If α_s of the far fiber is greater than that of the near one, the return signal is increased, making the apparent joint loss smaller. In fact, negative apparent losses have been observed. The way to overcome this is to take an OTDR measurement from both ends of the fiber pair being jointed and average the two measured joint losses [84], which can be shown to be equivalent to taking the discontinuity in the curve for the quotient $F(z)$ of (8) and therefore follows from that technique. Just as in the discussion about parameter-variation effects in OTDR loss measurement, this substantially complicates the joint measurement for field use. An alternate, but also inconvenient, means of correcting for α_s differences is to initially splice each fiber with a reference fiber to predetermine α_s differences; this also handles directional effects on the capture fraction S [84, 85].

FIG. 15. Backscattering paths before and after joint.

Another ambiguity in doing joint loss measurement via OTDR involves modal effects. We have seen above that loss in a joint is sensitive to modal excitation. In an OTDR measurement, even if the first traversal of the joint were done with the proper modal pattern, e.g., the steady-state distribution, the return beam, based on the scattering distribution, would not have the proper distribution. Another problem occurs if there is a mismatch of parameters at the joining (e.g., of NA), so that the splice loss in one direction would be different than in the other due to a modal distribution effect. The OTDR results would contain ambiguities due to this. The parameter differences, even if measurable by the above mentioned techniques, could not be applied to the joint loss in a simple way. Transient losses due to joints would also not be included in this measurement. We have already mentioned, further, that mode-conversion effects will produce apparent errors in OTDR results [86].

In summary, although splice optimization and splice-loss measurements are by-products of OTDR techniques requiring access to only one end of the fiber at a time, loss measurements done this way have many potential ambiguities and inaccuracies which require time-consuming techniques to be correctly treated.

Bandwidth Measurements

It has been shown in a previous chapter that there are two major components to pulse broadening, and thus to bandwidth limiting via intersymbol interference, in multimode fibers. The predominant one of these is usually modal dispersion (i.e., velocity differences between the modes), and the other is chromatic dispersion (variation of velocity with wavelength). Multimode-fiber bandwidth measurements generally involve measurements of pulse distortions (time-domain measurements) or of the baseband frequency response of the fiber, and are covered in the next two sections. For high-bandwidth (small intermodal velocity difference) fibers, chromatic-dispersion measurements are relevant and are also discussed below.

Time-Domain Techniques—In time-domain techniques, which are reviewed in [1, 87] and standardized in the EIA RS-455-51, a narrow pulse of light having a shape $P_{in}(t)$ is launched into the fiber. This is broadened in the fiber to emerge as an output pulse having a shape $P_{out}(t)$. Both pulse shapes are captured on a fast oscilloscope and stored, for example, in a digital processing oscilloscope; the frequency response is then found by taking the Fourier transform of each and dividing, i.e.,

$$h(f) = FT[P_{out}(t)] / FT[P_{in}(t)] , \qquad (11)$$

where FT symbolizes the Fourier transform of the bracketed function. It is noted that the numerator and denominator each have an amplitude and phase part. Knowing these parts, the overall frequency response $h(f)$ can be determined. If $h(f)$ is retransformed to the time domain, one obtains the fiber impulse response, which is the output pulse with the input pulse deconvolved from it.

A typical time-domain measurement apparatus is shown schematically in Fig. 16. A diode laser provides the short pulse launched into the fiber. Pulses in the 30- to 40-ps–width range have been produced by driving the laser with a comb generator driven, in turn, by a stable sinusoidal oscillator [88]. The pulse is launched into the fiber by means of relay lenses (as shown) or, for more stability, a pigtail from the laser. A mode scrambler is used to provide proper launching conditions, and it is discussed further below.

The output pulse, in Fig. 16, is collected at the fast detector, which can be a photodiode or an

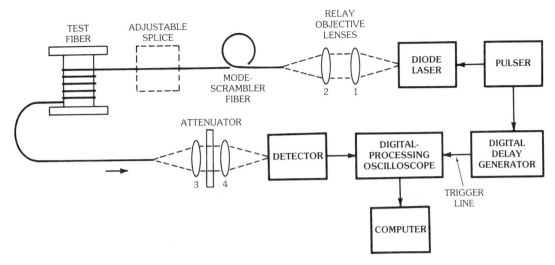

FIG. 16. Schematic diagram of time-domain bandwidth measurement.

APD, and then stored. Detectors as fast as 30 ps have been used in this measurement [88]. To get the reference measurement the fiber is cut back to a reference length and $P_{in}(t)$ is obtained; alternately the measurement can be done directly at the laser. If the source characteristics do not vary much with time, this does not have to be repeated with every run. Care must be exercised that the light is all captured by the detector within an area that has a uniform response. Otherwise, the output of some modes will be deemphasized, producing a distorted output pulse. Also, detector linearity should be verified. This is maximized by keeping the pulse peak power approximately the same for the $P_{out}(t)$ and $P_{in}(t)$ measurement, by means of an attenuator such as that shown in the figure.

In Fig. 16 the digital delay generator is put into the trigger circuit to make up for the delay in the fiber (about 5 μs/km). This would be made unnecessary if the repetition rate were sufficiently large, for then a later pulse could be used as trigger. A typical amplitude-response curve is shown in Fig. 17. The frequency bandwidth is usually taken to be the point at which the detected optical power is down by 3 dB, call it f_{3dB}. The electrical power, proportional to the square of detector current, would then be down by 6 dB. Bumpy, nonmonotonic frequency responses, often due to prepulses occurring from index-profile dips at the

fiber center, can give ambiguous values of f_{3dB}. One approach for reducing the ambiguity is to take the 3-dB frequency for a fitted Gaussian curve [89].

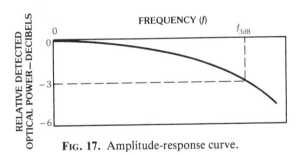

FIG. 17. Amplitude-response curve.

The launching conditions are quite important, since launching into a restricted number of modes can produce overlarge measured bandwidths. The most common launching technique involves use of a mode scrambler between the source and the test fiber (see "Introduction and Launching Techniques" under the subsection on attenuation measurements). One useful one is formed by 1-m–long step-, graded-, and step-index fibers fusion spliced together [49]; others are described in [55]. The mode scrambler approximates an overfilled launch and thus greatly decreases the sensitivity to source misalignment. It also compensates for the strong underfilling of the fiber by laser sources, which produces the overlarge-bandwidth problem mentioned above. Overexcitation of the fiber by the mode scrambler, of course, produces extra modes which act to produce extra pulse broadening, but the key transient loss mechanisms usually occur within the first few kilometers, leaving a modal power distribution and bandwidth close to that for steady state [17].

The changes in measured bandwidth produced by restricting the launch by various degrees below the overfilled launch case, using a mode scrambler followed by a mode filter, have been studied and found to be significant [9] over shorter distances (up to several kilometers) and, of course, are fiber-dependent, e.g., via differential modal attenuation. Specification of the launch conditions giving the most meaningful results in different cases is under study by standards committees. The EIA has standardized an overfilled launch, for the most part because it has good reproducibility, a factor that is very important for field applications.

The ideal launch would be that yielding predictability of link bandwidth from individual fiber bandwidths. Unlike for the case of attenuation, use of the steady-state launch has not yielded such predictability [17], nor has any other launch conditions. This is because of other factors involved, and we discuss this further under "Length Dependence and Concatenations" below.

Overall fiber bandwidths that can be measured by the time-domain technique are limited by the measurement-system bandwidth. Using appropriate microwave components, light sources, and detectors, bandwidths of several gigahertz or higher are possible. Further comments on the technique are given in the reviews cited above.

Frequency-Domain Techniques—In the frequency-domain technique, reviewed in [1, 90] and standardized in the EIA RS-455-30, the frequency response is found directly by launching light into the fiber that is modulated at a swept frequency f. The amount of signal transversing the fiber is measured as a function of f. A schematic diagram of the equipment is shown in Fig. 18. The source typically is a diode laser and a spectrum analyzer is used to measure the output modulation. A mode scrambler is again used to get the proper launching conditions. A run taken with a reference length of fiber gives the system amplitude response, which should be divided out from the long-fiber response data; with a sufficiently stable system this does not have to be repeated with every run. The reference measurement is shown as a dotted path.

Use of a tracking generator [14] permits the spectrum-analyzer frequency to track that of the modulation, permitting detection of the fiber out-

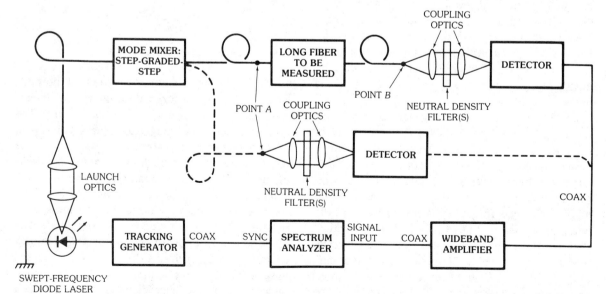

FIG. 18. Frequency-domain bandwidth measurement. (*After Chipman [14]*)

put with a narrow-band filter; manual tracking can also be done, giving the same advantage. In either case the narrow receiver bandwidth yields an improvement in signal-to-noise ratio.

The frequency- and time-domain techniques can be compared. There is very good agreement between the results of the two techniques when the same launching conditions are used. The frequency-domain technique generally has a much larger dynamic range because of the narrow-band filtering referred to above. Another advantage is that the data is obtained directly, with no Fourier transforming required, making data processing simpler and introducing less noise. Also, unlike the time-domain technique, which requires a trigger connection between the source and oscilloscope, versions of the frequency-domain technique exist which do not require any connection between input and output [14]. These all make the frequency-domain technique more applicable to field use than the time-domain technique. However, the frequency-domain technique has the disadvantage that the causes of fiber bandwidth limitations that can show up in the output pulse shape are not observed, making it less useful for fiber research. Also, the phase response is not obtained, except through more complicated measurements with phase-sensitive detection or network analyzers. The phase response, however, is only important in getting the detailed shape of the impulse response. From a system designer's perspective, digital systems are not sensitive to such shapes, which can be a secondary effect. If systems have such strict bandwidth needs that shape details become very important, it probably is more economical to go to single-mode fiber.

Chromatic-Dispersion Measurements—For fibers with index profiles close to optimum, contributions of chromatic dispersion to the fiber bandwidth can become important [10]. This is particularly so if the system wavelength is not near the minimum-dispersion wavelength, typically around 1.3 μm. In this case measurements of chromatic dispersion become useful. Techniques for this include pulse-delay techniques, where the delay of a narrow pulse of variable wavelength is measured versus wavelength. This is discussed in detail in the single-mode–fiber measurement section, under "Dispersion." Further,

by measuring the pulse broadening versus wavelength the spectral bandwidth can be obtained, which gives information on the frequency for which the index grading is optimum. In another class of techniques, a spectrally broad light source, such as an LED, is modulated and then is filtered by a monochromator or filter wheel, giving a scannable wavelength. The phase shift of the light in the fiber is measured versus wavelength, yielding the chromatic dispersion. This will also be described in detail in the single-mode–fiber "Dispersion" section, but to be applicable to multimode fibers the modulation frequency would have to be within the fiber bandwidth. The chromatic-dispersion and intermodal broadening components, can then be added together in a sum-of-squares method [10].

Length Dependence and Concatenations—Fibers can be characterized by a length-dependence exponent γ describing the growth of pulse width (and reduction of bandwidth) with length, as in

$$\Delta T = (\Delta T)_0 z^\gamma, \qquad (12)$$

where ΔT is pulse width at length z and $(\Delta T)_0$ is that at length zero, i.e., at the start of the fiber. The exponent factor γ depends on several factors. Modal dispersion, i.e., the effect of the index profile, yields a linear growth ($\gamma = 1$) but coupling between the modes (see Section 2 under "Modal Properties and Steady State) can reduce γ to as low as 0.5. As mentioned in Section 2, the fiber perturbation contribution to mode coupling can be very small in modern fibers and microbending can also be made to be very small with proper buffering and cabling, pushing γ to 1. However, perturbations in parameters such as index that vary randomly along the fiber length but too slowly to cause mode coupling, are important in modern fibers and can also reduce γ towards 0.5 [91] because of local compensation between broadenings at different rates, caused by the varying index profile. The reduction on γ is largest at wavelengths closest to that for minimum modal dispersion (optimum index profile) so that the linear behavior does not predominate; however, the linear behavior will become predominant after a sufficiently long length [92].

Microbending causes measured bandwidths to

depend on the environment of the fiber, i.e., the way the fiber is deployed, its temperature, etc. The resultant decrease of γ can give an optimistic measurement of bandwidth. Ideally the measurement condition should simulate field conditions. Cooling the measurement spool is one way that has been used to reduce winding tension during measurement and thus minimize microbending, to simulate the cabled case.

Predicting bandwidths of links made up of several fibers with individually measured bandwidths presents difficulties for several reasons [9]. First, bandwidth compensation can occur among linked fibers. For example, one fiber might have slower-going outer modes due to its grading coefficient α being larger than optimum; if this is linked with one having faster-going outer modes coming from an α smaller than optimum, the resultant bandwidth would be less than might be anticipated from individual fiber-bandwidth measurements. (A differential modal delay measurement [93] would be needed to study such effects.) Also, the exponent factor γ is imprecisely known, since it depends on the cabling, the fiber properties, any jointing, and on the launched modal distribution. It thus varies from fiber to fiber as well as with length in a given fiber. In fiber specifications and link designs conservative predictions ($\gamma=1$) are often made.

3. Single-Mode–Fiber Measurements

Single-mode–fiber propagation and coupling properties differ from those of multimode fiber in many ways, and this affects the measurements required on them. In multimode fibers the multiplicity of modes combined with modal velocity differences generally provides the largest limitation to bandwidth. Coupling among these modes in the cables or at joints has large and complicated effects on overall link bandwidth. Differences in attenuation among the modes as well as coupling between them affects overall attenuation and its variation along the fiber's length. The fiber optogeometric properties (core diameter and NA) determine the maximum number of modes and therefore the propagation and jointing properties

as well as the specific distribution of modes that should be launched to get the most desirable properties.

For single-mode fibers, the situation is different and simpler. In this case only one mode (the LP_{01} mode) can propagate. Generally, for non-birefringent fibers the two polarization modes can be assumed to propagate in the same way and their presence does not complicate things. It is the geometric distribution of the LP_{01} mode, rather than fiber properties, that directly determines its propagation and coupling. Launching does not have an effect except in the overall power injected.

In single-mode fibers the cutoff properties of the second mode (LP_{11}) become important because the latter can affect apparent attenuation and lead to modal (bimodal) noise and bandwidth degradation. Another important factor is noise due to variations in partitioning of power among the light source's longitudinal modes (mode-partition noise). This can be a bigger system limitation than bandwidth at high (gigabits-per-second) bit rates and depends directly on chromatic dispersion. The fiber's influence on bandwidth arises from its chromatic dispersion—there is no modal dispersion. For this case, however, it becomes necessary to measure this dispersion directly, rather than through the bandwidth, because the latter becomes too high to measure in a practical way.

In the following sections techniques will be described for measuring the key parameters for single-mode fibers—attenuation of fibers and joints, cutoff wavelength, mode-field diameter and dispersion. Only recent work will be reviewed. Previous review papers include [4, 5, 94, 95, 96]. Those specifically covering standards work are [6, 17, 18].

Attenuation
The measurement of spectral attenuation of single-mode fibers is very similar to that of multimode fibers, and the cutback, substitution, and OTDR measurements are still applicable. In this case, as mentioned above, a great simplification occurs in that there is no concern about exciting or propagating a specific modal distribution.

Transmission Techniques—To launch light into the fiber for transmission techniques, generally, with incoherent sources, one simply overfills the core and NA of the fiber. Since the loss of single-mode fiber is usually less than that of multimode fiber, the accuracy and precision of the measurement apparatus must be better. In addition, since the amount of light actually launched into the fiber is less, the requirements on the stability of the light source as well as on noise reduction get more strict so that lock-in amplifiers and cooled detectors are often needed. Reference 59 uses temperature-controlled LEDs as light sources instead of white-light sources and obtains very high precisions and stabilities, which makes this method applicable to very limited cable lengths.

To discuss attenuation measurements properly it is useful to examine a single-mode fiber attenuation curve, shown in Fig. 19. At the upper wavelength end the loss increases rapidly. For matched-clad fibers this occurs due to increased sensitivity to microbending as the percentage of light in this mode that exists in the cladding becomes substantial and increases further with increasing wavelength. In moderately depressed cladding fibers (these and matched-clad fibers form what has been called first-generation or dispersion-unshifted single-mode fibers), this edge forms as the fundamental mode cuts off. There is no such cutoff in a matched-clad fiber.

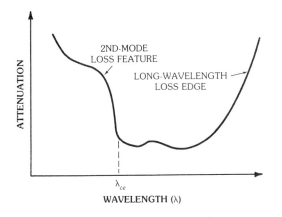

FIG. 19. Single-mode–fiber attenuation curve.

Concern must also be given to the effects of the second mode, the LP_{11}. In the region just below the cutoff wavelength of the LP_{11}, (λ_{ce} in Fig. 19), the percentage of light in the LP_{11} that is in the cladding becomes significant and continues to in-

crease with increasing wavelength. As a result of increased microbending and macrobending loss, as well as other loss components, the LP_{11} loss increases significantly in this region, and this is detected as a feature in the overall fiber attenuation curve. Once the cutoff wavelength is exceeded, however, the mode no longer propagates so that it does not contribute to the measured fiber loss. The cutoff thus appears as the long-wavelength edge of the feature. Measurement of the cutoff wavelength is discussed in the next subsection.

The second mode introduces an ambiguity in an attenuation measurement in the region of the feature because its loss is larger than that of the fundamental mode and, since the mode becomes insignificant in long lengths, its loss is not properly a part of the fiber attenuation. Nevertheless, since the LP_{11} mode can have measurable power in the reference length, it can show up in an attenuation curve as it did in Fig. 19. The apparent high measured loss can be a particular concern in first-generation, single-mode fibers used at 1.3 μm since this wavelength is likely to be in the neighborhood of the cutoff wavelength to permit alternate operation at 1.55 μm, i.e., to keep the high-wavelength–loss edge at wavelengths above 1.55 μm. If the LP_{11} mode produces such an ambiguity, a mode suppressor loop can be used to reduce the LP_{11} power sufficiently in the reference length. Mode-suppressor loops will be discussed in the next section.

The use of a mode stripper becomes a more critical issue for single-mode fiber than for multimode fiber because of the ease of exciting cladding modes in the former [4, 97]. Certain fiber buffers or coatings, however, can perform the job adequately. Any mode-stripping technique should be evaluated on a fiber type to ensure its effectiveness before the technique is used regularly. For reasons already discussed the need to avoid sharp bends and tight clamping, important for multimode fiber, is even more important for single-mode fiber because of the significant amount of the fundamental mode energy located in the cladding.

These considerations are also valid for the substitution technique. The error incurred in launching an improper light distribution at the joint between the source pigtail and the reference or test fiber is not a concern here, as it was for multimode fibers (Section 2, under "Optical Time-Domain Interferometry") but variations in joint loss are still important.

The cutback technique is the CCITT RTM and the EIA RS-455-78. The substitution method is not being considered by the CCITT as of this writing but an RS-455 is being developed for it by EIA [6].

Backscattering Techniques—As for multimode fiber, single-mode fiber OTDR has many applications in fiber and joint loss measurement, fiber fault location, fiber joint loss optimization, and fiber length measurement. It has the advantages of a nondestructive, portable technique which permits simultaneous determination of the length-dependent fiber loss and the presence and loss of joints and faults, all from one fiber end. The use of OTDR in single-mode fibers was reviewed in [4, 5, 96].

Theoretical studies [98, 99] show that an equation similar to (5) for the backscattered power holds in the case of single-mode fiber. It was pointed out in the discussion on multimode fiber OTDR that going from 0.8 μm to the longer wavelengths (1.3 and 1.55 μm) generally yields a decreased backscatter signal due to decreased scattering coefficient and decreased source power and detector sensitivity, although advances in the latter two have been made recently. The situation is further aggravated for single-mode fiber because the lower dopant levels used yield less scattering, because the lower number of modes decreases the backscatter capture fraction, and because there is less pulse power coupled into the fiber. Estimates indicate a 20- to 30-dB smaller backscatter signal for single-mode fiber than for multimode fiber [4], or a 10- to 15-dB smaller one-way range. However, because of decreased fiber loss, many tens of kilometers can still be viewed.

Many techniques have been reported for increasing the backscatter range, some similar to techniques described above for multimode fibers. The first single-mode fiber OTDR used a high-power laser, the 1.06-μm Nd:YAG laser [69], with coupled power limited so that nonlinear effects were insignificant. A loss measurement could be taken over a 15-dB range. Reference 100 used a 1.3-μm Nd:YAG laser to take advantage of the lower loss at that wavelength. An acousto-optic de-

flector was used as a directional coupler. The range in that case (as in the multimode fiber OTDR section, ranges imply one-way range to observe a nonreflecting break unless otherwise specified) is 21 dB, corresponding to a length of 34 km at the fiber attenuation of 0.63 dB/km; that for an accurate splice-loss measurement (within 0.1 dB) is 19 dB (30 km). Another group used Raman OTDR (see the multimode fiber OTDR section) in which the generated light signal at 1.5 to 1.6 μm extended the range to 52 km (25-dB range) [101]. Using a boxcar integrator, digital averaging and a liquid-nitrogen–cooled germanium APD, this was extended to a 102-km length, i.e., a 30-dB range [102, 103].

A calculation was made [104] of the maximum range for nonreflecting fault detection by Raman OTDR. Assuming the fiber loss is solely due to Rayleigh scattering, the result is 162 km, corresponding to an input pulse of 44 W from a Q-switched 1.06-μm Nd:YAG laser with the final generated and measuring wavelength at 1.61 μm and a fiber loss there of 0.16 dB/km. If a 0.15 dB/km component of wavelength-independent (e.g., microbending) loss also existed in the fiber, the maximum length would reduce to about 90 km. In this, the use of a germanium APD and an averaging correlator [67] was assumed, which was expected to give 50-dB enhancement. Experimental work involving a different high-power laser and not requiring nonlinear light generation in the fiber was reported in [105, 106]. The laser, a Q-switched Er:glass one, generated 1.55-μm light pulses, of which 2 to 4 W were coupled into a 100-km fiber. A TeO$_2$ acousto-optic light deflector was also used. The range was 30 dB (corresponding to 110 km of fiber with 0.26 dB/km loss including splice losses) if a germanium APD detector is used [105]. If, instead, liquid-nitrogen cooling of the detector (a germanium pin photodiode) and the amplifier's first stage and feedback element is used, this increases to a 34-dB range or about a 130-km length. In these the measurement time is about 15 min, in which averaging was only done for portions of the fiber above 59-km length. The slow repetition rate (5 pps) contributed to the long time.

While the above techniques permit testing long fibers because of the large laser powers combined with the longer wavelengths where fiber losses are low, the lasers are not suitable for field use. Several recent papers describe the use of semiconductor lasers for single-mode fiber OTDR. In [107] a 1.3-μm laser was used together with a liquid-nitrogen–cooled germanium APD detector operating in the photon-counting mode, with the receiver consisting of a digital correlator followed by a multichannel linear digital integrator. A 20-dB range was achieved for a 20-min measurement time, with a 10-dB improvement projected. Another work, with a laser at 1.55 μm, involved a similar detector but also with cooling of the feedback resistor in the receiver [108]. This decreases dark current and associated shot noise and suppresses the thermal noise in the front-end amplifier, extending the range to 42 km. However, the liquid-nitrogen cooling used in these methods is itself not field-suitable.

In order to eliminate the use of coolants an ultralow-noise receiver with an HgCdTe detector was designed and found to match the sensitivity of the liquid-nitrogen–cooled germanium APD detector used in the photon-counting mode [109]. Further improvements include the use of an acousto-optic beam deflector for maximum launch efficiency. Multichannel digital averagers handle 10^6 traces to produce a 30-dB (optical) noise reduction in 20 min. The result is a 30-dB range.

An alternative approach to these techniques is to use coherent detection, which can provide a quantum-noise limited sensitivity. Heterodyne detection [110] and homodyne detection [111] were utilized in OTDRs. A helium-neon laser at 1.55 μm and a pinFET receiver were used together with multichannel digital averaging. With launched power under 10 μW a range of 30 km (10 dB) was achieved. However, a fading phenomenon is observed [112], producing a jagged appearance in the trace, which can change from pulse to pulse. This arises from the use of the narrow–line–width light sources (gas lasers or injection-locked diode lasers) that are needed to maintain the advantage of heterodyne over direct detection. The jagged appearance is due to a combination of an optical mixing penalty caused by the random state of polarization of the backscattered wave (which mixes with the fixed-polarization wave from the local oscillator) and a specklelike phenomenon due to the summa-

tion of a very large number of scattered waves [113]. Much of the fading effect can be eliminated with an averaging technique [111], and by making the system as insensitive to polarization as possible. More recently [114], a 1.15-μm He-Ne laser was used, providing 120 μW into the fiber. A range of 64 dB was demonstrated using a measurement time of 8 min. The instrument was said to be robust and suitable for field use. It is noted that at 1.15 μm, losses can be 0.6 to 0.8 dB/km, so this range would correspond to an 80- to 105-km length. Sufficiently coherent and powerful sources at longer wavelengths are needed to extend this range.

The choice of the directional coupler is very important in single-mode fiber OTDR. Requirements include the lowest possible insertion loss, the maximum possible discrimination against the fiber's front-face reflection, the best isolation between transmitter and receiver, and the minimization of polarization discrimination. The last is of concern because of the polarization variations that can occur due to noise or to the birefringence effects in the fiber, which would show up as noise in the backscattered trace if there was discrimination at the coupler. In fact, polarization-sensitive OTDR can be used to measure birefringence beat-length information in fibers [114, 115] and also optical field distributions [116]. The polarization sensitivity gets much stronger for heterodyne schemes [110, 111, 114], and a polarization scrambler must often be used in that case.

One of the best coupler compromises with respect to the requirements listed above is the acousto-optic beam splitter cited in several of the above references. A very good TeO$_2$ version was described recently [117]. This also has the advantage of isolating specific parts of the return trace for detailed averaging and study. For high-power probes as in the Raman OTDR technique, however, this could be damaged; in that case a dichroic mirror could be used profitably [101] since the incident and returned beams have different wavelengths. The dichroic mirror scheme also gets rid of the front–fiber-face reflection but the mirror would have to be specially designed to minimize polarization discrimination. Further discussion on the splitter can be found in [4].

Just as for multimode fiber, fluctuations in

fiber parameters along the fiber length can confound OTDR measurements. Effects are similar to those for multimode fiber. They are simpler to predict, however, because modal effects presumably do not exist. The predictions of (8) and (9), using measurements from both ends of the fiber, are valid [99] so losses can be determined. An example is shown in [118]. In certain cases, even if several parameters are varying simultaneously, at least two of these can be determined from the forward and backward measurements [99], unlike the case of multimode fiber.

Joint Loss—The measurement of single-mode–fiber joint loss is simpler than that for multimode fiber since the concerns about modes excited and transient losses do not exist. However, in the region of the cutoff wavelength, care must be taken to avoid problems due to excitation of the LP_{11} mode, i.e., a mode-suppressor loop may be necessary. As described in the attenuation section on transmission techniques, care must also be taken to properly eliminate cladding modes, to gather all the light from the fiber, and, because the joint loss may be low, to keep the system properly stable.

The measurement of joint losses by OTDR is similar to that described above for multimode fiber (Section 2, "OTDR Measurements of Joint Losses"). Fluctuations in fiber properties can produce erroneous (even negative) loss results [4], that can be corrected by taking the mean value in decibels of the losses measured from the two fiber ends. The use of single-mode fiber eliminates the complexities due to modal effects such as transient losses and yields symmetrical losses (loss is the same from either direction), making the measurement more accurate. However, the inconvenience of a two-ended measurement still exists. One solution is to provide a reflector at the far fiber end, making visible the traversal through the joint from both directions in the same trace [119]. The OTDR's range must then equal or exceed the forward fiber loss plus that back to the joint, putting a more severe limit than usual on the fibers tested.

Cutoff Wavelength

The cutoff wavelength of a single-mode fiber generally refers to the wavelength at which the LP_{11} mode cuts off, i.e., can no longer propagate. It is

important to know this wavelength because, below it the presence of the LP_{11} mode can cause bimodal noise as well as a bandwidth decrease due to bimodal dispersion. The strict definition of cutoff wavelength corresponds to the theoretical cutoff wavelength, which is only appropriate for very short (millimeters) straight fibers. A more practical definition allows for the fact that in the region approaching cutoff, the growth of the spot size of the LP_{11} with increasing wavelength contributes to an increasing attenuation excess over the fundamental mode's attenuation (see Section 3, first subsection). Thus the actual propagation of the LP_{11} mode through a particular length of fiber depends on the distribution (length, radius) of bends in the fiber, the presence of fluctuations in diameter and index, and the fiber length [17, 120, 121, 122]. This leads to an effective cutoff wavelength, λ_{ce}, that is always less than the theoretical value since it takes account of the fact that the fiber in its "environment" tends to suppress the LP_{11} mode at wavelengths for which it would be present (not suppressed significantly) in a short, straight fiber. It is the cutoff wavelength that is observed experimentally (e.g., in Fig. 19) and is the wavelength at which the bimodal problems referred to above disappear. Since λ_{ce} is a function of the fiber length and environment (bends, etc.) any measurement of it must be made under standardized conditions. Perhaps the most precise method of defining effective cutoff wavelength is the wavelength at which a certain suppression (LP_{11} power a given amount of decibels below the LP_{01} power at the fiber end) occurs. This is discussed further below.

Any λ_{ce} measured on a test fiber (e.g., a 2-m–long fiber) must be related to that of a fiber in the field by determining relations between the fiber lengths and/or curvatures. Recent progress has been made in establishing such relations, and λ_{ce} is found to decrease with the logarithm of the length and, more crudely, with increasing curvature [121, 122].

The fact that cutoff wavelengths decrease with cable length can permit operating at wavelengths equal to or even below the short-length λ_{ce} [120]. Working in this region is advantageous from the point of view of minimizing microbending noise and pushing up the cutoff wavelength of the fundamental mode. Thus for a fiber to be used at both 1.3 and 1.55 μm, the short-length cutoff wavelength can be at 1.3 μm. For short fibers, however, such as in jumper cables, bimodal noise can be a problem [120, 123, 124] and, if so, must be eliminated by using jumper cables with shorter λ_{ce}s or by adding a mode-suppressor loop or other mode filter to the jumper.

It should be noted that highly birefringent fibers can have two slightly separated cutoff wavelengths corresponding to the two polarizations [5].

Several techniques have been reported for measuring λ_{ce}. One group of these is based on the spectral light transmission in a fiber. Because of the variabilities discussed above, these measurements are typically made on a fiber that has a standardized short length and is "straight," i.e., has a certain minimum radius of curvature. CCITT and the EIA have agreed upon 2 m as the length and 14 cm as the minimum radius, where the fiber should be loosely wound on its spool [6, 18, 125]. One popular technique, the single-bend attenuation technique [126], measures the spectral transmission of the fiber (the so-called straight fiber) and then does it again after a single bend is applied by loosely wrapping the fiber around a mandrel (typically 2 to 3 cm in diameter). Typical results are shown in Fig. 20a. The bend has the effect of reducing the cutoff wavelength. Taking the ratio of the power $P_b(\lambda)$ through the bent fiber to that, $P_s(\lambda)$, through the straight fiber, one gets the spectral attenuation of the bend (Fig. 20b). The long-wavelength edge of the curve is indicative of λ_{ce} of the straight fiber. This curve is more accurately measured than that for the straight fiber (Fig. 20a) because the measurement-system response (including the nonuniform light-source emission spectra or detector response) has been divided out. Wavelength λ_{ce} is taken to be the wavelength at which the bend loss is 0.1 dB; this specifies a particular suppression of the LP_{11} mode (19.3 dB, assuming the modes were equally excited originally [122]) and usually brings the measurement enough above background noise to be quite reproducible. This method was one of those tested in a recent EIA/NBS interlaboratory measurement comparison among several fiber manufacturers and users [97] in which it was found that the precision of this method was typically 2 nm. The man-

drel radius is not too critical; however, mandrels as small as 1.6-cm diameter caused increases in fundamental-mode attenuation for the fibers studied because of a shift in the microbending edge. Too large a bend radius yields an insufficient shift in λ_{ce} and gives poor reproducibility [97].

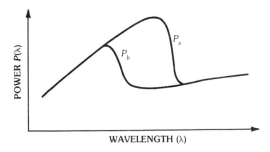

(a) Powers transmitted through bent (P_b) and straight (P_s) fibers vs wavelength.

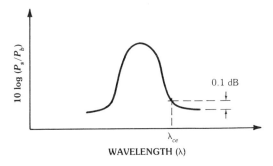

(b) Attenuation of bend and definition of effective cutoff wavelength λ_{ce}.

FIG. 20. Single-bend attenuation technique.

We have already referred to mode-suppressor loops, which are based on the above mandrel. The CCITT suggests a single loop of 2-cm diameter for the mode suppressor [121]. Other considerations on the loop are given in [97, 121].

Another way of removing the system response in a spectral power transmission measurement is to divide the attenuation of the single-mode fiber by that of a highly multimode fiber [33]. The interlaboratory comparison showed that if the same type of data analysis is used (e.g., the 0.1-dB technique) this method gives results insignificantly different from the single-bend method. This technique and the single-bend techniques have the advantage that they can work on the same setup used for measuring spectral attenuation. The techniques have been adopted as the CCITT RTM and the EIA RS-455-80 [6].

An alternate technique involves the spectral scanning of the mode-field diameter (MFD), which is the size of the light electric-field distribution (see next subsection). The technique [127, 128, 129] is shown in Fig. 21a. As wavelength increases from below cutoff, the diameter of the LP_{01} mode and the larger LP_{11} mode, and thus the overall MFD, increases at first. Then, as cutoff is approached, it decreases since the LP_{11} mode is eliminated. As wavelength is increased further, the MFD increases again. The effective cutoff wavelength λ_{ce} is taken to be at the intersection of straight lines fit to the decreasing and increasing portions of the curve (see figure). This technique, an alternate test method (ATM) of the CCITT, has the disadvantage that it depends on a spectral MFD measurement, which can be time consuming and prone to inaccuracy problems. A simpler technique [122], which is useful for long fibers (the single-bend technique is not useful for fibers longer than about 10 m [122]) is based on the sensitivity of splice loss to transverse offset. A single-mode fiber, selected to have a λ_{ce} well below the fiber being tested, is butt-jointed to that fiber, with index-matching fluid at the joint. Light is injected in the end of the low-λ_{ce} fiber, and the spectral throughput power is measured when the splice is adjusted for peak power and then again when the splice is transversely misaligned by a few micrometers. The ratio of the powers as a function of wavelength λ gives the spectrum of loss due to the offset splice (see Fig. 21b). This has three regions similar to those in Fig. 21a. In the region having wavelength above the cutoff wavelength, the fiber is single-mode and there is a characteristic splice loss. Well below λ_{ce} the fiber is strongly multimode, so the offset introduces a smaller loss. The third region is the transition region between these two. The effective cutoff wavelength λ_{ce} is taken to be the shortest wavelength at which there is no detectable power in the higher-order mode (i.e., no apparent increase in splice throughput) to within 0.1 dB.

Other techniques [130, 131, 132] have been described which allow the fraction of optical power in the fundamental mode with respect to total optical power, as well as λ_{ce}, to be measured. This permits the measurement of the LP_{11} mode sup-

pression and, particularly, the suppression caused by particular structures such as a mode-suppressor loop. Thus determinations of expected modal noise could be made [123].

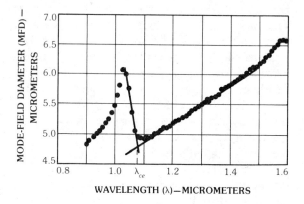

(a) *Scan of mode-field diameter.* (After Cannell et al. [129])

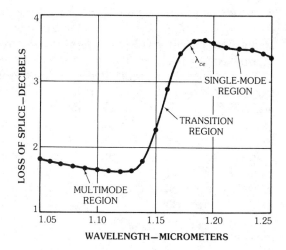

(b) *Scan of offset splice loss.* (After Anderson and Lenahan [122], ©1984 IEEE)

FIG. 21. Obtaining effective cutoff wavelength from spectral scans.

Techniques involving measuring the shift of λ_{ce} with changes in the radius of an applied bend or of variations in the near-field intensity pattern with wavelength have also been used, and these are compared with the single-bend and single-mode/multimode techniques in the interlaboratory comparison mentioned above [97]. There are discrepancies for certain fibers and there is a larger measurement spread, in addition to the disadvantages of greater complexity (see also comments on near-field techniques in the next subsection).

For the dispersion-shifted and dispersion-flattened fibers some of the techniques may have to be reexamined although it is expected that the simpler ones (e.g., the single-bend) would work in the same way and just as well.

Mode-Field Diameter

It has already been mentioned that for single-mode fiber the geometric distribution of light in the propagating mode, rather than the core diameter and NA, is what is important in predicting such operational properties as splice loss, microbending loss, dispersion, etc. In particular, a single parameter, the mode-field diameter (MFD), which is a measure of the width of the distribution of electric-field intensity, can be used in the prediction of many of these properties.

For first-generation single-mode fibers operated near cutoff so as to give minimum microbending losses, e.g., with λ_{ce} in the 1.2-μm range and the operating wavelength near the minimum-dispersion wavelength, i.e., near 1.3 μm, the electric-field distribution can be approximated by a Gaussian distribution [133, 134, 135]. The width of the distribution has been related to the sensitivity of splice loss to small offsets and tilts, to microbending loss, and, through its variation with wavelength, to the waveguide dispersion of the fiber [94]. The ability to characterize these properties depends on the exact way the width is defined, since the distribution is not exactly Gaussian. One technique [133] is to take the width (MFD), $2w_b$, to be twice the e^{-1} radius of the optical electric field (e^{-2} radius of optical power) of the Gaussian radial dependence

$$E_g(r) = E_0 \exp\left(-r^2/w_b^2\right), \qquad (13)$$

where r is the radius and E_0 the field at zero radius. The diameter $2w_b$ is chosen so this Gaussian distribution would produce a maximum launch efficiency (overlap) integral

$$I = \left[\int_0^\infty r E_g(r) E(r)\, dr\right]^2 \bigg/ \int_0^\infty r E_g^2(r)\, dr \int_0^\infty r E^2(r)\, dr \qquad (14)$$

into the fiber, where $E(r)$ is the actual field distribution.

The overlap integral has physical significance since it corresponds to the width of the Gaussian beam that would launch maximum power into the fiber. It will be discussed below that this gives good agreement between near and far-field MFD measurements based on this definition.

Another MFD definition, due to Petermann [136, 137], is

$$2w_j = 2\left\{2\int_0^\infty E^2(r)\, r\, dr \bigg/ \int_0^\infty [dE(r)/dr]^2\, r\, dr\right\}^{1/2}. \qquad (15)$$

It has been found that joint losses due to small lateral offsets and waveguide dispersion depend on this parameter [136, 137, 138], while microbending losses depend on the parameter, also due to Petermann [139],

$$2w_m = 2\left[2\int_0^\infty E^2(r)\, r^3\, dr \bigg/ \int_0^\infty E^2(r)\, r\, dr\right]^{1/2}, \qquad (16)$$

which is also the second moment of the field [140]. The splice loss due to small tilts also depends on this parameter [94].

It has been shown [94] that for first-generation single-mode fiber, $2w_j \leq 2w_b \leq 2w_m$, where the equalities hold only if the fields are truly Gaussian. Typically, differences do not exceed 5% if the fiber is operated near 1.3 μm, or 10% if it's operated near 1.5 μm, i.e., the fields are "quasi-Gaussian" in this range [17]. Thus it was proposed that if a single MFD definition is given for first-generation single-mode fiber, $2w_b$ can be a good compromise [94].

However, for dispersion-shifted fibers, such as those recently described [141], the distribution becomes less Gaussian and the Gaussian fits like that implicit in (13) and (14) give rise to big, noncorrelatable discrepancies between different measurement or fitting techniques [142, 143, 144], while (15) and (16), which do not require Gaussian fits, give more meaningful results.

The nomenclature "spot size" has also been used (e.g., in [134]) to describe the size of the field distribution. Since some ambiguities (field versus power distributions) have been associated with it, standards committees are favoring the mode-field diameter (MFD) nomenclature, which is what we use.

Several recent measurement procedures can be used to characterize the MFD. Earlier ones are discussed in the reviews [4, 5, 94]. We will discuss them in the light of the above considerations and make some comments on their ease of use. Complex, time-consuming techniques would lead to significant costs if used in quality control but might be favored for research if they have sufficient flexibility. Techniques which give meaningful answers correlatable with other techniques are the ones to favor.

Generally, the measurements are made on short (typically 2 or 3 m long) fibers because the property involved is a short-fiber property, in the same way as are index-profile measurements.

The most direct way of measuring the MFD is via measurement of the transmitted near-field intensity distribution [134, 145]. This is done in a manner similar to that described in Section 2, under "Near- and Far-Field Patterns." The result is then fit to a Gaussian distribution; the particular fitting technique used is quite important (see discussion below). The near-field technique suffers from inaccuracies due to lens distortion and from difficulties in locating and stably holding the image plane at the detector. Also, the dynamic range is limited, although solutions compensating for this have been proposed [94]. In addition, only a small portion of the fiber power reaches the detector, re-

quiring signal averaging techniques or large source powers coupled into the fiber.

Another technique, the transverse offset technique, involves the measurement of the variation in power transmitted through a mechanical butt splice as one of the fibers is swept transversely through the alignment position [134, 146] (Fig. 22). This technique makes use of the dependence of splice loss on spot size for Gaussian modes [147]. First, the fiber is either cut in two and spliced back together or the fiber is spliced to another with a known MFD. The variation of throughput power with offset is fit to the expected Gaussian dependence, which for the case of identical fibers having an MFD of $2w_a$ is

$$P(d) = P_0 \exp\left(-d^2/2\,w_a^2\right), \qquad (17)$$

where d is the offset and P_0 is the maximum transmitted power. Since the pattern departs from a Gaussian distribution, the means of fit is quite important and it is found that the use of an unweighted truncated fit (the truncation deemphasizes the tails) gives good agreement with near-field and one-dimensional far-field techniques (see below) for the first-generation single-mode fibers studied [144]. The transverse offset technique can be of special interest to those whose interest in the MFD is to determine splice losses. It is amenable to measurements over a range of wavelengths, which can be useful for determining the cutoff wavelength or dispersion (see next section). Another advantage is

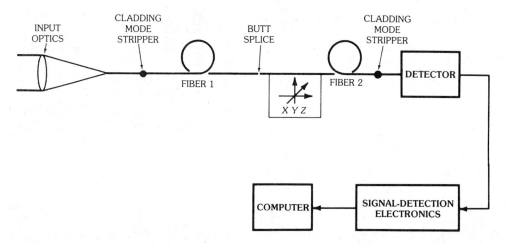

FIG. 22. Measurement of mode-field diameter by the transverse offset technique.

that the technique is efficient since most light is intercepted and transmitted, unlike for the near-field technique. However, it requires the ability to measure motions to submicrometer accuracy and is vibration sensitive. It also requires very good fiber-end cleaves so that the ends can be butted very close together to avoid an error due to beam spreading [134].

There are several techniques based on far-field measurements. One class of them involves making a one-dimensional linear scan with an apertured detector, similar to the measurement of NA (Section 2, under "Far-Field Pattern") (see [144, 134] and the latter's references). This technique is simpler than the two described above. Measurements to fractions of a micrometer, needed for the transverse offset technique, are not needed here. Also, the distortion and nonlinearity problems in the imaging and detection system that are encountered in the near-field technique are avoided here, so that a wider dynamic range is possible. Stability is also no longer so critical. As in the near-field technique, however, only a small fraction of the total source power is collected, limiting the selection of sources and detectors and making a wavelength-scanned approach difficult. Reference 134 describes details of one type of a far-field setup usable for this technique.

Interpretation of the far-field intensity data is done via one of two methods. In the first the data is transformed to the near field using the appropriate transformation, the inverse Hankel transform [5, 134, 144]. Then the resulting points are fitted to a Gaussian distribution. In the other approach, the far-field data are fitted to a Gaussian distribution and, from the width of this Gaussian, $2w_{ff}$, the near-field Gaussian width $2w_b$ is determined by the usual relation for Gaussian beams [134]. Because the field distribution is not exactly Gaussian, and

the departure from the Gaussian distribution increases with the operating wavelength above the cutoff wavelength, the fitting technique is quite important. Reference 140 shows that, if the overlap integral of the far-field pattern with a Gaussian is maximized (analogous to (14) for the near field), leading to an optimum far-field width $2w_{ff}$, the corresponding Gaussian near-field width will be that, $2w_b$, which maximizes the near-field overlap integral. Therefore, as long as far-field data is fitted to a Gaussian distribution via an overlap integral, the result will be the same as Hankel transforming the far-field data and fitting it via a near-field overlap integral. Thus near-field and far-field measurement results for MFD will agree using the overlap integral maximization.

Fitting the far-field data to a Gaussian distribution rather than transforming and then fitting has a big advantage—many fewer data points are needed in the former and the dynamic range required is less [140].

All the techniques described thus far involve fitting to Gaussian functions. While this fitting works well for first-generation single-mode fibers, MFDs determined this way for other single-mode fiber types with large departures from a Gaussian distribution may not have physical meaning, and major discrepancies in MFDs measured by different techniques can occur [144]. The following techniques have been adapted to more general fibers.

The variable-aperture technique, another far-field approach, is shown in Fig. 23. In a recent version [148, 149], each of a set of different-sized apertures is used sequentially to intercept the far-field pattern. Data of the relative powers transmitted through these known apertures is fit to a Gaussian far-field shape using a two-parameter nonlinear least-squares fit. The MFD is calculated from the determined far-field width. This works

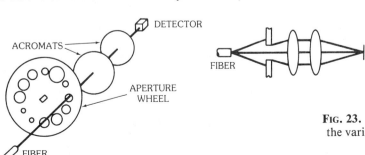

FIG. 23. Measurement of mode-field diameter by the variable-aperture far-field technique. (*After Westwig [149]*)

well for first-generation fibers. More recently, non-Gaussian fitting to variable-aperture data was performed [143] using the Petermann definition (15), which has a simple transformation from far-field data [137]. A three-parameter fit yields the MFD. This should be able to handle the other fiber types as well as first-generation fibers.

The variable-aperture technique shares with the transverse-offset method the advantages of a large efficiency, the ease of selecting sources, and the possibility of doing a wavelength scan so as to derive other parameters (MFD, cutoff wavelength). Also, like the other far-field techniques, it is not vibration sensitive. It can be done with a shared setup with other conventional measurements (e.g., attenuation). The aperture's transverse position can be scanned automatically to center on the pattern, making the fiber-end angle less critical. This technique is currently being used in manufacturing quality assurance [149].

Another far-field technique useful for general field distributions and utilizing Petermann's definition of (15) involves using a patterned mask in the far field [94, 150]. The power transmitted through the mask provides an evaluation of the integral forming the equation's numerator. The value of the denominator is the total power measured without a mask. The mask can either have an opaque zone bounded by an appropriate curve [94] (Fig. 24a) or a gray distribution (Fig. 24b). The first is easier to fabricate but requires a cylindrically symmetric far-field distribution. This symmetry is usually an excellent approximation and is easy to confirm by rotating the mask [94]. This technique has advantages of simplicity, good efficiency, and stability; moreover, it is tolerant of fiber-end quality, and it permits wavelength scans. Other mask techniques have been proposed (e.g., the Ronchi ruling grating mask technique [151]) but these are generally limited to Gaussian-beam assumptions.

The above techniques measure the MFD according to Gaussian-beam definitions or Petermann's first definition. These are useful for predicting the sensitivity of splice loss to small offsets, a major reason for determining MFD. However, if microbending loss and the splice loss due to small tilts are of interest, the MFD defined via Petermann's other definition (16) is more relevant. A mask tech-

nique similar to that described above has been proposed [94] for measuring this.

(a) Fermat's-spiral mask.

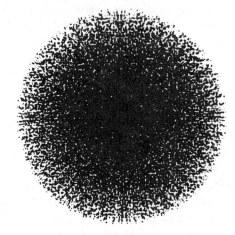

(b) Gray-distribution mask.

FIG. 24. Examples of masks used for mode-field diameter measurements. (*After Coppa et al. [94]*)

An interlaboratory comparison of different MFD measurement techniques on the same fiber was run by the EIA/NBS [125]. The techniques included the near-field, transverse offset, and the one-dimensional and variable-aperture far-field methods, all using Gaussian fitting. Results were obtained on first-generation single-mode fiber at both 1.3-μm and 1.5-μm wavelengths and, as expected, bigger variations occurred at the latter because of the less Gaussian mode shapes at that wavelength. Specific systematic offsets (up to 1 μm out of about 9 μm) were explainable by the difference between the techniques and the fitting procedures. In the standards area, CCITT has not resolved the MFD issue as of this writing, while EIA has issued a set of FOTPs covering several internally consistent procedures [6].

It is clear from the above discussions that a definition or definitions of MFD have to be identified for the various fiber types; this is particularly true for fibers other than first-generation ones. The definition(s) have to be physically meaningful and consistent and be measurable by relatively simple, straightforward, appropriately accurate, and precise techniques.

Dispersion

As mentioned in the beginning of Section 3, bandwidths for single-mode fibers are too high to measure except for very long fibers since direct-detection receivers cannot handle pulse widths less than 50 ps [95]. Also for single-mode fibers, bandwidth depends on source spectral properties and is thereby influenced by chirp and other complex effects. It thus becomes more practical to measure the chromatic dispersion, related closely to the variation of group delay with wavelength, from which the bandwidth of the fiber can be inferred once the source's spectral properties are known [152].

A knowledge of the dispersion also permits a calculation of mode-partition noise, arising from source multimoding, which can be a bigger limitation on system performance than is bandwidth at high modulation frequencies (gigahertz range) [152]. For lasers that emit largely in a single longitudinal mode under modulation, bandwidth and mode-partition noise can be negligible; in that case other dispersion-related phenomena which can limit system performance are laser chirping, chirp noise [153], and spectral sidebands due to modulation [154, 5].

Dispersion measurement techniques have previously been reviewed in [4, 5, 94, 95]. In the present treatment we do not consider polarization-mode dispersion, arising from a difference in velocity between the two polarization modes, since it is generally negligible (well under 1 ps/km) in the very circular low-birefringence fibers currently used for communications [155, 156]. Contributing to its low value are mixing between the polarization modes and random principal-axis orientation [5].

Time-Delay Techniques—The most direct measurement of dispersion is by time-delay techniques, i.e., by measuring the dependence of the time it takes for a pulse to traverse a fiber on the wavelength of the pulse. A very popular approach

is to use a fiber Raman laser [157], shown in Fig. 25. In this, high-intensity pulses from a mode-locked, Q-switched, 1.06-μm–wavelength Nd:YAG laser are injected into a single-mode fiber, producing short (approximately 150-ps) pulses containing a near continuum of wavelengths covering the range from 1.1 μm to above 1.6 μm, arising from a combination of stimulated Raman scattering and self-phase modulation. A monochromator is used to select particular wavelengths in this range. The filtered output is injected into the fiber being tested and the spectral dependence in pulse arrival time is determined with the help of a fast detector (e.g., a germanium APD). A low-jitter digital delay generator is needed to compensate for the fiber traversal time so as to yield proper triggering. If the spectral dependence of arrival times in a short reference is subtracted out to allow for delay variations in the laser and Raman fiber, the result is the spectral group delay, τ versus λ, of the test fiber (Fig. 26a). Generally, the group delay data is fit to an empirical expression, from which the dispersion parameters are obtained. For first-generation fibers the expressions usually used are a three-term Sellmeier one, e.g., $A_1\lambda^{-2}+A_2+A_3\lambda^2$, or a five-term one [95, 159]. The resulting curve of the spectral group delay τ per unit length can be differentiated to give the curve of dispersion $D=d\tau/d\lambda$ (Fig. 26b). For first-generation fibers it has been found that in the wavelength region of interest the dispersion performance can be well described by three parameters: the minimum-dispersion wavelength λ_0, the dispersion slope $dD/d\lambda$, and the dispersion curvature. Moreover, very near λ_0 only the first two suffice [5, 160]. More typically, however, fiber manufacturers specify dispersion in terms of a maximum value of D over a wavelength range. This description, although simpler, is not as precise (a disadvantage for high–bit-rate systems) and is made for the sake of fiber measurement time and yield [160]. Comparisons between using Sellmeier and non-Sellmeier expressions in terms of accuracy and number of measurements needed are given in [160]. For dispersion-shifted or dispersion-flattened designs, other types of expressions for fitting have to be used for maximum accuracy [135, 139].

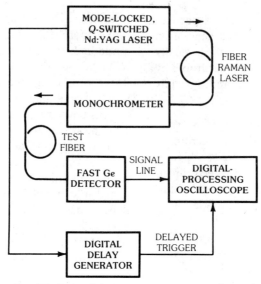

FIG. 25. Pulse-delay measurement of dispersion using a fiber Raman laser.

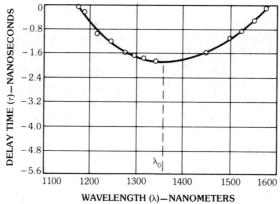

(a) Group delay.

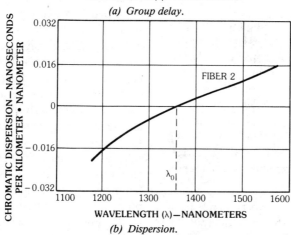

(b) Dispersion.

FIG. 26. Group delay and dispersion of a 1-km fiber. (*After Cohen et al. [158],* ©*1980 IEEE*)

The Raman laser measurement system involves a large, high-power, water-cooled laser that is not simple to maintain, and is not amenable to field or for most quality-control use. A more advantageous system for these applications would use a set of diode lasers emitting at different wavelengths as pulsed sources [159, 161]. A single-mode–fiber switch or manual alignment splice is needed to select between the lasers. Excellent agreement in dispersions measured by the two techniques (to within 0.35 ps/km·nm) occur for many types of single-mode fiber [161].

The discrete laser approach is much more compact than that based on the Raman laser, which is primarily a laboratory measurement. It does not have the dynamic range of the Raman laser method [161], although the difference is smaller with recently disclosed high-power lasers [162]. The wavelength coverage is considerably more sparse, so the technique would not be applicable for the detailed analysis involved in developing new fibers. Moreover, it places more importance on the proper selection of the analytical expression used in the fitting of the data.

Phase-Shift Techniques—A second class of dispersion measurements is based on phase-shift measurements. Usually in these the input light is sinusoidally modulated and the variation of the modulation phase shift over the fiber with wavelength is observed. In one version (shown in Fig. 27) an LED modulated at 30 MHz is used as a source; two LEDs are generally needed to cover the wavelength range 1.2 to 1.6 μm [165]. The light is coupled into the test fiber and the fiber's output light is filtered by a monochrometer and then detected by an APD followed by a low-noise, narrow-band amplifier. Finally, the phase is detected by a vector voltmeter, yielding the spectral phase shift. After the system phase shift is taken into account by repeating the above with a reference-length fiber, the dispersion can be determined. Measurements are found to agree well (to 1 ps/km·nm) with the results of the Raman pulse-delay technique.

The narrow-band filtering used in the phase-shift technique provides much more noise reduction than in pulse-delay techniques, similar to that described previously for frequency-domain tech-

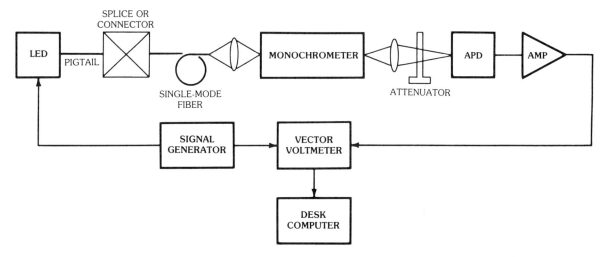

Fig. 27. Diagram of experimental setup of phase-shift dispersion measurement using a modulated LED as source. (*After Costa et al. [165],* ©*Institution of Electrical Engineers*)

For pulse-delay techniques, lengths have to be over about 0.5 km to measure group delay differences of ± 0.1 ns (the pulse width) in the neighborhood of λ_0, making them long-fiber techniques [95, 163, 164].

niques in the multimode-fiber bandwidth-measurement section. This permits the use of LED sources in spite of their low coupling efficiency into single-mode–fibers. The limiting resolution here is about 10 ps and fiber lengths are limited to about 1 km by the dynamic range, although the authors claim that improvements are possible [165].

Another phase shift technique, which couples much more light into the fiber uses a C^3 laser [166]. Light modulation is provided by 250-MHz modulation of the current of the front section of the laser; modulation of the current on the back section controls the laser wavelength by selecting which of the longitudinal modes lases. Dispersion can be measured down to 0.03 ps/km·nm resolution; this and the large coupled power permits measurements to be made on short lengths (as short as the 10-m range) as well as long lengths of fiber. Other work includes using a group of modulated lasers as phase-shift sources [167, 168]. Reference 168 describes an automated version using seven lasers in which the repeatability of λ_0 was within ±0.5 nm and that of the dispersion slope was 0.01 ps/km·nm^2 (this was for first-generation single-mode fibers). There was excellent agreement with the Raman technique, implying excellent accuracy. Spans of up to 50 km were said to be accommodated by this technique.

Recently, a multilaser version was described [169] in which the modulation frequency was swept rather than left constant. A pair of lasers was used, and their output powers were added. The lasers emitted at different wavelengths but were modulated at the same (scanned) frequency. The time position of nulls seen in the output are used in the evaluation of the dispersion parameters. Several laser pairs may be used to increase accuracy. The advantage of this technique is that it eliminates the need for a trigger or reference signal connection, which enhances its suitability for field use.

The multilaser phase-shift techniques make up in dynamic range what they lose in completeness of wavelength coverage (as compared with LED techniques) and are more suitable for quality control, where the coverage is not needed. The phase-shift techniques, in general, have better resolution, larger dynamic range and lower cost than pulse-delay techniques. They also have the advantage that they can make use of a setup very much like that used to measure multimode-fiber bandwidth in the frequency domain, as described previously.

Both phase-shift and time-delay methods have

been chosen as CCITT RTMs, and EIA will issue FOTPs on them [6].

Interference Techniques—A third class of dispersion measurements, which are very useful for short fibers, use interference of light. Recent ones (reviewed in [95]) work in the 1.3 to 1.6-μm wavelength region. In one type [164] a fiber Raman laser provides a tunable-wavelength source. In a Mach-Zehnder interferometer arrangement (Fig. 28a) the source beam is split into two beams by a beam splitter (labeled BS_1 in the figure). One beam goes through the test fiber in series with an optical delay line having variable delay. The other goes through a reference fiber having a length comparable to that of the test fiber (several meters) and having a known dispersion. The beams are combined on a beam splitter (BS_2) and detected. For each monochromator-selected wavelength a motor scans the delay in the variable-delay line and the detected signal is recorded (Fig. 28b). The delay value giving maximum fringe visibility, i.e., maximum cross correlation of the two beams and thus maximum detected signal envelope (shown by arrows in the figure), is noted and corresponds to equal interferometer arms. From the changes in delay with wavelength the relative group delay versus wavelength of the two fibers is determined. Since the dispersion of the reference fiber is known, the spectral group delay of the test fiber is found and, with the usual curve fitting, its dispersion characteristics can be obtained. A 0.3-ps temporal resolution, corresponding to 0.075 ns/km based on a 4-m fiber length, is obtainable and the agreement with time-delay data is quite good. Reference 170 gives a simple numerical method to extract the group delays from the measured visibility curves; this is said to eliminate human bias and error, enhance resolution, and facilitate automation of the measurement procedure.

These interference techniques are differentiated from most of the above techniques in that the fiber lengths can be short, with the advantage that one could pretest short fibers before drawing an entire preform into a fiber. This helps in optimizing fiber parameters to achieve low dispersion for high–bit-rate or wavelength–division-multiplexed system applications within desired wave-

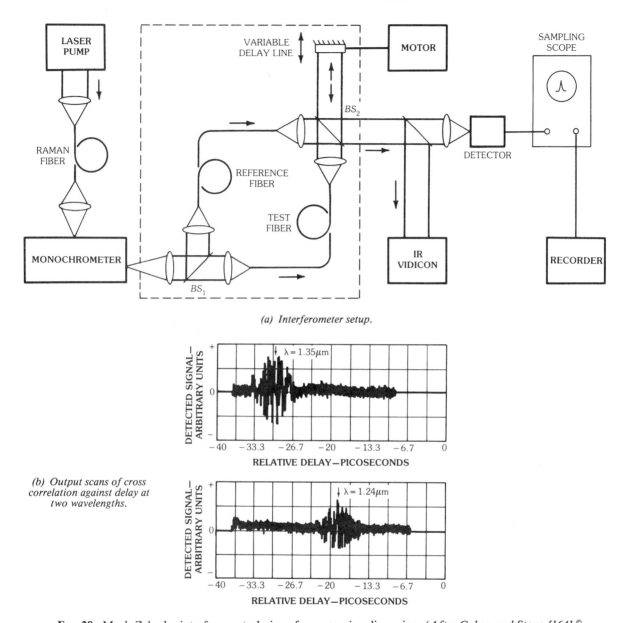

(a) Interferometer setup.

(b) Output scans of cross correlation against delay at two wavelengths.

FIG. 28. Mach-Zehnder interference technique for measuring dispersion. (*After Cohen and Stone [164],*© *Institution of Electrical Engineers*)

length ranges [171]. Also, they can check for axial variations in dispersion over the fiber length, due to variations in index or diameter. These variations can affect the total dispersion via the waveguide and profile dispersion components. This is particularly important in dispersion-shifted fibers or others in which the waveguide dispersion plays a major role. This subject is discussed further below.

The interference techniques have the disadvantage of sensitivity to vibrations and thermal effects; moreover, these use a light source that has already been described as limited to research uses. One improvement [172] utilizes a diode-laser source and permits a direct determination [173] of the dispersion D at the laser wavelength, which can be selected to be near the dispersion minimum. The technique, evaluated in [172], still requires stability of the apparatus. A method which substantially reduces the stability requirements is that of [174], which uses a white-light source (e.g., a cw halogen lamp) and a Mach-Zehnder interferometer similar to that of Fig. 28a. The lamp is coupled to a single-mode fiber which acts as a spatially coherent source for the interferometer. One of a set of interference filters after the interferometer output combiner acts as a wavelength selector. Once again a variable optical delay line in an interferometer arm is scanned and the equal-arm condition is seen via the increase in the envelope of detected light. As before, close agreement with the pulse-delay technique is seen.

Another version of this technique [175] uses a variable air path in one arm instead of a reference fiber; this was used to measure test fibers having lengths as small as 8 and 16 cm. Large standard deviations (16 and 5 nm for the 8- and 16-cm fiber lengths, respectively) were observed so the technique needs to be studied further; one possible problem is the presence of the LP_{11} and higher-order modes in the short lengths used. In yet another version the wavelength is scanned, permitting simultaneous determination of the second and third derivatives of propagation constant with respect to angular frequency, which are simply related to D and $dD/d\lambda$ [173].

The interference techniques have the advantage of 0.1-ps resolution, and thus can be used on short fiber lengths. As mentioned this permits studies of axial uniformity and assists in optimizing the fiber. It also has application as a quality-control technique but for this the fiber would have to be sufficiently uniform so that the short lengths are representative. A study of this was made in [171] using the white-light cross-correlation technique discussed above. Dispersion-shifted or dispersion-flattened fibers are quite sensitive to variations in index or diameter since waveguide dispersion must be precisely balanced against material dispersion. The study of [171] measured samples of these two types plus that of first-generation fibers and found that the uniformity was such that if dispersion is measured on short lengths taken from both ends of a long fiber, the measurement results agree very well with the long-fiber time-delay results in all cases if only fibers tapered in one direction are

considered and if the results on the two fiber ends are averaged. The results for any given fiber would depend on its type and manufacturing conditions and this should be checked (for example, with a short-fiber technique such as this) before conclusions on the validity of short-fiber results are made [142].

Interference techniques may not be suitable for most field applications since they are limited to short fiber lengths, and have vibration and thermal sensitivities that still have to be evaluated.

Spectral Mode-Field–Diameter Method—It is possible to determine the waveguide dispersion from a measurement of the dependence of mode-field diameter (MFD) on wavelength [142, 150, 176, 177] or, less directly, from the wavelength dependence of the near-field pattern [178]. To get the total dispersion the material and profile dispersion must be added to this. If one measured the MFD according to the Petermann definition of [136] (see (15) in the preceding subsection), the profile dispersion can be determined together with the waveguide dispersion [142]; however, a wavelength-dependent profile-dispersion parameter must also be known. For this and for an accurate knowledge of the material dispersion the index profile of the fiber and the dopant concentrations should be known at least approximately. Also, the MFD has to be measured quite accurately. Reference 143 determined the waveguide dispersion from MFD data obtained using the variable-aperture far-field technique. Material dispersion data obtained from [179] was added and the results for a first-generation fiber showed excellent agreement with pulse-delay data. Reference 142 used the transverse-offset technique to determine the MFD, adding in the Petermann profile-dispersion term and obtaining material dispersion by weighting material-dispersion data for each of the dopants. Differences from the pulse-delay data were attributed to the transverse-offset technique's inability to evaluate non-Gaussian field distributions. Also, concern about axial variations of dispersion with length was expressed since the MFD measurement is a short-fiber method.

The spectral MFD technique for determining dispersion has advantages of requiring only short lengths, applicability of the MFD-measurement apparatus, and the avoidance of time-dependent techniques. It has found use in quality control. The need to properly determine material and possibly profile dispersion are disadvantages except for well-characterized fiber.

4. Conclusions

The above discussions indicate that much progress has been made in devising and refining measurement techniques, although a state of flux is still evident. The latter is due especially to the development of new fiber types, the required identification of the important characterization parameters and the development of appropriate measurements. Much progress in standardization is also being made, although several issues still have to be resolved and, of course, new fibers are creating new standardization questions. Several alternative techniques have been described for each measurement. The optimum one to use depends on many factors, including standardization, the application (e.g., field use), cost, space, speed, sensitivity, and required precision and accuracy.

Acknowledgments

The author is grateful to B. Basch, W. J. Carlsen, E. Eichen, F. Kapron (ITT), P. R. Reitz (Corning), M. P. Singh, and R. A. Westwig (Corning) for helpful discussions. He also thanks B. Basch, R. Klein and M. P. Singh for a critical reading of the manuscript.

6. References

[1] G. CANCELLIERI and U. RAVAIOLI, "Measurement of Optical Fibers and Devices: Theory and Experiments," Dedham: Artech House, 1984.

[2] A. H. CHERIN et al., "Measurement of the Core Diameter of Multimode Graded-Index Fibers: A Comparison of Transmitted Near-Field and Index-Profiling Techniques," *IEEE J. Lightwave Tech.,* Vol. LT-1, p. 302, 1983.

[3] W. J. STEWART, "Optical Fiber and Preform Profiling Technology," *IEEE J. Quantum Electron.,* Vol. QE-18, p. 1451, 1982.

[4] L. B. JEUNHOMME, *Single-Mode Fiber Optics— Principles and Applications,* Chapter 5, New York: Marcel Dekker, 1983.

[5] F. P. KAPRON and P. D. LAZAY, "Monomode-Fiber Measurement Techniques and Standards," *Proc. SPIE Symp. Single-Mode Optical Fibers,* Vol. 425, p. 40, San Diego, 1983.

[6] F. P. KAPRON, "Issues in Single-Mode Fiber Standardization," *Proc. SPIE Symp. Fiber Optics: Short-Haul and Long-Haul Measurements and Applications II,* Vol. 500, p. 2, San Diego: SPIE, 1984.

[7] G. W. DAY, "Birefringence Measurements in Single-Mode Optical Fiber," *Proc. SPIE Symp. Single-Mode Fibers,* Vol. 425, p. 72, San Diego, 1983.

[8] S. C. RASHLEIGH, "Origins and Control of Polarization Effects in Single-Mode Fibers," *IEEE J. Lighwave Tech.,* p. 312, 1983; S. C. RASHLEIGH and R. H. STOLEN, "Preservation of Polarization in Single-Mode Fibers," *Laser Focus,* p. 155, May 1983.

[9] A. H. CHERIN, "Multimode Fiber Measurements— Present and Future," *Tech. Dig. Symp. Optical-Fiber Measurements,* (*NBS Special Publication 683*), p. 67, Boulder, 1984.

[10] J. E. MATTHEWS and P. REITZ, "Measurement and Characterization of Optical Fiber Designed for Communications Systems With LED Sources," *Proc. SPIE Conf. Fiber Optics: Short-Haul and Long-Haul Measurements,* Vol. 500, p. 82, San Diego, 1984.

[11] A. H. CHERIN, *An Introduction to Optical Fibers,* Chapter 8, New York: McGraw-Hill Book Co., 1981.

[12] P. S. LOVELY, "Measuring Fiber Performance Parameters," *Laser Focus,* p. 98, May 1984.

[13] A. H. CHERIN and W. B. GARDINER, "Standardization of Optical-Fiber Transmission Measurements," *Laser Focus,* Vol. 16, p. 60, August 1980.

[14] J. D. CHIPMAN, "Measurement Standards for Optical Fibers in Field Installations," *Laser Focus,* p. 99, July 1982.

[15] D. L. FRANZEN et al., "Standardizing Test Conditions for Characterizing Fibers," *Laser Focus,* p. 103, August 1981.

[16] R. L. GALLAWA and D. L. FRANZEN, "Progress in Fiber Test Standards," *Photonics Spectra,* p. 55, April 1983.

[17] P. R. REITZ, "Compatibility of National and International Standards for Optical Fibers," *Tech. Dig. Symp. Optical-Fiber Measurements, (NBS Special Publication 683)*, p. 41, Boulder, 1984.

[18] G. BONAVENTURA and U. ROSSI, "Standardization Within CCITT of Optical Fibres for Telecommunication Systems," *Proc. Globe Commun. Conf.,* Paper 31.1.1, p. 1030, Atlanta, 1984.

[19] R. OLSHANSKY, "Differential Modal Attenuation in Graded-Index Optical Waveguides," *Electro-Optical Syst. Des. Mag.,* Vol. 9, p. 56, November 1977; R. OLSHANSKY and S. M. OAKS, "Differential Modal Attenuation Measurements in Graded-Index Fibers," *Appl. Opt.,* Vol. 17, p. 1830, 1978.

[20] D. GLOGE, "Optical Power Flow in Multimode Fibers," *Bell Syst. Tech. J.,* Vol. 51, p. 1767, 1972.

[21] S. ZEMON and D. FELLOWS, "Characterization of the Approach to Steady State and the Steady-State Properties of Multimode Optical Fibers Using LED Excitation," *Opt. Commun.,* Vol. 13, p. 198, 1975.

[22] P. VELLA et al., "Precise Measurement of Steady-State Fiber Attenuation," *Tech. Dig. Symp. Optical Fiber Measurements, (NBS Special Publication 641)*, p. 55, Boulder, 1982.

[23] P. R. REITZ, "Prediction of Length Performance of Multimode Graded-Index Fiber," *Tech. Dig. Symp. Optical-Fiber Measurements (NBS Special Publication 641)*, p. 1, Boulder, 1982.

[24] P. R. REITZ, Corning Glass Works, private communication.

[25] M. L. DAKSS and B. KIM, "Effects of Fiber Propagation Loss on Diode-Laser-Fiber Coupling," *Electron. Lett.,* Vol. 16, p. 421, 1980.

[26] M. V. KLEIN, *Optics,* Chapter 7, New York: John Wiley & Sons, 1970.

[27] E. M. KIM and D. FRANZEN, "Measurement of Far-Field and Near-Field Radiation Patterns From Optical Fibers," *NBS Tech. Note 1032,* Washington: U.S. Government Printing Office, 1981.

[28] S. MAZZOLA and G. DEMARCHIS, "Analytical Relations Between Modal Power Distribution and Near-Field Intensity in Graded-Index Fibers," *Electron. Lett.,* Vol. 15, p. 721, 1979.

[29] G. K. GRAU and O. G. LEMINGER, "Relations Between Near-Field and Far-Field Intensities, Radiance and Modal Power Distribution of Multimode Graded-Index Fibers," *Appl. Opt.,* Vol. 20, p. 457, 1981.

[30] M. FOX and M. R. DENNIS, "Near-Field Intensity and Modal Power Distribution in Multimode Graded-Index Fibers," *Electron. Lett.,* Vol. 16, p. 678, 1980.

[31] F. M. E. SLADEN et al., "Determination of Optical-Fiber Refractive-Index Profiles by Near-Field Scanning Technique," *Appl. Phys. Lett.,* Vol. 28, p. 255, 1976.

[32] E. M. KIM and D. L. FRANZEN, "Measurement of the Core Diameter of Graded-Index Fibers by an Interlaboratory Comparison," *Appl. Opt.,* Vol. 21, p. 3443, 1982.

[33] Electronics Industries Association, 12001 Eye St., NW, Washington, DC, 20006.

[34] A long length can give a different answer because of steady-state effects. This is discussed in Section 2 under "Attenuation Measurements."

[35] R. L. GALLAWA, "On the Definition of Fiber Numerical Aperture," *Electro-Optic Syst. Des.,* p. 46, April 1982.

[36] A. H. CHERIN and E. D. HEAD, "Comparison of Numerical Apertures Obtained from Far-Field and Refractive-Index Profile Measurements," *Tech. Dig. Optical-Fiber Commun. Conf.,* Paper TuQ8, p. 68, San Diego, 1985.

[37] G. WORTHINGTON, "Acceptance-Angle Measurement of Multimode Fibers: a Comparison of Techniques," *Appl. Opt.,* Vol. 21, p. 3515, 1982.

[38] R. M. HAWK, "Multimode Waveguide Attenuation Measurements," *Tech. Dig. Symp. Optical-Fiber Measurements, (NBS Special Publication 597)*, p. 1, Boulder, 1980.

[39] P. R. REITZ, "Measuring Optical Waveguide Attenuation: The LPS Method," *Opt. Spectra,* p. 48, August 1981.

[40] G. W. DAY et al., "Measurement of Multimode Optical Fiber Attenuation," *NBS Tech. Note 1060,* Washington: U.S. Government Printing Office, 1983.

[41] A. B. SHARMA et al., "Effect of Source Near-Field and Far-Field Variations on the Accuracy of Multimode Fiber Attenuation Measurements," *Electron. Lett.,* Vol. 18, p. 49, 1982.

[42] M. DAKSS and D. YOUNG, GTE Laboratories, unpublished work.

[43] A. B. SHARMA et al., "Technique for Reducing Spatial and Wavelength Variations of Launch Spot in Fibre Attenuation Measurements," *Electron. Lett.,* Vol. 18, p. 244, 1982.

[44] P. LOVELY, Photon Kinetics Co., private communication.

[45] P. R. REITZ, "A Quality Check for Fiber End Faces," *Opt. Spectra,* Vol. 13, p. 39, December 1979.

[46] R. P. NOVAK et al., "Compact and Efficient Improvement of the 70 percent Beam Optics," *J. Opt. Commun.,* Vol. 3, p. 49, 1981.

[47] A. H. CHERIN et al., "Selection of Mandrel-Wrap Mode Filters for Optical Loss Measurements," *Fiber and Integrated Optics,* Vol. 4, p. 49, 1982.

[48] A. H. CHERIN, "A Fiber Concatenation Experiment Using a Standardized Loss Measurement Method," *Tech. Dig. Symp. Optical-Fiber Measurement,* (*NBS Special Publication 597*), p. 19, Boulder, 1980.

[49] W. F. LOVE, "Novel Mode Scrambler for Use in Optical Time-Domain Bandwidth Measurements," *Tech. Dig. Conf. Optical-Fiber Commun.,* (OFC), Paper ThG, p. 118, Washington, 1979.

[50] D. L. FRANZEN and E. M. KIM, "Interlaboratory Comparisons on Graded-Index Optical Fibers using standard measurement conditions," (*NBS Special Publication 637*): *Optical-Fiber Characterization,* Vol. 1, p. 141, 1982.

[51] M. TETEDA et al., "Optical Loss Measurement in Graded-Index Fiber Using a Dummy Fiber," *Appl. Opt.,* Vol. 18, p. 3272, 1979.

[52] M. IKEDA et al., "Multimode Optical Fibers: Steady-State Mode Exciter," *Appl. Opt.,* Vol. 15, p. 2116, 1976.

[53] T. LEHIEP and R. TH. KERSTEN, "A Combined Mode Filter/Mixer to Determine Spectral Attenuation of Graded-Index Fibers," *Opt. Commun.,* Vol. 40, p. 111, 1981.

[54] J. W. VERSLUIS and J. G. J. PEELEN, "Optical Communication Fibers—Manufacture and Properties," *Phillips Telecommun. Rev.,* Vol. 37, p. 215, 1979.

[55] A. K. AGARWAL and U. UNRAU, "Comparative Study of Methods to Produce Stationary Mode Power Distributions for Optical-Fiber Measurements," *J. Optical Commun.,* Vol. 4, p. 126, 1983.

[56] W. J. CARLSEN and P. MELMAN, "Elastic Tube Splice Performance With Single-Mode and Multimode Fibers," *Electron. Lett.,* Vol. 18, p. 320, 1982.

[57] S. SUMIDA et al., "A New Method of Optical-Fiber Loss Measurement by the Side-Illumination Technique," *IEEE J. Lightwave Tech.,* Vol. LT-2, p. 642, 1984.

[58] H. DOYLE and R. A. WEY, "Design and Operation of a Stable Attenuation Measurement System," *Laser Focus,* p. 117, May 1982.

[59] Y. NAMAHIRA et al., "Long-Term High-Stable Optical-Fiber Loss-Measuring Equipment," *Tech. Dig. Symp. Optical-Fiber Measurements,* (*NBS Special Publication 683*), p. 99, Boulder, 1984.

[60] L. S. SHORT and R. B. KUMMER, "Improved Automated Loss Set for Optical Cables," *Tech. Dig. Symp. Optical-Fiber Measurements,* (*NBS Special Publication 641*), p. 43, Boulder, 1982.

[61] J. GENTILE, "Characterizing Optical Fibers With an Optical Time-Domain Reflectometer," *Electro-Optic Syst. Des.,* p. 47, April 1981.

[62] M. K. BARNOSKI et al., "Optical Time-Domain Reflectometer," *Appl. Opt.,* Vol. 16, p. 2375, 1977; P. DIVITA and U. ROSSI, "The Backscattering Technique: Its Field of Application in Fiber Diagnostics and Attenuation Measurements," *Opt. Quantum Electron.,* Vol. 11, p. 17, 1980.

[63] S. D. PERSONICK, "Photon Probe—An Optical-Fiber Time-Domain Reflectometer," *Bell Syst., Tech. J.,* Vol. 56, p. 355, 1977.

[64] E. G. NEUMANN, "Optical Time-Domain Reflectometry: Comment," *Appl. Opt.,* Vol. 17, p. 1675, 1978.

[65] J. GENTILE, Laser Precision Co., Private Communication.

[66] Orionics Co., specifications.

[67] K. OKADA et al., "Optical Cable Fault Location Using Correlation Technique," *Electron. Lett.,* Vol. 16, p. 629, 1980.

[68] K. I. AOYAMA et al., "Optical Time-Domain Reflectometry in a Single-Mode Fiber," *IEEE J. Quantum Electron.,* Vol. QE-17, p. 862, 1981.

[69] D. L. PHILEN, "Optical Time-Domain Reflectometry Using a Q-Switched YAG Laser," *Tech. Dig. Symp. Optical-Fiber Measurements,* (*NBS Special Publication 597*), p. 97, Boulder, 1980.

[70] A. J. CONDUIT et al., "An Optimized Technique for Backscatter Attenuation Measurement in Optical Fibers," *Optical and Quantum Electron.,* Vol. 12, p. 169, 1980.

[71] J. L. HULLETT and R. D. JEFFREY, "Long-Range Optical-Fiber Backscatter Loss Signatures Using Two-Point Processing," *Optical and Quantum Electron.,* Vol. 14, p. 41, 1982.

[72] J. L. HULLETT and R. D. JEFFREY, "Noise in Optical-Fibre Backscatter Measurements," *Optical and Quantum Electron.,* Vol. 13, p. 117, 1981.

[73] P. HEALEY, "Optical Time-Domain Reflectometry by Photon Counting," *Tech. Dig. Eur. Conf. Opt. Commun.,* York, UK, p. 156, 1980; P. HEALEY and P. HENSEL, "Optical Time-Domain Reflectometry by Photon Counting," *Electron. Lett.,* Vol. 16, p. 631, 1980.

[74] P. HEALEY, "Pulse Compression Coding in Optical Time-Domain Reflectometry," *Tech. Dig. Eur. Conf. Optical Commun.,* p. 5.2-1, Copenhagen, 1981.

[75] M. NAKAZAWA et al., "OTDR at Wavelength of 1.5-μm Using Stimulated Raman Scattering in Multimode Graded-Index Optical Fiber," *J. Appl. Phys.,* Vol. 53, p. 1391, 1982.

[76] M. NAKAZAWA and M. TOKUDA, "Measurement of the Fiber Loss Spectrum Using Fiber-Raman Optical Time-Domain Reflectometry, *Appl. Opt.,* Vol. 22, p. 1910, 1983; K. NOGUCHI et al., "Ultralong Multimode Optical Fiber Fault Location Using Raman OTDR," *Electron. Lett.,* Vol. 18, p. 542, 1982.

[77] P. DIVITA and U. ROSSI, "Backscattering Measurement in Optical Fibers: Separation of Power Decay from Imperfection Contributions," *Electron. Lett.,* Vol. 15, p. 467, 1979.

[78] A. R. MICHELSON and M. ERIKSUD, "Theory of the Backscattering Process in Multimode Optical Fibers," *Appl. Opt.,* Vol. 21, p. 1898, 1982.

[79] N. SHIBATA et al., "Measurements of Waveguide Structure Fluctuations in a Multimode Fiber by Backscattering Technique," *IEEE J. Quantum Electron.,* Vol. QE-17, p. 39, 1981.

[80] A. J. CONDUIT et al., "Fiber Diameter Variations and Their Effect on Backscatter Loss Measurements," *Tech. Dig. Integrated Optics and Optical Commun. (IOOC) Conf.,* Paper TuK3, San Francisco, 1981.

[81] B. COSTA et al., "Power Distribution of Backward Scattered Radiation in Optical Fibers," *Tech. Dig. Integrated Optics and Optical Commun. (IOOC) Conf.,* Paper TuK2, San Francisco, 1981.

[82] M. ERIKSUD et al., "Backscattering Signatures From Optical Fibers With Differential Mode Attenuation," *IEEE J. Lightwave Tech.,* Vol. LT-2, p. 76, 1984.

[83] A. H. CHERIN and P. J. RICH, "Measurement of Loss and Output Numerical Aperture of Optical Fiber Splices," *Appl. Opt.,* Vol. 17, p. 642, 1978.

[84] P. MATTHIJSSE and C. M. DEBLOK, "Field Measurements of Splice-Loss Applying the Backscattering Method," *Electron. Lett.,* Vol. 15, p. 795, 1979.

[85] M. NAKAHIRA et al., "Measurement of Optical Fiber Loss and Splice Loss," *Trans. IECE of Japan,* Vol. E63, p. 762, 1980.

[86] M. ERIKSUD et al., "Backscatter Signatures From Graded-Index Fibers With Diameter Variations," *Electron. Lett.,* Vol. 17, p. 200, 1981.

[87] D. L. FRANZEN and G. W. DAY, "Measurement of Optical-Fiber Bandwidth in the Time Domain," *NBS Special Publication 637,* Vol. 1, p. 47, Boulder, July 1982.

[88] C. LIN, "Measuring High-Bandwidth Fibers in the 1.3-μm Region With Picosecond InGaAsP Injection Lasers and Ultrafast InGaAs-Detectors," *Electron. Lett.,* Vol. 17, p. 438, 1981.

[89] F. T. STONE and A. L. INGLES, "Improved Bandwidth-Measurement Reliability Using a Gaussian-Fit Technique," *Proc. Eur. Conf. Optical Commun.,* p. 379, Geneva, 1983.

[90] G. W. DAY, "Measurement of Optical-Fiber Bandwidth in the Frequency Domain," *NBS Special Publication 637,* Vol. 2, Chapter 2, Boulder, 1983.

[91] W. F. LOVE and D. A. NOLAN, "Wavelength Dependence of the Bandwidth Concatenation Exponent in Multimode Optical Fibers," *Proc. Euro. Conf. Optical Commun.,* Paper 12A5, p. 214, Stuttgart, 1984.

[92] W. F. LOVE, "Time Dispersion in Optical Fibers," *Laser Focus,* p. 113, June 1982.

[93] M. J. BUCKLER, "Optimization of Concatenated Fiber Bandwidth via Differential Modal Delay," *Tech. Dig. Symp. Optical-Fiber Measurements, (NBS Special Publication 597),* p. 59, Boulder, 1980; F. T. STONE et al., "Use of a Quantitative Differential Modal Delay Technique to Improve Fiber Bandwidth," *IEEE J. Lightwave Tech.,* Vol. LT-1, p. 585, 1983.

[94] G. COPPA et al., "Characterization of Single-Mode Optical Fibers," *Proc. SPIE Conf. Fiber Optics: Short-Haul and Long-Haul Measurements and Applications,* Vol. 500, p. 7, San Diego, 1984.

[95] L. G. COHEN, "Comparison of Single-Mode Fibre Dispersion Measurement Techniques," *Tech. Dig. Conf. Optical Fiber Commun. Conf. (OFC),* Paper TuF2, p. 36, San Diego, 1985.

[96] A. H. HARTOG, "Advances in Optical Time-Domain Reflectometry," *Tech. Dig. Symp. Optical Fiber Measurements, (NBS Special Publication 683),* p. 89, Boulder, 1984.

[97] D. L. FRANZEN, "Determining the Effective Cut-off Wavelength of Single-Mode Fibers: An Interlaboratory Comparison," *IEEE J. Lightwave Tech.,* Vol. LT-3, p. 128, 1985.

[98] E. BRINKMAYER, "Analysis of the Backscattering Method for Single-Mode Fibers," *J. Opt. Soc. Amer.,* Vol. 70, p. 1010, 1980.

[99] A. R. HARTOG and M. P. GOLD, "On the Theory of Backscattering in Single-Mode Optical Fibers," *IEEE J. Lightwave Tech.,* Vol. LT-2, p. 76, 1984.

[100] M. NAKAZAWA et al., "Marked Extension of Diagnosis Length in Optical Time-Domain Reflectometry Using 1.32-μm YAG Laser," *Electron. Lett.,* Vol. 17, p. 783, 1981.

[101] K. NOGUCHI et al., "52 km Long Single-Mode Optical-Fiber Fault Location Using Stimulated Raman Scattering Effect," *Electron. Lett.,* Vol. 18, p. 41, 1982.

[102] Y. MURAKAMI et al., "Maximum Measurable Distance for Fiber Fault Locator Using Raman OTDR," *Tech. Dig. Integrated Optics and Optical Commun. Conf. (IOOC),* Paper 28A2, p. 44, Tokyo, 1983.

[103] K. NOGUCHI, "A 100-km Long Single-Mode Optical-Fiber Fault Location," *IEEE J. Lightwave Tech.,* Vol. LT-2, p. 1, 1984.

[104] Y. MURAKAMI et al., "Maximum Measurable Distances for a Single-Mode Optical-Fiber Fault Locator Using the Stimulated Raman Scattering (SRS) Effect," *IEEE Trans. Microwave Theory Tech.,* Vol. MTT-30, p. 1461, 1982.

[105] N. NAKAZAWA et al., "1.5-μm OTDR for Single-Mode Optical Fiber Longer Than 110 km," *Electron. Lett.,* Vol. 20, p. 323, 1984.

[106] M. NAKAZAWA et al., "130-km Fault Location for Single-Mode Optical Fiber Using 1.55-μm Q-Switched Er^{3+}: Glass Laser," *Opt. Lett.,* Vol. 9, p. 312, 1984.

[107] P. HEALEY, "Multichannel Photon Counting Backscatter Measurement in Monomode Fiber," *Electron. Lett.,* Vol. 17, p. 741, 1981.

[108] M. FUJISE and M. KUWAZURU, "Rayleigh Scattering Measurement of a 42-km Single-Mode Fiber at 1.55-μm Wavelength Using a Laser Diode and a Liquid-Nitrogen–Cooled Detector," *Electron. Lett.,* Vol. 20, p. 232, 1984.

[109] M. P. GOLD and A. N. HARTOG, "Improved Dynamic-Range Single-Mode OTDR at 1.3 μm," *Electron. Lett.,* Vol. 20, p. 285, 1984.

[110] P. HEALEY and D. J. MAYLON, "Optical Time-Domain Reflectometry in Single-Mode Fibers at 1.5-μm Using Heterodyne Detection," *Electron. Lett.,* Vol. 18, p. 862, 1982.

[111] P. HEALEY et al., "OTDR in Single-Mode Fiber at 1.5-μm Using Homodyne Detection," *Electron. Lett.,* Vol. 20, p. 360, 1984.

[112] P. HEALEY, "Fading in Heterodyne Optical Time-Domain Reflectometry," *Electron. Lett.,* Vol. 20, p. 30, 1984.

[113] M. P. GOLD and A. N. HARTOG, "Ultra-Long Range OTDR in Single-Mode Fibers at 1.3 μm," *Electron. Lett.,* Vol. 19, p. 463, 1983.

[114] S. WRIGHT et al., "High-Dynamic-Range Coherent Reflectometer for Fault Location in Monomode Fibers," *Proc. Eur. Conf. Optical Commun.,* p. 175, Geneva, 1983.

[115] M. NAKAZAWA et al., "Measurement and Analysis on Polarization Properties on Backward Rayleigh Scattering for Single-Mode Optical Fibers," *IEEE J. Quantum Electron.,* Vol. QE-17, p. 2326, 1981; M. NAKAZAWA, Theory of Backward Rayleigh Scattering in Polarization-Maintaining Single-Mode Fibers and its Application to Polarization Optical Time-Domain Reflectometry," *IEEE J. Quantum Electron.,* Vol. QE-19, p. 854, 1983.

[116] A. J. ROGERS, "Polarization Optical Time-Domain Reflectometry: A Technique for the Measurement of Field Distributions," *Appl. Opt.,* Vol. 20, p. 1060, 1981.

[117] T. HORIGUCHI et al., "An Acousto-Optic Directional Coupler for an Optical Time-Domain Reflectometer," *IEEE J. Lightwave Tech.,* Vol. LT-2, p. 108, 1984.

[118] M. P. GOLD and A. H. HARTOG, "Determination of Structural Parameter Variations in Single-Mode Optical Fibers by Optical Time-Domain Reflectometry," *Electron. Lett.,* Vol. 18, p. 489, 1982.

[119] M. P. GOLD et al., "New Approach to Splice-Loss Monitoring Using Long-Range OTDR," *Electron. Lett.,* Vol. 20, p. 338, 1984.

[120] N. K. CHEUNG and P. KAISER, "Cutoff Wavelength and Modal Noise in Single-Mode Fiber Systems," *Tech. Dig. Symp. Optical-Fiber Measurements,* (*NBS Special Publication 683*), p. 15, Boulder, 1984.

[121] H. T. NIJNUIS and K. A. H. vanLEEUWEN, "Length and Curvature Dependence of Effective Cutoff Wavelength and LP_{11}-Mode Attenuation in Single-Mode Fiber," *Tech. Dig. Symp. Optical-Fiber Measurements,* (*NBS Special Publication 683*), p. 11, Boulder, 1984; V. SHAH, "Effective Cutoff Wavelength of Single-Mode Fibers—The Combined Effect of Curvature and Index Profile," ibid., p. 7.

[122] W. T. ANDERSON and T. A. LENAHAN, "Length Dependence of the Effective Cutoff Wavelength in Single-Mode Fibers," *IEEE J. Lightwave Tech.,* Vol. LT-2, p. 238, 1984.

[123] D. G. DUFF et al., "Measurements of Modal Noise in Single-Mode Lightwave Systems," *Tech. Dig. Optical Fiber Commun. Conf.,* Paper Tu01, p. 52, San Diego, 1985.

[124] N. K. CHEUNG et al., "Observation of Modal Noise in Single-Mode–Fiber Transmission Systems," *Electron. Lett.,* Vol. 21, p. 5, 1985.

[125] D. L. FRANZEN, "Interlaboratory Measurement Comparison Among Fiber Manufacturers to Determine Effective Cutoff Wavelength and Mode-Field Diameter of Single-Mode Fiber," *Tech. Dig. Optical-Fiber Commun. Conf.,* Paper TuF2, p. 36, San Diego, 1985.

[126] O. Y. KATSUYAMA et al., "New Method for Measuring *V*-Value of a Single-Mode Optical Fiber," *Electron. Lett.,* Vol. 12, p. 69, 1976; P. D. NICHOLS, "Comparison Between Techniques for Measuring the LP_{11}-Mode Cutoff Wavelength in Monomode Fibers," *Electron. Lett.,* Vol. 18, p. 1008, 1982.

[127] C. A. MILLAR, "Direct Method of Determining Equivalent Step-Index Profiles for Monomode Fibers," *Electron. Lett.,* Vol. 17, p. 458, 1981.

[128] C. A. MILLAR, "Application of Automated Equivalent Step-Index Profiling to a 31.6-km Monomode Fiber System," *Tech. Dig. Top. Mtg. Optical Fiber Commun.,* p. 62, Phoenix, 1982.

[129] G. J. CANNELL et al., "Measurement Repeatability and Comparison of Real and Equivalent Step-Index (ESI) Fibre Profiles," *Proc. Eur. Conf. Optical Commun.,* Paper A IV 5, Cannes, 1982.

[130] G. GROSSO et al., "A Novel Technique for Cutoff Wavelength Measurement in Single-Mode Fibres," *Proc. Eur. Conf. Optical Commun.,* p. 98, Cannes, 1982.

[131] G. COPPA and P. DIVITA, "Polarization Measurement of Cutoff Wavelength in Monomode Fibres," *Proc. Eur. Conf. Optical Commun.,* p. 193, Geneva, 1983.

[132] D. COPPA et al., "New Method for Cutoff Wavelength in Monomode Fibres," *Proc. Eur. Conf. Optical Commun.,* p. 120, Stuttgart, 1984.

[133] D. MARCUSE, "Gaussian Approximation of the Fundamental Modes of Graded-Index Fibers," *J. Opt. Soc. Am.,* Vol. 68, p. 103, 1978.

[134] W. T. ANDERSON and D. L. PHILEN, "Spot Size Measurements for Single-Mode Fibers—A Comparison of Four Techniques," *IEEE J. Lightwave Tech.,* Vol. LT-1, p. 20, 1983.

[135] P. R. REITZ, "New Single-Mode Fiber Measurement Techniques and Issues," *Tech. Dig. Optical-Fiber Commun. Conf.,* Paper TuB7, p. 28, San Diego, 1985.

[136] K. PETERMANN, "Constraints for Fundamental-Mode Spot-Size for Broadband Dispersion-Compensated Single-Mode Fibers," *Electron. Lett.,* Vol. 19, p. 712, 1983.

[137] C. PASK, "Physical Interpretation of Petermann's Strange Spot Size for Single-Mode Fibers," *Electron. Lett.,* Vol. 20, p. 144, 1984.

[138] G. COPPA et al., "Near-Field Measurements in Monomode Fibers: Determination of Chromatic Dispersion," *Electron. Lett.,* Vol. 19, p. 731, 1983.

[139] K. PETERMANN, "Fundamental-Mode Microbending Loss in Graded-Index and W-Fibers," *Opt. Quantum Electron.,* Vol. 9, p. 167, 1977.

[140] W. T. ANDERSON, "Consistency of Measurement Methods for the Mode-Field Radius in a Single-Mode Fiber," *IEEE J. Lightwave Tech.,* Vol. LT-2, p. 191, 1984.

[141] T. D. CROFT et al., "Low-Loss Dispersion-Shifted Fiber Manufactured by the OVD Process," *Tech. Dig. Conf. Optical-Fiber Commun.,* Paper WD2, p. 94, San Diego, 1985.

[142] H. KARSTENSEN and L. WETENKAMP, "Comparison of Chromatic-Dispersion Measurements of Single-Mode Optical Fibers by Spot-Size and Pulse-Delay Methods," *Tech. Dig. Symp. Optical-Fiber Measurements,* (*NBS Special Publication 683*), p. 131, Boulder, 1984.

[143] D. K. SMITH and R. A. WESTWIG, "Non-Gaussian Fitting of Variable-Aperture Far-Field Data and Waveguide Dispersion for Multiwavelength Measurements of Mode Radius," *Tech. Dig. Optical-Fiber Commun. Conf.,* Paper TuB2, p. 28, San Diego, 1985.

[144] A. T. KLEMAS et al., "Analysis of Mode-Field Radius Calculated From Far-Field Radiation," *Tech. Dig. Optical-Fiber Commun. Conf.,* Paper TuB3, p. 30, San Diego, 1985.

[145] G. COPPA et al., "Characterization of Single-Mode Fibers by Near-Field Measurement," *Electron. Lett.,* Vol. 19, p. 293, 1983.

[146] J. STRECKERT, "New Method for Measuring the Spot Size of Single-Mode Fibers," *Opt. Lett.,* Vol. 5, p. 505, 1980.

[147] D. MARCUSE, "Loss Analysis of Single-Mode Fiber Splices," *Bell Syst. Tech. J.,* Vol. 56, p. 703, 1977.

[148] J. M. DICK et al., "Automated Mode Radius Measurements Using the Variable-Aperture Method in the Far-Field," *Tech. Dig. Conf. Optical-Fiber Commun.,* Paper WB3, New Orleans, 1984.

[149] R. A. WESTWIG, "Mode-Field Diameter: An Important Single-Mode Fiber Parameter," *Laser Focus,* p. 80, May 1984.

[150] W. J. STEWART et al., "Waveguide Dispersion Measurements in Monomode Fibers from Spot-Size," *Proc. Eur. Conf. Optical-Commun.,* Paper 7A4, p. 122, Stuttgart, 1984.

[151] J. P. POCHOLLE and J. AUGE, "New Simple Method for Measuring the Mode Spot Size in Monomode Fibers," *Electron. Lett.,* Vol. 19, p. 191, 1983.

[152] K. OGAWA, "Considerations for Single-Mode Fiber Systems," *Bell Syst. Tech. J.,* Vol. 61, p. 1919, 1982.

[153] K. OGAWA and V. J. MAZORCZYK, "Chirping Noise in Single-Frequency Laser," *Tech. Dig. Optical-Fiber Commun. Conf.,* Paper W12, p. 102, San Diego, 1985.

[154] F. P. KAPRON, "Source and Modulation Effects in Monomode Fibers," *Tech. Dig. Sixth Eur. Conf. Optical Commun.,* IEE Conf. Publ. 190, p. 129, York, 1980.

[155] N. IMOTO and M. IKEDA, "Polarization Dispersion Measurements in Long Single-Mode Fibers With Zero Dispersion Wavelength at 1.5 μm," *IEEE J. Quantum Electron.,* Vol. QE-17, p. 542, 1981.

[156] K. MOCHIZUKI et al., "Polarization Mode Dispersion and Polarization Stability in an Optical-Fiber Submarine Cable," *Tech. Dig. Integrated Optics and Optical Commun. Conf. (IOOC),* Paper 29A4-3, Tokyo, 1983.

[157] L. G. COHEN and C. LIN, "A Universal Fiber-Optic (UFO) Measurement System Based on a Near-IR Fiber Raman Laser," *IEEE J. Quantum Electron.,* Vol. QE-14, p. 855, 1978.

[158] L. G. COHEN et al., "Experimental Technique for Evaluation of Fiber Transmission Loss and Dispersion," *Proc. IEEE,* Vol. 68, p. 1203, 1980.

[159] R. A. MODAVIS and W. F. LOVE, "Multiple-Wavelength System for Characterization of Dispersion in Single-Mode Optical Fibers," *Tech. Dig. Symp. Optical-Fiber Measurements,* (*NBS Special Publication 683*), p. 115, Boulder, 1984.

[160] F. P. KAPRON and T. C. OLSON, "Accurate Specifications of Single-Mode Dispersion Measurements," *Tech. Dig. Symp. Optical-Fiber Measurements,* (*NBS Special Publication 683*), p. 111, Boulder, 1984.

[161] C. LIN et al., "Chromatic Dispersion Measurements in Single-Mode Fibers Using Picosecond InGaAsP Injection Lasers in the 1.2- to 1.5-μm Spectral Region," *Bell System Tech. J.,* Vol. 62, p. 457, 1983.

[162] M. YAMAJUCHI et al., "High-Power CW Operation Over 100 mW at 1.3-μm in a DC-PBH Laser Diode with Reflectivity-Optimized Mirror Facets," *Conf. on Lasers and Electro-Optics (CLEO),* Paper Thl1, Baltimore, 1985.

[163] L. G. COHEN et al., "Experimental Techniques for Evaluation of Fiber Transmission Loss and Dispersion," *Proc. IEEE,* Vol. 68, p. 1203, 1980.

[164] L. G. COHEN and J. STONE, "Interferometric Measurements of Minimum Dispersion Spectra in Short Lengths of Single-Mode Fiber," *Electron. Lett.,* Vol. 18, p. 564, 1982.

[165] B. COSTA et al., "Phase-Shift Technique for the Measurement of Chromatic Dispersion in Fibers Using LEDs," *Electron. Lett.,* Vol. 19, p. 1074, 1983.

[166] N. A. OLSSON et al., "Fiber-Dispersion and Propagation-Delay Measurements with Frequency- and Amplitude-Modulated Cleaved–Coupled-Cavity Semiconductor Lasers," *Opt. Lett.,* Vol. 9, p. 180, 1984.

[167] P. J. VELLA et al., "Measurement of Chromatic Dispersion of Long Spans of Single-Mode Fibre: A Factory and Field Test Method," *Electron. Lett.,* Vol. 20, p. 67, 1984.

[168] K. TATEKURA et al., "High-Accuracy Measurement Equipment for Chromatic Dispersion, Making Use of the Phase-Shift Technique With Laser Diodes," *Tech. Dig. Symp. Optical-Fiber Measurements,* (*NBS Special Publication 683*), p. 119, Boulder, 1984.

[169] R. RAO, "Field Dispersion Measurement—A Swept-Frequency Technique," *Tech. Dig. Symp. Optical-Fiber Measurements,* (*NBS Special Publication 683*), p. 135, Boulder, 1984.

[170] L. OKSANEN and S. J. HALME, "Interferometric Dispersion Measurement in Single-Mode Fibers With a Numerical Method to Extract the Group Delays from the Measurement Visibility Curves," *Tech. Dig Symp. Optical-Fiber Measurements,* (*NBS Special Publication 683*), p. 127, Boulder, 1984.

[171] F. M. SEARS et al., "Interferometric Measurements of Dispersion Spectrum Variations in a Single-Mode Fiber," *IEEE J. Lightwave Tech.,* Vol. LT-2, p. 181, 1984.

[172] F. MENGEL, "Interferometric Monomode Fibre Measurements: Influence of Source Spectrum and Second-Order Dispersion," *Electron. Lett.,* Vol. 20, p. 66, 1984.

[173] D. MARCUSE and J. STONE, "Experimental Comparison of the Bandwidths of Standard and Dispersion-Shifted Fibers Near Their 'Zero-Dispersion' Wavelengths," *Opt. Lett.,* Vol. 10, p. 163, 1985; J. STONE and D. MARCUSE, "Direct Measurement of Second-Order Dispersion in Short Optical Fibers Using White-Light Interferometry," *Electron. Lett.,* Vol. 20, p. 741, 1984.

[174] J. STONE and L. G. COHEN, "Minimum Dispersion Spectra of Single-Mode Fibers Measured with Subpicosecond Resolution by White-Light Cross-Correlation," *Electron. Lett.,* Vol. 18, p. 716, 1982.

[175] M. J. SAUNDERS and W. B. GARDNER, "Precision Interferometric Measurement of Dispersion in Short Single-Mode Fibers," *Tech. Dig. Symp. Optical-Fiber Measurements,* (*NBS Special Publication 683*), p. 123, Boulder, 1984.

[176] P. SANSONETTI, "Prediction of Modal Dispersion in Single-Mode Fibres from Spectral Behavior of Mode Spot Size," *Electron. Lett.,* Vol. 18, p. 136, 1982.

[177] P. SANSONETTI, "Modal Dispersion in Single-Mode Fibers: Simple Approximation Issued from Mode Spot Size Spectral Behavior," *Electron. Lett.,* Vol. 18, p. 647, 1982.

[178] G. COPPA et al., "Near-Field Measurements in Monomode Fibers: Determination of Chromatic Dispersion," *Electron. Lett.,* Vol. 19, p. 731, 1983.

[179] S. KOBAYASHI et al., "Refractive-Index Dispersion of Doped Fused Silica," *Tech. Dig. Conf. on Integrated Optics and Optical Commun. (IOOC),* Paper B8-3, p. 309, Tokyo, 1977.

Nonlinear Optical Phenomena in Single-Mode Fibers

H. G. WINFUL

GTE Laboratories, Inc.

1. Introduction

The study of nonlinear optical effects in dielectric media began in the early 1960s when Franken, Hill, Peters, and Weinreich [1] observed the frequency doubling of a ruby laser beam by a quartz crystal. This was soon followed by the observation of a plethora of other nonlinear phenomena, including third-harmonic generation, optical rectification, parametric amplification and oscillation, self-focusing, and self-phase modulation. Because optical nonlinearities are intrinsically weak (at least far from material resonances), these developments depended critically on the availability of high-power lasers capable of output peak powers in excess of a kilowatt. The efficiency of a nonlinear optical process depends on the size of the nonlinear coefficient, the optical intensity, and the distance over which this intensity can effectively interact with the medium. Prior to the early 1970s, nonlinear optical interactions were studied primarily in bulk media, and the maximum efficiencies were limited by the achievable interaction lengths of much less than a meter. In bulk nonlinear optics the maximum effective interaction length is the Rayleigh length of the focusing optics. If one focuses a beam of P watts to a spot of radius w_0 centimeters, the inten-

sity at the focus can be approximately maintained over a distance given by the Rayleigh length $z_R = \pi w_0^2/\lambda$, where λ is the light wavelength in the medium. The product of this interaction length and the focal intensity determines the maximum efficiency of the nonlinear interaction and is given by [2]

$$E = (P/\pi w_0^2) \cdot (\pi w_0^2/\lambda) = P/\lambda . \tag{1}$$

This quantity is independent of the length of the nonlinear medium, and thus in bulk nonlinear optics for a given input power one cannot increase the maximum efficiency by using longer samples. The situation in an optical fiber is quite different. Because of the waveguiding properties of the fiber, the intensity in a small focal spot can be maintained over distances in excess of several kilometers. This leads to orders of magnitude increases in the yield of a nonlinear process in a fiber over that of the bulk material.

The first experiments in fiber nonlinear optics used relatively large core multimode fibers with rather high losses. Recently, nonlinear optical effects in fibers have become very important with the development of extremely low loss, single-mode fibers. The small core diameter, coupled with the long propagation distances attainable in these fibers, makes it possible to observe non-linear optical phenomena in fibers at power levels orders of magnitude smaller than required for bulk media. Some of these effects can be observed in fibers at power levels of a few milli-watts, well within the output capability of semi-conductor lasers.

Nonlinear effects in single-mode fibers have many important applications. Among these are the generation of new frequencies through Raman processes, the compression of optical pulses through self-phase modulation, and the all-optical processing of light signals through the optical Kerr effect. However, there are also several del-eterious effects that arise from these nonlinear phenomena. Raman and Brillouin scattering processes can limit the power-handling capability of fibers and lead to crosstalk in wavelength-division–multiplexed optical communications sys-tems. The intensity-dependent refractive index can lead to phase noise, while a poorly under-stood photosensitivity at visible wavelengths can lead to gradual decreases in the throughput of a fiber. In this chapter we review some of the underlying physics, point out the key experi-ments, and describe some applications of these nonlinear optical phenomena in fibers. Because of their importance for optical communications, we emphasize single-mode fibers and experiments done in the 1.0- to 1.6-μm spectral region. The reader will find some previous reviews of fiber nonlinear optics in [2–8].

2. Stimulated Scattering Processes

In this section we discuss nonlinear processes in which power launched into a fiber at a given wavelength is transferred to a set of longer wavelengths determined by certain characteristic vibrations of the fiber material. Some of these phenomena have sufficiently low thresholds in single-mode fibers that they will constrain the design of long-haul optical communications systems.

Background

When light is incident on a material it undergoes various scattering processes. Most of the scatter-ing is elastic, and the scattered wave has the same frequency as the incident wave. However, this scattered light is, in general, at some arbitrary angle to the forward direction of propagation. Hence, if one measures the transmitted light in the forward direction, there is a reduction in intensity as a result of the scattering into other directions. This loss is known as Rayleigh scatter-ing loss. The intensity of the scattered light is inversely proportional to the fourth power of the light wavelength. In good optical fibers the pri-mary source of loss is Rayleigh scattering due to compositional fluctuations. Because of the λ^{-4} dependence of the scattered intensity, propaga-tion losses can be minimized by operating at the longer wavelengths. Today's single-mode fibers have residual losses of order 0.1 dB/km in the wavelength region of 1.0 to 1.6 μm.

In addition to the elastically scattered compo-nent, a small fraction (about 1 in 10^6) of the inci-dent photons undergo inelastic scattering. The scattered photon emerges with a frequency shifted below or above the incident photon frequency. The difference in energy between the incident and scattered photons is deposited in, or extracted from, the scattering medium. The frequency shifts* can be small (approximately 1 cm^{-1}) or large (greater than 100 cm^{-1}). When the fre-quency shift is small, the process is known as Brillouin scattering. It is scattering from the acoustic vibrational modes of the material. Wave vector selection rules limit the Brillouin interac-tion in an optical fiber to backward-propagating scattered light. The larger frequency shifts char-acterize the regime of Raman scattering. In Raman scattering, the photon is scattered by local molecular vibrations or by optical phonons. Unlike Brillouin scattering, the Raman scattering process is isotropic and it can occur in both the forward and the backward directions in an optical fiber.

In general, the inelastic scattering produces photons that are either downshifted or upshifted in frequency. An upshifted scattered photon is

---

*Physicists tend to use centimeters^{-1} ("wave numbers") when describing frequency shifts. To convert to hertz, multi-ply the wave number shift $\Delta(1/\lambda)$ by the speed of light in centimeters per second. Thus, 1 cm^{-1}=30 GHz. Also, at any wavelength λ, the magnitude of the shift in wavelength units is $\Delta\lambda=\lambda^2\Delta(1/\lambda)$. Thus, at 1 μm, 1 cm^{-1}=1 Å=0.1 nm.

only possible if the material gives up a quantum of energy equal to the difference in energy between incident and scattered photon. The material must therefore be in an excited state before the incident photon arrives. Since the probability of a molecule being in an excited state is governed by Boltzman statistics, the flux of photons that are shifted upward in frequency (the anti-Stokes component) at room temperature is much smaller than those shifted downward (Stokes component).

Even if the incident beam is strictly monochromatic, the scattered light will exhibit a finite spectral width. The spectrum of the scattered radiation contains useful information concerning the lifetime of the material excitation that gave rise to the scattering. In the absence of other broadening mechanisms, the width of the scattered spectrum is inversely related to the lifetime of the excitation. There is a scattering cross section that relates incident intensity to scattered intensity. This cross section is measured in spontaneous scattering experiments [9]. Fig. 1 summarizes the relation between the incident laser frequency ω_p, also known as the pump frequency, and the Stokes (ω_s) and anti-Stokes (ω_a) components of the inelastically scattered radiation.

The scattering processes described are spontaneous scattering processes. The frequency-shifted photons are not present originally but are created when the incident photon scatters. When large numbers of Stokes photons are also present initially, they can combine with the incident laser photons to drive the material excitation, which in turn causes more photons to scatter inelastically. The spontaneous scattering process becomes stimulated and the Stokes photons experience gain. The spatial rate of growth of the number of Stokes photons is proportional to the existing number N_s of Stokes photons and is described by [10]

$$dN_s/dz = A\,N_p\,(N_s+1)\,, \qquad (2)$$

where A characterizes the strength of the scattering process, and N_p is the photon number at the incident laser frequency. The 1 in (2) describes the spontaneous scattering process. The stimulated effect will only be comparable to, or larger than, the spontaneous scattering if the number of Stokes photons N_s exceeds the number of vacuum electromagnetic modes contained in the frequency width of the scattered spectrum. In stimulated scattering ($N_s \gg 1$) the flux of scattered photons grows exponentially with propagation distance so that after an interaction length of L, the output flux is

$$N_s(L) = N_s(0)\exp{(g\,I_p\,L)}\,, \qquad (3)$$

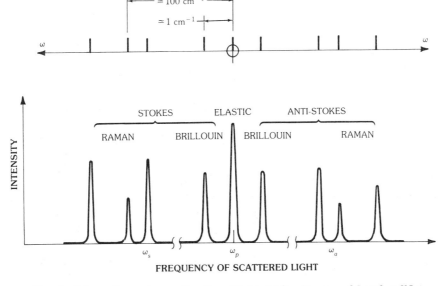

FIG. 1. Schematic spectrum of scattered light. (*After Hayes and Loudon [9]*)

where g is the gain coefficient of the individual scattering process, I_p is the incident laser intensity (proportional to the laser photon number N_p), and $N_s(0)$ is the initial flux of Stokes photons (due to spontaneous scattering).

In what follows we discuss Raman gain, its measurement and its optimization in optical fibers. Stimulated Raman scattering (SRS) is then considered as a source of new frequencies, as a means of amplifying weak signals, and as a source of crosstalk in wavelength-division-multiplexed systems. We then treat stimulated Brillouin scattering (SBS) and discuss ways in which it may be enhanced or suppressed. A potential application of SBS to coherent optical communications is also discussed. For a detailed review of these nonlinear scattering processes in bulk media, the articles by Kaiser and Maier [10] and by Bloembergen [11] are recommended.

Raman Processes

Raman Gain and Stimulated Raman Scattering—

In stimulated Raman scattering, the power in the Stokes wave grows exponentially with propagation distance at a rate proportional to the Raman gain coefficient and the intensity of the pump wave. When the Stokes power grows sufficiently large, it begins to deplete the pump wave, which decays at a rate proportional to the gain coefficient and the Stokes intensity. Both pump and Stokes waves also suffer attenuation as a result of linear fiber losses. The evolution of the pump (P_p) and Stokes (P_s) waves is governed by a pair of coupled-power equations [10, 12]:

$$(\partial/\partial z + v_p^{-1}\,\partial/\partial t)P_p = -\alpha_p P_p - g\,(\lambda_s/\lambda_p)$$

$$\times P_p P_s / K A_{\text{eff}}, \quad (4)$$

$$(\pm\partial/\partial z + v_s^{-1}\,\partial/\partial t)P_s = -\alpha_s P_s + g$$

$$\times P_p P_s / K A_{\text{eff}}, \quad (5)$$

where g is the Raman gain coefficient, A_{eff} is the effective area* of the fiber core, and α_s and α_p are

---

*The effective area is defined in terms of an overlap integral between the interacting waveguide modes. For a single-mode fiber, if a Gaussian approximation $\exp(-r^2/w^2)$ is used for the mode intensity, then the effective area is given by $A_{\text{eff}} = 2\pi w^2$ [6].

the loss coefficients at the Stokes and pump wavelengths (λ_s and λ_p), respectively. To resolve the ambiguity sign in (5), note that the plus sign governs the forward scattering process while the minus sign describes backward Raman scattering. Because of dispersion, the group velocities v_p and v_s of the pump and Stokes wave differ. This leads to the walk-off phenomenon in which pump and Stokes pulses can separate from each other during the forward interaction. Finally, the coefficient K is a factor whose value lies between 1 and 2, and which takes into account the relative polarizations of the pump and Stokes waves. It has a value of 1 if the two waves are in the same polarization state and 2 if the polarizations are totally scrambled [13].

In steady state ($\partial/\partial t = 0$) equations 4 and 5 can be solved exactly if $\alpha_s = \alpha_p$ [14]. A great deal of physical insight can be gained, however, by considering approximate solutions which hold in the limit of a nondepleted pump wave. In that limit the pump loses energy only through linear absorption and the steady-state solution for the pump wave is

$$P_p(z) = P_p(0)\exp(-\alpha_p z). \quad (6)$$

Use of (6) in (5) at steady state yields

$$dP_s/dz = \{-\alpha_s + [g\,P_p(0)/K A_{\text{eff}}]$$

$$\times \exp(-\alpha_p z)\} P_s, \quad (7)$$

which can be integrated to give the Stokes power

$$P_s(L) = P_s(0)\exp(-\alpha_s L + g\,P_p(0)\,L'/K A_{\text{eff}}), \quad (8)$$

where L is the length of the fiber and L' is the effective length defined as

$$L' = [1 - \exp(-\alpha_p L)]/\alpha_p. \quad (9)$$

The amplification factor for the Stokes wave is $\exp(g\,P_p(0)\,L'/K A_{\text{eff}})$.

In the absence of any input Stokes signal $P_s(0)$, the generated Stokes wave grows from the low-level spontaneous Raman scattering that occurs throughout the length of the fiber. This evolution from noise is usually treated theoretically by inserting one noise photon per mode (longitudinal and transverse) at the input of the fiber [15]. For a single-mode fiber the equivalent input signal $P_s(0)$ is equal to the Stokes photon energy multiplied by the number of longitudinal modes within the effective bandwidth of the Raman gain. With sufficient amplification the Raman Stokes signal can grow to a magnitude comparable to that of the pump. This is the regime usually referred to as stimulated Raman scattering. While there is no precise threshold for stimulated Raman scattering, it is possible to derive an approximate value for a critical pump power at which the spontaneous scattering process becomes stimulated [15]. For a single-mode fiber of effective area A_{eff} cm^2, loss coefficient of α_p cm^{-1} at the pump wavelength, and Raman coefficient of g cm/W, the critical power in watts for forward stimulated Raman scattering, is given by

$$P_{cr} = 16 K A_{eff}/g L'. \tag{10}$$

The threshold for backward SRS is about 25% higher. Note that the critical power is independent of the fiber length if $\alpha_p L \gg 1$, which is generally the case in long-span fibers. The analysis of Smith [15] neglects pump depletion. This is valid for predicting the threshold pump power since at threshold the signal amplitude builds up from the noise level. The final signal amplitude is limited by pump depletion and hence a fully nonlinear theory is necessary for finding this amplitude [16]. To estimate the critical power, consider, for example a single-mode fiber of length 10 km, core area 7.5×10^{-7} cm^2, and a loss at 1.3 μm of 0.6 dB/km (=0.138/km). If we take $g = 1.1 \times 10^{-11}$ cm/W and assume that the polarization is completely scrambled, we obtain a critical power of about 4.5 W. Stimulated Raman scattering thus limits the power-handling capability of long-haul, low-loss, optical-fiber transmission systems. The conversion of optical power from the pump wave to the Stokes signal appears as a nonlinear loss to a narrow-band detector placed at the receiving end of the fiber transmission system. This is one of the deleterious effects of stimulated Raman scattering in fibers. The other one is crosstalk in wavelength-division–multiplexed systems, a subject that will be treated in a subsequent section.

The gain constant g is related to the differential cross section $(d\sigma/d\Omega)$ for spontaneous Raman scattering and also depends on the frequency separation between the pump and Stokes waves. For ordinary silica fibers the Raman gain peaks at a frequency separation of about 440 cm^{-1}, where its value is of order 1.0×10^{-11} cm/W [17]. It can be shown (see, for example, Yariv [18]) that the peak value of the gain varies as

$$g \propto (d\sigma/d\Omega)/\omega_s^3 \Delta\omega, \tag{11}$$

where $\Delta\omega$ is the line width of the scattered radiation. Furthermore, the differential cross section is proportional to $\omega_s^3 \omega_p$ [9] and thus g varies linearly with pump frequency. Fig. 2 shows the Raman gain coefficient of pure silica as a function of frequency separation for a pump wavelength of 1.0 μm. For a different pump wavelength the gain coefficient scales as $1/\lambda$. There are actually two main peaks in the Raman gain spectrum—a broad one at 440 cm^{-1} and a narrower feature at 490 cm^{-1}. The large width of the Raman gain results from the fact that glass is an amorphous material, and hence the vibrational excitations have very short coherence lengths. Thus there are no wave-vector selection rules to contend with and the entire density of states can contribute to the observed Raman signal [19]. For comparison, one may note that the Raman line in crystalline quartz at room temperature is only a few wave numbers wide [20].

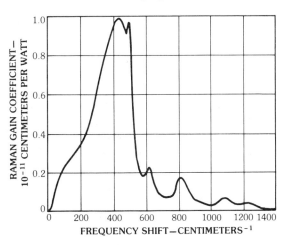

FIG. 2. Raman gain for fused silica. (*After Stolen [5]*)

The size of the Raman gain coefficient depends on the particular dopant material used in fiber fabrication. Germania (GeO_2) has a much larger gain coefficient than pure silica and is often used as a dopant in silica fibers intended for stimulated scattering applications [21]. In fact, pure germania multimode fibers have been fabricated and used in Raman generation experiments [22, 25]. However, the increased gain of germania is partially offset by the increased losses due to absorption in that material and by the large core diameter of the multimode fiber. Low-loss, single-mode GeO_2 fibers are not yet available. Other more exotic dopants have been used to tailor the Raman gain. Recently, it has been demonstrated that gaseous dopants can be dissolved in a fiber that has already been drawn, thus creating new Raman-gain media with certain desired properties [23]. For example, molecular deuterium (D_2) at room temperature and at pressures in excess of 10^5 torr will diffuse into a silica fiber and remain interstitially in molecular form. The long path length provided by the fiber yields much larger amplification factors compared to conventional pressurized gas cells [24]. The Raman shift of D_2 in silica is 2988 cm^{-1}, much larger than the 400-cm^{-1} shift of the virgin silica fiber. When pumped with the 1.06-μm output of a Nd:YAG laser, the D_2-charged fiber shows a peak at 1.56-μm, a wavelength of great importance for optical communications. A 100-m length of this ''gas-in-glass'' Raman medium has been used to achieve a 4000-fold amplification of the 1.56-μm output of a semiconductor laser [24].

Single-Pass Generation of New Frequencies—The low losses and relatively low amounts of dispersion that exist in silica fibers in the near infrared (1.0–1.6 μm) have spurred the development of fiber communications systems that operate in that wavelength region. There is thus a growing need for more optical sources in this spectral range. These sources may be used both for the transmission of information and for the characterization of optical fibers. Stimulated Raman scattering provides an efficient and versatile means of generating these new frequencies.

As already discussed, when the power in the pump wave launched into a fiber exceeds the threshold for stimulated Raman scattering, coherent light (known as the Stokes signal) is generated at a frequency shifted from the pump frequency by a vibrational frequency. When the Stokes signal grows sufficiently large, it can act as a pump for a higher-order Stokes line also displaced by the vibrational frequency. That in turn can drive other Stokes lines leading to the generation of a number of new frequencies from a single pump.

The first demonstration of single-pass Raman generation in the near-infrared in optical fibers was by Lin, Cohen, Stolen, Tasker, and French [25]. The fiber used in that experiment was fairly short (175 m) and quite lossy (approximately 5 dB/km or more). Thus the critical power for stimulated Raman scattering was of order 100 W, which required the use of a pulsed pump source. The experimental arrangement is shown in Fig. 3. With approximately 70 W of 1.064-μm radiation of duration 150 ns coupled into the 6-μm-diameter fiber, the first Stokes line at 1.12 μm is observed. By increasing the pump power to 250 W, higher-order Stokes components were observed at 1.18 μm, 1.23 μm, and 1.3 μm. In later experiments with up to 1 kW peak power in the pump, five Stokes orders were generated [26]. Fig. 4 shows the multiple-order Stokes spectrum obtained in a single pass through a silica fiber [26]. The limitation on the achievement of even higher orders of Stokes emission is the presence of self-phase modulation (see Section 5) and four-wave mixing processes (Section 3) which lead to a broadening of the individual spectral lines until they merge into a quasi-continuum between 1.3 μm and 1.7 μm [26–28]. This spectral broadening plays an important role in the spectral evolution of the Stokes signals at high pump power in the near-infrared where, because of the low fiber dispersion, four-wave mixing processes can be efficiently phase matched. In the visible the spectral dynamics are dominated by the peaks at 440 cm^{-1} and 490 cm^{-1} in the Raman gain spectrum. Studies at relatively low power show that the 490 cm^{-1} peak can grow at the expense of the one at 440 cm^{-1} [29]. In general, however, the spectral development of the Stokes emission is quite complicated and is subject to such effects as group-velocity dispersion, pulse walk-off, self-phase modulation, and pump depletion.

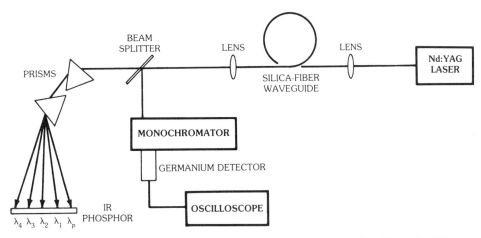

FIG. 3. Near-infrared Raman generation: Experimental setup. (*After Lin et al. [25]*)

The temporal dynamics of the pump and Stokes pulses in single-mode fibers is a subject of great current interest [30–33]. When a short pump pulse is used to generate a Stokes signal, the difference in group velocity between the two pulses means that they can physically separate from each other ("walk off") after propagating a distance known as the walkoff distance. For a pump pulse of duration T, this walkoff distance is given approximately by

$$L \simeq v_p\, v_s\, T/|v_p - v_s|\,. \qquad (12)$$

When the fiber length exceeds this walkoff distance the Stokes pulse will emerge ahead of the pump pulse if the pump wavelength lies in the region of positive group-velocity dispersion. Fig. 5 shows the intensity profiles of the transmitted pump and Stokes pulses in an experiment carried out with 75-ps, 1.06-μm pulses from a Nd:YAG laser focused into a 400-m long fiber [31]. The weaker pulse at the left is the Raman shifted pulse. As shown in this and other experiments, the combination of walkoff and pump depletion

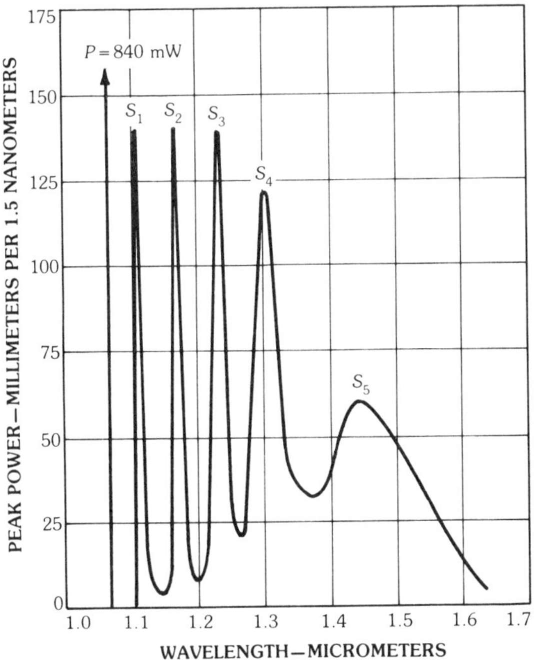

FIG. 4. Spectrum of multiple Stokes emission. (*After Cohen and Lin [26]*)

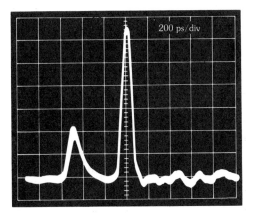

FIG. 5. Pulse walkoff in stimulated Raman scattering. (*After Weiner, Heritage, and Stolen [31]*)

leads to asymmetric shaping of the fundamental and Stokes pulses [30, 32, 33]. The Stokes pulse is also generated with a strong frequency chirp [30]. In the high conversion regime, the central and most intense part of the pump can be completely depleted, resulting in a temporal hole in the transmitted pump [25, 28]. The generated Stokes pulse is considerably shorter than the pump since it is only produced by those parts of the pump that are above the critical power for SRS.

One of the effects of pulse walkoff is to reduce the amount of conversion from the pump to the Stokes signal. The Raman interaction can be maximized by choosing the pump so that its wavelength and that of the Stokes pulse are symmetrically disposed about the zero-dispersion point of the fiber [34]. This equalizes the group velocities of the pump and Stokes waves. Such group-velocity matching has been used to obtain efficient generation of Stokes light at 1.56 μm in a D_2-in-glass fiber pumped by a Q-switched or mode-locked 1.06-μm Nd:YAG laser [35]. In multimode fibers group-velocity matching can be accomplished by using the dispersion of waveguide modes [36, 37].

Recent advances in fiber fabrication by the vapor axial deposition (VAD) technique have made it possible to draw extremely low loss fibers, with lengths of several kilometers, in a single span. Since the threshold for stimulated Raman scattering scales inversely as the fiber effective length, cw generation of Raman light should be possible in these long fibers. Sasaki and coworkers [38] have obtained continuous-wave, single-pass Raman generation by using a 35-km single-span VAD fiber with a loss of 1.13 dB/km at the pump wavelength of 1.064 μm. Generation of the first and second Stokes radiation was observed with 0.6 W and 4.8 W of pump power, respectively.

The coherent light generated by means of multiple Stokes scattering is useful for characterizing various transmission properties of fibers. By measuring the propagation delays of the different Stokes wavelengths through a length of fiber it is possible to determine the chromatic dispersion of the fiber and the wavelength of minimum dispersion [39, 26, 22]. The loss spectrum of the fiber is also readily determined by measuring the trans-

mission of the different wavelengths through the fiber [26]. Another application of single-pass Raman generation is in optical time-domain reflectometry (OTDR) for locating faults in long optical fibers [40–41]. In this technique a Stokes wave generated within the fiber undergoes Rayleigh backscattering when it encounters a fault. The arrival time of this scattered light at the input end is related to distance to determine the location of the fault. Use of the Stokes pulse instead of the fundamental improves the signal-to-noise ratio of the measurement technique.

Fiber Raman Oscillators—While single-pass Raman generation is a simple means of obtaining new frequencies, its usefulness as a source is somewhat limited because it lacks tunability. Furthermore, unless very long lengths of fiber are used, the threshold power is fairly high. By placing the fiber in a multipass optical cavity, a Raman oscillator is obtained which has both tunability and a reduced threshold. The cavity provides resonant, wavelength-selective feedback for the Stokes radiation, which therefore undergoes repetitive amplification.

Raman oscillation in an optical fiber was first observed by Stolen, Ippen, and Tynes [42]. That initial experiment was carried out at visible wavelengths (532-nm pump), used a short length of fiber (190 cm) and required 500 W of power in the fiber for oscillation. Hill, Kawasaki, and Johnson [43] were able to reduce the threshold to 5 W by using longer fibers (60 m), thus making cw operation possible. A cw argon ion laser operating at 488 nm was used as the pump source. A device of this sort oscillates at several wavelengths within the broad Raman gain bandwidth of silica. Continuously tunable oscillation over the 300-cm^{-1} band can be achieved by inserting an intracavity prism to spatially disperse the various wavelengths and turning the output mirror to select which wavelength resonates with the cavity [44, 45]. An even greater tuning range is achievable by generating multiple Stokes radiation and using a separate mirror to tune each order of Raman emission [46]. In a different approach, a ring resonator geometry has been used to obtain five orders of tunable Stokes radiation [47].

The oscillators described above were all demonstrated at visible wavelengths. Of greater interest for optical communications are fiber Raman oscillators that operate in the 1.0- to 1.6-μm region of the spectrum. A number of cw and pulsed sources in this spectral range have been demonstrated [48–54]. To cover the 1.07- to 1.32-μm range, Lin and French [50] use a cw mode-locked Nd:YAG laser operating at 1.064 μm to synchronously pump an 800-m–long fiber in a resonant cavity. In a synchronously pumped Raman oscillator the effective fiber resonator round-trip time for the Stokes pulses has to be an integral multiple of the pump pulse-repetition time in order to achieve oscillation. The length of the Stokes resonator is thus a critical quantity that has to be carefully adjusted. Because of fiber dispersion, different Stokes wavelengths experience different round-trip delays. Thus, by adjusting the resonator length properly, one can select the desired Stokes oscillation wavelength. This method of tuning is known as time-dispersion tuning, a technique first demonstrated by Stolen, Lin, and Jain [55]. The synchronously pumped, time-dispersion tuned Raman oscillator is shown in Fig. 6. Four orders of Stokes oscillations, peaked near 1.12, 1.18, 1.24, and 1.31 μm are obtained by pumping the fiber with about 5 W of average power. These four Stokes orders are then simultaneously tuned by translating the four mirrors which are mounted on rails. By this means a tuning range as broad as 250 nm is achieved. The pulse synchronization technique can also be used to measure the dispersion of the fiber since the

displacement of the tuning mirror is the spatial equivalent of the pulse delay experienced by the different oscillating wavelengths [56].

The 1.32- to 1.41-μm range has been accessed by using a cw mode-locked Nd:YAG laser operating at 1.32 μm as the pump source [51]. The extension of tunable Raman oscillation into the 1.4-μm spectral region had been difficult prior to that work because of an OH$^-$ absorption peak of greater than 10 dB/km that occurs at 1.39 μm in otherwise low-loss single-mode fibers. By using a low OH$^-$ single-mode fiber with a loss of 0.32 dB/km at 1.32 μm and 1 dB/km at 1.39 μm, Lin and Glodis [51] obtained more than 60% conversion to the Stokes component with an average pump power of 700 mW (60 W peak).

The formation of pulses in a synchronously pumped Raman oscillator is subject to the influence of group velocity dispersion and self-phase modulation. While these effects generally limit the attainable pulse widths, if the dispersion is properly chosen pulse compression within the cavity can result. For compression to occur the sign of the group velocity dispersion should be negative. Kafka and coworkers [54] have generated Raman shifted pulses as short as 0.8 ps from a dispersion-compensated oscillator pumped by a Nd:YAG laser emitting 80-ps pulses at 1.06 μm. The negative dispersion within the cavity is supplied by a grating-pair delay line to compensate for the positive dispersion seen by the 1.10-μm Stokes pulse within the fiber. There have been other experiments in which the wavelength of the

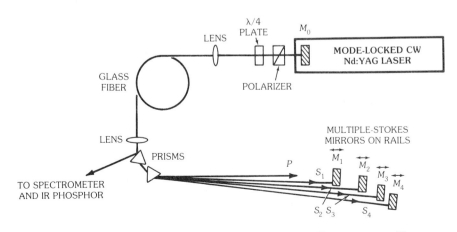

Fig. 6. A synchronously pumped, time-dispersion tuned fiber Raman oscillator. (*After Lin and French [50]*)

Stokes pulse lies in the negative-dispersion region (>1.3 μm) of the fiber. In that case no extra delay line is needed. The Raman pulse self-compresses as it circulates around the cavity. A D_2 gas-in-glass fiber Raman oscillator synchronously pumped by 120-ps pulses at 1.06 μm produces pulses as short as 15 ps at the Stokes wavelength of 1.56 μm [53].

The tunable sources in the near-infrared made possible by stimulated Raman scattering should prove useful for spectral and temporal studies of optical fibers and optical devices. Other suggested applications include use in fiber Raman gyroscopes [57, 58].

Fiber Raman Amplifiers—Stimulated Raman scattering in fibers holds great promise for the implementation of high-gain, high-speed, low-noise amplifiers and repeaters for use in optical communications. Since the same fiber transmission medium is used for amplification, Raman amplifiers have the advantage of simplicity over semiconductor laser amplifiers. In addition to the achievable gain and signal-to-noise ratio, some of the important considerations that will determine the utility of fiber Raman amplifiers are the possibility of cw operation and the availability of convenient pump sources in spectral regions of interest. Since the initial experiments by Ikeda on Raman amplification in the near-infrared [59], there have been numerous investigations by other workers in that same spectral region (1.0–1.6 μm) [60–67, 69–72, 74–76]. Here we review some of the key results of those experiments.

A schematic of the Raman amplification process is shown in Fig. 7. In forward Raman amplification the input pump $P_p(0)$ and the signal to be amplified $P_s(0)$ are injected at the same end of the fiber. In the backward amplification geometry the two beams counterpropagate. The output of the Raman amplifier consists of the amplified signal $P_s(L)$, the transmitted pump $P_p(L)$, and the amplified spontaneous Raman scattering P_{as}, which exists even in the absence of an input signal. The power P_{ss} measured by a detector tuned to the signal chanel consists of both the amplified signal and the amplified spontaneous scattering which is a source of noise.

Fig. 8 shows an experimental realization of a Raman amplifier in the backward pumped geometry [60]. Here the pump source is a cw 1.45-μm color-center laser and the signal is generated by a semiconductor laser operating at 1.5 μm. The gratings in the figure serve to separate pump and signal beams. The pump power is typically 100 mW for which the signal experiences 3.5 dB of Raman gain. Such an amplifier can be used to enhance receiver sensitivity (minimum number of photons required to detect a bit) in an optical-fiber communications system [61].

For signal levels much less than the pump power, the fiber Raman amplifier is characterized by a power gain or amplification factor given by

$$G = P_s(L)/P_s(0) \exp(-\alpha L)$$

$$= \exp[g\,P_p(0)\,L'/K\,A_{\mathrm{eff}}]. \qquad (13)$$

Operationally, this gain is determined by measuring the transmitted signal power with the pump on and dividing by the signal power with the pump off. Since the measured "signal" contains both the amplified input signal and the amplified spontaneous scattering, it is useful to define another quantity which measures the relative contributions of these two. This is the contrast ratio

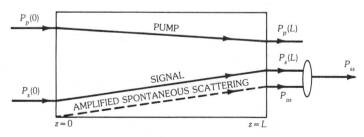

FIG. 7. Schematic of the forward Raman amplification process.

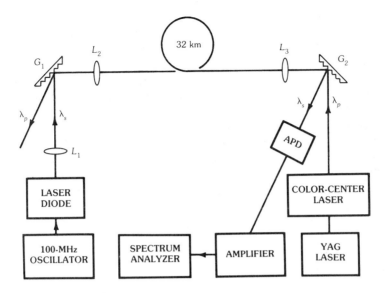

FIG. 8. Experimental setup for backward Raman amplification. (*After Olsson and Hegarty [60]*)

(or ON/OFF ratio) defined as the measured signal power with the pump on divided by the amplified spontaneous power (with no input signal):

$$\text{ON/OFF ratio} = P_{ss}/P_{as}. \qquad (14)$$

Note that this is not the same as the signal-to-noise ratio, which is a more complicated quantity that depends on the optical and electrical bandwidth of the receiver, the quantum efficiency of the detector, and other factors. Since the spectrum of the amplified spontaneous Raman scattering (noise) is much broader than that of the amplified signal, the contrast ratio may be improved by using narrow-band filters at the exit of the fiber.

In the small-signal amplification regime characterized by (13) the gain is independent of the input signal level. When the signal grows to a level that is no longer small compared to the pump, the gain begins to saturate. In the saturation regime the exact solution of the coupled power equations at steady state yields

$$G = [1 + P'_s(0)/P_p(0)]/\Big(P'_s(0)/P_p(0)$$

$$+ \exp\{-[1 + P'_s(0)/P_p(0)] \times g\, P_p(0)\, L'/K A_{\text{eff}}\}\Big),$$

$$(15)$$

where $P'_s(0) = \lambda_s P'_s(0)/\lambda_p$. The saturation characteristics of a fiber Raman amplifier are shown in Fig. 9 as universal curves of nonlinear gain versus pump power for several different input signal levels. The saturation power is on the order of the pump power, which is much higher than the saturation power of a semiconductor laser amplifier. This high saturation power of Raman amplifiers is a useful feature in applications that require the cascading of amplifiers [60].

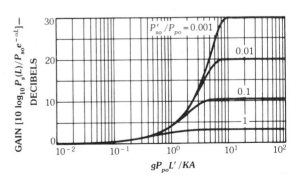

FIG. 9. Saturation characteristics of fiber Raman amplifier.

A linear gain of 40 dB and a saturated gain as high as 45 dB has been obtained by Desurvire and coworkers [62, 63]. The signal was an InGaAsP diode laser operating at 1.24 μm while the pump source was the 1.18-μm second Stokes radiation from an auxiliary fiber Raman generator excited at 1.06 μm. The pump power at 1.18 μm that yielded the 40-dB gain was about 1 W in a 200-ns pulse. The possibility of such large gains makes Raman amplification attractive in optical-fiber communications systems.

The Raman gain depends on the relative polarizations of the pump and signal waves, being maximum for parallel polarizations and nearly zero for orthogonal polarizations [13]. To optimize the power gain in long lengths of fiber the use of polarization-preserving fibers is necessary [64–66]. Nakazawa and colleagues [64] have recently used a 100-m–long birefringent, polarization-maintaining fiber to investigate the polarization dependence of the gain in backward stimulated Raman scattering. They find that the angular dependence of the gain obeys the relation

$$G(\theta) = (G_{\parallel} - 1)\cos^2\theta + 1, \qquad (16)$$

where θ is the angle between the pump polarization direction (oriented along a principal axis) and that of the signal. The term $G_{\parallel}$ is the gain when pump and signal have the same polarization direction. Fig. 10 shows the measured polarization dependence of the gain in a backward SRS geometry for a birefringent fiber with polarization extinction coefficient of 30 dB. (In this figure the circles and solid curves are experimental results and theoretical curves, respectively, and $\parallel$ denotes that polarization directions between the pump and signal pulses are the same, and $\perp$ shows that they are mutually orthogonal.) In subsequent experiments with longer fibers (about 1 km), the effect of the linear extinction ratio on the polarization dependence of the net Raman gain has been investigated [65]. The results show that the gain for an orthogonally polarized signal increases as the extinction ratio is decreased. This is to be expected since a lower extinction ratio means that a larger fraction of the power is transferred to the orthogonal axis as a result of random fiber imperfections. The surprising result is that the power dependence of the gain for an orthogonal signal cannot be explained on the basis of linear coupling between the fiber axes. There is an additional power-dependent coupling

which may be related to the intensity-dependent birefringence discussed in Section 4.

As mentioned in the introduction, the Raman process is isotropic, and hence the gain should be equal for both forward and backward scattering geometries. This equality of forward and backward gains has been confirmed in several experiments [64, 67]. The amplification factor measured in a particular experiment will of course depend on the dynamics of the interaction. When both pump and signal are short pulses, the achievable gain is limited by the extent of the temporal overlap between the pulses. In a backward interaction the effective length of the interaction region is given by $L_{\text{int}} = W v_g/2$, where W is the pulse width and v_g is the group velocity. For a forward interaction the effective length is the walkoff distance defined in (12). The backward pumping geometry has certain features which make it preferable to the forward geometry in certain applications. One practical advantage is that

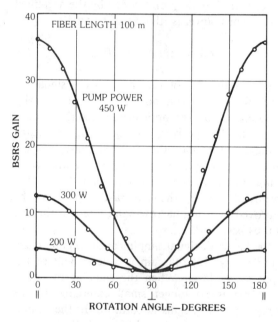

FIG. 10. Polarization dependence of backward stimulated Raman gain. (*After Nakazawa et al. [64]*)

it is much easier to separate the amplified signal from the pump beam if the two waves are counterpropagating. The backward scattering geometry can also lead to intense Stokes pulses much shorter that the pump pulse because essentially all the pump energy can be extracted by the leading edge of the Stokes pulse [68]. Finally, if the pump consists of a train of short pulses, as from a mode-locked laser, a counterpropagating signal pulse will encounter numerous pump pulses along the way and thus experience an integrated gain close to the cw gain [69]. This situation also averages out pump fluctuations and leads to stable amplification [60].

In a silica-fiber Raman amplifier maximum gain is obtained when the frequency separation between the pump and signal waves is about 440 cm^{-1}. This places restrictions on the pump and signal sources that can be used for efficient amplification. The most commonly used pump source in the near-infrared is the Nd:YAG laser operating at either 1.06 or 1.32 μm. This laser is capable of delivering high output power in a single longitudinal mode in either cw, Q-switched, or mode-locked operation. For this pump source the wavelengths of maximum gain, corresponding to the first Stokes lines, are around 1.12 μm and 1.40 μm. In order to extend the range of signal wavelengths that can be efficiently amplified with a Nd:YAG pump, it is possible to use the Raman gain at higher-order Stokes lines. For example, the 1.3-μm output wavelength of an InGaAsP diode laser coincides with the fourth Stokes line of a silica fiber pumped by a 1.06-μm Nd:YAG laser. The power in the third Stokes line can thus act as a pump for 1.3-μm radiation [64]. The amplification at higher-order Stokes lines is not as large as that obtained at the first Stokes. This is because of the $1/\lambda$ dependence of the Raman gain on the pump wavelength and the fact that not all the input power is transferred to the higher-order Stokes lines which pumps the adjacent signal line. There are situations however, in which the net gain at a higher-order Stokes line may be comparable to, or greater than, the net gain at the first Stokes line. This is the case for a 1.32-μm pump, whose first Stokes wavelength at 1.40 μm coincides with an absorption peak in silica. The second Stokes wavelength of 1.50 μm falls in a region of minimum loss. Nakamura and group [70] have obtained gains in excess of 24 dB by using a 1.5-μm InGaAsP laser at the second Stokes line of a Q-switched Nd:YAG laser operating at 1.32 μm. The maximum pump power was about 9 W. Nakazawa [66] has also used a 1.34-μm pump to obtain a 20-dB gain at the second Stokes wavelength of 1.52 μm. The combination of long length (1 km) and a high-germania core permitted a reduction in the required pump power to only 3.7 W. The ON/OFF ratio for higher-order Stokes amplification is not as high as that for the first Stokes.

Another attractive feature of fiber Raman amplifiers is the extremely broad gain bandwidth associated with the Raman process in silica. In principle, this broad bandwidth (about 100 cm^{-1}) means that subpicosecond pulses can be amplified without distortion at multigigabit rates. The only limitation on such an amplifier is the group-velocity dispersion which tends to broaden pulses. Raman amplification of subpicosecond pulses in fibers has not yet been reported. Kishida and colleagues [71] have studied the amplification of 100-ps pulses from a comb generator driven 1.4-μm InGaAsP diode laser. With 1.5 W from a cw 1.32-μm Nd:YAG laser pumping a 3.6-km–long fiber, a gain of about 20 dB was obtained. No distortion of either the pulse shape or spectrum was observed. To fully assess the high-speed capability of Raman amplifiers, much shorter signal pulses would be needed.

For practical reasons it would be desirable to use a low-power cw source to pump a Raman amplifier. Ideally, that source would be a semiconductor laser. The power required for cw Raman gain is of order 1 W. Experimental results on cw Raman amplification, however, have been sparse. Koepf, Kalen, and Greene [72] used a cw Nd:YAG laser operating at 1.064 μm as a pump to amplify the signal from an InGaAsP light-emitting diode at 1.21 μm. In that experiment it was observed that the onset of stimulated Brillouin scattering (SBS) caused a depletion of the pump power and a reduction of the Raman gain. The effect of SBS in a Raman amplifier is an important consideration when cw pumps are used. Stimulated Brillouin scattering may be suppressed by using pump pulses whose duration is less than the acoustic phonon lifetime (approximately 15 ns). In a cw experiment the coherence time of the pump must be less than this lifetime in order to suppress SBS. Aoki, Kishida, Honmou,

Washio, and Sugimoto [67] obtained efficient cw Raman amplification of InGaAsP diode laser light with a 1.32-μm Nd:YAG laser. No depletion due to stimulated Brillouin scattering was observed. The reason for the suppression of SBS is thought to be the large spectral broadening (and, hence, reduced coherence time) of the pump laser that occurs as a result of degenerate four-wave mixing. Since the pump wavelength lies in the minimum-dispersion region of the fiber, four-wave mixing interactions can be efficiently phase matched. Hegarty, Olsson, and Goldner [61] were able to suppress SBS in their cw Raman amplification experiment by using a pump laser that operates on two longitudinal modes.

The primary sources of noise in a Raman amplifier are pump fluctuations and amplified spontaneous Raman scattering. The effect of pump fluctuations has been measured by Aoki and coworkers [69]. They found that the power penalty due to gain fluctuations is only about 1 dB even at a gain as high as 14 dB. Hegarty, Olsson, and Goldner [61] have measured the improvement of receiver sensitivity that can be obtained by use of a Raman amplifier. They measured a net increase in receiver sensitivity of 3 dB/ 100 mW of pump power for input signal levels down to −30 dBm. The power penalty due to amplified spontaneous Raman scattering was only about 1 dB. Hegarty and Olsson have measured the noise properties of a fiber Raman amplifier [60]. They found that the dominant contribution to the degradation of the signal-to-noise ratio is the shot noise generated in the receiver due to spontaneous Raman scattering. There is no excess noise in the amplification process and generally the noise properties of a Raman amplifier compare favorably with those of a semiconductor laser amplifier. Theoretical calculations of the evolution of the amplified spontaneous Raman spectrum have been presented by Dakss and Melman [73]. Not included in that analysis is the role of four-wave mixing, group-velocity dispersion, and pump depletion. All these effects will profoundly influence the spontaneous Raman spectrum.

A number of other interesting results have recently been reported in connection with Raman amplification. Okamoto and Noda have incorpo-

rated a Solc-filter into their Raman fiber and used it to filter out the pump and other unwanted Stokes lines [76]. Fluoride glasses offer the possibility of extremely low loss transmission in the infrared. Raman gain in excess of 40 dB has been obtained in a fluoride glass fiber by Durteste and coworkers [77]. The Raman shift is 590 cm^{-1}. Desurvire and colleagues have used Raman amplification to compensate for optical losses in a reentrant single-mode fiber loop [78–80].

Raman Crosstalk in Optical Communications— The information capacity of an optical communications system can be increased by multiplexing several information channels at different wavelengths on a single fiber. This technique of wavelength-division multiplexing (wdm) has recently been used to demonstrate the transmission of 3 Gb/s from two distributed-feedback lasers with a channel spacing of only 2.9 nm [81]. The feasibility of a terabit-per-second system that employs 30 wavelength channels has also been demonstrated [82]. One of the potential limitations that needs to be seriously considered in the design of wdm systems is the effect of nonlinear crosstalk due to Raman gain [59]. Stimulated Raman scattering can convert energy from the short-wavelength channel to the longer-wavelength channels. This leads to a reduction in the signal-to-noise ratio for the short-wavelength channels and an increase in the bit error rate.

To quantify the effect of Raman crosstalk, first consider a two-channel system in which the short-wavelength channel (called the pump channel) is injected with P_{po} watts and the long-wavelength channel has a launched power of P_{so} watts. Assuming equal losses for both channels, the received power in the pump channel, taking into account the Raman interaction, is given by

$$P_p(L) = P_{po} e^{-\alpha L} (1 + P'_{so}/P_{po})$$

$$/\{1 + (P'_{so}/P_{po}) \exp [(1 + P'_{so}/P_{po})$$

$$\times g P_{po} L'/K A_{\text{eff}}]\}, \quad (17)$$

where $P'_{so} = \lambda_s P_{so}/\lambda_p$, λ_s and λ_p are the signal and pump wavelengths, respectively, L is the fiber length, and $L' = (1 - e^{-\alpha L})/\alpha$. The Raman gain coefficient is g, the fiber area is A_{eff}, and K is a polarization factor ($1 \leq K \leq 2$). The power in the pump channel is attenuated while the signal in the Stokes channel undergoes amplification with a gain given by (15). These results are shown in Fig. 11 as universal curves of nonlinear attenuation (or gain) versus the product of fiber effective length and launched power in the pump channel. For sufficiently large input power or effective length, essentially all the pump power is transferred to the longer-wavelength channel.

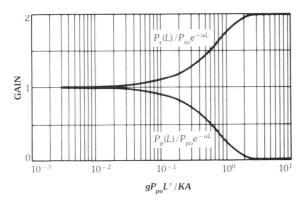

FIG. 11. Nonlinear attenuation (gain) at the pump (Stokes) wavelengths.

The amount of crosstalk due to Raman gain in a communications system will depend on the modulation scheme, the wavelength separation between channels, the number of channels, the launched power per channel, the group-velocity dispersion of the fiber, and many other factors. For a worst-case analysis of crosstalk, assume that the launched power in the pump channel equals the effective power P'_{so} in the Stokes channel. The fractional nonlinear attenuation of the pump due to the presence of a signal at the Stokes wavelength is given by

$$\beta \equiv 1 - P_p(L)/P_{po}\,e^{-\alpha L}. \qquad (18)$$

If this nonlinear attenuation or crosstalk is required to be less than a maximum β_M, then the launched power in either channel should be less than a critical value

$$P = (K A_{\text{eff}}/2g L') \ln\left[(1+\beta_M)/(1-\beta_M)\right]. \qquad (19)$$

For $\beta_M \ll 1$, one finds that the critical Stokes

power for a fractional attenuation of β_M is

$$P_{so} \simeq (K A_{\text{eff}}/g L')(\lambda_p/\lambda_s)\,\beta_M. \qquad (20)$$

Using typical values of $K=2$ (scrambled polarization), $\lambda_p/\lambda_s \simeq 1$, $A_{\text{eff}} = 5 \times 10^{-11}$ m^2, $g=7\times10^{-14}$ m/W, $\alpha=5\times10^{-5}$ m^{-1} and $L > 80$ km, we find that the input Stokes power that results in a 20% nonlinear loss is about 14 mW.

The performance degradation that results from stimulated Raman scattering in fibers has been analyzed by several workers [12, 83–86]. However, there are few experiments that directly measure the effects of this nonlinear crosstalk. Tomita [87] used two cw laser diodes operating at 1.26 and 1.34 μm to measure the level of crosstalk in a 21-km single-mode fiber. At a pump power level of 1.0 mW he found the stimulated Raman crosstalk to be -25.2 dB. More recently, Hegarty, Olsson and McGlashan-Powell [88] have measured the Raman crosstalk between two channels in a 45-km fiber transmission system as a function of channel separation and power. Bit-error-rate measurements were carried out on a signal channel modulated at 1 Gb/s at 1.50 μm and transmitted along with a second cw channel (pump) whose wavelength was tunable out to 1.57 μm. The power penalty at the receiver was found to be equal to the measured depletion of the signal. The power penalty on the shortest-wavelength channel depends on the number of longer-wavelength channels present. This dependence is shown in Fig. 12 for channels spaced 70 nm apart. It is seen that the power penalty increases rapidly with the power carried per channel and with the number of channels. The results shown here indicate that stimulated Raman crosstalk should be taken seriously if a wdm system is to transmit signals in excess of about 5 mW per channel.

Brillouin Processes

Stimulated Brillouin Scattering—Of the many nonlinear optical processes that can occur in single-mode fibers, stimulated Brillouin scattering (SBS) is the one most likely to limit the design and performance of lightwave communications systems [89]. This is especially true for coherent transmission systems, which will require sub-

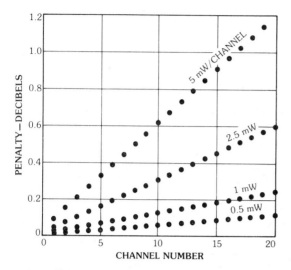

FIG. 12. Power penalty on shortest-wavelength channel as a function of channel number for different but equal powers per channel. (*After Hegarty, Olsson, and McGlashan-Powell [88]*)

megahertz laser line widths and will operate in the 1.5- to 1.6-μm wavelength region where fiber losses are a minimum. The reason why SBS looms so large over these systems is that its gain is two to three orders of magnitude larger than that of stimulated Raman scattering. Notwithstanding its large gain, the bandwidth associated with SBS is so narrow that ordinary wideband lasers do not efficiently pump the Brillouin process. When such lasers are used, stimulated Raman scattering usually dominates over SBS because the Raman gain bandwidth is much broader than even the laser line. When the laser line width is less than a few megahertz, the SBS process is greatly favored and can be observed with input powers of about 5 mW or less. The effect of stimulated Brillouin scattering on a communications link is to introduce strong nonlinear attenuation in the forward direction and to send intense Stokes-shifted light back towards the transmitter.

The process of stimulated Brillouin scattering may be described as a parametric interaction between a pump wave E_p, a Stokes-shifted wave E_s, and an acoustic wave ρ whose complex amplitudes are of the form

$$E_p = A_p\, e^{i(k_p \cdot r - \omega_p t)}, \tag{21a}$$

$$E_s = A_s\, e^{i(k_s \cdot r - \omega_s t)}, \tag{21b}$$

$$\rho = R\, e^{i(q \cdot r - \Omega t)}. \tag{21c}$$

The strong light fields generate acoustic waves through the process of electrostriction. These acoustic waves cause a periodic variation of the dielectric constant which then scatters the light wave. Because the sound-driven perturbation of the dielectric constant is a moving excitation, the scattered light is Doppler-shifted, with a frequency shift proportional to the velocity of sound in the medium. The frequencies of the incident (ω_p) and scattered (ω_s) light waves and that of the acoustic wave (Ω) satisfy the conservation law

$$\Omega = \omega_p - \omega_s. \tag{22}$$

Conservation of momentum (wave vector) also requires that for efficient interaction, the wave vectors must satisfy

$$q = k_p - k_s. \tag{23}$$

Because acoustic frequencies are several orders of magnitude lower than optical frequencies, the relation between the wave vectors reduces to

$$q = 2\,k_p \sin(\theta/2) = \Omega/V, \tag{24}$$

where θ is the angle between the incident and scattered waves and V is the velocity of sound in the medium. Thus the frequency shift in Brillouin scattering depends on the scattering direction. For a single-mode optical fiber the only relevant directions are the forward ($\theta = 0°$) and backward ($\theta = \pi$) directions. Frequency-shifted light is only generated in the backward direction,* and the size of the Stokes shift is given by

$$\nu = \frac{\Omega}{2\pi} = \frac{2\,n\,V}{\lambda}, \tag{25}$$

*This is strictly true for bulk media, where the acoustic excitations are plane waves. In a single-mode fiber the acoustic waves are guided and this leads to a relaxation of the wave vector selection rules. A very small amount of Stokes-shifted light is generated in the forward direction. The frequency shift is very small and is difficult to resolve. This phenomenon has been observed recently and christened GAWBS (guided acoustic wave Brillouin scattering) by Shelby et al. [90, 91].

where λ is the wavelength of the incident light and n is the refractive index of the fiber core. For a silica fiber where the acoustic velocity is 5.96×10^3 m·s^{-1}, and the refractive index is 1.45 at $\lambda = 1.5$ μm, the frequency shift is about 11.5 GHz.

At low input intensity the Stokes-shifted light is due to spontaneous Brillouin scattering from fluctuations in dielectric constant due to thermally excited acoustic waves. The spectrum of the scattered light yields valuable information concerning the lifetime of the acoustic waves. Spontaneous Brillouin scattering measurements on single-mode fibers yield line widths (FWHM) of order 150 MHz at $\lambda \simeq 0.5$ μm [92, 93]. The lifetime of the acoustic excitations is given by the reciprocal of the line width.

The backscattered waves in SBS can be quite intense, often containing more than 50% of the incident energy. In some experiments the backscattered light (or the acoustic shock associated with SBS [6]) has been intense enough to damage the input end of the fiber [2]. As in the Raman process it is possible to calculate a rough threshold value of input power at which SBS becomes significant [15]. To determine the gain and threshold of stimulated Brillouin scattering, it is valid to assume that the power P_p in the pump wave is undepleted by the nonlinear interaction and is attenuated only through the linear loss α:

$$P_p(z) = P_p(0) e^{-\alpha z}. \qquad (26)$$

In steady state the evolution of the backscattered wave is then described by

$$dP_s/dz = -g P_p P_s / K A_{\text{eff}} + \alpha P_s, \qquad (27)$$

where [94]

$$g = 2 \pi n^7 p_{12}{}^2 / \lambda^2 \rho_0 V \Delta\nu_B, \qquad (28)$$

A_{eff} is the fiber effective area, p_{12} is the longitudinal elasto-optic coefficient, ρ_0 is the material density, and $\Delta\nu_B$ is the Brillouin line width. The solution of (27) for the backscattered light at the input is

$$P_s(0) = P_s(L) \exp [G L' - \alpha L], \qquad (29)$$

where L' is the usual effective length and the small-signal gain coefficient is

$$G = g P_p(0) / K A_{\text{eff}}. \qquad (30)$$

Evaluation of the backscattered signal $P_s(0)$ requires knowledge of the value of this signal at $z = L$. In a fiber transmission link there is, of course, no input Brillouin-shifted light at the exit. What is usually done, in calculating thresholds, is to assume a fictitious input of one Stokes photon at a point in the fiber where the nonlinear gain exactly equals the natural loss of the fiber. The critical power for SBS is then somewhat arbitrarily defined as that input power (at $z = 0$) for which the backward stimulated Brillouin power equals the pump power. This threshold power is given approximately by [15]

$$P_{cr} \simeq 21 K A_{\text{eff}} / g L', \qquad (31)$$

where g is the peak Brillouin gain. For multi-kilometer lengths of single-mode fiber in the low-loss region the predicted threshold is only a few milliwatts.

Several experimental studies have confirmed the low threshold power required for SBS and have pointed out its potential significance for optical communication systems. The early work of Ippen and Stolen [95] used a pulsed xenon laser source and relatively short lengths (5.8 and 20 m) of fiber with linear loss of about 20 dB/km at the pump wavelength of 535.5 nm. An intracavity etalon was required to restrict the laser line width to less than the Brillouin line width of roughly 103 MHz. The measured threshold power was 2.3 W for the short fiber and less than 1 W for the longer fiber. One interesting feature of this experiment is that the SBS signal and the transmitted pump wave were observed to exhibit relaxation oscillations with a period equal to the round-trip transit time of the fiber. Such behavior, predicted earlier by Johnson and Marburger [96], is due to the finite length of the interaction region, and can be explained as follows. The growth of the backward Stokes-shifted wave causes a depletion of the pump wave near the input of the fiber. This reduces the gain in the fiber until the depleted portion of the pump wave passes out of the fiber. The gain then builds up again and the process continues. These finite-cell relaxation oscillations occur in the absence of reflective feedback. More recently, Bar-Joseph, Friesem, Lichtman, and Waarts [97] have shown that if external feedback is supplied, these relaxation oscillations become steady oscillations.

The threshold for stimulated Brillouin scattering can be quite low in long, low-loss fibers. Even at a pump wavelength of 0.71 μm, where the fiber loss is in excess of 4.0 dB/km, Uesugi, Ikeda, and Sasaki [98] measured substantial backward stimulated Brillouin power at power levels of 30 mW in a 4-km–long fiber. Beyond that input power level the transmitted power saturated at a value of about 9 mW and more than 56% of the input was converted into backscattered Brillouin-shifted light. In the low-loss region between 1.2 and 1.6 μm the effect of SBS should be observable at even lower input power. Cotter [99] has measured a threshold for SBS of only 5 mW in a 13.6-km–long single-mode fiber whose linear loss was 0.41 dB/km at the pump wavelength of 1.32 μm. The pump used was a cw single-frequency Nd:YAG whose line width of less than 1.6 MHz is an order of magnitude smaller than the Brillouin line width. Fig. 13 is a schematic of the experimental arrangement. The isolator serves to prevent the intense backscattered light from destabilizing the laser. The other instruments monitor the power and spectra of the transmitted and backscattered light. Fig. 14 shows the transmitted and backscattered power as a function of the launched power. At low input power the backscattered light is simply due to the 4% Fresnel reflection from the fiber end. Beyond an input of about 5 mW, the Brillouin scattering light reaches 65% of the input. Further increase in

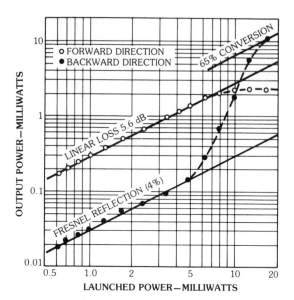

Fig. 14. Transmitted and reflected power in stimulated Brillouin scattering. (*After Cotter [99]*)

launched power leads to a saturation of the transmitted power of about 2 mW.

It is clear that the onset of SBS imposes a limitation on the power that can be launched in an optical-fiber communications system. For the extremely low loss fluoride fibers currently under development (with projected losses of less than 0.001 dB/km) the threshold power of SBS could be in the microwatt range. It is thus necessary to consider ways in which SBS may be suppressed in optical-fiber communications. Cotter [100] has proposed the use of certain modulation schemes to suppress the onset of stimulated Brillouin scattering. In systems that employ phase or frequency-shift keying (psk or fsk), it is possible to choose modulation parameters such that the

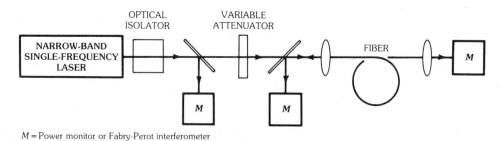

M = Power monitor or Fabry-Perot interferometer

Fig. 13. Experimental arrangement for observing stimulated Brillouin scattering. (*After Cotter [99]*)

threshold of SBS is raised. The scheme depends on the fact that SBS requires the buildup of a strong coherent acoustic wave within the decay time of an acoustic vibration. If the phase of the light pulses is continually changing, successive pulses do not add up in their effects and thus a coherent acoustic wave cannot build up. The same is true in fsk if the frequency shifts are large enough. Amplitude shift keying (ask), however, is additive in its effects and does not lead to much reduction of the Brillouin gain.

Cotter [100] has demonstrated the suppression of SBS by applying a phase modulation to the input optical field. The phase modulation arises from the mode beating that occurs when a laser operates on two longitudinal modes.

The experimental results of Cotter are shown in Fig. 15. When the laser operated in a single-frequency configuration, SBS was observed at power levels of about 6 mW. In the double-frequency configuration, no SBS was observed up to 90 mW, the maximum power available in this experiment. More recently experimenters [101] have used an acousto-optic modulator to suppress SBS. Aoki, Tajima, and Mituo have used fsk modulation to achieve similar results [102]. Thus, practical schemes for suppressing SBS in optical-fiber communications systems appear feasible.

Brillouin Oscillators—As is the case with stimulated Raman scattering, the presence of a resonator can lead to drastic reductions in the threshold power for stimulated Brillouin scattering [103]. In the first demonstration of such an oscillator, the ring cavity consisted of a 9.5-m–long fiber with loss at 514.5 nm of 100 dB/km, plus bulk optical

elements—lenses, mirrors, and beam splitters. The round-trip loss of such a resonator is quite high (about 70%) and thus the threshold power for oscillation is of order 100 mW [103–106]. By using an all-fiber ring resonator consisting of a single-mode fiber and a directional coupler, the threshold power can be reduced to a mere 0.56

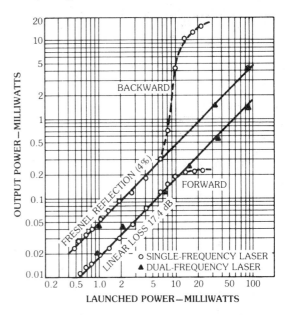

FIG. 15. Suppression of stimulated Brillouin scattering. (*After Cotter [100]*)

mW [107]. The all-fiber resonator has a round-trip loss of less than 3.5%, made possible by the use of a low-loss tunable evanescent-field directional coupler. A schematic of the fiber Brillouin oscillator is shown in Fig. 16, where P_i is the input pump power, and P_c the resonant circulat-

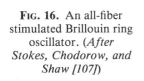
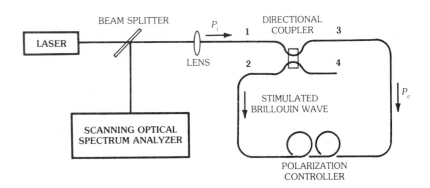

FIG. 16. An all-fiber stimulated Brillouin ring oscillator. (*After Stokes, Chodorow, and Shaw [107]*)

197

ing pump power. Possible applications of such low-threshold fiber ring oscillators include use in active inertial rotation sensing [108].

Application of Brillouin Amplification in Optical Communications—Coherent detection techniques (heterodyne or homodyne) have been used to enhance receiver sensitivity in optical communication systems [109]. Homodyne detection, while possessing some advantages over heterodyne detection, has the additional requirement that the local-oscillator laser be optoelectronically phase locked to the transmitter laser. To eliminate this need for optoelectronic phase locking, the use of Brillouin amplification has recently been suggested by Atkins and colleagues [110]. In that application the carrier component of the received optical signal is selectively amplified prior to photodetection [111]. If the power gain is G, the signal photocurrent after square-law detection is

$$i_s = C(P_s G)^{1/2}, \tag{32}$$

where P_s is the received signal power and C is a constant. If the carrier gain is large enough to overcome the noise of the electronic amplifier at the receiver, quantum-limited detection may be achieved.

This scheme requires an optical amplifier with a bandwidth sufficiently narrow that it can amplify the carrier without introducing gain at the signal sidebands. Brillouin amplification is ideally suited for this application because the gain bandwidth is only 15 to 25 MHz in silica fiber at a

pump wavelength of 1.5 μm. Fig. 17 shows an experimental arrangement used to demonstrate this scheme [110]. The transmitter is a 1.52-μm He-Ne laser followed by an external modulator while a color-center laser at the receiver end provides the backward pumping required for gain. The measured gain as a function of modulation frequency is shown in Fig. 18. The achievable

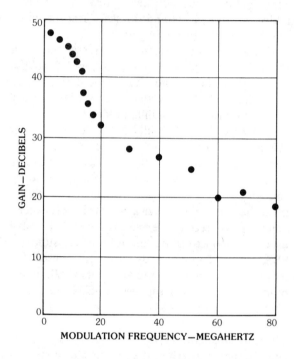

FIG. 18. Gain of Brillouin amplifier as a function of modulation frequency. (*After Atkins et al. [110]*)

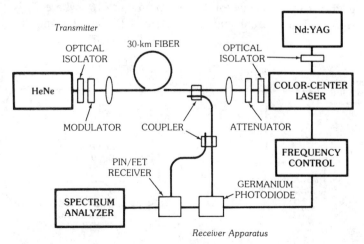

FIG. 17. Experimental realization of a Brillouin amplifier. (*After Atkins et al. [110]*)

gain is large enough to offer the possibility of quantum-limited detection.

There may be other applications, such as regenerative repeaters, that require amplifiers with greater bandwidth than the 15 MHz available in a fiber Brillouin amplifier. In that case it has been shown that the bandwidth can be increased by about a factor of ten by modulating the frequency of the pump laser [112]. The frequency modulation introduces a temporal variation in the absolute Stokes frequency seen by the probe laser. The counterpropagating nature of the Brillouin interaction and the long lengths of fiber involved cause an averaging of this temporal variation of the frequency. The net effect is that the probe sees a broadened Brillouin gain peak. This increase in bandwidth is, of course, accompanied by a decrease in net gain.

One possible limitation on the general use of Brillouin amplifiers is the presence of amplified spontaneous scattering from thermally excited phonons. The wider the bandwidth of the amplifier, the greater will be the contribution of noise to the signal. A general-purpose Brillouin amplifier may thus require cooling in order to improve its noise characteristics.

In a bidirectional, multiwavelength communication system, crosstalk due to Brillouin gain can occur if the frequency difference between the counterpropagating waves coincides with the Brillouin shift of about 20 GHz [113]. The power level at which significant crosstalk occurs is only of the order of 100 μW [114]. Fortunately, since the Brillouin gain bandwidth is so narrow, crosstalk can be avoided by a proper choice of the signal wavelengths without compromising the channel packing density of a wavelength-division–multiplexed system.

In practical applications to optical communications a fiber Brillouin amplifier would be pumped by a semiconductor laser. Typical semiconductor lasers, however, have line widths much too broad to effectively pump the Brillouin gain. Special line narrowing techniques, such as the use of external cavities or distributed feedback, are therefore necessary in order to use these lasers as efficient pumps. Recently, Olsson and van der Ziel have demonstrated a fiber Brillouin amplifier pumped by an external-cavity semiconductor laser with

line width less than 10 kHz [115]. The laser, tunable between 1.42 and 1.52 μm, can deliver up to 4 mW of power into the 37.5-km–long fiber. A probe laser of similar characteristics is tuned to the Stokes wavelength. Brillouin amplification of the probe laser was sufficient to cancel the fiber loss and supply 5 dB of net gain.

3. Parametric Processes

In the preceding section we considered nonlinear processes which involve the interaction of photons with material excitations such as molecular vibrations and acoustic phonons. The observed frequency shifts are determined by the frequencies of the material excitations involved. There is another class of phenomena, known as parametric processes, which originate from the nonlinear motions of instantaneously responding bound electrons. No net energy is deposited in the medium, which is passive and merely serves to mediate the interaction between photons. The frequency shifts are variable, ranging from about 1 cm^{-1} to greater than 1000 cm^{-1}, depending on phase-matching considerations. Processes that fall into this class include four-photon mixing, parametric amplification and oscillation, frequency conversion, and the generation of higher harmonics.

Physically, the presence of intense light waves in a dielectric can induce a nonlinear polarization which serves as a source of new waves. A phenomenological description of parametric processes usually begins with an expansion of the macroscopic polarization density $\mathbf{P}$ in powers of the electric field $\mathbf{E}$ [116]:

$$\mathbf{P} = \chi^{(1)}\mathbf{E} + \chi^{(2)}\mathbf{E}\mathbf{E} + \chi^{(3)}\mathbf{E}\mathbf{E}\mathbf{E} + \cdots + \text{electric quadrupole terms} + \cdots + \text{magnetic dipole terms} + \cdots . \tag{33}$$

Here $\chi^{(1)}$ is the usual linear susceptibility related to the linear refractive index through $n^2 = 1 + 4\pi\chi^{(1)}$. The second-order susceptibility $\chi^{(2)}$ leads to such effects as second-harmonic generation and two-wave sum-frequency generation. It vanishes in media which possess a center of inversion symmetry, such as glass. In optical fibers the first important nonlinear susceptibility is $\chi^{(3)}$, which gives rise to third-harmonic generation, four-photon mixing, and an intensity-dependent refractive index (Section 4).

The most studied parametric process in single-mode optical fibers is four-photon mixing, also known as three-wave mixing.* In this process an intense pump beam of frequency ω_p provides gain at two symmetrically displaced frequencies ω_s (Stokes) and ω_a (anti-Stokes). By analogy with microwave parametric amplification these two frequencies are also known as the signal and the idler, respectively. The relation between the interacting frequencies is shown in Fig. 19. As the pump beam propagates it acquires sidebands at ω_s and ω_a which grow at a rate proportional to $\chi^{(3)}$ and the pump intensity. Quantum mechanically, the process may be described as the absorption of

plies that the propagation constants $k_p = n_p \omega_p / c$, $k_s = n_s \omega_s / c$, and $k_a = n_a \omega_a / c$ must satisfy

$$2 k_p - k_s - k_a = 0 , \tag{35}$$

where n_p, n_s, and n_a are the effective indices at the pump, Stokes, and anti-Stokes wavelengths, respectively. Because of the dispersion in the refractive index, the phase-matching constraint is not automatically satisfied when (34) holds, There is generally a phase mismatch

$$\Delta k = k_a + k_s - 2 k_p = (n_a \omega_a + n_s \omega_s - 2 n_p \omega_p)/c \tag{36}$$

between the driving polarization at any frequency and the free-running wave at that frequency. One can define a coherence length

$$L_c = 2\pi / |\Delta k| \tag{37}$$

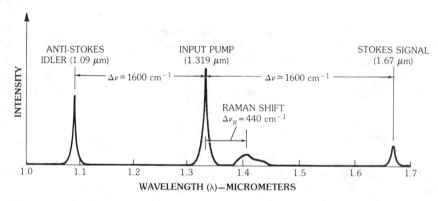

FIG. 19. Signal-idler pair generated by phase-matched four-photon mixing in a single-mode fiber pumped at 1.319 μm.(*After Lin et al. [129]*)

two pump photons to a virtual level and the emission of a signal photon and an idler photon. Conservation of energy requires that the three frequencies satisfy the relation

$$2\omega_p = \omega_s + \omega_a . \tag{34}$$

There is an additional constraint of momentum conservation or phase matching which im-

such that after a distance L_c the polarization and the radiated field are 180° out of phase. When that occurs energy is converted back from the signal to the pump. Efficient parametric signal generation thus requires $\Delta k = 0$ or be as small as possible. Phase matching can be approximately satisfied if the frequency shift is chosen sufficiently small that the coherence length is longer than the fiber length [117]. For multimode fibers waveguide mode dispersion can be used to compensate for material dispersion and thereby achieve phase matching [118–120]. In single-mode fibers two methods for achieving phase matching in parametric mixing with substantial frequency shifts involve the use of birefringent fibers or tun-

---

*The same process has been called three-wave mixing, four-wave mixing, and four-photon mixing in the literature. Here we will use "four-photon mixing" exclusively and reserve "three-wave mixing" for the process, mediated by $\chi^{(2)}$, in which two waves couple to generate a third.

ing to the minimum-dispersion wavelength of ordinary fibers. Both methods are described here.

Phase Matching in Birefringent Fibers

Stolen, Bosch, and Lin [121] first showed that phase matching of four-photon parametric interactions can be achieved by using the index difference between the two principal axes of a birefringent optical fiber. In this approach the pump beam is polarized along one of the principal axes and is orthogonal to the polarization of the signal and idler waves. For wavelengths in the visible region the refractive-index dispersion is such that in bulk silica the phase mismatch $\Delta k = k_s + k_a - 2 k_p$ is positive. In order to achieve phase matching therefore the pump should be oriented along the slow axis (i.e., the axis with the higher index). The frequency shift depends on the magnitude of the fiber birefringence.

Far from the minimum-dispersion wavelength, the phase mismatch can be approximated as a quadratic function of the frequency shift [117]:

$$\Delta k = k_s + k_a - 2 k_p = \beta \, \Omega^2 , \qquad (38)$$

where $\beta = \beta_m + \beta_w$ contains the material and waveguide contributions to the dispersion, and Ω is the frequency shift in centimeters^{-1}, defined as $2 \pi c \Omega = \omega_p - \omega_s$. For wavelengths far from the minimum-dispersion region, the material contribution dominates and one can write

$$\beta_m = 2 \pi \lambda D(\lambda) , \qquad (39)$$

where $D(\lambda)$ is the group-velocity dispersion defined as $\lambda^2 \, d^2 n / d\lambda^2$.

If the fiber has a birefringence δn, the contribution of this birefringence to the phase mismatch is $4 \pi \delta n / \lambda$. This contribution will compensate for the material dispersion if

$$\Omega = (4 \pi \delta n / \beta \lambda)^{1/2} . \qquad (40)$$

This relation shows that it is possible to tune the frequency shift by varying the fiber birefringence. Conversely, a measurement of the frequency shift gives an indication of the amount of birefringence in the fiber.

Stolen's group [121] obtained frequency shifts

between 340 and 1000 cm^{-1} in fibers with birefringence ranging from 1.5×10^{-5} to 9.0×10^{-5}. The four-photon parametric emission was only observed with the pump polarized along the major axis of the stress-birefringent fibers cladding ellipse. This unambiguously determines that axis as the slow axis of the fiber. The threshold input power from a doubled Nd:YAG laser ($\lambda_p = 532$ nm) was on the order of 500 W.

Fiber birefringence is a function of environmental factors such as stress and temperature. This makes it possible to tune the wavelength of the stimulated four-photon emission over a wide range. By applying an external stress of 0.3 kg/cm, Kitayama, Seikai, and Uchida [122] have been able to tune the emission frequency by up to 140 cm^{-1}. This force was applied by pressing the fiber with a flat plate. A more convenient way of applying stress to a fiber is to wrap it around a cylindrical rod. The bending-induced birefringence that results has been used to tune the Stokes emission by about 100 cm^{-1} in the spectral region around 1 μm [123]. Tuning of the phase-matched frequency by temperature-induced birefringence changes has also been reported [124]. The emission frequency changed by about 0.1 cm^{-1}/°C.

The four-photon mixing process can be used to measure the birefringence induced in a fiber by bending it. It also permits an unambiguous determination of the principal axes of the birefringent fiber. Shibata and team [125] have determined from a measured frequency shift of 1060 cm^{-1} that a birefringence of 7×10^{-5} is introduced by bending a fiber into a coil of radius 2.8 mm without axial strain. The contribution due to any extra axial strain has also been measured by Nakashima and colleagues [126] using the same four-photon mixing process.

The phase-matching process in which the pump is polarized along the slow axis and the signal and idler are along the fast axis is not the only possibility in birefringent fibers. Jain and Stenersen [127] have listed several additional phase matching configurations, some of which involve splitting the pump excitation between the fast and slow axes of the fiber. In their calculations they also include the dispersion of the fiber birefringence which leads to a correction of about 10% in the expression of Stolen's group [121] for the frequency shift. Sternesen and Jain [128] have demonstrated a phase-matching process in which the pump beam excites polarization modes along the two principal axes equally. These modes interact with a Stokes wave along the slow axis and an anti-Stokes wave along the fast axis. The frequency shifts obtainable by this method are small, on the order of 100 cm^{-1}.

Phase Matching in the Minimum-Dispersion Region

The magnitude of the chromatic dispersion of silica reaches a minimum at a wavelength λ_0 in the vicinity of 1.3 μm. This minimum in the dispersion characteristics permits phase matching of parametric interactions with frequency shifts of order 1000 cm^{-1} if the pump wavelength is tuned to a value slightly longer than λ_0 [129–131]. By making a Taylor series expansion of the refractive index it can be shown that in the vicinity of λ_0 the variation in the slope of $n(\lambda)$ makes it possible to obtain $\Delta k=0$ for a properly chosen pump wavelength. Such phase matching can occur even in bulk silica since it depends primarily on the material dispersion. Use of an optical fiber makes it possible to tune the wavelength of the emission by varying fiber parameters such as diameter and index difference between core and cladding [130]. The range of wavelengths that can be phase matched by this method is rather small. A typical output spectrum for phase-matched parametric mixing in the minimum-dispersion region is shown in Fig. 19. Note that the Raman band is also generated simultaneously. In fibers longer than a certain critical length ℓ_c the Raman output generally dominates the four-photon parametric signal. This is due to the spectral broadening of the pump owing to self-phase modulation. After a distance ℓ_c the pump spectrum is so broad that it cannot efficiently pump the narrow parametric gain (whose width decreases with distance) while it can still drive the broad Raman line [132]. As noted previously, the combination of multiple-Stokes generation, self-phase modulation, and four-photon mixing can also lead to the formation of a broad continuum [133]. These broad nanosecond continua have many useful spectroscopic applications [27, 133].

Four-Photon Parametric Amplification and Its Applications

The third-order nonlinear susceptibility $\chi^{(3)}$ mixes pump, signal and idler waves in the parametric interaction. In the limit of negligible pump depletion the evolution of the three waves is described by [134, 138]

$$dE_p/dz=i\gamma\,|E_p|^2\,E_p,\qquad(41\text{a})$$

$$dE_s^*/dz=-i\,2\,\gamma\,|E_p|^2\,E_s^*-i\,\gamma\,E_p^{*2}\,E_a,\quad(41\text{b})$$

$$dE_a/dz=i\,2\,\gamma\,|E_p|^2\,E_a+i\,\gamma\,E_p^2\,E_s^*,\quad(41\text{c})$$

where E_p, E_s, and E_a are the slowly varying amplitudes of the pump, signal and idler waves and $\gamma=(12\,\pi\,\omega/n\,c)\,f\,\chi^{(3)}$. The quantity f is a mode-overlap factor [6]. These equations are obtained by using the third-order nonlinear polarization in the wave equation, making the slowly varying envelope approximation, and assuming that the phase-matching condition holds. They can be readily solved for the values of the three fields at the exit of the fiber [134, 138].

The nature of the interaction depends on the relative phase between the three waves. If the signal and idler are generated from noise, the relative phase picks a value that optimizes the gain of the signal and idler. If all three waves are injected into the fiber, then depending on the relative phase the signal and idler will be amplified or attenuated [132, 134]. An experiment that clearly demonstrates this phase dependence has recently been carried out by Bar Joseph and coworkers [134]. In that experiment sidebands corresponding to injected signal and idler waves were generated by modulating a pump beam with an acousto-optic modulator. All three beams were coupled into a length of single-mode fiber and means were supplied for varying the relative phase between the waves. Fig. 20 shows the measured dependence of gain and attenuation on the relative phase.

The deamplification that can occur in the parametric process is the mechanism for the generation of the so-called squeezed states of light in which noise sidebands are attenuated. Squeezed states are states of the radiation field in which

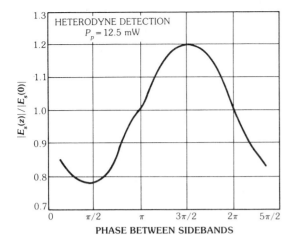

FIG. 20. Dependence of parametric gain on relative phase between signal and idler. (*After Bar-Joseph et al. [134]*)

noise fluctuations in one quadrature component are reduced below the quantum noise level. There has been much recent interest in such squeezing since it offers the possibility of greatly enhanced signal-to-noise ratios [135]. Levenson and colleagues have recently observed the squeezing of classical [136] and quantum [137] noise through four-photon mixing in a single-mode fiber. The observed quantum squeezing was about 12.5% below the quantum noise level. This observation represents an experimental *tour de force* since there are myriad other effects that compete with and usually dominate the delicate squeezing effect [138]. Foremost among these are guided acoustic wave Brillouin scattering and stimulated Brillouin scattering. The former was suppressed by cooling the fiber to liquid helium temperatures while the latter was eliminated by modulating the frequency of the pump beam.

Useful parametric amplification of weak signals had been demonstrated in a number of experiments [139, 63, 131]. Because of the need for phase matching, fiber parametric amplifiers are more difficult to implement than Raman amplifiers. The phase-matching condition is easily altered by environmental conditions such as temperature or stress. Ohashi and coworkers [139] have used externally applied stress to tune the phase-matching wavelength and thereby obtain gain in excess of 6000 for a semiconductor laser signal at 1.292 μm. The birefringent fiber was

pumped at a wavelength of 1.064 μm. Pocholle and group have achieved 37-dB gain for the 1.57-μm signal from a distributed-feedback semiconductor laser [63]. The pump was a Nd:YAG laser operating a 1.319 μm. In the absence of the injected signal, stimulated four-photon mixing induced by the pump beam leads to the generation of weak signal and idler waves. When the signal is injected it undergoes amplification and simultaneously the idler is also enhanced.

When a number of closely spaced frequency channels are multiplexed onto an optical fiber, crosstalk may occur as a result of four-photon parametric mixing [140, 141]. For reasonably long fibers this crosstalk will only be important for very closely spaced channels (about 10 GHz) since they are the only ones for which the four-photon interaction can be approximately phase matched.

Other Parametric Processes

A number of other parametric effects have also been observed in fibers [142–148]. In many cases the origin of the observed signal and the phase-matching process involved are not well understood. For example, third-harmonic generation has been observed in elliptical-core fibers [142]. While the third-harmonic signal is due to $\chi^{(3)}$ the phase-matching mechanism involved has not been elucidated.

Even more curious is the observation in optical fibers of several phenomena such as second-harmonic generation and two-wave sum-frequency generation [143–148] which are ordinarily due to the second-order susceptibility $\chi^{(2)}$. This susceptibility vanishes in isotropic media, such as glass. It is conjectured that the observed sum-frequency and second-harmonic radiation in single-mode fibers may be due to the quadrupole term in the polarization expansion. This is an area that requires more investigation. Particularly intriguing is the recent observation of copious amounts of second-harmonic generation in a photosensitized fiber [148]. The second harmonic, generated with 3% efficiency, was intense enough to pump a dye laser. The fiber was photosensitized by exposing it for several hours to the output of a cw mode-locked and Q-switched Nd:YAG laser. Initially no second-harmonic radiation was seen. After about 2 hours of continuous illumination, the second-harmonic output grew exponentially with exposure time. Fig. 21 shows the output second-harmonic power as a function of time. The physical mechanism responsible for frequency doubling in optical fibers is not yet understood.

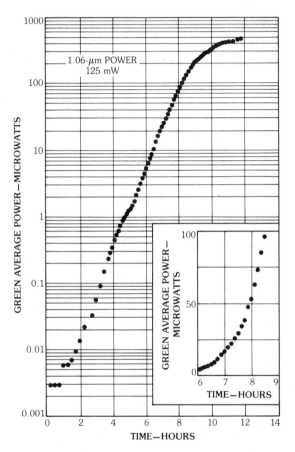

FIG. 21. Output power of second-harmonic light generated in a glass fiber as a function of time. (*After Österberg and Margulis [148]*)

4. Effects Due to an Intensity-Dependent Refractive Index

So far we have considered nonlinear processes in which photons from an intense pump beam scatter inelastically to generate frequency-shifted photons. The study of such inelastic scattering processes in bulk materials is often complicated by the onset of another nonlinear process known as self-focusing, which leads to the formation

of hot spots and causes the beam to break up into filaments. This effect has its origin in an intensity-dependent refractive index.

The Intensity-Dependent Refractive Index

There are several mechanisms which can lead to an intensity-dependent refractive index. Perhaps the most mundane is a thermal index change in which the heating of a weakly absorbing medium leads to changes in density and an accompanying change in index. In liquids, molecular reorientation by the applied field can be the primary cause of an intensity-dependent refractive index. In glasses the origin of a nonlinear index is thought to be the anharmonic response of bound electrons, though there is a nonnegligible component (about 20%) due to the displacement of the nuclei from their average positions in the absence of an applied field [149]. The nuclear motions occur on a time scale of order 10^{-12} s, and are the same vibrations that also give rise to the previously described Raman scattering. The electronic response occurs within 10^{-16} s. Since our approach to nonlinear effects in fibers is primarily phenomenological, we will not dwell on the microscopic origins of the nonlinear index and will assume that it can respond instantaneously to changes in the light intensity.

The intensity-dependent refractive index can be written

$$n(I) = n + n_2 I(r,t), \tag{42}$$

where $n_2 = 12 \pi \chi^{(3)}/n$ is the nonlinear index coefficient, $\chi^{(3)}$ is the third-order nonlinear susceptibility, and $I = \frac{1}{2}\langle E \rangle^2$ is the time-averaged intensity. In glasses the size of n_2 is on the order of 10^{-13} esu. By comparison, CS_2 has an n_2 of 10^{-11} esu, while GaAs at 1.06 μm has an n_2 of 10^{-9} esu. For glasses of interest in fiber optics, n_2 shows very little dispersion as one tunes from the infrared to the ultraviolet [150]. The value n_2 is usually quoted for linearly polarized light. For a purely electronic mechanism it has been shown that the nonlinear index for circularly polarized light is only 2/3 that for linearly polarized light [151] while in the case of random polarization its value is 5/6 that for linear polarization [152]. The nonlinear index coefficient is also often quoted in units of square centimeters per watt. The relation between the two units is

$$n_2(\text{cm}^2/\text{W}) = 4\pi n_2(\text{esu}) \times 10^7/nc, \qquad (43)$$

where n is the linear refractive index and c is the speed of light in centimeters per second. Thus, for example, n_2 for SiO_2 glass is 1.1×10^{-13} esu, or 3.2×10^{-16} cm^2/W [153]. It is useful to note that if n_2 is expressed in square centimeters per kilowatt, it has the same order of magnitude as n_2 in esu.

If the intensity $I(r,t)$ exhibits spatial and temporal variations, the nonlinear index will also be space and time dependent. Consider, for example, an incident beam with a Gaussian intensity profile. The induced index change will follow the Gaussian profile of the beam. This inhomogeneous distribution of refractive index acts as a nonlinear lens and leads to self-focusing in bulk materials. In single-mode fibers, however, the beam spot size is already of the order of a wavelength. Any additional confinement due to self-focusing is negligible. This is one of the features that makes single-mode fibers so attractive for nonlinear optical measurements. The beam profile is stable, well characterized, completely determined by the waveguide properties, and free from the deleterious effects of self-focusing.

We now consider several important effects in optical fibers that owe their origin to the intensity-dependent refractive index. These include the optical Kerr effect or light-induced birefringence, self-induced polarization changes, self-phase modulation, optical bistability and soliton propagation.

The Optical Kerr Effect

John Kerr (1824–1907) showed that a dc electric field could induce birefringence in an initially isotropic dielectric medium. Here we discuss a class of effects due to steady-state optical fields that alter the refractive index of a medium.

Light-Induced Birefringence—In the presence of an intense, linearly polarized light beam, an isotropic medium will become birefringent. A weaker probe beam will thus see a refractive-index change that differs whether the probe is polarized parallel to or normal to the strong pump. This optically-induced birefringence is known as the optical Kerr effect [154]. In a typical Kerr measurement an isotropic material is placed between a polarizer and analyzer (crossed polarizer). A strong linearly polarized pump beam induces a birefringence which is probed by a weaker probe beam, also linearly polarized at 45° to the pump. In the absence of the pump beam the probe transmission is blocked by the analyzer. When the pump beam is turned on, the medium becomes birefringent and a portion of the probe is transmitted. Clearly, this phenomenon has potential application as a shutter that can be operated at high repetition rates with picosecond pulses [155, 156].

To quantify the magnitude of the Kerr effect we consider the phase difference between the x and y components of the probe beam. After propagating a distance L through the medium in the presence of the strong beam, this phase shift is

$$\Delta\phi = 2\pi (\delta n_\parallel - \delta n_\perp) L/\lambda, \qquad (44)$$

where $\delta n_\parallel$ and $\delta n_\perp$ are the refractive-index changes parallel and perpendicular to the electric field of the strong beam. (In the presence of significant loss L should be replaced by the effective length $L' = [1 - \exp(-\alpha L)]/\alpha$.) The birefringence is related to the applied field through the following relation [155]:

$$\delta n_\parallel - \delta n_\perp = \tfrac{1}{2} n_{2B} |E|^2, \qquad (45)$$

where $n_{2B} = (3/2)n_2$ is the Kerr coefficient. The fraction of incident probe transmitted by the analyzer is given by

$$T = \sin^2 (\tfrac{1}{2}\Delta\phi). \qquad (46)$$

Maximum transmission occurs for a phase difference of π and a 100% modulation is achieved with a phase shift of 2π.

The long interaction lengths and high power densities attainable in fibers makes it possible to observe large phase shifts between the orthogonal components of the probe beam in Kerr measurements. In such measurements it is essential that the linear polarization state of the pump beam be maintained throughout the length of the interaction region. The pump beam is thus usually oriented along one of the principal axes of a birefringent fiber, and one measures the change in birefringence induced by this beam. The effect of the intrinsic birefringence on the probe beam is then compensated for by means of a phase plate located at the exit of the fiber.

Stolen and Ashkin [157] were the first to observe the optical Kerr effect in an optical fiber. Using a fiber length of about 6 m, a pulsed xenon laser as the pump and a cw HeNe laser as the probe, they obtained a maximum modulation of 5% with a pump power of 10 W. By going to much longer lengths of fiber, Dziedzic, Stolen, and Ashkin [158] were able to obtain Kerr modulations in excess of 100% (corresponding to phase differences greater than 2π). For a 50-m-long fiber an input power of order 1 W from an argon laser was sufficient to give a probe phase shift of π. More recently, 100% modulation depths have been observed with only 390 mW launched into a 500-m-long fiber [159]. The pump source was a Q-switched Nd:YAG laser operating at 1.064 μm and the probe was provided by a 1.15-μm cw HeNe laser. Stimulated Raman scattering was a factor in limiting the observed phase shifts to less than 2π.

The fast response of the nonlinearity responsible for the Kerr effect in optical fibers should, in principle, enable the operation of very high speed optical shutters. There are, however, other factors that will limit the performance of such devices. These include the group-velocity differ-

ence between the pump and probe wavelengths, fiber birefringence, and pulse spreading. Dziedzic and colleagues [158] have examined the limitations on Kerr shutters imposed by dispersion and fiber birefringence. They find that resolution times shorter than 1 ps are possible in principle but will not be easy to achieve. Kitayama's group [160] has used a fiber Kerr shutter to demonstrate optical sampling. A time-varying signal from a laser diode at 0.84 μm was sampled into a train of pulses by the mode-locked output of a 1.06-μm Nd:YAG laser. The mode-locked pulses repetitively open and close the shutter and thus perform a sampling operation on the signal. A schematic of the fiber Kerr shutter is shown in Fig. 22. In this experiment the sampling pulses used were fairly long (300 ps). Picosecond and subpicosecond operation of Kerr gates in optical fibers at high rates and with good modulation depth still remains to be demonstrated.

Light-Induced Nonreciprocity—The above discussion of the Kerr effect was limited to the case of a strong pump altering the refractive index seen by a weak copropagating probe beam of a different wavelength and polarization. Consider now two strong pump waves of the same frequency and linear polarization but counterpropagating with respect to each other. When these waves interact in a nonlinear medium each will modify the phase of the other through the intensity-dependent refractive index. Furthermore, the interference between the pump beams results in a nonreciprocal phase shift if the two

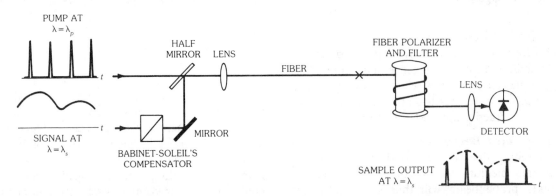

FIG. 22. Optical sampling by means of a fiber-optic Kerr shutter. (*After Kitayama et al. [160]*)

beams have unequal intensities. To calculate this phase shift, the nonlinear polarization induced by the applied fields is written as

$$P^{\text{NL}} = (n_0 n_2 / 4 \pi) |E|^2 E , \qquad (47)$$

where $E = (A_+ e^{ikz} + A_- e^{-ikz}) e^{-i\omega t}$ and $k = n_0 \omega / c$. The plus and minus subscripts refer to forward and backward propagating waves, respectively. By retaining only the terms with spatial variation exp (ikz) and exp $(-ikz)$, we find that the forward and backward nonlinear polarization components are

$$P_+^{\text{NL}} = (n_0 n_2 / 4 \pi)[|A_+|^2 + 2|A_-|^2] A_+ e^{ikz}, \qquad (48)$$

$$P_-^{\text{NL}} = (n_0 n_2 / 4 \pi)[|A_-|^2 + 2|A_+|^2] A_- e^{-ikz}. \qquad (49)$$

The factor of two is the result of interference between the counterpropagating waves which creates a refractive-index grating within the medium. Maxwell's equations with the above nonlinear source terms lead to the following results for the phase shifts seen by the two waves after propagating a distance L:

$$\Delta\phi^+ = 2\pi n_2 [|A_+|^2 + 2|A_-|^2] L / \lambda , \qquad (50)$$

$$\Delta\phi^- = 2\pi n_2 [|A_-|^2 + 2|A_+|^2] L / \lambda . \qquad (51)$$

Clearly these phase shifts are equal only if the counterpropagating intensities are equal. In the general case of unequal intensities the resulting nonreciprocal phase shifts have important consequences for fiber optic devices that use counterpropagating waves.

An example of the effect of this light-induced nonreciprocity is found in the operation of fiber-optic ring gyroscopes. Such ring gyroscopes are based on the Sagnac effect in which two counterpropagating waves in a rotating ring interferometer travel different optical path lengths depending on whether they are propagating with or against the rotation direction. This difference in path lengths is converted to a difference in amplitudes through the usual dependence of interferometer transmission on phase shift. Kaplan and Meystre [161] have suggested that by placing a nonlinear medium within the ring interferometer, this nor-

mal Sagnac effect can be enhanced by the nonlinear nonreciprocity. Ezekiel, Davis, and Hellwarth [162, 163] independently predicted and observed the effect of this intensity-induced nonreciprocity in a fiber-optic Sagnac interferometer. In their experiments they find that a power difference of only 1 μW between oppositely propagating light beams leads to a nonreciprocal phase shift of 1.4×10^{-6} rad. For their geometry, which consists of a 200-m-long fiber coiled around a 19-cm-diameter spool, this phase shift is equivalent to a rotation rate of $0.2°/\text{hr}$. Thus, in order to avoid rotation rate errors in precision fiber gyroscopes, strict control of the intensities of the counterpropagating waves would appear necessary. Fortunately, since the light-induced nonreciprocity owes its origin to interference effects, various temporal properties of the source can be modified in order to suppress this nonreciprocity. Bergh, Lefevre, and Shaw [164] suggest intensity modulation of the source in order to reduce the cross interaction between waves that results in phase nonreciprocity. In particular, for a square wave modulation, it is seen that the cross-phase effect is present only when the intensities of the two waves are coincident. The factor of two in the expressions for the phase shift (resulting from cross interaction) is reduced to an average value of unity. The nonreciprocity is thus effectively cancelled. Such compensation of the Kerr nonreciprocity in a fiber-optic gyroscope through source amplitude modulation has been demonstrated by Bergh and coworkers [164]. Other schemes for compensating the Kerr effect involve the use of broadband sources [165, 166]. Polarized thermal light, such as generated by a superluminescent diode, has the statistics required to completely cancel the Kerr nonreciprocity. Errors due to the Kerr effect have been eliminated by using a highly multimode laser to approximate such a source [167].

For completeness we note that Kravtsov and Serkin [168] have analyzed the effect of nonlinear nonreciprocity on a ring fiber Raman oscillator. They find that this nonreciprocity can lead to a frequency difference between the opposite traveling waves in the oscillator. This frequency splitting could be on the order of several hundred megahertz.

Optical Bistability and Chaos—Optical bistability is a phenomenon in which an optical system displays two stable output intensity states for the same input intensity. The two states are linked by a hysteresis loop, and transitions between them can be obtained by varying the input intensity adiabatically. There are many potential applications of this phenomenon in the areas of optical communications and optical signal processing. Among these applications are binary logic operations, differential amplification, pulse shaping and regeneration, optical limiting and ultrafast switching [169].

The physical requirements for bistability are (*a*) a medium whose refractive index or absorption is intensity-dependent and (*b*) some means of feedback such that the medium's transmission depends on the output intensity. The feedback is usually in the form of a Fabry-Perot resonator or a ring resonator that encloses the nonlinear medium. An intense input beam can change the refractive index of the intracavity medium, thereby altering the round-trip phase change and shifting the cavity on or off resonance with the input light wavelength. This leads to optical switching, hysteresis, and optical bistability.

Nakatsuka's team [170] has observed optical hysteresis in the transmission of a ring cavity containing a single-mode fiber. Mode-locked pulses from a Nd:YAG laser were incident on the cavity. The observed hysteresis loop is shown in Fig. 23. Above a certain input intensity the output exhibits highly irregular or chaotic behavior. The possibility of such optical chaos was first predicted by Ikeda [171] and is related to an instability of a multiply resonant four-photon parametric oscillator.

Self-Induced Polarization Changes

Strictly speaking, the Kerr effect refers to the birefringence induced by an intense linearly polarized beam in a nonlinear medium. This birefringence is probed through polarization changes suffered by a weaker copropagating beam. The polarization of the strong beam is preserved. There is also a class of "self-effects" in which the interaction of a single elliptically polarized beam with a nonlinear medium results in changes in the polarization state of the beam itself. If the nonlin-

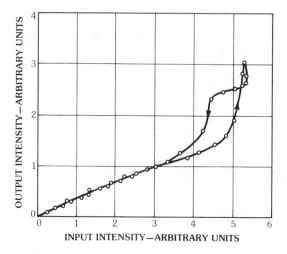

FIG. 23. Optical hysteresis in the transmission of a fiber-optic ring resonator. (*After Nakatsuka et al. [170]*)

ear medium is isotropic, the polarization ellipse merely rotates with propagation distance (and intensity) without changing shape [172]. If the nonlinear medium is anisotropic, as in a birefringent optical fiber, both the shape and orientation of the polarization ellipse will change with intensity and with propagation distance [173]. For anisotropic media, self-induced polarization changes are possible even with linearly polarized inputs so long as the electric-field vector does not lie along a principal axis of the medium. (See, however, [174] on the fast-axis instability.) Clearly, an intensity-dependent polarization state from an optical fiber is undesirable in optical communications systems that have polarization-sensitive elements. On the other hand, useful pulse shaping, intensity discrimination, and optical logic functions can be achieved when these same intensity-dependent polarization changes are converted to nonlinear transmission through a polarizer. Before reviewing these applications we outline a coupled-mode formalism that permits a convenient analysis of self-induced polarization changes in both birefringent and isotropic single-mode fibers.

A general elliptical state of polarization can be described in terms of orthogonal right and left circularly polarized basis fields c_+ and c_-. The polarization ellipse is characterized by its ellipticity $e = (|c_+| - |c_-|)/(|c_+| + |c_-|)$ and azimuth

$\theta = \frac{1}{2}(\phi_+ - \phi_-)$, where ϕ_+ and ϕ_- are the phases of c_+ and c_-. The spatial evolution of the polarization state in a birefringent fiber is governed by a pair of coupled wave equations for the circular modes [173, 174]:

$$dc_+/dz = i\kappa c_- + i\beta |c_-|^2 c_+ , \qquad (52)$$

$$dc_-/dz = i\kappa c_+ + i\beta |c_+|^2 c_- . \qquad (53)$$

The circular mode amplitudes are coupled because of the linear birefringence δn through $\kappa = \pi \delta n/\lambda$, and thus periodically exchange energy as they propagate. In the absence of the nonlinear terms the terms in κ describe the well-known sinusoidal evolution of the polarization state with propagation distance in a birefringent medium. The nonlinear coefficient β is proportional to n_2 and is given by

$$\beta = 4\pi\chi/3\lambda \quad (\text{W cm})^{-1} , \qquad (54)$$

where $\chi = 4\pi n_2 \times 10^7 n c A_{\text{eff}}$ and A_{eff} is the fiber effective area. It is the nonlinear terms in β that lead to an intensity-dependent phase difference between the circular mode amplitudes and, hence, an intensity-dependent rotation of the polarization ellipse. To see this, set $\kappa = 0$ (isotropic medium) and write $c_\pm = |c_\pm| \exp (i\phi_\pm)$. Then from (52) and (53) one finds that the amplitudes $|c_\pm|$ do not change and that the relative phase $\phi = \phi_+ - \phi_-$ evolves as

$$d\phi/dz = \beta (|c_-|^2 - |c_+|^2) . \qquad (55)$$

Since the angle between the major axis of the polarization ellipse and the x axis is given by $\phi/2$, it is clear that after propagating a distance L in an isotropic nonlinear medium, the polarization has been rotated through an angle

$$\theta = \frac{1}{2} \beta L (|c_-|^2 - |c_+|^2) . \qquad (56)$$

Thus, ellipse rotation in an isotropic medium is the result of the difference in intensity-dependent phase shifts suffered by the unequal left and right circular components of an elliptically polarized beam [172].

In a birefringent fiber the output polarization state is determined by competition between the existing linear birefringence and the ellipse rotation component. Equations 52 and 53 can be solved analytically in terms of elliptic functions for the output polarization [173]. Fig. 24a shows the dependence of the output polarization (ellipticity and azimuth) on incident power for an input beam linearly polarized at 45° to the principal axes of a birefringent fiber. For applications to pulse shaping and optical logic a polarizer is placed at the fiber exit and oriented such that it rejects the output polarization at low intensity. The calculated nonlinear transmission through this polarizer is shown in Fig. 24b. When the light-induced birefringence is weak compared to the existing fiber birefringence the transmission

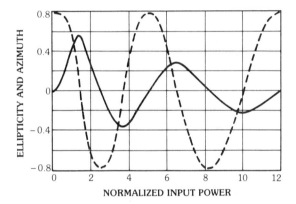

(a) Polarization of transmitted beam versus input power. (After Winful [173])

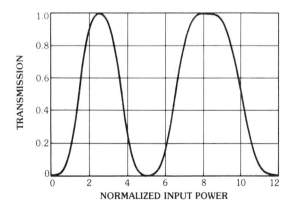

(b) Transmission of a birefringent fiber and crossed polarizer versus input power.

FIG. 24. Power-dependent transmission properties of a birefringent fiber.

through the crossed polarizer is approximately quadratic in input power.

The use of birefringent single-mode fibers for pulse shaping and intensity discrimination was first described by Stolen, Botineau, and Ashkin [175]. In that application the low-intensity parts of an incident pulse are blocked by the output polarizer while the high-intensity part, which undergoes nonlinear polarization rotation, is transmitted. Subsequent experiments by Niko-laus, Grischkowsky, and Balant [176] also showed dramatic reshaping and clipping of pulses through nonlinear polarization rotation. It appears that this technique will be useful in removing the extensive wings that often accompany pulses compressed through self-phase modulation in optical fibers (this section, under "Pulse Compression"). Mollenauer's group [177] has successfully used it to suppress the broad pedestal associated with solitons in optical fibers (see this section, under "Soliton Propagation"). An example of the kind of suppression achievable by this technique is shown in Fig. 25. More recently, Halas and Grischkowsky [178] have used self-phase modulation and polarization rotation simultaneously to obtain exceptionally clean compressed pulses in a fiber-grating compression scheme. In this application to pulse shaping it has been observed experimentally [176] and shown theoretically [179] that the circular birefringence

induced by axial twists of the fiber can be used to enhance the nonlinear transmission of the fiber-polarizer combination. The nonlinear transmission of a birefringent fiber and crossed polarizer has also been used to demonstrate all-optical logic gates [180, 181].

Ellipse rotation can be used to measure the nonlinear index of fibers. Some measurements of this sort have been made by Al'tshuler and colleagues [182] and by Crosignani and colleagues [183]. Their results are in fair agreement with those obtained by Owyoung, Hellwarth, and George [153] for bulk fused quartz.

There has been much recent interest in polarization instabilities in birefringent optical fibers [173, 184–186]. These instabilities result from the intensity-dependent refractive index and are such that small changes in the polarization or intensity of an input beam lead to large changes in the output polarization. For example, an intense beam oriented along the fast axis of a birefringent fiber reduces the fiber birefringence as a result of the intensity-dependent refractive index. At a critical input power the light-induced birefringence can cancel the existing fiber birefringence. The intensity-dependent beat length associated with these birefringence changes is shown in Fig. 26. Recently an experimental observation of polarization instabilities in a birefringent fiber has been reported [188].

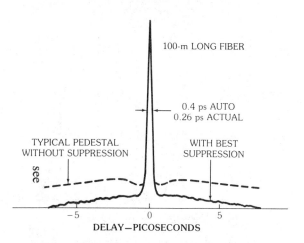

FIG. 25. Suppression of the "wings" of a pulse by means of nonlinear polarization rotation in an optical fiber and transmission through a crossed polarizer. (*After Mollenauer et al. [177]*)

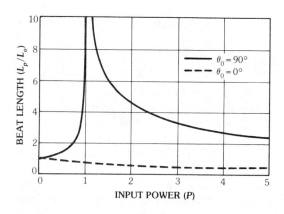

FIG. 26. Change of beat length of a birefringent fiber owing to an intense beam oriented along the fast axis (solid line) or along the slow axis (dashed line). (*After Winful [174]*)

The power required for these self-effects is fairly high. From (52) and (53) we find a critical power $P_c = 2\kappa/\beta$ watts at which the ellipse rotation component becomes comparable to the linear birefringence term. Even for a weakly birefringent fiber of $\delta n = 3 \times 10^{-6}$, this power is on the order of 150 W. For nonbirefringent fibers the effect of random mode coupling will overshadow the intensity-dependent polarization component for the power levels available from semiconductor lasers. Thus it is unlikely that these self-effects will have any serious consequences for optical communications. It has been suggested by Kitayama's group [187] that the strength of the random mode coupling, which depends on fiber birefringence, can be tuned by altering the fiber birefringence with an intense beam. The power required, however, is rather high (10^5 W).

Self-Phase Modulation

The nonlinear processes of self-focusing and the optical Kerr effect can occur even with cw beams. We now consider an effect which depends on the temporal variation of the intensity of an input beam for its observation. This effect, known as self-phase modulation (spm), is the temporal analog of self-focusing and leads to spectral broadening, frequency chirping, and, depending on the sign of the group-velocity dispersion, pulse compression or pulse broadening. The effect can also be described as a four-photon mixing process with small frequency shift [152].

Pure Self-Phase Modulation—A light wave propagating through a single-mode fiber of length L and refractive index n will acquire a phase shift of $\phi = (2\pi L/\lambda) n$. (As usual, for fibers with appreciable loss, L should be replaced by the effective length $L' = [1 - \exp(-\alpha L)]/\alpha$.) If the refractive index is intensity-dependent and the input beam has an amplitude modulation, the transmitted beam will exhibit an intensity-dependent and temporally varying phase given by

$$\Delta\phi(t) = (2\pi L/\lambda)\,\delta n(t),\qquad (57)$$

where $\delta n(t) = \frac{1}{2} n_2 |E(t)|^2$, and the electric field in esu is related to the power in watts through

$$|E(t)|^2 = 8\pi P(t) \times 10^7 / n c A_{\text{eff}}.\qquad (58)$$

It is this conversion of an amplitude modulation to a phase modulation through the nonlinear index that is known as self-phase modulation (spm). It modifies the frequency spectrum of an intense short pulse by generating new frequency components [188]. The instantaneous frequency shift at any point in the pulse is given by the time derivative of the phase modulation:

$$\omega - \omega_0 = \partial(\Delta\phi)/\partial t.\qquad (59)$$

In a dispersive medium the different frequency components in a pulse will, in general, travel at different speeds. Ignoring for the moment the effect of dispersion, let us consider in greater detail the effect of self-phase modulation on the spectrum of an initially Gaussian pulse.

The spectral content of a band-limited plane wave of carrier frequency ω_0 is obtained by taking the Fourier transform of the complex field E:

$$S(\omega) = \left| (1/2\pi)^{1/2} \int_{-\infty}^{\infty} |E(t)|\, e^{i[\phi_0(t) + \Delta\phi(t)]}\, e^{i(\omega - \omega_0)t}\, dt \right|^2.\qquad (60)$$

Both the amplitude and phase contribute to the Fourier spectrum. In a transform-limited pulse the spectral information is contained entirely in the amplitude function and there is no modulation of the phase. For example, the spectrum of a Gaussian pulse in the absence of phase modulation is also Gaussian, with a bandwidth inversely related to the temporal width of the pulse. The presence of a time-varying phase will result in excess bandwidth. The intensity-dependent phase factor $\Delta\phi(t) = 2\pi n_2 I(t) L/\lambda$ is shown in Fig. 27a for a Gaussian pulse in a nonlinear medium. The instantaneous frequency shift $\partial(\Delta\phi)/\partial t$ is also shown in Fig. 27b. It is seen that at the leading edge of the pulse the instantaneous frequency is lowered, since $\partial(\Delta\phi)/\partial t > 0$ in that region. Conversely, the trailing edge is elevated in frequency. This variation in frequency across the pulse is known as a chirp and plays an important role in the pulse compression schemes to be discussed later. Because n_2 in fibers is positive, the chirp induced by self-phase modulation is also positive (i.e., the frequency increases from the front to the rear of the pulse). Two points are worthy of note concerning the instantaneous frequency. The first is that the maximum frequency shifts originate from the inflection points of the pulse temporal profile. Secondly, it is possible to find pairs of points in time that have equal instantaneous frequencies. In the Fourier integral that yields the frequency spectrum, the spectral amplitude at a particular frequency will therefore contain the contributions from two different points in time. These contributions, being complex quantities, may add up in phase or out of phase. If the phase difference exceeds π, cancellations will occur leading to an oscillatory frequency spectrum. The maximum phase shift is related to the total number of peaks N in the oscillatory spectrum through [189]

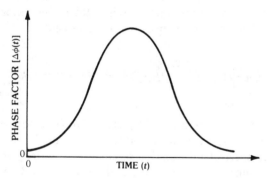

(a) Intensity-dependent phase factor for a Gaussian pulse in a nonlinear medium.

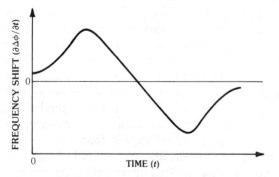

(b) The instantaneous frequency shift.

FIG. 27. Phase shift and instantaneous frequency of a Gaussian pulse undergoing self-phase modulation.

$$\Delta\phi_{max} = \tfrac{1}{2}\pi(2N-1). \tag{61}$$

This relation follows from an approximate evaluation of the Fourier integral in (60) by the method of stationary phase. Knowing the maximum phase shift, one can estimate the amount of frequency broadening by using the relation [190]

$$\delta\omega = 0.86\,\Delta\omega\,\Delta\phi_{max}, \tag{62}$$

where $\Delta\omega$ is the spectral width (FW $1/e$) of the input pulse. This is an excellent approximation if $\Delta\phi_{max} \gg 1$. For intermediate values of $\Delta\phi_{max}$ a small correction to this expression may be necessary [191].

Calculated frequency spectra of a Gaussian pulse undergoing self-phase modulation are shown in Fig. 28 for several different peak powers [192]. In this calculation it is assumed that the input pulses are transform limited ($\phi_0 = 0$) and symmetric.

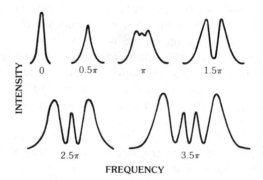

FIG. 28. Calculated frequency spectra for a Gaussian pulse undergoing self-phase modulation. The spectra are labeled by maximum phase shift at the peak of the pulse. (*After Stolen and Lin [192]*)

The definitive experiment on pure self-phase modulation in single-mode fibers is the work of Stolen and Lin [192]. Pulses of width 180 ps (FW $1/e$), from a mode-locked argon-ion laser, were propagated down a 99-m–long fiber with a core diameter of 3.35 μm and a loss at 514.5 nm of 17 dB/km. For a range of input power levels the spectra of the pulses at the input and output of the fiber were measured with scanning Fabry-Perot interferometers. The input pulse shape and output spectra are shown in Fig. 29. For a peak

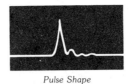

Pulse Shape

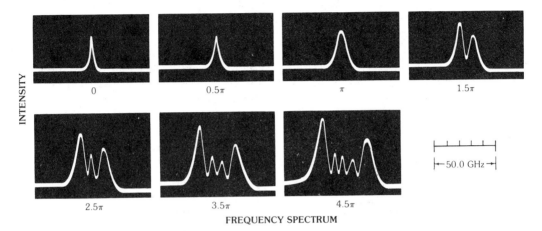

FREQUENCY SPECTRUM

FIG. 29. Input pulse shape and the output spectrum from a 3.35-μm silica-core fiber. The spectra are labeled by the maximum phase shift, which is proportional to peak power. (*After Stolen and Lin [192]*)

power of 3 W (average power 80 mW), the frequency bandwidth was increased a factor of ten as a result of self-phase modulation. The peak phase shift at the center of the pulse relative to the edges was as large as $4.5\,\pi$. Good agreement was found between the experimentally measured spectra and those calculated from a Fourier transform of the pulse amplitude.

One noteworthy feature of the Stolen and Lin experiment is the fact that the output pulse showed no temporal broadening. For a pure phase modulation a temporal broadening is obtained if group-velocity dispersion plays an important role. A measure of the importance of group-velocity dispersion is a critical length [193]

$$Z_c = c\,T_p/[D(\lambda)\,\delta n_{max}]^{1/2}\,, \qquad (63)$$

at which the pulse width approximately doubles. Here $D(\lambda) = \lambda^2\,d^2n/d\lambda^2$ is the dispersion, δn_{max} is the maximum index change, and $c\,T_p$ is the spatial extent of the pulse. Using the measured pulse and fiber parameters Z_c is found to be 1.5 km for these experimental conditions. Negligible pulse broadening is therefore expected for a fiber of

only 99 m. The self-phase modulation experiment also yields a value for n_2 of 1.14×10^{-13} esu.

Finally it should be noted that the observed spectra in Fig. 29 show an asymmetry not present in the simulations of Fig. 28. This asymmetry has been ascribed to a possible asymmetry in the input pulse shape [192, 194]. It could also originate from an initial nonlinear chirp on the output of the laser. The effect of a quadratic chirp can be modeled by using an initial time-dependent phase of the form

$$\phi_0 = \beta\,t^3/6\,. \qquad (64)$$

The output of many passively mode-locked lasers exhibits a nonlinear chirp [195]. Use of such pulses in self-phase modulation experiments will result in asymmetrically broadened spectra. In other self-phase modulation experiments it has been shown that pulse distortion due to stimulated Raman scattering can lead to asymmetric broadened spectra [31]. Other effects that can lead to asymmetric spectra include pulse self-steepening owing to a derivative nonlinear term usually neglected in the pulse propagation equations [196–198].

Dispersive Self-Phase Modulation—In a single-mode optical fiber the refractive index of the core material depends weakly on the light wavelength. This dispersion in the refractive index leads to group-velocity dispersion in which Fourier components in different wavelength intervals travel with different group velocities. The group velocity dispersion (gvd) is characterized by the derivative of the group delay (inverse group velocity) with respect to wavelength:

$$D(\lambda)=d(1/v_g)/d\lambda=\lambda^2 d^2n/d\lambda^2 . \qquad (65)$$

The group-velocity dispersion may be positive ("normal") or negative ("anomalous") depending on fiber composition, waveguide geometry, and light wavelength. However, regardless of its sign, gvd alone always leads to the broadening of a transform-limited input pulse. In particular, a Gaussian pulse of duration τ picoseconds (FWHM) will broaden, after propagating a distance L kilometers in a fiber, to a new width [199]

$$\tau'=\tau [1+(1.47 L D \lambda^2/\tau^2)^2]^{1/2} , \qquad (66)$$

where D is the dispersion parameter in picoseconds per nanometer per kilometer (ps/nm/km) and λ is the vacuum wavelength in micrometers. This dispersion parameter is related to the dimensionless group-velocity dispersion $D(\lambda)$ through $D=D(\lambda)/\lambda c$. As an example, a 5-ps input pulse will double in width after propagating a distance of about 800 m if the value of the dispersion parameter is -16 ps/s /nm/km at a wavelength of 1.5 μm. In typical fibers the dispersion is positive for wavelengths shorter than 1.3 μm and becomes negative for longer wavelengths. Recent advances in fiber design, however, have led to a new class of dispersion-shifted fibers where the zero-dispersion point coincides with the minimum-loss wavelength of 1.55 μm. These dispersion-shifted fibers should play an important role in future nonlinear optics experiments.

In addition to broadening a transform-limited pulse, group-velocity dispersion also imposes a linear chirp on the pulse since the slower frequency components end up at the rear end of the pulse. If one writes the instantaneous frequency of the pulse as

$$\omega=\omega_0+\beta t , \qquad (67)$$

then the chirp parameter is given by

$$\beta = D L/[|\tau^4+(D L)^2 |]. \qquad (68)$$

For fibers that are long enough for propagation effects to be important, the interplay between group-velocity dispersion and self-phase modulation must be considered. The coupling between these two effects can lead to substantial changes in both the shape and spectrum of an intense input pulse.

Nonlinear pulse propagation in a dispersive medium is quite accurately described by the nonlinear Schroedinger equation [200]:

$$i \partial u/\partial z+|u|^2 u=\pm\partial^2 u/\partial t^2 , \qquad (69)$$

where u is the dimensionless electric-field envelope of the pulse. The term on the right-hand side accounts for the effects of gvd, and is positive for normal dispersion and negative otherwise. For positive gvd this equation predicts large nonlinear spectral and temporal broadening of short pulses in long lengths of fiber. Numerical studies of the interplay between self-phase modulation and group-velocity dispersion have been carried out by Fisher and Bischel [193, 201]. They used a physically motivated algorithm in which the sample is broken up into a large number of segments which are alternately purely nonlinear or purely dispersive. The technique is also known as the split-step Fourier method. In propagating through a nonlinear segment of length dz the electric field acquires a phase shift $\delta\phi=\delta n \, \omega \, dz/c$, where, for a medium with instantaneous response, $\delta n=\frac{1}{2} n_2|E|^2$. The phase-shifted pulse from the nonlinear segment is then Fourier transformed, multiplied by the transfer function of the dispersive segment, and then transformed back to the time domain. If the segments are chosen thin enough, this scheme is highly stable and efficient.

The calculations of Fisher and Bischel [193, 201] and of Grischkowsky and Balant [202] show how the interaction between self-phase modulation (spm) and group-velocity dispersion (gvd) modifies the temporal shape and frequency spectrum of an intense pulse in a nonlinear medium.

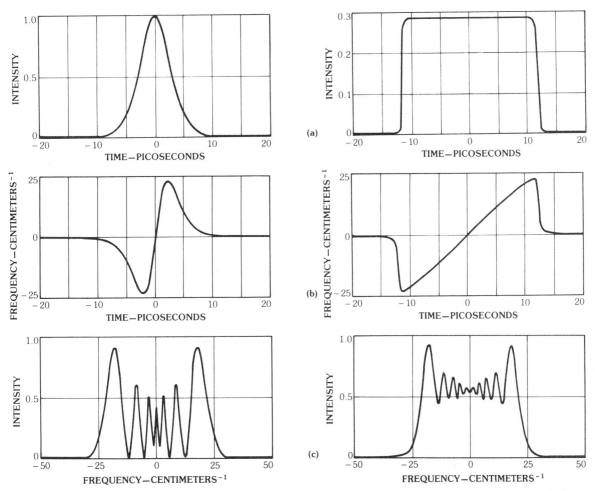

FIG. 30. Comparison of the transmitted pulse shapes (a), frequency chirp (b), and frequency spectra (c), in pure self-phase modulation (left) and in dispersive self-phase modulation (right). (*After Grischkowsky and Balant [202]*)

In the absence of dispersion, self-phase modulation by itself imparts a frequency chirp and increases the spectral width of a pulse. The temporal shape and width do not change. In the absence of nonlinearity, dispersion alone causes a Gaussian input pulse to spread temporally while maintaining its Gaussian shape. In the presence of both spm and positive gvd drastic pulse shaping occurs. An initially Gaussian pulse emerges with a flattened top, steepened sides, and greatly enhanced width. The reason for this pulse stretching is that the chirp due to spm depresses frequencies near the front and elevates frequencies near the rear of the pulse. With positive gvd the front will travel faster than the rear, thus stretching out the pulse. The spectral broadening due to spm

permits gvd to further disperse these frequencies, thus causing enhanced temporal broadening. Fig. 30 shows the pulse shape, frequency chirp, and frequency spectrum of a pulse that undergoes self-phase modulation with and without gvd [202]. Note that in the absence of dispersion the frequency chirp due to spm is linear only over the central part of the pulse. The presence of gvd greatly expands the region over which the chirp is linear [202, 193].

Experiments on the interplay between spm and positive gvd in a fiber were first done by Nakasuka and coworkers [203]. They observed the temporal and spectral broadening of picosecond input pulses and confirmed that the output pulses had a positive chirp. Nelson's team [204] has observed the broadening of 150-ps pulses from a Nd:YAG laser to as much as 2.5 ns in a 20-km length of single-mode fiber. At low power levels (less than 400 mW) the output pulse width was roughly the same as that of the input. At higher power levels dramatic pulse shaping and broadening was observed. At an input of 40 W a 16-fold broadening was obtained.

Another result of the interplay between spm and gvd is a phenomenon that has been called optical wave breaking by Tomlinson and colleagues [205] in analogy to the breaking of water waves. This effect manifests itself in the creation of side lobes in the spm spectra. Spectra showing this effect have been observed by Johnson's group [206] and by Gomes' group [207]. In the temporal domain, optical wave breaking results in fine structure in the pulse tails [205, 208]. No experiments have yet been reported which have the requisite temporal resolution to observe this fine structure.

Pulse Compression

The presence of group-velocity dispersion, regardless of its sign, will always lead to the broadening of a transform-limited pulse. If the pulse is chirped, however, then pulse compression can occur wherever the dispersion has the right sign to reverse the chirp. This idea is the basis of chirp radar technology and has now been used in the optical domain to obtain pulses as short as 8-fs. In a typical pulse compression experiment the pulse to be compressed is propagated down a length of optical fiber. The emergent pulse will have a positive frequency sweep, or chirp, impressed on it as a result of self-phase modulation. The process also leads to a spectral broadening of the pulse. The chirped and spectrally broadened pulse is then propagated along a delay line that has negative group-velocity dispersion. On such a delay line the high-frequency components, which are in the rear of the chirped pulse, travel faster than the low-frequency components, which are located near the front. The rear of the pulse thus catches up with the front, and pulse compression occurs. The initial suggestion that pulses could be compressed through self-phase modulation in a Kerr medium and the subsequent action of a delay line was by Fisher, Kelley, and Gustafson [209]. Because of various transverse effects that occur in bulk media, this compression technique did not become practical until it was demonstrated with an optical fiber several years later.

The first experimental observation of pulse compression in an optical fiber and external delay line combination was by Nakatsuka, Grischkowsky, and Balant [201]. Input pulses of duration 5.5-ps (FWHM) and peak power 10 W from a mode-locked dye laser operating at 587.4 nm were propagated through a 70-m single-mode fiber. The emergent pulses were temporally and spectrally broadened and linearly chirped as a result of dispersive self-phase modulation within the fiber. By passing these output pulses through a delay line of atomic sodium vapor, compression to a final pulse width of 1.5-ps was obtained. These 1.5-ps pulses were then used to probe the output pulses of the fiber in a cross-correlation scheme. The measurements showed impressive agreement with the predictions of the nonlinear Schroedinger equation. To achieve optimum compression it is important that the entire input pulse be subjected to a linear frequency chirp. Self-phase modulation alone leads to a positive linear chirp over only the central portion of the pulse. Grischkowsky and Balant [202] have shown through numerical simulations that inclusion of a positive group-velocity dispersion leads to an enhanced frequency chirp that is positive and linear over essentially the entire pulse width. In previous simulations on pulse compression in CS_2 and in glass, Fisher and Bischel [193] also noted that positive gvd assists in the preparation of chirped pulses for subsequent grating compression. Some analytic results for the optimum chirp have been obtained by Meinel [210].

The next advance in optical-fiber pulse compression was by Shank, Fork, Yen, Stolen, and Tomlinson [211], who extended this technique to the femtosecond (10^{-15} s) time domain. Short pulses of 90-fs duration from a colliding pulse mode-locked dye laser operating at 619 nm were amplified by a four-stage dye amplifier. A few nanojoules of the amplified pulse energy was then coupled into a 15-cm–long polarization-preserving optical fiber. The choice of 15 cm was dictated by a balance between self-phase modulation and linear dispersion. The spectrally broadened output pulses were then compressed to 30-fs by using an external delay line in the form of a grating pair. Treacy [212] has shown that a grating pair yields negative group-velocity dispersion, and it is now the most widely used delay line for pulse compression purposes.

Improvements in the basic optical-fiber grating-pair pulse compression technique have led to the achievement of very large compression ratios and the generation of the shortest optical pulses yet measured. Nikolaus and Grischkowsky [213] obtained a factor of 12 compression of the 5.4-ps, 1-kW pulses from a mode-locked dye laser. The compressed pulses could be tuned across the wavelength region of 570 to 600 nm by varying the dye laser wavelength and making slight angular adjustments to the grating pair. In that experiment the compression factor exceeded the loss factor of the pulse compressor, and thus the peak power of the output pulses (3 kW) was higher than that of the initial pulses. This raises the possibility of further compression of those output pulses by passage through a second compressor. Indeed, Nikolaus and Grischkowsky [214] have obtained a factor of 65 compression by using such a two-stage scheme. Starting with the 2-kW, 5.9-ps pulses from a commercial dye laser, they obtained 20-kW, 200-fs pulses from the first compressor and 10-kW, 90-fs pulses from the second compressor. Insertion of an optical amplifier

after the first compressor has made it possible to obtain even greater compression from the second stage [215]. By far the largest single-stage compression factor of 80 has been achieved by Johnson, Stolen, and Simpson [216, 217]. Their experimental arrangement is shown in Fig. 31. The 0.532-μm, 33-ps pulses from a frequency doubled Nd:YAG laser were compressed to 0.41-ps in a single stage, while their peak power was increased by more than a factor of five. Shown in Fig. 32a are the spectral profiles of the input pulse and of the spectrally broadened output pulse after propagating down the optical fiber. The autocorrelation functions of the input and output pulses are shown in Fig. 32b. A key factor in choosing the optimum fiber lengths and grating separations for pulse compression has been the numerical calculations of Tomlinson, Stolen, and Shank [218].

In the march toward shorter and shorter pulses, Fujimoto, Weiner, and Ippen [219] followed on the heels of Shank's team [211] by compressing 70-fs pulses to 16-fs using a short (8-mm) length of fiber. Halbout and Grischkowsky [220] subsequently obtained 12-fs pulses by compressing 110-fs pulses in a 15-mm–long fiber. More recently, Knox, Fork, Downer, Stolen, Shank, and Valdmanis [221] have compressed 40-fs pulses to 8-fs duration at a 5-kHz repetition rate. These pulses, whose center wavelength is 620 nm,

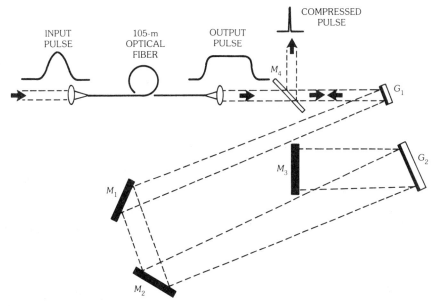

FIG. 31. An optical pulse compressor, with the dispersive delay line consisting of gratings G_1 and G_2 and mirrors M_1, M_2, and M_3. The compressed pulse is deflected out by mirror M_4.

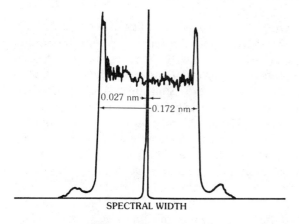

SPECTRAL WIDTH

(a) Spectral width of the 235-W, 0.532-μm input pulses and the spectrally broadened output pulses after propagation down a 93.5-m length of the 4.1-μm core-diameter polarization-preserving fiber. The side lobes on the broadened spectrum are due to the wave-breaking phenomenon. (After Johnson and Simpson [206])

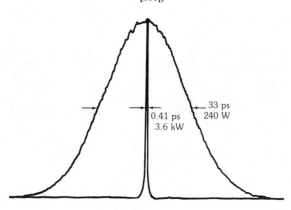

(b) The autocorrelation function of the input and compressed pulses displayed on the same scale. This represents a compression of 80× in a single stage. (After Johnson, Stolen, and Simpson [217])

FIG. 32. Spectral and temporal profiles of the input and output pulses in fiber/grating pulse compression.

contain only four optical periods! The spectral width of the compressed pulses exceed 70 nm. Here too the frequency chirp and spectral broadening were obtained in a fiber as short as 7 mm. One can now envision the use of other nonlinear materials, such as the polydiacetylenes, with large, fast-responding nonlinearities to obtain even greater self-phase modulation at lower input intensities.

Pulse compression techniques have now been

extended to the near-infrared region of the spectrum and to relatively long pulses (on the order of 100-ps) [222–230]. The low amounts of gvd in this wavelength region makes it difficult to operate in the dispersive self-phase modulation regime unless very long lengths of fiber are used. The absence of significant amounts of gvd means that the compressed pulses emerge with significant amounts of energy in the uncompressed wings of the pulse. These wings originate from the high- and low-frequency ends of the spm pulse spectrum which are not linearly chirped. Heritage and colleagues [230–232] have used a technique known as spectral windowing to remove these frequency components and thereby obtain compressed pulses with reduced wings. The best compression ratios obtained to date in this spectral region are on the order of 50 at 1.06 μm [225, 227] and at 1.319 μm [223]. In a two-stage compressor, a 450× compression of mode-locked Nd:YAG pulses at 1.06 μm from 90-ps to 200-fs has been reported [228].

The fiber grating-pair pulse compression technique is now so well developed that commercial pulse compressors are now available (e.g., Spectra-Physics Models 3610, 3690). This is probably the first use of nonlinear optical effects in fibers in an off-the-shelf, commercially available item. The short pulses obtained through these compression techniques have already proved invaluable in probing fast molecular processes, phase transitions, and other rapid events of scientific and technological importance.

Cross-Phase Modulation

When two or more waves propagate in a nonlinear medium, an amplitude modulation of one wave will result in a phase modulation of the others through the same mechanism that leads to self-phase modulation [233]. Consider the interaction between two waves $E_1 = A_1(t) \exp [i(k_1 z - \omega_1 t)]$ and $E_2 = A_2(t) \exp [i(k_2 z - \omega_2 t)]$ through the nonlinear index n_2. In the slowly varying envelope approximation the interaction between the waves is described by

$$\partial A_1 / \partial \tau = i(\pi n_2 / \lambda)(|A_1|^2 + 2|A_2|^2)A_1, \quad \text{(70a)}$$

$$\partial A_2 / \partial \tau = i(\pi n_2 / \lambda)(|A_2|^2 + 2|A_1|^2)A_2, \quad \text{(70b)}$$

where $\tau = t - z/c$, a retarded time. The first term on the right-hand side of each of these equations describes self-phase modulation, while the second term describes cross-phase modulation. It is seen that a given wave is twice as effective at modulating the phase of another wave as it is at modulating its own. The phase modulation of A_2 due to the presence of A_1 is given by

$$\phi(t) = 2\pi n_2 |A_1|^2 L/\lambda . \tag{71}$$

The effect of cross-phase modulation can be large in wavelength-division multiplexed systems, where several optical channels propagate simultaneously within the same fiber [234]. If an angle modulation scheme is used, the digital message is impressed on the phase angle of the optical carrier while the amplitude is kept constant. Any fluctuations in the phase will therefore seriously affect system performance. Since cross-phase modulation converts power fluctuations in one channel to phase fluctuations in another, stringent requirements are placed on the amplitude stability of the carrier. Chraplyvy and coworkers [234] calculate that in wdm systems containing as few as four channels an rms power fluctuation of 1 mW leads to phase noise in excess of 0.15 rad and hence, a power penalty greater than 0.5 dB. For amplitude-modulated systems the phase noise is much larger due to the intentionally large amplitude changes. This phase noise will have important consequences if a phase-sensitive detection scheme (coherent detection) is used.

Measurements by Chraplyvy and Stone [235] have shown the importance of cross-phase modulation at diode laser power levels. Light from two InGaAsP lasers at wavelengths of 1.3 μm and 1.54 μm was multiplexed into a 15-km–long single-mode fiber. By using a novel interferometric technique, Chraplyvy and Stone found that a 1-mW power change in one channel was sufficient to produce a 1.4° phase shift in the other channel.

Soliton Propagation

One of the factors that limits the bandwidth of an optical communication system is the pulse distortion that arises from dispersive spreading. The effect of dispersion can be reduced by operating at the minimum-dispersion wavelength (approximately 1.3 μm). However, such an approach lacks flexibility since it requires precise tuning to a specific wavelength. Furthermore, higher-order dispersion terms become important at that wavelength and these can lead to pulse distortion especially when the effect of a nonlinear index is included. Several years ago, Hasegawa and Tappert [236] proposed an alternative approach based on the fiber's intensity-dependent refractive index and the negative group-velocity dispersion that occurs at wavelengths longer than 1.3 μm. They showed that the effect of nonlinearity can be made to balance the effect of dispersion, thus leading to the formation of solitons: pulses that maintain their shapes, or whose shapes evolve periodically with propagation distance. Solitons also have the property that they can collide with each other and emerge unscathed except for a possible phase shift. In this section we discuss some of the properties of solitons, their experimental observation, and their applications in optical communications and in novel laser systems. A highly readable account of soliton effects in fibers has been given by Mollenauer and Stolen [237].

Solitons in Single-Mode Fibers—The envelope $A(z,t)$ of the electric field $E(z,t) = A(z,t) \exp [i(k_0 z - \omega_0 t)]$ in an optical fiber satisfies the wave equation

$$i(\partial A/\partial z + k_1 \partial A/\partial t) = -\tfrac{1}{2} k_2 \partial^2 A/\partial t^2 + \kappa |A|^2 A , \tag{72}$$

where $k_1 = \partial k/\partial \omega$, $k_2 = \partial^2 k/\partial \omega^2$, and $\kappa = \tfrac{1}{2} k_0 n_2 n_0$. By making the transformation [238]

$$s = |t - k_1 z|/t_0 , \tag{73a}$$

$$\zeta = |k_2| z/t_0^2 , \tag{73b}$$

$$u = t_0 (\kappa/|k_2|)^{1/2} A , \tag{73c}$$

where t_0 is an arbitrary parameter usually taken as the pulse width, (72) can be converted to the dimensionless form:

$$-i \partial A/\partial \zeta = \tfrac{1}{2} \partial^2 A/\partial s^2 + |A|^2 A , \tag{74}$$

which is known as the nonlinear Schroedinger equation. The first term on the right-hand side describes the effect of dispersion, while the second term describes the effect of nonlinearity. In the absence of these two terms the left-hand term alone would describe the distortionless propagation of the pulse envelope. Thus one can imagine that when the non-linearity balances the dispersion, similar distortion-less propagation might occur. In fact, it can be shown that in the absence of loss, a pulse whose form at the input is

$$V(A=0,t)=N \text{sech } (t/t_0) \qquad (75)$$

will maintain its shape and width if $N=1$. This is known as the fundamental soliton. Other integer values of N lead to higher-order solitons whose shapes evolve periodically with propagation distance. The period of the spatial evolution is given by [238]

$$Z_0=\pi \tau^2/2 |k_2|=0.322 \pi^2 c^2 t^2/D(\lambda) \lambda_0 . \qquad (76)$$

Here τ is the pulse width (FWHM of intensity) in seconds, $D(\lambda)=\lambda^2 d^2n/d\lambda^2$ is the dimensionless group-velocity dispersion and λ is the vacuum wavelength in centimeters. The pulse width τ is related to the parameter t_0 through

$$t_0=0.568 \tau .$$

From (73c) one can determine the critical power for the fundamental soliton as

$$P_1=n c \lambda_0 A_{\text{eff}}/16 \pi Z_0 n_2 . \qquad (77)$$

For a fiber of effective core area 1.0×10^{-6} cm^2, $n_2=1.1 \times 10^{13}$ esu, $\lambda_0=1.55$ μm, the critical power is $P_1=1.0$ W.

Fig. 33 shows the evolution of the first-, second-, and third-order solitons as numerically calculated from the nonlinear Schroedinger equation. The power required for the Nth-order soliton is N^2 times the power for the fundamental soliton. For all the higher-order solitons the initial evolution is characterized by a narrowing of the pulse, with the point of minimum width occurring closer to the fiber input end as the input power is increased. Subse-

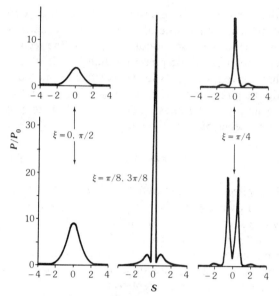

FIG. 33. Computer-generated solutions to the nonlinear Schroedinger equation. Above, the $N=2$ soliton; below, the $N=3$ soliton. (*After Mollenauer, Stolen, and Gordon [238]*)

quent evolution involves complicated behavior that includes periodic pulse splitting, broadening, and narrowing. The pulse spectrum, of course, also evolves periodically.

The formation of solitons is a relatively forgiving process. The input pulse need not be an exact soliton. It is essential, however, that the pulse have no excess bandwidth, as is the case with chirped input pulses. If one defines a pulse area as $R=\int A(t) dt$, then any reasonably shaped pulse whose area falls within the limits $R_0/2<R<3 R_0/2$ will evolve into a fundamental soliton whose area is R_0. The nonsoliton parts of the pulse are stripped off and propagate in the form of dispersive waves. The soliton is also remarkably robust and stable to perturbations such as loss.

All the features of solitons described above have been confirmed experimentally. Mollenauer, Stolen, and Gordon [238] first reported the observation of soliton behavior in single mode fibers. Pulses of 7-ps duration from a mode-locked color-center laser were propagated down a 700-m–long single-mode optical fiber. This length of fiber is roughly one-half the soliton period for the and large negative dispersion ($D=-16$ ps/nm/

km). At low input power the transmitted pulses were seen to broaden as a result of group-velocity dispersion. As the input power was increased, the output pulse narrowed continuously until at a power of 1.2 W it had narrowed to the input pulse width. Further increases in input power led to further narrowings and splittings consistent with soliton behavior. Fig. 34 shows the measured autocorrelation traces of the output pulse shapes for increasing values of pump power.

have demonstrated a laser that incorporates a length of optical fiber and, through soliton effects, produces clean, well-characterized picosecond pulses. The pulse shapes are accurately described by a sech^2 intensity function and the pulse widths are controllable by varying the fiber length. A schematic of the soliton laser is shown in Fig. 35. It consists of a mode-locked color-center laser synchronously pumped with 5 W of average power from a Nd:YAG laser operating at 1.06

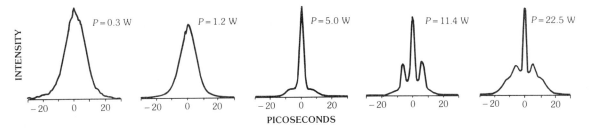

FIG. 34. Observation of solitons in an optical fiber. Autocorrelation traces of the fiber output as a function of power. (*After Mollenauer, Stolen, and Gordon [238]*)

The combination of nonlinearity and negative gvd can result in the production of extremely narrow pulses through soliton effects. The amount of achievable pulse narrowing increases with the order of the soliton. Mollenauer's group [177] has obtained a $27\times$ compression factor for an $N=13$ soliton in a 100-m–long fiber. Dianov and coworkers [239] have obtained a $100\times$ compression of a 30-ps pulse from a parametric oscillator tunable over the range 1.5 to 1.65 μm. The input pulse corresponds to an $N=100$ soliton in the 250-m–long fiber. Tai and Tomita [240] have combined fiber-grating pulse compression with higher-order soliton compression to achieve a total compression factor of $1100\times$. One problem with soliton pulse compression is that the resulting pulses are accompanied by a broad pedestal which contains a substantial fraction of the pulse energy. One method of suppressing this pedestal uses the intensity-dependent polarization state described in this section under "The Optical Kerr Effect: Self-Induced Polarization Changes." The restoration of the input pulse shape and spectrum at the soliton period has also been observed by Stolen and colleagues [241].

The Soliton Laser—Mollenauer and Stolen [242]

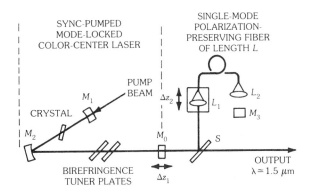

FIG. 35. Schematic of the soliton laser. Typical reflectivities: $M_0 \approx 70\%$, $S \approx 50\%$; M_1, M_2, $M_3 \approx 100\%$. (*After Mollenauer and Stolen [242]*)

μm. The output of the color-center laser is tunable between 1.4 μm and 1.6 μm by means of birefringent tuner plates. The mode-locked color-center laser by itself is capable of producing pulses no shorter than about 8-ps. A 50% beam splitter directs part of the color-center laser output through lens L_1, into a single-mode, polarization-preserving fiber of length L. A cat's-eye retroreflector comprising lens L_2 and mirror M_3 reflects the radiation back through the fiber and into the laser cavity.

The principle of operation of the soliton laser

is as follows. As the output of the color-center laser builds up from noise, the initially broad pulses are narrowed upon double passage through the fiber (by means of self-phase modulation and negative gvd). The narrowed pulses are reinjected into the laser cavity and force the laser to produce yet narrower pulses which undergo further reshaping by the fiber. The process continues until the pulses attain a stable steady-state shape and duration corresponding to solitons of period twice the length of the fiber. In steady state, pulses that exit the fiber have the same shape with which they entered. The pulses reinjected into the laser must be in phase with those already in the main cavity.

The soliton laser has yielded stable pulses as short as 210-fs. The pulse width scales as the square root of the fiber length. Experimentally, it is observed that the laser prefers to operate on the $N=2$ soliton. The reason for this behavior is not known. Both the fundamental and $N=2$ solitons have been shown theoretically to be stable in the soliton laser configuration [243, 244]. In their analyses Haus and Islam [243] use an equivalent single-cavity approach, while Blow and Wood use a model that considers the coupling between the color-center cavity and the optical-fiber cavity [244]. Although the latter approach appears more realistic neither theory completely explains the experimental results.

Solitons and the Stimulated Raman Effect—The interplay between soliton propagation and stimulated Raman scattering is a subject of great current interest. On the one hand there is interest in using Raman gain to amplify and reshape solitons that lose energy as a result of finite fiber losses. On the other hand there is the possibility that an intense pump pulse can generate solitons at a wavelength Stokes-shifted from that of the pump.

Consider first the effect of loss on a fundamental soliton. In the presence of loss the nonlinear Schroedinger equation can be written

$$-i\,\partial u/\partial z = \tfrac{1}{2}\,\partial^2 u/\partial t^2 + |u|^2\,u + i\,\Gamma u,\qquad (78)$$

where Γ is the loss factor of the pulse amplitude envelope. If $\Gamma z \ll 1$, perturbation theory shows

that the envelope of the $N=1$ soliton is given by [245]

$$u(z,t) = \exp\{i[1-\exp(-4\,\Gamma z)]/4\,\Gamma\}$$

$$\times \operatorname{sech}(t\,e^{2\Gamma z}).\qquad (79)$$

This is a pulse whose width grows exponentially at a rate $e^{2\Gamma z}$ while its amplitude decreases exponentially at a rate $e^{-2\Gamma z}$. The pulse area (amplitude×width) thus remains constant, and the soliton maintains its shape in the presence of finite (but small) loss. The pulse spreading due to fiber loss results in a reduction of the achievable bit rate in a soliton-based communication system. Since the soliton retains its shape, Hasegawa and Kodama have suggested, in a series of papers, that some form of gain be introduced in the fiber transmission to amplify the attenuated pulses and restore the pulse widths [246–250]. The most promising amplification and reshaping scheme is one that uses the intrinsic Raman gain of the optical-fiber medium itself [249]. Here pump light, at a shorter wavelength than that of the soliton signal, is injected periodically at a spacing determined by fiber loss and pump depletion. The soliton is reshaped and amplified adiabatically and the generation of dispersive waves (the nonsoliton component) is minimal. Hasegawa [250] has numerically studied the stability of such an amplification scheme for long transmission lengths which require many Raman amplification steps. With careful choice of parameters, stable transmission of a 10-Gb/s soliton train with peak amplitude of 30 mW is feasible for a distance of approximately 5000 km when pump power is injected every 34.4 km. The pump power is injected in the form of two 40-mW beams propagating both co- and counter-directionally with the soliton train. For different parameter values chaotic behavior of the propagating solitons is possible. More extensive design calculations by Mollenauer, Gordon, and Islam [251] show that this scheme has a number of advantages over one that uses conventional repeaters.

An experimental verification of the use of Raman gain to amplify and reshape a fundamental soliton has been reported by Mollenauer, Stolen, and Islam [252]. In that work, Raman gain supplied by a cw color-center laser at 1.46 μm was used to achieve distortionless propagation of 10-ps fundamental soliton pulses ($\lambda=1.56$ μm) over a 10-km length of fiber. The results are shown in Fig. 36. Note that in the absence of Raman gain the soliton broadened considerably.

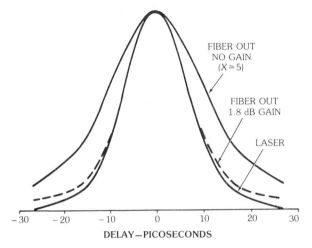

FIG. 36. Autocorrelation traces of pulses at the output end of a 10-km fiber, with and without gain, and of the laser output pulses. Note that the height of the no-gain curve has been considerably magnified to facilitate comparison of pulse widths. (*After Mollenauer, Stolen, and Islam [252]*)

The generation of solitons whose wavelength is Stokes-shifted from that of the pump is an exciting possibility [253]. The dynamics of the process involves both an "active" component and a "reactive" component. The active component transfers energy from the pump wave to the Stokes-shifted frequency through the usual Raman susceptibility. In the reactive component any frequency modulation or chirp acquired by the pump through self-phase modulation is transferred to the Stokes wave by means of cross-phase modulation [254]. If the Stokes wavelength lies in the region of negative group-velocity dispersion, self-contraction of the Stokes pulse results in the formation of Raman solitons. Thus the pump wavelength need not lie in the negative dispersion region for the generation of these solitons.

When the pump wavelength lies in the negative dispersion region, both pump and Stokes pulses can propagate as solitons. A pump pulse launched as a high-order soliton can be converted into a

fundamental soliton at a Stokes-shifted wavelength. Because only the most intense parts of the pump pulse contribute to the generation of the Stokes pulse, the Stokes soliton is created with much reduced wings. This conversion of a high-order soliton at the pump wavelength to an intense fundamental soliton at the Stokes wavelength has been observed by Dianov's team [255]. Dianov and coworkers [256] have also investigated the picosecond structure of the pump pulse in SRS. In the strong pump depletion regime the pump fragments end up with a chirp opposite to that of the Stokes pulse, which is generated by the central part of the pump pulse.

The combination of Raman generation and soliton propagation leads to some very interesting intracavity dynamics [257, 58, 53]. A fiber Raman amplification soliton laser (FRASL) has been demonstrated that produces Stokes pulses as short as 240-fs when pumped with 10-ps pulses from a color-center laser [258]. It is simply a synchronously pumped fiber Raman oscillator whose output Stokes wavelength lies in the negative dispersion region of the fiber. The Stokes pulses undergo self-compression as a result of chirp transfer from the pump, self-phase modulation, and the negative group velocity dispersion. Chraplyvy and Stone [53] were actually the first to demonstrate a FRASL, though they did not call it such. Their D_2 gas-in-glass fiber Raman laser operated at 1.56 μm and yielded 15-ps Stokes pulses from 100-ps pump pulses at 1.06 μm.

Modulational Instability—An intense continuous light wave in a single-mode fiber is unstable to amplitude and phase modulation if the light wavelength lies in the region of negative gvd and if the frequency of modulation is less than a certain critical value [259]. This modulational instability transfers energy from the pump wave to a set of sidebands symmetrically disposed about the pump frequency. The sidebands thus experience exponential gain and will grow to an amplitude limited by pump depletion. Use of this instability as a source of new frequencies in the infrared has been suggested by Hasegawa and Brinkman [260]. Similar instabilities have been studied in fluid dynamics [261] and in plasma physics [262]. The modulational instability can convert an initial

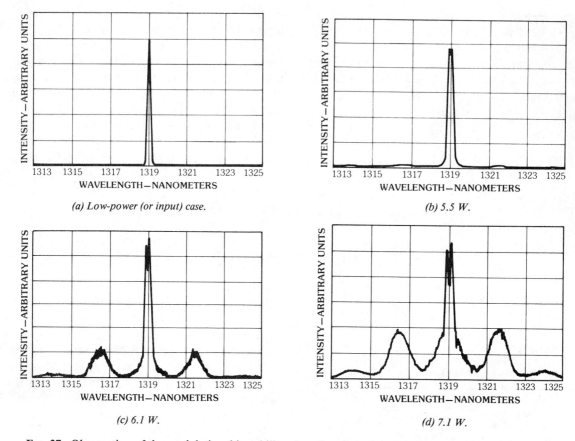

FIG. 37. Observation of the modulational instability. A series of power spectra measured at the output end as a function of the peak power. The modulation-frequency sidebands are clearly shown in (b), (c), and (d). (*After Tai et al. [266]*)

sinusoidal modulation on an input beam into a train of solitonlike pulses [263, 264]. The modulational instability has been observed by Tai and coworkers [265] (Fig. 37) and has been used to produce picosecond pulse trains at very high repetition rates [266] (Fig. 38). The possible effect of this instability on coherent communications systems has been considered [267–269].

Solitons in Optical Communications—An ideal, isolated fundamental soliton in a lossless optical fiber will propagate indefinitely without changing its shape or width. This distortionless transmission at first glance offers the possibility of high–bit-rate communication over unrepeatered spans in excess of thousands of kilometers. There are, however, several effects that limit the potential of soliton-based transmission systems.

Numerous theoretical papers have addressed the effects of loss, higher-order dispersion, soliton interactions, and intensity and phase fluctuations on the ultimate bandwidth of communication systems that use nonlinear effects to overcome linear dispersion [270–291].

In a linear transmission system the signal at the receiver can always be enhanced by increasing the launched power. The ultimate bandwidth of the system is limited by the pulse spreading that results from group-velocity dispersion. In a soliton-based system, however, this linear dispersion is canceled out and pulse spreading is a consequence of fiber loss. As already discussed, for a small enough loss rate, perturbation theory shows that the soliton retains its area while spreading. Use of distributed gain, such as that provided by Raman amplification can then restore the pulse height and width.

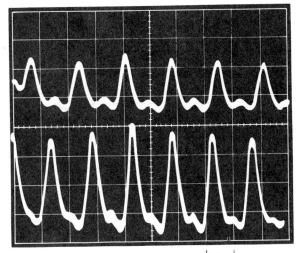

→|1.8 ps|←

FIG. 38. Generation of picosecond pulse trains by means of the induced modulational instability. The two traces correspond to two different initial modulation frequencies. (*After Tai et al. [266]*)

For long fibers the validity of the perturbative solution is questionable. Numerical solutions of the NLSE with loss by Blow and Doran [271–273] have revealed the true evolution of lossy solitons for large propagation distances. The exponential increase in pulse width is seen to hold only for a distance less than $z = 1/\Gamma$, where Γ is the loss rate. Beyond this distance the pulse spreads at a linear rate. This asymptotic spreading occurs at a rate slower than that of a low-intensity pulse launched at $z = 0$ with the same pulse width. The asymptotic spreading rate is determined by the width of the pulse at the point at which the intensity of the pulse has dropped to a level where nonlinear effects are unimportant. Another interesting result of the numerical calculations is that in the presence of loss the periodic oscillations with distance of higher-order solitons are strongly damped. The higher-order solitons evolve into a broad single hump which continues to spread but does not oscillate. Many features of these calculations have been reproduced by approximate analytical expressions obtained by Anderson (1983) and Anderson and Lisak [273, 274].

Another factor that reduces the ultimate bandwidth of nonlinear transmission systems is the mutual interaction between solitons that results from the intensity-dependent refractive index [277–283]. Depending on their relative phase, solitons that are spaced less than about ten pulsewidths apart will attract each other, merge, and then separate as they propagate. The interaction between solitons is much reduced if there is a phase shift of 180° between them.

In their original experiments on soliton propagation, Mollenauer, Stolen, and Gordon (1980) stressed the need for input pulses that are as nearly as possible transform-limited. Any excess bandwidth tends to counteract the process of soliton formation. Recent computer studies by Lassen's group (1985) have further elucidated the nature of the propagation of non-transform-limited pulses in nonlinear media. Solutions of the nonlinear Schroedinger equation showed that the pulse broadening resulting from chirp greatly exceeds the broadening due to fiber loss. Positively chirped pulses undergo an initial compression after which they broaden at a rate close to that of chirped pulses in linear media. Negatively chirped pulses only experience enhanced broadening. Since the output of directly modulated semiconductor lasers exhibits a frequency chirp, the enhanced broadening due to this chirp should be considered in designing soliton optical communication systems.

A recent design study by Mollenauer and colleagues [251] takes into account the effects of fiber loss and soliton interaction on a soliton-based optical communication system. It concludes that the achievable bit rate of such a system is at least an order of magnitude greater than that contemplated for the most advanced conventional system. Studies such as these, coupled with experiments, will be necessary to fully assess the potential utility of solitons in optical communications.

5. Photosensitivity in Fibers

In the last section we examined effects such as self-phase modulation and soliton propagation which have their origin in a nonlinear index that responds instantaneously to changes in light intensity. We now consider a rather poorly understood effect which depends on the time integrated laser power, responds in minutes, and leads to permanent changes in the refractive index of germania-doped silica fibers. The effect was first reported by Hill, Fujii, Johnson and Kawasaki [296] and is of great interest because it offers the possibility of fabricating tunable, high-quality optical-waveguide filters with low scattering losses and high frequency selectivity. Narrow-band waveguide filters can play an important role in separating the individual channels in high-capacity wavelength-division–multiplexed (wdm) systems. In what follows we discuss the fabrication of such filters and some of their proposed applications.

Fabrication of Optical-Fiber Filters

The microscopic origin of the photosensitivity that leads to grating formation in fibers is not known. It does appear, however, that germania (GeO_2) is a necessary ingredient in fibers that exhibit photosensitivity. The observed photosensitivity is greater the larger the germanium dopant concentration. (Typical germanium concentration in fibers is about 4%.) In fact, gratings have been created with high efficiency in thin-film waveguides of pure GeO_2 [297]. The photosensitivity is also higher the shorter the wavelength of the exposing laser. For example, exposure times are an order of magnitude longer using red light (647.1 nm) compared to exposure times for light in the blue-green region.

A schematic of the fiber-filter fabrication apparatus is shown in Fig. 39 [298]. The exposing laser is a cw single–longitudinal-mode argon-ion laser operating typically on the 514.5-nm or 488.0-nm lines. To prevent back-reflected light from destabilizing the laser an attenuator and beam-splitter combination or a Faraday isolator is used. The launched power in the fiber is around 50 mW. The grating formation process is initiated by the interference pattern set up within the fiber by the primary beam and the Fresnel-reflected beam from the output end of the fiber. This interference pattern exposes the fiber core and creates a periodic perturbation of the refractive index. The grating thus created reflects more of the incident light, which leads to a deeper modulation of the interference pattern and in turn a stronger periodic perturbation of the refractive index and a higher reflection coefficient. The process builds on itself. The reflectivity of the fiber can grow

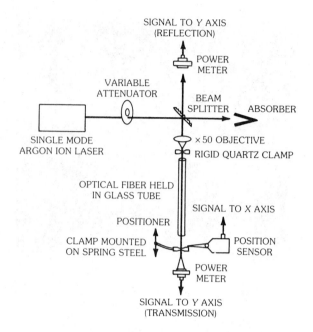

FIG. 39. Schematic of apparatus used for recording reflection filters in optical fibers. (*After Kawasaki et al. [298]*)

from the initial level of about 4% to more than 60% in the course of a few minutes. Fig. 40 shows the growth of the reflection coefficient as a function of time in a typical filter fabrication process. As in holography, the stability requirements are stringent. Shielding of the fiber from thermal effects and minimization of tension changes is important since these affect the filter response. The length of the filter is limited primarily by the coherence length of the exposing laser radiation. A typical coherence length for an argon-ion laser with an intracavity etalon is 10 m, while an actively stabilized, single-mode ring dye laser can yield coherence lengths more than 100 m. Kawasaki and team [298] have reported making fiber filters whose lengths vary from 1 cm to more than 1 m. Other aspects of the fabrication and characterization of these fiber filters can be found in [299–305].

The phase gratings created by the process described above function as narrow-band Bragg reflection filters. The properties of such filters are conveniently described by the coupled-mode theory of Kogelnik. From that theory the bandwidth of a Bragg filter is given by

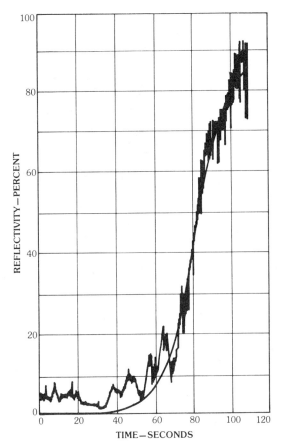

FIG. 40. Growth of the reflectivity as a function of time during fabrication of a fiber filter. The smooth curve is calculated using a dynamic coupled-wave theory. (*After Bures et al. [299]*)

$$\Delta v = (c/2\,n)\,[(\delta n/\lambda)^2 + (1/L)^2]^{1/2},$$

where L is the length of the filter, λ is the wavelength of the radiation used to form the grating, σn is the amplitude of the index perturbation, (typically 10^{-6}), n is the unperturbed refractive index, and c is the speed of light. Filter bandwidths as narrow as 306 MHz have been obtained in a 33.5-cm–long fiber. Most of the demonstrated fiber filters operate in the length-limited regime, where $\Delta v \simeq c/2\,nL$.

These optical fiber filters have been used as reflectors in an argon-ion laser [296]. Because of its wavelength-selective reflection properties, when such a filter replaces a regular mirror in a laser it forces the laser to operate in a single longitudinal mode. Other suggested applications make use of the group-velocity dispersion of these filters. The filter gvd varies from positive to negative in an oscillatory manner as one tunes the wavelength of the incident light in the vicinity of the stop band. Lam's group [306] has suggested that the negative dispersion characteristic of the filter can be used to equalize the material dispersion in an optical-fiber communication link. Finally, Winful [307] has proposed that the negative dispersion of the fiber filter can be used in conjunction with the Kerr nonlinearity of the fiber to achieve pulse compression without an external delay line. This scheme may also make possible the observation of soliton propagation at visible wavelengths where the unperturbed fiber has positive gvd.

6. Conclusion

In optical communications, the optical fiber is usually regarded as a completely passive, linear medium whose only effect on the light signal is to attenuate and disperse it by a small amount. In this chapter we have seen that the fiber transmission medium is inherently nonlinear and that many interesting and potentially useful phenomena arise from this nonlinearity. The small core diameters, long path lengths, and extremely low losses available with today's single-mode fibers have led to greatly reduced threshold power levels for nonlinear interactions in fibers. One can now envisage using an injection-locked diode laser array, operating in the fundamental Gaussian mode, to drive such effects as soliton propagation in fibers. The extremely short response time of the nonlinear index of silica fibers has made it possible to generate the shortest pulses ever. The single-mode property of the fiber makes it easy to perform clean, well-characterized nonlinear experiments that show impressive agreement with theoretical predictions. New fibers, such as the fluoride-glass fibers, have potential for providing new Raman lines and enhanced gain coefficients. Single-crystal fibers currently under development offer possibilities for efficient harmonic generation [308]. Silica fibers doped with specific gain media provide laser action when pumped by diode lasers [309]. The soliton laser, based on nonlinear interactions in fibers, is a source of clean, stable, well-characterized pulses.

While there are some deletrious effects that arise from a nonlinear transmitting medium, the writer believes that the useful effects far outweigh the undesirable ones. The undesirable effects can often be suppressed. For example, by modulating the phase of an input laser signal, one can suppress effects such as stimulated Brillouin scattering and the rotation error due to Kerr nonreciprocity in fiber gyroscopes. By choosing the proper axis for the polarization of an intense beam in a birefringent fiber, four-photon mixing can be suppressed. Conversely, one can enhance nonlinear interactions when their presence is desired. Nonlinear optical phenomena will play an increasing role in optical-fiber communications.

7. References

[1] P. A. FRANKEN, A. E. HILL, C. W. PETERS and G. WEINREICH, "Generation of Optical Harmonics," *Phys. Rev. Lett.,* Vol. 7, p. 118, 1961.

[2] E. P. IPPEN, "Nonlinear Effects in Optical Fibers," in *Laser Applications to Optics and Spectroscopy,* ed. by S. F. JACOBS, M. SARGENT III, J. F. SCOTT, and M. O. SCULLY, Eds. Reading: Addison-Wesley Publishing Co., 1975.

[3] R. H. STOLEN, "Active Fibers," in *New Directions in Guided Wave Optics,* Vol. 1, (NATO ASI series E: Applied Science N.78), ed. by D. B. OSTROWSKY and E. SPITZ, Boston: Martinus Nijhoff, 1984.

[4] R. H. STOLEN, "Nonlinearity in Fiber Transmission," *Proc. IEEE,* Vol. 68, p. 1232, 1980.

[5] R. H. STOLEN, "Fiber Raman Lasers," in *Fiber and Integrated Optics,* ed. by D. B. OSTROWSKY, New York: Plenum Press, 1979.

[6] R. H. STOLEN, "Nonlinear Properties of Optical Fibers," in *Optical Fiber Telecommunications,* ed. by S. E. MILLER and A. G. CHYNOWETH, New York: Academic Press, 1979.

[7] L. JEUNHOMME, *Single-Mode Fiber Optics,* New York: Marcel Dekker, 1983, Ch. 9.

[8] K. O. HILL, B. S. KAWASAKI, D. C. JOHNSON, and Y. FUJII, "Nonlinear Effects in Optical Fibers: Application to the Fabrication of Active and Passive Devices," in *Fiber Optics—Advances in Research and Development,* ed. by B. BENDOW and S. S. MITRA, New York: Plenum Press, 1979.

[9] W. R. HAYES and R. LOUDON, *Scattering of Light by Crystals,* New York: John Wiley & Sons, 1978, p. 25.

[10] W. KAISER and M. MAIER, "Stimulated Rayleigh, Brillouin and Raman Spectroscopy," in *Laser Handbook,* Vol. 2, ed. by F. T. ARECCHI and E. O. SCHULZ-DUBOIS, Amsterdam: North-Holland Publishing Co., 1972.

[11] N. BLOEMBERGEN, "The Stimulated Raman Effect," *Am. Jr. Phys.,* Vol. 35, p. 989, 1967.

[12] D. COTTER and A. M. HILL, "Stimulated Raman Crosstalk in Optical Transmission: Effects of Group Velocity Dispersion," *Electron. Lett.,* Vol. 20, p. 185, 1984.

[13] R. H. STOLEN, "Polarization Effects in Fiber Raman and Brillouin Lasers," *IEEE J. Quantum Electron.,* Vol. QE-15, p. 1157, 1979.

[14] R. H. ENNS and I. P. BATRA, "Saturation and Depletion in Stimulated Light Scattering," *Phys. Lett.,* Vol. 28A, p. 591, 1969.

[15] R. G. SMITH, "Optical Power Handling Capacity of Low Loss Optical Fibers as Determined by Stimulated Raman and Brillouin Scattering," *Appl. Opt.,* Vol. 11, p. 2489, 1972.

[16] J. AU YEUNG and A. YARIV, "Spontaneous and Stimulated Raman Scattering in Long Low Loss Fibers," *IEEE J. Quantum Electron.,* Vol. QE-14, p. 347, 1978.

[17] R. H. STOLEN and E. P. IPPEN, "Raman Gain in Glass Optical Waveguides," *Appl. Phys. Lett.,* Vol. 22, p. 276, 1973.

[18] A. YARIV, *Quantum Electronics,* 2nd ed., New York: John Wiley & Sons, 1975, p. 479.

[19] R. SHUKER and R. W. GAMMON, "Raman-Scattering Selection-Rule Breaking and the Density of States in Amorphous Materials," *Phys. Rev. Lett.,* Vol. 25, p. 222, 1970.

[20] P. E. TANNENWALD, "Mode Pulling in a Stimulated Raman Oscillator," *J. Appl. Phys.,* Vol. 38, p. 4788, 1967.

[21] F. L. GALEENER, J. C. MIKKELSEN, R. H. GEILS, and W. J. MOSBY, "The Relative Raman Cross Sections of Vitreous SiO_2, GeO_2, B_2O_3, and P_2O_5," *Appl. Phys. Lett.,* Vol. 32, p. 34, 1978.

[22] H. TAKAHASHI, J. CHANG, K. NAKAMURA, I. SUGIMOTO, T. TAKABAYASHI, A. OYOBE, and Y. FUJII, "Efficient Single-Pass Raman Generation in a GeO_2 Optical Fiber and its Application to Measurement of Chromatic Dispersion," *Opt. Lett.,* Vol. 11, p. 383, 1986.

[23] J. STONE, A. R. CHRAPLYVY, and C. A. BURRUS, "Gas-in-Glass—A New Raman-Gain Medium: Molecular Hydrogen in Solid-Silica Optical Fibers," *Opt. Lett.,* Vol. 7, p. 297, 1982.

[24] A. R. CHRAPLYVY, J. STONE, and C. A. BURRUS, "Optical Gain Exceeding 35 dB at 1.56 μm due to Stimulated Raman Scattering by Molecular D_2 in a Solid Silica Optical Fiber," *Opt. Lett.,* Vol. 8, p. 415, 1983.

[25] C. LIN, L. G. COHEN, R. H. STOLEN, G. W. TASKER, and W. G. FRENCH, "Near-Infrared Sources in the 1–1.3 μm Region by Efficient Stimulated Raman Emission in Glass Fibers," *Opt. Commun.,* Vol. 20, p. 426, 1977.

[26] L. G. COHEN and C. LIN, "A Universal Fiber-Optic (UFO) Measurement System Based on a Near-IR Fiber Raman Laser," *IEEE J. Quantum Electron.,* Vol. QE-14, p. 855, 1978.

[27] C. LIN and R. H. STOLEN, "New Nanosecond Continuum for Excited-State Spectroscopy," *Appl. Phys. Lett.,* Vol. 28, p. 216, 1976.

[28] Y. OHMORI, Y. SASAKI, and T. EDAHIRO, "Stimulated Raman Scattering in Optical Fibers," *Trans. IECE Japan,* Vol. E66, p. 146, 1983.

[29] R. H. STOLEN, C. LEE, and R. K. JAIN, "Development of the Stimulated Raman Spectrum in Single-Mode Silica Fibers," *J. Opt. Soc. Am. B,* Vol. 1, p. 652, 1984.

[30] R. H. STOLEN and A. M. JOHNSON, "Effect of Pulse Walkoff on Stimulated Raman Scattering in Fibers," in *Tech. Dig. Conf. on Lasers and Electro-Optics,* San Francisco, 1986, Paper THC3.

[31] A. M. WEINER, J. P. HERITAGE, and R. H. STOLEN, "Effect of Stimulated Raman Scattering and Pulse Walkoff on Self-Phase Modulation in Fibers," in *Tech. Dig. Conf. on Lasers and Electro-Optics,* San Francisco, 1986, Paper THC4.

[32] B. STOLZ, U. OSTERBERG, A. S. L. GOMES, W. SIBBET, and J. R. TAYLOR, "Streak Camera Investigation of Raman Pulse Generation and Propagation in an Optical Fiber," *IEEE J. Lightwave Tech.,* Vol. LT-4, p. 55, 1986.

[33] B. VALK, and H. P. WEBER, "Stimulated Raman Spectra Generated by ps-Pulses in a Single-Mode Fiber," *Opt. Commun.,* Vol. 54, p. 363, 1985.

[34] C. LIN, "Designing Optical Fibers for Frequency Conversion and Optical Amplification by Stimulated Raman Scattering and Phase-Matched Four-Photon Mixing," *J. Opt. Commun.,* Vol. 4, p. 2, 1983.

[35] A. R. CHRAPLYVY and J. STONE, "Single-Pass Mode-Locked or Q-Switched Pump Operation of D_2 Gas-in-Glass Fiber Raman Lasers Operating at 1.56-μm Wavelength," *Opt. Lett.,* Vol. 10, p. 344, 1985.

[36] C. LIN, R. H. STOLEN, and R. K. JAIN, "Group Velocity Matching in Optical Fibers," *Opt. Lett.,* Vol. 1, p. 205, 1977.

[37] Y. OHMORI, Y. SASAKI, M. KAWACHI, and T. EDAHIRO, "Single-Pass Raman Generation Pumped by a Mode-Locked Laser," *Electron. Lett.,* Vol. 17, p. 594, 1981.

[38] Y. SASAKI, Y. OHMORI, M. KAWACHI, and T. EDAHIRO, "CW Single-Pass Raman Generation in Optical Fibres," *Electron. Lett.,* Vol. 17, p. 315, 1981.

[39] L. G. COHEN and C. LIN, "Pulse Delay Measurements in the Zero Material Dispersion Wavelength Region for Optical Fibers," *Appl. Opt.,* Vol. 16, p. 3136, 1977.

[40] K. NOGUCHI, Y. MURAKAMI, K. YAMASHITA, and F. ASHIYA, "52 km-Long Single-Mode Optical Fibre Fault Location Using the Stimulated Raman Scattering Effect," *Electron. Lett.,* Vol. 18, p. 41, 1982.

[41] K. NOGUCHI, "A 100-km-Long Single-Mode Optical-Fiber Fault Location," *IEEE J. Lightwave Tech.,* Vol. LT-2, p. 1, 1984.

[42] R. H. STOLEN, E. P. IPPEN, and A. R. TYNES, "Raman Oscillation in Glass Optical Waveguide," *Appl. Phys. Lett.,* Vol. 20, p. 62, 1972.

[43] K. O. HILL, B. S. KAWASAKI, and D. C. JOHNSON, "Low-Threshold CW Raman Laser," *Appl. Phys. Lett.,* Vol. 29, p. 181, 1976.

[44] D. C. JOHNSON, K. O. HILL, B. S. KAWASAKI, and D. KATO, "Tunable Raman Fibre-Optic Laser," *Electron. Lett.,* Vol. 13, p. 53, 1977.

[45] R. K. JAIN, C. LIN, R. H. STOLEN, W. PLEIBEL, and P. KAISER, "A High-Efficiency Tunable CW Raman Oscillator," *Appl. Phys. Lett.,* Vol. 30, p. 162, 1977.

[46] R. K. JAIN, C. LIN, R. H. STOLEN, and A. ASHKIN, "A Tunable Multiple Stokes CW Fiber Raman Oscillator," *Appl. Phys. Lett.,* Vol. 31, p. 89, 1977.

[47] R. H. STOLEN, C. LIN, J. SHAH, and R. F. LEHENY, "A Fiber Raman Ring Laser," *IEEE J. Quantum Electron.,* Vol. 14, p. 860, 1978.

[48] C. LIN, R. H. STOLEN, and L. G. COHEN, "A Tunable 1.1 μm Fiber Raman Oscillator," *Appl. Phys. Lett.,* Vol. 31, p. 97, 1977.

[49] C. LIN, R. H. STOLEN, W. G. FRENCH, and T. G. MALONE, "A CW Tunable Near-Infrared (1.085-1.175 μm) Raman Oscillator," *Opt. Lett.,* Vol. 1, p. 96, 1977.

[50] C. LIN and W. G. FRENCH, "A Near-Infrared Fiber Raman Oscillator Tunable From 1.07 to 1.32 μm," *Appl. Phys. Lett.,* Vol. 34, p. 10, 1979.

[51] C. LIN and P. F. GLODIS, "Tunable Fibre Raman Oscillator in the 1.32–1.41 μm Spectral Region Using a Low-Loss, Low OH—Single-Mode Fibre," *Electron. Lett.,* Vol. 18, p. 696, 1982.

[52] M. NAKAZAWA, T. MASAMITSU, and N. ICHIDA, "Continuous-Wave Raman Oscillation for a Nd^{3+}:YAG Intracavity Fiber Laser," *J. Opt. Soc. Am. B,* Vol. 1, p. 86, 1984.

[53] A. R. CHRAPLYVY and J. STONE, "Synchronously Pumped D$_2$ Gas-in-Glass Fiber Raman Laser Operating at 1.56 μm," *Opt. Lett.,* Vol. 9, p. 241, 1984.

[54] J. D. KAFKA, D. F. HEAD, and T. BAER, "Dispersion Compensated Fiber Raman Oscillator," in *Tech. Dig. Conf. on Lasers and Electro-Optics,* San Francisco, 1986, Post-deadline Paper ThU6.

[55] R. H. STOLEN, C. LIN, and R. K. JAIN, "A Time-Dispersion-Tuned Fiber Raman Oscillator," *Appl. Phys. Lett.,* Vol. 30, p. 340, 1977.

[56] C. LIN, L. G. COHEN, W. G. FRENCH, and H. M. PRESBY, "Measuring Dispersion in Single-Mode Fibers in the 1.1–1.3 μm Spectral Region—A Pulse Synchronization Technique," *IEEE J. Quantum Electron.,* Vol. QE-16, p. 33, 1980.

[57] M. NAKAZAWA, "Synchronously Pumped Fiber Raman Gyroscope," *Opt. Lett.,* Vol. 10, p. 193, 1985.

[58] E. DESURVIRE, B. Y. KIM, K. A. FESLER and H. J. SHAW, "Reentrant Fiber Raman Gyroscope," in *Tech. Dig. Conf. on Lasers and Electro-Optics,* San Francisco, 1986, Post-deadline Paper ThU5.

[59] M. IKEDA, "Stimulated Raman Amplification Characteristics in Long Span Single-Mode Silica Fibers," *Opt. Commun.,* Vol. 39, p. 148, 1981.

[60] N. A. OLSSON and J. HEGARTY, "Noise Properties of a Raman Amplifier," *IEEE J. Lightwave Tech.,* Vol. LT-4, p. 396, 1986.

[61] J. HEGARTY, N. A. OLSSON, and L. GOLDNER, "CW Pumped Raman Preamplifier in a 45 km-Long Fibre Transmission System Operating at 1.5 μm and 1 Gbit/s," *Electron. Lett.,* Vol. 21, p. 290, 1985.

[62] E. DESURVIRE, M. PAPUCHON, J. P. POCHOLLE, and J. RAFFY, "High-Gain Optical Amplification of Laser Diode Signal by Raman Scattering in Single-Mode Fibres," *Electron. Lett.,* Vol. 19, p. 751, 1983.

[63] J. P. POCHOLLE, J. RAFFY, M. PAPUCHON, and E. DESURVIRE, "Raman and Four Photon Mixing Amplification in Single Mode Fibers," *Opt. Eng.,* Vol. 24, p. 600, 1985.

[64] M. NAKAZAWA, M. TOKUDA, Y. NEGISHI, and N. UCHIDA, "Active Transmission Line: Light Amplification by Backward-Stimulated Raman Scattering in Polarization–Maintaining Optical Fiber," *J. Opt. Soc. Am. B,* Vol. 1, p. 80, 1984.

[65] M. NAKAZAWA, T. NAKASHIMA, and S. SEIKAI, "Raman Amplification in 1.4–1.5 μm Spectral Region in Polarization-Preserving Optical Fibers," *J. Opt. Soc. Am. B,* Vol. 2, p. 515, 1985.

[66] M. NAKAZAWA, "Highly Efficient Raman Amplification in a Polarization-Preserving Optical Fiber," *Appl. Phys. Lett.,* Vol. 46, p. 628, 1985.

[67] Y. AOKI, S. KISHIDA, H. HONMOU, K. WASHIO, and M. SUGIMOTO, "Efficient Backward and Forward Pumping CW Raman Amplification for InGaAsP Laser Light in Silica Fibres," *Electron. Lett.,* Vol. 19, p. 620, 1983.

[68] C. LIN and R. H. STOLEN, "Backward Raman Amplification and Pulse Steepening in Silica FIbers," *Appl. Phys. Lett.,* Vol. 29, p. 428, 1976.

[69] Y. AOKI, S. KISHIDA, K. WASHIO, and K. MINEMURA, "Bit Error Rate Evaluation of Optical Signals Ammplified via Stimulated Raman Process in an Optical Fibre," *Electron. Lett.,* Vol. 21, p. 191, 1985.

[70] K. NAKAMURA, M. KIMURA, S. YOSHIDA, T. HIKADA, and Y. MITSUHASHI, "Raman Amplification of 1.50-μm Laser Diode Light in a Low Fiber Loss Region," *IEEE J. Lightwave Tech.,* Vol. LT-2, p. 379, 1984.

[71] S. KISHIDA, Y. AOKI, H. HONMOU, K. WASHIO, and M. SUGIMOTO, "An Active Fiber for Raman Amplification of Picosecond Light Pulses," in *Tech. Dig. Fourth Intl. Conf. IOOC,* Tokyo, 1983, Paper 29C3.

[72] G. A. KOEPF, D. M. KALEN, and K. H. GREENE, "Raman Amplification at 1.118 μm in Single-Mode fiber and its limitation by Brillouin scattering," *Electron. Lett.,* Vol. 18, p. 942, 1982.

[73] M. L. DAKSS and P. MELMAN, "Amplified Spontaneous Raman Scattering and Gain in Fiber Raman Amplifiers," *IEEE J. Lightwave Tech.*, Vol. LT-3, p. 806, 1985.

[74] T. NAKASHIMA, S. SEIKAI, and M. NAKAZAWA, "Dependence of Raman Gain on Relative Index Difference for GeO$_2$-Doped Single-Mode Fibers," *Opt. Lett.*, Vol. 10, p. 420, 1985.

[75] T. NAKASHIMA, S. SEIKAI, and M. NAKAZAWA, "Configuration of the Optical Transmission Line Using Stimulated Raman Scattering for Signal Light Amplification," *IEEE J. Lightwave Tech.*, Vol. LT-4, p. 569, 1986.

[76] K. OKAMOTO, J. NODA, and H. MIYAZAWA, "Fibre-Optic Solc Filter for Use in Raman Amplification of Light," *Electron. Lett.*, Vol. 21, p. 90, 1985.

[77] Y. DURTESTE, M. MONERIE, and P. LAMOULER, "Raman Amplification in Flouride Glass Fibers," *Electron. Lett.*, Vol. 21, p. 723, 1985.

[78] E. DESURVIRE, M. DIGONNET, and H. J. SHAW, "Raman Amplification of Recirculating Pulses in a Reentrant Fiber Loop." *Opt. Lett.*, Vol. 10, p. 83, 1985.

[79] E. DESURVIRE, M. J. F. DIGONNET, and H. J. SHAW, "Theory and Implementation of a Raman Active Fiber Delay Line," *IEEE J. Lightwave Tech.*, Vol. LT-4, p. 426, 1986.

[80] E. DESURVIRE, M. TUR, and H. J. SHAW, "Signal-to-Noise Ratio in Raman Active Fiber Systems: Application to Recirculating Delay Lines," *IEEE J. Lightwave Tech.*, Vol. LT-4, p. 560, 1986.

[81] N. A. OLSSON, R. A. LOGAN, and L. F. JOHNSON, "Transmission Experiment at 3 Gbit/s With Close-Spaced Wavelength-Division-Multiplexed Single-Frequency Lasers at 1.5 μm," *Electron. Lett.*, Vol. 20, p. 673, 1984.

[82] N. A. OLSSON, J. HEGARTY, R. A. LOGAN, L. F. JOHNSON, K. L. WALKER, L. G. COHEN, B. L. KASPER, and J. C. CAMPBELL, "68.3 km Transmission With 1.37 Tbit-km/s Capacity Using Wavelength Division Multiplexing of Ten Single-Frequency Lasers at 1.5μm," *Electron. Lett.*, Vol. 21, p. 105, 1985.

[83] A. R. CHRAPLYVY and P. S. HENRY, "Performance Degradation due to Stimulated Raman Scattering in Wavelength-Division-Multiplexed Optical-Fibre systems," *Electron. Lett.*, Vol. 19, p. 641, 1983.

[84] A. R. CHRAPLYVY "Optical Power Limits in Multichannel Wavelength-Division Multiplexed Systems due to Stimulated Raman Scattering," *Electron. Lett.*, Vol. 20, p. 58, 1984.

[85] A. M. HILL, D. COTTER, and J. V. WRIGHT, "Nonlinear Crosstalk due to Stimulated Raman Scattering in a Two-Channel Wavelength-Division-Multiplexed System," *Electron. Lett.*, Vol. 20, p. 247, 1984.

[86] H. F. MAHLEIN, "Crosstalk due to Stimulated Raman Scattering in Single-Mode Fibres for Optical Communication in Wavelength Division Multiplex Systems," *Optical and Quantum Electron.*, Vol. 16, p. 409, 1984.

[87] A. TOMITA, "Crosstalk Caused by Stimulated Raman Scattering in Single-Mode Wavelength-Division Multiplexed Systems," *Opt. Lett.*, Vol. 8, p. 412, 1983.

[88] J. HEGARTY, N. A. OLSSON, and M. McGLASHAN-POWELL, "Measurement of the Raman Crosstalk at 1.5 μm in a Wavelength-Division-Multiplexed Transmission System," *Electron. Lett.*, Vol. 21, p. 395, 1985.

[89] D. COTTER, "Stimulated Brillouin Scattering in Monomode Optical Fiber," *J. Opt. Commun.*, Vol. 4, p. 10, 1983.

[90] R. M. SHELBY, M. D. LEVENSON, and P. W. BAYER, "Resolved Forward Brillouin Scattering in Optical Fibers," *Phys. Rev. Lett.*, Vol. 54, p. 939, 1985.

[91] R. M. SHELBY, M. D. LEVENSON, and P. W. BAYER, "Guided Acoustic-Wave Brillouin Scattering," *Phys. Rev. B*, Vol. 31, p. 5244, 1985.

[92] N. L. ROWELL, P. J. THOMAS, H. M. VAN DRIEL, and G. I. STEGEMAN, "Brillouin Spectrum of Single-Mode Optical Fibers," *Appl. Phys. Lett.*, Vol. 34, p. 139, 1979.

[93] P. J. THOMAS, N. L. ROWELL, H. M. VAN DRIEL, and G. I. STEGEMAN, "Normal Acoustic Modes and Brillouin Scattering in Single-Mode Optical Fibers," *Phys. Rev. B*, Vol. 19, p. 4986, 1979.

[94] C. L. TANG, "Saturation and Spectral Characteristics of the Stokes Emission in the Simulated Brillouin Process," *J. Appl. Phys.*, Vol. 37, p. 2945, 1966.

[95] E. P. IPPEN and R. H. STOLEN, "Stimulated Brillouin Scattering in Optical Fibers," *Appl. Phys. Lett.*, Vol. 21, p. 539, 1972.

[96] R. V. JOHNSON and J. H. MARBURGER, "Relaxation Oscillations in Stimulated Raman Scattering," *Phys. Rev. A,* Vol. 4, p. 1175, 1971.

[97] I. BAR-JOSEPH, A. A. FRIESEM, E. LICHTMAN, and R. G. WAARTS, "Steady and Relaxation Oscillations of Stimulated Brillouin Scattering in Single-Mode Optical Fibers," *J. Opt. Soc. Am. B,* Vol. 2, p. 1606, 1985.

[98] N. UESUGI, M. IKEDA, and Y. SASAKI, "Maximum Single Frequency Input Power in a Long Optical Fibre Determined by Stimulated Brillouin Scattering," *Electron. Lett.,* Vol. 17, p. 379, 1981.

[99] D. COTTER, "Observation of Stimulated Brillouin Scattering in Low-Loss Silica Fibre at 1.3 μm," *Electron. Lett.,* Vol. 18, p. 495, 1982.

[100] D. COTTER, "Suppression of Stimulated Brillouin Scattering During Transmission of High-Power Narrowband Laser Light in Monomode Fibre," *Electron. Lett.,* Vol. 18, p. 638, 1982.

[101] M. TSUBOKAWA, S. SEIKAI, T. NAKASHIMA, and N. SHIBATA, "Suppression of Stimulated Brillouin Scattering in a Single-Mode Fibre by an Acousto-Optic Modulator," *Electron. Lett.,* Vol. 22, p. 473, 1986.

[102] Y. AOKI, K. TAJIMA, and I. MITUO, "Observation of Stimulated Brillouin Scattering in Single-Mode Fibers With DFB-LD Pumping and its Suppression by FSK Modulation," in *Tech. Dig. Conf. on Lasers and Electro-Optics,* San Francisco, 1986, Post-deadline Paper ThU4.

[103] K. O. HILL, B. S. KAWASAKI, and D. C. JOHNSON, "CW Brillouin Laser," *Appl. Phys. Lett.,* Vol. 26, p. 608, 1976.

[104] K. O. HILL, D. C. JOHNSON, and B. S. KAWASAKI, "CW Generation of Multiple Stokes and Anti-Stokes Brillouin Shifted Frequencies," *Appl. Phys. Lett.,* Vol. 29, p. 185, 1976.

[105] B. S. KAWASAKI, D. C. JOHNSON, Y. FUJII, and K. O. HILL, "Bandwidth-Limited Operation of a Mode-Locked Brillouin Parametric Oscillator," *Appl. Phys. Lett.,* Vol. 32, p. 429, 1978.

[106] D. R. PONIKVAR and S. EZEKIEL, "Stabilized Single-Frequency Stimulated Brillouin Fiber Ring Laser," *Opt. Lett.,* Vol. 6, p. 398, 1981.

[107] L. F. STOKES, M. CHODOROW, and H. J. SHAW, "All-Fiber Stimulated Brillouin Ring Laser With Submilliwatt Pump Threshold," *Opt. Lett.,* Vol. 7, p. 509, 1982.

[108] P. J. THOMAS, H. M. VAN DRIEL, and G. I. A. STEGEMAN, "Possibility of Using an Optical Fiber Brillouin Ring Laser for Inertial Sensing," *Appl. Opt.,* Vol. 19, p. 1906, 1980.

[109] T. G. HODGKINSON, D. W. SMITH, R. WYATT, and D. J. MALYON, "Coherent Optical Fibre Transmission Systems," *Br. Telecom Tech. J.,* Vol. 3, p. 5, 1985.

[110] C. G. ATKINS, D. COTTER, D. W. SMITH, and R. WYATT, "Application of Brillouin Amplification in Coherent Optical Transmission," *Electron. Lett.,* Vol. 22, p. 556, 1986.

[111] J. A. ARNAUD, "Enhancement of Optical Receiver Sensitivies by Amplification of the Carrier," *IEEE J. Quantum Electron.,* Vol. QE-4, p. 893, 1968.

[112] N. A. OLSSON and J. P. VAN DER ZIEL, "Fibre Brillouin amplifier with electronically controlled bandwidth," *Electron. Lett.,* Vol. 22, p. 488, 1986.

[113] R. G. WAARTS and R. P. BRAUN, "Crosstalk due to Stimulated Brillouin Scattering in Monomode Fibre," *Electron. Lett.,* Vol. 21, p. 1114, 1985.

[114] E. J. BACHUS, R. P. BRAUN, W. EUTIN, E. GROSSMAN, H. FOISEL, K. HEIMES, and B. STREBEL, "Coherent Optical-Fibre Subscriber Line," *Electron. Lett.,* Vol. 21, p. 1203, 1985.

[115] N. A. OLSSON and J. P. VAN DER ZIEL, "Cancellation of Fiber Loss by Semiconductor Laser Pumped Brillouin Amplification at 1.5 μm," *Appl. Phys. Lett.,* Vol. 48, p. 1329, 1986.

[116] Y. R. SHEN, *The Principles of Nonlinear Optics,* New York: John Wiley & Sons, 1984.

[117] K. O. HILL, D. C. JOHNSON, B. S. KAWASAKI, and R. I. MACDONALD, "CW Three-Wave Mixing in Single-Mode Optical Fibers," *J. Appl. Phys.,* Vol. 49, p. 5098, 1978.

[118] R. H. STOLEN, J. E. BJORKHOLM, and A. ASHKIN, "Phase-Matched Three-Wave Mixing in Silica Fiber Optical Waveguides," *Appl. Phys. Lett.,* Vol. 24, p. 308, 1974.

[119] R. H. STOLEN, "Phase-Matched Stimulated Four-Photon Mixing in Silica-Fiber Waveguides," *IEEE J. Quantum Electron,* Vol. QE-11, p. 100, 1975.

[120] C. LIN and M. A. BOSCH, "Large-Stokes-Shift Stimulated Four-Photon Mixing in Optical Fibers," *Appl. Phys. Lett.,* Vol. 38, p. 479, 1981.

[121] R. H. STOLEN, M. A. BOSCH, and C. LIN, "Phase Matching in Birefringent Fibers," *Opt. Lett.,* Vol. 6, p. 213, 1981.

[122] K. KITAYAMA, S. SEIKAI, and N. UCHIDA, "Stress-Induced Frequency Tuning for Stimulated Four-Photon Mixing in a Birefringent Single-Mode Fiber," *Appl. Phys. Lett.,* Vol. 41, p. 322, 1982.

[123] K. KITAYAMA and M. OHASHI, "Frequency Tuning for Stimulated Four-Photon Mixing by Bending-Induced Birefringence in a Single-Mode Fiber," *Appl. Phys. Lett.,* Vol. 41, p. 619, 1982.

[124] M. OHASHI, K. KITAYAMA, N. SHIBATA, and S. SEIKAI, "Frequency Tuning of a Stokes Wave for Stimulated Four-Photon Mixing by Temperature-Induced Birefringence Change," *Opt. Lett.,* Vol. 10, p. 77, 1985.

[125] N. SHIBATA, M. OHASHI, K. KITAYAMA, and S. SEIKAI, "Evaluation of Bending-Induced Birefringence Based on Stimulated Four-Photon Mixing," *Opt. Lett.,* Vol. 10, p. 154, 1985.

[126] T. NAKASHIMA, N. SHIBATA, M. OHASHI, and S. SEIKAI, "Evaluation of Axial Fibre Strain Based on the Stimulated Four-Photon Mixing Process," *Electron Lett.,* Vol. 21, p. 935, 1985.

[127] R. K. JAIN and K. STENERSEN, "Phase-Matched Four Photon Mixing Processes in Birefringent Fibers," *Appl. Phys. B,* Vol. 35, p. 49, 1984.

[128] K. STENERSEN and R. K. JAIN, "Small-Stokes-Shift Frequency Conversion in Single-Mode Birefringent Fibers," *Opt. Commun.,* Vol. 51, p. 121, 1984.

[129] C. LIN, W. A. REED, A. D. PEARSON, and H.-T. SHANG, "Phase Matching in the Minimum-Chromatic-Dispersion Region of Single-Mode Fibers for Stimulated Four-Photon Mixing," *Opt. Lett.,* Vol. 6, p. 493, 1981.

[130] C. LIN, W. A. REED, A. D. PEARSON, H.-T. SHANG, and P. F. GLODIS, "Designing Single-Mode Fibres for Near-IR (1.1–1.7 μm) Frequency Generation by Phase-Matched Four-Photon Mixing in the Minimum Chromatic Dispersion Region," *Electron. Lett.,* Vol. 18, p. 87, 1982.

[131] K. WASHIO, K. INOUE, and S. KISHIDA, "Efficient Large-Frequency-Shifted Three-Wave Mixing in Low Dispersion Wavelength Region in Single-Mode Optical Fibre," *Electron. Lett.,* Vol. 16, p. 658, 1980.

[132] R. H. STOLEN and J. E. BJORKHOLM, "Parametric Amplification and Frequency Conversion in Optical Fibers," *IEEE J. Quantum Electron.,* Vol. QE–18, p. 1062, 1982.

[133] C. LIN, V. T. NGUYEN, and W. G. FRENCH, "Wideband Near-IR Continuum (0.7–2.1 μm) Generated in Low-Loss Optical Fibres," *Electron. Lett.,* Vol. 14, p. 822, 1978.

[134] I. BAR-JOSEPH, A. A. FRIESEM, R. G. WAARTS, and H. H. YAFFE, "Parametric Interaction of a Modulated Wave in a Single-Mode Fiber," *Opt. Lett.,* Vol. 11, p. 534, 1986.

[135] D. F. WALLS, "Squeezed States of Light," *Nature,* Vol. 306, p. 141, 1983.

[136] M. D. LEVENSON, R. M. SHELBY, and S. H. PERLMUTTER, "Squeezing of Classical Noise by Nondegenerate Four-Wave Mixing in an Optical Fiber," *Opt. Lett.,* Vol. 10, p. 514, 1985.

[137] R. M. SHELBY, M. D. LEVENSON, S. H. PERLMUTTER, R. G. DEVOE, and D. F. WALLS, "Broad-Band Parametric Deamplification of Quantum Noise in an Optical Fiber," *Phys. Rev. Lett.,* Vol. 57, p. 691, 1986.

[138] M. D. LEVENSON, R. M. SHELBY, A. ASPECT, M. REID, and D. F. WALLS, "Generation and Detection of Squeezed States of Light by Nondegenerate Four-Wave Mixing in an Optical Fiber," *Phys. Rev. A,* Vol. 32, p. 1550, 1985.

[139] M. OHASHI, K. KITAYAMA, Y. ISHIDA, and N. UCHIDA, "Phase-Matched Light Amplification by Three-Wave Mixing Process in a Birefringent Fiber due to Externally Applied Stress," *Appl. Phys. Lett.,* Vol. 41, p. 1111, 1982.

[140] N. SHIBATA, R. P. BRAUN, and R. G. WAARTS, "Crosstalk due to Three-Wave Mixing Process in a Coherent Single-Mode Transmission Line," *Electron. Lett.,* Vol. 22, p. 675, 1986.

[141] R. WAARTS and R. P. BRAUN, "System Limitations due to Four-Wave Mixing in Single-Mode Optical Fibres," *Electron. Lett.,* Vol. 22, p. 873, 1986.

[142] J. M. GABRIAGUES, "Third-Harmonic and Three-Wave Sum-Frequency Light Generation in an Elliptical-Core Optical Fiber," *Opt. Lett.,* Vol. 8, p. 183, 1983.

[143] Y. FUJII, B. S. KAWASAKI, K. O. HILL, and D. C. JOHNSON, "Sum-Frequency Light Generation in Optical Fibers," *Opt. Lett.,* Vol. 5, p. 48, 1980.

[144] Y. SASAKI and Y. OHMORI, "Phase-Matched Sum-Frequency Light Generation in Optical Fibers," *Appl. Phys. Lett.,* Vol. 39, p. 466, 1981.

[145] Y. OHMORI and Y. SASAKI, "Two-Wave Sum-Frequency Light Generation in Optical Fibers," *IEEE J. Quantum Electron.,* Vol. QE–18, p. 758, 1982.

[146] Y. SASAKI and Y. OHMORI, "Sum-Frequency Wave Generation in Optical Fibers," *J. Opt. Commun.,* Vol. 4, p. 3, 1983.

[147] M. NAKAZAWA, T. NAKASHIMA, and S. SEIKAI, "Efficient Multiple Visible Light Generation in a Polarization-Preserving Optical Fiber Pumped by a 1.064-μm Yttrium Aluminum Garnet Laser," *Appl. Phys. Lett.,* Vol. 45, p. 823, 1984.

[148] U. OSTERBERG and W. MARGULIS, "Dye Laser Pumped by Nd:YAG Laser Pulses Frequency Doubled in a Glass Optical Fiber," *Opt. Lett.,* Vol. 11, p. 516, 1986.

[149] R. W. HELLWARTH, J. CHERLOW, and T. T. YANG, "Origin and Frequency Dependence of Nonlinear Optical Susceptibilities of Glasses," *Phys. Rev. B,* Vol. 11, p. 964, 1975.

[150] W. T. WHITE III, W. L. SMITH, and D. MILAN, "Direct Measurement of the Nonlinear Refractive Index Coefficient γ at 355 nm in Fused Silica and BK-10 Glass," *Opt. Lett.,* Vol. 9, p. 10, 1984.

[151] R. W. HELLWARTH, A. OWYOUNG and N. GEORGE, "Origin of the Nonlinear Refractive Index in Liquid CCl_4," *Phys. Rev. A,* Vol. 4, p. 2342, 1971.

[152] J. BOTINEAU and R. H. STOLEN, "Effect of Polarization on Spectral Broadening in Optical Fibers," *J. Opt. Soc. Am.,* Vol. 72, p. 1592, 1982.

[153] A. OWYOUNG, R. W. HELLWARTH, and N. GEORGE, "Intensity-Induced Changes in Optical Polarizations in Glasses," *Phys. Rev. B,* Vol. 5, p. 628, 1972.

[154] G. MEYER and F. GIRES, "Action d'une Onde Lumineuse Intense Sur L'indice de Refraction des Liquides," *C. R. Acad. Sc. Paris,* Vol. 158, Groupe 6, p. 2039, 1964.

[155] M. A. DUGUAY and J. W. HANSEN, "An Ultrafast Light Gate," *Appl. Phys. Lett.,* Vol. 15, p. 192, 1969.

[156] E. P. IPPEN and C. V. SHANK, "Picosecond Response of a High-Repetition-Rate CS_2 Optical Kerr Gate," *Appl. Phys. Lett.,* Vol. 26, p. 92, 1975.

[157] R. H. STOLEN and A. ASHKIN, "Optical Kerr Effect in Glass Waveguide," *Appl. Phys. Lett.,* Vol. 22, p. 294, 1972.

[158] J. M. DZIEDZIC, R. H. STOLEN, and A. ASHKIN, "Optical Kerr Effect in Long Fibers," *Appl. Opt.,* Vol. 20, p. 1403, 1981.

[159] J. L. AYRAL, J. P. POCHOLLE, J. RAFFY, and M. PAPUCHON, "Optical Kerr Coefficient measurement at 1.15 μm in Single-Mode Optical Fibers," *Opt. Commun.,* Vol. 49, p. 405, 1984.

[160] K. KITAYAMA, Y. KIMURA, K. OKAMOTO, and S. SEIKAI, "Optical Sampling Using an All-Fiber Optical Kerr Shutter," *Appl. Phys. Lett.,* Vol. 46, p. 623, 1985.

[161] A. E. KAPLAN and P. MEYSTRE, "Enhancement of the Sagnac Effect due to Nonlinearly Induced Nonreciprocity," *Opt. Lett.,* Vol. 6, p. 590, 1981.

[162] S. EZEKIEL, J. L. DAVIS, and R. W. HELLWARTH, "Intensity Dependent Nonreciprocal Phase Shift in a Fiberoptic Gyroscope," in *Fiber-Optic Rotation Sensors and Related Technologies: Proceedings of the First International Conference,* Cambridge, Massachusetts, 1981.

[163] S. EZEKIEL, J. L. DAVIS, and R. W. HELLWARTH, "Observation of Intensity-Induced Nonreciprocity in a Fiber-Optic Gyroscope," *Opt. Lett.,* Vol. 7, p. 457, 1982.

[164] R. A. BERGH, H. C. LEFEVRE, and H. J. SHAW, "Compensation of the Optical Kerr Effect in Fiber-Optic Gyroscopes," *Opt. Lett.,* Vol. 7, p. 282, 1982.

[165] N. J. FRIGO, H. F. TAYLOR, L. GOLDBERG, J. F. WELLER, and S. C. RASHLEIGH, "Optical Kerr Effect in Fiber-Gyroscopes: Effects of Nonmonochromatic Sources," *Opt. Lett.,* Vol. 8, p. 119, 1983.

[166] B. CROSIGNANI and A. YARIV, "Kerr Effect and Chromatic Dispersion in Fiber-Optic Gyroscopes," *IEEE J. Lightwave Technol.,* Vol. LT-3, p. 914, 1985.

[167] R. A. BERGH, B. CULSHAW, C. C. CUTLER, H. C. LEFEVRE, and H. J. SHAW, "Source Statistics and the Kerr Effect in Fiber-Optic Gyroscopes," *Opt. Lett.,* Vol. 7, p. 563, 1983.

[168] N. V. KRAVTSOV and V. N. SERKIN, "Optical Nonreciprocity in a Ring Raman Fiber Laser," *Sov. J. Quantum Electron.,* Vol. 13, p. 111, 1983.

[169] H. M. GIBBS, *Optical Bistability: Controlling Light with Light,* New York: Academic Press, 1985.

[170] H. NAKATSUKA, S. ASAKA, H. ITOH, K. IKEDA, and M. MATSUOKA, "Observation of Bifurcation to Chaos in an All-Optical Bistable System," *Phys. Rev. Lett.,* Vol. 50, p. 109, 1983.

[171] K. IKEDA, H. DAIDO, and O. AKIMOTO, "Optical Turbulence: Chaotic Behavior of Transmitted Light From a Ring Cavity," *Phys. Rev. Lett.,* Vol. 45, p. 709, 1980.

[172] P. D. MAKER, R. W. TERHUNE, and C. M. SAVAGE, "Intensity-Dependent Changes in the Refractive Index of Liquids," *Phys. Rev. Lett.,* Vol. 12, p. 507, 1964.

[173] H. G. WINFUL, "Self-Induced Polarization Changes in Birefringent Optical Fibers," *Appl. Phys. Lett.,* Vol. 47, p. 213, 1985.

[174] H. G. WINFUL, "Polarization Instabilities in Birefringent Nonlinear Media: Application to Fiber-Optic Devices," *Opt. Lett.,* Vol. 11, p. 33, 1986.

[175] R. H. STOLEN, J. BOTINEAU, and A. ASHKIN, "Intensity Discrimination of Optical Pulses With Birefringent Fibers," *Opt. Lett.,* Vol. 7, p. 512, 1982.

[176] B. NIKOLAUS, D. GRISCHKOWSKY, and A. C. BALANT, "Optical Pulse Reshaping Based on the Nonlinear Birefringence of Single-Mode Optical Fibers," *Opt. Lett.,* Vol. 8, p. 189, 1983.

[177] L. F. MOLLENAUER, R. H. STOLEN, J. P. GORDON, and W. J. TOMLINSON, "Extreme Picosecond Pulse Narrowing by Means of Soliton Effect in Single-Mode Optical Fibers," *Opt. Lett.,* Vol. 8, p. 289, 1983.

[178] N. J. HALAS and D. GRISCHKOWSKY, "Simultaneous Optical Pulse Compression and Wing Reduction," *Appl. Phys. Lett.,* Vol. 48, p. 823, 1986.

[179] H. G. WINFUL and A. HU, "Intensity Discrimination With Twisted Birefringent Optical Fibers," *Opt. Lett.,* October, 1986.

[180] K. KITAYAMA, Y. KIMURA, and S. SEIKAI, "Fiber-Optic Logic Gate," *Appl. Phys. Lett.,* Vol. 46, p. 317, 1985.

[181] Y. KIMURA, K. KITAYAMA, N. SHIBATA, and S. SEIKAI, "All-Fibre-Optic Logic 'AND' Gate," *Electron. Lett.,* Vol. 22, p. 277, 1986.

[182] G. B. AL'TSHULER, K. I. KRYLOV, S. M. MERSADYKOV, V. G. ROMANOV, and L. B. YANUSHANETS, "Self-Rotation of the Polarization Ellipse in Optical Fibers,"*Izvestiya Akad. Nauk SSSR. Ser. Fizich.,* Vol. 45, p. 2222, 1981.

[183] B. CROSIGNANI, S. PLAZZOLA, P. SPANO, and P. DIPORTO, "Direct Measurement of the Nonlinear Phase Shift Between the Orthogonally Polarized States of a Single-Mode Fiber," *Opt. Lett.,* Vol. 10, p. 89, 1985.

[184] B. DAINO, G. GREGORI, and S. WABNITZ, "New All-Optical Devices Based on Third-Order Nonlinearity of Birefringent Fibers," *Opt. Lett.,* Vol. 11, p. 42, 1986.

[185] F. MATERA and S. WABNITZ, "Nonlinear Evolution and Instability in a Twisted Birefringent Fiber," *Opt. Lett.,* Vol. 11, p. 467, 1986.

[186] S. TRILLO, S. WABNITZ, R. H. STOLEN, G. ASSANTO, C. T. SEATON, and G. I. STEGEMAN, "Experimental Observation of Polarization Instability in a Birefringent Optical Fiber," *Appl. Phys. Lett.,* to be published, 1986.

[187] K. KITAYAMA, Y. KIMURA, and S. SEIKAI, "Nonlinear Mode Coupling in Birefringent Fiber: Application to Optical Pulse Reshaping," *Appl. Phys. Lett.,* Vol. 45, p. 838, 1984.

[188] F. SHIMIZU, "Frequency Broadening in Liquids by a Short Light Pulse," *Phys. Rev. Lett.,* Vol. 19, p. 1097, 1967.

[189] R. CUBEDDU, R. POLLONI, C. S. SACCHI, and O. SVELTO, "Self-Phase Modulation and 'Rocking' of Molecules in Trapped Filaments of Light With Picosecond Pulses," *Phys. Rev. A,* Vol. 2, p. 1955, 1970.

[190] C. H. LIN and T. K. GUSTAFSON, "Optical Pulse Width Measurement Using Self-Phase Modulation," *IEEE J. Quantum Electron.,* Vol. QE–8, p. 429, 1972.

[191] S. C. PINAULT and M. J. POTASEK, "Frequency Broadening by Self-Phase Modulation in Optical Fibers," *J. Opt. Soc. Am. B,* Vol. 2, p. 1318, 1985.

[192] R. H. STOLEN and C. LIN, "Self-Phase Modulation in Silica Optical Fibers," *Phys. Rev. A,* Vol. 17, p. 1448, 1978.

[193] R. A. FISHER and W. K. BISCHEL, "Numerical Studies of the Interplay Between Self-Phase Modulation and Dispersion for Intense Plane-Wave Laser Pulses," *J. Appl. Phys.,* Vol. 46, p. 4921, 1975.

[194] B. VALK, W. HODEL, and H. P. WEBER, "High Intensity Picosecond Pulse Transmission Through Optical Fibers," *Opt. Commun.,* Vol. 50, p. 63, 1984.

[195] J. C. DIELS, J. J. FONTAINE, I. C. MCMICHAEL, and F. SIMONI, "Control and Measurement of Ultrashort Pulse Shapes (in Amplitude and Phase) With Femtosecond Accuracy," *Appl. Opt.,* Vol. 24, p. 1270, 1985.

[196] N. TZOAR and M. JAIN, "Self-Phase Modulation in Long-Geometry Optical Waveguides," *Phys. Rev. A,* Vol. 23, p. 1266, 1981.

[197] D. ANDERSON and M. LISAK, "Nonlinear Asymmetric Pulse Distortion in Long Optical Fibers," *Opt. Lett.,* Vol. 7, p. 394, 1982.

[198] D. ANDERSON and M. LISAK, "Nonlinear Asymmetric Self-Phase Modulation and Self-Steepening of Pulses in Long Optical Waveguides," *Phys. Rev. A,* Vol. 27, p. 1393, 1983.

[199] A. YARIV and P. YEH, *Optical Waves in Crystals,* New York: John Wiley & Sons, p. 48, 1984.

[200] G. B. WHITHAM, *Linear and Nonlinear Waves,* New York: John Wiley & Sons, Ch. 17, 1974.

[201] R. A. FISHER and W. BISCHEL, "The Role of Linear Dispersion in Plane-Wave Self-Phase Modulation," *Appl. Phys. Lett.,* Vol. 23, p. 661, 1973.

[202] D. GRISCHKOWSKY and A. C. BALANT, "Optical Pulse Compression Based on Enhanced Frequency Chirping," *Appl. Phys. Lett.,* Vol. 41, p. 1, 1982.

[203] H. NAKATSUKA, D. GRISCHKOWSKY, and A. C. BALANT, "Nonlinear Picosecond Pulse Propagation Through Optical Fibers With Positive Group Velocity Dispersion," *Phys. Rev. Lett.,* Vol. 47, p. 910, 1981.

[204] B. P. NELSON, D. COTTER, K. J. BLOW, and N. J. DORAN, "Large Nonlinear Pulse Broadening in Long Lengths of Monomode Fibre," *Opt. Commun.,* Vol. 48, p. 292, 1983.

[205] W. J. TOMLINSON, R. H. STOLEN, and A. M. JOHNSON, "Optical Wave Breaking of Pulses in Nonlinear Optical Fibers," *Opt. Lett.,* Vol. 10, p. 457, 1985.

[206] A. M. JOHNSON and W. M. SIMPSON, "Tunable Femtosecond Dye Laser Synchronously Pumped by the Compressed Second Harmonic of Nd:YAG," *J. Opt. Soc. Am. B,* Vol. 2, p. 619, 1985.

[207] A. S. L. GOMES, A. S. GOUVEIA-NETO, and J. R. TAYLOR, "Direct Measurement of Chirped Optical Pulses with Picosecond Resolution," *Electron. Lett.,* Vol. 22, p. 41, 1986.

[208] H. E. LASSEN, F. MENGEL, B. TROMBORG, N. C. ALBERTSEN, and P. L. CHRISTIANSEN, "Evolution of Chirped Pulses in Nonlinear Single-Mode Fibers," *Opt. Lett.,* Vol. 10, p. 34, 1985.

[209] R. A. FISHER, P. L. KELLEY, and T. K. GUSTAFSON, "Subpicosecond Pulse Generation Using the Optical Kerr Effect," *Appl. Phys. Lett.,* Vol. 14, p. 140, 1969.

[210] R. MEINEL, "Generation of Chirped Pulses in Optical Fibers Suitable for an Effective Pulse Compression," *Opt. Commun.,* Vol. 47, p. 343, 1983.

[211] C. V. SHANK, R. L. FORK, R. YEN, R. H. STOLEN, and W. J. TOMLINSON, "Compression of Femtosecond Optical Pulses," *Appl. Phys. Lett.,* Vol. 40, p. 761, 1982.

[212] E. B. TREACY, "Optical Pulse Compression With Diffraction Gratings," *IEEE J. Quantum Electron.,* Vol. QE–5, p. 454, 1969.

[213] B. NIKOLAUS and D. GRISCHKOWSKY, "12 × Pulse Compression Using Optical Fibers," *Appl. Phys. Lett.,* Vol. 42, p. 1, 1983.

[214] B. NIKOLAUS and D. GRISCHKOWSKY, "90-fs Tunable Optical Pulses Obtained by Two-Stage Pulse Compression," *Appl. Phys. Lett.,* Vol. 43, p. 228, 1983.

[215] S. L. PALFREY and D. GRISCHKOWSKY, "Generation of 16-fsec Frequency-Tunable Pulses by Optical Pulse Compression," *Opt. Lett.,* Vol. 10, p. 562, 1985.

[216] A. M. JOHNSON, R. H. STOLEN, and W. M. SIMPSON, "Generation of 0.41-picosecond Pulses by the Single-Stage Compression of Frequency Doubled Nd:YAG Laser Pulses," in *Ultrafast Phenomena IV,* ed. by D. H. AUSTON and K. B. EISENTHAL, New York: Springer-Verlag, 1984.

[217] A. M. JOHNSON, R. H. STOLEN, and W. M. SIMPSON, "80 × Single-Stage Compression of Frequency Doubled Nd:yttrium Aluminum Garnet Laser Pulses," *Appl. Phys. Lett.,* Vol. 44, p. 729, 1984.

[218] W. J. TOMLINSON, R. H. STOLEN, and C. V. SHANK, "Compression of Optical Pulses Chirped by Self-Phase Modulation in Fibers," *J. Opt. Soc. Am. B,* Vol. 1, p. 139, 1984.

[219] J. G. FUJIMOTO, A. M. WEINER, and E. P. IPPEN, "Generation and Measurement of Optical Pulses as Short as 16 fs," *Appl. Phys. Lett.,* Vol. 44, p. 832, 1984.

[220] J. M. HALBOUT and D. GRISCHKOWSKY, "12-fs Ultrashort Optical Pulse Compression at a High Repetition Rate," *Appl. Phys. Lett.,* Vol. 45, p. 1281, 1984.

[221] W. H. KNOX, R. L. FORK, M. C. DOWNER, R. H. STOLEN, and C. V. SHANK, "Optical Pulse Compression to 8 fs at a 5-kHz Repetition Rate," *Appl. Phys. Lett.,* Vol. 46, p. 1120, 1985.

[222] E. M. DIANOV, A. YA. KARASIK, P. V. MAMYSHEV, G. I. ONISHCHUKOV, A. M. PROKHOROV, M. F. STEL'MAKH, and A. A. FORMICHEV, "Effective Shortening of Picosecond Pulses Emitted by a YAG:nD^{3+} Laser," *Sov. J. Quantum Electron.,* Vol. 14, p. 726, 1984.

[223] K. TAI and A. TOMITA, "50x Optical Fiber Pulse Compression at 1.319 μm," *Appl. Phys. Lett.,* Vol. 48, p. 309, 1986.

[224] K. J. BLOW, N. J. DORAN, and B. P. NELSON, "All-Fiber Pulse Compression at 1.32 µm," *Opt. Lett.,* Vol. 10, p. 393, 1985.

[225] A. S. L. GOMES, U. OSTERBERG, W. SIBBETT, and J. R. TAYLOR, "An Experimental Study of the Primary Parameters That Determine the Temporal Compression of CW Nd:YAG Laser Pulses," *Opt. Commun.,* Vol. 54, p. 377, 1985.

[226] A. S. L. GOMES, W. SIBBETT, and J. R. TAYLOR, "Generation of Subpicosecond Pulses From a Continuous-Wave Mode-Locked Nd:YAG Laser Using a Two-Stage Optical Compression Technique," *Opt. Lett.,* Vol. 10, p. 338, 1985.

[227] J. D. KAFKA, B. H. KOLNER, T. BAER, and D. M. BLOOM, "Compression of Pulses From a Continuous-Wave Mode-Locked Nd:YAG Laser," *Opt. Lett.,* Vol. 9, p. 505, 1984.

[228] B. ZYSSET, W. HODEL, P. BEAUD, and H. P. WEBER, "200-Femtosecond Pulses at 1.06 µm Generated with a Double-Stage Pulse Compressor," *Opt. Lett.,* Vol. 11, p. 156, 1986.

[229] D. STRICKLAND and G. MOUROU, "Compression of Amplified Chirped Optical Pulses," *Opt. Commun.,* Vol. 55, p. 447, 1985.

[230] J. P. HERITAGE, R. N. THURSTON, W. J. TOMLINSON, A. M. WEINER, and R. H. STOLEN, "Spectral Windowing of Frequency-Modulated Optical Pulses in a Grating Compressor," *Appl. Phys. Lett.,* Vol. 47, p. 87, 1985.

[231] J. P. HERITAGE, A. M. WEINER, and R. M. THURSTON, "Picosecond Pulse Shaping by Spectral Phase and Amplitude Manipulation," *Opt. Lett.,* Vol. 10, p. 609, 1985.

[232] A. M. WEINER, J. P. HERITAGE, and R. N. THURSTON, "Synthesis of Phase-Coherent, Picosecond Optical Square Pulses," *Opt. Lett.,* Vol. 11, p. 153, 1986.

[233] S. A. AKHMANOV, R. V. KHOKHLOV, and A. P. SUKHORUKOV, "Self-Focusing, Self-Defocusing, and Self-Modulation of Laser Beams," in *Laser Handbook,* Vol. 2, ed. by F. T. ARECCHI and E. O. SCHULZ-DUBOIS, Amsterdam: North-Holland Publishing Co., 1972.

[234] A. R. CHRAPLYVY, D. MARCUSE, and P. S. HENRY, "Carrier-Induced Phase Noise in Angle-Modulated Optical-Fiber Systems," *IEEE J. Lightwave Tech.,* Vol. LT-2, p. 6, 1984.

[235] A. R. CHRAPLYVY and J. STONE, "Measurement of Crossphase Modulation in Coherent Wavelength-Division Multiplexing Using Injection Lasers," *Electron. Lett.,* Vol. 20, p. 996, 1984.

[236] A. HASEGAWA and F. TAPPERT, "Transmission of Stationary Optical Pulses in Dispersive Dielectric Fibers: 1. Anomalous Dispersion," *Appl. Phys. Lett.,* Vol. 23, p. 142, 1973.

[237] L. F. MOLLENAUER and R. H. STOLEN, "Solitons in Optical Fibers," *Laser Focus,* Vol. 18, No. 4, p. 193, 1982.

[238] L. F. MOLLENAUER, R. H. STOLEN, and J. P. GORDON, "Experimental Observation of Picosecond Pulse Narrowing and Solitons in Optical Fibers," *Phys. Rev. Lett.,* Vol. 45, p. 1095, 1980.

[239] E. M. DIANOV, A. YA. KARASIK, P. V. MAMYSHEV, G. I. ONISHCHUKOV, A. M. PROKHOROV, M. F. STEL'MAKH, and A. A. FOMICHEV, "100-fold Compression of Picosecond Pulses From a Parametric Light Source in Single-Mode Optical Fibers at Wavelengths 1.5–1.65 µm," *Pis'ma Zh. Eksp. Teor. Fiz.,* Vol. 40, p. 148, 1984. *JETP Lett.,* Vol. 40, p. 903, 1984.

[240] K. TAI and A. TOMITA, "1100x Optical Fiber Pulse Compression Using Grating Pair and Soliton Effect at 1.319 µm," *Appl. Phys. Lett.,* Vol. 48, p. 1033, 1986.

[241] R. H. STOLEN, L. F. MOLLENAUER, and W. J. TOMLINSON, "Observation of Pulse Restoration at the Soliton Period in Optical Fibers," *Opt. Lett.,* Vol. 8, p. 186, 1983.

[242] L. F. MOLLENAUER and R. H. STOLEN, "The Soliton Laser," *Opt. Lett.,* Vol. 9, p. 13, 1984.

[243] H. A. HAUS, and M. N. ISLAM, "Theory of the Soliton Laser," *IEEE J. Quantum Electron.,* Vol. QE-21, p. 1172, 1985.

[244] K. J. BLOW and D. WOOD, "Stability and Compression of Pulses in the Soliton Laser," *IEEE J. Quantum Electron.,* Vol. QE-22, p. 1109, 1986.

[245] A. HASEGAWA and Y. KODAMA, "Signal Transmission by Optical Solitons in Monomode Fiber," *Proc. IEEE,* Vol. 69, p. 1145, 1981.

[246] A. HASEGAWA and Y. KODAMA, "Amplification and Reshaping of Optical Solitons in a Glass Fiber—I," *Opt. Lett.,* Vol. 7, p. 285, 1982.

[247] Y. KODAMA and A. HASEGAWA, "Amplification and Reshaping of Optical Solitons in Glass Fiber—II," *Opt. Lett.,* Vol. 7, p. 339, 1982.

[248] Y. KODAMA and A. HASEGAWA, "Amplification and Reshaping of Optical Solitons in Glass Fiber—III. Amplifiers With Random Gain," *Opt. Lett.,* Vol. 8, p. 342, 1983.

[249] A. HASEGAWA, "Amplification and Reshaping of Optical Solitons in a Glass Fiber—IV: Use of the Stimulated Raman Process," *Opt. Lett.,* Vol. 8, p. 650, 1983.

[250] A. HASEGAWA, "Numerical Study of Optical Soliton Transmission Amplified Periodically by the Stimulated Raman Process," *Appl. Opt.,* Vol. 23, p. 3302, 1984.

[251] L. F. MOLLENAUER, J. P. GORDON, and M. N. ISLAM, "Soliton Propagation in Long Fibers With Periodically Compensated Loss," *IEEE J. Quantum Electron.,* Vol. QE–22, p. 157, 1986.

[252] L. F. MOLLENAUER, R. H. STOLEN, and M. N. ISLAM, "Experimental Demonstration of Soliton Propagation in Long Fibers: Loss Compensated by Raman Gain," *Opt. Lett.,* Vol. 10, p. 229, 1985.

[253] V. A. YVSLOUKH and V. N. SERKIN, "Generation of High-Energy Solitons of Stimulated Raman Radiation in Fiber Light Guides," *Pis'ma Zh. Eksp. Teor. Fiz.,* Vol. 38, p. 170, 1983. [*JETP Lett.,* Vol. 38, p. 199, 1983.]

[254] V. N. LUGOVOI, "Stimulated Raman Emission and Frequency Scanning in an Optical Waveguide," *Zh. Eksp. Teor. Fiz.,* Vol. 71, p. 1307, 1967. [*Sov. Phys. JETP,* Vol. 44, p. 683, 1976.]

[255] E. M. DIANOV, A. YA. KARASIK, P. V. MAMYSHEV, A. M. PROKHOROV, V. N. SERKIN, M. F. STEL'MAKH, and A. A. FOMICHEV, "Stimulated Raman Conversion of Multisoliton Pulses in Quartz Optical Fibers," *Pis'ma Zh. Eksp. Teor. Fiz.,* Vol. 41, p. 242, 1985. [*JETP Lett.,* Vol. 41, p. 294, 1985.]

[256] E. M. DIANOV, A. YA. KARASIK, P. V. MAMYSHEV, G. I. ONISHCHUKOV, A. M. PROKHOROV, M. F. STEL'MAKH, and A. A. FOMICHEV, "Picosecond Structure of the Pump Pulse in Stimulated Raman Scattering in a Single-Mode Optical Fiber," *Pis'ma Zh. Eksp. Teor. Fiz.,* Vol. 39, p. 564, 1984. [*JETP Lett.,* Vol. 39, p. 691, 1984.]

[257] E. M. DIANOV, A. M. PROKHOROV, and V. N. SERKIN, "Dynamics of Ultrashort-Pulse Generation by Raman Fiber Lasers: Cascade Self-Mode Locking, Optical Pulsons, and Solitons," *Opt. Lett.,* Vol. 11, p. 158, 1986.

[258] M. N. ISLAM and L. F. MOLLENAUER, "Fiber Raman Amplification Soliton Laser," in *Tech. Dig. Int. Quantum Electron. Conf.,* Pap. TuHH1, San Francisco, 1986.

[259] L. A. OSTROVSKII, "Propagation of Wave Packets and Space-Time Self-Focusing in a Nonlinear Medium," *Sov. Phys. JETP,* Vol. 24, p. 797, 1967.

[260] A. HASEGAWA and W. F. BRINKMAN, "Tunable Coherent IR and FIR Sources utilizing Modulational Instability," *IEEE J. Quantum Electron.,* Vol. QE–16, p. 694, 1980.

[261] T. BROOKE BENJAMIN and J. E. FEIR, "The Disintegration of Wave Trains on Deep Water," *J. Fluid Mech.,* Vol. 27, Pt. 3, p. 417, 1967.

[262] A. HASEGAWA, *Plasma Instabilities and Nonlinear Effects,* Heidelberg: Springer-Veriag, 1975, p. 201.

[263] A. HASEGAWA, "Generation of a Train of Soliton Pulses by Induced Modulational Instability In Optical Fibers," *Opt. Lett.,* Vol. 9, p. 288, 1984.

[264] N. N. AKHMEDIEVA, V. M. ELEONSKII, and N. E. KULAGIN, "Generation of Periodic Trains of Picosecond Pulses in an Optical Fiber: Exact Solutions," *ZH. Eksp. Teor. Fiz.,* Vol. 89, p. 1542, 1985 [*Sov. Phys. JETP,* Vol. 62, p. 894, 1985.]

[265] K. TAI, A. HASEGAWA, and A. TOMITA, "Observation of Modulational Instability in Optical Fibers," *Phys. Rev. Lett.,* Vol. 56, p. 135, 1986.

[266] K. TAI, A. TOMITA, J. L. JEWELL, and A. HASEGAWA, "Generation of Subpicosecond Soliton-like Optical Pulses at 0.3 THz Repetition Rate by Induced Modulational Instability," *Appl. Phys. Lett.,* Vol. 49, p. 236, 1986.

[267] P. K. SHUKLA and J. J. RASMUSSEN, "Modulational Instability of Short Pulses in Long Optical Fibers," *Opt. Lett.,* Vol. 11, p. 171, 1986.

[268] D. ANDERSON and M. LISAK, "Modulational Instability of Coherent Optical-Fiber Transmission Signals," *Opt. Lett.,* Vol. 9, p. 468, 1984.

[269] B. HERMANSSON and D. YEVICK, "Modulational Instability Effects in PSK Modulated Coherent Fiber Systems and Their Reduction by Optical Loss," *Opt. Commun.,* Vol. 52, p. 99, 1984.

[270] N. J. DORAN and K. J. BLOW, "Solitons in Optical Communications," *IEEE J. Quantum Electron.,* Vol. QE–19, p. 1883, 1983.

[271] K. J. BLOW and N. J. DORAN, "High Bit Rate Communication Systems Using Nonlinear Effects," *Opt. Commun.,* Vol. 42, p. 403, 1982.

[272] K. J. BLOW and N. J. DORAN, "The Asymptotic Dispersion of Soliton Pulses in Lossy Fibres," *Opt. Commun.,* Vol. 52, p. 367, 1985.

[273] D. ANDERSON and M. LISAK, "Asymptotic Linear Dispersion of Optical Pulses in the Presence of Fiber Nonlinearity and Loss," *Opt. Lett.,* Vol. 10, p. 390, 1985.

[274] D. ANDERSON, "High Transmission Rate Communication Systems Using Lossy Optical Solitons," *Opt. Commun.,* Vol. 48, p. 107, 1983.

[275] D. YEVICK and B. HERMANSSON, "Soliton Analysis With the Propagating Beam Method," *Opt. Commun.,* Vol. 47, p. 101, 1983.

[276] K. J. BLOW and N. J. DORAN, "Bandwidth Limits of Nonlinear (Soliton) Optical Communication Systems," *Electron Lett.,* Vol. 19, p. 429, 1983.

[277] J. P. GORDON, "Interaction Forces Among Solitons in Optical Fibers," *Opt. Lett.,* Vol. 8, p. 596, 1983.

[278] P. L. CHU and C. DESEM, "Gaussian Pulse Propagation in Nonlinear Optical Fiber," *Electron. Lett.,* Vol. 19, p. 956, 1983.

[279] P. L. CHU and C. DESEM, "Optical Fibre Communication Using Solitons," *Technical Digest, IOOC '83,* Tokyo, June 27–30, pp. 52–53, 1983.

[280] P. L. CHU and C. DESEM, "Mutual Interaction Between Solitons of Unequal Amplitudes in Optical Fibre," *Electron. Lett.,* Vol. 21, p. 1133, 1985.

[281] B. HERMANSSON and D. YEVICK, "Numerical Investigation of Soliton Interaction." *Electron. Lett.,* Vol. 19, p. 570, 1983.

[282] D. ANDERSON and M. LISAK, "Bandwidth Limits due to Mutual Pulse Interaction in Optical Soliton Communication Systems," *Opt. Lett.,* Vol. 11, p. 174, 1986.

[283] E. SHIOJIRI and Y. FUJII, "Transmission Capability of an Optical Fiber Communication System Using Index Nonlinearity," *Appl. Opt.,* Vol. 24, p. 358, 1985.

[284] K. J. BLOW, N. J. DORAN, and E. CUMMINS, "Nonlinear Limits on Bandwidth at the Minimum Dispersion in Optical Fibres," *Opt. Commun.,* Vol. 48, p. 181, 1983.

[285] P. L. CHU and C. DESEM, "Effect of Third-Order Dispersion of Optical Fibre on Soliton Interaction," *Electron. Lett.,* Vol. 21, p. 228, 1985.

[286] V. A. VYSLOUKH, "Propagation of Pulses in Optical Fibers in the Region of a Dispersion Minimum: Role of Nonlinearity and Higher-Order Dispersion," *Sov. J. Quantum Electron.,* Vol. 13, p. 1113, 1983.

[287] P. K. A. WAI, C. R. MENYUK, Y. C. LEE, and H. H. CHEN, "Nonlinear Pulse Propagation in the Neighborhood of the Zero-Dispersion Wavelength of Monomode Optical Fibers," *Opt. Lett.,* Vol. 11, p. 464, 1986.

[288] A. M. FATTAKHOV and A. S. CHIRKIN, "Nonlinear Propagation of Phase-Modulated Optical Pulses," *Sov. J. Quantum Electron.,* Vol. 14, p. 1556, 1984.

[289] D. ANDERSON, M. LISAK, and P. ANDERSON, "Nonlinearly Enhanced Chirp Pulse Compression in Single-Mode Fibers," *Opt. Lett.,* Vol. 10, p. 134, 1985.

[290] G. P. BAVA, G. GHIONE, and I. MAIO, "Influence of Laser Fluctuations on Soliton Propagation in Optical Fibers," *Electron. Lett.,* Vol. 20, p. 1002, 1984.

[291] B. CROSIGNANI, C. H. PAPAS, and P. DIPORTO, "Role of Intensity Fluctuations in Nonlinear Pulse Propagation," *Opt. Lett.,* Vol. 5, p. 467, 1980.

[292] A. HASEGAWA and F. TAPPERT, "Transmission of Stationary Nonlinear Optical Pulses in Dispersive Dielectric Fibers. II. Normal Dispersion," *Appl. Phys. Lett.,* Vol. 23, p. 171, 1973.

[293] K. J. BLOW and N. J. DORAN, "Multiple Dark Soliton Solutions of the Nonlinear Schrodinger Equation," *Phys. Lett.,* Vol. 107A, p. 55, 1985.

[294] D. N. CHRISTODOULIDES and R. I. JOSEPH, "Dark Solitary Waves in Optical Fibers," *Opt. Lett.,* Vol. 9, p. 408, 1984.

[295] B. BENDOW and P. D. GIANINO, "Theory of Nonlinear Pulse Propagation in Inhomogeneous Waveguides," *Opt. Lett.,* Vol. 4, p. 164, 1979.

[296] K. O. HILL, Y. FUJII, D. C. JOHNSON, and B. S. KAWASAKI, "Photosensitivity in Optical Fiber Waveguides: Application to Reflection Filter Fabrication," *Appl. Phys. Lett.,* Vol. 32, p. 647, 1978.

[297] ZHONG-YI YIN, PAUL E. JESSOP, and BRIAN K. GARSIDE, "Photoinduced Grating Filters in GeO$_2$ Thin-Film Waveguides," *Appl. Opt.,* Vol. 22, p. 4088, 1983.

[298] B. S. KAWASAKI, K. O. HILL, D. C. JOHNSON, and Y. FUJII, "Narrow-Band Bragg Reflectors in Optical Fibers," *Opt. Lett.,* Vol. 3, p. 66, 1978.

[299] J. BURES, J. LAPIERRE, and D. PASCALE, "Photosensitivity Effect in Optical Fibers: A Model for the Growth of an Interference Filter," *Appl. Phys. Lett.,* Vol. 37, p. 860, 1980.

[300] J. LAPIERRE, J. BURES, and D. PASCALE, "Photosensitive Phenomena in Optical Fibers," in *Physics of Fiber Optics,* ed. by B. BENDOW and S. S. MITRA, American Ceramic Society, 1981.

[301] J. LAPIERRE, J. BURES, and D. PASCALE, "Mesure de la Reponse en Frequence de Filtres de Bragg dans une Fibre Optique par Balayage Thermique," *Opt. Commun.,* Vol. 40, p. 95, 1981.

[302] J. LAPIERRE, J. BURES, and G. CHEVALIER, "Fiber-Optic Integrated Interference Filters," *Opt. Lett.,* Vol. 7, p. 37, 1982.

[303] J. BURES, S. LACROIX, and J. LAPIERRE, "Reflecteur de Bragg Induit par Photosensibilite dans une Fibre Optique: Modele de Croissance et Reponse en Frequence," *Appl. Opt.,* Vol. 21, p. 3502, 1982.

[304] M. PARENT, J. BURES, S. LACROIX, and J. LAPIERRE, "Proprietes de Polarisation des Reflecteurs de Bragg Induits par Photosensibilite dans les Fibres Optiques Monomodes," *Appl. Opt.,* Vol. 24, p. 354, 1985.

[305] D. K. W. LAM and B. K. GARSIDE, "Character-ization of Single-Mode Optical Fiber Filters," *Appl. Opt.,* Vol. 20, p. 440, 1981.

[306] D. K. W. LAM, B. K. GARSIDE, and K. O. HILL, "Dispersion Cancellation Using Optical-Fiber Fil-ters," *Opt. Lett.,* Vol. 7, p. 291, 1982.

[307] H. G. WINFUL, "Pulse Compression in Optical Fiber Filters," *Appl. Phys.,* Vol. 46, p. 527, 1985.

[308] M. M. FEJER and R. L. BYER, "Nonlinear Inter-actions in Crystal Fibers," in *Tech. Dig. Conf. on Lasers and Electro-Optics,* Paper WR1, San Fran-cisco, 1986.

[309] D. N. PAYNE, L. REEKIE, R. J. MEARS, S. B. POOLE, I. M. JAUNCEY, and J. T. LIN, "Rare-Earth Doped Single-Mode Fiber Lasers, Amplifiers, and Devices," *Tech. Dig. Conf. on Lasers and Electro-Optics,* Paper FN1, San Francisco, 1986.

Passive Optical Components

J. STRAUS & B. KAWASAKI
Northern Telecom Canada, Ltd.

1. Introduction

Developments in the structure and performance of passive components for use in optical-fiber–based systems have complemented the recent advances in performance of the various types of optical fiber and of the optoelectronic interfaces to the electrical signals with the aim of making optimum use of the available spectrum. The processing of optical signals through passive components permits many alternatives to the simple use of optical fiber as a one-channel point-to-point transmission medium between optoelectronic transducers. Common examples of such alternatives are (*a*) the use of multiport branching devices for the broadcast of signals to many receivers when individual transmission paths are not limited by optical attenuation and (*b*) the use of wavelength-division- multiplexing (wdm) components to permit a greater channel capacity when a higher bit rate is not warranted. Passive components such as isolators, attenuators, mode conditioners (cladding mode stripper, mode scrambler), and polarizers permit more predictable and optimized performance of optical links.

For all of these components the advances have necessarily followed the requirements imposed by the fiber performance. With fiber attenuation now very low, passive components have been developed to have low insertion loss in order not to limit unduly the system performance. The use of single-mode fiber is now common, and single-mode passive components are being developed. Together with single-mode operation has come the use of the longer-wavelength windows of 1.3 and 1.55 μm and passive components that operate efficiently in these ranges are being developed.

The use of single-mode fiber has also added a further dimension to the possible passive components in that interference-based devices utilizing the coherent guiding properties of the fiber are being developed to process signals in optical form.

In this chapter the recent advances in the various categories of passive fiber-optic components are described from the points of view of structure, performance, packaging, and applications.

2. Multiport Optical Couplers

Multiport optical couplers are the basic interconnection elements for assembling a variety of passive distribution networks that employ optical fiber. These networks can be unidirectional for signal broadcast from a single source to multiple receivers or bidirectional as in various databus or local-area network structures. In the great majority of networks the performance of the coupling elements, rather than the transmission characteristics of the fiber lines themselves, limits the network performance and hence determines the optimum network configurations. Ideal coupling elements should distribute light among the branches of the network with no scattering loss or generation of noise and they should do so with complete insensitivity to such factors as the distribution of light among the fiber modes and the state of polarization of the light. Also, couplers designed to operate in a particular fiber wavelength window should perform independently of source wavelength in that range. Finite scattering loss at the coupler will limit optimum use of available light power and reduce the number of terminals that can be connected and/or the span of the network, whereas the generation of noise at the coupler points will restrict the bandwidth-span factor of the network. Ideal couplers should also be flexible in function to permit the layout of different network configurations for optimum usage of cable or available launched optical power. They should also be able to distribute the light into or from an arbitrary number of branches of varying fiber types with or without directional characteristics, and with selectable branching ratios.

No family of fiber-optical couplers possess all of these attributes, and in practice the deviation from the ideal exhibited by existing fiber couplers necessitates many compromises in their applications. In this section the more common types of multiport optical couplers are described with respect to coupling principle, performance, and limitations. There are a great many types of coupling structures that could be described, but for reasons of brevity we will restrict ourselves to those that are or apparently will be in common use. We will also restrict ourselves in this section to components which are "fiber-to-fiber" structures and not hybrid components combining fiber lines with light sources or detectors.

Micro-Optic Beam-Splitter Couplers

These devices perform branching of light from one fiber input port to two or more fiber output ports by wavefront division. In one form the output from a fiber is expanded and collimated by a lens; the wavefront is divided into two or more parts by partial reflection or spatial division; and the separated light beams are refocused into the output fibers by means of additional lenses. Typical of this type of coupler is the Selfoc lens splitter [1] shown in Fig. 1. In this device 1/4-pitch Selfoc lenses are used as the collimating elements. Division of the wavefront is by means of a partial mirror and separation of the input path from the output path corresponding to the reflected beam is achieved by offsetting the fibers from the axis of the lenses. With conventional lenses this type of offset would result in significant distortion of the beam through the imaging system but, as shown by Tomlinson [2] the imaging properties of Selfoc lenses are such as to minimize this distortion for parallel input and output fibers and thus permit efficient coupling.

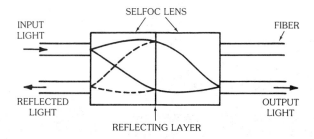

Fig. 1. Selfoc lens coupler. (*After Kobayashi et al. [1]*)

The typical excess loss for this type of structure is 0.5 dB with multimode fiber and is only slightly higher with single-mode fiber. Couplers of this type for single-mode fiber have been developed particularly for wdm applications [3]. As is apparent from the beam paths shown in Fig. 1, these devices are directional couplers; light entering from the input port emerges only from the two output ports and does not return (except by scattering) to the input port. The typical directional isolation is 30 dB in optical power. This method of coupling is extendable to a larger number of output ports. As shown in Fig. 2, Ishikawa and colleagues [4] demonstrated a power-divider structure with high coupling efficiency, by using wedge-shaped lenses. The performance of couplers using Selfoc lenses has improved steadily, the progress being aided by recent improvements in the lenses themselves [5], and the devices are widely used. A problem still to be overcome for this structure is the high cost resulting from the precision alignments involved in the construction and the stable packaging schemes required for practical applications. Other forms of lenses than Selfoc are also used for this principle of coupling. A ball-lens arrangement to make a compact single-mode directional coupler was demonstrated [6]. This device showed particularly good performance with respect to directivity and return loss.

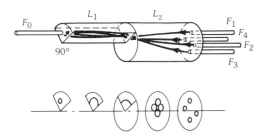

FIG. 2. Four-port micro-optic star coupler, together with beam patterns. (*After Ishikawa et al. [4]. Acknowledgment is made of the prior publication of this material by the IEE*)

Another form of micro-optic beam-splitter coupler is based on the spatial division of the output from the input fiber and the coupling to output fibers through a spatial filter or to the output fibers directly without lensing. A typical example of this coupling principle is shown in Fig. 3, a multimode fiber tapping element developed by Witte [7]. In this device a photolithographic technique is used to produce a mold in which an offset butt-joint is produced between the input fiber and the output fiber. Part of the light which is not coupled to this output fiber is guided via a plastic waveguide to an additional output fiber, the tap. Advantages of this structure are its potential low cost and reproducibility. Disadvantages include its low coupling efficiency to the tapping branch and incompatibility with single-mode optical fiber. Other couplers based on this principle were described by Fujita and coworkers [8] and Suzuki and Kashiwagi [9].

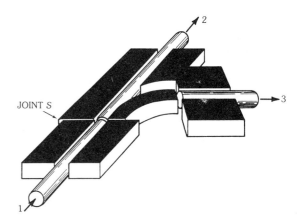

FIG. 3. Multimode-fiber tapping element. (*After Witte [7]*)

Fused Biconical Taper Couplers

The fused biconical taper technique can be used to produce a family of functionally different coupling structures. The basic structure is shown in Fig. 4. It is a four-port symmetric directional coupler [10]. The device is fabricated by spot fusing, side by side, two sections of optical fiber from which the jacketing material has been removed, and then elongating the fused portion to form a twin biconically tapered structure. It operates by converting a portion of the input light in one port into radiation or cladding modes in the narrowing taper region of the device and then recapturing this portion of light as core-guided light divides approximately equally between the two output ports. This recapture of light is accomplished in the expanding taper region of the device. The splitting ratio of the coupler is related to the narrowness of the waist region produced; the narrower this region, the larger the portion of light that is shared approximately equally between the two output ports and the closer the splitting ratio will approach the limit of 1:1. The device is symmetric in that the portion of light which is cou-

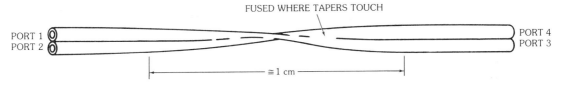

FIG. 4. Fused biconical taper coupler. (*After Kawasaki and Hill [10]*)

pled from a throughpath, port 1 to port 3 or port 2 to port 4, to the coupled output port, port 4 or port 3, respectively, is ideally independent of which port is used as the input port.

Advantages of this structure are its relatively low scattering loss, typically less than 0.5 dB (ten times the common logarithm of the ratio of the total output optical power to the input optical power) and high directional isolation, typically greater than 50 dB (ten times the common logarithm of the ratio of the power at the isolated port to the input power). This device can be made from many different fiber types and sizes, but in practice the performance of the couplers obtained depends somewhat on the fiber structure.

The primary disadvantage of the fused biconical taper coupler is related to the nature of the coupling action. Since the coupling occurs by separating the input light into two portions, a portion associated with low-order modes which passes directly from the input port to one output port, and a second portion associated with higher-order modes which are shared between the two output ports, the coupling action and splitting ratio are dependent on the mode-fill distribution of the input light. This fiber-mode dependence of the coupler leads to two difficulties for the system designer. Firstly, since each coupler in a network can change the distribution of light in the fiber modes as the light signal passes along a given path and also since each coupler has a splitting ratio that is a function of this mode-fill condition, then the couplers in the network cannot be treated as individual elements with known parameters. The system designer does not have sufficient information about the distribution of light at each point in the network to evaluate precisely the performance of the whole network. Consequently he or she must overdesign the network by providing a large margin (difference between available optical power budget and maximum optical loss) to ensure that adequate power is received at each terminal. Secondly, the mode-dependent splitting by a coupler used in conjunction with coherent light sources can lead to the generation of modal noise [11].

Symmetric devices other than the directional coupler can be produced by increasing the number of fibers that are fused together and tapered, and by forming return paths in some arms of the

structure. Examples of the three functionally distinct types of devices so formed are shown in Fig. 5. Fig. 5a is an 8×8-port transmission star coupler formed by fusing and tapering eight strands of fiber. This is a multiway signal distribution device. Light which is input at any one port is divided among the eight ports on the opposite side of the taper region. The remaining ports on the input side are isolated from the light signal. By forming loops on one side of the device as in Fig. 5b, a reflection star coupler is formed.

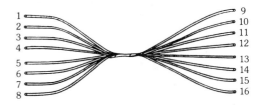

(a) 8-way transmission star coupler.

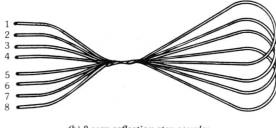

(b) 8-way reflection star coupler.

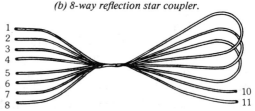

(c) 2 × 8 hybrid transmission-reflection star coupler.

FIG. 5. Fused biconical taper star couplers.

Here light which enters any port crosses the taper region to the loops which return the light back into the taper region. The light then emerges from all of the ports, including the input port. Fig. 5c shows the third form of star, which combines the actions of the other two. This coupler operates such that if light is input at one subset of the ports, those on the left-hand side of the diagram, the device will operate as a reflection star coupler, but, if light is input in one of the ports on the right-hand side, the device acts as a transmission star coupler and light emerges only from those ports on the opposite side of the tapered region. In practice this combination of properties proves useful in the avoidance of multiple signal paths in networks which combine more than one star coupler. These star devices can be made with an arbitrary number of ports. Devices with up to 200 ports have been demonstrated [12]. On the negative side is the nonuniformity which is observed in the distribution of light to the output ports. Variation of output power can be as great as an order of magnitude difference between maximum and minimum output powers. More recent developments of this technique have, however, realized couplers with a fair degree of uniformity in the coupled output power [13].

Asymmetric couplers can be of use in applications such as a linear data bus. Here one would ideally wish to couple all of the input light at a port onto the trunk fiber and remove only a small fraction of the light propagating along the trunk fiber. Asymmetry of coupling can be introduced into the fused biconical taper structure through the use of fibers with differing initial parameters, such as core or cladding dimensions or numerical aperture. Asymmetrical coupling will also result from unequal processing of the fibers in the fabrication of the coupler. Examples of unequal processing in the formation of a four-port directional coupler are the initial etching of one fiber to reduce its cladding diameter and the stretching of a fiber to reduce its cross-sectional dimension before fusing to the second fiber and tapering. The use of asymmetrical coupling structures can produce an additional benefit in the optical behavior of the coupler with respect to mode-dependent coupling behavior. These asymmetric devices can be made with relatively mode-independent splitting ratios [14].

The principle of operation described above for biconical taper couplers, the radiation of core-guided light to cladding modes and recapture of the cladding light, would seem to preclude operation of these devices with single-mode fiber in which the recapture of light would be very inefficient. Notwithstanding this, very efficient single-mode operation of four-port devices has been demonstrated [15–17]. The light-coupling mechanism in these devices is a beat phenomenon between two normal modes of the glass-air interface waveguide formed at the taper waist [18]. These devices have been demonstrated to have a tunable coupling ratio with the tuning mechanism being a mechanical-motion [19] or a refractive-index change [20].

Evanescent-Wave Coupling Devices (Single Mode)

The earliest demonstration of an efficient single-mode–fiber coupler was the "bottle coupler" [21] as shown in Fig. 6. This coupler was formed by a

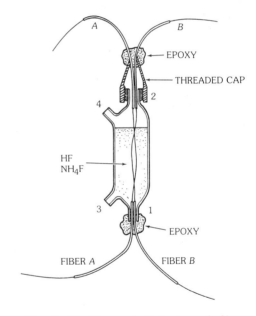

FIG. 6. "Bottle coupler" single-mode fiber coupler. (*After Sheem and Giallorenzi [21]*)

twist-etching technique, in which the plastic jackets were stripped from the fibers which were then twisted together. The fibers were then etched in the bottle almost to the core-cladding interface until the evanescent fields of the optical beams guided in the primary and secondary fibers overlapped to produce optical coupling. The twist-etched fibers were subsequently immersed in oil having an index of refraction close to, but smaller than, the index of the core. Control of tension of the twisted fibers, the number of twists, and the selection of the oil were parameters used to control the coupling ratio. Later developments in this technique aimed at improving the mechanical and environmental stability of these devices included fusion of the coupling region [22]. By using more than two fibers etched and twisted together, a transmission star device can be produced in similar fashion [23]. In these devices a complex braiding technique is used to achieve a measure of uniformity in the power coupled to the output fibers. In a 10×10 fiber coupler a uniformity of greater than 2 dB among all fibers was achieved and greater than 1 dB over a selection of eight of the ten fibers. Throughput loss was less than 1 dB.

Another form of evanescent-wave single-mode coupler with more rugged characteristics is the quartz-block coupler [24] shown schematically in Fig. 7. This technique is straightforward although requiring considerable precision. A fiber is bonded into a slot in a quartz block. The block and fiber are then ground and polished to within a few micrometers of the fiber core. Placing two such units in close proximity produces envanescent coupling of the guided light between the core regions. Refractive-index matching oil is employed between the two blocks. An important advantage of this coupling structure is that the coupling ratio can be controlled by small relative motions of the blocks with little effect on insertion loss or directivity. Also, the structure is potentially rugged. This coupling structure has been used to produce an adjustable single-mode–fiber multiterminal directional coupler that exhibits efficient coupling and uniform coupling over a range of coupling coefficients [25]. The device is made by forming multiple loops of fiber which meet in the same quartz blocks to be polished simultaneously.

Still another form of evanescent-wave coupler employs a specially prepared fiber to produce an all-fiber coupler [26]. The fiber from which this device is made is drawn from a preform ground parallel to the core such that the core is close to the flattened surface of the fiber. To form a directional coupler from this fiber two sections of the fiber are arranged flat-side together in a capillary tube so that the distance between their cores is as small as possible. The fibers are given a common polysiloxane coating inside the capillary tube over a length of 1 to 2 cm. The capillary tube can be removed after this process and the mechanical stability of the resultant coupler increased by embedding it in hard plastic. Like the similar fused biconical taper coupler, this device can be tuned in coupling ratio by bending.

3. Wavelength-Division Multiplexing Couplers

Wavelength-division multiplexing (wdm) is the use of light from two or more optical sources of differing wavelengths propagating in the same optical fiber. The light at different wavelengths can be propagating in the same or opposite directions in the fiber. Wavelength-division multiplexing couplers are the passive optical components that perform the efficient multiplexing and demultiplexing of the light sources into and out of the trunk fiber.

For fiber-optic communications the use of wdm has become a realistic option for the more efficient use of the medium because of the coincidence of several factors. Firstly, in contrast to metallic waveguides, the wavelength windows of operation in optical fibers from the viewpoint of low attenuation and usably low chromatic dispersion are many orders of magnitude wider than a single modulation band of say 1 GHz [27]. In fact, for most systems the unmodulated line width of the light source primarily determines the amount of spectral bandwidth being used, and this line width is orders of magnitude narrower than the windows. Secondly, light sources can be stabilized in spectral width so that several wdm channels can be spaced in the operating windows of optical fiber. And finally, many fiber links are limited by the length-band-

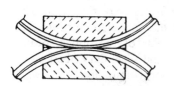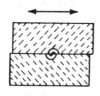

FIG. 7. "Quartz-block" coupler. (*After Bergh, Kotler, and Shaw [24]. Acknowledgment is made of the prior publication of this material by the IEE*)

width product rather than by optical attenuation, and hence the additional optical loss associated with the insertion of wdm components can be tolerated. In practice, several other factors must also be considered. Link capacity can be increased by raising the bit rate for digital systems or by frequency-division multiplexing and also by spatial multiplexing in the sense of adding additional fiber lines. Economic considerations would thus necessitate a comparison between the wdm option and the availability and cost of different electronic terminal equipments for higher electrical bandwidth. The cost of additional fiber lines would be compared with the cost of the wdm components and the possibility of crosstalk introduced by these components. In the past the wdm option has often been rejected because of lack of availability, poor performance, and cost of the components. More recently, however, these components are being mass produced in structures that have excellent performance and at costs competitive with the alternatives.

This section describes wdm structures typifying the different forms that have been developed and discusses some possibilities for future developments. The two most common means of separating or combining light paths in wdm components are the interference filter in high-pass, low-pass, or bandpass form and the diffraction grating in various configurations. Generally speaking, practical diffraction gratings can separate or combine more closely spaced wavelength channels than can available interference filters. Also, single gratings can operate on several channels simultaneously, unlike interference filters. However, when gratings are coupled to optical fibers, the passbands of operation are highly dependent on the fiber position and fiber structure, whereas interference filters have more stable spectral characteristics. With this mix of advantages and disadvantages for these two dispersive elements, it is understandable that we find both types used and sometimes used in combination.

General Principles

The important optical parameters for a wdm component are identified in Fig. 8, which depicts the transmission characteristics of a three-channel device with nominal operating wavelengths of λ_1,

λ_2, and λ_3. In this figure L_1 is the attenuation of light wavelength λ_1 in the λ_1 channel. Also, L_{12} is the crosstalk of light at wavelength λ_2 into the λ_1 channel. Ideally, the component should have a low loss in the transmission window corresponding to a given wavelength, i.e., low L_1 at λ_1. It should have high interband isolation, that is L_{12}, the attenuation of λ_2 leaking into channel 1, should be high. The channel separation, $\lambda_2 - \lambda_1$, say, should be as small as permitted by light source availability and stability. Low transmission-band attenuation will minimize system margin degradation in comparison with a single-channel system. High interchannel isolation will minimize crosstalk. In practice, high interchannel isolation is required only at the receiver (demultiplexer) end of the link (both ends for a two-way system).

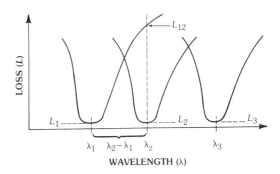

FIG. 8. Transmission characteristics of a three-channel wdm device.

Interference-Filter–Based WDM Couplers

The optical performance of this type of device is determined primarily by the characteristics of the dielectric interference filters incorporated into its structure. In order to achieve a wavelength channel separation of 30 nm, corresponding to a reasonable requirement for light source availability and stability, quite severe requirements are imposed on the filters. They must have very steep cuton and/or cutoff slopes to achieve sufficient interchannel isolation, say 30 dB, while maintaining high inband transmittance. Such filters have been developed for this application [28] in high-pass, low-pass, and bandpass form. These filters have been designed to be sufficiently stable against moisture, aging, and temperature as not to affect the window of operation for the laser. Since by nature this type of filter can operate to combine or separate only two wavelength channels or two groups of wavelength channels at a time, an n-channel wdm device will necessarily contain at least $n-1$ interference filters.

The most simple structures for wdm components based on interference filters contain no optical parts other than a fiber junction and interference filters formed on the fiber end faces. Such an all-fiber structure [29] is shown in Fig. 9. While this type of structure is of highly compact design, there are penalties of insertion loss since the spaces between the fiber end faces are without waveguiding properties. Also with the edge filter, F, that can be applied to a fiber end face, it is necessary to have additional interference filters, F_1 and F_2, applied to the end faces of the receive channels of the demultiplexer in order to obtain a high level of crosstalk isolation.

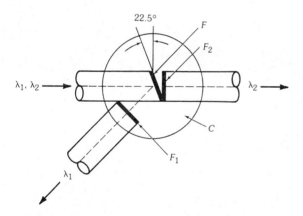

FIG. 9. All-fiber wdm coupler. (*After Miyauchi et al. [29]*)

In a demultiplexing device of this form one can take advantage of different fiber parameters to overcome some of the insertion loss. For example, in the case of a demultiplexer for a single-mode fiber line, one of the receive lines can be a multimode fiber [30] as shown in Fig. 10, since this receive line is presumably short and will not constitute a limit in bandwidth or attenuation. Also, since the coupling from this fiber is to a detector, no mismatch will occur at this point. In this structure the spacing between fibers 1 and 2 can be minimized to reduce insertion loss at this

interface. This type of fiber mix cannot be utilized for a multiplexer, however.

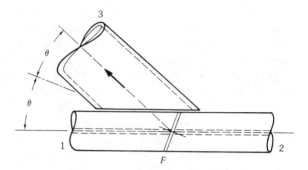

FIG. 10. Single-mode to multimode wdm all-fiber coupler. (*After Winzer, Mahlein, and Reichelt [30]*)

The next level of complexity is to use lenses to minimize geometrical losses at the fiber junction. The lenses can be in the form of a shaping of the fiber end faces [31] as shown in Fig. 11. In this device, using typical multimode graded-index fiber, losses of 1 to 1.5 dB and greater than 35-dB crosstalk attenuation were achieved.

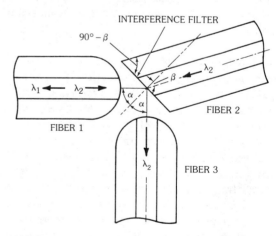

FIG. 11. Fiber-lens wdm coupler. (*After Rode and Weidel [31]. Acknowledgment is made of the prior publication of this material by the IEE*)

A more sophisticated lensing system uses 1/4-pitch graded-index–rod lenses to collimate the beam from the input fiber and then refocus the beams to the output fibers after the operation of the interference filter. Such a structure is shown in Fig. 12. The collimation and expansion of the input beam by the graded-index–rod lens permits use of a large and high-performance filter. Typical performance of this type of structure for single-mode or multimode–fiber devices is less than 0.5-dB insertion loss and crosstalk which can be as low as −35 dB, depending on channel spacing. Fig. 13 shows how several such devices can be cascaded into a multichannel wdm coupler [32].

The number of graded-index–rod lenses can be reduced by the use of tilted mirrors or filters [33], as in Fig. 14. Here λ_1 is reflected at interference filter F, and λ_2 is reflected at mirror R. The tilt permits the focus of these two wavelengths and hence the output fiber positions to be offset.

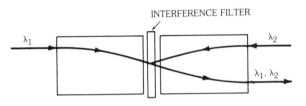

(a) λ_1 and λ_2 output together.

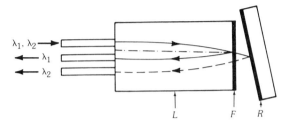

FIG. 14. Wavelength-division–rod lens and tilted mirror. (*After Tanaka, Serizawa, and Tsujimoto [33]. Acknowledgment is made of the prior publication of this material by the IEE*)

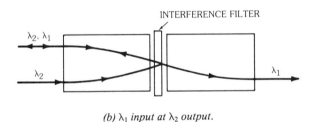

(b) λ_1 input at λ_2 output.

FIG. 12. Wavelength-division–multiplexing coupler with graded-index–rod lenses for collimation.

Another form of multiport wdm multiplexer/demultiplexer using graded-index–rod lenses [34] is shown in Fig. 15. This device uses a polygonal prism structure to minimize interlens spacing in a compact multireflection structure. It is optimized for ease of assembly, low insertion loss, and high mechanical stability.

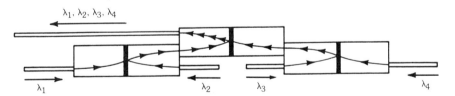

(a) Four wavelengths output at one port.

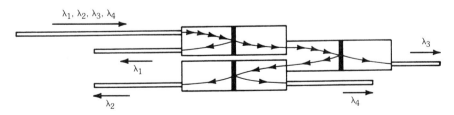

(b) One wavelength output at four ports.

FIG. 13. Two cascaded-lens multichannel wdm coupler. (*After Kobayashi et al. [32]*)

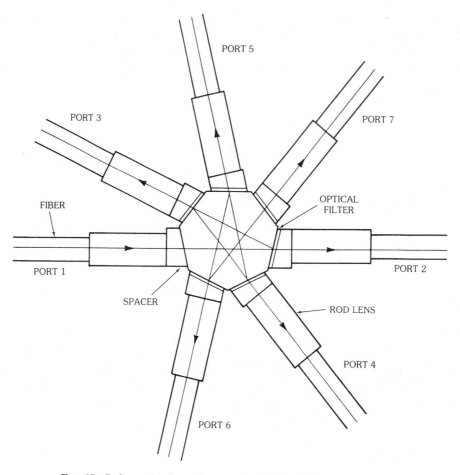

FIG. 15. Polygonal-prism wdm coupler. (*After Ishikawa et al. [34]*)

Diffraction-Grating–Based WDM Couplers

The main advantage of diffraction-grating–based couplers in comparison with interference-filter–based devices is the intrinsic ability of the grating to process all the wavelength channels in parallel and hence to avoid a multiplicity of wavelength-selection elements as the number of wdm channels is increased. This insertion loss does not necessarily increase with number of channels and the loss can be uniform for all channels. This advantage comes at the cost of a structure-dependent passband; the bandshape will depend on the positioning of the input and output fibers in the structure and on such parameters as the core dimensions and the fiber numerical aperture. Also, since effective operation of a grating depends on having a major portion of its surface area illuminated, the throw

distance between input fiber and the individual output fibers is usually substantial, necessitating highly accurate focusing elements and/or a trade-off in terms of the fiber parameters. In general, diffraction-grating–based devices are preferred in demultiplexers where such trade-offs can be made and in cases where a large number of channels are to be demultiplexed.

Because of the long path lengths between input and output fibers, focusing optics are mandatory in these devices. The different forms that this focusing takes distinguish the basic forms of diffraction-based wdm couplers that have been developed.

The earliest practical demonstrations of grating wdm multiplexers used a graded-index–rod lens as the focusing element [35, 36] with the plane grating in Littrow mount configuration. The form of these devices is shown in Fig. 16. The grating and graded-index–rod lens were separate elements with an air space between them. Insertion loss was approximately 2.4 dB with approximately −30-dB crosstalk at channel spacings of 27 nm. Later improvements on this configuration incorporated a larger-diameter graded-index–rod lens [37] in order to increase the allowable channel number to 12 and maintain similar per-channel optical performance. These later devices also made use of wedge-shaped grating blocks in order to fill in the air gap between grating and lens.

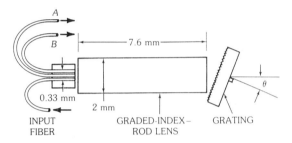

FIG. 16. Diffraction-grating wdm multiplexer.

Grating-based demultiplexers using conventional lenses have also been demonstrated [38]. This demonstration was noteworthy for a high optical efficiency, environmentally stable echelette grating using an anisotropic etching of a silicon substrate. This development of gratings aimed specifically for this application rather than the use of commercial pressed gratings reflects the recognition of the necessity to develop wdm components capable of surviving in field environments. A more recent development in this direction has been the elimination of a separate focusing element by the use of specially developed curved gratings [39–41]. An example of this form of wdm component is shown in Fig. 17. This device consists of a concave diffraction grating, a multimode slab waveguide, and a fiber array. The concave diffraction grating is made with a constant groove angle by cylindrically bending a silicon plane diffraction grating. Placement of the input and output fibers can be arranged to obtain high diffraction efficiency for such a curved grating. The slab waveguide is all-glass formed by a

hot-press method and when bonded to the grating and fiber array forms a mechanically stable structure. In this particular device use was made of output fibers larger than the input fiber in order to reduce the insertion loss.

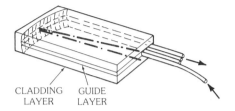

FIG. 17. Wavelength-division multiplexer using a focusing grating. (*After Fujii and Minowa [39], ©1980 IEEE, Yen et al. [40], and Nosu and Watanabe [41]. Acknowledgment is made of the prior publication of this material by the IEE*)

Single-Mode Guided-Wave Interference- Based WDM Couplers

Single-mode coupling structures can show wdm coupler behavior without the use of wavelength-selective elements, such as inteference filters or diffraction gratings. The wavelength-selective behavior results from wavelength-dependent coupling in single coupling regions between two light guides or by interference effects between two or more coupling regions joined by multiple paths. The former type of wdm device was studied theoretically and experimentally by Digonnet and Shaw [42] using their previously developed quartz-block fiber coupler. In this type of coupler the interaction length and strength could be chosen through the fiber curvature and the proximity of the two fiber cores. Wavelength dependence of the coupling ratio of a given coupler derives from the wavelength dependence of the electric-field distributions of the waveguide modes of the fibers forming the coupler. These field distributions determine the light coupling between the arms of the device. In a perfect coupler having no loss, polarization dispersion, or mismatch of propagation constant between the two light guiding regions, the output at each arm of the coupler will show a sine-squared characteristic against the wavelength of light. The pitch and phase of the characteristic is determined by the parameters of interaction length and proximity of the core regions at the waist of the coupler. Thus, with this coupler, one signal wavelength can be totally coupled while a signal wavelength 90° apart along this characteristic will be totally uncoupled. These two signals can accordingly be multiplexed or demultiplexed by the coupler by feeding them in on separate or the same input arms. Similar behavior to this has been observed and analyzed in a single-mode–fiber coupler based on a two-core fiber [43].

The single coupling region in these devices causes the coupling interaction length and coupling strength to be interlinked and therefore control of the pitch of the coupling characteristic while maintaining a high extinction ratio is difficult. Separate control of these parameters can be achieved by taking advantage of guided-wave interference between spatially separated coupling regions in an interferometer structure. This principle of wdm coupling in an optical-fiber structure was first demonstrated in a modification of the bottle coupler in a Mach-Zehnder configuration [44] as shown in Fig. 18. Here the structure consists of two 3-dB couplers in the two twist regions and the asymmetric wrapping separating these two regions creates a differential optical path length between them. The tuning of the

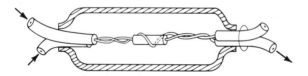

FIG. 18. Single-mode–fiber wavelength multiplexer. (*After Sheem and Moeller [44]*)

couplers to a 3-dB split for the wavelength of operation determines the extinction ratio, while the path difference between these couplers determines the pitch of the overall coupling characteristic. In the device demonstrated, a path difference of 3.5 nm at a wavelength of approximately 600 nm was achieved. No information about the extinction ratio was given. Ideal operation of this type of device would require several features. The individual 3-dB couplers should be dimensioned to have high coupling strength so that in a given spectral working range they func-

tion as independently of wavelength as possible. Also, the couplers should not perturb the polarization of the light; the insertion loss should be low; and the paths between the couplers should be arranged to minimize differential polarization shifts. More nearly ideal behavior can probably be achieved by using more recently developed coupling structures such as the quartz-block coupler [24] or the fused biconical taper single-mode coupler [15], particularly in a tunable, polarization-maintaining form [45, 46].

This type of wdm coupler is not limited to two-channel wdm. When individual two-channel devices have been tailored they can be cascaded (Fig. 19) to form four-channel wdm devices. In this assembly the couplers have two forms; the upstream couplers in a multiplexing structure have a pitch twice that of the following downstream coupler.

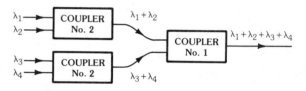

FIG. 19. Cascaded two-channel wdm couplers to form four-channel wdm coupling structure.

4. Optical-Fiber Switches

The optical switching of light guided in fiber offers an alternative to electrical switching in which the light is detected to produce an electrical signal which is switched or otherwise processed and subsequently converted to an optical signal again in one or more fibers. The choice of optical switching depends not only on the availability of suitable optical switches but on the mixture of advantages and disadvantages it offers compared with electrical switching. On the plus side, the optical switch offers a simple crosspoint or matrix of crosspoints which will not limit the signal bandwidth and can offer high reliability with the absence of optoelectronic conversions. On the negative side, this direct routing of an optical signal does not provide the opportunity to process the signal as does electrical switching. Such processing could include amplification, pulse shaping, retiming, and a variety of forms of signal multiplexing. Perhaps because of this limited applicability the development and availability of fiber-optic switches is not extensive. There have, however, been a large number of mechanisms demonstrated for optical switching.

The desirable characteristics of a switch include high switching speed, low insertion loss, low crosstalk and high directivity, and wide spectral bandwidth. The extent to which these characteristics can be achieved in a given structure depends to a large extent on the switching mechanism.

The most straightforward mechanism for switching light guided in optical fiber is mechanical motion of the fibers to make and break optical paths [47]. An early demonstration of this switching principle is shown in Fig. 20. In this device,

alignment is accomplished by reference to the fiber cladding surface. A similar structure has been used to fabricate single-mode–fiber switches in 1×4 form by making use of all four corners of the square tubing [48]. In this device the actuation mechanism was a transverse deflection of the single fiber using a rotating section. The average insertion loss of the device was less than 0.5 dB with crosstalk better than −70 dB. A further advance in mechanically actuated switches makes use of preferentially etched silicon chips for the fiber alignment mechanism [49] as shown in Fig. 21. In this device, arrays of fibers are accurately positioned relative to one another by the etched grooves in the chips and the arrays are moved relative to one another by controlled movement of the chips. A positive switching action can thus be made to occur

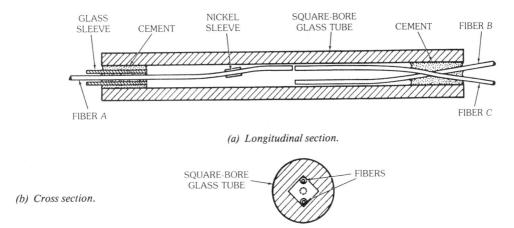

(a) Longitudinal section.

(b) Cross section.

FIG. 20. Optical switch based on movement of optical fibers in tube enclosure. (*After Hale and Kompfer [47]. Acknowledgment is made of the prior publication of this material by the IEE*)

optical contact is made or broken between a single fiber, *C,* and two other fibers, *A* and *B,* by mechanically moving fiber *C* to opposite corners of a square bore tube in which fibers *A* and *B* are fixed. In the actual demonstration, motion of fiber *C* was accomplished by pushing it transversely from outside the tube, but, as shown in the illustration, the use of a magnetic sleeve would permit electromagnetic control of this motion. This form of switching takes advantage of the small cross section of fibers which permits the switching to be accomplished by small transverse motions. Also it depends on accurate control of the dimensions and core/cladding concentricity of the fibers since

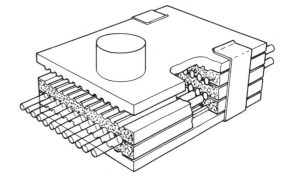

FIG. 21. Cascaded multipole optical switch. (*After Young and Curtis [49]. Acknowledgment is made of the prior publication of this material by the IEE*)

in a multiple-pole form. In this configuration, high-capacity switching can occur with fewer switching stages than would be required with single-pole switches. In single-mode–fiber form, this switch showed an insertion loss of less than 1 dB. Other forms of mechanically actuated switched switches utilize change in beam paths between fibers rather than fiber motion. Fig. 22 shows an optical switch used as an element to bypass a failed terminal in a bus network [50]. Here a prism can be rotated in two orientations to redirect the optical beam paths. This type of switch has also been demonstrated in a matrix form. [51].

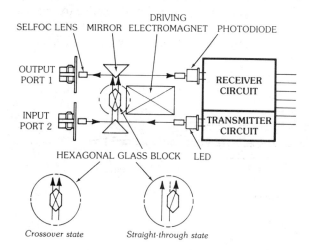

FIG. 22. Prism optical switch used as bypass element. (*After Uda, Aoyama, and Kikuo [50]*)

Switches based on mechanical motion can have high performance with respect to insertion loss and crosstalk, and, by careful design, factors such as chatter can be minimized. Also, this mechanism intrinsically does not affect the bandwidth of the switched signal. However, it is unlikely that high switching speed can be achieved in this type of switch and mechanical wear and subsequent degradation of performance or failure is also a possibility. Furthermore, substantial energy is required to perform the switching action. These problems of switching speed, wear, and energy requirement can be overcome in switches using several other mechanisms.

Nonmechanical switching of light in structures compatible with optical fiber has been demon-

strated in devices based on liquid crystals [52]. This technique offers potential for low-voltage, low-power control, high reliability, low insertion loss, and low optical crosstalk. Some of this potential has been realized. Many different configurations for this type of switch have been demonstrated. The double-pass switch [52] of Fig. 23 illustrates the principle of operation. This is a 2×2 switch for unpolarized light with multimode fiber forming the input and output ports. Two prisms of high-index glass enclose a transparent layer of field-effect liquid crystal. The light is obliquely incident on the liquid-crystal film at two points in the device. At the first crossing the originally unpolarized light is divided into two orthogonally polarized beams. At the second crossing the state of the liquid crystal can be changed electrically to permit recombination of the two beams in either of two directions as illustrated. Graded-index–rod lenses are used to collimate the light coming out of the input fibers and to focus the light into the output fibers. This device had a directivity (signal-to-crosstalk) ratio of 13 dB in both the on and off states. In it the insertion loss is comprised of intrinsic scattering loss in the liquid-crystal layer, reflection loss, and optical coupling loss. The insertion loss of this double-pass device was approximately 1 dB.

More elaborate liquid-crystal switch structures

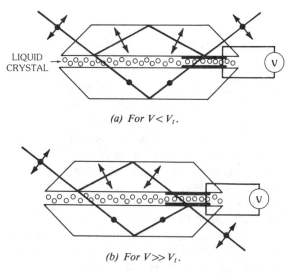

(a) For $V < V_t$.

(b) For $V \gg V_t$.

FIG. 23. Liquid-crystal switch. (*After Soref and Kestigan [52]*)

have been demonstrated to improve crosstalk performance and to provide matrix switch functions. Fig. 24B shows a 4×4 switching array [53]. In this two-stage device any of the four inputs can be switched to any of the four outputs, thus giving 24 different configurations of the switch. Furthermore, the use of two stages of switching greatly improves the crosstalk performance since most of the leakage from the first stage is rejected by the second stage. In the device crosstalk between outputs was of the order of −40 dB. The switching time was 5 ns with drive voltage at 40 V rms. Switching in matrix form of unpolarized light using a more complex prism configuration [54] has also been demonstrated.

polarization of an incident beam and hence the ratio of power in the two orthogonally polarized output beams defined by ports 3 and 4. The relationship of the parameters involved in this effect is

$$V = \delta \lambda d / 2 \pi \eta_0^3 r_{22} L,$$

where

V = the voltage across the crystal,
d = the thickness of the crystal,
L = the length of the crystal,
r_{22} = the electro-optic coefficient,
η_0 = ordinary refractive index,
λ = the free-space wavelength,
δ = the phase difference caused by the electro-optic effect.

Thus, incident light can be split into any proportion at the output ports. In a demonstration using a multimode-fiber input collimated by a graded-index-rod lens, an extinction ratio of 33 dB between the two outputs was obtained with an insertion loss of 3.5 dB.

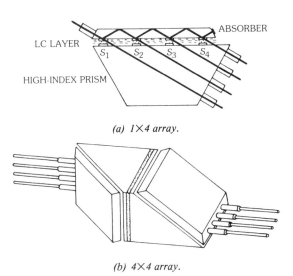

(a) 1×4 array.

(b) 4×4 array.

FIG. 24. Liquid-crystal switch. (*After Soref [53]*)

There are nonmechanical switches based on polarization rotation in crystals. These devices have potential for high switching speed and the reliability associated with no moving parts. One type shown in Fig. 25 is based on an electro-optic effect [55]. An LiNbO₃ crystal is located between two Wollaston prisms (WP). Incident beams from the left-hand side following paths defined by ports 1 and 2 are transformed into two orthogonal linear polarizations by the first prism. Those portions of these incident beam corresponding to the wrong polarization, exit along the dotted paths. A voltage applied to the LiNbO₃ crystal serves to change the state of

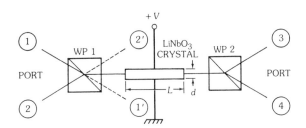

FIG. 25. Polarization-rotation optical switch. (*After Takahashi et al. [55], ©1982 IEEE*)

A polarization-based switch using a magneto-optic effect [56] is shown in Fig. 26. The operation is qualitatively similar to the electro-optic device previously described with an input polarizer for the incident beam and two orthogonally polarized output paths defined by the output polarizer. In this device, however, the switching is performed by using the Faraday effect in a single crystal of yttrium iron garnet (YIG). The device was optimized for use at 1.3-mm wavelength and for lens coupling to single-mode–fiber input and output ports. To minimize the required external magnetic field the YIG crystal was made in the form of a thin plate with the small cross section along the light beam path. This is a bistable device in that operation of the switch involves a pulse of current in either direction through the coil to reverse the direction of magnetization in the semihard magnetic material of the electromagnet. Thus a magnetic field can be applied to the crystal in either direction with no current flowing.

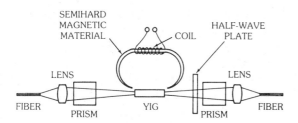

FIG. 26. Magneto-optic optical switch. (*After Shirasaki et al. [56]*)

A refinement of this configuration to permit operation with unpolarized input beams [56] is shown in Fig. 27. Prisms are used to create two beam paths internal to the switch corresponding to the two orthogonally polarized portions of light. This device operating at 1.3 nm had an insertion loss of 1.4 dB and −28 dB crosstalk. The state of the switch could be changed in 10 ms.

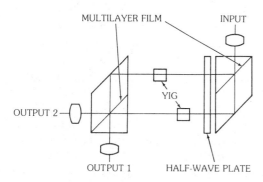

FIG. 27. Magneto-optic optical switch for unpolarized light. (*After Shirasaki et al. [56]*)

A switch developed for multimode fibers that has attributes and characteristics combining those of mechanical and nonmechanical switches is the electrowetting optical switch [57]. This device depends for its switching action on the motion of reflecting liquid (mercury) driven by the electro-

wetting effect. This effect is illustrated in Fig. 28. Platinum electrodes are placed at either end of a sealed capillary containing a liquid electrolyte and a small slug of mercury. A thin layer of electrolyte fills the region between the mercury and the walls of the capillary. This layer is sufficiently thin that a voltage applied between the electrodes drops mainly in this region, thus causing a surface tension gradient at this interface. Consequently an electrowetting pressure gradient occurs in this region which tends to cause the mercury slug to move. This motion is toward the negative electrode. The motion continues as long as the potential is applied or until a sufficient restoring force is present.

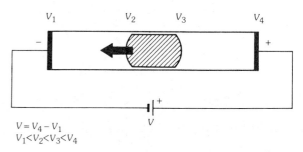

$$V = V_4 - V_1$$
$$V_1 < V_2 < V_3 < V_4$$

FIG. 28. Electrowetting effect. (*After Jackel, Hackwood, and Beni [57]*)

Fig. 29 shows the principle of operation of the switch based on this effect. The capillary is sandwiched between two 1/4-pitch graded-index–rod lenses which serve as collimating and focusing elements to couple light among the beam paths illustrated. An applied voltage moves the mercury slug into and out of the beam path to switch the light between the two states. The switch had a low crosstalk of −30 dB, low operating voltage of 1 V, power consumption of 1 mW, and a switching speed of 20 ms.

5. Optical Isolators

An optical isolator is a device that is intended to prevent return reflections along a transmission path. For optical communications the primary use of this device is to prevent return of reflected or scattered light into the semiconductor lasers, whose stability of oscillation is particularly sus-

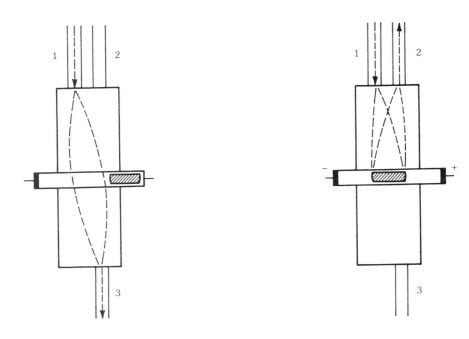

(a) With no voltage applied.   *(b) With voltage applied to electrodes.*

FIG. 29. Electrowetting-based optical switch. (*After Jackel, Hackwood, and Beni [57]*)

ceptible to this type of disturbance. These devices can also be used in fiber networks to suppress echos and thus the pulse distortion or false signals caused by the echos.

Desirable optical characteristic of isolators are (*a*) low insertion loss for the forward-going signal, (*b*) high isolation of the return signal, and (*c*) wide spectral bandwidth of operation. In addition, for practical use the devices should have certain physical attributes, such as compact size and environmentally stable performance. These characteristics have been achieved in isolators based on the Faraday effect. The basic form of this device is shown in Fig. 30. A Faraday rotation crystal which ro-

tates the plane of polarization of a passing light beam by 45° is located between two polarizers oriented at 45° to each other. Thus a light beam traveling to the right, the "forward" direction, will, if polarized in the orientation of the first polarizer, be rotated 45° clockwise and pass the second polarizer. On the other hand, a return reflection that is oriented with the second polarizer will be rotated anticlockwise 45° because of the nonreciprocal nature of Faraday rotation and will thus have a polarization crossed with the first polarizer at that point. Usually the degree of isolation of this backward beam is limited not by the extinction ratio of the polarizers but by imperfection of the Faraday

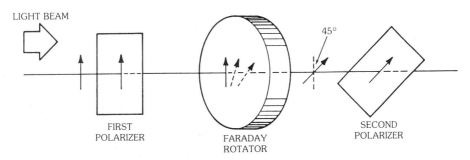

FIG. 30. Principle of operation of Faraday-effect isolator.

257

rotation. The material for the rotator is chosen to provide low optical absorption and strong Faraday rotation at the desired wavelength of operation. In the visible region, paramagnetic glasses [58] and terbium aluminum garnets [59] have been used. For the longer-wavelength windows used for fiber-optic communications the rotators are usually based on iron-garnet materials [60, 61]. In the following paragraphs particular isolator structures developed for optical-fiber communications are discussed.

Design for Compact Size

Compact optical isolators can be achieved by using Faraday rotator materials of type and dimensions that can be saturated with small permanent magnets. Fig. 31 shows such a design using a rod of $Y_3Fe_5O_{12}$ (YIG) crystal and a ring-shaped samarium cobalt magnet [62]. This device was intended for use at a wavelength of 1.153 mm. The rotator was 8.6 mm in diameter by 5.4 mm long with a 2-mm clear aperture and had antireflection coated faces. As shown the ring magnet was sized to this crystal. The polarizers were Glan-Thompson calcite prisms also antireflection coated. The forward loss of this device was 1.3 dB for linearly polarized light primarily due to loss in the rotator crystal. The backward loss was greater than 30 dB.

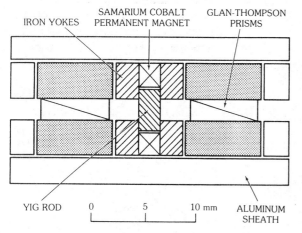

FIG. 31. Cross section of isolator. (*After Iwamura, Hayaski, and Iwasaki [62]*)

For the particular application of isolation of a laser diode coupling to a single-mode fiber, a very compact isolator structure has been demonstrated.

This structure integrates the function of coupling and isolation by using a YIG sphere as the coupling lens [63] as shown in Fig. 32. The 2.1-mm diameter YIG rotates the polarization direction of light by 45° under a field of 1700 G. This structure takes advantage of the fact that the light from a laser diode is highly polarized and eliminates one of the usual two polarizers. The isolation value is limited by the YIG sphere. There is a nonuniformity of the rotation angle of the sphere caused by inherent crystal nonuniformity and also by the fact that light traversing the sphere at different angles travels different distances through the material. Nevertheless, the isolation provided by this structure was greater than 30 dB and the coupling efficiency to a single-mode fiber was a very acceptable −5.2 dB.

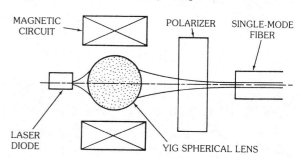

FIG. 32. Isolator for laser-diode coupling to optical fiber. (*After Sugie and Saruwatari [63]. Acknowledgment is made of the prior publication of this material by the IEE*)

Other efforts being made toward compactness of isolators involve the use of thick Gd-YIG films [64] for the rotator material. This material can use smaller magnets because of the lower field required for saturation in comparison with bulk YIG crystals. In addition, there is the potential for lower cost resulting from greater consistency of material growth.

Polarization-Independent Isolators

Polarization-independent operation of Faraday-effect optical isolators can be achieved by taking advantage of birefringence in the polarizers. Fig. 33 shows the layout of the isolator [65]. The polarizers are in the form of tilted wedge-shaped sections of a birefringent material, titanium-dioxide single crystal. If the forward direction is from left to right, then light originally in the fiber is coupled out through an antireflection-coated glass plate and then collimated by a lens. This originally unpolarized beam is then separated into two orthogonally polarized parts corresponding to the ordinary and extraordinary rays in the birefringent material. Because of the tilt and wedge shape of the polarizer, these two beams will propagate at a slight angle to each other. After rotation through the Faraday rotator these two beams are processed through the second wedge, which is oriented in its optical axis and its wedge direction with respect to the first polarizer so that the ordinary and extraordinary rays emerge parallel but slightly deviated with respect to each other. The two beams are then efficiently refocused into the downstream fiber. For light propagating in the reverse direction the ordinary and extraordinary beams formed by the first-encountered wedge will be separated in angle by this first wedge and then, after rotation by the Faraday crystal, will be deflected from each other even farther by the second-encountered wedge. This different behavior results from the nonreciprocal polarization rotation of the Faraday crystal. These two nonaxial beams are then focused to miss the downstream fiber and thus this return light has high attenuation through the isolator structure. The structure operating at 1.3-mm wavelength had a forward insertion loss of 0.8 dB and a 35-dB backward loss into multimode fiber.

Broadband Isolator Structures

Optical isolators based on Faraday rotation usually have a limited range of operation with respect to the wavelength of light because of the dispersion of the Faraday rotation. In a YIG crystal the Faraday rotation in the wavelength range of interest for optical-fiber communications decreases monotonically with increasing wavelength. The resulting differences in rotation angle cause a decrease in backward loss and an increase in forward loss.

In [62] an isolator based on YIG was made to operate in a broadband fashion by using the dispersion of optical activity in a section of quartz to cancel the dispersion from the YIG crystal. In this device the optically active element contributed a reciprocal rotation of polarization of 45°, thus producing a total rotation of the angle of polarization of 90° in one propagation direction and 0° in the other when combined with a 45° Faraday rotator. These two beams could be passed or isolated by polarizers set at 90° with respect to each other.

Broadband operation can also be obtained by selection of the Faraday rotator material to have low dispersion in the wavelength region of interest. Isolators using yttrium iron garnet have been shown [66] to have wavelength-independent operation over the range 1.3 to 1.7 mm.

Fiber-Based Isolator Structures

A Faraday rotator in the form of a single-mode fiber would considerably simplify the overall isolator structure since it could eliminate the need for collimating optics and permit the use of micropolarizers in a very stable structure. A single-mode fiber with a large Faraday effect [67] was made using Hoya FR-5 glass [68] and a rod-in-tube method. At a wavelength of 0.633 nm this fiber could produce a 45° Faraday rotation in a distance of a few centimeters. A proposed structure for the isolator [67] is shown in Fig. 34. The micropolarizers would be deposited as periodic metal-dielectric layers in grooves etched in the fiber.

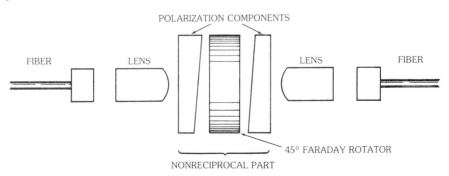

FIG. 33. Configuration of polarization-independent optical isolator. (*After Shirasaki and Asama [65]*)

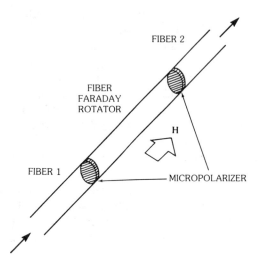

FIG. 34. Optical isolator in optical fiber. (*After Shiraishi, Sugaya, and Kawakami [67]*)

6. Optical Attenuators

An optical attenuator is a device that is inserted into a transmission path to cause a controlled insertion loss. The attenuator can provide a fixed insertion loss or may provide an adjustable insertion loss over a wide range of optical signals. Optical attenuators suitable for use in fiber-optic networks or associated with fiber-optic test equipment should have a number of attributes:

1. *Mode-independent optical characteristics:* The mechanism for attenuating light should operate equally on all guided modes in a multimode optical fiber so as not to change the mode distribution as light traverses the attenuator. Similarly, the attenuation should be independent of the state of polarization of the light. Mode-dependent behavior can lead to the generation of modal noise under various conditions of type of light source and mechanical stability of system parts and is ideally to be avoided.

2. *Low reflection coefficient:* If the mechanism for attenuation results in reflected light along the transmission path, this reflected light may degrade the properties of other components. In particular, reflected light can cause noisy behavior in laser diodes and can degrade the directionality of branching devices.

3. *Wide spectral range of operation:* The attenuation provided should be independent of the wavelength of light over the intended operation region.

4. *Environmentally stable properties:* Since attenuators may be used in rather harsh outside environments, the attenuation mechanism and packaging should be consistent with stable performance over wide ranges of environmental and mechanical conditions.

In addition, attenuators for particular uses may have to be capable of handling high optical-power levels or be very compact or accurately calibrated. A number of attenuator designs have been developed to provide these characteristics.

Fixed Optical Attenuators

These are usually small in-line components used to bring the optical power level into the dynamic range of the receiver. The simplest mechanism for attenuation that is consistent with attributes 1 and 2 above is an optical absorbing layer consisting of a medium which is transparent to the light but which is doped with one or more substances which have absorption bands in the wavelength range of operation. Examples of such layers are gelatine filters (Kodak Wratten) or inorganic filters (Specivex MTO). Figs. 35a and 35b show two versions of fixed attenuators. In Fig. 35a the absorbing layer lies between two fibers which are aligned axially. In this structure it is important that the absorbing layer be very thin in order to minimize attenuation associated with spreading of the light as it leaves the fiber and traverses the absorbing medium. This type of loss is mode dependent. A further deficiency of this structure is the small cross-sectional area occupied by the light between the fibers. This structure is thus not consistent with higher light levels as heat will be dissipated in a very small volume. In the structure shown in Fig. 35b these problems are largely overcome by the use of graded-index–rod lenses to collimate the light between the fibers and then refocus the light into the downstream fiber. The absorbing layer is sandwiched between the two lenses. Because the light is collimated at the absorbing layer, this layer can be thicker than in the former structure and also the power density of the light is lower for a given power level in the fiber because of the beam expansion caused by the lenses.

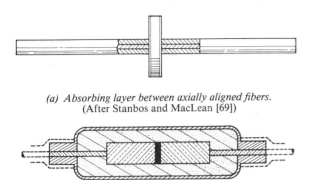

(a) Absorbing layer between axially aligned fibers.
(After Stanbos and MacLean [69])

(b) Use of graded-index–rod lenses to collimate light.
(After Barbaudy and Galaup [70])

Fig. 35. Fixed attenuators.

Variable Optical Attenuators

A variety of variable optical attenuators can be devised using polarization rotators located between polarizing elements. However, a variable attenuator is more versatile and easier to use if it operates independently of the state of polarization of the light. One form of polarization-independent attenuator [71] is shown schematically in Fig. 36. This device uses planar transmissive slabs of crystalline calcite to separate and recombine orthogonal polarizations and a field-effect twisted nematic liquid-crystal cell sandwiched between the calcite slabs to rotate the plane of polarization of the

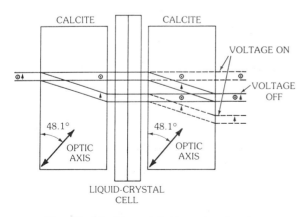

Fig. 36. Liquid-crystal attenuator showing polarization states (arrows) of transmitted light beams. (*After Hanson [71]*)

passing light beam. A 90° rotation of the polarization causes the output beam to change from the combined condition indicated by solid lines in the figure to the two separated beams shown by dotted lines. The output focusing optics (not shown in figure) can be arranged to focus only the combined beam into the downstream fiber. Thus the attenuator is varied by changing the relative portions of light in the combined and separated beams. In the device demonstrated, an insertion loss of 0.87 dB was measured with a dynamic range of 31.5 dB. The attenuation limit was related to the tolerance on the 90° twist angle by the liquid-crystal cell.

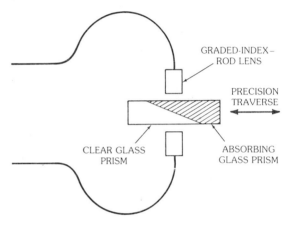

Fig. 37. Glass-prism variable optical attenuator. (*After JDS Optics Co. [72]*)

A simple mechanically operated variable optical attenuator is based on the use of a graded absorbing element [72] as shown in Fig. 37. A slab of glass with absorption varying linearly along one axis is formed by two prisms, one absorbing and one clear. Variation of attenuation is performed by linear translation of this element across the path of the light beam collimated by graded-index–rod lenses.

7. References

[1] K. Kobayashi, R. Ishikawa, K. Minemura, and S. Sugimoto, "Micro-Optic Devices for Branching Coupling, Multiplexing and Demultiplexing," *Proc. IOOC,* Paper B11-4, Tokyo, July 1977.

[2] W. J. Tomlinson, "Applications of GRIN-Rod Lenses in Optical Fiber Communication Systems," *Appl. Opt.,* Vol. 19, p. 1127, 1980.

[3] R. J. S. BATES, J. D. SPALINK, S. J. BUTTERFIELD, J. LIPSON, C. A. BURRUS, T. P. LEE, and R. A. LOGAN, "1.3-mm/1.5-mm Bidirectional WDM Optical Fibre Transmission System Experiment at 144 Mbits," *Electron. Lett.*, Vol. 19, p. 458, 1983.

[4] R. ISHIKAWA, K. KAEDA, H. NISHIMATA, K. MINEMURA, and S. MATSUSHITA, "Kaleidoscope Micro-Optic Star Coupler," *Electron. Lett.* Vol. 16, p. 248, 1980.

[5] K. THYAGARAJAN, A. ROHRA, and A. K. GHATAK, "Aberration Losses of the Microoptic Directional Coupler," *Appl. Opt.*, Vol. 19, p. 266, 1980.

[6] S. MASUDA and T. IWAMA, "Single-Mode Fiber-Optic Directional Coupler," *Appl. Opt.*, Vol. 21, p. 3484, 1982.

[7] H. H. WITTE, "Optical Tapping Element for Multimode Fibers," *Opt. Commun.*, Vol. 18, p. 559, 1976.

[8] H. JUJITA, Y. SUZAKI, and A. TACHIBANA, "Optical Fiber Wave Splitting Coupler," *Appl. Opt.*, Vol. 15, p. 2031, 1976.

[9] Y. SUZUKI and H. KASHIWAGI, "Concentrated-Type Directional Coupler for Optical Fibers," *Appl. Opt.*, Vol. 15, p. 2032, 1976.

[10] B. S. KAWASAKI and K. O. HILL, "Low-Loss Access Coupler for Multimode Optical Fiber Distribution Networks," *Appl. Opt.*, Vol. 16, p. 1794, 1977.

[11] B. S. KAWASAKI, K. O. HILL, and Y. TREMBLAY, "Modal-Noise Generation in Biconical Taper Couplers," *Opt. Lett.*, Vol. 6, p. 499, 1981.

[12] E. G. RAWSON and M. D. BAILEY, "Bitaper Star Couplers With up to 100 Fibre Channels," *Electron. Lett.*, Vol. 15, p. 432, 1979.

[13] J. C. WILLIAMS, S. E. GOODMAN, and R. L. COON, "Fiber-Optic Subsystem Considerations of Multimode Star Coupler Performance," *Proc. Conf. Optical Fiber Commun.*, Paper WC6, New Orleans, January 23–25, 1984.

[14] Y. TREMBLAY, B. S. KAWASAKI, and F. BILODEAU, "Modal-Insensitive Biconical-Taper Couplex," *Opt. Lett.*, Vol. 1, p. 506, 1982.

[15] B. S. KAWASAKI, K. O. HILL, and R. G. LAMONT, "Biconical-Taper Single-Mode Fiber Coupler," *Opt. Lett.*, Vol. 6, p. 327, 1981.

[16] M. H. SLONECKER, "Single-Mode Fused Biconical Taper Coupler," *Tech. Dig. Top. Meeting Optical Fiber Commun.*, Paper WBB7, Phoenix, 1982.

[17] C. A. VILLARRUEL and A. P. MOELLER, "Optimized Single-Mode Tapped Tee Data Bus," *Tech. Dig. Top. Meeting Optical Fiber Commun.*, Paper TuFF3, Phoenix, 1982.

[18] J. BURES, S. LACROIX, and J. LAPIERRE, "Analyse d'un Coupleur Bidirectionnel a Fibres Optime Monomodes Fusionnees," *Appl. Opt.*, Vol. 22, p. 1918, 1983.

[19] B. S. KAWASAKI, M. KAWACHI, K. O. HILL, and D. C. JOHNSON, "Single-Mode Fiber Coupler With Variable Coupling Ratio," *IEEE J. Lightwave Tech.* Vol. LT-1, p. 126, 1983.

[20] K. O. HILL, D. C. JOHNSON, and R. G. LAMONT, "Efficient Coupling-Ratio Control in Single Mode Fiber Biconical-Taper Couplers," *Proc. Conf. Optical Fiber Commun.*, Paper WE2, New Orleans, January 23–25, 1984.

[21] S. K. SHEEM and T. G. GIALLORENZI, "Single-Mode Fiber-Optical Power Divider Encapulated Etching Technique," *Opt. Lett.*, Vol. 4, p. 29, 1979.

[22] C. A. VILLARRUEL and R. P. MOELLER, "Fused Single Mode Fibre Access Couplers," *Electron. Lett.*, Vol. 17, p. 243, 1981.

[23] S. K. SHEEM and T. G. GIALLORENZI, "Single-Mode Fiber Multiterminal Star Directional Coupler," *Appl. Phys. Lett.*, Vol., 35, p. 131, 1979.

[24] R. A. BERGH, G. KOTLER, and H. J. SHAW, "Single-Mode Fiber Optic Directional Coupler," *Electron. Lett.*, Vol. 16, p. 260, 1980.

[25] S. A. NEWTON, J. E. BOWERS, G. KOTLER, and H. J. SHAW, "Single-Mode Fiber $1 \times N$ Directional Coupler," *Opt. Lett.*, Vol. 8, p. 6, 1983.

[26] G. SCHONER, E. KLEMENT, G. SCHIFFNER, and N. DOUKLIAS, "Novel Method for Making Single Mode Optical Fibre Directional Couplers," *Electron. Lett.*, Vol. 18, p. 517, 1982.

[27] H. F. MAHLEIN, "Fiber-Optic Communication in the Wavelength Division Multiplex Mode," *Fiber and Integrated Optics*, Vol. 4, p. 339, 1983.

[28] S. P. BANDETTINI, "Optical Filters for Wavelength Division Multiplexing," *Proc. SPIE: Fiber Optics Multiplexing and Modulation*, p. 67, Arlington, April 7, 1983.

[29] E. MIYAUCHI, T. IWAMA, H. NAKAJIMA, N. TOKOYO, and K. TERAI, "Compact Wavelength Multiplexer Using Optical Fiber Pieces," *Opt. Lett.*, Vol. 5, p. 321, 1980.

[30] G. Winzer, H. F. Mahlein, and A. Reichelt, "Single-Mode and Multimode All Fiber Directional Couplers for WDM," *Appl. Opt.,* Vol. 20, p. 3128, 1981.

[31] M. Rode and E. Weidel, "Compact and Rugged All-Fibre Coupler for Wavelength Division Multiplexing," *Electron. Lett.,* Vol. 18, p. 898, 1982.

[32] K. Kobayashi, R. Ishikawa, K. Minemura, and S. Sugimoto, "Microoptic Devices for Fiber-Optic Communications," *Fiber and Integrated Optics,* Vol. 2, p. 1, 1979.

[33] T. Tanaka, H. Serizawa, and Y. Tsujimoto, "Simple High Isolation Multi/Demultiplexer," *Electron. Lett.,* Vol. 16, p. 869, 1980.

[34] S Ishikawa, F. Matsumura, K. Takahashi, and K. Okuno, "High Stability Wavelength Division Multi/Demultiplexer With Polygonal Structure," *Proc. SPIE: Fiber Optics Multiplexing and Modulation,* p. 44, April 7, 1983.

[35] W. J. Tomlinson and G. D. Aumiller, "Optical Multiplexer for Multimode Fiber Transmission Systems," *Appl. Phys. Lett.,* Vol. 31, p. 179, 1977.

[36] W. J. Tomlinson and C. Lin, "Optical Wavelength-Division Multiplexer for the 1-1.4 μm Spectral Region," *Electron. Lett.,* Vol. 14, p. 345, 1978.

[37] B. D. Metcalf and J. F. Providakes, "High Capacity Wavelength Demultiplexer With a Large Diameter GRIN-Rod Lens," *Appl. Opt.,* Vol. 21, p. 794, 1982.

[38] Y. Fujii, K. I. Aoyama, and J. I. Minowa, "Optical Demultiplexer Using a Silicon Echelette Grating," *IEEE J. Quantum Electron,* Vol. QE-16, p. 165, 1980.

[39] Y. Fujii and J. Minowa, "Optical Demultiplexer Using a Silicon Concave Diffraction Grating," *Appl. Opt.,* Vol. 22, p. 974, 1983.

[40] H. W. Yen, H. R. Friedrich, R. J. Morrison, and G. L. Tangonan, "Planar Rowland Spectrometer for Fiber-Optic Wavelength Demultiplexing," *Opt. Lett.,* Vol. 6, p. 639, 1981.

[41] K. Nosu and R. Watanabe, "Slab Waveguide Star Coupler for Multimode Optical Fibers," *Electron. Lett.,* Vol. 16, p. 608, 1980.

[42] M. Digonnet and H. J. Shaw, "Wavelength Multiplexing in Single-Mode Fiber Couplers," *Appl. Opt.,* Vol. 22 p. 484, 1983.

[43] Y. Murakami and S. Sudo, "Coupling Characteristics Measurements Between Curved Waveguides Using a Two-Core Fiber Coupler," *Appl. Opt.,* Vol. 20, p. 417, 1981.

[44] S. K. Sheem and R. P. Moeller, "Single-Mode Fiber Wavelength Multiplexer," *J. Appl. Phys.,* Vol. 51, p. 4050, 1980.

[45] M. Kawachi, B. S. Kawasaki, K. O. Hill, and T. Edahiro, "Fabrication of Single-Polarization Single Mode Couplers," *Electron. Lett.,* Vol. 18, p. 963, 1982.

[46] C. A. Villarruel, M. Abebe, and W. K. Burns, "Polarization Perserving Single Mode Fibre Coupler," *Electron. Lett.,* Vol. 19, p. 17, 1983.

[47] P. G. Hale and R. Kompfner, "Mechanical Optical-Fibre Switch," *Electron. Lett.,* Vol. 12, p. 388, 1976.

[48] C. M. Miller, R. B. Kummer, S. C. Mettler, and D. N. Ridgway, "Single-Mode Optical Fibre Switch," *Electron. Lett.,* Vol. 16, p. 783, 1980.

[49] W. C. Young and L. Curtis, "Cascaded Multipole Switches for Single-Mode and Multimode Optical Fibres," *Electron. Lett.,* Vol. 17, p. 571, 1981.

[50] Y. Uda, T. Aoyama, and D. Kikuo, "Local Network Modules With No Restriction on Number of Bypassed Terminals," *Proc. Eur. Conf. Optical Commun.,* p. 371, Paper AXI-3, Cannes, 1982.

[51] M. Oda and M. Shiraishi, "Mechanically Operated Optical Matrix Switch," *Fujitsu Sci. Tech. J.,* p. 121, September 1981.

[52] R. A. Soref and M. Kestigian, "Development of Liquid Crystal and Magnetic Stripe Domain Multimode Optical Switches," *Sperry Research Center, Final Technical Report, RADC-TR-80-217,* p. 11, July 1980.

[53] R. A. Soref, "Electrooptic 4×4 Matrix Switch for Multimode Fiber-Optic Systems," *Appl. Opt.,* Vol. 21, p. 1386, 1982.

[54] R. F. Wagner and J. Cheng, "Electronically Controlled Optical Switch for Multimode Fiber Applications," *Appl. Opt.,* Vol. 19, p. 2921, 1980.

[55] H. Takahashi, C. Masuda, S. Satoh, K. Naiki, and K. Miyaji, "2×2 Optical Switch and its Applications," *IEEE J. Quantum Electron.,* Vol. QE-18, p. 210, 1982.

[56] M. Shirasaki, H. Nakajima, T. Obokata, and K. Asama, "Nonmechanical Optical Switch for Single-Mode Fibers," *Appl. Opt.,* Vol. 21, p. 4229, 1982.

[57] J. L. Jackel, S. Hackwood, and G. Beni, "Electrowetting Optical Switch," *Appl. Phys. Lett.,* Vol. 40, p. 4, 1982.

[58] J. Aplet and J. W. Carson, "A Faraday Effect Optical Isolator," *Appl. Opt.,* Vol. 3, p. 544, 1964.

[59] F. J. SANSALONE, "Compact Optical Isolator," *Appl. Opt.,* Vol. 10, p. 2329, 1971.

[60] B. JOHNSON, "The Faraday Effect at Near Infrared Wavelength in Rare-Earth Garnets," *Brit. J. Appl. Phys.,* Vol. 17, p. 1441, 1966.

[61] T. KIMURA and K. KAZUHIRO, "A Proposal for an Optical Fibre Transmission Systems in a Low-Loss 1.0–1.4 μm Wavelength Region," *Optical and Quantum Electron.,* London: Chapman & Hall, Vol. 9, p. 33, 1977.

[62] H. IWAMURA, S. HAYASHI, and H. IWASAKI, "A compact Optical Isolator Using a $Y_3Fe_5O_{12}$ Crystal for Near Infrared Radiation," *Optical and Quantum Electron.,* London: Chapman & Hall, Vol. 10, p. 393, 1978.

[63] T. SUGIE and M. SARUWATARI, "Nonreciprocal Circuit for Laser-Diode-to-Single-Mode-Fibre Coupling Employing a YIG Sphere," *Electron. Lett.,* Vol. 18, p. 1026, 1982.

[64] T. AOYAMA, K. DOI, H. UCHIDA, T. HIBIYA, Y. OHTA, and T. MATSUMI, "A Low Cost, Compact Optical Isolator Using a Thick Ad:YIG Film Grown by Liquid Phase Epitaxy," *Proc. Eur. Conf. Optical Commun.,* Paper 8.2-2, 1982.

[65] M. SHIRASAKI and K. ASAMA, "Compact Optical Isolator for Fibers Using Birefringent Wedges," *Appl. Opt.,* Vol. 21, p. 4296, 1982.

[66] R. C. BOOTH, D. COTTER, and E. A. D. WHITE, "Optical Isolators for Long-Wavelength Fiber-Optic Communication Systems," *Proc. Conf. Optical Fiber Commun.,* Paper MG6, New Orleans, February 1983.

[67] K. SHIRAISHI, S. SUGAYA, and S. KAWAKAMI, "Fiber Faraday Rotator," *Appl. Opt.,* Vol. 23, p. 1103, 1984.

[68] HOYA CORP., *Hoya Faraday Rotator Glass Report,* Nishi-Shinjuku-Showa Bld. 13-12, 1 Chome Nishi-Shinjuku, Shinjuku-ku Tokyo, 160 Japan.

[69] W. C. STANBOS and J. R. MACLEAN, "In-Line Optic Attenuators for Optical Fibers," U.S. Patent 4,261,640.

[70] A. BARBAUDY and J. GALAUP, "Fixed Optical Attenuator for Light Rays Guided by Optical Fibres," U.S. Patent 4,257,671.

[71] E. G. HANSON, "Polarization-Independent Liquid-Crystal Optical Attenuator for Fiber-Optics Applications," *Appl. Opt.,* Vol. 21, p. 1342, 1982.

[72] Private Communication, JDS Optics Co., Ottawa, Ontario, Canada.

9

Optical Sources for Lightwave System Applications

N. K. DUTTA

AT&T Bell Laboratories

1. Introduction

The lightwave transmission systems that are being installed throughout the world have been made possible by two major technological advances. One of them is the development of low-loss silica (glass) fibers which act as a transmission medium and the other is the development of high-performance, reliable, semiconductor lasers. Many lightwave systems installed so far use multimode fibers and an operating wavelength of about 0.85 μm. AlGaAs semiconductor lasers are used as sources for these systems. Third-generation systems using single-mode fibers and an operating wavelength near 1.3 μm offer longer repeater spacings because of the lower silica-fiber loss near 1.3 μm than at approximately 0.85 μm. Even larger repeater spacing is allowed for an operating wavelength near 1.55 μm, where the fiber loss is minimum. Semiconductor lasers fabricated using the InGaAsP material system are used as sources for commercial high–data-rate, long-haul lightwave systems operating near 1.3 μm and 1.55 μm. For short-haul transmission systems, where the performance requirements are not as critical, light-emitting diodes (LEDs) operating near 0.85 μm or 1.3 μm are used as sources.

The principles of operation, fabrication, and performance characteristics of different types of semiconductor lasers are described in this chapter. Emphasis is on lasers fabricated using the InGaAsP material system and operating in the wavelength ranges of 1.3 μm and 1.55 μm since these are regions of interest for modern fiber-optic transmission systems.

Basic Laser Diode

The semiconductor injection laser was discovered in 1962 [1–4]. Since then it has emerged as an important component in many optoelectronic systems, such as optical recording, high-speed data transmission through fibers, optical sensors, high-speed printing, and guided-wave signal processing. Perhaps, the most important impact of semiconductor lasers is in the area of lightwave transmission systems where the information is sent through encoded light beams propagating in glass fibers.

The concept of a semiconductor laser diode is a unique blend of semiconductor-device physics and quantum electronics. The basic laser-diode chip consists of two parallel cleaved facets which form the optical cavity (Fig. 1). The other two edges are saw cut. The device has a pn junction near the light-emitting region (active region) for effective current injection. The typical cavity length is about 200 to 400 μm.

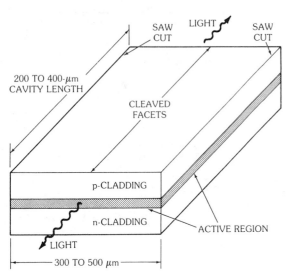

FIG. 1. Schematic of the basic semiconductor laser diode.

The light/current characteristic of a laser diode is shown in Fig. 2. This device is a proton-stripe gallium-arsenide laser. As the current injected is increased beyond a certain value (threshold current), the light output from the facet increases dramatically. Associated with this is a narrowing of the

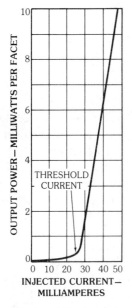

FIG. 2. Light emitted from the facet versus the injected current for a semiconductor injection laser. (*Courtesy R.L. Hartman*)

spectral emission from approximately 60 nm in the below-threshold region to about 3 to 4 nm in the above-threshold region. The light below the threshold is emitted spontaneously as the injected electrons and holes recombine in the active region. Above the threshold, stimulated emission into a low-divergence beam characteristic of laser action is observed. For digital lightwave systems the light signal is encoded by modulating the injection current, which generally switches the laser between the "on" state (stimulated-emission regime) and the "off" state (spontaneous-emission regime).

Early work in the area of semiconductor lasers have been extensively reviewed in three books [5–7]. Much of the material presented in this chapter are recent developments which complements previous work [5–7].

Light-Emitting Diode

A close relative of the injection laser is the light-emitting diode (LED) [8]. The LED is a spontaneous-emission device, and its emission characteristics are similar to that of lasers below threshold. The LEDs generally emit lower power, and have a broader spectral emission and lower modulation speed than lasers.

The LEDs are often divided into two categories: the edge-emitting LED [9] and the surface-emitting LED [8]. The edge-emitting LED structure is similar to that of a laser except that the device is operated in the spontaneous-emission regime (below threshold). For the surface-emitting LED the light is emitted normal to the surface of the wafer. The schematic of a Burrus-type surface-emitting LED is shown in Fig. 3. The edge-emitting LEDs exhibit narrower spectral width and beam divergence than surface-emitting LEDs.

LEDs are generally used as sources for short-distance, multimode fiber transmission systems. However, the small divergence of the output from edge-emitting LEDs makes them a potential candidate as sources for short-haul, single-mode fiber, transmission systems [10].

Materials Growth

The fabrication of high-quality, reliable, semiconductor lasers has been made possible by the advances made in the materials-growth technology over the last two decades. The laser and LED

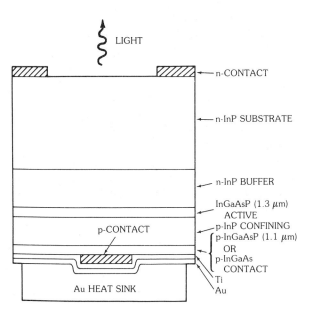

FIG. 3. Schematic of a surface-emitting LED.

2. Operating Principles

The principles of operation of a semiconductor injection laser are described in this section.

The Dielectric Waveguide

Conceptually, the stimulated emission in a semiconductor laser arises from the radiative recombination of electrons and holes in the active region, and the light generated is confined and guided by a dielectric waveguide. The active region has a slightly higher index than the p- and n-type cladding layers, and the three layers form a dielectric waveguide as shown in Fig. 4. In this figure n_2 is the refractive index of the cladding layers and n_1 that of the active region, so that $n_1 > n_2$. The cladding layers also have a higher-band gap than the active region; this confines the recombining electrons and holes to the active region.

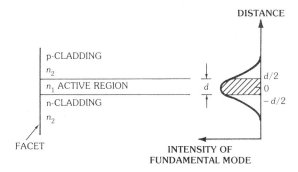

FIG. 4. The dielectric waveguide of the semiconductor laser.

structures shown in Figs. 1 and 3 are fabricated using the epitaxial-growth technology, where one semiconductor is grown over another, preserving the crystal structure across the interface. The commonly used epitaxial-growth techniques are liquid-phase epitaxy (LPE) [11], vapor-phase epitaxy [12], and the molecular-beam epitaxy (MBE) [13].

In LPE the epitaxial layer is grown by cooling a saturated solution of the components of the layer to be grown while the solution is in contact with the substrate. The epitaxial layers of both AlGaAs and InGaAsP materials were first grown by LPE.

In VPE the epitaxial layer is grown by the reaction of gaseous elements or compounds at the surface of a heated substrate. The VPE technique has also been called chemical vapor deposition (CVD), depending on the constituents of the reactants. A variant of the same technique is the metal-organic chemical vapor deposition (MOCVD) [14], which has been very successful for lasers, where metal alkyls are used as the compound source.

In MBE the epitaxial layer growth is achieved by the reaction of atomic or molecular beams of the constituent elements (of the layer to be grown) with a crystalline substrate held at high temperature in ultrahigh vacuum.

The propagation characteristics of the optical modes in a dielectric waveguide has been extensively studied [15, 16]. Two types of fundamental transverse modes can propagate in the waveguide; they are the transverse-electric (TE) and the transverse-magnetic (TM) modes. The intensity or energy distribution of the fundamental mode of the waveguide is also sketched in Fig. 4. A fraction of the optical mode, shown as crosshatched, is confined to the active region. The confinement factor Γ, i.e., the fraction of the mode in the active region is a quantity of practical interest for the design of low-threshold–current lasers, and it can be calculated by solving Maxwell's equations in the active and cladding layers with appropriate boundary conditions. Fig. 5 shows the calculated Γ as a function of active-layer thickness for the TE and TM modes for $\lambda=1.3$-μm InGaAsP double heterostructure with p-InP and n-InP cladding layers. The refractive indexes of the active and cladding regions are 3.51 and 3.22, respectively. The active-layer thickness of a InGaAsP laser is typically in the range 0.15 to 0.25 μm.

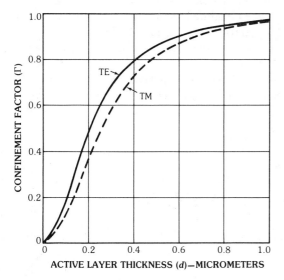

FIG. 5. Confinement factor of the fundamental TE and TM modes for a waveguide with an InGaAsP active layer and InP cladding layers as a function of the thickness of the active region.

The evaluation of Γ is in general quite tedious. For the fundamental mode, however, a remarkably simple expression,

$$\Gamma \simeq D^2/(2+D^2), \tag{1}$$

with

$$D=k_0(n_a^2-n_c^2)^{1/2}d,$$

is found [17] to be accurate to within 1.5%. In the above expression, d is the active-layer thickness, $k_0=2\pi/\lambda_0$ where λ_0 is the wavelength in free space and n_a and n_c are the refractive indexes of the active layer and the cladding layer, respectively.

Radiative Recombination

The basis of light emission in semiconductors is the recombination of an electron in the conduction band with a hole from the valence band, with the excess energy emitted as a photon (light quantum). The process is called *radiative recombination*. The energy versus wave-vector diagram of the electrons and holes in a cubic (zinc-blend type) semiconductor is shown in Fig. 6. For direct-gap semiconductors the bottom of the conduction band and the top of the valence band is at the same point in momentum space or k space ($k=0$ in Fig. 6). This allows both energy and momentum conservation in the process of photon emission by electron-hole recombination. For indirect-gap semiconductors (e.g. silicon) the momentum conservation can be achieved with the assistance of a phonon (lattice vibration), which significantly decreases the probability of radiative recombination.

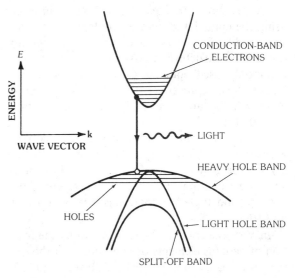

FIG. 6. Energy versus wave vector of the four major energy bands for a zinc-blend–type, direct-gap semiconductor.

The valence band in many III–V semiconductors is represented by three major subbands [18]. These are the heavy hole band, light hole band, and the spin split-off band. The radiative transitions occurring near band-gap energies are due to the recombination of electrons with heavy holes and light holes. The heavy hole, as the name implies, has larger effective mass than the light hole, which makes the density of states (and also the available number of heavy holes for a given Fermi level) larger than that for the light holes.

Condition for Stimulated Emission—Sufficient number of electrons and holes must be excited in the semiconductor for stimulated emission or net optical gain. The condition for net gain at photon energy E is given by [19].

$$E_{fc} + E_{fv} = E - E_g , \qquad (2)$$

where E_{fc} and E_{fv} are the quasi-Fermi levels of electrons and holes, respectively, measured from the respective band edges (positive into the band) and E_g is the band gap of the semiconductor. For zero net gain (transparency) the above condition becomes $E_{fc} + E_{fv} = 0$. Electrons and holes are injected into the active region by the injected current, and the quasi Fermi energies increase as the electron and hole density increases. Optical gain occurs when (2) is satisfied.

Gain Calculation—The quantities associated with a radiative recombination are the absorption spectrum, emission spectrum, gain spectrum, and the total radiative emission rate. The optical absorption or gain for a transition between the valence band and the conduction band at an energy E is given by [6] (Pt. A, p. 129):

$$a(E) = (e^2 h / 2 \epsilon_0 m_0^2 c n E) \int_{-\infty}^{\infty} \rho_c(E') \rho_v(E'')$$

$$\times |M(E', E'')|^2 [f(E'' = E' - E) - f(E')] \, dE' , \qquad (3)$$

where m_0 is the free electron mass, e is the electron charge, ϵ_0 is the permittivity of free space, E is the photon energy, n is the refractive index at energy E, ρ_c and ρ_v are the densities of state for a unit volume, per unit energy in the conduction and valence bands, respectively, $f(E')$ is the probability that a state of energy E' is occupied by an electron, and M is the effective matrix element between the conduction-band state of energy E' and the valence-band state of energy E''. In addition to the states in the parabolic band the contribution of the band-tail impurity states to optical gain can be significant for photon energies near the band edges. Several models for the density of

states and matrix element for transition between band tail states exist (see [6], Pt. A, Ch. 3). The latest of such models which take into account the contributions of the parabolic bands and that of the impurities is a Gaussian fit of the Kane form to the Halperin-Lax model of band tails. This was first proposed by Stern [20] and used to calculate the gain and recombination rate in gallium arsenide.

The calculated spectral dependence of absorption or gain at various injected carrier densities is shown in Fig. 7 for undoped InGaAsP ($\lambda \approx 1.3$ μm). The calculation was done using Gaussian Halperin-Lax band tails and Stern's matrix element. The material parameters used and the method of calculation are described in detail in Dutta [21]. Fig. 7 shows that the gain peak, i.e., the spectral position for maximum gain shifts to higher energies with increasing injection. Fig. 8 shows the

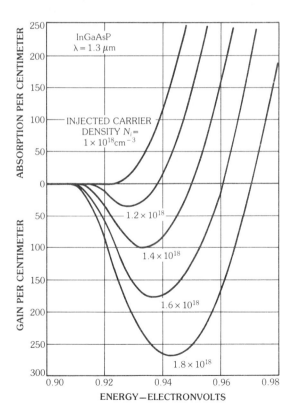

FIG. 7. Calculated gain or absorption as a function of photon energy for $\lambda = 1.3$-μm ($E_g = 0.96$ eV) InGaAsP at various injected carrier densities. (*After Dutta [21]*)

maximum gain g as a function of injected-carrier density at different temperatures. Note that considerably lower injected carrier density is needed at a lower temperature to achieve the same gain. This is the origin of the lower threshold current at low temperature. The carrier density at which $g=0$ in Fig. 8 is the injected-carrier density needed for transparency.

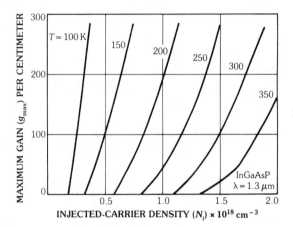

FIG. 8. The maximum gain as a function of injected carrier density for undoped InGaAsP ($\lambda=1.3$ μm) at different temperatures.

Spontaneous Emission Rate—At unity quantum efficiency the total spontaneous radiative recombination rate R equals the excitation rate. The latter is usually expressed in terms of the nominal current density J_n [20, 21]:

$$J_n(\text{A/cm}^2 \cdot \mu\text{m})=eR,\qquad(4)$$

with

$$R=\int r_{\text{spon}}(E)\,dE,$$

where e is the electron charge and the thickness of the active region is assumed to be 1 μm, and $r_{\text{spon}}(E)$ is the spontaneous emission rate at a photon energy E. It is given by [20, 21]

$$r_{\text{spon}}(E)=(4\pi\,n\,e^2\,E\,/\,m_0^2\,\epsilon_0\,h^2\,c^3)\int_{-\infty}^{\infty}\rho_c(E')\,\rho_v(E'')$$

$$\times|M(E',E'')|^2 f(E')[1-f(E'')]dE'.\quad(5)$$

The optical gain g as a function of nominal cur-

rent density has been calculated both for GaAs and InGaAsP lasers. To a good approximation the gain is found to vary linearly with J_n. For GaAs, at a temperature of 300 K, the relation is [22]

$$g=0.045\,(J_n-4222)\qquad(6)$$

per centimeter, where J_n is expressed in amperes per micrometer·square centimeter. The total spontaneous radiative recombination rate R can be approximated by

$$R=BNP,\qquad(7)$$

where B is the radiative recombination coefficient, and N and P are the electron and hole densities, respectively. For undoped semiconductors (7) becomes $R=BN^2$. For GaAs, the measured $B=1\times10^{-10}\text{cm}^3\text{ s}^{-1}$. Calculation of the radiative recombination rate shows that B decreases with increasing carrier density [22]. This has been recently confirmed by Olshansky and coworkers [23] using carrier lifetime measurements.

Nonradiative Recombination
An electron-hole pair can recombine nonradiatively, which means that the recombination can occur through any process that does not emit a photon. In many semiconductors, for example pure germanium or silicon, the nonradiative recombination dominates radiative recombination.

The effect of nonradiative recombination on the performance of injection lasers is to increase the threshold current. If τ_{nr} is the carrier lifetime associated with the nonradiative process, the increase in threshold-current density is approximately given by

$$J_{nr}=e\,n_{th}\,d/\tau_{nr},\qquad(8)$$

where N_{th} is the carrier density at threshold, d is the active-layer thickness, and e is the electron charge.

The nonradiative recombination processes described in this section are (*a*) the Auger effect, (*b*) surface recombination, and (*c*) recombination at defects or traps.

The Auger Effect—Since the pioneering work by Beattie and Landsburg [25], it is generally accepted that Auger recombination can be a major nonradiative mechanism in narrow-gap semiconductors. Recent attention to the Auger effect has been in connection with the observed higher temperature dependence of threshold current of long-wavelength InGaAsP lasers compared to short-wavelength AlGaAs lasers [25, 26–29]. It is generally believed that the Auger effect plays a sig-

nificant role in determining the observed high-temperature sensitivity of the threshold current of InGaAsP lasers emitting near 1.3 μm and 1.55 μm. A detailed discussion of this subject, however, is beyond the scope of this chapter.

There are several different types of Auger recombination processes. The three major types are band-to-band processes, phonon-assisted Auger processes, and the trap-assisted Auger processes.

The band-to-band Auger processes in direct-

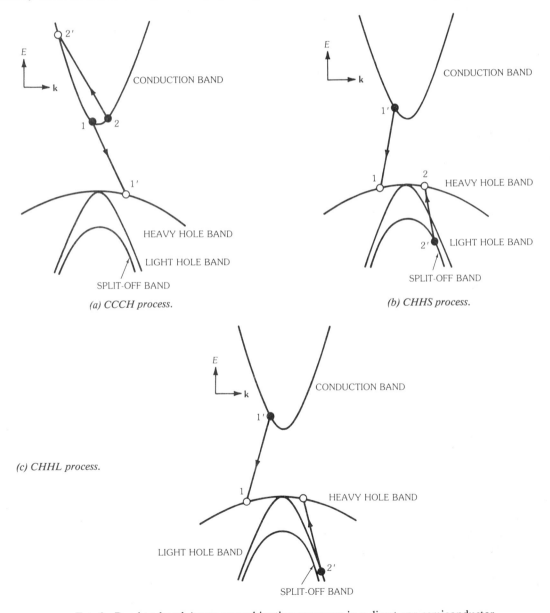

(a) CCCH process.

(b) CHHS process.

(c) CHHL process.

FIG. 9. Band-to-band Auger recombination processes in a direct-gap semiconductor.

gap semiconductors are shown in Fig. 9. The three processes are labeled as CCCH, CHHS, and CHHL, where C stands for conduction band, H for heavy hole band, L for light hole band, and S for spin split-off band. The CCCH mechanism involves three electrons and a heavy hole and is dominant in n-type material. This process was first considered by Beattie and Landsberg [25]. The CHHS process involves one electron, two heavy holes, and a split-off–band hole. The CHHL process is similar to the CHHS process except that it involves a light hole. The CHHS and the CHHL mechanism are dominant in p-type material [30, 31]. Under high-injection conditions as is present in lasers, all the above mechanisms must be considered.

The band-to-band Auger processes were characterized by a strong temperature dependence and band-gap dependence, the Auger rate decreasing rapidly either at low temperature or for high–band-gap materials. Such strong temperature and band-gap dependencies are not expected for phonon-assisted and trap-assisted Auger processes.

The active region of an InGaAsP laser is nominally undoped. Under high injection in undoped material the Auger rate R_a varies approximately as

$$R_a = C_A N_i^3, \qquad (9)$$

where N_i is the injected carrier density and C_A is called the Auger coefficient. The current lost to Auger recombination at threshold is given by

$$J_{nr} = e \, d \, C_A N_{th}^3. \qquad (10)$$

Calculations of the Auger coefficient using the Kane-band model [18] yield a value of $1 \times 10^{-28} \mathrm{cm}^6 \mathrm{s}^{-1}$ for $\lambda \simeq 1.3$ μm InGaAsP [24]. Using the Chelikowsky and Cohen model [32] for the band structure, Haug [33] has recently calculated a value of approximately $2.5 \times 10^{-29} \mathrm{cm}^2 \mathrm{s}^{-1}$ for $\lambda \simeq 1.3$ μm InGaAsP. The experimental values of the Auger coefficient for this material are in the range $2-7 \times 10^{-29} \mathrm{cm}^6 \mathrm{s}^{-1}$ [34–38]. For GaAs, $C_A \simeq 10^{-31} \mathrm{cm}^6 \mathrm{s}^{-1}$. Thus the effect of Auger recombination on the threshold current of GaAs lasers is small compared with that for InGaAsP lasers be-

cause the Auger coefficient in GaAs is smaller by two orders of magnitude.

Surface Recombination—In an injection laser the cleaved facets are surfaces exposed to the ambient. In addition, in many index-guided laser structures the edges of the active region can be in contact with curved surfaces which may not be a perfect lattice. A surface, in general, is a strong perturbation of the lattice, creating many dangling bonds which can absorb impurities from the ambient. Hence a high concentration of defects can occur which can act as nonradiative recombination centers. Such localized nonradiative centers, in addition to increasing the threshold current, can cause other performance problems (e.g., sustained oscillations) in lasers.

Recombination at Defects—Defects in the active region of an injection laser can be formed in several different ways. In many cases they are grown during the epitaxial growth process. They can also be generated, multiply, or propagate during a stress aging test [39]. Defects can propagate along a specific crystal axis in a strained lattice. The well-known dark-line defect, DLD (dark region of linear aspect), is generally believed to be responsible for high degradation rate (short life span) of early AlGaAs lasers.

Defects produce a continuum of states in a localized region. Electrons or holes which are within a diffusion length from the edge of the defect recombine nonradiatively via these continum of states.

Threshold-Current Calculation

At threshold the optical gain equals the total optical loss in the cavity. The condition for threshold is

$$\Gamma g_{th} = \alpha_a \Gamma + (1 - \Gamma) \alpha_c + L^{-1} \ln (1/R_m), \quad (11)$$

where g_{th} is the threshold gain in the active region, α_a and α_c are the absorption losses in the active and cladding region, respectively, L is the cavity length, and R_m is the mirror reflectivity. Typically $L \simeq 300$ μm, $R_m \simeq 0.3$, and values of Γ are in the range of 0.3 to 0.7, depending on the thickness of the active region. For a 0.2-μm-thick $\lambda \simeq 1.3$-μm InGaAsP-InP double-heterostructure layer Γ(TE) is approximately 0.47 (from Fig. 5), and using $\alpha_c = \alpha_a = 30$ cm^{-1}, the calculated $g_{th} \simeq 150$ cm^{-1}. The threshold gains for semiconductor lasers are generally in the range 100 to 200 cm^{-1}. The threshold current of a 250-μm–long GaAs/AlGaAs double-heterostructure laser with 2-μm×0.2-μm active region can now be calculated using (6) and (11). The calculated $\Gamma \simeq 0.6$ for 0.2-μm–thick GaAsAl$_{0.36}$GA$_{0.64}$As double heterostructure. Assuming $R_m = 0.3$ and $\alpha_a = \alpha_c = 20$ cm^{-1}, we get $g_{th} = 114$ cm^{-1}, $J_{th} = 1.35$ kA/cm^2 and a threshold current of 6.7 mA. This compares well with the experimentally observed threshold currents in the range 5 to 10 mA for GaAs/AlGaAs double-heterostructure lasers with good current confinement [40]. Fig. 8 shows that carrier density N_c needed to achieve a certain optical gain increases with increasing temperature and since $J_n \simeq e B N^2$ it follows that the threshold current density of a laser is expected to increase with increasing temperature. It is experimentally observed that the threshold-current density J_{th} of a laser varies with the temperature T as $J_{th} \simeq J_0 \exp (T/T_0)$, where T_0 is a parameter determining the temperature sensitivity [26–28]. For $g_{th} \simeq 114$ cm^{-1}, the calculated $T_0 \simeq 210$ K in the temperature range 300 K to 350 K. This agrees well with the measured values for GaAs lasers.

For InGaAsP lasers a significant amount of nonradiative recombination is believed to be present at carrier densities comparable to the lasing threshold near room temperature. The nonradiative recombination arises from the Auger effect, which is considerably larger in InGaAsP material than in GaAs.

Since the threshold carrier density increases with increasing temperature, the carrier loss to the nonradiative Auger effect increases with increasing temperature, which results in a more rapid increase of threshold current with increasing temperature (low T_0) for long-wavelength InGaAsP lasers than that for GaAs lasers. The measured variation of threshold current density as a function of temperature for InGaAsP lasers can be expressed by the relation $J_{th} \simeq J_0 \exp (T/T_0)$ with $T_0 \simeq 50$ to 70 K in the temperature range 300 to 350 K [26–29]. It is generally believed that Auger recombination plays a significant role in determining the smaller T_0 values of InGaAsP ($\lambda \simeq$ 1.3 μm, 1.55 μm) lasers compared with shorter-wavelength ($\lambda \simeq 0.85$ μm) AlGaAs lasers [26–29, 40].

3. Laser Structures and Their Performance

Most of the initial work on semiconductor lasers were carried out using the AlGaAs material system and these lasers operating near 0.85 μm were also used in many initial optical-fiber transmission systems. Semiconductor lasers operating in the wavelength range of 1.3 μm and 1.55 μm, which are the regions of low loss for commercial silica fibers, are fabricated using the InGaAsP quaternary material, which can be grown lattice matched over an InP substrate. In this section different types of semiconductor laser structures are described. Although these structures are generic, in principle, to all semiconductor lasers, we use the InGaAsP material system for the discussion of each type.

In a double-heterostructure laser the lower refractive index of the cladding layers than that of the active region confines the optical mode perpendicular to the junction plane, as discussed in Section 2. For stable fundamental-mode operation with a low threshold current, additional confinement of the optical mode along the junction plane is required. A laser structure that does not have any lateral mode confinement and current confinement along the junction plane is called a *broad-area laser*. These lasers typically have threshold current density of approximately 1 kA/cm^2, i.e, a threshold current of about 1 A for a 250-μm× 380-μm laser. Such a high threshold current would limit the usefulness of an injection laser. The laser structures described in this chapter are essentially different ways of realizing lateral mode confinement with the eventual aim of obtaining a laser structure with low threshold current that can operate in the fundamental lateral mode up to high powers. The current state of the art InGaAsP lasers emitting near 1.3 μm generally have threshold currents in the range 10 to 15 mA at 30°C.

The laser structures are often classified into two groups: gain guided and index guided. In the gain-guided structure the width of the lasing optical mode along the junction plane is principally determined by the width of the optical gain region, which is determined by the width of the current pumped region (typically in the range 5 to 10 μm).

In index-guided lasers a narrow central region of relatively higher index in the junction plane confines the lasing mode to that region. The index-guided lasers can, in general, be divided into two groups: weakly index-guided and strongly index-

oxide-stripe device, originally fabricated by Dyment [42] and used to study the transverse-mode structure along the junction plane in AlGaAs injection lasers. Fig. 10 shows various gain-guided laser structures fabricated using the InGaAsP material systems. These are (a) the oxide-stripe laser where an SiO$_2$ layer on the p contact confines the injected current to a small region through an opening in the dielectric [43], (b) the proton-stripe [44] or the deuteron-stripe [45] laser, where the implanted

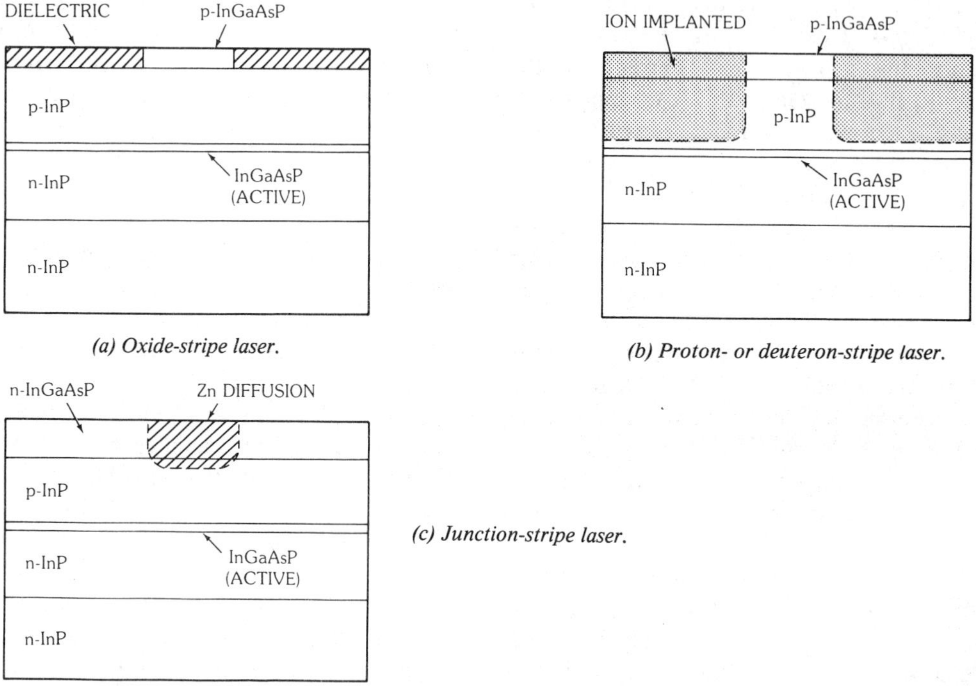

(a) Oxide-stripe laser.

(b) Proton- or deuteron-stripe laser.

(c) Junction-stripe laser.

FIG. 10. Schematic cross sections of different types of gain-guided laser structures.

guided. In weakly index-guided lasers the active region is generally continuous and the effective index discontinuity is provided by a cladding layer of varying thickness. The strongly index-guided lasers, by contrast, employ a buried heterostructure. In these lasers the active region is bounded by lower-index epitaxially grown layers both along and normal to the junction plane. The lateral index difference is approximately 0.01 to 0.03 for weakly index-guided lasers while it is approximately 0.2 for strongly index-guided buried-heterostructure lasers.

Gain-Guided Lasers

The simplest gain-guided laser structure is an

protons or deuterons create a high-resistivity region which restricts the current to an opening in the implanted region, and (c) the junction-stripe laser [46], where the top n-type layer is type converted to p type by zinc diffusion in a small region to provide a current path, while the reverse-biased junction provides current confinement. Variations of the above structures can also be found in the literature.

In all the above structures the active region is planar and continuous. The stimulated-emission characteristics of such a laser are determined by the carrier distribution (which provides optical gain) along the junction plane. Since the optical-mode distribution along the junction is determined by the optical gain, these lasers are called *gain-guided lasers*. The fabrication of an oxide-stripe, gain-guided laser involves the following steps. First, four epitaxial layers are grown by LPE on (100) oriented n-InP substrate. These layers are n-InP (cladding layer), undoped InGaAsP ($\lambda \approx 1.3$ μm, active layer, p-InP (cladding layer) and p-InGaAsP ($\lambda \approx 1.3$ μm, contact layer). The active-layer thickness is approximately 0.2 μm. A 300-nm–thick SiO_2 layer is then deposited on the wafer. Windows 6 to 9 μm wide are then opened on the SiO_2 using photolithographic techniques and plasma etching. The wafer, typically 250 to 400 μm thick, is then thinned down to a thickness of 75 to 100 μm and n-contact (typically AuSn or AuGe alloy) and p-contact (typically CrAu or AuZn alloy) are deposited on the substrate side and epitaxially grown side, respectively. The thinning of the wafer is necessary to facilitate cleaving along a crystallographic plane. The cleaved planes (typically (110)) form the mirrors of the laser cavity. The other sides are saw cut.

The typical light-power/current characteristic of a gain-guided ($\lambda = 1.3$-μm InGaAsP) laser is shown in Fig. 11a. The 250-μm–long lasers typically have threshold currents in the range 100 to 150 mA. Typical spectral-emission characteristics of these lasers are shown in Fig. 11b. The laser emits in a group of wavelengths each of which correspond to the longitudinal modes of the laser cavity.

As the current through the laser is increased, the light/current characteristics of a gain-guided laser often exhibits a "kink" or nonlinearity. The nonlinearity can be associated with a movement of the optical mode along the junction plane, the transition to higher-order modes or the transition from TE to TM mode. Such a nonlinear response can severely alter the amount of light coupled into an optical fiber and thus limit the usefulness of the laser in an optical communication system. The lasers with narrow stripe widths exhibit kinks at higher operating powers than lasers with wide stripe widths as observed by Dixons group [47] for AlGaAs proton-stripe lasers.

Weakly Index-Guided Lasers

Gain-guided InGaAsP lasers have undesirable characteristics such as a high threshold current,

low differential quantum efficiency, and occurrence of light/current kinks at relatively low output powers. These effects are principally caused by carrier-induced index reduction, which leads to the movement of the optical mode along the junction plane. The effective index depression caused by

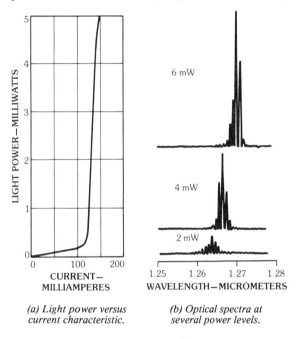

(a) Light power versus current characteristic.

(b) Optical spectra at several power levels.

FIG. 11. Measured cw light/current characteristics and optical spectra of a deuteron-stripe gain-guided laser. (*After Schwartz et al. [45], ©1984 IEEE*)

carriers at the waveguide center is approximately 5×10^{-3}. In weakly index-guided lasers the laser structure is modified so as to introduce an effective index step of approximately 10^{-2}, which is larger than the carrier-induced reduction. Hence the optical mode along the junction plane in a weakly index-guided laser is essentially determined by the device structure.

Several types of weakly index-guided laser structures known under various names, such as rib waveguide, ridge waveguide, plano-convex waveguide and channeled-substrate plane waveguide, have been reported, and some of them are shown schematically in Fig. 12. It is possible to group these structures into two categories, collectively called ridge-waveguide type and rib-waveguide type. In the ridge-waveguide type structure the use of a dielectric around the ridge (see the

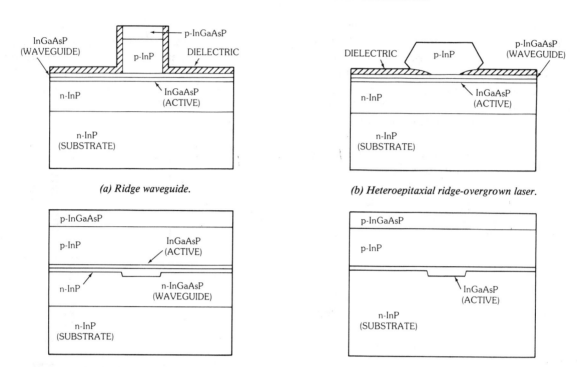

(a) Ridge waveguide.

(b) Heteroepitaxial ridge-overgrown laser.

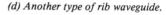

(c) Rib waveguide, also known as plano-convex waveguide.

(d) Another type of rib waveguide.

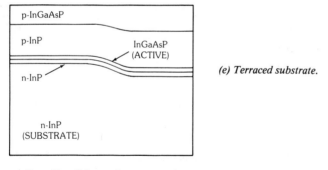

(e) Terraced substrate.

FIG. 12. Schematic cross sections of several types of weakly index-guided laser structures.

schemes (a) and (b) in Fig. 12) inhibits current spreading in the p-InP layer. By contrast, in the rib-waveguide type structure (see the schemes (c) and (d) in Fig. 12) current spreading in the p-InP layer can influence the threshold current. Both schemes are, however, capable of providing an effective index step of approximately 10^{-2} for the lateral mode under the ridge or rib region, and the mode is essentially index guided. The ridge-waveguide laser has been reported by several authors [48, 49]. The device fabrication involves the following steps: The planar epitaxial layers are successively grown as shown in Fig. 12 by either LPE, VPE, or MOCVD growth technique

on a (100) oriented N-InP substrate. The wafer is then etched to form a ridge of width of approximately 5 μm. The $\lambda=1.1$-μm InGaAsP waveguide layer acts as a stop etch layer. The wafer is then processed to produce lasers using standard dielectric deposition, photolithographic, and metallization techniques.

The active layer of a ridge-waveguide laser is planar and thus the threshold current is affected by carrier diffusion in the active region. Fig. 13 shows the typical light/current characteristics of a $\lambda=1.3$-μm InGaAsP ridge-waveguide laser. Also shown are emission patterns along and normal to the junction plane. Typical threshold currents are

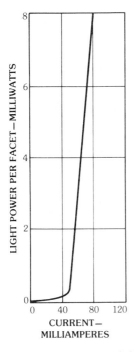

(a) Light/current characteristic.

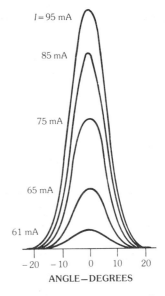

(b) Far fields parallel to junction plane.

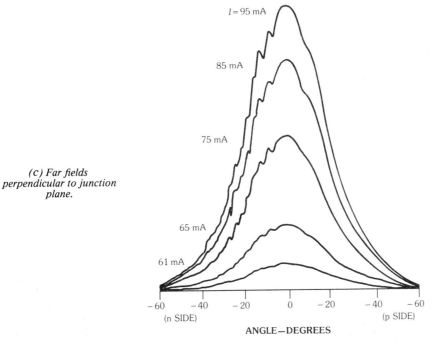

*(c) Far fields
perpendicular to junction
plane.*

FIG. 13. Typical light/current characteristic and far fields of a $\lambda=1.3$-μm InGaAsP ridge-waveguide laser with 5-μm ridge, $L=250$ μm, $T=20°$C, $I_t=47$ mA cw, and $\eta=26\%$ facet. (*After Kaminow et al. [47],* ©*1983 IEEE*)

in the range of 40 mA to 60 mA. The rib-wave-guide lasers have been fabricated by several investigators [50, 51]. The threshold current of these lasers is influenced by current spreading in the p-cladding layer, carrier diffusion in the active region, and the lateral index step. The measured threshold currents of these lasers are typically in the range of 60 mA to 80 mA.

Strongly Index-Guided Lasers

The transverse-mode control in injection lasers can be achieved using index guiding along the junction plane. The mode control is necessary for improving the light/current linearity and the modulation response of injection lasers. In strongly index-guided lasers the active region is buried in higher band-gap layers (e.g., InP) on all sides. For this reason these lasers are called *buried-heterostructure* (BH) lasers. The lateral index step along the junction plane is approximately 0.2 in these laser structures. This built-in index step is about two orders of magnitude larger than the carrier-induced effects, and thus the lasing characteristics of BH lasers are primarily determined by the dielectric waveguiding, which confines the mode inside the buried active region.

The buried heterostructure was first proposed for GaAs lasers [52]. Several types of InGaAsP buried-heterostructure lasers have been reported over the last few years [53–62]. These lasers can be, in general, divided into two categories: (*a*) planar active-layer structures and (*b*) nonplanar active-layer structures.

The schematic cross sections of three types of BH lasers is shown in Fig. 14. They are (*a*) the buried heterostructure [53] (BH), which we shall call the "etched-mesa buried heterostructure" in order to distinguish it from other BH lasers (EMBH), (*b*) the double-channel planar buried-heterostructure (DCPBH) [55] and (*c*) the channeled-substrate buried-heterostructure (CSBH) [58, 59]. Variations of the above laser structures and other types of BH lasers have been reported in the literature. The EMBH and DCPBH lasers have planar active regions and the CSBH laser has nonplanar active regions. Laser structures with planar active regions are suitable for distributed-feedback type semiconductor lasers.

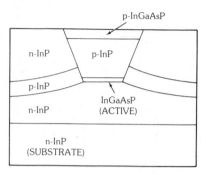

(a) Etched-mesa type.

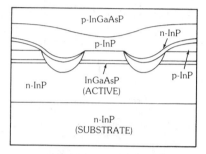

(b) Double-channel planar type.

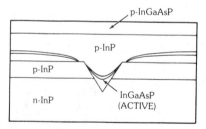

(c) Channeled-substrate type.

FIG. 14. Schematic cross section of three types of buried-heterostructure lasers.

In buried-heterostructure lasers of the types shown in Fig. 14 the confinement of the injected current to the active region is achieved through reverse-biased junctions or forward-biased junctions of a higher band-gap material (e.g., InP). The difference between the total current I injected into the laser and the current I_a passing through the active region is the leakage current I_L, i.e.,

$$I_L = I - I_a . \tag{12}$$

The magnitude of I_L and its variation with I depends on the thickness and the carrier concentration of the current-confining layers [63]. A large leakage may lead to high threshold current and also limit the maximum output power. For low-threshold operation the carrier concentration and the layer thicknesses of a BH laser must be optimized to reduce the leakage current. The fabrication and performance characteristics of a planar active-region (DCPBH) laser and a nonplanar active-region (CSBH) laser are described below.

Double-Channel Planar Buried-Heterostructure (DCPBH) Lasers—An SEM photomicrograph of a DCPBH laser is shown in Fig. 15. The active region in this device consists of a 1.3-μm InGaAsP layer. The device fabrication involves two epitaxial growth steps. In the first growth, four layers are grown by the LPE technique. They can also be grown by the VPE, MBE, or MOCVD technique. The top InGaAsP layer is removed using an etching mixture of $10H_2SO_4:1H_2O_2:1H_2O$. Two channels, about 3 μm deep and 5 μm wide, are then etched on the wafer using a plasma-deposited SiO_2

as etching mask. The mask is then removed and four epitaxial layers, p-InP, n-InP, p-InP and p-InGaAsP ($\lambda=1.3$ μm), are grown on the wafer using a second LPE growth step. The last quaternary layer serves as a contact layer. The wafer is then processed to produce laser chips using a procedure similar to that for gain-guided lasers except that the contact stripe is aligned over the active region using photolithographic techniques.

The light/current characteristics at different temperatures of a 1.3-μm InGaAsP DCPBH laser are shown in Fig. 16. The device can operate cw up to 130°C. DCPBH lasers typically exhibit room-temperature threshold currents in the range 10 to 15 mA. The lateral-mode control is achieved by reducing the dimension of the active region. For fundamental-mode operation the cross-sectional area of the active region must be less than 0.3 μm^2.

InGaAsP lasers emitting near 1.55 μm are also of considerable interest as a source in fiber transmission systems. Fig. 17 shows the light/current characteristics at different temperatures of a DCPBH laser emitting near 1.55 μm. The laser operates cw up to 110°C. The lasers emitting near

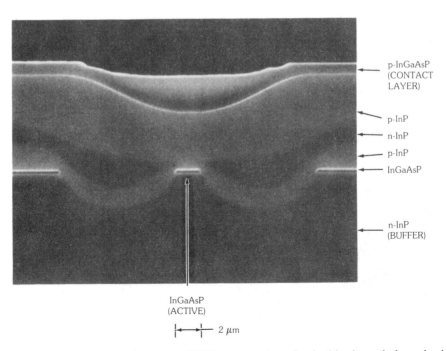

FIG. 15. Scanning electron microscope (SEM) cross section of a double-channel planar buried-heterostructure (DCPBH) laser.

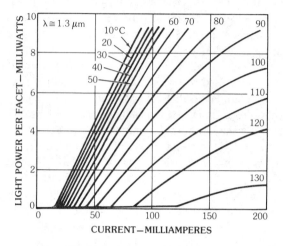

FIG. 16. Light/current characteristics at different temperatures of a λ≅1.3-μm InGaAsP DCPBH laser.

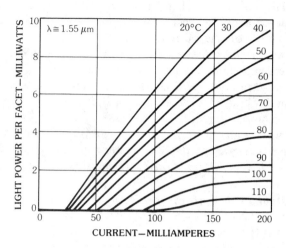

FIG. 17. Light/current characteristics at different temperatures of a λ≅1.55-μm InGaAsP DCPBH laser.

1.55 μm exhibit lower differential quantum efficiency than lasers emitting near 1.3 μm. This may be due to larger intervalence-band absorption at longer wavelengths as proposed by Adams and colleagues [64].

Channeled-Substrate Buried-Heterostructure (CSBH) Laser—A SEM photomicrograph of a CSBH laser is shown in Fig. 18. The p-InP current-blocking layer shown in Fig. 18 can be grown by LPE or VPE on a (100) oriented n-InP

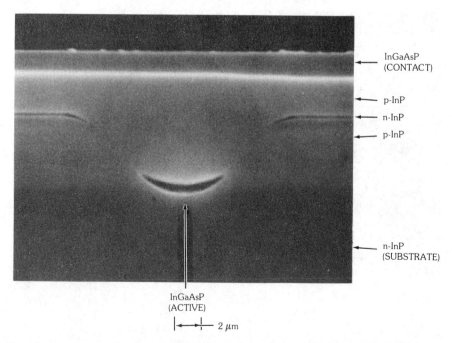

FIG. 18. Scanning electron microscope (SEM) cross section of a channeled-substrate buried-heterostructure (CSBH) laser.

substrate. A sealed-ampule cadmium-diffusion using Cd_3P_2 source at 600° to 650°C can also be used to type convert a thin layer (about 2 μm) on the surface on n-InP substrate [65]. This method allows the fabrication of the device using one LPE growth step. Vee grooves are then etched on the wafer parallel to (110) using $3HC1:1H_3PO_4$ and a SiO_2 etching mask. Four epitaxial layers are then grown by LPE. They are the n-InP buffer layer, undoped ($\lambda = 1.3$ μm) InGaAsP active layer, p-InP cladding layer, and p-InGaAs contact layer. The wafer is then processed to produce CSBH lasers using a procedure very similar to that used for oxide-stripe, gain-guided lasers.

The cw light/current characteristics at different temperatures of a CSBH laser emitting at 1.3 μm are shown in Fig. 19. The threshold current of these lasers are typically in the range of 10 to 15 mA at room temperature, similar to that for EMBH and DCPBH lasers. Lasers emitting near 1.55 μm have also been fabricated using the CSBH structure with performance characteristics similar to those obtained for DCPBH lasers. The lateral-mode control, needed for kink-free operation up to high powers can be achieved by controlling the dimensions for the active region.

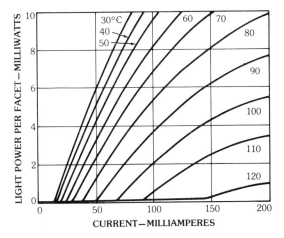

FIG. 19. Typical light/current characteristics at different temperatures of a $\lambda = 1.3$-μm InGaAsP CSBH laser.

One of the technological issues associated with the fabrication of high-performance, index-guided lasers is the control of leakage current. The strong-ly index-guided laser structures described above utilize junctions between different semiconductors for confinement of the current to the active region. Many other laser structures exist in the literature, some of which are variations of the above structures, that use semi-insulating layers (e.g., Fe-doped InP) or highly resistive layers (undoped InP) for current confinement.

Gain-guided and weakly index-guided lasers are easier to fabricate than index-guided lasers. However, they exhibit higher threshold currents than index-guided lasers. Typical threshold currents of InGaAsP gain-guided lasers are in the range 100 to 150 mA and those of weakly index-guided lasers are in the range 40 to 60 mA, compared with 10 to 15 mA for buried-heterostructure lasers. The fabrication of strongly index-guided lasers generally requires epitaxial growth over nonplanar surfaces or two epitaxial growths and, in addition, careful attention to processing. In spite of these difficulties in fabrication the superior performance characteristics, viz., low threshold current, stable fundamental-mode operation at high powers, and good high-speed modulation characteristics, make the strongly index-guided lasers primary candidates for high-performance applications. In particular, their use as a source in a lightwave transmission system is highly attractive.

Miscellaneous Laser Structures

InGaAsP semiconductor laser structures described so far have been developed for low-power (5 to 10 mW/facet) applications, such as that of a source in a lightwave communication system. These lasers can generally emit continuously up to powers in the range 30 to 60 mW/facet near room temperature. This limitation in output power mainly arises from the leakage currents which increases with increasing injection [63]. For higher-power applications laser arrays may be useful. The laser array has closely spaced multiple active regions, each of which emits light. The stimulated emission from the neighboring emitters in a laser array overlap, establishing a definite phase relation between them and hence the array emits a high-power coherent beam [66]. Power outputs of greater than 1 W have been reported from 40-element AlGaAs laser arrays [67] and 600 mW from a 10-element InGaAsP laser array [68].

Another type of laser structure is a surface-emitting laser where the laser output is normal to the wafer surface. Iga and coworkers [69] have fabricated such surface-emitting lasers. Their device structure is shown in Fig. 20. The surfaces of the wafer form the Fabry-Perot cavity of the laser. The lasers have threshold current density of approximately 11 kA/cm^2 at 77 K and have been operated at output powers of several milliwatts.

nantly in a single longitudinal mode. Such single-frequency lasers [71] are of much interest in view of their potential application in optical communication at the 1.55-μm wavelength, where the loss of silica fibers is minimum.

In conventional Fabry-Perot (FP) type semiconductor lasers the feedback is provided by facet reflections and its magnitude remains the same for all longitudinal modes. The only longitudinal-

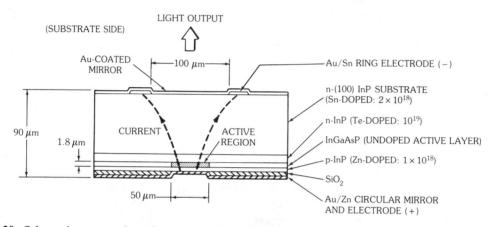

FIG. 20. Schematic cross section of a surface-emitting laser where the surfaces of the wafer form the Fabry-Perot cavity. (*After Soda et al. [69]*)

Liau and Walpole [70] have reported a buried-heterostructure laser whose light output is deflected perpendicular to the wafer surface, so that the laser in effect is a surface-emitting laser. The mirrors of the laser, which are perpendicular to the wafer surface, are formed by vapor-phase transport and the beam is deflected normal to the surface using a monolithically integrated 45° parabolic mirror. The lasers have room-temperature threshold currents in the range 12 to 18 mA and exhibit performance characteristics comparable to that of conventional cleaved-facet lasers.

4. Single-Frequency Lasers

A conventional semiconductor laser does not emit light in a single longitudinal mode. In general, few modes near the gain peak contain most of the optical power. In the presence of fiber chromatic dispersion the unwanted side modes limit the information-transmission rate by reducing the fiber bandwidth. It is therefore desirable to have semiconductor lasers that emit light predomi-

mode discrimination in such a laser is provided by the gain spectrum itself. Since the gain spectrum is usually much wider than the longitudinal-mode spacing the resulting mode discrimination is poor. One way of improving the mode selectivity is to have frequency-selective feedback so that the cavity loss is different for different longitudinal modes. Two mechanisms have been found useful in this respect; they are known as the distributed-feedback and the coupled-cavity mechanisms.

Distributed-Feedback Lasers

In a distributed-feedback (DFB) laser the feedback necessary for the lasing action is not localized at the cavity facets but is distributed throughout the cavity length. This is achieved through the use of a grating etched along the cavity length. Mode selectivity of the DFB mechanism results from Bragg's law, which states that coherent coupling between forward and backward traveling waves occur for wavelengths such that the grating period $\Lambda = m\lambda_m/2$, where λ_m is the wavelength of the light inside the laser medium and the integer m is the order of Bragg diffraction induced by the grating. By choosing Λ appropriately, such a device can be made to provide distributed feedback only at selective wavelengths within the gain spectrum of the active layer.

Kogelnik and Shank [72] were the first to observe the lasing action in a periodic structure that utilized the DFB mechanism. Although most of the early work [73–75] was related to GaAs lasers the need of a single-frequency semiconductor laser operating at the fiber-loss–minimum wavelength of 1.55 μm has resulted in the development of DFB type InGaAsP lasers [76–78]. From the viewpoint of device operation, semiconductor lasers employing the DFB mechanism can be classified into two broad categories, shown schematically in Fig. 21 and referred to as DFB lasers and distributed–Bragg-reflector (DBR) lasers [79]. In DBR lasers the grating is etched near the cavity ends and distributed feedback does not take place in the central active region. The unpumped corrugated end-regions act as an effective mirror whose reflectivity is of DFB origin and is therefore dependent on wavelength.

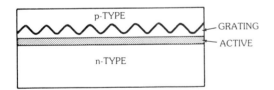

(a) DFB laser.

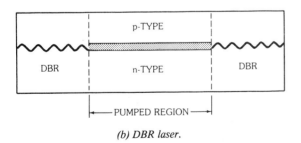

(b) DBR laser.

FIG. 21. The distributed-feedback and the distributed–Bragg-reflector laser structures.

In Section 3 we have discussed various semiconductor-laser structures that are commonly used. Any of these structures can be employed to make a DFB semiconductor laser after etching a grating into one of the layers. The direct etching of the active layer is generally not preferred since it can increase the rate of nonradiative recombination by introducing defects in the active region. This would effect the device performance through a higher threshold current. One of the cladding layers is therefore used to etch the grating. Since only the evanescent field associated with the fundamental transverse mode interacts with the grating, it is clear that the exact location of the grating with respect to the active layer and the corrugation depth are critical in determining the effectiveness of the grating. The grating period Λ is determined by the laser wavelength in the medium and the order of Bragg diffraction used for the distributed feedback. The Bragg condition for the mth-order coupling between the forward and backward propagation waves is

$$\Lambda = m\lambda/2\bar{n}, \qquad (13)$$

where $\bar{n}$ is the effective-mode index and $\lambda/\bar{n}$ is the wavelength inside the medium. For a 1.55-μm InGaAsP laser $\Lambda \simeq 0.23$ μm, if we use $m=1$ (first-order grating) and a typical value $\bar{n} \simeq 3.4$. This value doubles if a second-order grating is used. Both first-order and second-order gratings have been employed to fabricate InGaAsP DFB lasers [80].

Two techniques have been used for the formation of a grating with submicrometer periodicity. In the holographic technique [80], optical interference is used to form a fringe pattern on the photoresist deposited on the wafer surface. In the electron-beam lithographic technique an electron beam scans the wafer surface and "writes" the desirable pattern on the electron-beam resist. In both cases chemical etching is generally used to produce the grating corrugations, with the patterned resist acting as a mask. Once the grating has been etched on the substrate or on an epitaxial layer, the wafer can be processed in the usual way to obtain a specific laser structure.

Fig. 22 shows the scanning electron microscope cross section of a 1.55-μm InGaAsP double-heterostructure with a grating on the n-cladding layer. In the case of Fig. 22 the second-order grating with a periodicity of about 470 nm was formed on

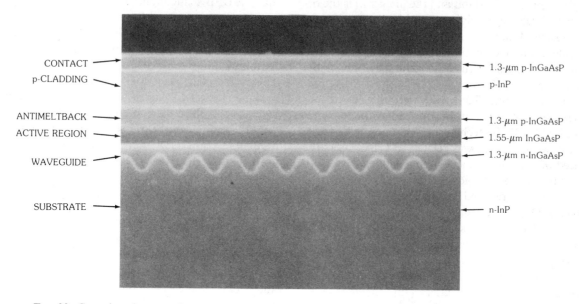

CONTACT — ← 1.3-μm p-InGaAsP
p-CLADDING — ← p-InP
ANTIMELTBACK — ← 1.3-μm p-InGaAsP
ACTIVE REGION — ← 1.55-μm InGaAsP
WAVEGUIDE — ← 1.3-μm n-InGaAsP
SUBSTRATE — ← n-InP

FIG. 22. Scanning electron microscope cross section of $\lambda=1.55=\mu$m InGaAsP double heterostructure fabricated with a second-order grating on the substrate.

the n-InP substrate using a holographic technique. These five layers were grown by the LPE growth technique. These layers are (*a*) n-InGaAsP ($\lambda \simeq 1.3\ \mu$m) waveguide layer (*b*) undoped InGaAsP ($\lambda \simeq 1.55\ \mu$m) active layer (*c*) p-InGaAsP ($\lambda \simeq 1.3\ \mu$m) antimeltback layer, (*d*) p-InP cladding layer, and (*e*) p-InGaAsP ($\lambda \simeq 1.3\mu$m) contact layer. This layer structure can be used to fabricate the DCPBH type lasers described in Section 3.

Fig. 23 shows the light/current characteristics of a 1.55-μm DFB laser of the DCPBH type [81]. This device had a first-order grating ($\Lambda=240$ nm) and the corrugation depth after regrowth was estimated to be 30 nm. The cavity length was about 300 μm with both facets cleaved. At room temperature the threshold current is about 30 mA and the slope is nearly constant up to a power level of 10 mW from each facet. The efficiency η_d at room temperature is typically 30 to 40% for these devices [81]. From the temperature dependence of the threshold current the characteristic temperature T_0 (see Section 2) is estimated to be 67 K at room temperature. The performance of these DFB lasers is comparable to DCPBH Fabry-Perot (FP) lasers as far as light/current curves are concerned. Longitudinal-mode spectra at $I=1.5\ I_{th}$ are shown at several temperatures in Fig. 24 for the device of

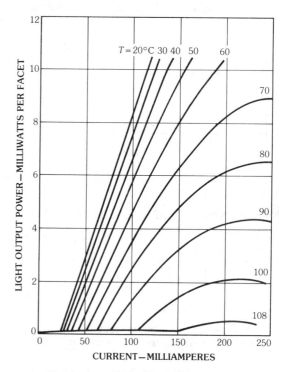

FIG. 23. Light/current characteristics of a DFB DCPBH laser at various temperatures. (*After Kitamura et al. [80], ©1984 IEEE*)

Fig. 23. The DFB laser maintains the same longitudinal mode in the entire temperature range of 20°C to 108°C. By contrast, a Fabry-Perot semiconductor laser would exhibit several mode jumps over this temperature range because of the temperature-induced shift of the gain peak. The stability of the DFB longitudinal mode is due to the built-in grating, the period of which determines the lasing wavelength.

Fig. 24 shows that the wavelength changes slightly with temperature at a rate of $d\lambda/dT \simeq$ 0.09 nm/°C. This happens because the mode index $\bar{n}$ in (13) varies with temperature. The wavelength shift for a DFB laser is, however, considerably smaller compared with an FP laser where $d\lambda/dT \simeq$ 0.5 nm/°C.

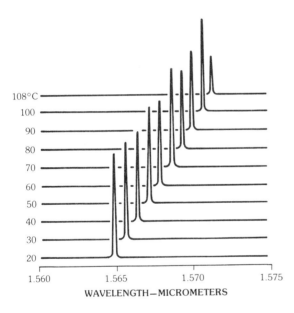

FIG. 24. Emission wavelength as a function of temperature at $I = 1.5I_{th}$ for the laser of Fig. 23. (*After Kitamura et al. [80], ©1984 IEEE*)

The light-output modulation using a pulsed modulation current leads to wavelength chirping, i.e., a variation of the laser wavelength over a small range during each modulation cycle. This wavelength range is usually referred to as the *chirp width* and is typically about 0.1 nm under practical conditions, i.e., for a typical modulation current of 50 mA [80–82]. It has been shown [84] that the wavelength chirp leads to a significant dispersion

penalty in optical communication systems and is likely to be the limiting factor in their performance. The origin of the chirp is related to the carrier-induced index change accompanying the modulation-induced carrier-density variations.

Coupled-Cavity Laser

The basic concept of longitudinal-mode selection by the coupled-cavity scheme is known from the early work on gas lasers and has been widely used for mode selection. Recently, semiconductor coupled-cavity devices have been extensively studied for optical communication in the 1.55-μm wavelength region [85–87].

The mechanism of mode selectivity in coupled-cavity lasers can be easily understood by referring to Fig. 25. Each Fabry-Perot (FP) cavity has its own set of longitudinal modes, which in general do not coincide because of different cavity lengths. Periodically, however, the longitudinal modes of each cavity nearly coincide and constitute the longitudinal modes of the coupled system for which both cavities are in resonance. One such pair closest to the gain-spectrum peak has the lowest threshold and the lasing occurs at this particular wavelength. Other pairs of coincident modes are

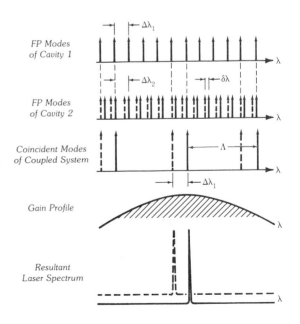

FIG. 25. Mechanism of mode selectivity in coupled-cavity laser.

effectively discriminated by the gain rolloff because of their large separation from each other. Tunability of coupled-cavity lasers arises from the fact that in semiconductor lasers the longitudinal-mode wavelength depends on the refractive index, which can be changed by varying the current or temperature. If the longitudinal modes of one cavity are made to shift by external means, a different pair of modes becomes resonant with respect to both cavities and the device wavelength jumps to this new wavelength. In this way the wavelength can be tuned discretely over several single-cavity modes in the vicinity of the gain peak.

Coupled-cavity semiconductor lasers can be classified into two broad categories. We shall refer to them as *active-passive* and *active-active* schemes, depending on whether the second cavity remains unpumped or can be pumped to provide gain.

In the active-passive scheme the semiconductor laser is coupled to an external cavity that is unpumped and plays a passive role. In the simplest design a plane or a spherical mirror is placed at a short distance from one of the facets which may be antireflection-coated to increase the coupling between the two cavity sections.

Such a scheme which employs a short (approximately 100- to 200-μm) rod of the graded-index (GRIN) multimode fiber with one end gold-coated [88] is shown in Fig. 26. Single-frequency lasers of this type have been used in transmission system experiments near 1.55 μm.

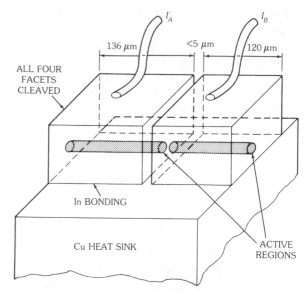

FIG. 27. Schematic of a C^3 laser. (*After Tsang, Olsson, and Logan [87]*)

In the active-active scheme both sections can be independently pumped, giving an additional degree of freedom that can be used to control device performance, e.g., tunability. Cleaving [87] and etching [86] techniques have been used to fabricate this type of coupled-cavity laser.

Coupled-cavity lasers made by the cleaving techniques are usually referred to a *cleaved–coupled cavity* or C^3 lasers. The schematic of a C^3 laser is shown in Fig. 27. A C^3 laser can be fabricated using any of the multimode laser structures discussed in Section 3. Fig. 28 shows the longitudinal-mode spectra of C^3 laser. The single-frequency emission can be tuned discretely over a range of 26 nm, by varying the current through one section. The tuning occurs by mode jumps of about 2 nm each. The ability to tune the emission wavelength simply by changing the current makes C^3 lasers very attractive for a variety of applications. Many of these applications have been reviewed by Tsang [89].

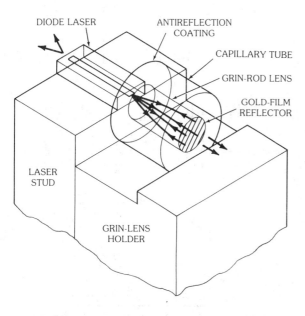

FIG. 26. Schematic of a short external-cavity laser using a graded-index fiber. (*After Liou et al. [88]*)

width decreases with increasing cavity length. Continuous-wave line widths of 2 MHz have been reported for long-wavelength DFB lasers at high output power [91]. The line width of C^3 lasers is found to be smaller (by a factor of 3 to 4) than that of DFB lasers at the same output power [92].

For very narrow line-width operation, i.e., in the range 10 to 50 kHz, which is required for high–bit-rate coherent transmission systems, it is necessary to have a laser coupled to a long (approximately 10 cm) external cavity [93]. Single-frequency operation with narrow line width is obtained using a grating for frequency-selective feedback and an antireflection coated laser. The schematic diagram of an external cavity laser used in a coherent transmission system experiment is shown in Fig. 29. Fujita and coworkers [94] have recently reported an integrated external cavity laser with a line width of 0.9 MHz.

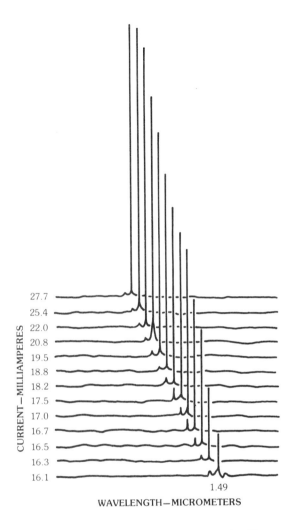

FIG. 28. Longitudinal-mode spectra of a C^3 laser. (*After Tsang, Olsson, and Logan [87]*)

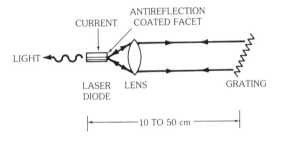

FIG. 29. Schematic of an external-cavity, single-frequency laser with narrow line width.

CW Laser Line Width

The line width of an injection laser under cw operation is an important parameter when the laser used is used as a source in coherent transmission system. The line width arises from noise-induced phase fluctuations occurring inside the laser cavity. The cw line width of a typical 250-μm single-frequency laser, such as a DFB laser, typically lies in the range 60 to 100 MHz at 1 mW output power and is found to vary inversely with the output power in agreement with the modified Schllow-Townes formula [90]. The laser line

5. Quantum-Well Lasers

A double-heterostructure laser consists of an active layer sandwiched between two higher-gap cladding layers. The active-layer thickness is typically in the range 0.1 to 0.3 μm. In the last few years double-heterostructure lasers with an active-layer thickness of about 10 nm have been fabricated. The carrier (electron or hole) motion normal to the active layer in these structures is restricted. As a result the kinetic energy of the carriers moving in that direction are quantized into discrete energy levels similar to the well-known quantum-mechanical problem of the one-dimensional potential well, and hence these lasers are called "quantum-well lasers."

Energy Levels

When the thickness L_z of a narrow-gap semiconductor layer confined between two wide-gap semiconductors becomes comparable to the deBroglie wavelength ($\lambda = h/p \simeq L_z$), the quantum-mechanical effects are expected to occur. The energy levels of the carriers confined in the narrow-gap semiconductor can be determined by separating the Hamiltonian into a component normal to the layer (confined) and into the usual (unconfined) Bloch function components (x, y) in the plane of the layer. The resulting energy eigenvalues are

$$E(n, k_x, k_y) = E_n + (\hbar^2/2 m_n^*)(k_x^2 + k_y^2), \quad (14)$$

where E_n is the nth confined-particle energy level for carrier motion normal to the well, m_n^* is the effective mass of the nth level, $\hbar$ is the Planck's constant ($h/2\pi$), and k_x and k_y are the usual Bloch function wave vectors in the x and y directions.

Fig. 30 shows schematically the energy levels E_n of the electrons and holes confined in a quantum well. The confined-particle energy levels E_n are denoted by E_{1c}, E_{2c}, and E_{3c} for electrons, E_{1hh}, E_{2hh}, and E_{3hh} for heavy holes, and E_{1lh} and E_{2lh} for the light holes. The calculation of these quantities is a standard problem in quantum mechanics for a given potential barrier (ΔE_c, ΔE_v). For an infinite potential well the following simple result is obtained

$$E_n = h^2 n^2/8 L_z^2 m_n^*. \quad (15)$$

The electron-hole recombination in a quantum well follows the selection rule $\Delta n = 0$, i.e., the electrons in states E_{1c} (E_{2c}, E_{3c}, etc.) can combine the heavy holes E_{1hh} (E_{2hh}, E_{3hh}, etc.) and with light holes E_{1lh} (E_{2lh}, E_{3lh}, etc.). Note, however, that since $E_{1lh} > E_{1hh}$ the light hole transitions are at a higher energy than the heavy hold transitions. Since the separation between the lowest conduction-band level and the highest valence-band level is given by

$$E_q = E_g + E_{1c} + E_{1hh}$$
$$\simeq E_g + (h^2/8 L_z^2)(1/m_c + 1/m_{hh}), \quad (16)$$

it follows that in a quantum-well structure the energy of the emitted photons can be varied simply by varying the well width L_z. Fig. 31 shows the experimental results of Temkin and colleagues for InGaAs quantum-well lasers with different well thicknesses bounded by InP cladding layers. For smaller well thicknesses the laser emission shifts to higher energies.

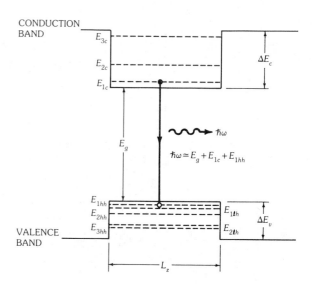

FIG. 30. Energy levels in a quantum-well structure. (*After Holonyak et al. [100], ©1980 IEEE*)

Gain and Radiative Recombination

The optical gain and spontaneous emission rate in a quantum-well structure can be calculated using (3) and (5) with an appropriate modification of the density of the states. If the bands in the x, y direction are assumed parabolic, the density of states (per unit area) for a given subband (nth level) is given by [95, 96]

$$\rho = m_n^*/\pi \hbar^{-2} L_z, \quad (17)$$

where m_n^* is the effective mass of the subband. Since the density of states is independent of energy in quantum-well structures, a group of electrons with nearly the same energy (e.g., E_{1c}) can recombine with a group of holes also with nearly the same energy (e.g., E_{1hh}). In bulk semi-

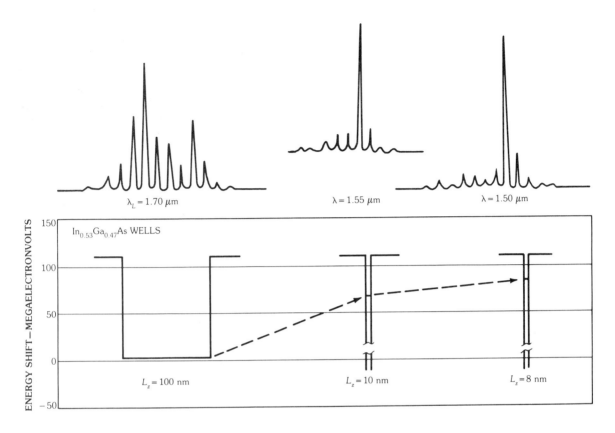

FIG. 31. The lasing spectrum for different well widths for an InGaAs quantum-well structure, showing that the emission shifts to higher energies as the well width is decreased. (*Courtesy H. Temkin*)

conductors, $\rho \simeq E^{1/2}$, i.e., the recombining electrons and holes are distributed over a wide energy range with smaller densities near the band edges, where the Fermi factors allow more occupancy. Thus in quantum-well structures the optical gain at a given injected carrier density can be larger than that for bulk semiconductors [96].

Single–Quantum-Well and Multiquantum-Well Lasers

Quantum-well injection lasers with both single and multiple active layers have been fabricated. Quantum-well lasers with one active region are called *single–quantum-well* (SQW) *lasers* and those with multiple active regions are called *multiquantum-well* (MQW) *lasers*. The layers separating the active layers in a MQW structure are called *barrier layers*. The energy-band diagrams

of these active region structures are schematically shown in Fig. 32. The MQW lasers where the band gap of the barrier layers are different from that of the cladding layers are sometimes referred to as *modified multiquantum-well lasers* [97].

One of the main differences between the SQW and the MQW lasers is that the confinement factor Γ of the optical mode is significantly smaller for the former than that for the latter. This can result in higher threshold carrier density and higher threshold current density for SQW lasers when compared with MQW lasers. The confinement factor of a SQW heterostructure can be significantly increased using a graded-index cladding layer (see Fig. 32b). This allows the use of the intrinsic advantage of the quantum-well structure (high gain at low carrier density) without the penalty of a small mode-confinement factor.

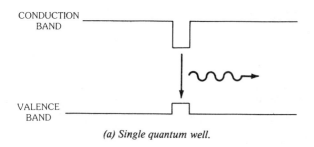

(a) Single quantum well.

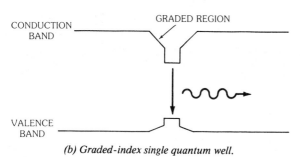

(b) Graded-index single quantum well.

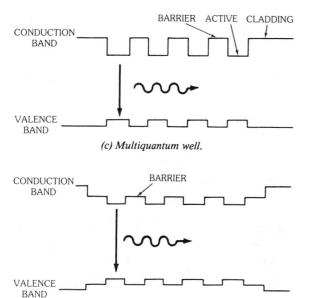

(c) Multiquantum well.

(d) Modified multiquantum well.

FIG. 32. Schematic energy-band diagram of different types of quantum-well structure.

Threshold current densities as low as 200 A/cm² have been reported for a GaAs graded-index SQW laser [98].

Tsang [97] has reported measurements of threshold current densities of AlGaAs multiquan-

tum-well lasers with different barrier heights. His results are shown in Fig. 33. The cladding layers were $Al_{0.35}Ga_{0.65}As$. The results show that there is an optimum value for barrier height ($Al_xGa_{1-x}As$ with $x \approx 0.19$) for lowest threshold current. For larger x the barrier height is too large for sufficient current injection, and for very small x the effect of the potential well which gives rise to confined two-dimensionallike states is reduced.

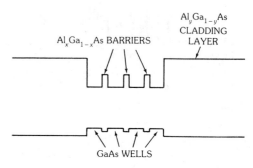

(a) Multiquantum-well structure.

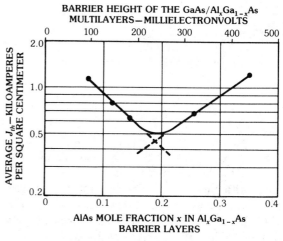

(b) Threshold current density as a function of barrier energy.

FIG. 33. An AlGaAs multiquantum-well laser structure and threshold-current/barrier characteristic. (*After Tsang [97]*)

AlGaAs Quantum-Well Lasers

A considerable amount of experimental work has been reported on quantum-well lasers fabricated using the AlGaAs material system. The epitaxial layers are grown using molecular-beam epitaxy (MBE) or organometallic vapor-phase epitaxy

(OMVPE or MOCVD) growth technique. The performances of AlGaAs lasers grown by MBE and MOCVD techniques have been reviewed by Tsang [99] and Holonyak and coworkers [100]. These results show that AlGaAs quantum-well lasers exhibit lower threshold current, somewhat higher differential quantum efficiency, and a weaker temperature dependence of threshold current when compared with regular double-heterostructure lasers.

Some of the results that demonstrate the high performance of AlGaAs quantum-well lasers are fabrication of graded-index SQW lasers with the lowest reported threshold current density [98] of 200 A/cm^2 (this compares with values in the range 0.6 to 1 kA/cm^2 for regular DH lasers), fabrication of modified MQW lasers with a threshold-current density [97] of 250 A/cm^2, and fabrication of laser arrays with an MQW active region which have been operated to highest reported cw power [101] of 2 W.

Tsang and colleagues [102] have reported a BH laser with graded-index SQW active region that operated with the lowest reported threshold current of 2.5 mA. The laser structure is shown in Fig. 34a and was fabricated by hybrid epitaxial growth. The graded-index SQW active region with the cladding layers was grown by MBE. Mesas were then etched using standard photolithographic techniques and the current-confining layers were grown by LPE. The cw and pulsed light/current characteristics of the laser are shown in Fig. 34b.

InGaAsP Quantum-Well Lasers

Compared with the literature on quantum-well lasers using AlGaAs material system, there is considerably less amount of work reported using the InGaAsP material system. It is the author's perception that as the materials growth technology matures, long-wavelength quantum-well lasers (e.g., InGaAsP) will be extensively studied. The results reported below are observations to date

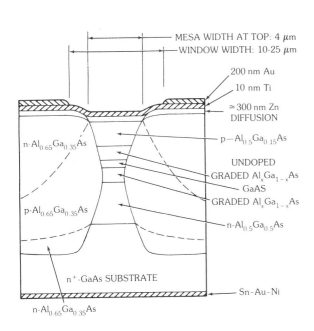

(a) Schematic of structure.

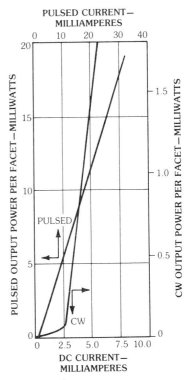

(b) Pulsed and cw light versus current characteristics of the laser shown in (a).

FIG. 34. A buried-heterostructure laser with graded-index single–quantum-well active region. (*After Tsang, Logan, and Ditzenberger [102], © Institution of Electrical Engineers*)

which may improve as the technology develops.

Rezek and coworkers [103, 104] first reported the growth of InGaAsP quantum-well structures by liquid-phase epitaxy (LPE). Quantum-well lasers with a room-temperature threshold-current density of approximately 5.2 kA/cm² were reported [104].

Yanase and colleagues [105] have reported InGaAsP-InP MQW lasers emitting near 1.3 μm and fabricated using the hybrid transport VPE growth technique. Threshold-current densities were in the range 2 to 3 kA/cm² for well thick-

nesses in the range 10 to 30 nm.

Dutta and coworkers [106, 107] have reported the fabrication and performance characteristics of InGaAsP double-channel planar buried-heterostructure (DCPBH) laser with a multiquantum-well active layer. The MQW structure was grown by LPE. Two types of MQW structures were studied: (a) with InP barrier layers [106] and (b) with InGaAsP ($\lambda \approx 1.03 \mu m$) barrier layers [107]. The MQW active region had four active wells ($\lambda \approx 1.3$-μm InGaAsP) and three barrier layers. The schematic cross section of the DCPBH structure and the MQW structure are shown in Fig. 35. The DCPBH lasers with InP barrier layers had room-temperature threshold currents in the range 40 to 50 mA, and for InGaAsP ($\lambda \approx 1.03$ μm) barrier layers the threshold currents were in the range 20 to 25 mA. The lower threshold current of devices with lower barrier height may be due to more uniform current injection. Similar observation was reported by Tsang [97] for AlGaAs MQW lasers as shown in Fig. 33. The cw light/current characteristics at different temperatures of an InGaAsP DCPBH MQW laser is shown in Fig. 36. The temperature dependence of

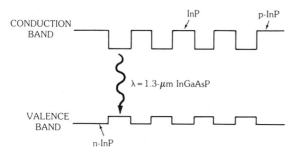

(a) Active layers, with InP barrier layers.

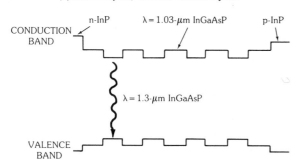

(b) Active layers, with InGaAsP barrier layers.

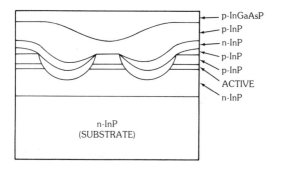

(c) DCPBH laser schematic.

FIG. 35. The MQW and the DCPBH structures.

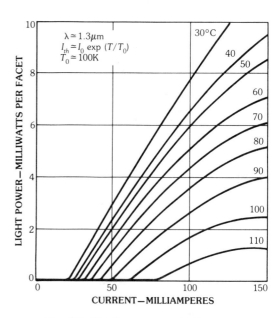

FIG. 36. Continuous-wave light versus characteristics at different temperatures of a DCPBH MQW laser with InGaAsP ($\lambda \approx 1.03$-μm) barrier layers.

threshold current of these MQW lasers are lower than that for regular DH lasers.

Quantum-well lasers emitting near 1.55 μm, which is the region of lowest optical loss in silica fibers, have been reported by several investigators [108–111]. InGaAs/AlInAs MQW structures emitting in the range 1.5 to 1.6 μm have been fabricated by MBE growth techniques [108]. The broad-area threshold current is approximately 2.4 kA/cm^2 near room temperature. Similar results for lasers with InGaAs active wells and InP barriers have been reported by Tsang [109].

InGaAsP-InP MQW injection lasers emitting near 1.55 μm and fabricated by the LPE growth technique have been reported by Dutta and co-workers [111]. Both the threshold current and the temperature dependence of threshold are comparable to those for regular DH lasers. Single-frequency, distributed-feedback type InGaAsP MQW lasers emitting near 1.55 μm with cw performance characteristics comparable to regular DH lasers have also been reported [111].

For lightwave system applications the laser light is modulated by modulating the current injected into the laser. The current modulates the injected carrier density and hence the refractive index of the guided wave, which leads to a wavelength chirp as mentioned previously. Low chirp width is desirable for high–data-rate optical communication systems. Dutta and coworkers [106] have reported that the measured dynamic line width of quantum-well lasers is smaller (by a factor of about 2) than that for regular double-heterostructure lasers. This is consistent with the theoretical results of Burt [112] and Arakawa and coworkers [113], who showed that the modification of the density of states in a quantum-well structure reduces the free-carrier contribution to the refractive index. Thus the advantages of quantum-well lasers over regular DH lasers lie in the possibility of lower threshold current, lower temperature dependence of threshold, and lower chirp.

6. Degradation and Reliability

The performance characteristic of injection lasers can degrade during their operation. This degrada-

tion is usually characterized by an increase in the threshold current which is often accompanied by a decrease in the external differential quantum efficiency. The mechanisms responsible for this degradation are determined by any or all of the several fabrication processes, including epitaxial growth, wafer quality, device processing, and bonding. In addition, the degradation rate of lasers processed from a given wafer depends on the operating conditions, viz., the operating temperature and the injection current. Although many of the degradation mechanisms are not fully understood, extensive amounts of empirical observations exist in the literature, which has allowed the fabrication of InGaAsP laser diodes with extrapolated median lifetimes in excess of 25 years [114] at an operating temperature of 10°C.

The detailed studies of degradation mechanisms of injection laser diodes have been motivated by the desire to have a reasonably accurate estimate of the operating lifetime before they are used in practical systems. Since for many applications the laser sources are expected to operate reliably over a period in excess of 10 years, an appropriate reliability assurance procedure becomes necessary, especially for applications such as an undersea lightwave transmission system, where the laser replacement cost is very high.

Aside from catastrophic damage, which is a sudden failure mechanism caused by unforeseen events (such as power supply failure), the degradation mechanisms can be separated into two categories. They are (a) defect formation in the active region and (b) degradation of current-confining junctions. It is important to mention at the outset that although reliable semiconductor lasers have been fabricated using both the AlGaAs and InGaAsP alloy system, the study of degradation modes has been descriptive, and the results in many cases are tentatively interpretative.

Defect Formation in the Active Region

The high density of recombining electrons and holes and a possible presence of strain and thermal gradients can promote defect formation in the active region of the laser. The defect structures that are generally observed are the dark-spot defect (DSD) and the dark-line defect (DLD).

The dark-line defect, as the name suggests, is a

region of greatly reduced radiative efficiency of roughly linear form. The DLD was first reported by DeLoach and colleagues [39] in the active region of an aged AlGaAs proton-stripe double-heterostructure laser. The DLD appears as a linear dark feature crossing the luminescent stripe at 45° in degraded lasers. Since the active stripe is oriented along the (110) direction, the DLDs are oriented along the (100) direction.

The observation of nonluminescent regions or regions of reduced radiative efficiency, such as a DLD inside the active region of a semiconductor laser, requires the fabrication of a special laser structure. This laser structure, commonly known as the *window laser,* is shown in Fig. 37. A "window" typically 20 to 25 μm wide is formed using photolithographic techniques on the substrate side. Since the InGaAsP laser emits at energies smaller than the band gap of the n-InP substrate, the spontaneous emission from the active region can be directly observed through the window. The window-laser structure can allow continuous monitoring of luminescence efficiency of the active region and is also compatible with the normal "p-down" bonding configuration. Very often, degraded lasers show dark regions in the active-strip luminescence when observed through the window. DLDs and DSDs have been observed in both AlGaAs and InGaAsP lasers and LEDs. Detailed study of these defect structures requires the use of techniques such as transmission electron microscopy (TEM) [115] deep-level transient spectroscopy (DLTS) [116], scanning electron microscopy (SEM), electron-beam–induced current (EBIC) mode of SEM, cathodoluminescence [117] and scanning photoluminescence [118].

Accelerated aging techniques, which include high-temperature and high-power operation, are generally used to estimate the usable lifetime of injection lasers under normal operating conditions. The generation rate of DSDs and DLDs is enhanced under accelerated aging. When the accelerated aging is done at high temperature an activation energy can be defined for the generation rate of defects in the active region. The activation energy is defined using

$$t_d = t_0 \exp\left(E_a / k\, T\right), \qquad (18)$$

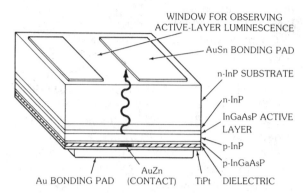

FIG. 37. Schematic of a "window" laser structure.

where t_d is the generation time for the first defect, t_0 is a constant, E_a is the activation energy, k is Boltzmann's constant, and T is the operating temperature. Fukuda and coworkers [119] have measured the generation time of DSDs and DLDs in InGaAsP gain-guided lasers operating at 1.3-μm and 1.55-μm wavelengths. The generation of DSDs and DLDs was observed by electroluminescence using a window laser configuration. The lasers that did not exhibit DSDs and DLDs operated for a long time without degradation. The measured pulsed threshold current normalized to the initial value is shown in Fig. 38. The increase in threshold current was associated with an increase in the number of DSDs and (100) DLDs. Gallium- and arsenic-rich regions in the active layer were correlated with the location of the DSDs. Fukuda and colleagues [119] observed a saturation in the number of DSDs and DLDs in about 50 hrs of aging at 250°C; beyond that time the increase in threshold current caused by further aging also showed a tendency to saturate. This observation is very important for the purpose of reliability assurance because it shows a possible presence of a saturable mode of degradation, i.e., after a certain period of aging the degradation rate may be very small.

Degradation of Current-Confining Junctions

As discussed in Section 3, many index-guided laser structures utilize current restriction layers so that most of the injected current will flow through the active region. Effective current injection to the active region is necessary in order to obtain low

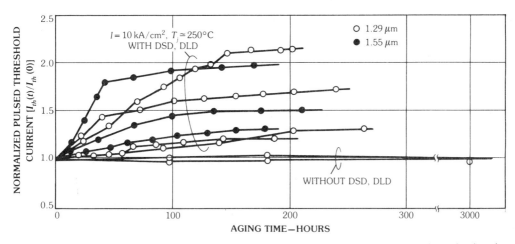

FIG. 38. Change in pulsed threshold current normalized to its initial value as a function of aging time. (*After Mitzuishi et al. [120], ©1983 IEEE*)

threshold current and high output powers. The current outside the active region in buried-heterostructure (BH) type lasers is called the *leakage current*. A mode of degradation that is associated with BH lasers is an increase in the leakage current (which increases the device threshold and decreases the external differential quantum efficiency) under accelerated aging. An increase in leakage current not only increases the device threshold but also generally decreases the external differential quantum efficiency. Increased leakage current usually appears as a "soft turn-on" in the current/voltage characteristics of the laser. Electron-beam–induced current (EBIC) observation of laser facets is a useful technique in detecting defects at current-confining junctions [120].

Reliability Assurance

In some semiconductor laser applications the system design lifetimes are long (about 20 to 25 years) and the replacement of components (e.g., lasers) can be too expensive. An example of such an application is provided by the repeaters of an undersea lightwave transmission system, where the failure of just a few lasers can cripple the system. Thus it is important to have a strategy by means of which it is possible to establish the expected operating lifetime of a laser.

Some lasers exhibit an initial rapid degradation, after which the operating characteristics of the la-

sers are very stable. Given a population of lasers it is possible to quickly identify the "stable" lasers by a high stress test (also known as the purge test) [121]. The stress test implies that operating the laser under a set of high-stress conditions (e.g., high current, high temperature, high power) would cause the weak lasers to fail and stabilize the possible winners. Observations on the operating current after stress aging have been reported by Nash and colleagues [114]. Fig. 39 shows their measured data. The operating current required for an output power of 3 mW/facet at 60°C is plotted as a function of stress-aging time. The stress conditions required the cw operation of the laser at 100°C with a 250-mA current. The lasers were of the buried-heterostructure type. Some lasers exhibit an increase in operating current before they stabilize similar to the observations of Fukuda and coworkers [119], while others exhibit stable characteristics without a significant increase in the operating current. Figs. 38 and 39 show that the high-stress test can be used as a screening procedure to identify robust devices. It is important to point out that the determination of the duration and the specific conditions for stress aging are critical to the success of this screening procedure.

Central to the determination of the expected operating lifetime is the concept of thermally accelerated aging, of which the validity for AlGaAs injection lasers was shown by Hartman

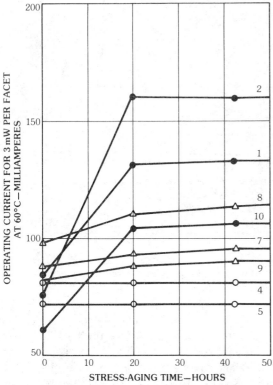

FIG. 39. The operating current required for 3 mW/facet at 60°C as a function of stress-aging time. (*After Nash et al. [114]*)

and Dixon [122]. The lifetime τ at a temperature T varies as

$$\tau = \tau_0 \exp\left(-E_a/kT\right),\qquad(19)$$

where E_a is the activation energy and k is Boltzmann's constant. Measurements of activation energy of InGaAsP-InP BH lasers have been reported by Mizuishi and colleagues [123] and Hakki and colleagues [124] by measuring the degradation rate at various temperatures. From their measurements Mizushi's group obtained $E_a = 0.9$ eV and Hakki's group obtained $E_a = 1 \pm 0.13$ eV.

The data of Nash and coworkers [114] on the reliability of BH lasers are shown in Fig. 40. The normalized operating currents (I/I_0) for 14 lasers aged for approximately 7000 hrs at 60°C at 3mW operating conditions are shown. The maximum degradation rate observed is 3.1% per 1000 hr. The measured degradation rate can be used to obtain an expected operating life using a 50% change in I/I_0 as a failure criterion. The extrapolated life at an operating temperature of 10°C (which is the ocean-bottom temperature) is then obtained using (19) and an activation energy of 0.9 eV. Then 1200 hr of aging time at 60°C at

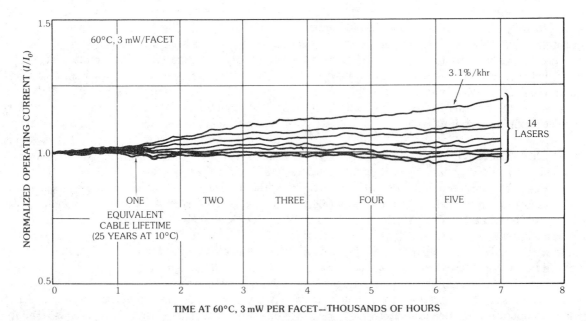

FIG. 40. Accelerated aging rate at 60°C, at 3 mW/facet of a group of 14 prescreened lasers. (*After Nash et al. [114]*)

3 mW/facet is equivalent to 25 years of operating time at 10°C. The arrows in Fig. 40 represent 25 years of equivalent operating life at 10°C, which is the expected cable lifetime of the transatlantic lightwave system (TAT8) that is currently being deployed [125].

Acknowledgments

The author gratefully acknowledges the support of the semiconductor laser development department in preparation of this manuscript. He thanks G. P. Agrawal for assistance in writing the section on single-frequency lasers.

7. References

[1] R. N. HALL, G. E. FENNER, J. D. KINGSLEY, T. J. SOLTYS, and R. O. CARLSON, "Coherent Light Emission From GaAs Junctions" *Phys. Rev. Lett.,* Vol. 9, pp. 336–368, November 1962.

[2] M. I. NATHAN, W. P. DUMKE, G. BURNS, F. H. DILL, JR., and G. LASHER, "Stimulated Emission of Radiation From GaAs pn Junctions" *Appl. Phys. Lett.,* Vol. 1, pp. 63–64, November 1962.

[3] T. M. QUIST, R. H. REDIKER, R. J. KEYES, W. E. KRAG, B. LAZ, A. L. MCWHORTER, and H. J. ZEIGLER, "Stimulated Maser of GaAs" *Appl. Phys. Lett.,* Vol. 1, pp. 91–92, December 1962.

[4] N. HOLONYAK, JR. and S. F. BEVACQUA, "Coherent Visible Light Emission From $GaAs_{1-x}P_x$ Junctions," *Appl. Phys. Lett.,* Vol. 1, pp. 82–83, December 1962.

[5] H. KRESSEL, and J. K. BUTLER, *Semiconductor Lasers and Heterojunction LEDs,* New York: Academic Press, 1977.

[6] H. C. CASEY, JR., and M. B. PANISH, *Heterostructure Lasers,* New York: Academic Press, 1978.

[7] G. H. B. THOMPSON, *Physics of Semiconductor Laser Devices,* New York: John Wiley & Sons, 1981.

[8] C. A. BURRUS and R. W. DAWSON, "High-Radiance Light Emitting Diode," *Appl. Phys. Lett.,* Vol. 17, p. 97, 1970.

[9] H. KRESSEL, M. ETTENBERG, J. P. WILTKE and I. LADANY, Chapter 2 in *Topics in Applied Physics,* Vol. 39, *Semiconductor Devices for Optical Communications,* ed. by H. KRESSEL, New York: Springer-Verlag, 1982.

[10] G. ARNOLD et al., "1.3-μm Edge-Emitting Diodes Launching 250 μW Into a Single Mode Fibre at 100 mA," *Electron. Lett.,* Vol. 21, p. 993, 1985.

[11] H. NELSON, "Liquid Phase Epitaxy of GaAs," *RCA Rev.,* Vol. 24, p. 603, 1963.

[12] J. J. TIETJEN and J. A. AMICK, "The Preparation and Properties of Vapor Deposited Epitaxial GaAsP Using Arsine and Phosphine," *J. Electrochem. Soc.,* Vol. 113, p. 724, 1966.

[13] A. Y. CHO, "Recent Developments in Molecular Beam Epitaxy," *J. Vac. Sci. Tech.,* Vol. 16, p. 275, 1979.

[14] R. D. DUPUIS, "MOCVD Growth of GaAs," *J. Crystal Growth,* Vol. 55, p. 213, 1981.

[15] D. MARCUSE, *Light Transmission Optics,* New York: Van Nostrand Reinhold Co., 1982.

[16] L. M. MAGID, *Electromagnetic Fields, Energy and Waves,* New York: John Wiley & Sons, 1972.

[17] D. BOTEZ, "Effective Refractive Index and First-Order Mode Cutoff Conditions in InGaAsP/InP DH Laser Structures," *IEEE J. Quantum Electron.,* Vol. QE-17, p. 178, 1981; Vol. QE-18, p. 865, 1982.

[18] E. O. KANE, "Band Structure of InSb," *J. Phys. Chem. of Solids,* Vol. 1, p. 249, 1957.

[19] M. G. A. BERNARD and G. DURAFFOURG, "Laser Conditions in Semiconductors," *Phys. Status Solidi,* Vol. 1, p. 699, 1961.

[20] F. STERN, "Gain-Current Relation for GaAs Lasers With n-Type and Undoped Active Layers," *IEEE J Quantum Electron.,* Vol. QE-9, p. 290, 1973, and references therein.

[21] N. K. DUTTA, "Calculated Absorption, Emission and Gain in $In_{0.72}Ga_{0.28}As_{0.6}P_{0.4}$," *J. Appl. Phys.,* Vol. 51, pp. 6095–6100, 1980; "Gain-Current Relation for InGaAsP Lasers," *J. Appl. Phys.,* Vol. 52, pp. 55–56, 1981.

[22] F. SERN, "Calculated Spectral Dependence of Gain in Excited GaAs," *J. Appl. Phys.,* Vol. 47, pp. 5382–5386, 1976.

[23] R. OLSHANSKY, C. B. SU, J. MANNING, and W. POWAZNIK, "Measurement of Radiative and Nonradiative Recombination Rates in InGaAsP and AlGaAs Light Sources," *IEEE J. Quantum Electron.,* Vol. QE-20, p. 838, 1984.

[24] N. K. DUTTA and R. J. NELSON, "The Case for Auger Recombination in InGaAsP," *J. Appl. Phys.,* Vol. 53, pp. 74–92, 1982.

[25] A. R. BEATTIE and P. T. LANDSBERG, "Auger Effect in Semiconductors," *Proc. Royal Soc. London,* Vol. 249, pp. 16–28, 1959.

[26] Y. HORIKOSHI and Y. FURUKAWA, "Temperature-Sensitive Threshold Current of InGaAsP-InP Double-Heterostructure Lasers," *Jap. J. Appl. Phys.,* Vol. 18, p. 809, 1979.

[27] G. H. B. THOMPSON and G. D. HENSHALL, "Non-radiative Carrier Loss and Temperature Sensitivity of Threshold in 1.27-μm InGaAsP DH Lasers," *Electron. Lett.,* Vol. 16, pp. 42–44, 1980.

[28] N. K. DUTTA and R. J. NELSON, "Temperature Dependence of Threshold of InGaAsP-InP DH Lasers and Auger Recombination," *Appl. Phys. Lett.,* Vol. 38, pp. 407–409, 1981.

[29] A. SUGIMURA, "Band-to-Band Auger Recombination Effect on InGaAsP Laser Threshold," *IEEE J. Quantum Electron.,* Vol. QE-17, pp. 627–635, 1981.

[30] A. R. BEATTIE and G. SMITH, "Recombination in Semiconductors by a Light Hole Transition," *Phys. Status Solidi,* Vol. 19, p. 577, 1967.

[31] M. TAKISHIMA, "Auger Recombination in InAs, GaSb, InP and GaAs," *J. Appl. Phys.,* Vol. 43, p. 4114, 1972; *J. Appl. Phys.,* Vol. 44, p. 4717, 1973.

[32] J. R. CHELIKOWSKY and M. L. COHEN, "Nonlocal Pseudo-Potential Calculation for the Electronic Structure of Eleven Diamond and Zinc-Blend Semiconductors," *Phys. Rev.,* Vol. 1314, pp. 556–582, 1976.

[33] A. HAUG, "Auger Recombination in InGaAsP," *Appl. Phys. Lett.,* Vol. 42, pp. 512–514, 1983.

[34] B. SERMAGE, H. J. EICHLER, J. HERITAGE, R. J. NELSON, and N. K. DUTTA, "Photoexcited Carrier Lifetime and Auger Recombination in 1.3-μm InGaAsP," *Appl. Phys. Lett.,* Vol. 42, pp. 259–261, 1983.

[35] C. H. HENRY, B. F. LEVINE, R. A. LOGAN, and C. G. BETHEA, "Minority Carrier Lifetime and Luminescence Efficiency of 1.3-μm InGaAsP-InP DH Lasers," *IEEE J. Quantum Electron.,* Vol. QE-19, pp. 905–913, 1983.

[36] E. WINTNER and E. P. IPPEN, "Nonlinear Carrier Dynamics in GaInAsP Compounds," *Appl. Phys. Lett.,* Vol. 44, p. 999, 1984.

[37] A. MOZER, K. M. ROMANEK, O. HILDEBRAND, W. SCHMID, and M. H. PILKHUN, "Losses in GaInAsP/InP and GaAlSbAs-GaSb Lasers—The Influence of Split-off Band," *IEEE J. Quantun Electron.,* Vol. QE-19, pp. 913–916, 1983.

[38] C. B. SU, J. SCHLAFER, J. MANNING, and R. OLSHANSKY, "Measurement of Radiative and Auger Recombination Rates in p-Type InGaAsP Diode Lasers," *Electron. Lett.,* Vol. 18, p. 595, 1982a; Vol. 18, p. 1108, 1982b.

[39] B. C. DELOACH, JR., B. W. HAKKI, R. L. HARTMAN, and L. A. D'ASARO, " Degradation of CW GaAs Double-Heterojunction Lasers at 300 K," *Proc. IEEE,* Vol. 61, p. 1042, 1973.

[40] C. H. HENRY, R. A. LOGAN, F. R. MERRITT, and J. P. LUONGO, "The Effect of Intervalence-Band Absorption on the Thermal Behavior of InGaAsP Lasers," *IEEE J. Quantum Electron.,* Vol. QE-19, pp. 947–53, 1983.

[41] H. C. CASEY, JR., "Temperature Dependence of Threshold Current of InGaAsP Lasers," *J. Appl. Phys.,* Vol. 56, p. 1959, 1984.

[42] J. C. DYMENT, "Hermite-Gaussian Mode Patterns in GaAs Junction Lasers, " *Appl. Phys. Lett.,* Vol. 10, p. 84, 1967.

[43] J. J. HSIEH, J. A. ROSSI, and J. P. DONNELLY, "Room-Temperature CW Operation of GaInAsP/InP DH Diode Lasers Emitting at 1.1 μm," *Appl. Phys. Lett.,* Vol. 28, p. 709, 1976.

[44] G. D. HENSHALL, G. H. B. TTHOMPSON, J. E. A. WHITEWAY, P. R. SELWLAY, and M. BROOMFIELD, *Solid State and Electron Devices,* Vol. 3, No. 1, 1979.

[45] B. SCHWARTZ, M. W. FOCHT, N. K. DUTTA, R. J. NELSON, and P. BESOMI, "Stripe Geometry InP/InGaAsP Lasers Fabricated with Deuteron Bombardment," *IEEE Trans. Electron. Devices,* Vol. ED-31, p. 841, 1984.

[46] K. OE and K. SUGIYAMA, "GaInAsP/InP Planar Stripe Lasers Prepared by Using Sputtered SiO_2 film as a Zn Diffusion Mask," *J. Appl. Phys.,* Vol. 51, p. 43, 1980.

[47] R. W. DIXON, F. R. NASH, R. L. HARTMAN, and R. T. HEPPLEWHITE, "Nonlinearization in the Light Current Characteristics of GaAs Lasers," *Appl. Phys. Lett.,* Vol. 29, p. 372, 1976.

[48] I. P. KAMINOW, L. W. STULZ, J. S. KO, A. G. DENTAI, R. E. NAHORY, J. C. DEWINTER, and R. L. HARTMAN, "Low-Threshold InGaAsP Ridge-Waveguide Injection Lasers at 1.3 μm," *IEEE J. Quantum Electron.,* Vol. QE-19, p. 1312, 1983.

[49] W. T. TSANG, and R. A. LOGAN, "A New High-Power Transverse-Mode Stabilized Semiconductor Laser at 1.5 μm: the Heteroepitaxial Ridge-Overgrown Laser," *Appl. Phys. Lett.,* Vol. 45, p. 1025, 1984.

[50] K. AIKI, M. NAKAMURA, T. KURODA, and J. UMEDA, "Channeled-Substrate Planar Structure AlGaAs Injection Lasers," *Appl. Phys. Lett.,* Vol. 48, p. 649, 1977.

[51] S. E. H. TURLEY, G. D. HENSHALL, P. D. GREENE, V. P. KNIGHT, D. M. MOULE, and S. A. WHEELER, "Properties of Inverted Rib-Waveguide Lasers Operating at 1.3 μm Wavelength," *Electron. Lett.,* Vol. 17, p. 868, 1981.

[52] T. TASUKADA, "GaAs-GaAlAs Buried-Heterostructure Injection Lasers," *J. Appl. Phys.,* Vol. 45, p. 4899, 1974.

[53] M. HIRAO, S. TSUJI, K. MIZUSHI, A. DOI, and M. NAKAMURA, "Long-Wavelength InGaAsP/InP Lasers for Optical-Fiber Communication Systems," *J. Opt. Commun.,* Vol. 1, p. 10, 1980.

[54] Z. L. LIAU and J. N. WALPOLE, "A Novel Technique for GaInAsP/InP Buried-Heterostructure Laser Fabrication," *Appl. Phys. Lett.,* Vol. 40, p. 568, 1982.

[55] I. MITO, M. KITAMURA, K. KOBAYASHI, S. MURATA, M. SEKI, Y. ODAGIRI, H. NISHIMOTO, M. YAMAGUCHI, and K. KOBAYASHI, "InGaAsP Double-Channel Planar Buried-Heterostructure Laser Diode (DCPBH LD) With Effective Current Confinement," *IEEE J. Lightwave Tech.,* Vol. LT-1, pp. 195–202, 1983.

[56] I. MITO, M. KITAMURA, K. KAEDA, Y. ODAGIRI, M. SEKI, M. SUGIMOTO, and K. KOBAYASHI, *Electron. Lett.,* p. 2, January 1982.

[57] R. J. NELSON, P. D. WRIGHT, P. A. BARNES, R. L. BROWN, T. CELLA, and R. G. SOBERS, "High Output Power InGaAsP (λ = 1.3 μm) Strip-Buried Heterostructure Lasers," *Appl. Phys. Lett.,* Vol. 36, p. 358, 1980.

[58] H. ISHIKAWA, H. IMAI, T. TANAHASHI, Y. NISHITANI, M. TAKASAGAWA, and K. TAKAHI, "V-Grooved Substrate Buried-Heterostructure InGaAsP/InP Laser," *Electron. Lett.,* Vol. 17, p. 415, 1981.

[59] N. K. DUTTA, D. P. WILT, P. BESOMI, W. D. DAUTREMONT-SMITH, P. D. WRIGHT, and R. J. NELSON, "Improved Linearity and Kink Criteria for 1.3-μm InGaAsP-InP Channeled-Substrate Buried-Heterostructure Lasers," *Appl. Phys. Lett.,* Vol. 44, p. 483, 1984.

[60] R. A. LOGAN, J. P. VAN DER ZIEL, and H. TEMKIN, "V-Groove Substrate BH Lasers," *Proc. SPIE,* Vol. 380, p. 181–185, 1982.

[61] M. ORON, N. TAMARI, H. SHTRIKMAN, and C. A. BURRUS, "Lasing Properties of InGaAsP Buried-Heterojunction Lasers Grown on a Mesa Substrate," *Appl. Phys. Lett.,* Vol. 41, p. 609, 1982.

[62] T. MURTANI, E. OOMURA, H. HIGUCHI, H. NAMIZAKI, and W. SUSAKI, "InGaAsP/InP Buried Crescent Laser Emitting at 1.3 μm With Very Low Threshold Current," *Electron. Lett.,* Vol. 16, p. 566, 1980.

[63] N. K. DUTTA, D. P. WILT, and R. K. NELSON, "Analysis of Leakage Currents in 1.3-μm InGaAsP Real Index-Guided Lasers," *IEEE J. Lightwave Tech.,* Vol. LT-2, pp. 201–208, 1984.

[64] A. R. ADAMS, M. ASADA, Y. SUEMATSU, and S. ARAI, "The Temperature Dependence of Efficiency and Threshold Current of InGaAsP Lasers Related to Intervalence Absorption," *Jap. J. Appl. Phys.,* Vol. 19, p. L621, 1980.

[65] H. ISHIKAWA, H. IMAI, I. UMEBU, K. HORI, and M. TAKUSAGAWA, "V-Grooved Substrate BH InGaAsP/InP Lasers by One-Step Epitaxy," *J. Appl. Phys.,* Vol. 53, p. 2851, 1982.

[66] D. R. SCIFRES, R. D. BURNHAM, and W. STREIFER, *Appl. Phys. Lett.,* Vol. 34, p. 259, 1979.

[67] W. STREIFER, R. D. BURNHAM, T. L. PAOLI, and D. R. SCIFRES, "High-Power Laser Diodes," *Laser Focus,* June 1984.

[68] N. K. DUTTA, L. A. KOSZI, B. P. SEGNER, S. G. NAPHOLTZ, and D. C. CRAFT, "High-Power Index-Guided Multiridge-Waveguide Laser Array," *Appl. Phys. Lett.,* Vol. 46, p. 803, 1985.

[69] H. SODA, K. IGA, C. KITAHARA and Y. SUEMATSU, *Jap. J. Appl. Phys.,* Vol. 18, p. 2329, 1979.

[70] Z. L. LIAU and J. N. WALPOLE, "Surface-Emitting GaInAsP/InP Laser With Low Threshold Current and High Efficiency," *Appl. Phys. Lett.,* Vol. 46, p. 115, 1985.

[71] T. E. BELL, "Single-Frequency Semiconductor Lasers," *IEEE Spectrum,* Vol. 20, p. 38, 1983.

[72] H. KOGELNIK and C. V. SHANK, "Stimulated Emission in a Periodic Structure," *Appl. Phys. Lett.,* Vol. 18, p. 152, 1971.

[73] M. NAKAMURA, A. YARIV, H. W. YEN, S. SOMEKH, and H. L. GARVIN, *Appl. Phys. Lett.,* Vol. 22, p. 515, 1973; Vol. 23, p. 224, 1973.

[74] C. V. SHANK, R. V. SCHMIDT, and B. I. MILLER, "Double-Heterostructure GaAs Distributed-Feedback Laser," *Appl. Phys. Lett.,* Vol. 25, p. 200, 1974.

[75] D. R. SCIFRES, R. D. BURNHAM, and W. STREIFER, "Distributed-Feedback Single-Heterojunction GaAs Diode Laser," *Appl. Phys. Lett.,* Vol. 25, p. 203, 1974.

[76] K. UUAKA, S. AKIBA, K. SAKAI, and Y. MATSUSHIMA, "Room-Temperature CW Operation of Distributed-Feedback InGaAsP/InP Laser, *Electron. Lett.,* Vol. 17, p. 961, 1981.

[77] S. AKIBA, K. UTAKA, K. SAKAI, and Y. MATSUSHIMA, "Distributed-Feedback InGaAsP/InP Lasers With Window Region Emitting at 1.5 μm Range," *IEEE J. Quantum Electron.,* Vol. QE-19, p. 1052, 1983.

[78] M. KITAMURA, M. SEKI, M. YAMAGUCHI, I. MITO, K. E. KOBAYASHI, and K. O. KOBAYSHI, "High-Power Single-Mode Operation of 1.3-μm DFB DCP BH LD," *Electron. Lett.,* Vol. 19, p. 840, 1983.

[79] W. T. TSANG and S. WANG, "GaAs-AlGaAs Double-Heterostructure Injection Laser With Distributed Bragg Reflector," *Appl. Phys. Lett.,* Vol. 28, p. 596, 1976.

[80] M. KITAMURA, M. YAMAGUCHI, S. MURATA, I. MITO, K. KOBAYASHI, *IEEE J. Lightwave Tech.,* Vol. LT-2, p. 363, 1984.

[81] N. K. DUTTA, T. WESSEL T. CELLA, and R. L. BROWEN, "Continuously Tunable Distributed-Feedback Laser Diode," *Appl. Phys. Lett.,* Vol. 47, p. 981, 1985.

[82] L. D. WESTBROOK, A. W. NELSON, P. J. FIDDYMENT, and J. S. EVANS, "Continuous-Wave Operation of 1.5-μm DFB Ridge-Waveguide Lasers," *Electron. Lett.,* Vol. 20, p. 225, 1984.

[83] H. TEMKIN, G. J. DOLAN, N. A. OLSSON, C. H. HENRY, R. A. LOGAN, R. F. KAZARINOV, and L. F. JOHNSON, "1.55-μm InGaAsP Ridge-Waveguide Distributed-Feedback Laser," *Appl. Phys. Lett.,* Vol. 45, p. 1178, 1984.

[84] R. A. LINKE, "Transient Chirping in Single-Frequency Lasers," *Electron. Lett.,* Vol. 20, p. 472, 1984.

[85] M. B. CHANG and E. GARMIRE, *IEEE J. Quantum Electron.,* Vol. QE-16, p. 997, 1980.

[86] L. A. COLDREN, B. I. MILLER, K. IGA, and J. RENTSCHLER, "Monolithic Two-Section GaInAsP/InP Active-Optical-Resonator Devices Formed by Reactive Ion Etching," *Appl. Phys. Lett.,* Vol. 38, p. 315, 1981.

[87] W. T. TSANG, N. A. OLSSON, and R. A. LOGAN, "High-Speed Direct Single-Frequency Modulation With Large Tuning Rate in Cleaved–Coupled-Cavity Lasers," *Appl. Phys. Lett.,* Vol. 42, p. 650, 1983; *Electron. Lett.,* Vol. 19, p. 438, 1983.

[88] K-Y. LIOU, C. A. BURRUS, R. A. LINKE, I. P. KAMINOW, S. W. GRANLUND, C. B. SWAN, and P. BESOMI, "Single Longitudinal-Mode Stabilized Graded-Index-Rod External Coupled-Cavity Laser," *Appl. Phys. Lett.,* Vol. 45, p. 729, 1984.

[89] W. T. TSANG, "Lightwave Communications Technology, Vol. 22, Pt. B of *Semiconductors and Semimetals,* ed. by W. T. TSANG, New York: Academic Press, 1985.

[90] C. H. HENRY, "Theory of Line Width of Semiconductor Lasers," *IEEE J. Quantum Electron.,* Vol. QE-18, p. 259, 1982.

[91] K-Y. LIOU, N. K. DUTTA, and C. A. BURRUS, paper presented at Conference on Lasers and Electro-Optics (CLEO), San Francisco, 1986.

[92] T. P. LEE, C. A. BURRUS, K. Y. LIOU, N. A. OLSSON, R. A. LOGAN, and D. P. WILT, "CW Line Width of Single-Frequency Lasers," *Electron. Lett.,* Vol. 20, p. 1011, 1984.

[93] G. P. AGRAWAL, "Line Narrowing in a Single-Mode Injection Laser due to External Optical Feedback," *IEEE J. Quantum Electron.,* Vol. QE-20, p. 468, 1984.

[94] T. FUJITA, J. OHYA, K. MATSUDA, M. ISHINO, H. SATO, and H. SERIZAWA, "Narrow Spectral Line Width Characteristics of Monolithic Integrated–Passive-Cavity InGaAsP/InP Semiconductor Lasers," *Electron. Lett.,* Vol. 21, p. 374, 1985.

[95] K. HESS, B. A. VOJAK, N. HOLONYAK, JR., R. CHIN, and P. D. DAPKUS, "Temperature Dependence of Threshold Current for a Quantum-Well Heterostructure Laser," *Solid-State Electron.,* Vol. 23, p. 585, 1980.

[96] N. K. DUTTA, *Electron. Lett.,* Vol. 18, p. 451, 1982; "Calculated Threshold Current of GaAs Quantum-Well Lasers," *J. Appl. Phys.,* Vol. 53, p. 7211, 1982.

[97] W. T. TSANG, "Extremely Low Threshold AlGaAs Modified Multiquantum-Well Heterostructure Lasers Grown by MBE," *Appl. Phys. Lett.,* Vol. 39, p. 786, 1981.

[98] S. D. Hersee, B. de Cremoux, and J. P. Duchemin, "Some Characteristics of GaAs/GaAlAs Graded-Index Separate-Confinement Heterostructure Quantum-Well Laser," *Appl. Phys. Lett.,* Vol. 44, p. 476, 1984.

[99] W. T. Tsang, "Heterostructure Semiconductor Lasers Prepared by MBE," *IEEE J. Quantum Electron.,* Vol. QE-20, p. 1119, 1984.

[100] N. Holonyak, Jr., R. M. Kolbas, R. D. Dupuis, P. D. Dapkus, "Quantum-Well Heterostructure Lasers," *IEEE J. Quantum Electron.,* Vol. QE-16, p. 170, 1980.

[101] D. R. Scifres, R. D. Burnham, C. Lindstrom, W. Streifer, and T. L. Paoli, *Appl. Phys. Lett.,* Vol. 42, p. 645, 1983.

[102] W. T. Tsang, R. A. Logan, and J. A. Ditzenberger, *Electron. Lett.,* Vol. 18, p. 845, 1982.

[103] E. A. Rezek, R. Chin, N. Holonyak, Jr., S. W. Kirchofer, and R. M. Kolbas, "Quantum-Well InP-InGaAsP Heterostructure Lasers Grown by LPE," *J. Electron. Mat.,* Vol. 9, p. 1, 1980.

[104] E. A. Rezek, N. Holonyak, Jr., and B. K. Fuller, "Temperature Dependence of Threshold of InGaAsP Quantum-Well Lasers," *J. Appl. Phys.,* Vol. 51, p. 2402, 1980.

[105] T. Yanase, Y. Kato, I. Mito, M. Yamoguchi, K. Nishi, K. Kobayashi and R. Lang, "1.3-μm InGaAsP/InP Multiquantum-Well Lasers Grown by Vapour Phase Epitaxy," *Electron. Lett.,* Vol. 14, p. 700, 1983.

[106] N. K. Dutta, S. G. Napholtz, R. Yen, R. L. Brown, T. M. Shen, N. A. Olsson, and D. C. Craft, "Fabrication and Performance Characteristics of InGaAsP Multiquantum-Well Double-Chanel Planar BH Lasers," *Appl. Phys. Lett.,* Vol. 46, p. 19, 1985.

[107] N. K. Dutta, S. G. Napholtz, R. Yen, T. Wessel, N. A. Olsson, "Long-Wavelength InGaAsP Modified Multiquantum-Well Laser," *Appl. Phys. Lett.,* Vol. 46, p. 1036, 1985.

[108] H. Temkin, K. Alavi, W. R. Wagner, T. P. Pearsall, and A. Y. Cho, *Appl. Phys. Lett.,* Vol. 42, p. 845, 1983.

[109] W. T. Tsang, "GaInAs/InP Multiquantum-Well Heterostructure Lasers Grown by MBE," *Appl. Phys. Lett.,* Vo.. 44, p. 288, 1984.

[110] N. K. Dutta, T. Wessel, N. A. Olsson, R. A. Logan, and R. Yen, "Fabrication and Performance Characteristics of 1.55-μm InGaAsP Multiquantum-Well Ridge-Guide Lasers," *Appl. Phys. Lett.,* Vol. 46, p. 525, 1985.

[111] N. K. Dutta, T. Wessel, N. A. Olsson, R. A. Logan, R. Yen, and P. J. Anthony , "Fabrication and Performance Characteristics of InGaAsP Ridge-Guide Distributed-Feedback Multiquantum-Well Lasers," *Electron. Lett.,* Vol. 21, p. 571, 1985.

[112] M. G. Burt, "Line Width Enhancement Factor for Quantum-Well Lasers," *Electron. Lett.,* Vol. 20, p. 27, 1984.

[113] Y. Arakawa, K. Vahala, and A. Yariv, "Quantum Noise and Dynamics in Quantum-Well and Quantum-Wire Lasers," *Appl. Phys. Lett.,* Vol. 45, p. 950, 1984.

[114] F. R. Nash, W. J. Sundberg, R. L. Hartman, J. R. Pawlik, D. A. Ackerman, N. K. Dutta, and R. W. Dixon, "Implementation of the Proposed Reliability Assurance Strategy for an InGaAsP/InP Planar Mesa BH Laser for Use in a Submarine Cable," *AT&T Tech. J.,* Vol. 64, p. 809, 1985.

[115] S. Mahajan, W. D. Johnston, Jr., M. A. Pollack, and R. E. Nahory, "The Mechanism of Optically Induced Degradation in InP/InGaAsP Heterostructures," *Appl. Phys. Lett.,* Vol. 34, p. 717, 1979.

[116] P. M. Petroff and D. V. Lang, "A New Spectroscopic Technique for Imaging the Spatial Distribution of Nonradiative Defects," *Appl. Phys. Lett.,* Vol. 31, p. 60, 1977.

[117] S. Mahagan, A. K. Chin, C. L. Zipfel, D. Brasen, B. H. Chin, R. T. Tung, and S. Nakahara, *Materials Lett.,* Vol. 2, p. 184, 1984.

[118] W. D. Johnston, Jr., G. Y. Epps, R. E. Nahory, and M. A. Pollack, "Spatially Resolved Photoluminsence Characterization and Optically Induced Degradation of InGaAsP DH Laser Material," *Appl. Phys. Lett.,* Vol. 33, p. 992, 1978.

[119] M. Fukuda, K. Wakita, and G. Iwane, "Dark Defects in InGaAsP/InP DH Lasers Under Accelerated Aging," *J. Appl. Phys.,* Vol. 54, p. 1246, 1983.

[120] K. Mizuishi, M. Sawai, S. Todoroki, S. Tsuji, M. Hirao, and M. Nakamura, "Reliability of InGaAsP/InP Buried-Heterostructure 1.3-μm Lasers," *IEEE J. Quantum Electron.,* Vol. QE-19, p. 1294, 1983.

[121] E. I. GORDON, F. R. NASH, and R. L. HARTMAN, "Purging: A Reliability Assurance Technique for New Technology Semiconductor Devices," *IEEE Electron. Dev. Lett.,* Vol. EDL-4, p. 465, 1983.

[122] R. L. HARTMAN, and R. W. DIXON, "Reliability of DH GaAs Lasers at Elevated Temperatures," *Appl. Phys. Lett.,* Vol. 26, p. 239, 1975.

[123] K. MIZUISHI, M. HIRAO, S. TSUJI, H. SATO, and M. NAKAMURA, "Reliability of InGaAsP/InP Buried-Heterostructure Lasers, *Jap. J. Appl. Phys.,* Vol. 21, p. 359, 1982.

[124] B. W. HAKKI, P. E. FRALEY, and T. ELTRINGHAM, "1.3-μm Laser Reliability Determination for Submarine Cable Systems," *AT&T Tech. J.,* Vol. 64, p. 771, 1985.

[125] P. K. RUNGE, and P. R. TRISCHITTA, "The SL Undersea Lightwave System," *IEEE J. Lightwave Tech.,* Vol. LT-2, p. 744, 1984.

10

Modulation of Optical Sources

S. M. STONE

GTE Laboratories, Inc.

1. Introduction

To transmit information in an optical communication system the source of optical power must be modified in response to the information to be transmitted. The modification of the source could be in intensity, frequency, or phase. Most commonly the source intensity is caused to vary in accordance to an analog or digital electrical signal. When semiconductor laser diodes (LDs) or light-emitting diodes (LEDs) are used, their emission intensity can be modulated simply by varying the current injected into these devices. In the case of a laser diode emitting in a single longitudinal mode, varying the injected current also results in a varying emission frequency. If this laser is injection locked to a stable frequency source, the phase of the optical frequency can be modulated in response to an injected signal current. Information can be transmitted by directly modulating the source with a subcarrier signal which has been modulated by the baseband signal. By using a subcarrier signal, information can be transmitted by frequency modulation using an LED source.

Modulation of an optical source can also be produced with an external modulator through which the light is caused to transit. The most common devices use electro-optic or acousto-optic effects, while some less frequently used devices employ electroabsorption or magneto-optic phenomena. Electro-optic devices are used to modulate the intensity or phase of laser emission, and acousto-optic devices modulate its intensity or frequency.

Present-day optical communication systems predominantly make use of current modulation of the source intensity because of its simplicity and adequacy in satisfying the system requirements. Digital signals are preferred to analog signals because of the lower signal-to-noise ratio required to achieve the desired signal quality. The smaller signal-to-noise ratio requirement is offset by a requirement for wider-bandwidth modulated sources, detectors, and fibers, which can be readily met with present-day components.

The fact that current modulation of a laser diode modulates its frequency, as well as its intensity, provides a simple means of achieving frequency modulation of analog signals and frequency-shift keying (fsk) of digital signals.

2. Direct Modulation of Semiconductor Optical Sources

Light-Emitting Diodes
The number of photons (light intensity) emitted

when electrons and holes recombine in the active region of a semiconductor pn junction is a function of the diode current as illustrated in Fig. 1, which is a plot of optical output power versus current for a representative surface-emitting InGaAsP/InP LED coupled to a 50-μm–core graded-index fiber.

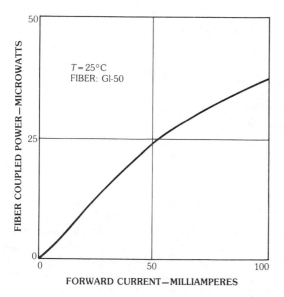

FIG. 1. Optical output power versus current for a representative surface-emitting 1.3-μm InGaAsP LED coupled to a 50-μm–core graded-index fiber.

By varying the current about a fixed bias level the output of an LED can be modulated with an upper frequency limit essentially governed by the carrier lifetime (τ). The modulation bandwidth (3 dB electrical or 1.5 dB optical) is approximately equal to

$$\Delta f_{3dB} \simeq 1/2 \pi \tau . \qquad (1)$$

In practical LEDs the optical response to an electrical current pulse is limited by the *RC* time constants of the device and drive circuitry. The capacitance of the LED device includes a space-charge capacitance of the entire pn junction area and a diffusion capacitance in the area associated with the region in which the injected carriers combine. Resistance can be determined from the *I-V* characteristics of the diode. Calculated response

time and experimental data have been reported in the literature [1, 2, 3, 4]. The results of these investigations indicate that because the current is injected into the pn junction through a very small area electrode as compared to the entire junction area, the resistance of the pn junction decreases with increasing drive current whereas the spreading resistance which charges the rest of the pn junction area is much higher and remains relatively constant, and most of the current is therefore crowded into the low resistance and small capacitance of the emitting area. As a result, at drive currents normally encountered in practice the rise time is essentially governed by the lifetime (recombination time) τ of the injected carriers. The effects of the diffusion capacitance and the peripheral parasitic capacitance of the non-emitting region can be further reduced by the application of a bias current in addition to the current pulses. The space-charge capacitance can be minimized by etching away, as much as possible, the peripheral junction surrounding the emitting area.

An equivalent circuit, shown in Fig. 2 has been proposed [4] to predict the pulse behavior of a surface-emitting LED having a cross section illustrated in Fig. 3. The LED structure is assumed to be disklike in which the peripheral area surrounding the emitting area is divided into eight concentric areas, the size of each selected to have an equal spreading resistance. Space-charge capacitances are designated by C_{Si} ($i=E, 1, 2, 3, \ldots$) and diffusion capacitances, which are dependent on the carrier lifetime, are designated by C_{Di}. Calculated optical pulse shapes and experimental results shown in Fig. 4 indicate reasonably good agreement. It should be noted that the fall time of the optical emission is shorter than the rise time.

Carrier lifetime is determined by nonradiative as well as radiative recombination:

$$1/\tau = 1/\tau_r + 1/\tau_{nr}, \qquad (2)$$

where τ_r is the radiative lifetime and τ_{nr} is the nonradiative lifetime.

The internal quantum efficiency η of an LED is determined by the ratio

$$\eta = \tau/\tau_r . \qquad (3)$$

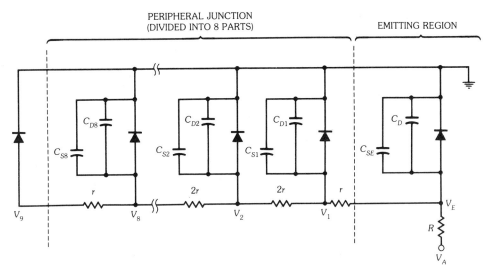

FIG. 2. Equivalent circuit proposed to predict pulse response of a surface-emitting LED.
(*After Hino and Iwamoto [4], ©1979 IEEE*)

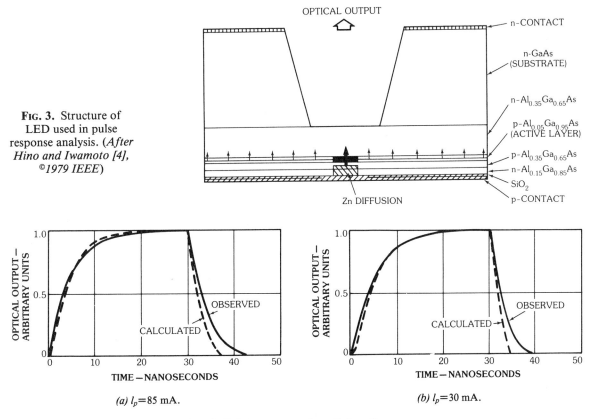

FIG. 3. Structure of LED used in pulse response analysis. (*After Hino and Iwamoto [4], ©1979 IEEE*)

(a) l_p=85 mA.

(b) l_p=30 mA.

FIG. 4. Pulse response of small–junction–area LED. (*After Hino and Iwamoto [4], © 1979 IEEE*)

Nonradiative recombination occurs at impurity sites and crystal structure defects as well as at the junction interfaces. In present-day commercial LEDs the rate of nonradiative recombination is very much less than radiative recombination.

For very low doping concentration (less than the injected electron concentration) [5]

$$\tau_r \propto (d/J)^{1/2}, \qquad (4)$$

where d is the active-layer thickness and J is the electron density.

At high doping levels [6]

$$\tau_r \propto 1/(n_0 + p_0), \qquad (5)$$

where n_0 is the donor impurity doping concentration and p_0 is the acceptor impurity doping concentration.

At doping concentrations of less than $10^{18}/cm^3$ the nonradiative recombinations occur mainly at the junction interfaces [7] and

$$\tau_{nr} \propto d/v_{sr}, \qquad (6)$$

where v_{sr} is the surface recombination velocity.

Rise time can be minimized by minimizing the active-layer thickness d or by maximizing the dop-

ing level $(n_0 + p_0)$, either of which results in reduced quantum efficiency. Therefore, in general, increased modulation bandwidth can be obtained at the expense of output intensity.

Germanium has been used as the dopant in AlGaAs LEDs because of the relatively high levels of concentration which can be attained [2, 8]. Bandwidth in excess of 100 MHz has been achieved at germanium concentrations of greater than $10^{19}/cm^3$. Fig. 5 is a plot of bandwidth obtained at various levels of germanium concentration [3]. Also shown is the calculated lifetime for each concentration.

InGaAsP LEDs have been made with high-bitrate modulation capabilities. One such surface-emitting device [9], the structure of which is shown in Fig. 6, has achieved 1.6 Gb/s nrz modulation. This was accomplished with a zinc doping concentration of $1.3 \times 10^{19}/cm^3$ and a 13-μm-diameter emitting area. The peak power launched into a 50-μm core, 0.2-NA fiber was −17.0 dBm, with a 6-dB on/off optical pulse ratio. This same device when operated at 1.0 Gb/s, coupled −17.5-dBm peak optical power into the same fiber with a 9-dB on/off ratio.

Modulation bandwidth versus output power obtained by various researchers for InGaAsP 1.3-μm LEDs are plotted in Fig. 7 [10]. The solid

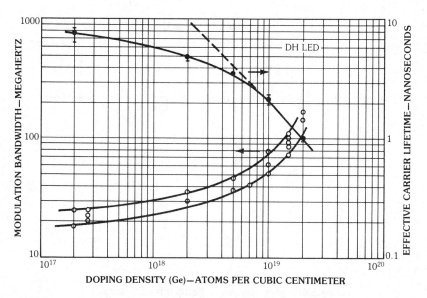

FIG. 5. Modulation bandwidth and carrier lifetime at 100-mA drive current as a function of the doping concentration in the active region. (*After Lee and Dentai [3], ©1978 IEEE*)

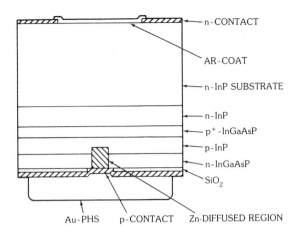

FIG. 6. A 1.3-μm high-speed double heterojunction LED structure. (*After Suzuki et al. [9], ©Institution of Electrical Engineers*)

line indicates the best results for AlGaAs LEDs. For comparison the dashed line represents the AlGaAs results shifted by the ratio of photon energy at 0.85 μm to the photon energy at 1.3 μm.

Laser Diodes

As can be seen from Fig. 8, which is a plot of the power output versus input current for a fiber-coupled InGaAsP/InP buried-heterostructure laser, the relationship is quasi-linear above lasing threshold. It should be mentioned that lasers of early design (gain-guided stripe width greater than 10μm) had kinks in their output power versus input current characteristics which were due to transitions in their transverse-mode configurations. Present-day lasers with index-guided active layers with widths less than 10 μm do not generally exhibit kinks in their normal operating current ranges.

Fig. 8 represents the expected performance of an unmodulated laser operating at various bias levels, and reveals very little about the laser's performance when the driving current is modulated, except that the output would be expected to respond to the current. In the first place, semiconductor lasers exhibit an initial delay [11] of 1 to 2 ns before lasing commences after the driving current is suddenly switched on, which is due to the time required for electron density in the active region to build up to the lasing threshold level.

After this initial delay the speed of response is determined by the recombination lifetime of the injected electrons, which is much shorter for laser diodes (less than 10^{-8} s) than it is for LEDs due to

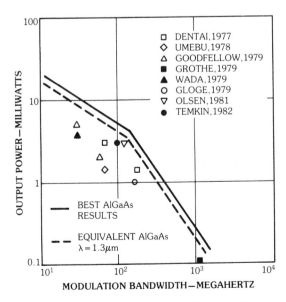

FIG. 7. Reported output power and bandwidth for 1.3-μm InGaAsP LEDs. (*After Saul [10], ©1983 IEEE*)

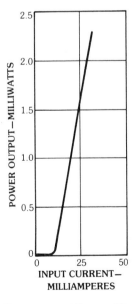

FIG. 8. Power output from a 50-μm–core graded-index fiber coupled to a 1.3-μm InGaAsP/InP laser.

the effect of stimulated emission [12, 13, 14, 15]. Since the lifetime of photons in the optical cavity is very short (less than 10^{-11} s) it is possible to modulate semiconductor laser diodes at extremely high rates. At present, direct intensity modulation is the method predominantly used in optical pulse-code-modulation (pcm) communication systems in conjunction with direct-detection receivers, and it has also been employed in experimental systems with heterodyne detection receivers [16, 17].

Semiconductor lasers exhibit a dynamic transient behavior in response to a current pulse sud-denly applied. This response is due to the quantum-mechanical process in which the injected electrons interact with the generated photons. When the current driving the laser is applied from zero to above threshold level with a short rise time, there is the previously mentioned delay of 1 to 2 ns before laser output occurs [11], the length of the delay being affected by the presence or absence of previous current pulses (pattern effect), which affects the stored charge [18, 19]. Then an oscillation develops after the leading edge of the pulse [18] as shown in Fig. 9, which usually damps out after a few cycles and the laser output comes to equilibrium. In this

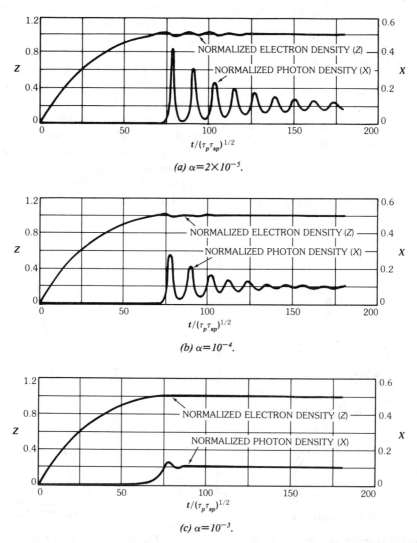

FIG. 9. Calculated transient response of an injection laser to a current pulse, where α is proportional to the spontaneous emission received into the lasing modes. (*After Arnold and Russer [18]*)

figure τ_p is the photon lifetime in the optical cavity with zero excitation current, τ_{sp} is the differential lifetime of the electrons at threshold, and α is the portion of spontaneous emission received into the lasing modes. This behavior can be explained in the following manner. When a large current pulse is suddenly applied, there is a delay before the electron density in the active layer builds up sufficiently for laser output to occur and the photon density rises quickly. While the photon density is below the steady-state value the electron density continues to increase. When the photon density rises above the steady-state value, the electron density rapidly decreases due to the rapidly increasing recombination process, but the photon density increases further until the electron density is depleted at which time the photon density rapidly reduces. The electron density then rises again and the process is repeated except that there is a higher photon density in the active region at each upturn in the electron density so that each successive cycle has a reduced amplitude until a steady-state condition is reached. If the laser is biased above threshold before the current pulse is applied, the optical pulse rise time and the amplitude of the relaxation oscillation will be reduced due to minimized electron density excursions.

The behavior of semiconductor lasers is governed by the quantum-mechanical rate equations which result in information concerning the generation of photons as a function of time. When a small time-varying signal is included in the analysis, a resonance is predicted [20], the frequency f_r of which is given by

$$f_r = (1/2\,\pi)[(1/\tau_{sp}\,\tau_p)(I/I_{th}-1)]^{1/2}, \qquad (7)$$

where I is the excitation current, I_{th} is the threshold current, τ_{sp} is the differential lifetime of the electrons at threshold and τ_p is the photon lifetime in the optical cavity when the excitation current is zero. Fig. 10 shows a normalized response curve as a function of modulation frequency obtained from such an analysis [12, 21, 22], taking into account the fraction of the spontaneous emission coupling into the lasing modes. Laser structure details greatly influence both τ_{sp} and τ_p and hence f_r is different for each type of laser. The height and width of the resonance peak

is influenced by damping mechanisms. In narrow-stripe lasers, lateral electron diffusion has been proposed as the major damping factor. A theoretical analysis [23] indicates that the height of the resonance peak is at a minimum when the width of the active layer is comparable to the electron diffusion length. An oscillogram of the response of a particular laser to a current pulse is shown in Fig. 11. More recent theoretical work

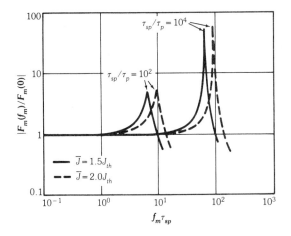

FIG. 10. Theoretical variation of injection-laser modulation depth, $F_m\,(f_m)$, normalized to its value at $f_m=0$, with τ_{sp}/τ_p and $\bar{J}/J_{th}$ as parameters. (After Paoli and Ripper [12], ©1970 IEEE)

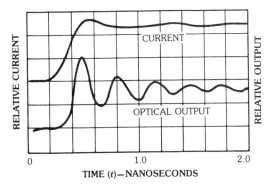

FIG. 11. Oscillogram of laser transient response to a current pulse.

[24] proposes that nonlinear gain is the dominant factor in damping relaxation oscillations. The calculated and observed dynamic responses of a buried-crescent InGaAsP laser just below thresh-

old, with and without the nonlinear gain contribution, are shown in Fig. 12.

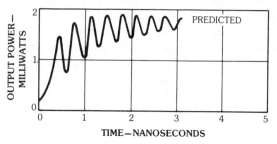

(a) Without symmetric nonlinear gain.

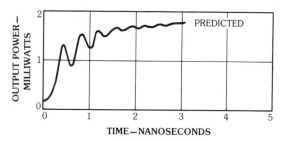

(b) With symmetric nonlinear gain.

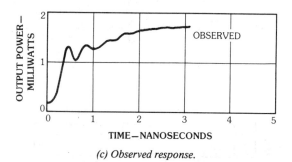

(c) Observed response.

FIG. 12. Observed and calculated dynamic responses of a 1.3-μm buried-crescent laser just below threshold. (*After Manning et al. [24], ©Institution of Electrical Engineers*)

When modulated with large signals the resonant frequency is lower than for small-signal modulation [25] and, in addition, a train of mode-locked pulses can be generated if the semiconductor laser is coupled to an external cavity and the large signal driving current is at the fundamental frequency of the optical-cavity resonance or at its harmonics.

In one such pulse generator [26] the laser had one facet antireflection coated and was coupled to a graded-index fiber several centimeters long with its cleaved far end goldplated for high reflectivity. Transform limited pulses 30 ps at FWHM were generated at 3.33, 6.66, and 10 GHz. It has been reported that pulse rates of 13 GHz have been achieved in the same type of external cavity laser [27].

An examination of the modulation response curve of a semiconductor laser as shown in Fig. 10 indicates that the resonant frequency limits its bandwidth capabilities. For distortionless modulation of small signals the frequency content of the signal should be kept in the flat portion of the response. For frequencies beyond resonance the response falls off sharply. Consequently, considerable effort has been expended to extend the frequency of resonance.

A judicious modification of the rate equation combined with small-signal analysis leads to an expression for f_r using different laser parameters [28]:

$$f_r = (1/2\,\pi)(A\,p_{ss}/\tau_p)^{1/2}, \qquad (8)$$

where A is the differential-gain constant, which is governed by the properties of the laser material, p_{ss} is the steady-state photon density in the active region, and τ_p is the photon lifetime in the cavity, which is governed by the device structure. From this equation it can be seen that the resonant frequency could be increased by individually, or in combination, increasing A and p_{ss}, and decreasing τ_p.

The photon lifetime τ_p can be reduced by shortening the laser cavity, and photon density p_{ss} can be increased by raising the bias current, with an upper limit set by the output power density at which facet damage occurs. Another limiting factor is the higher junction temperature generated by the increased current density resulting from a shorter cavity. The facet damage threshold can be raised by fabricating the laser with a window structure. The differential-gain constant A could be increased by operating the laser at a low temperature (approximately −50°C) but, for practical reasons, room temperature is most desirable. Making use of these considerations a GaAlAs/GaAs windowed, buried-heterostructure laser has been fabricated on a semi-insulating substrate (to reduce the parasitic capacitance) which has a bandwidth of 11 GHz [29]. Modulation characteristics of this laser are shown in Fig. 13, where curves 1 through 5 correspond to bias optical powers of 1.7, 3.6, 6.7, 8.4, and 16 mW. The absence of strong relaxation peaks is attributed to the effect of superluminescent damping [30] resulting from the presence of the window.

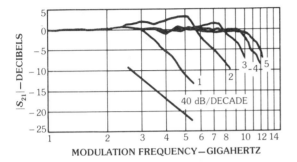

FIG. 13. Modulation characteristics of a window, buried-heterostructure laser on a semi-insulating substrate at various bias optical power levels at room temperature. (*After Lau, Bar-Chaim, and Ury [29]*)

When modulating at these high frequencies consideration must be given to the deleterious effect of parasitic circuit elements, those that are part of the laser chip structure and those that are introduced by packaging. For the laser just referred to, the 3-dB cutoff frequency of the drive current was approximately 12 GHz. To utilize any further increase of bandwidth in the active region of the laser, the parasitic capacitance would have to be reduced.

The differential-gain constant A can be substantially increased by raising the doping level in the active layer [31]. Exploiting this effect, an InGaAsP vapor-phase-regrown buried-heterostructure laser was fabricated [32] which exhibited a small-signal bandwidth of 15 GHz. This laser structure resulted in a parasitic chip capacitance of less than 2 pf, which did not limit its bandwidth. Fig. 14 is a plot of its modulation characteristics at various bias optical power levels. In this case bandwidth was limited by the temperature of the junction, which could be lowered with improved heat transfer.

Most of the analysis and experimentation that has been performed and reported has been

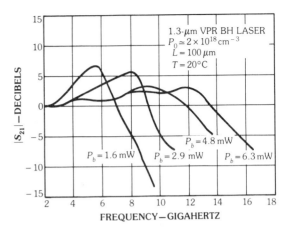

FIG. 14. Small-signal modulation characteristics versus frequency at different bias optical powers P_b. (*After Su et al. [32], ©Institution of Electrical Engineers*)

done with small-signal modulation. "Small-signal" usually means that the modulation depth of the optical output power P is considerably less than one, where modulation depth is given by $(P_{max} - P_{min})/P_{max}$. Large-signal modulation where the depth approaches one usually results in an optical response that is degraded. This can be explained by the fact that the laser output cannot be quickly raised from a low optical output level to a higher level because of the low rate of stimulated emission at the low output state [30]. Small-signal response can be expected up to a modulation depth of approximately 0.7; however, some specific types of lasers have given good optical response with large signals at modulation rates exceeding 8 Gb/s [34]. Fig. 15 shows the optical response to a drive current with a quasi-random 8.2-Gb/s pattern with a modulation depth of approximately 0.6.

In addition to the effects associated with modulation on the output intensity of semiconductor lasers caused by various phenomena, there is also an effect on the optical frequencies generated. Below threshold the emission spectrum is very broad but narrows as the current is increased above threshold, as illustrated in Fig. 16 for a mesa-stripe GaAlAs double-heterostructure laser [18]. Spectral changes from a buried-heterostructure laser 120 μm long as its current is increased to various

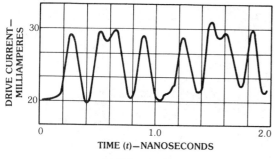

(a) Drive current.

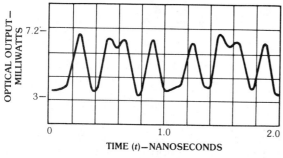

(b) Optical current.

FIG. 15. Large-signal laser response at 8.2 Gb/s with approximately 0.6 modulation depth. (*After Lau and Yariv [33]*)

levels above threshold [34] are shown in Fig. 17. Fig. 18 shows the spectra for a similar laser except that its cavity length is 250 μm. This indicates the effect that the increased mode selectivity of the shorter laser has on the suppression of adjacent modes. These are time-averaged spectra. If the entire spectrum were observed instant to instant, it would be seen that the relative intensities of the modes were constantly changing although the total intensity remained quite constant. This mode-partitioning effect would be observed to be stronger with modulation. It leads to intensity fluctuations in the receiver of long-length optical-fiber systems, where the fiber has some dispersion, which degrades its bit-error rate performance [35].

Modal fluctuations also occur during a modulating pulse [36], which is illustrated in Fig. 19 for two different types of lasers showing time resolved dynamic spectra at different points of a single 1-ns-wide pulse.

With modulation the width of the spectrum also increases [34], as shown in Fig. 20, which further

reduces system performance due to broadening of the received pulses. Ideally, a laser emitting in a single longitudinal mode and lowest-order transverse mode is most desirable for high–bit-rate, long–link-length, fiber-optic communication systems, but except for distributed-feedback lasers, which are not readily available commercially at present, most lasers emit in several longitudinal modes, while some lasers exhibit quasi-single-mode cw emission. To suppress modes adjacent to the dominant mode, lasers have been made in laboratories which are coupled to passive external cavities [37] or active external cavities such as the C^3 laser [38]. This laser [39] has been modulated at a bit rate of 2 Gb/s, with on/off ratios greater than 10:1 with side-mode suppression of greater than 1200:1.

Dynamic line-width broadening, referred to as *chirping,* is another effect resulting from modulating the laser current [40, 41, 42]. It is caused by variation of the refractive index of the active layer

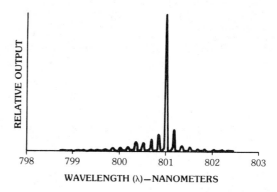

(a) At 3% above threshold.

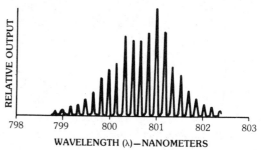

(b) Close to threshold.

FIG. 16. Emission spectrum of a low mesa-stripe AlGaAs DHS injection laser. (*After Arnold and Russer [18]*)

312

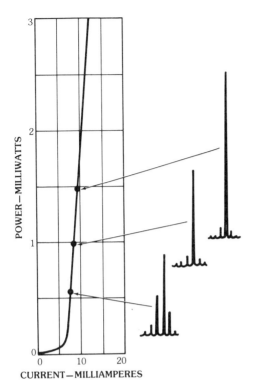

FIG. 17. Continuous-wave light versus current and spectral characteristics of a GaAs laser whose cavity length is 120 μm. (*After Lau, Harder, and Yariv [34], ©1984 IEEE*)

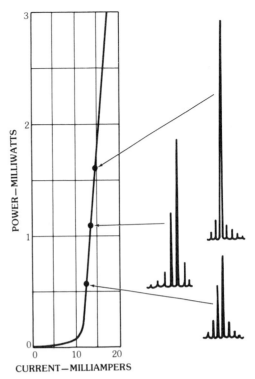

FIG. 18. Continuous-wave light versus current and spectral characteristics of a GaAs laser identical with that in Fig. 17 except that its cavity length is 250 μm. (*After Lau, Harder, and Yariv [34], ©1984 IEEE*)

due to the varying electron density during modulation, which varies the optical length of the laser cavity and hence shifts its resonant wavelength [40] as much as 1 nm. One method used to significantly reduce the chirp utilized a cleaved–coupled-cavity laser in which the relative amplitudes and phases of the modulating currents to each section of the laser were carefully adjusted [41]. With a modulation depth of 0.9 at 1.7 GHz the chirp was reduced to 0.02 nm for a 1.3-μm InGaAsP laser. Another method reported recently [42] used injection locking of a distributed-feedback laser which resulted in chirp reduction to below the measurement-resolution limit of 0.02 nm.

This same phenomenon, which may be deleterious to some direct-detection optical communication systems, can be effectively utilized in coherent detection system where frequency-shift keyed (fsk) and phase-shift keyed (psk) modulation results in greatly increased receiver sensitivity [43].

The maximum frequency deviation (Δf) as a function of current has been experimentally determined for various types of lasers at several modulation frequencies (f_m). Fig. 21 is a plot of the data obtained on an AlGaAs channeled-substrate planar (CSP) laser at a bias current 1.4 times the threshold current I_{th}. Frequency deviation normalized by modulation current ($\Delta f/\Delta i_m$) versus modulation frequency is shown in Fig. 22 with the ratio of bias current to threshold current as a parameter. The solid lines represent calculated values obtained from rate-equation analysis which includes lateral carrier diffusion [44]. Data from a 1.5-μm InGaAsP crescent buried-heterostructure cleaved-coupled-cavity (C^3) laser [45] are shown in Fig. 23 as a function of bias current. The normalized frequency deviation $\Delta f/\Delta i_m$ for two types of C^3 lasers is plotted in Fig. 24 as a function of the modulating frequency, for two schemes of modulation: one modulating the modulator section and the other modulating the laser section.

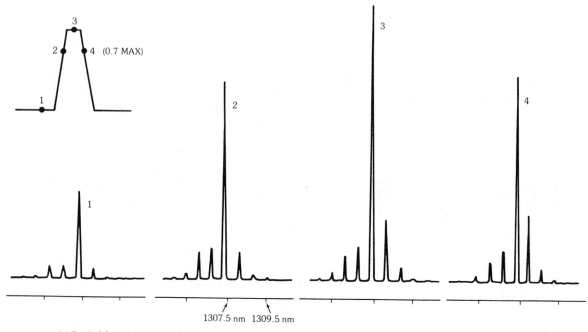

(a) Buried-heterojunction laser, HLP 5400/9365, with T=20 C, I_{th}=31.5 mA, I_B=38.3 mA, and I_m=15 mA.

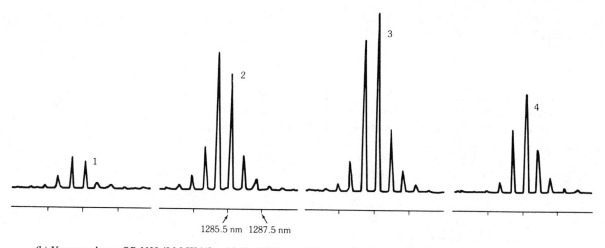

(b) V-groove laser, QB 113L/LM III/4S, with T=20°C, I_{th}=123 mA, I_B=131.9 mA, I_m=15 mA, and pulse width=1 ns.

FIG. 19. Time-resolved dynamic spectra at different points of a single pulse with 1-ns halfwidth. (*After Wenke and Enning [36]*)

Direct fsk modulation of a double-channel planar buried-heterostructure (DCPBH) 1.5-μm laser was used successfully in an experimental 140-Mb/s, 200-km, fiber-optic system using heterodyne detection where the bit-error-rate was 10^{-9} at a signal level of −55.3 dBm at the receiver [46]. The laser was biased at 95 mA and modulated with a 32-mA peak-to-peak current, which resulted in a frequency deviation of 83 MHz for a frequency-modulation index ($\Delta f/f_m$) of 0.6.

If the emission frequency of a single longitudinal-mode semiconductor laser is locked by

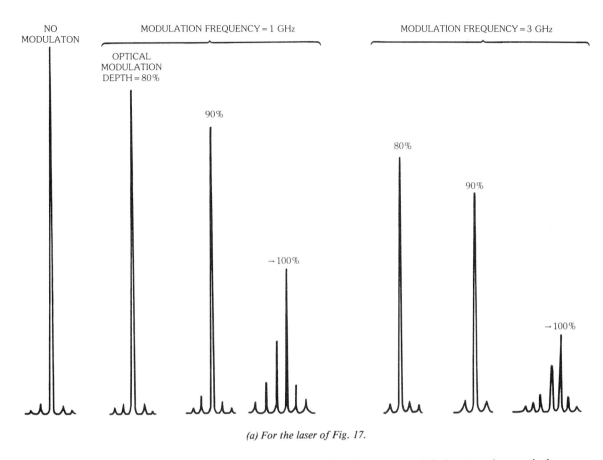

(a) For the laser of Fig. 17.

FIG. 20. Observed time-averaged spectrum of a laser under microwave modulation at various optical modulation depths, at modulation frequencies of 1 GHz and 3 GHz, with the laser biased at a dc optical power of 1.5 mW. (*After Lau, Harder, and Yariv [34], ©1984 IEEE*)

injection to the frequency of a narrow–line-width, single-frequency laser, the phase of the optical emission can be shifted by changing the driving current. Phase-shift data obtained with a channeled-substrate planar AlGaAs laser [47] by varying its bias current is shown in Fig. 25. A static phase shift of π radians resulted from a bias change of 0.48 mA. When modulated with sinusoidal current the optical emission was effectively phase modulated at frequencies up to 1 GHz. Fig. 26 is a plot of data showing the effect of injected power on the maximum phase deviation per milliampere of rf drive current.

Fortuitously, direct modulation can be used in optical communication systems requiring any one of a variety of modulation and detection meth-

ods. Some of the deleterious effects caused by the quantum mechanics of the lasing process can be avoided by operating the laser at a constant power level and modulating its emission with an external device at the expense of cost, complexity, and optical loss.

3. External Modulators for Optical Sources

Emission from an optical source can be modulated by an external device whose operation is based on one of several phenomena. Modulators which have been most highly developed and are commercially available make use of the acousto-optic effect or electro-optic effect, while some

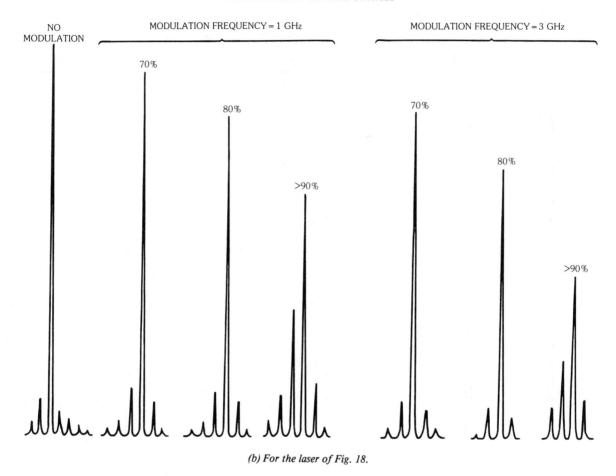

(b) For the laser of Fig. 18.

Fig. 20—cont.

experimental devices have utilized the magneto-optic effect and electroabsorption as well as other phenomena. In addition, external modulators are classified into two other categories, namely bulk and thin-film waveguide devices.

Acousto-Optic Modulators

The interaction of sound and light was predicted by Brillouin [48] in 1922. Sound waves cause periodic compression and rarefaction of the medium in which they are propagating, resulting in periodic variations of the medium's refractive index, which acts as a phase grating that diffracts part or all of the incident light. This phenomenon, called *Bragg diffraction,* is illustrated in Fig. 27, which shows light beams incident on a transparent medium in which an ultrasonic wave is propagating. At incident angles of $\pm\theta_B$ the diffracted

beam exists at $2\theta_B$ [49]. The Bragg angle is

$$\theta_B = \sin^{-1}(\lambda/2\Lambda),\qquad(9)$$

where λ is the vacuum wavelength of the light beam and Λ is the acoustic wavelength. A necessary condition for obtaining a single-diffraction-order beam [49] is

$$(2\pi\lambda L)/\Lambda^2 \gg 1,\qquad(10)$$

where L is the length of the interaction region. The fraction I/I_0 of the incident light intensity I_0 diffracted is related to the amplitude of the phase grating by [50]

$$I/I_0 = \sin^2(\Delta\phi/2),\qquad(11)$$

316

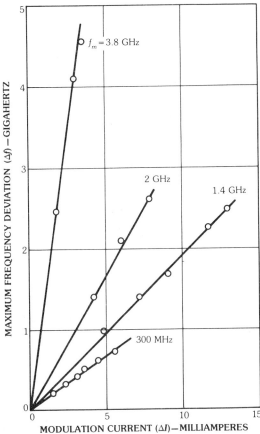

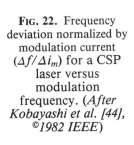

FIG. 21. Maximum frequency deviation for a CSP laser versus modulation current. (*After Kobayashi et al. [44], ©1982 IEEE*)

FIG. 22. Frequency deviation normalized by modulation current ($\Delta f / \Delta i_m$) for a CSP laser versus modulation frequency. (*After Kobayashi et al. [44], ©1982 IEEE*)

where $\Delta\phi$ is the peak-to-peak phase difference of the grating and I is the intensity of the diffracted light. Complete diffraction occurs when $\Delta\phi = \pi$.

Because the sound wave is moving, the diffracted light is frequency shifted by an amount equal to the acoustic frequency f_{ac}. If the light and acoustic-wave directions are chosen as shown at the top of Fig. 27, the optical frequency will be down shifted, and for the directions shown at the bottom of Fig. 27 an upshift results [49].

The photoelastic effect is observable in all forms of matter. The choice of material is dependent on the properties the material possesses. The acousto-optic phase change $\Delta\phi$ is given by [50]

$$\Delta\phi = \pi[(2/\lambda^2)(L/H)(n^6 p_c^{\,2}/\rho \, v_{ac}^{\,3})P_{ac}]^{1/2}, \quad (12)$$

where L is the interaction length, H is the height of the acoustic wave, n is the refractive index, p_c is the photoelastic coefficient, v_{ac} is the acoustic velocity, ρ is the material density, and P_{ac} is the acoustic power.

In order to compare different materials for use in acousto-optic devices, a figure of merit frequently used is

$$M_2 \equiv (n^6 p_c^{\,2}/\rho \, v_{ac}^{\,3}). \quad (13)$$

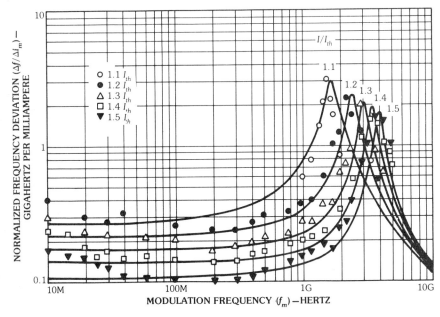

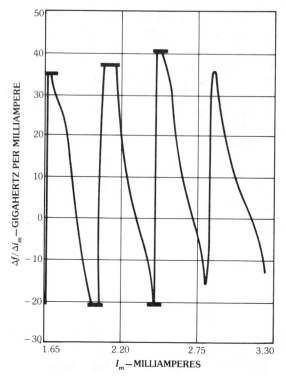

FIG. 23. Typical recording of frequency deviation per unit modulation current amplitude ($\Delta f/\Delta i_m$) at $f_m = 100$ kHz as a function of the dc bias current I_m of the modulator section. (*After Tsang and Olsson [45]*)

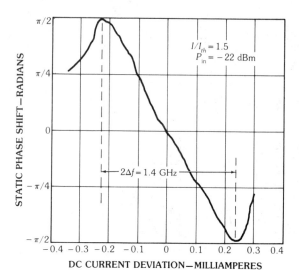

FIG. 25. Static phase shift obtained by varying the bias current of the injection-locked laser. (*After Kobayashi and Kimura [47], ©1982 IEEE*)

One other important material characteristic must be considered, namely intrinsic acoustic attenuation. Table 1 lists several properties of materials which are used in some commercial modulators and experimental devices. The great-

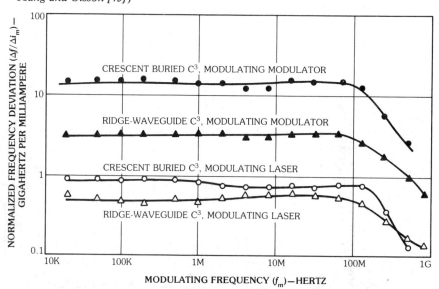

FIG. 24. Frequency deviation per unit modulation current ($\Delta f/\Delta i_m$) as a function of modulating frequency for 1.5-μm buried-crescent and ridge-waveguide C^3 InGaAsP lasers. (*After Tsang and Olsson [45]*)

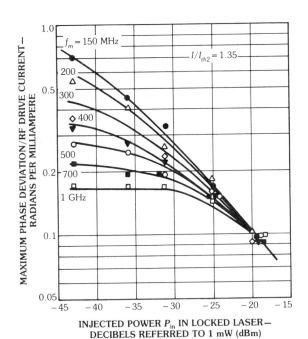

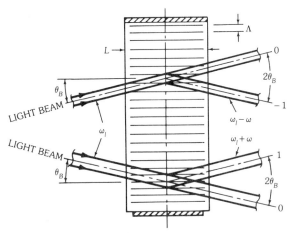

FIG. 27. Diffraction in the Bragg region showing downshifted (top) and upshifted (bottom) interaction. (*After Korpel [49], ©1982 IEEE*)

FIG. 26. Maximum phase deviation as a function of power injected into locked laser with modulating frequency as a parameter. (*After Kobayashi and Kimura [47], ©1982 IEEE*)

est variation in M_2 results from differences of refractive index n and acoustic velocity v_{ac}. Liquids have high figures of merit because of their relatively low acoustic velocities. They are, however, not very useful because of their excessive acoustic attenuation. In general, a trade-off is required between figure of merit and acoustic attenuation, because materials with high figures of merit usually have high acoustic attenuations, while materials with low attenuations usually have low figures of merit. The materials most commonly used in commercially available modulators for the near-infrared and visible wavelengths are $PbMoO_4$, TeO_2, SiO_2, GaP, and various glasses. The performance of present-day modulators is limited by available materials.

The modulation-frequency limit $f_{m(\max)}$ is reached when the acoustic wavelength Λ equals the diameter d of the light beam [49]:

$$f_{m(\max)} = v_{ac}/d = 1/\tau,\qquad(14)$$

where τ is the transit time of the sound wave through the light beam. In order to achieve a high modulating frequency the light beam must be focused into the interactive medium [55] to decrease d and therefore decrease τ, as shown in Fig. 28. However, since a focused light beam first converges and then diverges, good diffraction efficiency requires that the acoustic beam must have a related angular spread. If the optical angular spread $\Delta\theta_{op}$ is much larger than the acoustic angular spread $\Delta\theta_{ac}$, the diffracted beam will have a very elliptical cross section. Reduced bandwidth is the result if $\Delta\theta_{ac} \gg \Delta\theta_{op}$. To achieve a diffracted beam with minimal ellipticity and minimally reduced bandwidth, the ratio $R = \Delta\theta_{ac}/\Delta\theta_{op}$ should be kept in the range of 1.5 to 2.0.

The optical-intensity rise time t_r (10% to 90%) for a Gaussian input beam is [55]

$$t_r = \tau/1.5,\qquad(15)$$

where τ is the time required for the acoustic wavelength Λ to pass through the $1/e^2$ points of the beam waist. Fig. 29 shows the frequency response of an acousto-optic modulator as a function of the product of the acoustic frequency and the transit time.

In order to launch an acoustic wave into the

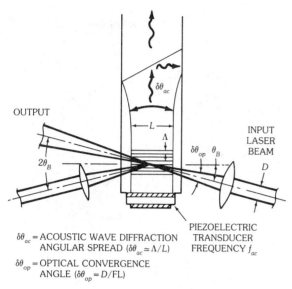

OUTPUT

$2\theta_B$

$\delta\theta_{ac}$ = ACOUSTIC WAVE DIFFRACTION ANGULAR SPREAD ($\delta\theta_{ac} \simeq \Lambda/L$)

$\delta\theta_{op}$ = OPTICAL CONVERGENCE ANGLE ($\delta\theta_{op} = D/FL$)

INPUT LASER BEAM

PIEZOELECTRIC TRANSDUCER FREQUENCY f_{ac}

(a) Top view.

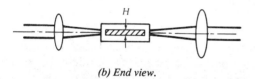

(b) End view.

FIG. 28. Acousto-optic modulator configuration. (*After Young and Yao [55], ©1981 IEEE*)

modulator material, mechanically coupled piezoelectric ultrasonic transducers are employed. These transducers are generally made by one of two methods [56]. In one technique a piezoelectric crystal wafer is bonded to the acousto-optic material which had a previously deposited metallic film electrode applied. The crystal is then ground and lapped to a thickness of approximately 10 to 15 μm and then a metallic film electrode is applied to this new surface of the crystal.

In the second method a film of piezoelectric material is evaporated or sputtered onto a previously electroded acousto-optic material to a thickness in the range of 3 to 5 μm and a second electrode is then deposited.

Single-crystal transducers are usually made with LiNbO$_3$, and thin-film transducers with ZnO [55]. The sonic coupling and bandwidth of a transducer are strongly influenced by the electrode material used and its thickness, as well as

the bonding material and its thickness [57]. Great care must be taken in the transducer design and fabrication in order to achieve good performance.

One device [58] designed to modulate a 1.06-μm laser beam had a LiNbO$_3$ transducer bonded to a vitreous As$_2$Se$_3$ acousto-optic material. It required 45 mW of electrical drive power at a frequency of 200 MHz to achieve 70% modulation of a 50-Mb/s bit stream. Since the modulated beam was the deflected beam, the on/off ratio was 100%, although the modulation efficiency was 70% at the reported drive power. This performance, which is representative of a well-designed modulator, had a power-to-bandwidth ratio of approximately 1 mW/MHz. This ratio is considered high when it is contemplated that the objective for fiber-optic communications is bandwidths in the gigahertz region.

Besides this drawback the bandwidth of acousto-optic modulators is usually a small fraction of the acoustic frequency, generally approximately 25% [59]. Bandwidths greater than 25% of the acoustic frequency can be achieved by trading off diffraction efficiency [59]. One commercially available modulator has a specified bandwidth of 1 GHz with a center frequency of 1.75 GHz, but with a diffraction efficiency of 8%/W at 0.633 μm. The actual realizable diffraction efficiency of this device is lower than 8%

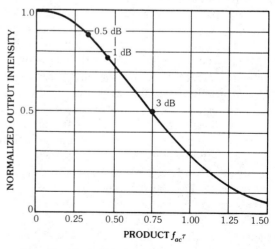

FIG. 29. Modulation frequency response for analog acousto-optic modulation. (*After Young and Yao [55], ©1981 IEEE*)

because at 1 W of drive power, nonlinear effects become very pronounced. Further, diffraction efficiency [60] is proportional to $1/\lambda^2$, which further limits the performance of these devices for fiber-optic communications which utilize optical sources at 0.85, 1.3 and 1.55 μm.

Considerable research effort has been expended on thin-film acousto-optic devices which utilize surface acoustic waves with guided optic waves. This research has been motivated by the potential of improved performance due to the confinement of the acoustic and optic waves, and the exploitation of fabrication processes which can lead to miniaturized integrated devices with reduced manufacturing costs. These devices also provide the opportunity for a wide range of possibilities for transducer designs to maximize acousto-optic interaction over a large bandwidth.

A basic configuration [61] for a thin-film acousto-optic modulator is shown in Fig. 30. Surface acoustic waves are generated by an array of interdigital electrodes which are deposited directly onto the modulator substrate if it is piezoelectric, such as Y-cut LiNbO$_3$. Optic wave-guiding can be accomplished by diffusing titanium into the LiNbO$_3$ surface [62]. If the substrate is not piezoelectric, a thin film of piezoelectric material such as ZnO must first be deposited on the thin-film

waveguide and the interdigital electrode array then applied. An acousto-optic modulator making use of the piezoelectric effect and optical waveguiding of titanium-diffused LiNbO$_3$ is shown schematically [63] in Fig. 31.

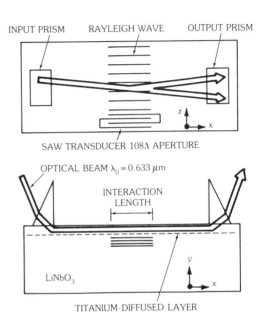

FIG. 31. Experimental arrangement for acousto-optical interaction between guided optical waves and surface acoustic waves in LiNbO$_3$. (*After Lean, White, and Wilkinson [63], ©1976 IEEE*)

Basically, the interaction between the acoustic wave and light beam is the same as in bulk devices. However, if the material is piezoelectric, the electric fields generated by the acoustic waves influence the refractive-index variations, and surface ripples are also formed, both of which affect the diffraction efficiency [61, 63].

The bandwidth of thin-film acousto-optic modulators is limited either by the bandwidth of the transducer, or bandwidth resulting from the Bragg phase-matching requirements and the confinement of the surface acoustic wave [61, 63]. For a given acoustic power the product of Bragg bandwidth and diffraction efficiency (with a basic transducer) is a constant which is independent of the center frequency or the length of the interaction region [61]. This results in the necessity for a trade-off of bandwidth for diffraction efficiency, which is a condition similar for bulk modulators.

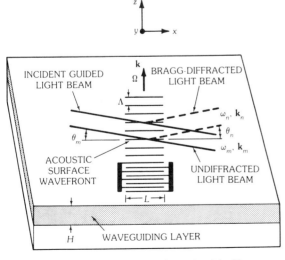

FIG. 30. Schematic drawing of a thin-film guided-wave acousto-optic modulator showing Bragg diffraction. (*After Tsai [61], ©1979 IEEE*)

A requirement for high diffraction efficiency with large bandwidth necessitates high acoustic drive power which may exceed the capability of the transducer before it is destroyed. Also, at high acoustic power, nonlinear phenomena introduce adverse acousto-optic effects [61].

Many types of transducers have been designed and tested to extend the bandwidth and diffraction efficiency of thin-film acousto-optic modulators. A few of these designs [61] are shown in Fig. 32, some of which require electronic phase shifters and others which require separate drivers with different center frequencies.

The largest bandwidth achieved for any of these designs was 680 MHz with a diffraction efficiency of 8%, at 0.633 µm, with 800 mW of drive power for a transducer of the type shown in Fig. 32c. A 50% diffraction efficiency was obtained at a bandwidth of 112 MHz with 68 mW of drive power using the transducer of Fig. 32d. It has been proposed that with improvements in design, 1 GHz of bandwidth with 50% diffraction efficiency with 1 W of drive power could be achieved [61]. All of the reported results and projections were based on an optical wavelength of 0.633 µm. Because of the factor of $1/\lambda^2$ the figure of merit in the order of 1 mW/MHz at 0.633 µm would be reduced [60] by a factor of approximately 6 if the optical wavelength were 1.55 µm.

Performance achieved and projected for thin-film acousto-optic modulators is, in general, comparable to that of bulk modulators. The bandwidth required for systems being developed with bit rates above 1 Gb/s has not been achieved. Both types require drive power orders of magnitude larger than directly driven light sources for practical present and future optical-fiber communication systems, and as compared with electro-optic modulators described in the following section.

Electro-Optic Modulators

Certain crystals exhibit the phenomenon of linear changes in their indexes of refraction when an electric field is applied. The field must be applied in a particular direction, and the light must be polarized in a specific direction for maximum effect. In uniaxial crystals such as potassium dihydrogen phosphate (KDP), potassium dideuterium phosphate (KD*P), ammonium dihydrogen phosphate (ADP) and other crystals of the form XH_2PO_4 [64], the application of an electric field in the direction of the optic (Z) axis causes the crystal to become biaxial. Two axes of birefringence are induced which are inclined at 45° to the X and Y crystalographic axes. The refractive index increases on one induced axis (X') and decreases on the orthogonal axis (Y'), and the change (Δn) being linearly proportional to the

(a) Multiple tilted transducers of staggered center frequency.

(b) Curved transducer evolved from large number of tilted transducers of staggered center frequency.

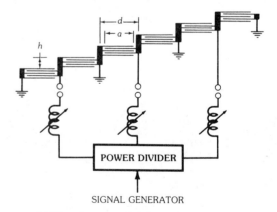

(c) Phased-array transducer with matching and driving circuits.

(d) Multiple tilted phased-array transducer.

FIG. 32. Various types of surface acoustic wave transducers. *(After Tsai [61], ©1979 IEEE)*

TABLE 1. Acousto-Optic Properties of Selected Materials [50–54]

Material	Range of Transmission (μm)	Wavelength of Measurement (μm)	Density ρ (g/cm^3)	Acoustic Velocity v_{ac} (km/s)	Refractive Index n	Photoelastic Coefficient p_c	Acoustic Attenuation (dB/cm·GHz)	Figure of Merit M_2 (10^{-18} s^3/g)
LiNbO$_3$	0.4–4	0.633	4.64	6.57	2.20	0.20	0.15	7.0
LiTaO$_3$	0.4–5	0.633	7.45	6.19	2.18	0.15	0.1	1.37
TiO$_2$	0.45–6	0.633	4.23	10.3	2.58	0.17	0.55	1.52
GaP	0.6–10	0.633	4.13	6.32	3.31	0.15	6.25	44.6
PbMoO$_4$	0.42–5.5	0.633	6.95	3.63	2.26	0.28	15	36.3
TeO$_2$	0.35–5	0.633	6.00	4.20	2.26	0.34	15	34.5
				0.62*	2.26	0.091	176	793
Al$_2$O$_3$	0.15–6.5	0.633	4.0	11.0	1.76	0.25	0.2	0.36
SiO$_2$ (fused)	0.2–4.5	0.633	2.2	5.96	1.46	0.27	12	1.56
As$_2$S$_3$ (glass)	0.6–11	0.633	3.2	2.6	2.61	0.3	170	433
As$_8$Se$_3$ (glass)	0.9–11	1.153	4.64	2.25	2.89	0.31	280	1090
H$_2$O	0.2–0.9	0.633	1.0	1.49	1.33	0.31	2400	126

*Shear wave, all others longitudinal wave.

applied field. If polarized laser light is coupled into the crystal with its polarization plane coincident with the crystalographic axis (X or Y), two orthogonal components of the light propagate along the planes containing the induced axes and

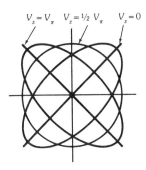

FIG. 33. Polarization states produced by electro-optic modulator. (*After Stone [65]. Permission for reprint courtesy Society for Information Display*)

a relative phase difference develops along the length L of the crystal. With zero electric field the polarization state of the beam exiting the crystal is linearly polarized in the same plane as the entering beam. As the field is monotonically increased, the polarization becomes eliptical, then circular, then eliptical, and then linear again, but polarized at 90° to the entrance beam [65], which is illustrated in Fig. 33. If a polarization analyzer is oriented at 90° to the input polarization at the output end of the crystal, the light intensity I would vary with the phase difference Γ as

$$I = I_0 \sin^2(\Gamma/2), \qquad (16)$$

where I_0 is the intensity of the beam entering the crystal. This modulation technique [65] is illustrated in Fig. 34.

The difference in phase (Γ) between the two waves propagating along the planes of the

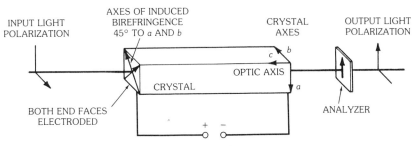

FIG. 34. Longitudinal electro-optic modulator. (*After Stone [65]. Permission for reprint courtesy Society for Information Display*)

induced axes is given by the following:

$$\Gamma = (2\pi L/\lambda)(n_{x'} - n_{y'}) = (2\pi L/\lambda)(\Delta n) \quad (17)$$

and

$$\Delta n = (n_0^3 r_{63} V_z)/L, \quad (18)$$

where n_0 is the index of refraction at zero field, and r_{63} is the linear electro-optic coefficient for the electric field (E_z) in the direction of the optic axis. The term V_z is the voltage applied across the crystal of length L. Then

$$\Gamma = (2\pi n_0^3/\lambda) r_{63} V_z. \quad (19)$$

It can be seen that the retardation is independent of the crystal length, and to achieve $\lambda/2$ retardation at the lowest V_z, crystals must be chosen with the highest values of $n_0^3 r_{63}$. Retardation can also be expressed as

$$\Gamma = \pi V_z/V_\pi, \quad (20)$$

where V_π is the voltage required to produce π radians retardation and is referred to as the *half-wave voltage*. The intensity of the light out of the modulator is then

$$I = I_0 \sin^2 (\pi V_z/2 V_\pi) \quad (21)$$

as shown in Fig. 35.

If the linear polarized light entering the crystal along the optic axis is aligned at 45° to the crystal axes, it will then be aligned with one of the induced axes of birefringence and the light will be phase modulated. The voltage required for a phase shift of π radians will be twice the value of V_π required for a phase difference of π radians for maximum intensity modulation.

Values of V_π for several XH_2PO_4 crystals at 0.633 μm are given in Table 2.

It is obvious that modulation voltages of this magnitude would be impractical for optical communication. One method used to reduce the voltage requirement is to propagate the light transverse to the electric field and to reduce the crystal thickness d in the direction of the field, which reduces the half-wave voltage by the ratio

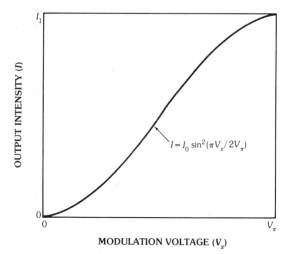

FIG. 35. Intensity variation versus applied voltage produced by electro-optic modulator followed by polarizer. (*After Stone [65]. Permission for reprint courtesy Society for Information Display*)

of 2 d/L. In this crystal orientation, intrinsic birefringence is present, which can be compensated by mounting two similar crystals of equal physical dimension with the optic axis of one crystal rotated 90° with respect to the other [65] as illustrated in Fig. 36.

Another more common form of compensated-birefringence modulator is illustrated in Fig. 37. Two similar crystals of equal dimensions are arranged with their axes as shown with a $\lambda/2$ retardation plate between them. The purpose of the $\lambda/2$ plate is to switch the direction of polarization of each component by 90°. This type of modulator has been preferred because the electrodes are coplanar and therefore more amenable to fabrication and the crystal alignment is less critical than for the arrangement shown in Fig. 36. However, the $\lambda/2$ retardation plate is specific

TABLE 2. Half-Wave Voltage (V_π) for Several Electro-Optical Crystals

Crystal	Half-Wave Voltage V_π (kV)	Wavelength λ (μm)
(KDP) KH_2PO_4	$\simeq 7.5$	0.633
(KD*P) KD_2PO_4	$\simeq 3.4$	0.633
(ADP) $NH_4H_2PO_4$	$\simeq 9.6$	0.633

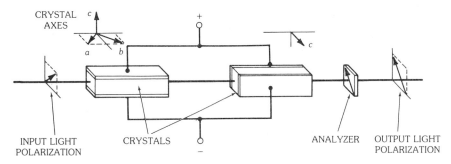

FIG. 36. A compensated-birefringence electro-optic modulator. (*After Stone [65]. Permission for reprint courtesy Society for Information Display*)

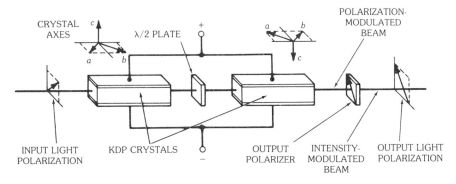

FIG. 37. A preferred type of compensated birefrigence electro-optic modulator. (*After Stone [65]. Permission for reprint courtesy Society for Information Display*)

for each wavelength, whereas the rotated crystal modulator is not wavelength-selective.

Assuming the optical beam to be circular in cross section, the thickness and width of the crystal can be equal, and the minimum thickness of the modulator crystals is governed by diffraction [66]. Assuming a Gaussian intensity distribution in the optical beam, referring to Fig. 38

$$d = S (2 \lambda L / n \pi)^{1/2} , \qquad (22)$$

where S is a safety factor which practically should be between 3 and 6 to avoid excessive beam clipping. Compensated-birefringence modulators [67] have been made with V_π of less than 100 V.

For pulse-code modulation the nonlinearity at the extremes of the transfer function, shown in Fig. 35, is acceptable, but for analog modulation the region around the midpoint of the curve is fairly linear and can be further corrected by compensating the electrical driving signal, using a

feedback loop. Either electrical or optical biasing can be used to determine the operating point.

In addition to the $X_2H_2PO_4$ crystals considerable development work has been done on lithium niobate ($LiNbO_3$) and lithium tantalate ($LiTaO_3$) crystals. These crystals are grown from a melt with resulting high optical quality. They are hard,

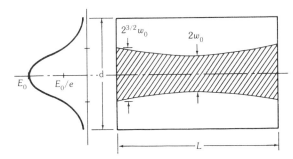

FIG. 38. A beam of Gaussian cross section passing through a rod of diameter d and length L. (*After Kaminow and Turner [66], ©1966 IEEE*)

325

nonhygroscopic, and readily cut, ground, and polished. By contrast the $X_2H_2PO_4$ crystals are grown from a water solution, are hygroscopic, soft, and more difficult to polish to good optical surfaces.

LiNbO$_3$ and LiTaO$_3$ crystals have very large electro-optic coefficients when the electric field is applied in the direction of the optic (Z) axis and light is propagated orthogonally along the X or Y axis. Phase modulation results if the light is polarized coincident with the X or Y crystal axes. Intensity modulation is produced if the light is polarized at 45° to the Z and X (or Y) axes, and a properly oriented $\lambda/4$ retardation plate precedes a polarization analyzer plate oriented at 90° to the original polarization [68]. In this orientation the crystals exhibit natural birefringence which is temperature dependent, and for a usable modulator the temperature must be closely controlled [69]. At 0.633 μm the half-wave voltage for LiNbO$_3$ is approximately 3.0 kV and for LiTaO$_3$ it is approximately 2.8 kV.

In a manner similar to the transverse $X_2H_2PO_4$ modulators, two LiNbO$_3$ or LiTaO$_3$ crystals in tandem can be oriented to compensate for the natural birefringence with and without the addition of a $\lambda/2$ plate as shown schematically in Fig. 39.

The power per unit bandwidth required to drive an electro-optic modulator [66] is proportional to d^2/L. In an optical waveguide this ratio can be made exceedingly small because diffraction does not limit the small cross-sectional dimension, d. As a result modulators based on optical waveguides require much lower drive power than bulk modulators.

Considerable research and development effort has been expended on waveguide modulators based upon titanium diffused into LiNbO$_3$ crystal substrates because of the relatively low propagation loss (absorption plus scattering) that is achievable, and the high electro-optic coefficients which can be effectively utilized. Propagation loss as low as 0.3 dB/cm at 1.32 μm has been measured [70]. However, coupling losses due to misfit of the near-field mode patterns and Fresnel reflections result in an insertion loss which is larger. The total insertion loss reported at 1.32 μm [70], excluding Fresnel reflection loss, which can be

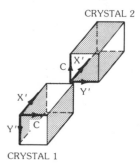

(a) Without $\lambda/2$ plate.

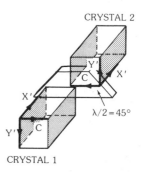

(b) With $\lambda/2$ plate.

FIG. 39. Lithium-tantalate modulator crystal orientation. (*After Biazzo [69]*)

minimized by using an interface liquid with a refractive index of 1.65 between the waveguide and the fibers, was 1.0 dB for a Z-cut Ti:LiNbO$_3$ waveguide 1.3 cm long.

In addition to the reduced drive-power required, the waveguide dimensions are compatible with single-mode fibers and potential optical integrated devices.

Modulators have been developed which use the electro-optic effect in waveguides with various configurations. Figs. 40a and 40b show the electrode arrangement for a single-strip–waveguide phase modulator which has a simple design concept for a very efficient device. This configuration results in a horizontal electric field, which produces the largest electro-optic effect in a Y-cut LiNbO$_3$ crystal when light is polarized in the plane of the substrate in the Z direction and propagating in the X direction or in an X-cut crystal with light propagating in the Y direction. In both cases the r_{33} coefficient, which is the largest, is operative. For an electrode configuration as shown in Fig. 40c a component of the electric field is vertical and the r_{33} coefficient is employed if the crystal is Z cut, and the light is polarized perpendicular to the substrate and propagating in either the X or Y direction.

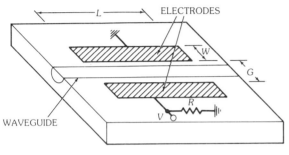

(a) Modulator and parameters.

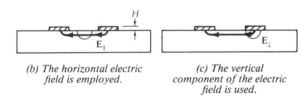

| *(b) The horizontal electric field is employed.* | *(c) The vertical component of the electric field is used.* |

FIG. 40. Single-strip waveguide modulator configurations. (*After Alferness [72],* *©1982 IEEE*)

Response time of the electro-optic effect is in the range of 10^{-12} s and therefore it does not limit the modulation bandwidth for communications. The limiting factors are the circuitry driving the modulator and the transit time of the light propagating in the crystal. In the modulators described, the electric field is applied to a lumped capacitor (modulator electrodes) and it completes a polarity reversal in one modulation-frequency period, $1/f_m$. If this period is equal to the transit time of the light through the modulator, the net modulation is zero. At 3 GHz this would occur in a crystal with a refractive index of 1.5 and a length of 6.7 cm. Maximum modulation for a given applied voltage at 3 GHz would be obtained with a modulator of one-half the length [71].

The bandwidth of this type of modulator is usually limited by circuit parameters. The required driving voltage as well as bandwidth varies inversely with electrode length, if the electrode resistance

is ignored, and therefore bandwidth can be increased at the expense of voltage [72]. For a modulator of practical length, reasonable parameter values, and reasonable drive voltage requirements, a bandwidth of approximately 2 GHz can be achieved at 0.633 μm.

A figure of merit [72] which has been used for electro-optic modulators is $V_\pi/\Delta f$, where V_π is the drive voltage required for π radians phase shift and Δf is the bandwidth. The ratio $V_\pi/\Delta f$ is a function of λ^x, where x falls between 1 and 2. Approximate values of $V_\pi/\Delta f$ for optimized modulators are approximately 0.5 V/GHz at 0.633 μm and 1.5 V/GHz at 1.32 μm. Using the figure of merit accepted for acousto-optic modulator, the power per unit bandwidth ($P/\Delta f$) would be approximately 0.6 μw/MHz at 0.633 μm and approximately 5.6 μW/MHz at 1.32 μm for a modulator with a 50-Ω termination.

To overcome the bandwidth-voltage requirement limitation of the lumped circuit modulator, traveling-wave modulators have been developed in which the electrodes are designed to act as a transmission line with the same characteristic impedance as the cable connecting it to the driving circuit, and with the velocity of propagation of the electric wave along the electrodes matched as closely as possible to the velocity of light in the waveguide [73, 74]. A modulator of this type is shown schematically in Fig. 41, in which the asymmetric electrode structure was chosen because it couples well to a coaxial cable and has low propagation loss.

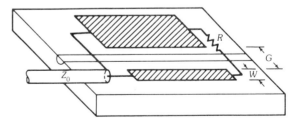

FIG. 41. Asymmetric coplanar strip traveling-wave electrode. (*After Alferness [72],* *©1982 IEEE*)

Phase modulation of guided waves can be transformed into intensity modulation in a switchable directional-coupler device [72] shown in Fig. 42. In

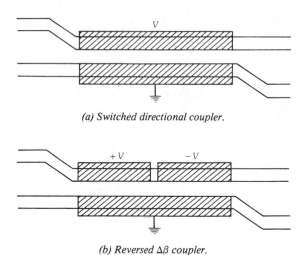

(a) Switched directional coupler.

(b) Reversed $\Delta\beta$ coupler.

FIG. 42. Electrode configuration of two types of directional-coupler modulators. (*After Alferness [72], ©1982 IEEE*)

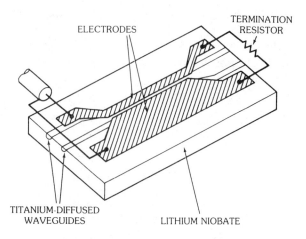

FIG. 43. Schematic drawing of directional-coupler traveling-wave switch/modulator. (*After Alferness et al. [75], ©1983 IEEE*)

this type of modulator two identical waveguides are fabricated by titanium diffusion into an LiNbO₃ substrate with very close spacing between them, and electrodes are applied. Light is coupled between the two waveguides by means of the evanescent optical fields. At a specific length of interaction, which depends on the waveguide parameters, the separation between them, and the wavelength, all the light from one waveguide is coupled into the adjacent one. If a sufficient index difference between the two waveguides is produced electro-optically, the net coupled energy can be made zero. A directional-coupler modulator of this type [75] shown schematically in Fig. 43 has been made with a bandwidth of 7.2 GHz and a $P/\Delta f$ of 7.6 μW/MHz at 1.32 μm with an extinction ratio of approximately 14 dB. However, the total fiber-coupled insertion loss was about 7.5 dB. Another modulator of this type [76] designed for lower optical insertion loss had a bandwidth of 5.5 GHz, with a $P/\Delta f$ of 10.9 μW/MHz at 1.32 μm with an extinction ratio of 10 dB. The total coupled optical insertion loss was 2.3 dB.

Phase modulation can also be transformed into intensity modulation [77] by interferometrically combining two otherwise identical waves. In the most efficient implementation, polarized light is divided equally into two waveguides, as illus-

trated in Fig. 44, both of which are phase modulated by applying a varying voltage to the electrodes indicated. If the light is then recombined, the output will be intensity modulated, in accordance with

$$I = (1 - \alpha_0)\, I_0 \cos^2\left(\Gamma/2\right), \qquad (23)$$

where I is the output intensity, I_0 is the input intensity, Γ is the total phase difference, and α_0 is the total optical loss.

A traveling-wave modulator of this type [77], often referred to as a Mach-Zehnder interferometric modulator, had a bandwidth of 17 GHz and required ± 3.5 V at a drive power of 120 mW for

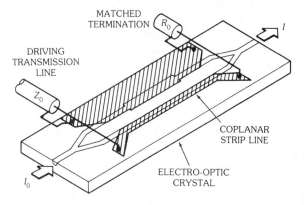

FIG. 44. Illustration of a Mach-Zehnder interferometric traveling-wave modulator. (*After Gee, Thurmond, and Yen [77]*)

maximum modulation at 0.83 μm with an extinction ratio of 15 dB. The ratio of $P/\Delta f$ for this modulator was 7 μW/MHz.

Research and development on other types of electro-optic waveguide modulators has been reported, including devices made on semiconductor substrates [78] which, in the future, may be used for integrated laser source and modulator devices. To date, however, the switchable coupler and interferometric modulators have received the greatest attention. Recently, a Ti:LiNbO$_3$ traveling-wave directional-coupler waveguide switch was used to demonstrate 4-Gb/s transmission over 117 km of optical fiber [79].

Experimental modulators have been made utilizing other than the electro-optic and acousto-optic phenomena. One of them is the electro-absorption (Franz-Keldysh) effect, where the absorption edge of a semiconductor material is caused to shift by varying the back-biased electric field applied across it. A modulator has been made in the form of an Al$_y$Ga$_{1-y}$As -Al$_x$Ga$_{1-x}$As double-heterostructure device [80], with which 90% modulation was obtained at 0.9 μm with 4 V applied and a $P/\Delta f$ of 0.2 mW/MHz.

Another experimental modulator was based on the magneto-optic (Faraday rotation) effect in an iron-garnet thin film grown on a gallium-garnet substrate, using a serpentine electric circuit to supply the magnetic field [81]. In this modulator a TE wave is converted to a TM wave due to Far-aday rotation. The two modes are angularly separated at the output by a birefringent coupler as illustrated in Fig. 45. As the current in the serpentine circuit is switched on and off, the light beam emerging from the output prism coupler switches between two different directions. With this device 100% modulation can be achieved at very low magnetic fields (less than 6 Oe) at frequencies up to 300 MHz with potential frequencies greater than 1 GHz.

4. Summary

It is clear that direct modulation of optical sources is the simplest way to impress information on an optical carrier. Bit rates of 1.6 Gb/s have been achieved [9] with LEDs, but because of their lower fiber-coupled power and their large spectral width acting on fiber dispersion, their use will be limited to shorter optical-fiber link lengths.

Laser diodes are capable of much higher bit rates (greater than 8 Gb/s) [33]. For multimode lasers and quasi–single-mode lasers, mode-partition noise and chirping limit the bit-rate–distance product. The frequency shift with current, however, may be used to advantage in conjunction with a frequency discriminator at the receiver in a fsk direct-detection system and in coherent detection systems where psk modulation of injection-locked lasers may be employed.

For external modulators, acousto-optic devices are very limited in bandwidth where high diffraction efficiency is required for low optical insertion loss. In addition, these modulators require a driving frequency approximately four times the required bandwidth, and the electrical drive power per unit bandwidth is orders of magnitude higher than that required for direct modulation.

Of all the external types, waveguide electro-optic modulators appear to offer the greatest compatibility with optical-fiber systems because the optical waveguide-mode dimensions are compatible with single-mode fiber. They have wide-bandwidth and high–bit-rate capabilities, they require low power per unit bandwidth, and they can be produced using integrated-circuit techniques, which should minimize cost.

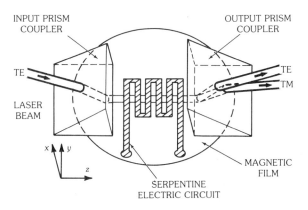

FIG. 45. Schematic drawing of the film-waveguide magneto-optical switch in operation. (*After Tien, Schinke, and Blank [81]*)

5. References

[1] T. P. LEE, "Effect of Junction Capacitance on the Rise Time of LEDs and on the Turn-On Delay of Injection Lasers," *Bell Sys. Tech. J.,* Vol. 54, No. 1, pp. 53–69, January 1975.

[2] J. ZUCKER, R. LAUER, and J. SCHLAFER, "Response Time of Ge-Doped (AlGa) As-GaAs Double Heterostructure LEDs," *J. Appl. Phys.,* Vol. 47, No. 5, pp. 2082–2084, May 1976.

[3] T. P. LEE, and A. G. DENTAI, "Power and Modulation Bandwidth of GaAs-AlGaAs High-Radiance LEDs for Optical Communication Systems," *IEEE J. Quantum Electron.,* Vol. QE-14, No. 3, pp. 150–159, March 1978.

[4] I. HINO, and K. IWAMOTO, "LED Pulse Response Analysis Considering the Distributed CR Constant in the Peripheral Junction," *IEEE Trans. Electron Dev.,* Vol. ED-26, No. 8, pp. 1238–1242, August 1979.

[5] H. NAMIZAKI, H. KAN, M. ISHII, and A. ITO, "Current Dependence of Spontaneous Carrier Lifetimes in GaAs-Ga$_{1-x}$Al$_x$As Double Heterostructure Lasers," *Appl. Phys. Lett.,* Vol. 24, No. 10, pp. 486–487, May 15, 1974.

[6] R. N. HALL, "Recombination Processes in Semiconductors," *Proc. IEE,* Vol. 106, Supp. 17, Pt. B, pp. 923–931, March 1960.

[7] M. ETTENBERG, and H. KRESSEL, "Interfacial Recombination at AlGaAs/GaAs Heterojunction Structures," *J. Appl. Phys.,* Vol. 47, No. 4, pp. 1538–1544, April 1976.

[8] M. ETTENBERG, H. KRESSEL, and S. GILBERT, "Minority Carrier Diffusion Length and Recombination Lifetime in GaAs:Ge Prepared by Liquid-Phase Epitaxy," *J. Appl. Phys.,* Vol. 44, No. 2, pp. 827–831, February 1973.

[9] A. SUZUKI, Y. INOMOTO, J. HAYASHI, Y. ISODA, T. UJI, and H. NOMURA, "Gbit/s Modulation of Heavily Zn-Doped Surface Emitting InGaAsP/InP DH LED," *Electron. Lett.,* Vol. 20, No. 7, pp. 273–274, March 29, 1984.

[10] R. H. SAUL, "Recent Advances in the Performance and Reliability of InGaAsP LEDs for Lightwave Communication Systems," *IEEE Trans. Electron Devices,* Vol. ED-30, No. 4, pp. 285–295, April 1983.

[11] K. KONNERTH and C. LANZA, "Delay Between Current Pulse and Light Emission of a Gallium-Arsenide Injection Laser," *Appl. Phys. Lett.,* Vol. 4, No. 7, pp. 120–121, April 1, 1964.

[12] T. L. PAOLI and J. E. RIPPER, "Direct Modulation of Semiconductor Lasers," *Proc. IEEE,* Vol. 58, No. 10, pp. 1457–1465, October 1970.

[13] M. TAKUSAGAWA, T. FUNAYAMA, J. NISHIZAWA, K. DEMIZU, and T. NAKANO, "Measurement of Lifetime in GaAs Diodes," *J. Appl. Phys.,* Vol. 38, pp. 4084–4086, September 1967.

[14] J. I. NISHIZAWA, "Recombination Lifetime in a Semiconductor Laser Diode," *IEEE J. Quantum Electron.,* Vol. QE-4, No. 4, pp. 143–147, April 1968.

[15] H. RIECK, "The Effective Lifetime of Stimulated and Spontaneous Emission in Semiconductor Laser Diodes," *Solid-State Electron.,* Vol. 8, pp. 83–85, January 1965.

[16] T. G. HODGKINSON, D. W. SMITH, and R. WYATT, "1.5-μm Optical Heterodyne System Operating Over 30 km of Monomode Fiber," *Electron. Lett.,* Vol. 18, No. 21, pp. 929–930, October 14, 1982.

[17] M. SHIKADA, K. EMURA, S. FUJITA, M. KITAMURA, M. ARAI, M. KONDO, and K. MINEMURA, "100 Mbit/s ASK Heterodyne Detection Experiment Using 1.3-μm DFB-Laser Diodes," *Electron. Lett.,* Vol. 20, No. 4, pp. 164–165, February 16, 1984.

[18] G. ARNOLD and P. RUSSER, "Modulation Behavior of Semiconductor Injection Lasers," *Appl. Phys.,* Vol. 14, pp. 255–265, 1977.

[19] T. OZEKI and T. ITO, "A New Method for Reducing Pattern Effect in PCM Current Modulation of DH-GaAlAs Lasers," *IEEE J. Quantum Electron.,* Vol. QE-9, No. 11, pp. 1098–1101, November 1973.

[20] T. L. PAOLI, "Magnitude of the Intrinsic Resonant Frequency in a Semiconductor Laser," *Appl. Phys. Lett.,* Vol. 39, No. 7, pp. 522–524, October 1, 1981.

[21] T. IKEGAMI and Y. SUEMATSU, "Direct Modulation of Semiconductor Junction Laser," *Electronics and Communications in Japan,* Vol. 51-B, No. 2, pp. 51–58, February 1968.

[22] P. M. BOERS, M. T. VLAARDINGERBROEK, and M. DANIELSEN, "Dynamic Behavior of Semiconductor Lasers," *Electron. Lett.,* Vol. 11, No. 10, pp. 206–208. May 15, 1975.

[23] K. FURUYA, Y. SUEMATSU, and T. HONG, "Reduction of Resonance-like Peak in Direct Modulation Due to Carrier Diffusion in Injection Laser," *Appl. Opt.,* Vol. 17, No. 12, pp. 1949–1952, June 15, 1978.

[24] J. MANNING, R. OLSHANSKY, D. M. FYE, and W. POWAZINIK, "Strong Influence of Nonlinear Gain on Spectral and Dynamic Characteristics of InGaAsP Lasers," *Electron. Lett.*, Vol. 21, No. 11, pp. 496–497, May 23, 1985.

[25] T. IKEGAMI and Y. SUEMATSU, "Large-Signal Characteristics of Directly Modulated Semiconductor Injection Lasers," *Electronics and Communications in Japan*, Vol. 53B, No. 9, pp. 69–75, September 1970.

[26] R. S. TUCKER, G. EISENSTEIN and I. P. KAMINOW, "10-GHz Active Mode Locking of a 1.3-μm Ridge-Waveguide Laser in an Optical-Fiber Cavity," *Electron. Lett.*, Vol. 19, No. 14, pp. 552–553, July 7, 1983.

[27] I. P. KAMINOW, "High-Speed Modulation of Semiconductor Lasers," *Conf. Opt. Fiber Commun.*, MJI, p. 28, January 23, 1984.

[28] K. Y. LAU, N. BAR-CHAIM, and I. URY, "Direct Amplitude Modulation of Short-Cavity GaAs Lasers up to X-Band Frequencies," *Appl. Phys. Lett.*, Vol. 43, No. 1, pp. 1–3, July 1, 1983.

[29] K. Y. LAU, N. BAR-CHAIM, and I. URY, "11-GHz Direct Modulation Bandwidth GaAlAs Window Laser on Semi-Insulating Substrate Operating at Room Temperature," *Appl. Phys. Lett.*, Vol. 45, No. 4, pp. 316–318, August 15, 1984.

[30] K. Y. LAU and A. YARIV, "Ultrahigh-Speed Semiconductor Lasers," *IEEE J. Quantum Electron.*, Vol. QE-21, No. 2, pp. 121–138, February 1985.

[31] C. B. SU and V. LANZISERA, "Effect of Doping Level on the Gain Constant and Modulation Bandwidth of InGaAsP Semiconductor Lasers," *Appl. Phys. Lett.*, Vol. 45, No. 12, pp. 1302–1304, December 15, 1984.

[32] C. B. SU, V. LANZISERA, W. POWAZINIK, E. MELAND, J. SCHLAFER, R. OLSHANSKY and R. B. LAUER, "15-GHz Direct-Modulation Bandwidth of Vapor-Phase-Regrown 1.3-μm InGaAsP Buried-Heterostructure Lasers Under CW Operation at Room Temperature," *Electron. Lett.*, Vol. 21, No. 13, pp. 577–579, June 20, 1985.

[33] K. Y. LAU and A. YARIV, "Large-Signal Dynamics of an Ultrafast Semiconductor Laser at Digital Modulation Rates Approaching 10 Gbit/s," *Appl. Phys. Lett.*, Vol. 47, No. 2, pp. 84–86, July 15, 1985.

[34] K. Y. LAU, C. HARDER, and A. YARIV, "Longitudinal Mode Spectrum of Semiconductor Lasers Under High-Speed Modulation," *IEEE J. Quantum Electron.*, Vol. QE-20, No. 1, pp. 71–79, January 1984.

[35] K. OGAWA, "Analysis of Mode-Partition Noise in Laser Transmission Systems," *IEEE J. Quantum Electron.*, Vol. QE-18, No. 5, pp. 849–855, May 1982.

[36] G. WENKE and B. ENNING, "Spectral Behavior of InGaAsP/InP 1.3-μm Lasers and Implications on the Transmission Performance of Broadband Gbit/s Signals," *J. Opt. Commun.*, Vol. 3, No. 4, pp. 122–128, 1982.

[37] C. LIN, C. A. BURRUS, JR. and L. A. COLDREN, "Characteristics of Single Longitudinal-Mode Selection in Short-Coupled-Cavity (SCC) Injection Lasers," *IEEE J. Lightwave Tech.*, Vol. LT-2, No. 4, pp. 544–549, August 1984.

[38] W. T. TSANG, N. A. OLSSON and R. A. LOGAN, "High-Speed Direct Single-Frequency Modulation with Large Tuning Rate and Frequency Excursion in Cleaved-Coupled-Cavity Semiconductor Lasers," *Appl. Phys. Lett.*, Vol. 42, No. 8, pp. 650–652, April 15, 1983.

[39] I. P. KAMINOW, J. S. KO, R. A. LINKE, and L. W. STULZ, "High-Speed 1.55-μm Single-Longitudinal-Mode Ridge Waveguide C^3 Laser," *Electron. Lett.*, Vol. 19, No. 19, pp. 784–785, September 15, 1983.

[40] R. A. LINKE, "Transient Chirping in Single-Frequency Lasers: Light Wave System Consequences," *Electron. Lett.*, Vol. 20, No. 11, pp. 472–474, May 24, 1984.

[41] L. A. COLDREN, G. D. BOYD, J. E. BOWERS, and C. A. BURRUS, "Reduced Dynamic Linewidth in Three-Terminal Two-Section Diode Lasers," *Appl. Phys. Lett.*, Vol. 46, No. 2, pp. 125–127, January 15, 1985.

[42] N. A. OLSSON, H. TEMKIN, R. A. LOGAN, L. F. JOHNSON, G. DOLAN, J. P. VAN DER ZIEL, and J. C. CAMPBELL, "Chirp-Free Transmission over 82.5 km of Single-Mode Fiber at 2 Gbit/s With Injection-Locked DFB Semiconductor Lasers," *IEEE J. Lightwave Tech.*, Vol. LT-3, No. 1, pp. 63–67, February 1985.

[43] Y. YAMAMOTO, "Receiver Performance Evaluation of Various Digital Optical Modulation-Demodulation Systems in the 0.5-10 μm Wavelength Region," *IEEE J. Quantum Electron.*, Vol. QE-16, No. 11, pp. 1251–1259, November 1980.

[44] S. Kobayashi, Y. Yamamoto, M. Ito, and T. Kimura, "Direct Frequency Modulation in AlGaAs Semiconductor Lasers," *IEEE J. Quantum Electron.,* Vol. QE-18, No. 4, pp. 582–595, April 1982.

[45] W. T. Tsang and N. A. Olsson, "Enhanced Frequency Modulation in Cleaved–Coupled-Cavity Semiconductor Lasers with Reduced Spurious Intensity Modulation," *Appl. Phys. Lett.,* Vol. 43, No. 6, pp. 527–529, September 15, 1983.

[46] R. Wyatt, D. W. Smith, T. G. Hodgkinson, R. A. Harmon, and W. J. Devlin, "140 Mbit/s Optical-Fiber Heterodyne Experiment at 1.54 μm," *Electron. Lett.,* Vol. 20, No. 22, pp. 912–913, October 25, 1984.

[47] S. Kobayashi and T. Kimura, "Optical Phase Modulation in an Injection Locked AlGaAs Semiconductor Laser, *IEEE J. Quantum Electron.,* Vol. QE-18, No. 10, pp. 1662–1669, October 1982.

[48] I. Brillouin, "Diffusion De La Lumiere Et Des Rayons X par Un Corps Transparent Homogene," *Ann Phys* (France), 9th Ser., Vol. 17, pp. 88–122, 1922.

[49] A. Korpel, "Acousto-Optics—A Review of Fundamentals," *Proc. IEEE,* Vol. 69, No. 1, pp. 48–53, January 1981.

[50] D. A. Pinnow, "Guide Lines for the Selection of Acousto-Optic Materials," *IEEE J. Quantum Electron.* Vol. QE-6, No. 4, pp. 223–238, April 1974.

[51] R. W. Dixon, "Photoelastic Properties of Selected Materials and Their Relevance for Applications to Acoustic Light Modulators and Scanners," *J. Appl. Phys.,* Vol. 38, No. 13, pp. 5149–5153, December 1967.

[52] N. Uchida and N. Niizeki, "Acousto-Optic Deflection Materials and Techniques," *Proc. IEEE,* Vol. 61, No. 8, pp. 1073–1092, August 1973.

[53] Y. Ohmachi and N. Uchida, "Vitreous As_2Se_3: Investigation of Acousto-Optical Properties and Application to Infrared Modulator," *J. Appl. Phys.,* Vol. 43, No. 4, pp. 1709–1712, April 1972.

[54] N. Uchida and Y. Ohmachi, "Elastic and Photoelastic Properties of TeO_2 Single Crystal," *J. Appl. Phys.,* Vol. 40, No. 12, pp. 4692–4695, November 1969.

[55] E. H. Young, Jr., and S-K Yao, "Design Considerations for Acousto-Optic Devices," *Proc. IEEE,* Vol. 69, No. 1, pp. 54–64, January 1981.

[56] A. H. Meitzler and E. K. Sittig, "Characterization of Piezoelectric Transducers Used in Ultrasonic Devices Operating Above 0.1 GHz," *J. Appl. Phys.,* Vol. 40, No. 11, pp. 4341–4352, October 1969.

[57] E. K. Sittig, "Effects of Bonding and Electrode Layers on Transmission Parameters of Piezoelectric Transducers Used in Ultrasonic Digital Delay Lines," *IEEE Trans. Sonic and Ultrasonics,* Vol. SU-16, No. 1, pp. 2–10, January 1969.

[58] A. W. Warner and D. A. Pinnow, "Miniature Acousto-Optic Modulators for Optical Communications," *IEEE J. Quantum Electron.,* Vol. QE-9, No. 12, pp. 1155–1157, December 1973.

[59] I. C. Chang, "Acousto-Optic Devices and Applications," *IEEE Trans. Sonic and Ultrasonics,* Vol. SU-23, No. 1, pp. 2–22, January 1976.

[60] R. V. Schmidt and I. P. Kaminow, "Acousto-Optic Bragg Deflection in $LiNbO_3$ Ti-Diffused Waveguides," *IEEE J. Quantum Electron.,* Vol. QE-11, No. 1, pp. 57–59, January 1975.

[61] C. S. Tsai, "Guided-Wave Acousto-Optic Bragg Modulators for Wide-Band Integrated Optic Communications and Signal Processing," *IEEE Trans. Circuits and Systems,* Vol. CAS-26, No. 12, pp. 1072–1098, December 1979.

[62] R. V. Schmidt and I. P. Kaminow, "Metal-Diffused Optical Waveguides in $LiNbO_3$," *Appl. Phys. Lett.,* Vol. 25, No. 8, pp. 458–460, October 15, 1974.

[63] E. G. H. Lean, J. M. White, and C. D. W. Wilkinson, "Thin-Film Acousto-Optic Devices," *Proc. IEEE,* Vol. 64, No. 5, pp. 779–788, May 1976.

[64] B. H. Billings, "The Electro-Optical Effect in Uniaxial Crystals of the Type XH_2PO_4," *J. Opt. Soc. Am.,* Vol. 39, No. 10, pp. 797–808. October 1949.

[65] S. M. Stone, "Experimental Multicolor Real-Time Laser Display System," *Proc. 8th Natl. Symp. on Information Display,* San Francisco, pp.161–167, May 24–26, 1967.

[66] I. P. Kaminow, and E. H. Turner, "Electro-Optic Light Modulators," *Proc. IEEE,* Vol. 54, No. 10, pp. 1374–1390, October 1966.

[67] J. Schlafer and V. J. Fowler, "A Low-Voltage Multiple-Wavelength Electro-Optic Modulator," *Proc. SID,* Vol. 12, No. 2, pp. 72–76, Second Quarter, 1971.

[68] E. G. Spencer, P. V. Lenzo, and A. A. Ballman, "Dielectric Materials for Electro-Optic Elasto-Optic and Ultrasonic Device Applications," *Proc. IEEE,* Vol. 55, No. 12, pp. 2074–2108, December 1967.

[69] M. R. Biazzo, "Fabrication of a Lithium-Tantalate Temperature-Stabilized Optical Modulator," *Appl. Opt.,* Vol. 10, No. 5, pp. 1016–1021, May 1971.

[70] R. C. Alferness, V. R. Ramaswamy, S K. Korotky, M. D. Divino, and L. L. Buhl, "Efficient Single-Mode Fiber to Titanium Diffused Lithium Niobate Waveguide Coupling for λ=1.32 μm," *IEEE J. Quantum Electron.,* Vol. QE-18, No. 10, pp. 1807–1813, October 1982.

[71] S. M. Stone, "A Microwave Electro-Optic Modulator Which Overcomes Transit-Time Limitations," *Proc. IEEE,* Vol. 52, No. 4, p. 409, April 1964.

[72] R. C. Alferness, "Waveguide Electro-Optic Modulators," *IEEE Trans. Microwave Theory Tech.,* Vol. MIT-30, No. 8, pp. 1121–1137, August 1982.

[73] C. J. Peters, "Gigacycle-Bandwidth Coherent-Light Traveling-Wave Amplitude Modulator," *Proc. IEEE,* Vol. 53, pp. 455–460, May 1965.

[74] F. S. Chen, "Modulators for Optical Communications," *Proc. IEEE,* Vol. 58, No. 10, pp. 1440–1457, October 1970.

[75] R. C. Alferness, C. H. Joyner, L. L. Buhl, and S. K. Korotky, "High-Speed Traveling-Wave Directional-Coupler Switch/Modulator for λ=1.32 μm," *IEEE J. Quantum Electron.,* Vol. QE-19, No. 9, pp. 1339–1341, September 1983.

[76] R. C. Alferness, S. K. Korotky, L. L. Buhl, and M. D. Divino, "High-Speed Low-Loss Low–Drive-Power Traveling-Wave Optical Modulator at 1.32 μm," *Electron. Lett.,* Vol. 20, No. 8, pp. 354–355, April 12, 1984.

[77] C. M. Gee, D. Thurmond and H. W. Yen, "17-GHz Bandwidth Electro-Optic Modulator," *Appl. Phys. Lett.,* Vol. 43, No. 11, pp. 998–1000, December 1983.

[78] J. P. Donnelly, N. L. Demeo, G. A. Ferrante and K. B. Nichols, "A High-Frequency GaAs Optical Guided-Wave Electro-Optic Interferometric Modulator," *IEEE J. Quantum Electron.,* Vol. QE-21, No. 1, pp. 18–21, January 1985.

[79] S. K. Korotky, G. Eisenstein, A. H. Gnauck, B. C. Kasper, J. J. Veselka, R. C. Alferness, L. L. Buhl, C. A. Burrus, T. C. D. Hud, L. W. Stultz, K. Ciemiecki Nelson, L. G. Cohen, R. W. Dawson, and J. C. Campbell, "4-Gb/s Transmission Experiment over 117 km of Optical Fiber Using a Ti:LiNbO₃ External Modulator," *Conf. Optical-Fiber Commun.,* (OFC '85) Post Deadline Paper PD1, February 11–13, 1985.

[80] F. K. Reinhart, "Electroabsorption in Al_yGa_{1-y}As-Al_xGa_{1-x}As Double Heterostructures," *Appl. Phys. Lett.,* Vol. 22, No. 8, pp. 372–374, April 15, 1973.

[81] P. K. Tien, D. P. Schinke and S. L. Blank, "Magneto-Optics and Motion of the Magnetization in a Film-Waveguide Optical Switch," *J. Appl. Phys.,* Vol. 45, No. 7, pp. 3059–3068, July 1974.

Detectors for Optical-Waveguide Communications

G. E. STILLMAN

Electrical Engineering Research Laboratory of the University of Illinois

1. Introduction

While much of the research emphasis on components for optical waveguide communication has been directed toward the fibers and optical sources as described in the previous chapters, detectors are equally important in determining the performance limitations of the total system. In the early work the near-infrared spectral region covered by AlGaAs-GaAs diode laser sources was of interest, and the already well developed silicon pin and avalanche photodiode (APD) detectors were well suited for detector applications in these systems. However, as low-loss optical fibers were extended to longer wavelengths where the advantages of zero fiber dispersion could also be capitalized on, it became necessary to examine new materials and schemes for detectors.

The requirements that a detector suitable for fiber-optical communications must satisfy include good sensitivity at the operating wavelength and the ability to detect high-frequency signals. These two requirements plus the requirement of small, durable, and inexpensive devices, limit the acceptable detection schemes to semiconductor photon detectors.

In this chapter we will first review the fundamentals of semiconductor photon detectors and then describe how these fundamentals limit the sensitivity of photon detectors. For wide-bandwidth detection the sensitivity is often limited by the thermal noise of the load resistor and amplifier, and in this case increased sensitivity can be obtained with detectors which have internal gain or carrier multiplication, such as photoconductive detectors or avalanche photodiodes. Because of the importance to future systems the structure and limitations of the performance of these devices will be discussed in detail.

2. Semiconductor Detector Fundamentals

Although there is considerable confusion in the published literature over the actual meaning of the terms, the most useful semiconductor photon detectors are generally classified as either *photoconductive* or *photovoltaic*. This discussion will be limited to the near-infrared spectral region, where the influence of background radiation can generally be neglected. In this chapter, and in most of the literature, the term "photoconductive detector" is used to describe those detectors which are homogeneous semiconductor samples with ohmic contacts in which, with a dc bias voltage applied to the detector element, an increased current flows when light is incident on the sample due to increased conductivity which results from the photogenerated carriers. The term "photovoltaic detector" includes all detectors which have a built-in electric field and produce a current due to the separation of photogenerated electron-hole pairs, whether the device is operated in the short-circuit (zero voltage), open-circuit (zero current), or reverse-biased (depletion) mode of operation. In some of the literature these three modes of pn junction detector operation are referred to as photodiode, photovoltaic, and photoconductive modes, respectively. The latter is particularly confusing since the more widely used meaning of the term "photoconductive detector" does not include pn junction devices.

Direct Photon Detection

In direct photon detection an incident photon is absorbed and directly excites an electron from a nonconducting state to a conducting state. In intrinsic direct photon detection an electron is excited from the valence band to the conduction band of a semiconductor, and in n-type extrinsic direct photon detection excitation, an electron is excited from the ground state of a neutral donor impurity to the conduction band. Another form of extrinsic photon detection occurs in free carrier photoconductors, when an electron is excited from a given state to one of higher or lower mobility. These three mechanisms are illustrated schematically in Fig. 1. The intrinsic and extrinsic excitation processes are characterized by a threshold energy which is required to excite the carrier (electron) from its bound state to a conducting state, while the free-electron absorption process does not have a low-energy or long-wavelength threshold but instead shows an increase with the square of the wavelength from short wavelength up to some constant value at very long wavelengths, the same as for free carrier absorption. This mechanism is related to the "heating" of the free carriers above the lattice temperature by the absorbed radiation, and this type of detector is often referred to as a *free-electron bolometer*. While this mechanism is important for some far-infrared detectors it is of no practical use for near-infrared fiber-optical communications. The

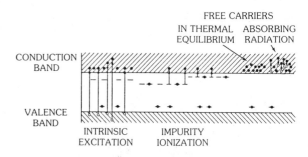

(a) Mechanisms of direct photon detection.

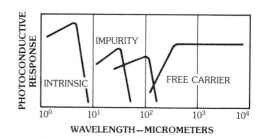

(b) Photoconductive responses.

FIG. 1. Schematic diagram of different photoconductive processes in semiconductors and corresponding spectral responses. (*After Putley [1]*)

long-wavelength threshold λ_g corresponds to the energy gap E_g in the case of intrinsic excitation and to the impurity ionization energy E_I in the case of extrinsic excitation. That is, a photon must have an energy E, in electronvolts, such that

$$E \geq E_g = h c/\lambda_g = 1.24/\lambda_g. \tag{1}$$

with λ_g in micrometers for intrinsic excitation and

$$E \geq E_I \tag{2}$$

for extrinsic excitation. For either intrinsic or extrinsic direct photon detection the photocurrent which is generated can be written as

$$I_p = q \eta \Phi M, \tag{3}$$

where q is the electronic charge, η is defined as the quantum efficiency of the direct photon detection process, Φ is the incident photon flux (photons per second), and M is the current gain or multiplication of the process—i.e., the number of electrons flowing through the external electrical circuit per absorbed photon. If the photon flux Φ in (3) consists of photons of wavelength λ, and energy $E_p = hc/\lambda$, we can write the responsivity R_λ in amperes per watt of the detection process at wavelength λ as

$$R_\lambda = I_p/\Phi E_p = I_p \lambda/\Phi hc = q\eta\lambda M/hc. \quad (4)$$

Thus the responsivity R_λ is zero for $\lambda > \lambda_{co}$, where λ_{co} is the threshold wavelength, and, for constant η, decreases linearly with decreasing λ from its maximum value at $\lambda = \lambda_{co}$. For a photoconductive gain or avalanche multiplication of unity the relationship between responsivity and quantum efficiency is given by

$$R = (q\lambda/hc)\eta = \eta\lambda/1.24 \quad (5)$$

amperes per watt, with λ expressed in micrometers. Thus, at a wavelength of 1.24 μm, a quantum efficiency of 100% corresponds to a responsivity of 1 A/W, while at a wavelength of 0.62 μm a quantum efficiency of 100% corresponds to a responsivity of only 0.5 A/W.

Photoconductive Detectors

For the intrinsic photoexcitation process in a semiconductor the excitation of an electron from the valence band to the conduction band occurs with the absorption of a photon having an energy greater than the energy gap. This generation process also creates a hole in the valence band, and both carriers contribute to the conductivity of the semiconductor sample. The electron-hole pair generation can be described in terms of the absorption coefficient of the semiconductor at the wavelength of the incident radiation. If a total steady optical signal power P_0 of wavelength λ is incident on a semiconductor sample, a part of the incident power will be reflected, and a part will be absorbed in the semiconductor sample. This process is shown schematically in Fig. 2 in which the optical generation rate of electron-hole pairs as a function of distance x into the sample is given by

$$g(x)\,dx = (P_0/h\nu)(1-R)\,\alpha e^{-\alpha x}\,dx, \quad (6)$$

where α is the absorption coefficient at the energy $h\nu$ or wavelength $\lambda = hc/E$. The reflection for normal incidence can be estimated from

$$R = (n_1 - n_2)^2/(n_1 + n_2)^2, \quad (7)$$

where n_1 is the index of refraction of air ($n_1 \simeq 1$) and n_2 is the index of refraction of the semiconductor at the wavelength of the incident radiation. For semiconductors used for near-infrared detectors, $n_2 \simeq 3.4$ so $R \simeq 30\%$—i.e., nearly one-third of the incident radiation is reflected. The reflectivity can be reduced to a few percent at a particular wavelength of interest through the use of antireflection coatings. For cases where αW is small it is necessary to take into account multiple internal reflections that have been neglected in (6) and (7). The dependence of the optical absorption coefficient on photon energy for several semiconductors of interest is shown in Fig. 3, along with the energy ranges of representative laser sources. From Fig. 2 it is clear that essentially all of the

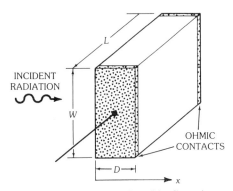

(a) Photoconductive sample and its dimensions.

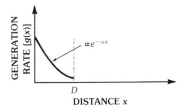

(b) Generation rate g (x) of carriers throughout sample thickness D.

FIG. 2. Photoconductive sample and the generation rate of carriers throughout the sample thickness.

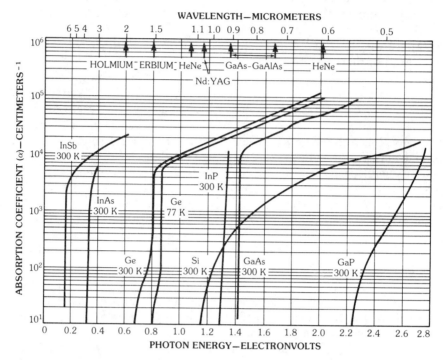

FIG. 3. Energy dependence of absorption coefficient for representative semiconductor materials. (*After Stillman and Wolfe [3]*)

radiation will be absorbed in a sample that has a thickness several times α^{-1}. For an intrinsic photoconductive detector the change in the *average* steady-state electron and hole concentrations can be written as

$$\Delta n = \Delta p = G\,\tau_c = (1/WLD) \int_0^D g(x)\,\tau_c\,dx$$

$$= (P_0/WLD\,h\nu)\,(1-R)\,(1-e^{-\alpha D})\,\tau_c\,, \qquad (8)$$

where G is the electron and hole generation rate per cubic centimeter and τ_c is the excess carrier lifetime assumed to be the same for electrons and holes (i.e., carrier trapping is negligible). The sample width, length, and thickness dimensions are W, L, and D, respectively. Then the change in conductivity of the sample due to the optical excitation of Δn and Δp can be written as

$$\Delta\sigma = q(\Delta n\,\mu_n + \Delta p\,\mu_p)$$

$$= (q\,G\,\tau_c/WDL)\,(\mu_n + \mu_p)\,. \qquad (9)$$

When a voltage V is applied to the photoconductor between the ohmic contacts a distance L apart, a photoinduced current Δi, given in amperes by

$$\Delta i = (\Delta\sigma\,V/L)\,WD \qquad (10)$$

or

$$\Delta i = (q\,G\,\tau_c\,V/L^2)\,(\mu_n + \mu_p)\,, \qquad (11)$$

will flow, where as above L is the length between the ohmic contacts and WD is the cross-sectional area of the sample. The usual definitions of quantum efficiency for a photoconductor is

η = number of electron-hole pairs generated per second ÷ number of incident photons per second. (12)

We can define a photoconductive gain M_{pc} for photoconductors such that

$\eta\, M_{pc}$=number of electron-hole pairs collected per second ÷ number of incident photons per second

$$= (\Delta i/q)/(G/\eta) = \eta\, \tau_c\, (V/L^2)\, (\mu_n + \mu_p)$$

$$= \tau_c/\tau_t, \tag{13}$$

where τ_t is the average carrier transit time, and the photoconductive quantum efficiency is

$$\eta = (1-R)(1-e^{-\alpha D}). \tag{14}$$

Thus, with a large excess carrier lifetime, large applied voltage, short length L, and high mobilities, M_{pc} and $\eta\, M_{pc}$ can be greater than unity. When we consider sensitivity of photodetectors below, we will see that this is a desirable situation for wide-bandwidth detection systems.

For optical-waveguide communications, although there has been some recent consideration of intrinsic photoconductive detectors, the photodiode or photovoltaic detector is the type of detector being most widely studied and utilized because of its small size, short response time, and high sensitivity. For the near-infrared spectral region the reverse-biased photodiode is the most commonly used detector. Because of the importance of this device the parameters which influence the quantum efficiency, response time, and noise properties of reverse-biased photodiodes will be examined in detail.

Reverse-Biased PN-Junction Detectors

Many of the same considerations important for photoconductive detectors apply to photovoltaic detectors, but there are a few significant differences. The definition of the quantum efficiency for pn-junction detectors is slightly revised, since in contrast to a photoconductive detector not every electron or hole that is generated contributes to the photocurrent in reverse-biased or short-circuit modes of operation, or to the diode voltage in open-circuit operation. The photocurrent in a pn-junction detector results from the separation, by the electric field in the junction, of the electron-hole pairs that are generated by the absorption of incident photons. Fig. 4 shows the structure of a

reverse-biased pn-junction detector and the carrier generation rate for radiation incident from the left on the semiconductor material with reflectivity R and absorption coefficient α at the energy $h\nu$ of the incident radiation. The same structure, generation rate, and energy-band diagram also apply qualitatively to both open-circuit and short-circuit photovoltaic operations as well. The thin p^+ and n^+ regions serve as ohmic contacts to the p and n regions that form the active part of the device. The depletion region extends a distance W_p into the p-type material and a distance W_n into the n-type material. When photons are incident on the detector as shown, electron-hole pairs are generated within the structure as described previously for intrinsic photoconductive detectors, and the spatial variation of the generation rate is given by $g(x)$. Those electron-hole pairs that are generated within the depletion region are separated by the electric field and contribute to the current directly. Outside of the depletion region some of the generated electron-hole pairs recombine, but minority carriers generated within a diffusion length (L_n or L_p) of the depletion region diffuse to the edge of the depletion region and are collected. These carriers contribute to the photocurrent at low frequencies, but if the modulation frequency is high relative to the time it takes for these carriers to diffuse to the edge of the depletion region, they are effectively lost. The fraction of carriers generated in the different regions of the detector is related to the device dimensions and to the value of the absorption coefficient α, and these must both be optimized for a particular device application. Thus the carriers generated by photons absorbed outside of the crosshatched and shaded areas are lost from the photocurrent generation process and thus do not contribute to the quantum efficiency of the pn-junction detector, in contrast to the case for the intrinsic photoconductive detector, where all of the optically generated carriers contribute to the photocurrent. Also, since the carriers generated within a diffusion length of the edges of the depletion region may or may not contribute to the photocurrent depending on the modulation frequency, the quantum efficiency can be a function of modulation frequency. The time required for the carriers to diffuse a distance d can be estimated from

$$\tau_{\text{diff}} = d^2/2 D_c, \qquad (15)$$

where D_c is the diffusion coefficient for the carrier in question. Fig. 5 shows the response of a detector, with the structure indicated in Fig. 4, to a rectangular light pulse. The wavelength of the light is assumed to be such that the radiation is completely absorbed before it reaches the bulk n-type material beyond the depletion region. The resulting detector pulse consists of a fast and a slow component and the shape is considerably distorted from that of the incident light pulse. The fast component, shown by the shaded area in

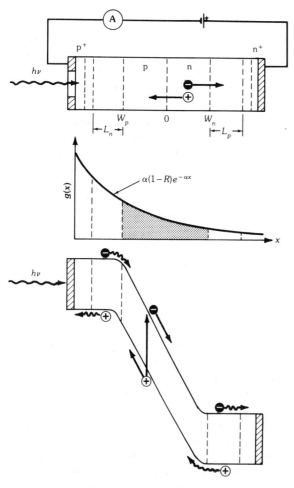

FIG. 4. Schematic diagram of a reverse-biased pn-junction detector showing the structure (top), carrier generation (center), and energy-band diagram (bottom). (*After Stillman and Wolfe [3]*)

Fig. 5, is due to the carriers that are generated in the depletion region. The rise time of this component can be limited by the transit time, by $R_{\text{eq}} C$ effects, or (if the device is operated with avalanche gain) by the gain-bandwidth product. The slow component is due to the diffusion of electrons that are generated within a distance L_n of the edge of the depletion region. This slow component would build up to a steady-state value if the light pulse were sufficiently long.

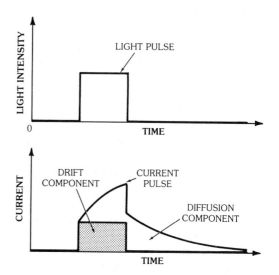

FIG. 5. Detector current pulse resulting from detection of a rectangular light pulse showing drift and diffusion components of the photogenerated current. (*After Stillman and Wolfe [3]*)

At the end of the light pulse the carriers in the depletion region are collected quickly, so that, neglecting $R_{\text{eq}} C$ effects, the detector photocurrent decreases by the same amount as the original increase due to the space-charge–generated current in a time comparable to the transit time as these carriers are swept out of the depletion region. This is the origin of the fast component of the detector response at the end of the light pulse. The diffusion of the carriers which are within a distance L_n of the edge of the depletion region at the end of the light pulse produces the slow response at the end of the pulse. Although the time constant describing the diffusion process is the same for both the rising and falling part of the pulse, the shapes of these two parts of the response can appear to be drastically different, depending on how close the diffusion current in the rising part of the pulse is to the steady-state value. The actual shape of the pulse will depend on the device structure, the absorption coefficient, the minority-carrier diffusion lengths, and the conditions under which the detector is operated (wavelength and avalanche gain).

The contribution of the diffusion current to the detector response can be reduced by increasing the depletion-layer width (for weakly absorbed radiation) and/or by reducing the diffusion length and the dimensions of the absorbing detector material outside of the depletion width. Increasing the depletion width will also increase the drift component for weakly absorbed radiation, but in addition will decrease the transit-time–limited bandwidth. Since the minority-carrier diffusion length generally decreases with increasing carrier concentration, the contribution of the diffusion current to the total detector current can be reduced by using epitaxial layers on substrates with higher carrier concentrations. For avalanche photodiodes as discussed below, the relative contribution of the diffusion current to that of the drift current can be further reduced by arranging the device structure so that the diffusion component of the current consists mainly of carriers with the lower ionization coefficient. In this manner the effective multiplication for the diffusion current is less than that for the drift current.

In the case of pn-junction or photovoltaic detectors a better definition of the quantum efficiency than (12) is

$\eta \equiv$ number of electron hole pairs collected per second $\div$ number of incident photons per second. (16)

For detectors in optical-waveguide communication systems the useful bit rates are generally quite high, so that it is desirable to have essentially all of the photons absorbed in the depletion region. If this is the case, the internal quantum efficiency (the ratio of collected electron-hole pairs to the absorbed photons) will be nearly unity, and the external quantum efficiency, the ratio of collected electron hole pairs to the *incident* photons, will be limited primarily by the

reflectivity of the detector and will be independent of frequency. To accomplish this the detector dimensions, doping, etc., must be adjusted for the absorption coefficient of the semiconductor material, at the operating wavelength. If the device dimensions are adjusted so that all of the nonreflected incident radiation is absorbed in the depletion region, the external quantum efficiency for wavelength λ is approximately

$$\eta(\lambda) = [1 - R(\lambda)] [1 - e^{-\alpha(\lambda) W}], \quad (17)$$

where W is the width of the depletion layer. In this case the quantum efficiency for a photovoltaic detector is the same as (14) for the extrinsic photoconductor with the sample thickness D replaced by the depletion width W for the diode. It should be noted that W cannot be increased indefinitely to increase the quantum efficiency. The material purity will limit the maximum attainable value of W. In order to obtain a high quantum efficiency for detectors where α is small, it is necessary to have a wide depletion region. The device structure which permits this most readily is the so-called pin structure. Diagrams of a p^+inn^+ structure and the electric-field variation in this structure are compared to an ordinary p^+n diode in Fig. 6. The pin structure permits a wide depletion width at relatively modest reverse bias voltages. If the lightly doped region, referred to as the i region, were truly intrinsic, the electric field in this region would be constant. With practical semiconductors for near-infrared detectors, however, it is not possible to obtain intrinsic material, and the maximum value for the width of the so-called i region will be also limited by the purity of the material. More importantly, the transit time of carriers through the depletion region,

$$\tau_t = W / v_s, \quad (18)$$

where v_s is the average drift velocity of the carrier in question and W is the width of the i region, can limit the high-frequency response. This becomes a limiting factor in long-wavelength silicon detectors because of the indirect band structure and low absorption coefficient shown in Fig. 3. Fig. 7 shows the internal quantum efficiencies attainable at several wavelengths with silicon detectors for various transit-time–limited bandwidths. Even for wavelengths as short as 0.8 μm, the internal quantum efficiency for a 1-GHz transit-time–limited base-bandwidth is less than 80%, and for 0.85 μm the maximum quantum efficiency attainable is only 60% or less.

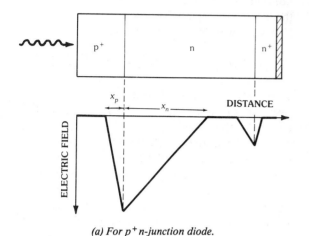

(a) For p^+n-junction diode.

(b) For pin-junction diode.

Fig. 6. Structures and electric-field distributions for junction diodes.

For a photovoltaic detector designed so that all the radiation is absorbed in the high-field region, the peak signal current is

$$i_s = q\,\eta\,p/h\,\nu\,, \qquad (19)$$

where *p is the peak incident optical power at frequency ν.* This expression can be used to calculate the peak power required in a pulse to give a peak

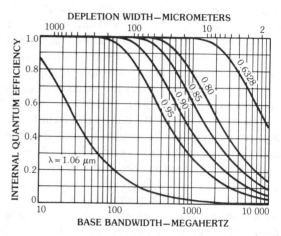

Fig. 7. Variation of internal quantum efficiency of silicon pin detectors with depletion width and transit-time-limited bandwidth for several wavelengths, assuming a saturated drift velocity of 10^7cm/s. *(After Stillman and Wolfe [3])*

signal current in a pulse-modulated optical-waveguide communication system. This relation will be used below to estimate the signal-to-noise ratio required to obtain a given bit error rate in such a system.

3. Optical-Waveguide Communication System Requirements

To evaluate the limitations of detectors it is necessary to know the performance required to satisfy certain system applications. Although receiver design is considered in detail in the next chapter, in this section a simple argument is given which relates the signal-to-noise ratio of a detector to the bit error rate in a digital communication system.

The received optical power in a digital modulation format in which the signal information is carried in a two-level binary form consists of a series of light pulses which can be represented by

$$P_0(t) = \sum_{k=-\infty}^{+\infty} a_k\,h_p(t-k\,T)\,, \qquad (20)$$

where $a_k = 0$ or 1 corresponding to the binary information being transmitted, $h_p(t)$ is the shape of a single pulse, and T is the pulse repetition time or pulse spacing. Fig. 8a shows an optical signal for the case where the pulse is square with a length T. The pulse length does not have to fill the entire time T (called the "bit period"), and Fig. 8c shows the optical signal for the case where $h_p(t)$ has a Gaussian shape with a width much less than the bit period T. The signal current is given by

$$i_s(t) = R P_p(t) = R \sum_{k=-\infty}^{+\infty} a_k h_p(t - kT), \qquad (21)$$

where R is the detector current responsivity in amperes per watt at the wavelength of the incident optical signal. This current is then amplified and filtered to produce an output current or voltage which can be represented by

$$i_{\text{out}}(t) = \sum_{k=-\infty}^{+\infty} a_k h_{i_{(\text{out})}}(t - kT)$$

or

$$v_{\text{out}}(t) = \sum_{k=-\infty}^{+\infty} a_k h_{v_{(\text{out})}}(t - kT), \qquad (22)$$

where $h_{i_{(\text{out})}}(t)$ and $h_{v_{(\text{out})}}(t)$ are the shapes of single amplified and filtered current or voltage output pulses, respectively.

The signal level at the output during each bit period is compared with a predetermined decision threshold level—if the signal exceeds the threshold level a pulse is assumed to be present, and if it is less than the threshold level it is assumed that no pulse is present. If the received power $P_0(t)$ is large and the receiver is well designed, the transmitted pulses will be faithfully reproduced as shown in Fig. 8c and there will be negligible uncertainty in determining whether or not a pulse is present in a given bit period. In practical systems, however, the signal will always be very weak because of signal attenuation in transmission, so that any noise generated in the detector or receiver becomes comparable to the signal and occasionally the true signal is either increased or decreased sufficiently by the noise that a wrong decision is made concerning the presence or absence of a pulse. The required fidelity or acceptable error rate of a particular application is usually specified in terms of a bit error rate, the probability of either a miss or a false alarm.

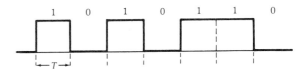

(a) Square pulse with pulse width equal to T.

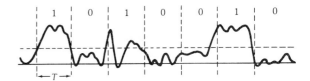

(b) Effect of Gaussian-noise current on (a).

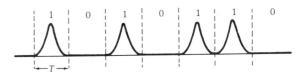

(c) Received optical power for binary signal with Gaussian pulse with width much less than T.

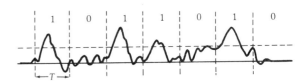

(d) Effect of Gaussian-noise current on (c).

FIG. 8. Possible optical and electrical signals in a fiber-optic system. (*After Stillman [40]*)

Even if we have an ideal detector with no dark current and sufficiently low thermal and amplifier noise that it can detect a single electron-hole pair or $1/\eta$ photons, where η is the detector quantum efficiency, because of the statistical nature of the incident photons there is still a limit to the incident power or to the minimum number of incident photons in a pulse that can be detected reliably. For an incident optical pulse $P_0(t)$ the average number of electron-hole pairs generated and collected in a photodiode is given by

$$\bar{N} = (\eta/h\nu) \int_T P_0(t)\, dt = (\eta/h\nu) E, \qquad (23)$$

where $h\nu$ is the photon energy and E is the energy in the individual pulse. Because of the statistical nature of the incident radiation the actual number of electron-hole pairs generated by a particular optical pulse deviates from the average value and the probability that n pairs will be produced is given by the Poisson distribution

$$p(n) = \bar{N}^n e^{-\bar{N}} / n! . \tag{24}$$

The uncertainty in the actual number of electron-hole pairs generated for a particular optical pulse places a lower limit on the optical power that can be detected even when no other noise sources are present. Since a single electron-hole pair can be detected with this ideal detector, if an optical pulse generates just one pair, the pulse will be detected. But if the power level is low, there is a finite probability that no electron-hole pairs will be generated by a given optical pulse, and thus it is possible that a pulse may not be detected when it is actually present. If the particular system application requires an error probability of less than 10^{-9}, this ideal detector requires sufficient optical energy per pulse that

$$p(0) = e^{-\bar{N}} \leq 10^{-9} , \tag{25}$$

or $\bar{N} \geq 21$ electrons. Thus the minimum number of photons per pulse that can be detected with a probability of error of less than 10^{-9} is $21/\eta$. Because this limit results from the statistical properties (i.e., fluctuation) in the arrival of photons, it is referred to as the quantum limit of detection, the limit due to fluctuation of photons in the signal itself.

In all real detectors for fiber-optic systems there will be other noise sources more important than the signal photon noise discussed above. Many of these noise sources are characterized by a Gaussian probability distribution, so that even when no signal is present the probability of detecting a current or voltage amplitude of x is given by

$$P(x) = [1/\sigma (2\pi)^{1/2}] \exp (-x^2 / 2 \sigma^2) , \tag{26}$$

where σ is the rms value of the current or voltage. Thus the noise current or voltage is just as likely to be positive as negative, and if it is sufficiently large, it can either increase a binary zero to a one or decrease a binary one to a zero. The influence

of a Gaussian-noise source on the received signals of Fig. 8a and 8c is shown in Fig. 8b and 8d. For the threshold detection limit i_t, indicated by the horizontal dashed lines in Figs. 8b and 8d, if at anytime during a mark bit (i.e., a bit when a binary one is transmitted) the noise current i_N is negative such that

$$i_N < -(i_s - i_t) , \tag{27}$$

the resulting current $i_s + i_N$ will be less than i_t and an error will result. The probability of this occurring can be written as

$$P(0,1) = \int_{-\infty}^{i_t} p(i, i_s) \, di , \tag{28}$$

in which

$$p(i, i_s) = (2\pi \overline{i_N^2})^{-1/2} \exp [-(i - i_s)^2 / 2 \overline{i_N^2}] \tag{29}$$

and i is the actual current, i_s is the peak signal current during a mark bit, and $\overline{i_N^2}$ is the mean-square noise current. Similarly, the probability that a one will be received when a zero is transmitted is just the total probability that the received current will be greater than i_t at some time during the "off" bit interval and is given by

$$P(i,0) = \int_{i_t}^{\infty} p(i,0) \, di . \tag{30}$$

Now, (28) and (30) can be written in terms of the error functions

$$\text{erf } x = (4/\pi)^{1/2} \int_0^x e^{-t^2} dt \tag{31}$$

and

$$\text{erfc } x = (4/\pi)^{1/2} \int_x^{\infty} e^{-t^2} dt = 1 - \text{erf } x \tag{32}$$

as

$$P(0,1) = (1/2) \{[1 - \text{erf } [|i_s - i_t| / (2 \overline{i_N^2})^{1/2}]\}$$

$$= (1/2) \text{erfc } [|i_s - i_t| / (2 \overline{i_N^2})^{1/2}] \tag{33}$$

and

$$P(1,0) = (1/2) \text{erfc } [|i_s - i_t| / (2 \overline{i_N^2})^{1/2}] , \tag{34}$$

where for a zero bit we assume $i_s = 0$. If we pick $i_t = i_s/2$, and assume that a binary code is chosen such that the number of "on" and "off" pulses are approximately equal, the total error probability (bit error rate) is given by

$$P_e = (1/2)[P(0,1) + P(1,0)]$$

$$= (1/2)\{\mathrm{erfc}\,[|i_s - i_t|/(2\,\overline{i_N^2})^{1/2}]$$

$$+ (1/2)\,\mathrm{erfc}\,[|0 - i_t|/(2\,\overline{i_N^2})^{1/2}]\}$$

$$= (1/2)\,\mathrm{erfc}\,[i_s/2\,(2\,\overline{i_N^2})^{1/2}] \qquad (35)$$

Thus the choice $i_t = i_s/2$ results in a simple expression which relates the analog signal-to-noise power ratio (snr) to the bit error rate by

$$P_e = (1/2)\,\mathrm{erfc}\,[(\mathrm{snr})^{1/2}/2^{3/2}] \qquad (36)$$

since $\mathrm{snr} = i_s^2/\overline{i_N^2}$. (In this expression snr is the ratio of the *peak* signal power to the rms noise power rather than the more common ratio of rms signal power to rms noise power.) Fig. 9 is a plot of (36) and from this figure it can be seen that in order to achieve a bit error less than 10^{-9} as discussed previously, the received signal power must be large enough that the peak signal-to-noise

power ratio is about 144, or that the peak signal current is about 12 times greater than the rms noise current. The number of photons that can be detected with this error rate is thus dependent on the magnitude and source of the noise currents, but in all cases of interest will be much greater than $21/\eta$, the minimum detectable number of photons when the only limitation is shot noise in the signal itself.

4. Sensitivity of Photovoltaic Detectors

Since the bit error rate can be directly estimated from the snr of a photodetector, it is important to understand the material and structural factors that limit this ratio. The generalized photodetection process for a pin photodiode [2, 3] is shown schematically in Fig. 10. The block diagram in Fig. 10a illustrates the separate elements of the detection process. The optical signal and background radiation are absorbed in the diode, where they produce electrical currents by the internal photoelectric effect as described above. The radiation gives up its energy to electrons in the valence band of the semiconductor, exciting the electrons into the conduction band and leaving holes in the valence band. These electrons and holes are then separated and collected by the electric field of the device to produce an electric current. The quantum efficiency of this conversion depends on the wavelength of the incident radiation, the properties of the material used to form the junction, and the geometry of the detector. It can also depend on the modulation frequency of the incident radiation. Some of these considerations will be discussed in detail later. From this diagram it is clear that the snr is sequentially degraded by the detection process from the snr present in the incident signal. The noise sources that must be considered to evaluate the snr are shown in the equivalent circuit of Fig. 10b and include the shot noise in the average signal, dark, and background currents, I_p, I_d, and I_b, respectively, and the thermal noise of the load resistor or bias resistor and the excess noise of the amplifier. Thus the total mean-square noise current can be represented as

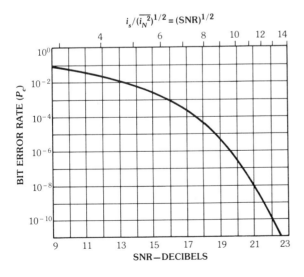

FIG. 9. Bit error rate as a function of power snr and current snr $i_s/(\overline{i_N^2})^{1/2}$, as calculated from (34). (*After Stillman [40]*)

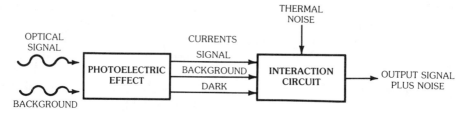

(a) Block diagram separating elements of detection process.

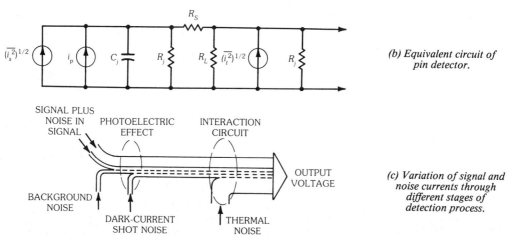

(b) Equivalent circuit of pin detector.

(c) Variation of signal and noise currents through different stages of detection process.

FIG. 10. Generalized photodetection process of pin detectors. (*After Anderson and McNurtry [4], ©1966 IEEE*)

$$\overline{i_N^2} = 2q(I_p + I_d + I_b)B + (4kT_{\text{eff}}/R_{\text{eq}})B, \quad (37)$$

where k is Boltzmann's constant, T_{eff} is the effective temperature including the amplifier noise, if any, R_{eq} the equivalent load resistance, and B the bandwidth. For the wavelengths of interest in optical waveguide transmission the background photocurrent is negligible, but this term can be used to describe the photocurrent resulting from dc biasing of a pulse-modulated semiconductor laser source close to threshold. Combining (19) and (37), the power snr for a reverse-biased pn junction detector is

$$\text{snr} = i_s^2 R_{\text{eq}}/\overline{i_N^2} R_{\text{eq}} = (q\eta p/h\nu)^2$$

$$/[2q(I_p + I_d + I_b)B + (4kT_{\text{eff}}/R_{\text{eq}})B], \quad (38)$$

and the minimum *peak* power necessary to give a peak signal-to-noise ratio (snr) can be estimated from

$$p = [2^{1/2} h\nu B/\eta](\text{snr})$$

$$\times \{1 + [1 + I_{\text{eq}}/qB(\text{snr})]^{1/2}\}, \quad (39)$$

in which

$$I_{\text{eq}} = I_d + I_b + (2kT_{\text{eff}}/qR_{\text{eq}}). \quad (40)$$

(The actual value of p required to give a specified rms snr will depend on the modulation format used.)

The thermal-noise contribution to I_{eq} is minimized by making R_{eq} large, but for wide bandwidths B, the equivalent load resistance R_{eq} cannot be made arbitrarily large. The base bandwidth, assuming the transit-time and diffusion-time limitations discussed above have been avoided, will be determined by the $R_{\text{eq}}C$ cutoff, where C is the sum of the pn junction capacitance and any stray package and/or wiring capacitance.

Fig. 11 shows the variation of the noise equiva-

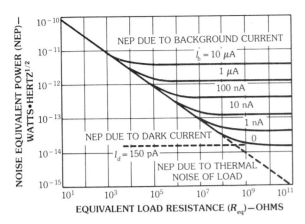

FIG. 11. Variation of NEP of pin diode (with $\eta = 75\%$ at $\lambda = 0.77\mu$m) with load resistance R_{eq} for 150-pA dark current and various background currents. (*After Stillman and Wolfe [3]*)

lent power (NEP), i.e., the peak signal power required to produce an rms of unity with a 1-Hz bandwidth) for a typical silicon pin detector at a wavelength of 0.77 μm calculated from (39). If the dark current is 150 pA, a dark-current–limited NEP of just slightly greater than 10^{-14} W·Hz$^{1/2}$ can be achieved for equivalent load resistances greater than about 10^9 Ω. For higher values of dark or background current, the NEP will be shot-noise limited for lower values of R_{eq} as shown, but at higher (poorer) values of NEP. Thus it is clear that R_{eq} should be large to make the thermal-noise contribution negligible. Fig. 12 shows the variation of the base bandwidth or cutoff frequency with R_{eq} for diodes with three different capacitance values of 0.5, 1.0, and 2.0 pF. Thus, a 10^9-Ω load resistor, as required if the detector is to be dark-current shot-noise limited with $I_d = 150$ pA, permits a bandwidth of only a few hundred hertz. Conversely, even for diodes with capacitances as low as 0.5 pF, to obtain a cutoff frequency of 10^9 Hz, R_{eq} must be no larger than a few hundred ohms. In this case the NEP is limited by the thermal noise of the load resistor even for dark currents up the 10 μA or greater. This conclusion is valid even if the following amplifier has zero noise (i.e., a noise figure of 1.0 or a noise equivalent temperature of 300 K) and zero input capacitance, as long as the detector and amplifier are operated at room temperature.

It has often been pointed out that the value of load resistance R_{eq} does not have to be restricted by the limitation that $R_{eq}C$ be less than the reciprocal of the bit rate, since the signal which is integrated by a large time constant can be restored through differentiation later in the signal processing. However, this high-impedance integrating amplifier technique is most useful when the signal current pulse utilizes a small part of the marking interval or bit period, and thus is limited in its application to high–bit-rate transmission systems. Another possibility is a feedback amplifier which develops a low impedance at the amplifier input through feedback, without affecting the amplifier noise properties. Both of these amplifier types are discussed in the next chapter.

The performance or sensitivity of reverse-biased pn junction detectors in high–bit-rate transmission systems is almost always limited by the thermal noise of R_{eq}, and for these devices it is important to optimize the quantum efficiency, minimize the capacitance, and develop low-noise amplifiers. With avalanche photodiodes, however, it is possible to obtain internal current gain which can make the thermal noise and amplifier noise insignificant and result in a substantial increase in the snr for a given incident optical power or a decrease in NEP. In these devices the excess noise of the avalanche gain process and the leakage current are important considerations, and these topics are discussed below.

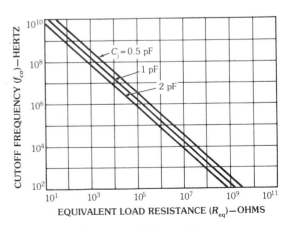

FIG. 12. Variation of *RC* cutoff frequency or base bandwidth with load resistance for three typical values of capacitance. (*After Stillman and Wolfe [3]*)

5. Sensitivity of Avalanche Photodiodes

The avalanche gain mechanism multiplies not only the photocurrent due to the signal and background, but also the dark current. In addition, the avalanche gain process also introduces extra noise due to the statistics of the carrier multiplication process. The minimum detectable peak power for an avalanche photodiode can be estimated from

$$p = 2^{1/2} h \nu B F_p \eta^{-1} (\text{snr})$$

$$\times \{1 + [1 + I_{eq}/q B F_p^2 (\text{snr})]^{1/2}\}, \quad (41)$$

in which

$$I_{eq} = I_b F_b + I_{dm} F_d + I_{du}/M^2 + 2 k T_{eff}/q R_{eq} M^2.$$

The factors F_p, F_b, and F_d are the excess-noise factors for the photocurrents, background currents, and dark currents, respectively, and are often considered to be equal and given by F. The dark current is divided into a bulk component I_{dm} that is multiplied and an unmultiplied surface leakage component, I_{du}.

Comparison of (41) with the similar result for a pn-junction detector without gain, (39) shows that if I_{eq} is negligible, the minimum detectable power will be larger (i.e., poorer) for the avalanche photodiode than for the reverse-biased pn-junction detector without gain by the factor F_p, the excess-noise factor. The internal gain mechanism reduces the importance of both the unmultiplied dark current and the thermal and excess amplifier noise. If the contribution of these terms is negligible (because of sufficient multiplication M) the relative values of background and bulk dark current and the bandwidth and snr will determine the minimum detectable power in the same way as for the photodiodes without gain. Thus, for wide-bandwidth applications such as those of interest in high–bit-rate fiber-optic transmission systems, avalanche photodiodes with small excess-noise factors and low bulk leakage currents permit a minimum detectable power considerably less than that required by thermal-noise–limited pin detectors.

6. Avalanche Gain Mechanism

The values of the excess-noise factors and the gain M in (41) are strongly influenced by the semiconductor material as well as by the device structure used. Some of these effects can be explained qualitatively by considering the avalanche gain process for two special cases: (1) only electrons have a finite impact ionization coefficient α, the probability per unit length that a carrier will generate an extra electron-hole pair by impact ionization; and (2) the electron and hole impact ionization coefficients, α and β, are equal: $\alpha = \beta$.

Fig. 13a shows the multiplication process graphically for electron injection where only the electrons can cause impact ionization. In this configuration the electric field is directed in the negative x direction and the currents are defined as shown in Fig. 13a. The injected electrons at $x=0$ are described by the current $J_n(0)$ shown in Fig. 13a. The electron multiplication for this case can be written as

$$\ln M_n = \int_0^w \alpha \, dx$$

or

$$M_n = \exp(\alpha W). \quad (42)$$

where α in this expression is the electron impact ionization coefficient.

The general features of the gain process can be seen by considering the high-field depletion region of an avalanche photodiode shown schematically in Fig. 13c. The spatial variation of the field in this region does not need to be specified, but the direction of the field is assumed to be as shown, so that electrons within this region travel in the positive x direction and holes travel in the negative x direction. Thus the direction of current, whether due to electrons or holes, is in the same direction as the electric field, but the electron current increases with increasing x while the hole current decreases with increasing x. Neglecting the spatial dependence of α and β described by Beni and Capasso [5] and Thornber [6] the variation of the electron current in the high-field depletion region can be written as

$$-dJ_p(x)/dx = \alpha(x)\,J_n(x) + \beta(x)\,J_p(x)\,. \qquad (44)$$

The total current J is the sum of the electron and hole currents, and under dc conditions $J = J_n(x) + J_p(x) =$ constant. The differential equations for the electron and hold currents can then be written in terms of the total current J as

$$dJ_n(x)/dx = [\alpha(x) - \beta(x)]\,J_n(x) + \beta(x)\,J\,, \qquad (45)$$

and

$$dJ_p(x)/dx = [\alpha(x) - \beta(x)]\,J_p(x) - \alpha(x)\,J\,. \qquad (46)$$

These expressions can be solved by using the integrating factor

$$\exp\left[-\int_0^x (\alpha - \beta)\,dx'\right] \equiv \exp\left[-\phi(x)\right]$$

and integrating from $x = 0$ to W to obtain

$$J = \left[J_p(W) + J_n(0)\,\exp[\phi(x)]\right] \Big/ \left\{1 - \int_0^W \beta(x)\,\exp\left[\int_x^W (\alpha - \beta)\,dx'\right]dx\right\} \qquad (47)$$

from (45), and

$$J = \left[J_p(W)\,\exp[\phi(x)] + J_n(0)\right] \Big/ \left\{1 - \int_0^W \alpha(x)\,\exp\left[-\int_0^x (\alpha - \beta)\,dx'\right]dx\right\} \qquad (48)$$

from (46). If hole injection at $x = W$ and electron injection at $x = 0$ are considered separately, the usual multiplication factors for electrons and holes can be obtained from (47) and (48). Thus

$$M_p = J/J_p(W) = 1 \Big/ \left\{1 - \int_0^W \beta\,\exp\left[\int_x^W (\alpha - \beta)\,dx'\right]dx\right\}$$

$$= \left\{\exp\left[-\int_0^W (\alpha - \beta)\,dx\right]\right\} \Big/ \left\{1 - \int_0^W \alpha\,\exp\left[-\int_0^x (\alpha - \beta)\,dx'\right]dx\right\} \qquad (49)$$

and

$$M_n = J/J_n(0) = \left\{\exp\left[\int_0^W (\alpha - \beta)\,dx\right]\right\} \Big/ \left\{1 - \int_0^W \beta\,\exp\left[\int_x^W (\alpha - \beta)\,dx'\right]dx\right\}$$

$$= 1 \Big/ \left\{1 - \int_0^W \alpha\,\exp\left[-\int_0^x (\alpha - \beta)\,dx'\right]dx\right\}\,. \qquad (50)$$

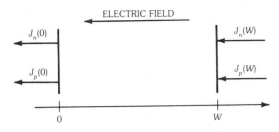

ELECTRIC FIELD

$J_n(0)$ $J_n(W)$

$J_p(0)$ $J_p(W)$

0 W

(a) Boundary conditions and electric field.

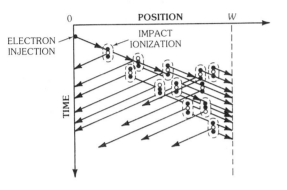

POSITION

ELECTRON INJECTION

IMPACT IONIZATION

TIME

(b) Avalanche gain process for $\beta = 0$.

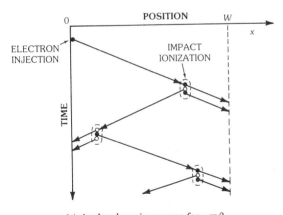

POSITION

ELECTRON INJECTION

IMPACT IONIZATION

TIME

(c) Avalanche gain process for $\alpha = \beta$.

FIG. 13. Schematic diagram of avalanche multiplication process. (*After Stillman and Wolfe [3]*)

$$dJ_n(x)/dx = \alpha(x)\,J_n(x) + \beta(x)\,J_p(x)\,, \qquad (43)$$

where q is the magnitude of the electronic charge. (The variation of the ionization coefficients with x can be obtained from the spatial dependence of the electric field as discussed below.) The variation of the hole current is given by

These relations are the same as those previously reviewed by Stillman and Wolfe [3] and originally derived by Howard [7] and Lee and colleagues [8] and differ from the equations derived by Moll [9] only because of the assumed direction of the electric field. The avalanche breakdown voltage is defined as the voltage for which the multiplication becomes infinite and thus is the voltage for which

$$\int_0^W \beta_p \exp\left[\int_x^W (\alpha-\beta)\,dx'\right] dx = 1 \qquad (51)$$

with hole injection at $x=W$, or for which

$$\int_0^W \alpha \exp\left[-\int_0^x (\alpha-\beta)\,dx'\right] dx = 1 \qquad (52)$$

with electron injection at $x=0$. These two conditions are identical. When both electrons and holes are injected simultaneously the breakdown voltage is that voltage for which the total current J in (47) and (48) becomes infinite. This voltage is the same as that given by (51) and (52). The breakdown voltage is not influenced by the *type* of carrier injection.

The equations presented above are, in general, quite complicated because of the variation of $\alpha(x)$ and $\beta(x)$ with position. However, effects of the ionization coefficients on the avalanche gain mechanism can be demonstrated by two special cases, both assuming a uniform field in the depletion region: (1) $\beta=0$ (or $\alpha=0$), which will later be shown to be the desired case for the highest performance avalanche photodiode; and (2) $\alpha=\beta$, the worst case for the performance of a properly designed avalanche photodiode.

For $\beta=0$ and a uniform field the multiplication factor for electrons injected at $x=0$ is, from (50) simply that given by (42) and repeated here,

$$M_n = \exp\left[\int_0^W \alpha\,dx\right] = e^{\alpha W}.$$

Under these conditions M_n just continues to increase exponentially with increasing values of αW. For this case there is no true avalanche breakdown, and for a uniform electric field the current increases exponentially with αW, where α is the electron impact ionization coefficient and W is the width of the high field depletion region.

The current pulse which results from the injec-

tion of a single electron builds up exponentially with time until the primary injected electron arrives at the anode, i.e., during the electron transit time, and then decreases to zero during the hole transit time. Thus the current pulse lasts about twice as long with avalanche gain as it does without, but since the pulse width is equal to approximately two transit times, independent of avalanche gain, there is no gain-bandwidth product limitation when $\beta=0$ and only one carrier can impact ionize. In addition, it is clear from the schematic diagram of Fig. 13b that high avalanche gain implies a large number of ionizing carriers in the depletion region. Thus the statistical fluctuation of the impact ionization process will produce only a small fluctuation in the total number of carriers, and therefore it is reasonable to expect that for single carrier impact ionization there is little excess noise introduced by the avalanche gain process.

The behavior just described, however, is significantly different from that for the case where both electrons and holes can cause carrier multiplication through impact ionization. Fig. 13c gives a pictorial representation of the time dependence of the avalanche multiplication process of an injected electron for the case where the electrons and holes have equal impact ionization probability, $\alpha=\beta$. When $\alpha=\beta$ and the field is uniform,

$$M_n = M_p = 1 \Big/ \left(1 - \int_0^W \alpha\,dx\right) = 1/(1-\alpha W), \quad (53)$$

and a true avalanche breakdown is obtained. The breakdown voltage corresponds to the situation where $\alpha W=1$, which occurs when each injected carrier on the average generates one electron-hole pair during its transit through the depletion region. Fig. 13c shows that when $\alpha=\beta$ the gain or multiplication can be very high, but when the gain is high the pulse width is also very long, so that there is a definite gain-bandwidth relationship. (When the multiplication is infinite, the current pulse length is infinite and we have avalanche breakdown). It can be seen from this diagram that even at high gain there are relatively few carriers in the depletion region at any given time. Thus statistical variations in the impact-ionization process can cause large fluctuations in the gain or multiplication and contribute considerable excess noise.

From a practical point of view, in most semiconductors both the electrons and holes contribute to the ionization process and the ionization coefficients are not equal, so that real multiplication processes are somewhere between the two extremes just considered. In the general case, given arbitrary values of α and β that are independent of position (such as for an ideal pin avalanche photodiode), the multiplication of electrons injected into the high-field region at $x=0$ is (see (50))

$$M_n = (1 - \beta/\alpha) \exp [\alpha W (1 - \beta/\alpha)]$$

$$/ \{ 1 - (\beta/\alpha) \exp [\alpha W (1 - \beta/\alpha)] \} . \quad (54)$$

For small values of β/α, the positive-feedback characteristic of the avalanche breakdown becomes apparent by rewriting (54) as

$$M_n \simeq M_1 / [1 - (\beta/\alpha) M_1] , \quad (55)$$

where $M_1 = \exp (\alpha W)$ is the electron multiplication (42) that would apply for $\beta = 0$ (the unilateral gain), and β/α is the positive-feedback factor.

The electron multiplication calculated from (54) as a function of αW for various ratios of β/α is shown in Fig. 14. These results are plotted in terms of the electric field in a 1-μm-wide "i" region of a pin avalanche photodiode for which the electron ionization coefficient is given by

$$\alpha = 3.36 \times 10^6 \exp (-1.75 \times 10^6 / |\mathbf{E}|) . \quad (56)$$

(This is the variation of α with electric field for silicon attributed to Lee and colleagues [8]). The corresponding reverse-bias voltage is 1×10^{-4} cm times the electric field and thus ranges from 20 to 45 V in this figure. It is clear from Fig. 14 that when the ratio β/α is near unity there will be, in addition to the lower gain-bandwidth product and higher excess noise mentioned above, severe practical limitations on the uniformity and stability of the avalanche gain process for high values of multiplication. For $\beta/\alpha = 0.01$, at a reverse bias of 100 V with a 0.5% variation in the electric field

would result in a 20% variation in multiplication, while for $\beta/\alpha = 1$ the same variation in electric field produces a variation in multiplication of over 300%. Since in real avalanche photodiodes such local variations in electric field could easily be caused by small inhomogeneities in the semiconductor doping level, these results show that there will be severe (probably insurmountable) technological problems in fabricating uniform *high-gain* avalanche photodiodes from semiconductors in which α and β are nearly equal. However, because of the excess-noise problem to be discussed below such devices are not useful at high avalanche gain for other reasons as well.

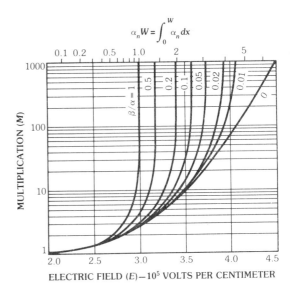

FIG. 14. Calculated electron multiplication as a function of electric field for a 1-μm-wide pin diode with $\alpha_n = 3.36 \times 10^6 \exp (1.75 \times 10^6 / |\mathbf{E}|)$ for various ratios β_p / α_n. *(After Webb et al. [13])*

For devices fabricated from material in which the ionization coefficients are significantly different, it is important that the device structure be designed to take advantage of this difference. That is, when α and β are not equal, the avalanche gain depends on where the optically excited (primary) carriers are generated in, or injected into, the high-field multiplication region. The multiplication or gain which results from the injection (or generation) of an electron-hole pair at position x_0 in the depletion region is given by

$$M(x_0) = \left\{ \exp \left[\int_{x_0}^{W} (\alpha - \beta)\, dx \right] \right\}$$

$$\Big/ \left\{ 1 - \int_0^W \beta \exp \left[\int_x^W (\alpha - \beta)\, dx' \right] dx \right\}$$

$$= \left\{ \exp \left[- \int_0^{x_0} (\alpha - \beta)\, dx \right] \right\}$$

$$\Big/ \left\{ 1 - \int_0^W \alpha \exp \left[- \int_0^x (\alpha - \beta)\, dx' \right] dx \right\}. \quad (57)$$

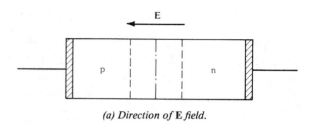

(a) Direction of E field.

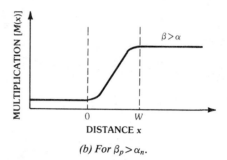

(b) For $\beta_p > \alpha_n$.

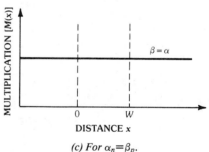

(c) For $\alpha_n = \beta_p$.

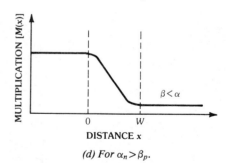

(d) For $\alpha_n > \beta_p$.

FIG. 15. Variation of multiplication with position on injection for different relative values of α and β.

Thus the actual multiplication depends on where the electron-hole pair is injected (that is, on the value of x_0) as well as on the spatial variation of α and β throughout the depletion region due to the variation of the electric field.

It is important that the primary photocurrent consists primarily of carriers with the higher ionization coefficient injected into the high-field region, if the lowest possible excess-noise factor is to be obtained. This dependence of the multiplication on the position at which the electron-hole pairs are injected in the high-field region can also produce a wavelength (λ) dependence of the average multiplication when $\alpha \neq \beta$.

From this discussion one might conclude, as is often done in the literature, that $\alpha = \beta$ represents the worst case from the standpoint of excess noise in avalanche photodiodes. This is not always true, however, because of the dependence of the multiplication on where an electron-hole pair is injected into the depletion region. A qualitative picture of the variation of multiplication with injection position is given in Fig. 15. The minority carriers that are injected into the p region for $x < 0$ and the n region for $x > W$ must first diffuse to the high-field region before they are accelerated by the electric field and multiplied. Thus, as long as the width of the undepleted p and n regions is less than a diffusion length, each minority carrier generated in these regions will be collected, and the minority carrier electrons will be multiplied by the factor M_n and the minority carrier holes will be multiplied by the factor M_p. If $\alpha = \beta$, M_n and M_p will be equal and the multiplication will also be the same for carriers injected at any position within the depletion region as shown in Fig. 15c. If $\alpha \neq \beta$, the multiplication will depend on where the carriers are injected into the depletion region, and if $\alpha > \beta$, electrons should be injected from the p side as in Fig. 15d for highest gain, and if $\beta > \alpha$, holes should be injected from the n side as in Fig. 15b.

It is qualitatively clear that this procedure will result in the largest number of ionizing carriers in the depletion region and thus the least noise or dependence on the statistical fluctuation of the impact ionization process and also the least degradation in the pulse width or frequency response.

These considerations can be described quantitatively, and Fig. 16 shows the results of a calculation by McIntyre [10] of the variation of the excess-noise factor with multiplication for either electron or hole injection, as a function of the ratio of the impact ionization coefficients. This figure makes it clear that for a low excess-noise factor the electron and hole impact ionization coefficients must be significantly different and, in addition, the device structure must be designed so that only the carrier with the highest ionization coefficient is injected into the high-field region. If the impact ionization coefficient ratio is considerably different from unity and predominantly carriers with the lower ionization coefficients are injected into the high-field region, the excess-noise factor can be much greater than M, the value for $\alpha = \beta$. Thus it is essential that the device structure is designed so that the primary photocurrent consists entirely of carriers with the higher ionization coefficient.

In the initial work on avalanche photodiodes it

tional material property that is essential for low-noise avalanche photodiodes, in addition to those properties required for pin detectors, is a large asymmetry in the electron and hole impact ionization coefficients. Silicon is so far the only semiconductor material in which a large asymmetry in α and β is well established experimentally. It is generally agreed that $\alpha \gg \beta$ for silicon, although the actual values and field dependence of α and β are not so well established. Nevertheless, ratios $\beta / \alpha < 0.006$ have been estimated from noise measurements, and high uniform gain, low-noise silicon avalanche photodiodes are produced more or less routinely.

7. Silicon Reach-Through APD Structure

The reach-through avalanche photodiode structure, which is shown in Fig. 17 and was first described by Ruegg [11], has several desirable

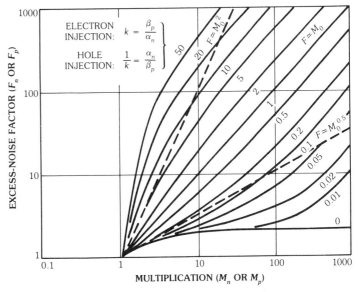

FIG. 16. Excess-noise–factor variation with multiplication for different ratios of the electron and hole impact ionization coefficients. (*After McIntyre [10], ©1966 IEEE*)

was realized that the excess-noise factor was nearly unity for single-carrier ionization and M for $\alpha = \beta$, so it was common to express F as $F = M^x$ where $0 \le x \le 1$. Fig. 16 shows that this is a poor approximation in general, and systems analyses which rely on this expression will probably not be reliable, at least over wide ranges of M.

The above discussion indicates that an addi-

characteristics not present in other device structures. The highest-performance silicon avalanche photodiodes available commercially utilize this structure. The reach-through avalanche photodiode is fabricated by starting with very high quality, high-resistivity, p-type (i.e., π-type) substrates with typical resistivities of about 5000 $\Omega \cdot$ cm. The active $n^+ p$ junction is formed by two diffusions:

the p-type using boron and the n-type using phosphorus. A thin p^+ diffusion is used to form the window and ohmic back contact to the π region. The diffusion times and temperatures are adjusted so that when a reverse-bias voltage is applied, the depletion region in the p^+ layer reaches through to the π region just before the n^+p^+ junction breaks down. Further increases in reverse-bias voltage then rapidly deplete the high-resistivity π region and increase the uniform field in this region while only increasing slightly the peak field at the np

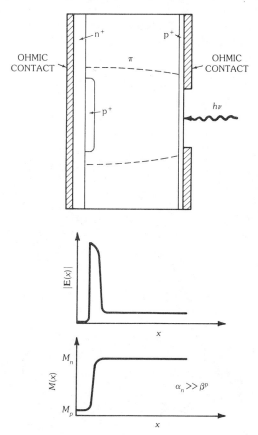

FIG. 17. Reach-through avalanche photodiode structure, with the variation of electric field and multiplication with position.

junction. This slower increase in the peak electric field results in the much slower variation of multiplication with bias voltage which is characteristic of this structure. The diffusion steps are critical to the performance of the device. If the phosphorus n-type diffusion is not deep enough, the np junction will break down before reach-through occurs.

In this case the gain will vary rapidly with voltage and the quantum efficiency for radiation incident on the p^+ contact will be low. If the phosphorus diffusion is too deep, an np^- or $n\pi$ junction will be formed and sufficient gain will be obtained only at extremely high voltages. Recently, to achieve better uniformity, ion implantation has been used to deposit the boron for the p-type diffusion [12].

The trade-off between the quantum efficiency at long wavelengths and the speed of response can be accurately controlled in this structure by changing the thickness of the intrinsic region. Since multiplication occurs only in the narrow high-field region, the avalanche gain characteristics are not affected. The wide low-field region in these devices produces a much more gradual change in the avalanche gain with bias voltage, and this is desirable in practical applications of these devices. In addition, the structure provides nearly pure electron injection regardless of wavelength (assuming the n^+ region is sufficiently thin or that the absorption coefficient is low as for indirect band-gap semiconductors), so that the noise performance of these devices should be essentially independent of wavelength and with the optimum electric field as good as can be obtained with silicon. Because of these advantages, McIntyre, Webb, Conradi, and coworkers at RCA Limited have expended considerable effort developing and characterizing this type of device [13–17], several versions of which are now available commercially.

The experimental variation of the electron and hole multiplication with bias voltage for a reach-through avalanche photodiode is shown [15] by the data points in Fig. 18. Pure hole injection was obtained by using the 404.7-nm line of a mercury lamp with a narrow-line filter. This radiation was incident on the n side of the device, where the junction was 8 to 10 μm deep. Pure electron injection was obtained by using 0.8-μm radiation incident through the p^+ window. As expected, the electron multiplication M_n was significantly larger than the hole multiplication M_p. The curves shown in Fig. 18 were calculated by Conradi [15] for the electric-field distribution estimated for the device from the ionization-coefficient data of Lee and colleagues [8] (solid curves) and van Overstraeten and DeMan [18] (dashed curves). It

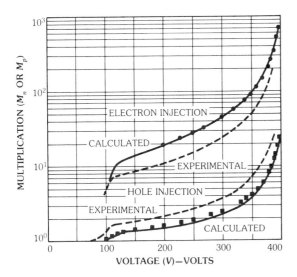

FIG. 18. Experimental and calculated variation of the electron and hole multiplication with bias voltage for a reach-through avalanche photodiode. (*After Conradi [15]*)

is clear that much better agreement with experimental data is obtained by using the ionization rate of Lee [8], although, as will be discussed below, the data from which these ionization rates were determined is not very reliable.

Low-frequency room-temperature noise measurements were made on these devices. Good agreement between the experimental and calculated excess-noise factors are only obtained when an effective ionization coefficient ratio k_{eff} obtained from the data of Lee [8] was used [13]. Noise measurements on similar devices at 77 K were previously analyzed in the constant-k approximation, and a best fit was obtained wth $k=\alpha/\beta=0.028$.

As discussed above, the quantum efficiency and speed of response for the reach-through avalanche photodiode are closely related. An increase in the width of the high-resistance π region increases the quantum efficiency for weakly absorbed radiation but decreases the response time in the transit-time–limited case. The impulse response depends somewhat on where the radiation is absorbed. For strongly absorbed radiation there is very little photocurrent until the injected electrons reach the high-field region (the electron transit time) and are multiplied. Then the multiplied current persists for essentially the transit

time of the holes back to the p$^+$contact. For weakly absorbed radiation some of the injected carriers are multiplied immediately and the resulting current pulse has a trapezoidal shape.

In the wavelength range of GaAs and Nd:YAG lasers, 0.8 to 1.06 μm, the response time of silicon avalanche photodiodes with reasonable quantum efficiencies will be greater than about 1 ns.

8. Germanium Avalanche Photodiodes

For detectors with faster response and/or higher quantum efficiency at long wavelengths than is

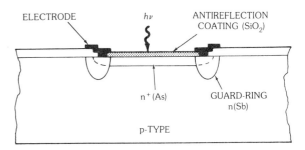

(a) Of n$^+$p type.

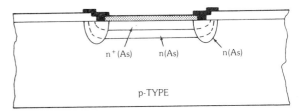

(b) Of n$^+$np type.

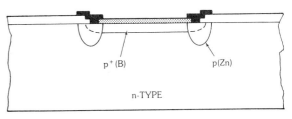

(c) Of p$^+$n type.

FIG. 19. Cross-sectional views of germanium APD structures. (*After Mikami et al. [19],* ©*1980 IEEE*)

obtainable with silicon, a material with a higher absorption coefficient and smaller band gap is required. The elemental semiconductor germanium has been used to fabricate sensitive and fast avalanche photodiodes that are useful for nearly the entire 1.8- to 1.6-μm wavelength range at present of primary interest for optical-waveguide communications [19]. These devices have conventionally used an n^+p structure as shown in Fig. 19a, for reasons of simplicity in fabrication [19]. However, when the wavelength of the optical signal approaches the cutoff wavelength for germanium the absorption coefficient becomes small (*cf.* Fig. 3), and the fractional part of the primary photocurrent that is due to electron injection increases. For wavelengths longer than 1 μm the fractional part of the primary photocurrent due to electrons is over 50% for a diode with a 0.4-μm-deep n^+p junction, and when the wavelength is longer than 1.5 μm the fractional part of the primary photo-

current due to electrons exceeds 80% [19]. This is undesirable for avalanche photodiode operation since for germanium the impact ionization coefficient for holes is slightly greater than that for electrons and, as discussed above, the contamination of the primary photocurrent with electrons will in this case lead to a larger excess-noise factor than could be obtained with a primary photocurrent consisting entirely of the carrier with the higher ionization coefficient, holes for the case of germanium. The structures shown in Fig. 19b and 19c can be designed to obtain nearly pure hole injection in the long-wavelength region. Fig. 20 shows the impurity and electric-field profile for the n^+np and p^+n structure and the weakly absorbed long-wavelength radiation. For the n^+np structure a sufficiently thin n^+ region ensures that most of the radiation is absorbed in the depletion region to maximize the quantum efficiencies and ensure high-frequency response. The structure is designed

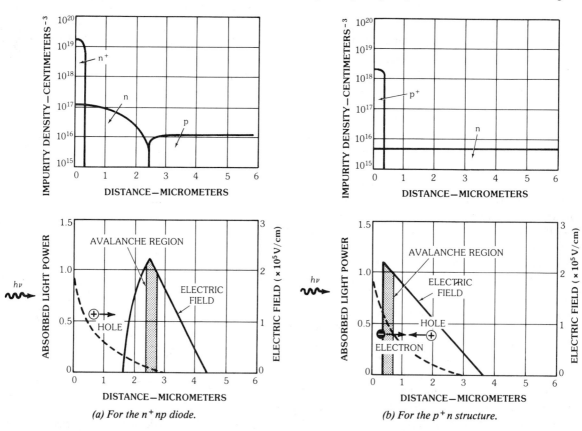

(a) For the n^+np diode.

(b) For the p^+n structure.

FIG. 20. Impurity and electric-field profiles, and spatial carrier injection. (*After Mikami et al. [19], ©1981 IEEE*).

so that all of the incident radiation is absorbed before it reaches the np junction, and thus mainly holes are injected into the avalanche region and low–excess-noise characteristics are obtained. A low-noise n^+np germanium APD operating at $\lambda = 1.3$ μm with an excess-noise factor of $F \simeq 7$ at a multiplication factor of $M \simeq 10$ and with a quantum efficiency of 80% and a cutoff frequency of 500 MHz has been reported by Mikawa and coworkers [20].

9. Impact Ionization Coefficients

The experimental measurement of the impact ionization coefficients is important for the evaluation of the ultimate performance capability of avalanche photodiodes in new materials. Recently there have been many advances in the theory of impact ionization phenomena and in the ability to calculate these effects, including the details of the energy-band structure. The experimental determination of the impact ionization coefficients is also important for the evaluation of the physical parameters used in these theoretical calculations, as well as to refine and evaluate the validity of the theoretical approach. The experimental determination of the electron and hole impact ionization coefficients is not an easy task, however, as is evident from the wide discrepancies of experimental results in the literature. In compound semiconductor materials there is not even general agreement on which carrier type has the higher ionization coefficient, and in silicon there is only poor agreement among the results of different investigators.

In this section we will review the experimental techniques which permit the reliable determination of the impact ionization coefficients, briefly review earlier results for silicon, germanium and gallium arsenide, and give the most recent and reliable experimental results for electron and hole impact ionization in gallium arsenide and indium phosphide.

The accurate determination of the impact ionization coefficients requires that certain experimental conditions be met. These experimental conditions have been known for some time [21, 22] but is only recently that it has been realized how difficult it is to actually achieve these conditions. The quantity that is usually experimentally

measured in order to determine the ionization coefficients is the variation with reverse bias of the photocurrent in a diode due to pure hole and pure electron injection. From this data and the current multiplication and ultimately the ionization coefficients are calculated. Careful design of the experiment can facilitate these calculations. Some of the most important experimental considerations are detailed below.

To calculate the ionization coefficients from the experimental photocurrent versus voltage data, the multiplication for both pure hole and pure electron injection must be determined. This requires that a device be fabricated which permits illumination of both sides of the junction separately. It is generally a simple matter to obtain one type of injection by shining strongly absorbed light on the top contact layer of the diode. There are several methods that have been used for obtaining the photocurrent for injection of the other type of carrier. One of these is to illuminate the side of the mesa with strongly absorbed light. This method, however, has several potential problems. One of these is that it is possible for stray light to hit the top layer and thereby cause mixed injection. Another problem is that the mesa edge is beveled and so the electric field at the edge is different from the field in the bulk. Therefore, if the carriers stay at the surface of the device, they will multiply in a way that is not typical of the bulk field. Another method for obtaining carrier injection is to illuminate the device from the backside or through the substrate. For direct band-gap compound semiconductors it has been shown that the material where the light is absorbed must be no more than a few diffusion lengths from the edge of the high-field region. This is because the subband-gap light generated by the radiative recombination of the carriers as they diffuse toward the depletion region can be absorbed through the Franz-Keldysh effect in the high-field region at the junction and produce mixed carrier injection [23]. This can be avoided by thinning the substrate underneath the junction to a few diffusion lengths. One way to do this reproducibly is to grow a thin layer of a lattice-matched material with different etching characteristics on top of the substrate to act as an etch-stop layer.

For the calculations of the electron and hole impact ionization coefficients, the actual carrier multiplication must be determined from the variation of the photocurrent for both types of injection. To do this the primary photocurrent in the absence of multiplication must be known. But since the multiplication process cannot be turned off, the primary photocurrent after the onset of multiplication must be calculated based on both the variation of photocurrent with bias voltages below that where multiplication occurs and on the theoretically expected primary photocurrent for the particular device structure under study. Several methods have been used to do this and many of the early methods and results have been reviewed by Stillman and Wolfe [3]. The simplest method is to assume that the primary photocurrent does not change with bias. This may be a reasonable assumption when the depletion-region edge is in heavily doped material and so does not move toward the illuminated surface as the bias is increased. For lightly doped material, however, the width of the depletion region will increase with bias and the collection efficiency will increase as the distance the carriers must diffuse is reduced. When such an increase in the primary photocurrent is observed at low bias voltages, a linear extrapolation of the low bias photocurrent to voltages higher than the onset of multiplication is often used. However, a more accurate procedure is to use a model that takes into account the physical processes that are responsible for the increase in the collection efficiency. This requires a knowledge of the doping levels in the device as well as the layer thicknesses. The movement of the depletion edge toward the illuminated surface with increasing voltage can be calculated, and by treating the minority carrier diffusion length as an adjustable parameter, a least-squares fit to the experimental data can be obtained. This model can then be extrapolated to calculate the primary photocurrent at voltages where the avalanche process is taking place.

After the multiplication has been calculated, this data is then used to calculate the ionization coefficients. If the spatial variation of the electric field is known at each reverse-bias voltage, the ionization coefficients can be calculated from the variation of the electron and hole multiplication with reverse-bias voltage for any type of structure by using numerical techniques. If, however, the electric-field variation has a simple form, such as a constant field or a linearly varying field, then analytical expressions can be used to calculate the ionization coefficients. Therefore it is advantageous to fabricate diodes that have abrupt junctions and constant doping profiles in the layers. Reach-through structures may be used if the reach-through occurs before the onset of multiplication. If the electric field has a rapid spatial variation, such as in a heavily doped depletion region, the carrier may travel through a significant part of the high-field region without reaching the threshold for ionization. This results in a "dead space" that must be taken into account in the calculations.

Even when all of the above-mentioned factors are taken into consideration, other errors can be introduced into the ionization coefficient determination. It is important that the device have a spatially uniform photoresponse so the measured photocurrent is characteristic of the structure and the electric-field profile of the device. This means that care must be taken to avoid illuminating the edge of a mesa diode where the electric field is not typical of the bulk field. Even if the multiplication has been carefully calculated using a physical model, the ionization coefficient values calculated from multiplication values very close to unity must be viewed with suspicion. Also, for values of voltage that are close to the breakdown voltage of a particular device, the multiplication may increase so rapidly with voltage that it is impossible to calculate accurately the derivative needed for the evaluation of the ionization coefficients. Therefore a number of devices that operate in overlapping electric-field ranges must be measured and the data from these devices combined to obtain ionization coefficients over a large range of electric fields.

After the ionization coefficients have been determined for a particular device structure or for a set of structures, it is common to calculate the breakdown voltage for those structures from the experimental ionization coefficients and then compare the calculated voltage to the experimentally measured breakdown voltage. While this serves as a check on the self-consistency of the calculations, it is not an independent verification of the correctness of the results. An independent check of the results for α and β can be obtained, however, from excess-noise measurements for both pure electron and pure hole injection on the same devices that were used for the photocurrent multiplication determination of the ionization coefficients. Comparison of experimentally determined excess-noise factors with the noise factors calculated from McIntyre's noise theory [10] using the ionization coefficients determined from the photocurrent multiplication measurements can indicate whether pure carrier injection was obtained and whether the relative magnitudes of the electron and hole ionization coefficients were correctly determined.

Accurate values for the impact ionization coefficients in (100) GaAs have recently been determined experimentally [24]. In this study, devices from eight wafers with different doping concentrations were studied. Typically, six devices per wafer were measured. Five of these wafers had heavily doped p^+ layers and the data from these devices therefore had to be corrected for the "dead space" mentioned previously. The other three wafers had more lightly doped p-type regions and did not require this correction. The combined results of all of these measurements are shown in Fig. 21. There is excellent agreement between the devices on a single wafer and between devices from different wafers. These results show that α is greater than β, with the two becoming almost equal at high fields. The least-squares fits to these data are given by

$$\alpha = 1.9 \times 10^5 \exp\left[-(5.75 \times 10^5 / E)^{1.82}\right] \quad (58)$$

and

$$\beta = 2.2 \times 10^5 \exp\left[-(6.57 \times 10^5 / E)^{1.75}\right] \quad (59)$$

in centimeters^{-1}. The breakdown voltages and electric-field values calculated using these fits, as well as the experimentally observed breakdown voltages are shown in Fig. 22, where the dashed curves are for n layer thicknesses shown in which the electric field reaches through the n region before breakdown occurs. Several of the same devices used for the ionization coefficient determination were also used for noise measurements, and the calculated and the experimental excess-noise factors were in excellent agreement with each other.

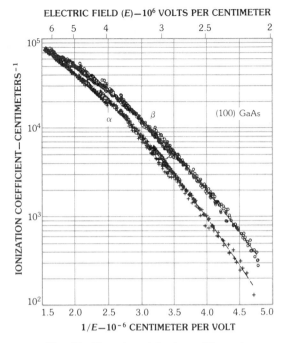

FIG. 21. Experimental values of impact ionization coefficients in (100) GaAs. (*After Bulman et al. [24], ©1983 IEEE*)

Impact ionization in (100) InP has also been studied [23]. In this case three wafers were measured to obtain the ionization coefficients over a wide range of electric fields. The more heavily doped side of the junction in these devices was not so highly doped that dead-space correction was necessary for these devices. The results of these measurements are given in Fig. 23. In this material β is greater than α over the range of electric fields measured, but again the two become nearly equal at very high electric fields. The solid lines are a parameterization obtained by a least-squares fit to these data, given by

$$\alpha = 3.5 \times 10^5 \exp\left[-(1.04 \times 10^6 / E)^{1.54}\right] \quad (60)$$

and

$$\beta = 3.8 \times 10^5 \exp\left[-(1.01 \times 10^6 / E)^{1.46}\right] \quad (61)$$

in centimeters^{-1}. Excess-noise measurements on the devices used for these measurements were also made with both types of carrier injection, and good agreement was found between the experimental and the calculated noise factors.

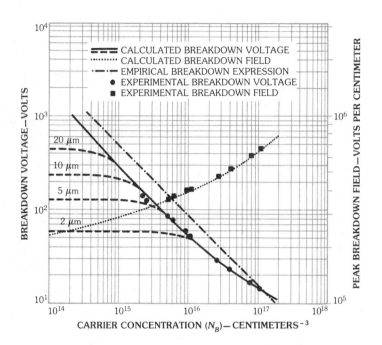

FIG. 22. Calculated and observed breakdown voltages in (100) p^+nn^+ GaAs devices.

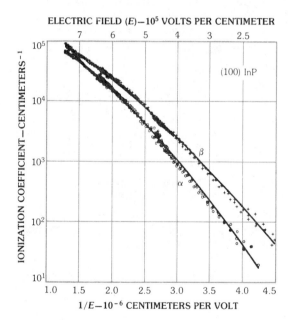

FIG. 23. Experimental values of impact ionization coefficients in (100) InP devices. (*After Cook, Bulman, and Stillman [25]*)

electric-field range of the (111) measurements is not as great because of the difficulties in device fabrication. The ionization rates for the two orientations differ only by a 5% uncertainty in the measured value of the peak electric field which causes a slight shift along the horizontal coordinate. There is essentially no difference in the ratio of α and β for these two orientations. In a separate study the (110) and the (100) orientations were compared and there was also no difference

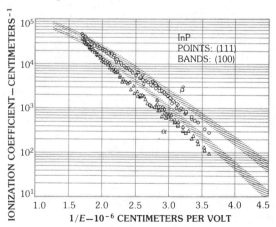

FIG. 24. Experimental values of impact ionization coefficients in (111) InP devices. (*After Tabatabaie et al. [26]*)

The ionization coefficients have also been measured in (111) InP [26]. These results are shown in Fig. 24 and are compared to the (100) InP results that are indicated by the shaded bands. The

between the ionization coefficients in these directions [27].

In addition to the experimental measurements, Monte Carlo calculations of the ionization coefficients have been performed and the results compared to the experimental data [28]. The calculated values agree well with the experimental values. The calculations also show no anisotropy in the electron ionization coefficients as a function of orientation for either gallium arsenide or indium phosphide, in agreement with the experimental results in indium phosphide but in contrast to the early experimental reports by other workers for gallium arsenide [10]. The Monte Carlo results also indicate that the reversal of the α to β ratio be-

10. Separate Absorption and Multiplication (SAM) APDs

For applications at longer wavelengths where silicon devices have too low quantum efficiency and/or too long response times as discussed above, the GaInAsP quaternary alloy system is being studied extensively. As can be seen in Fig. 25, this alloy system can be lattice matched to indium-phosphide substrates while still permitting the bandgap to be adjusted over a significant range of interest for long-wavelength (1.3–1.5 μm) fiberoptic applications. There has been much work on this alloy system [29], but although reasonably low leakage currents have been obtained at low

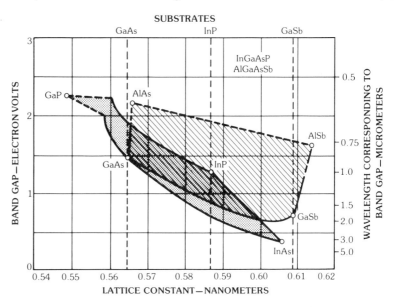

FIG. 25. Possible ranges of lattice constant and bandgap energy of the quaterary alloys GaInAsP and AlGaAsSb.

tween the two materials can be explained in terms of the ionization threshold energies and the densities of states in the two materials. The threshold energies for both holes and electrons are the same in gallium arsenide whereas in indium phosphide the hole threshold energy is smaller than the electron threshold energy. The densities of states in the valence and conduction bands of both materials would tend to make the electron rate greater than the hole rate. In gallium arsenide, since the threshold energies are equal, α is greater than β. However, in indium phosphide the smaller threshold energy of the holes overcomes the effects of the density of states and so the hole ionization coefficient is greater than that for electrons.

voltages, the leakage currents at high reverse-bias voltages become excessive due to tunneling. In addition, preliminary measurement of the impact ionization coefficients in this alloy indicate that α and β are nearly equal.

If suitable materials were available, it would be possible to separate the absorption region of the device from the avalanche gain region so that optimum parameters could be chosen for each. An example of such a device is shown in Fig. 26. The top part of Fig. 26 shows the equilibrium band structure of an InP-InGaAsP or InP-InGaAs heterostructure detector designed so that multiplication is obtained in the InP pn junction. Radiation is incident from the left in this structure and passes through the wide–band-gap indium phosphide to the alloy materials where it is absorbed. Under reverse-bias conditions shown in the bottom of Fig. 26 the carriers are generated in a relatively high electric-field region in the alloy where the holes gain sufficient energy from the field to either get over or tunnel through the hole barrier due to the valence-band discontinuity at the heterojunction. Once in the indium phosphide the holes continue to gain energy and avalanche gain occurs in the usual way. This structure can produce high quantum efficiency under reverse bias and pure hole injection, so if β were much greater than α in indium phosphide it would be an ideal structure for an avalanche photodiode. Unfortunately, recent measurements show that β is only a little larger than α in indium phosphide. This type structure, however, has been used successfully by several groups working on InGaAsP or InGaAs avalanche photodiodes to reduce the large dark tunneling currents that are frequently present in pn junctions in these materials, and further development may permit the fabrication of high-performance devices based on the idea of optimum materials for separate multiplication and absorption.

11. Performance Improvement With SAM APDs

Interest in long-wavelength (1.3–1.6 μm) detectors has been stimulated by the development of optical fibers with minimum loss and dispersion in this spectral range. Quaternary III-V semiconductor alloys are particularly suitable for infrared detector applications because they allow independent adjustment of the lattice constant and the band gap over a significant range, making it possible to build lattice-matched heterostructures. Lattice-matched conditions are very important in heterostructure devices to minimize strains and defects which otherwise would degrade device performance. The InGaAsP/InP quaternary alloy system, covering the 0.91- to 1.72-μm wavelength range, has been extensively studied. The growth and properties of this alloy system have recently been reviewed by Stillman and colleagues [29].

The ultimate performance of long-wavelength InGaAsP/InP or InGaAs/InP avalanche photodiodes is limited by fundamental material proper-

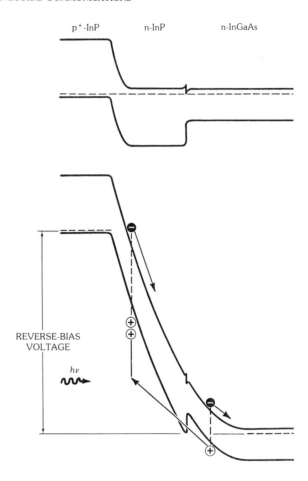

FIG. 26. Energy-band diagram of device structure with separate absorption and avalanche gain regions in equilibrium (top) and under reverse bias (bottom).

ties. A first limitation is related to the large tunneling currents associated with the narrow band gap required for long-wavelength optical absorption. Band-to-band or defect-assisted tunneling currents in the narrow–band-gap InGaAsP alloy become too large before the electric field is high enough to obtain any siginificant avalanche gain [29]. This problem can be substantially alleviated using a separate absorption and multiplication (SAM) device structure described above which was first proposed by Nishida and coworkers [30]. The basic features of SAM APDs have been described by Susa and colleagues [31] and are reviewed below.

The performance of SAM avalanche photodiodes is strongly determined by the impact ionization coefficient ratio in the wide–band-gap material. Thus the performance of InGaAsP/InP avalanche photodiodes is limited by the fact that, in indium phosphide the ionization coefficient ratio β/α is close to unity over the entire range of useful electric-field values. The measured values of α and β in (100) InP as a function of the electric field are given in Fig. 23, and although the dependence of the ionization coefficient ratio on the electric field is rather small, it still leaves some room to optimize device performance, using appropriate device configurations.

In this section we will examine the limits imposed by material properties and device configuration on the performance of InGaAsP/InP and InGaAs/InP APDs. The experimental values of α and β in Fig. 23 can be used to calculate multiplication and excess-noise characteristics, optimum avalanche gain, minimum detectable optical power, and maximum allowable dark current, for different device configurations. By comparing the performance of avalanche photodiodes with that of unity-gain pin photodiodes, the improvement that can be achieved through the use of APDs can be examined.

The SAM structures are an effective way to reduce the large tunneling currents which occur in pn junctions in narrow–band-gap semiconductor alloys. The incident radiation is absorbed in the narrow–band-gap ternary layer, where the electric field is kept low enough to minimize tunneling currents but high enough to obtain high quantum efficiency, while multiplication takes place in the high-field region of the wide–band-gap indium-phosphide material. Therefore the quantum efficiency of a SAM APD is determined by the properties of the ternary layer and of the InGaAsP/InP hetero-interface, while the multiplication and excess-noise properties are determined by the avalanche multiplication process in indium phosphide. Proper device operation critically depends on the thickness and doping level of the wide–band-gap semiconductor, so these two factors have to be adjusted in such a way that the electric field in the narrow–band-gap material never exceeds the value which causes significant tunneling current. The structure of the device shown in Fig. 26 will provide hole injection into the multiplication layer for wavelengths longer

than that corresponding to the band gap of indium phosphide, but shorter than that corresponding to the narrow–band-gap alloy.

High-speed operation of SAM APD devices can be limited by a slow component of the photoresponse which has been associated with a pileup of the photogenerated holes in the valence-band discontinuity at the InGaAs/InP interface [33]. The slow-response component results from the thermal emission of the holes trapped at the interface, when the electric field or the temperature is not sufficiently high. It was also suggested that this effect could be reduced using a compositionally graded region between InGaAs and InP over a length of 600 to 1000 μm. High-speed operation of SAM APD devices which incorporate an intermediate–band-gap InGaAsP grading layer between the InP multiplication layer and the InGaAs absorption layer has recently been demonstrated [34]. A properly graded InGaAs/InP interface will be assumed in this study, so that injection of holes in the multiplication region is not limited by the valence-band discontinuity, even at very low values of the electric field.

Another device structure with separate absorption and multiplication regions has recently been proposed [35]. Its operation has been demonstrated in the MBE-grown $Al_{0.48}In_{0.52}As/Ga_{0.47}In_{0.53}As$ material system [36], but the same structure could be implemented in the InGaAsP/InP system. This new structure, called HI-LO SAM APD by its authors, makes use of a steplike electric-field profile, defining a high-field zone and a low-field zone in the wide–band-gap material. The structure and electric-field profile for the conventional SAM APD and the HI-LO SAM APD are shown in Fig. 27a and 27b, respectively. The multiplication process in this device is confined to the high-field zone of the wide–band-gap material where the electric field has an uniform spatial distribution. The steplike field profile is achieved by introducing a thin doping spike in the ultralow-doped wide–band-gap layer. The doping spike is located a distance (about 0.2 μm) away from the narrow-band-gap absorbing layer to prevent the formation of defects at the hetero-interface by high doping densities. The width and doping of the spike are adjusted so that the electric field at the hetero-interface does not exceed the value which causes a significant tunneling current in the narrow–band-gap material. Since the magnitude of the electric field at the interface is adjusted by a doping spike rather than by the doping concentration over the entire indium-phosphide layer as in the conventional SAM structure, the HI-LO SAM APD should have better manufacturing tolerance to simultaneously obtaining high multiplication and low tunneling current. In addition, the HI-LO SAM structure is expected to produce lower excess-noise factors because of its uniform electric-field profile which can provide the required avalanche gain at lower electric fields where the ratio β/α is larger.

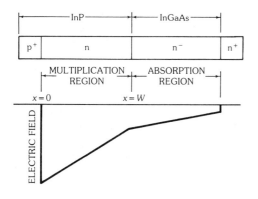

(a) For conventional SAM APD.

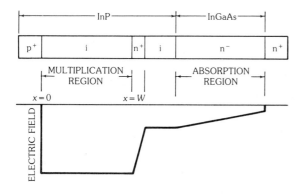

(b) For HI-LO SAM APD.

FIG. 27. Structure and electric-field profile.

Signal-to-Noise Ratio and Optimum Avalanche Gain

For a 100% sinusoidally modulated optical signal with average power P_0, the signal-to-noise ratio (snr) over a bandwidth B at the output of a photodetector consisting of an avalanche photodiode, with multiplication M, excess-noise factor F, and an interaction circuit characterized by a noise temperature T_{eff} and an equivalent load resistance R_{eq}, is given by [3]

$$\text{snr} = (1/2)\,(q\,\eta\,P_0/h\,\nu)^2\,M^2$$

$$/(2\,q\,I_p\,F\,M^2\,B + 4\,k\,T_{\text{eff}}\,B/R_{\text{eq}})\,, \quad (62)$$

where k is Boltzmann's constant, if the shot-noise contributions of the background and dark multiplied and unmultiplied (bulk and surface) currents can be neglected compared to the shot noise associated with the average photocurrent

$$I_p = q\,\eta\,P_0/h\,\nu\,, \qquad (63)$$

where q is the electron charge, η the quantum efficiency, h Planck's constant, and ν the frequency of the incident radiation.

From (58) it is apparent that the use of an avalanche photodiode instead of a unity-gain pin detector results in an improvement of the signal-to-noise ratio as long as the thermal noise of the interaction circuit dominates the detector shot noise. In this case the snr increases with M until the multiplied shot noise overrides the thermal noise. When that happens, further increase of the multiplication results in lower snr values, because the excess-noise factor F is a monotonically increasing function of M. In consequence, an optimum value of the avalanche gain exists which maximizes the snr for a given average incident power P_0.

Using (41) the following expression can be obtained for the minimum average optical power per unit bandwidth (P_0/B) required to achieve a given snr:

$$P_0/B = (2\,h\,\nu\,F/\eta)\,(\text{snr})\{1 + [1 + 4\,\pi\,k\,T_{\text{eff}}\,C_{\text{eq}}$$

$$/q^2\,F^2\,M^2\,(\text{snr})]^{1/2}\}\,. \quad (64)$$

Here, C_{eq} is the equivalent capacitance which, along with R_{eq}, determines the bandwidth B of the system as

$$B = 1/2\,\pi\,R_{eq}\,C_{eq}\,. \qquad (65)$$

Since the dark-current contribution to the shot noise has been neglected in these expressions, it is important to have a means to determine when the predicted performance will be degraded by the dark current in particular cases. One way of doing this is to determine the value of the bulk dark current per unit bandwidth (I_{dm}/B) which causes a 3-dB increase in the minimum optical power (P_0/B) required to obtain a given snr. Assuming that the excess-noise factor for the multiplied bulk dark current is the same as that for the photocurrent, this value of the dark current is given by

$$I_{dm}/B = q\,F(\text{snr})\,\Big(1 + 4\,\{1 + [4\,\pi\,k\,T_{eff}/q^2]$$

$$\times [C_{eq}/F^2\,M^2\,(\text{snr})]\}^{1/2} + 3\{1 + [4\,\pi\,k\,T_{eff}/q^2]$$

$$\times [C_{eq}/F^2\,M^2\,(\text{snr})]\}\Big)\,. \qquad (66)$$

Multiplication and the Excess-Noise Factor

While the preceding analysis applies to any type of avalanche photodiode, multiplication and excess-noise behavior are dependent on device geometry and material. To evaluate the potential improvement obtainable with SAM APDs, we will assume that the multiplication process is confined to the n-InP layer in conventional SAM structures, and to the high-field zone of the n-InP layer in HI-LO SAM devices (see Figs. 27a and 27b). The electric field varies linearly with position across the multiplication layer of the conventional SAM device, with a slope determined by the uniform doping concentration N_D-N_A. The magnitude of the electric field in the indium phosphide is a maximum at the pn junction plane $(x=0)$ and minimum at $x = W$, where the depletion region reaches through the hetero-interface. Because of this spatial variation of the electric field the ionization coefficients α and β are also functions of position across the multiplication layer of the SAM device. On the other hand, the electric field and α and β are nearly constant in the ultralow-doped multiplication region of the HI-LO SAM structure shown in Fig. 27b.

Assuming hole injection at $x = W$ in a multiplication region extending from $x=0$ to $x=W$,

where the ionization coefficients $\alpha(x)$ and $\beta(x)$ are position dependent in the most general case, the resulting avalanche gain M is given by (57). The corresponding excess-noise factor F is given by the following expression, derived by McIntyre [37]:

$$F = K'_{eff}\,M + (2 - 1/M)(1 - K'_{eff})\,, \qquad (67)$$

in which

$$K'_{eff} = K_{eff}/K_1{}^2\,,$$

$$K_{eff} = (K_2 - K_1{}^2)/(1 - K_2)\,, \qquad (68)$$

$$K_1 = \int_0^W \beta(x)\,M(x)\,dx \Big/ \int_0^W \alpha(x)\,M(x)\,dx\,, \qquad (69)$$

and

$$K_2 = \int_0^W \beta(x)\,M^2(x)\,dx \Big/ \int_0^W \alpha(x)\,M^2(x)\,dx\,. \qquad (70)$$

In order to evaluate M and F using (57) and (67)–(70) it is necessary to know the dependence of α and β on the magnitude of the electric field, so that $\alpha(x)$ and $\beta(x)$ can be determined for a given electric-field profile. Fig. 23 shows the measured variation of α and β with the electric field for (100) InP [32] and least-squares fits to this experimental data shown by the solid curves are given by (60) and (61), where E is the magnitude of the electric field in volts per centimeter.

Noise Performance of Conventional SAM APDs

Design of a SAM avalanche photodiode must consider that a suitable combination of width W and doping concentration N of the multiplication layer for proper device operation. There are in particular two important operating conditions to be met: (1) the electric field at the hetero-interface of the multiplication and absorption layers has to be smaller than 2.2×10^5 V/cm at breakdown [39] in order to avoid large tunneling currents in the narrow–band-gap absorption layer, and (2) the depletion region in the multiplication layer has to reach through to the absorption layer before breakdown, since failure to do so would result in poor collection efficiency and slow response time. Calculations were made to determine the doping concentration range satisfying the two conditions stated above for different multiplication layer widths and the results are shown in Table 1. This table shows the suitable doping concentration range in the multiplication region of conventional SAM structures with different multiplication layer widths. For $N < N_{min}$, device performance is limited by large tunneling currents originated in the narrow–band-gap absorption layer. For $N > N_{max}$, breakdown occurs before the depletion region reaches through the hetero-interface.

TABLE 1. Width and Doping Concentration of SAM APD

W (μm)	N_{min} (cm^{-3})	N_{max} (cm^{-3})
1	2.50×10^{16}	4.46×10^{16}
2	9.65×10^{15}	1.90×10^{16}
3	5.42×10^{15}	1.17×10^{16}
4	3.72×10^{15}	8.33×10^{15}
5	2.74×10^{15}	6.41×10^{15}
6	2.13×10^{15}	5.19×10^{15}
7	1.72×10^{15}	4.33×10^{15}

To study the noise performance of conventional SAM APDs once the doping range was determined, some particular configurations with different thicknesses W of the n-InP region shown in Fig. 27a were selected. Equations (57) through (70) were used to numerically calculate multiplication and excess-noise–factor values for a given configuration. Fig. 28 shows the calculated excess-noise factor as a function of the multiplication for four configurations having different multiplication layer widths. For each one of these configurations the doping concentration was chosen to be the average value of the corresponding N_{min} and N_{max}, so that the electric field at the InP-InGaAs interface when avalanche breakdown occurs is about 1×10^5 V/cm. It is observed that the excess-noise factor decreases as the multiplication-layer width increases. Little improvement, however, is achieved for the larger widths. The effect of changing the doping concentration with a fixed value of W was also studied. For the configuration with a 5-μm-wide multiplication layer, a 6% decrease in noise factor for $N = N_{min} = 2.74 \times 10^{15}$ cm^{-3} and a 4% increase as N was approaching $N_{max} = 6.41 \times 10^{15}$ cm^{-3} was found, with respect to noise-factor figures for $N = 4.58 \times 10^{15}$ cm^{-3}.

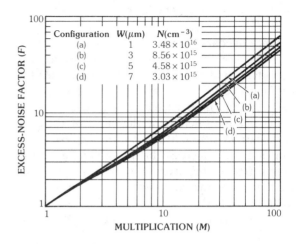

FIG. 28. Calculated excess-noise factor as a function of multiplication for conventional SAM APD configurations having different multiplication-layer widths, W.

The calculated excess-noise factors, along with (64), were used in a computer program to determine the optimum avalanche gain which results in the lowest average optical power required to obtain a given value of snr at three different capacitance values. A noise temperature of $T_{eff} = 300$ K was assumed in the computations. The results are shown in Fig. 29.

The average minimum optical power per unit bandwidth (P_0/B) required to obtain a given snr was calculated using the optimum avalanche gain results. The term P_0/B was also evaluated for a unity-gain pin photodiode. Both sets of curves are displayed in Fig. 30 for configuration (c). For the calculation of the results in Fig. 30 the following values were used: the wavelength of the optical signal, $\lambda = 1.3$ μm, the quantum efficiency $\eta = 0.7$ and the noise temperature $T_{eff} = 300$ K. Suitable rescaling will allow one to determine P_0/B for other values of λ and η. As expected, at low snr values a significant improvement in P_0/B is obtained by using an avalanche photodiode, while at high snr values there is little difference between the pin and APD detector performance. It has been shown [40] that in order to achieve a bit error rate of less than 10^{-9} in the presence of Gaussian noise, an snr of about 144 is required. These calculations indicate that more than a 10-dB improvement can be achieved in the

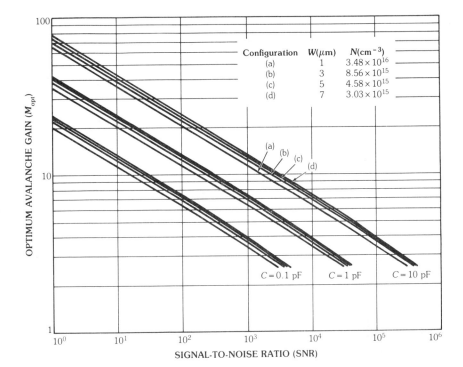

FIG. 29. Optimum avalanche gain resulting in the lowest average optical power required to obtain a given snr at three capacitances, with $T_{eff}=300$ K, for the conventional SAM APD.

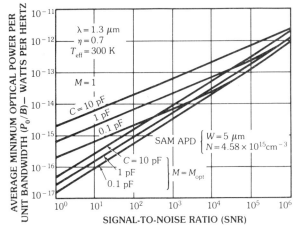

FIG. 30. Average minimum optical power per unit bandwidth required to obtain a given snr for optimum avalanche gain and for a unity-gain pin photodiode, for configuration (c) of the conventional SAM APD.

minimum optical power needed to obtain this particular value of snr when an APD is utilized.

Equation 66 was used to calculate (I_{dm}/B) for configuration (c) which when multiplied will cause a 3-dB increase in the minimum optical power required to obtain a given snr at $M=M_{opt}$. Calculations were also made for a pin ($M=1$)

photodetector and the results are shown in Fig. 31. For a 0.1-pF capacitance, 1-GHz bandwidth, and 144 snr the dark current can be as high as 0.1 mA and cause only a 3-dB increase in the minimum P_0/B required for a pin detector, while for the avalanche photodiode a bulk dark current

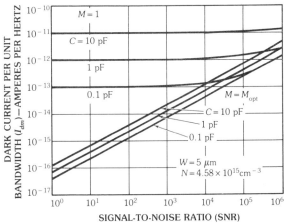

FIG. 31. Multiplied bulk dark current for configuration (c) of the conventional SAM APD with optimum avalanche gain and for pin photodetector with $M=1$, which when multiplied will cause a 3-dB degradation in the minimum optical power required to produce a given snr as given in Fig. 30.

of 1 μA will cause a similar degradation of P_0/B. With the SAM structure it is possible to obtain dark current significantly lower than these values, so that with careful design and fabrication leakage currents do not limit the performance of avalanche photodiodes for the 1.3- to 1.5-μm wavelength range.

Noise Performance of HI-LO SAM APDs

Excess-noise factor versus multiplication and optimum avalanche gain were also calculated for a number of HI-LO SAM configurations with different multiplication-layer widths. The results are shown in Figs. 32 and 33. For equal multiplication-layer widths the excess-noise factors of the HI-LO SAM structures are about 5 to 7% lower than the corresponding excess-noise values for the conventional SAM devices analyzed above.

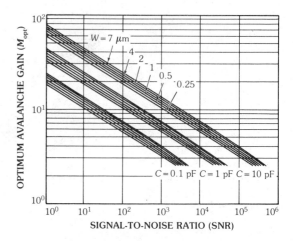

Fig. 33. Optimum avalanche gain as a function of snr for different multiplication-layer widths in the HI-LO SAM APD structure.

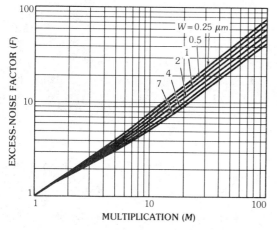

Fig. 32. Excess-noise factor as a function of multiplication for various multiplication-layer widths in the HI-LO SAM APD structure.

Multiplication Versus Voltage Characteristics

As discussed previously, besides limiting noise performance, the near-unity impact ionization coefficient ratio in indium phosphide is also responsible for the sharpness of multiplication versus bias voltage (M-V) characteristics in APD detectors. As a consequence, even moderate gain operation of avalanche photodiodes is limited by practical considerations, including stability requirements for the bias voltage source and a high probability for unstable or nonuniform avalanche processes taking place.

A way to obtain more slowly varying M-V characteristics is to use a thick, lightly doped absorption layer, such that the applied bias voltage is shared between the absorption and the multiplication layers, reducing the corresponding variation of the electric field in the multiplication layer and the sharpness of the M-V curve. This effect is illustrated in Fig. 34, where several M-V characteristics at different doping concentrations of the ternary layer have been plotted for the conventional SAM

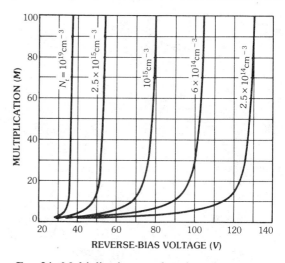

Fig. 34. Multiplication as a function of reverse-bias voltage at different doping concentrations of the ternary layer for conventional SAM configurations, in which there is negligible multiplication before the depletion region reaches through to the ternary absorption layer.

configurations with the width and doping of the multiplication region given by $W=1$ μm and $N=3.48\times10^{16}$ cm^{-3}, respectively. The lower the doping concentration of the ternary layer, the softer the M-V characteristics are, but also the larger the bias voltage required. For these structures there is no avalanche gain at the electric field where the depletion region reaches through the indium phosphide to the ternary.

The sharpness of the M-V characteristic can be reduced with a smaller voltage penalty by further modifying the structure. In considering a conventional SAM structure where the multiplication-layer width and doping concentration have been chosen in such a way that the depletion region reaches through the hetero-interface when the gain is already larger than unity, the electric field in the multiplication region, and hence the gain, will start increasing less rapidly with voltage, provided that the doping concentration in the ternary layer is low enough. This effect is shown in Fig. 35 for a SAM structure, where the required doping concentration of the 1-μm–wide multiplication layer was determined to be about 4.3×10^{16} cm^{-3}, for the depletion region to reach-through the InP-InGaAs interface when the gain is about 10. Each of the five plots corresponds to a different doping concentra-

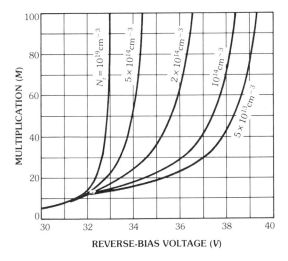

FIG. 35. Reduced sharpness of the multiplication/voltage characteristics for conventional SAM structures which have an avalanche gain of about 10 when the edge of the depletion region reaches through the ternary absorption layer with the doping level indicated.

tion of the absorption layer, which was assumed to be 5 μm wide. Obviously, the same method can be applied to the HI-LO SAM structure, the only difference being that the gain value at reach-through is controlled in this case by the width and doping concentration of the doping spike.

12. New APD Structures

The above discussion of the InGaAs/InP SAM avalanche photodiode indicates that significant improvements in wide-bandwidth system performance can be obtained with this device structure even though the excess-noise factor is large because of the near-unity ratio of the electron and hole impact ionization coefficients in indium phosphide. There is a large amount of development work around the world to enable the routine production of these devices for long-wavelength applications. There is also considerable interest, however, in developing new materials and/or new structures which have the potential for even higher performance.

The major limitation on the performance of the SAM avalanche photodiodes is the near-unity ratio of the electron and hole impact ionization coefficients in indium phosphide. This same limitation seems to apply to other III-V compounds that have potential for long-wavelength optical waveguide communications. Because of the limitations imposed by bulk material properties there has been considerable interest in using the properties of artificially structured materials or multiple-layer heterojunction structures to obtain devices in which there is a large difference in the electron and hole impact ionization coefficients, even though these quantities are nearly equal in the bulk material used to form the multilayered heterojunction structures. The first proposal for such a device structure utilized the influence of the band-edge discontinuities on the transport of electrons and holes through the heterostructure. The basic structure proposed in [41] used AlGaAs/GaAs as an example and is shown in Fig. 36. The basic physical idea in this structure involves the influence of the different band-edge discontinuities in the conduction and valence bands on the electrons and holes, respectively.

The impact ionization rate depends exponentially on the impact ionization threshold and on the energy from which the electron (hole) starts to be accelerated. Therefore a steplike band structure as shown in Fig. 36, where the discontinuity in the conduction band is greater than that in the valence band, will enhance the ionization rate for electrons. This can easily be seen by subtracting (adding) the conduction band-edge step to the ionization threshold in the Baraff theory [42]. A rigorous justification for this procedure can be given for special cases only, because in general the effect of the band-edge step on the ionization rate will depend on the energy which the electron already has when it approaches the junction.

The scattering rate in quantum wells is different from that in the bulk and, at low energies at least, holes are scattered more often and thus are collected more effectively in the quantum wells as shown by Holonyak and coworkers [43, 44]. An analysis of this effect is difficult because it must include the reflection and transmission through the (nonideal) heterojunction. However, inserting different mean free paths in the Baraff theory shows immediately that this effect can strongly change the ratio of a α and β if it prevails up to the energy of the impact ionization threshold. Based on a structure appropriate for GaAs-Al$_x$Ga$_{1-x}$As layers, with the material constants adjusted to give α/β when inserted in the Baraff theory for bulk material, calculations of the enhancement of α/β have been made. A polynomial fit to the Baraff theory is used in the

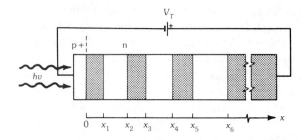

(a) Shaded regions, GaAs; unshaded regions, Al$_x$Ga$_{1-x}$As.

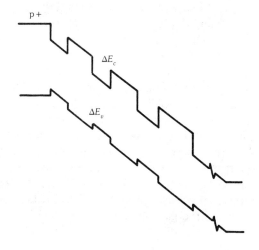

(b) Energy band diagram.

FIG. 36. Basic structure of APD with quantum-well layers. (*After Chin et al. [41], ©Institution of Electrical Engineers*)

TABLE 2. Material Constants

	Threshold Ionization Energy E_I (eV)	Optical Phonon Mean Free Path λ (nm)	Impact Ionization Mean Free Path d (nm)	Band-Edge Discontinuity ΔE (eV)
Electron	2.0	5	15	0.6
Hole	1.5	4	15	0.106

calculations [42]. The values of the material constants are listed in Table 2.

The p$^+$n junction is assumed to be steplike with uniform doping concentrations on each side. The electric field accelerating the electrons and causing impact ionization has three sources: the external voltage, the built-in voltage, and the band-edge step of the heterojunction. The first two are taken into account in the usual way by inserting space-dependent electric fields in the Baraff theory. As mentioned above, the band-edge discontinuity has to be treated in a different way because the band-edge energy changes over a distance which is much shorter than the mean free path between collisions. To assess the effect of the band-edge step on the ionization rate, consider an electron that travels from one $Al_xGa_{1-x}As$ layer into a gallium-arsenide layer. When it arrives in the gallium arsenide it "sees" not only the smaller band-gap of the gallium arsenide, but it also starts at an energy ΔE_c (conduction band edge step) above the gallium-arsenide band edge. To include this effect the "excess" energy ΔE_c was subtracted from the gallium-arsenide impact ionization threshold within a distance from the discontinuity equal to the impact ionization mean free path. This process increases α much more than β since, when the same procedure is followed for holes, the valence band edge step ΔE_v is much smaller [43, 44].

The thickness of the layers should be chosen so that after an electron impact ionizes in the gallium-arsenide layer it can gain sufficient energy in this layer to get out of the well, and after the electron enters the $Al_xGa_{1-x}As$ layer with small kinetic energy, it gains the effective threshold energy $E_1(GaAs) - \Delta E_c$ before arriving at the next gallium-arsenide layer. $E_1(GaAs)$ is the impact ionization threshold for bulk gallium arsenide. Note that the values of the layer width x_s vary for different design parameters, i.e., doping concentration N_D, the applied voltage and the position within the structure. However, the calculated value of the thickness is always close to 30 nm for GaAs layers, and 50 nm for the $Al_xGa_{1-x}As$ layers.

The above procedure gives the α/β ratio at a particular location in the layers. The value of α/β relevant to the multiplication factor in an APD is given by the spatial average over the whole structure. The results for the average effective α/β ratio are shown in Fig. 37. As can be seen the enhancement of the α/β ratio is substantial especially for low values of the donor density N_D. The multiplication factor is in an interesting range however (i.e., of the order of ten or larger) only for the two highest doping concentrations shown in Fig. 37. The reason for this are outlined below.

The second effect which enhances α/β, the higher phonon scattering rate of holes in the heterojunction structures, has not been included in the results of Fig. 37. It is clear that this effect once more increases α/β. The mean free path for phonon scattering enters more sensitively in the impact ionization rate than the ionization threshold. The amount of the reduction of mean free

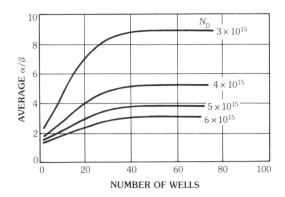

FIG. 37. Variations of α/β with number of quantum wells for an external voltage of 100 V and the net doping concentrations shown.

path is not known exactly, since the results of Holonyak and coworkers apply only for electrons with rather low energy. The hole mean free path was changed arbitrarily from 4 nm to 3 nm in the gallium-arsenide quantum wells to examine the effect. The results for the average α/β ratio are shown in Fig. 37 once more as a function of the number of wells. It is clear that the α/β ratio depends sensitively on the structure and doping concentration. This is due to the fact that α is considerably larger than β only for electric fields $E < 10^6$ V/cm. Therefore, if we choose to increase the multiplication factor, we automatically reduce the α/β ratio. Nevertheless, this analysis indicates that even for multiplication factors above 10 the α/β ratio can be substantially enhanced in quantum-well structures, especially because of the second enhancement effect, the reduction of the phonon mean free path.

There have been two experimental reports which show that indeed the ratio α/β can be enhanced by this technique. Capasso and colleagues [45] reported an enhancement of α/β to a value of 10 at an avalanche gain of 10. Juang and coworkers [46] reported value of $\alpha/\beta = 2$ to 5 for the structure described above and $\alpha/\beta > 10$ for a graded–band-gap superlattice. These early results are probably not reliable, however, because the structures used for the measurements of α and β had the same flaws as described previously for ionization coefficient measurements. The main motivation for developing structures with enhanced ratios of the impact ionization coefficients is to obtain avalanche

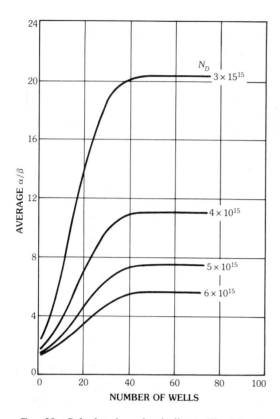

FIG. 38. Calculated results similar to Fig. 37, but with reduced optical phonon mean free paths for holes (3 nm instead of 4 nm).

photodiodes with lower excess-noise factors. Although there is one published report stating that "independent noise measurements conclusively confirm the high α/β," [47], the noise measurements have never been published. Before the effectiveness of the quantum-well heterostructure for enhanced α/β ratios can be reliably evaluated, excess-noise measurements must be correlated with the direct measurements of α and β. This is especially true in light of the sensitiveness of the calculated α/β ratio to the mean free path for phonon scattering of holes and the expected sensitivity to the abruptness and defect nature of the heterojunction interfaces, as well as other work which had indicated no enhancement of the α/β ratio in these structures [48].

The special structure described above [41] is still of considerable interest, but there have been many other suggestions for structures that will enhance the α/β ratio obtainable in bulk semicon-ductor materials. Many of these structures have been recently reviewed by Capasso [49], and the reader is referred to this reference for a review of these structures.

13. References

[1] E.H. Putley, "InSb Submillimeter Photoconductive Detectors," in *Semiconductors and Semimetals*, ed. by R. K. Willardson and A. C. Beer, Vol. 12, New York: Academic Press, 1966, pp. 143–158.

[2] L. K. Anderson, M. DiDomenico Jr., and M. B. Fisher, "High Speed Photodetectors for Microwave Demodulation of Light," *Advan. Microwaves*, Vol. 5, p. 1335, 1970.

[3] G. E. Stillman and C. M. Wolfe, "Avalanche Photodiodes," in *Semiconductors and Semimetals*, ed. by R. K. Willardson and A. C. Beer, Vol. 12, New York: Academic Press, 1977, pp. 291–393.

[4] L. K. Anderson and B. J. McNurtry, "High Speed Photodetectors," *Proc. IEEE*, Vol. 54, p. 1335, 1966.

[5] G. Beni and F. Capasso, "Effect of Drift Velocities on Measured Ionization Coefficients in Avalanching Semiconductors," *Phys. Rev.*, B, Vol. 19, p. 2197, 1979.

[6] K. K. Thornber, "Applications of Scaling to Problems in High Field Electronic Transport," *J. Appl. Phys.*, Vol. 52, p. 279, 1981.

[7] N. R. Howard, "Avalanche Multiplication in Silicon Junctions," *J. Electron. Contr.*, Vol. 13, p. 537, 1962.

[8] C. A. Lee, R. A. Logan, R. L. Batdorf, J. J. Kleimack and W. Wiegmann, "Ionization Rates of Holes and Electrons in Silicon," *Phys. Rev.*, Vol. 134, p. A761, 1964.

[9] J. L. Moll, *Physics of Semiconductors*, New York: McGraw-Hill Book Co., 1964, p. 225.

[10] R. J. McIntyre, "Multiplication Noise in Uniform Avalanche Diodes," *IEEE Trans. Electron. Dev.*, Vol. ED-13, p. 164, 1966.

[11] H. W. Ruegg, "A Fast High-Gain Silicon Diode," *IEEE Intl. Solid-State Circuits Conf. Dig.*, p. 56, 1966.

[12] W.N. Shaunfield and D. W. Boone, *1968 Solid State Sensors Symp., Minneapolis, Conf. Record*, p. 1281, Inst. of Electron. and Electron. Eng., New York, 1968.

[13] P. P. Webb, R. J. McIntyre, and J. Conradi, "Properties of Avalanche Photodiodes," *RCA Rev.,* Vol. 35, p. 234, 1974.

[14] R. J. McIntyre, IEEE Intl. Electron. Devices Meeting, Washington, D.C., 1973, *Tech. Digest,* p. 213, 1973.

[15] J. Conradi, "Temperature Effects in Silicon Avalanche Diodes," *Solid-State Electron.,* Vol. 17, p. 99, 1974.

[16] P. P. Webb and R. J. McIntyre, "A Silicon Avalanche Photodiode for 1.06 μm Radiation," *1970 Solid State Sensors Symp., Minneapolis, Conf. Record,* p. 82, Inst. of Electron. and Electron. Eng., New York, 1970.

[17] P. P. Webb and A. R. Jones, "Large Area Reach-Through Avalanche Diodes for Radiation Monitoring," *IEEE Trans. Nucl. Sci.,* Vol. NS-21, p. 151, 1974.

[18] R. van Overstraeten and H. DeMan, "Measurement of the Ionization Rates in Diffused Silicon p-n Junctions," *Solid-State Electron.,* Vol. 13, p. 583, 1970.

[19] O. Mikami, H. Ando, H. Kanbe, T. Mikawa, T. Kaneda, and Y. Toyam, "Improved Germanium Avalanche Photodiodes," *IEEE J. Quantum Electron.,* Vol. QE-16, p. 1002, 1980.

[20] T. Mikawa, S. Kagawa, T. Kaneda, T. Lakurai, H. Ando, and O. Mikami, "A Low-Noise $n^{+}np$ Germanium Avalanche Photodiode," *IEEE J. Quantum Electron.,* Vol. QE-17, p. 210, 1981.

[21] A. G. Gynoweth, "Charge Multiplication Phenomena," in *Semiconductors and Semimetals,* ed. by R. K. Willardson and A. C. Beer, Vol. 4, New York: Academic Press, 1968, pp. 263–323.

[22] G. E. Stillman, "Ionization Coefficients in GaAs and Related Compounds," *Intl. Symp. on Gallium Arsenide and Related Compounds, Edinburgh* (1976), *Inst. Phys., Conf. Ser. Vol. 33a,* London, 1977, pp. 185–209.

[23] G. E. Bulman, L. W. Cook, and G. E. Stillman, "The Effect of Electroabsorption on the Determination of Ionization Coefficients," *Appl. Phys. Lett.,* Vol. 39, p. 813, 1981.

[24] G. E. Bulman, V. M. Robbins, K. F. Brennan, K. Hess, and G. E. Stillman, "Experimental Determination of Impact Ionization Coefficients in (100) GaAs," *IEEE Electron. Dev. Lett.,* Vol. EDL-4, p. 181, 1983.

[25] L. W. Cook, G. E. Bulman, and G. E. Stillman, "Electron and Hole Impact Ionization Coefficients in InP Determined by Photomultiplication Measurements," *Appl. Phys. Lett.,* Vol. 40, p. 589, 1982.

[26] N. Tabatabaie, V. M. Robbins, N. Pan, and G. E. Stillman, "Impact Ionization Coefficients in (111) InP," *Appl. Phys. Lett.,* Vol. 46, p. 182, 1985.

[27] C. A. Armiento and S. H. Groves, "Impact Ionization in (100)-, (110)-, and (111)-Oriented InP Avalanche Photodiodes," *Appl. Phys. Lett.,* Vol. 43, p. 198, 1983.

[28] K. Brennan and K. Hess, "Theory of High-Field Transport of Holes in GaAs and InP," *Phys. Rev.,* B, Vol. 29, p. 5581, 1984.

[29] G. E. Stillman, L. W. Cook, N. Tabatabaie, G. E. Bulman, and V. M. Robbins, "InGaAsP Photodiodes," *IEEE Trans. Electron. Dev.,* Vol. ED-30, pp. 364–381, 1983.

[30] K. Nishida, K. Taguchi, and Y. Matsumoto, "InGaAsP Heterostructure Avalanche Photodiodes with High Avalanche Gain," *Appl. Phys. Lett.,* Vol. 35, pp. 251–253, 1979.

[31] N. Susa, H. Nakagome, O. Mikami, H. Ando, and H. Kanbe, "New InGaAs/InP Avalanche Photodiode Structure for the 1–1.6 μm Wavelength Region," *IEEE J. Quantum Electron.,* Vol. QE-14, pp. 864–870, 1980.

[32] L. W. Cook, G. E. Bulman, and G. E. Stillman, "Ionization Coefficient Determination in InP by Analysis of Avalanche Photomultiplication and Noise Measurements," in *Inst. Phys. Conf. Ser. No. 63:* Chapter 6, pp. 281–286, 1982.

[33] S. R. Forrest, O. K. Kim, and R. G. Smith, "Optical Response Time in $In_{0.53}GA_{0.47}As/InP$ Avalanche Photodiodes," *Appl. Phys. Lett.,* Vol. 41, pp. 95–98, 1982.

[34] J. C. Campbell, A. G. Dentai, W. S. Holder, and B. L. Kasper, "High Performance Avalanche Photodiode With Separate Absorption 'Grading' and Multiplication Regions," *Electron. Lett.,* Vol. 19, pp. 818–820, 1983.

[35] F. Capasso, K. Alavi, A. Y. Cho, P. W. Foy, and C. G. Bethea, "Long Wavelength, Wide Spectral Response (0.8–1.8 μm) $Al_{0.48}In_{0.52}As/Ga_{0.47}In_{0.53}As$ Avalanche Photodiodes and $Al_{0.48}In_{0.52}As$ Electroabsorption p-i-n Avalanche Detectors Grown by Molecular Beam Epitaxy," *Proc. IEEE Intl. Electron. Dev. Mtg.,* pp. 468–471, 1983.

[36] F. Capasso, A. Y. Cho, and P. W. Foy, "Low-Dark-Current Low-Voltage 1.3-1.6 μm Avalanche Photodiode With High-Low Electric Field Profile and Separate Absorption and Multiplication Regions by Molecular Beam Epitaxy," *Electron. Lett.*, Vol. 20, pp. 635–637, 1984.

[37] R. J. McIntyre, "The Distribution of Gains in Uniformly Multiplying Avalanche Photodiodes: Theory," *IEEE Trans. Electron. Dev.*, Vol. ED-19, pp. 703–713, 1972.

[38] G. E. Stillman, N. Tabatabaie, V. M. Robbins, and J. A. Aguilar, "III-V Compound Semiconductor Optical Detectors," Intl. Symp. on GaAs and Related Compounds, Biarritz, 1984.

[39] G. E. Stillman, L. W. Cook, G. E. Bulman, N. Tabatabaie, R. Chin, P. D. Dapkus, "Long-Wavelength (1.3 to 1/6 μm) Detectors for Fiber-Optical Communication," *IEEE Trans. Electron. Dev.*, Vol. ED-29, pp. 1355–1371, 1982.

[40] G. E. Stillman, "Design Considerations for Fiber-Optic Detectors," *Proc. SPIE*, Vol. 239, pp. 42–52, 1980.

[41] R. Chin, H. Holonyak Jr., G. E. Stillman, J. Y. Tang, and K. Hess, "Impact Ionisation in Multilayered Heterojunction Structures," *Electron. Lett.*, Vol. 16, pp. 467–468, 1980.

[42] C. R. Crowell and S. M. Sze, "Temperature Dependence of Avalanche Multiplication in Semiconductors," *Appl. Phys. Lett.*, Vol. 9, pp. 242–244, 1966.

[43] H. Shichijo, K. M. Kolbas, N. Holonyak, Jr., R. D. Dupuis, and P. D. Dapkus, "Carrier Collection in a Semiconductor Quantum Well," *Solid State Commun.*, Vol. 27, pp. 1029–1032, 1978.

[44] N. Holonyak, Jr., R. M. Kolbas, R. D. Dupuis, and P. D. Dapkus, "Quantum-Well Heterostructure Lasers," *IEEE J. Quantum Electron.*, Vol. QE-16, pp. 170–186, 1980.

[45] F. Capasso, W. T. Tsang, A. L. Hutchinson, and G. F. Williams, "Enhancement of Electron Impact Ionization in a Superlattice: A New Avalanche Photodiode With a Large Ionization Rate Ratio," *Appl. Phys. Lett.*, Vol. 40, pp. 38–40, 1982.

[46] F-Y. Juang, U. Das, Y. Nashimoto, and P. K. Bhattacharya, "Electron and Hole Impact Ionization Coefficients in GaAs-$Al_x Ga_{1-x}As$ Superlattices," *Appl. Phys. Lett.*, Vol. 47, pp. 972–974, 1985.

[47] F. Capasso, W. T. Tsang, A. Hutchinson, and G. F. Williams, "Enhancement of Electron Impact Ionization in a Superlattice: A New Avalanche Photodiode With a Large Ionization Rate Ratio," *Inst. Phys. Conf. Ser. No. 63*, Chapter 12, pp. 569–570, 1981.

[48] N. Susa and H. Okamoto, Musashino Electrical Communication Laboratory, NTT, unpublished, private communication.

[49] F. Capasso, "New Superlattice Structures and Heterojunction Devices by Band-Gap Engineering," in *Gallium Arsenide Technology*, ed. by David K. Ferry, Indianapolis: Howard W. Sams & Co., 1985, pp. 303–330.

Receiver Design
of
Optical-Fiber Systems

T. V. MUOI

PlessCor Optronics

1. Introduction

The receiver is a critical part of an optical-fiber communication system since it often dictates the overall system performance. The function of the receiver is to detect the incident optical power (by a photodetector) and extract from it the information that is being transmitted. In a digital communication system the receiver output consists of the regenerated data and normally the recovered clock as well. In an analog transmission system the receiver must also demodulate the detected signal to obtain the original transmitted message.

The receiver must achieve the above function under a number of requirements which relate to system performance. Of primary importance is the receiver sensitivity, which determines the minimum incident optical power required at the receiver in order to satisfy a specified value of bit error rate (for digital systems) or signal-to-noise ratio and signal-to-distortion ratio (for analog systems). Other practical requirements for the optical receiver include a wide input dynamic range, capability to accept unrestricted data format, fast acquisition time, capability of being adaptive to multiple or variable bit-rate operation, low power consumption, low cost, and so on. The above requirements are often conflicting.

A complete understanding of the optical receiver and system operation, its noise and its degradation sources is necessary to examine all the trade-offs involved in the receiver design because of the above conflicting requirements.

The photodetector used in optical-fiber communication systems can be either a pin or avalanche photodiode (APD). Avalanche photodiodes are more sensitive (i.e., can detect weaker light levels) than pin photodiodes because of their internal avalanche gain. Much effort has been devoted to the APD signal and noise characterization since it determines the ultimate limit on receiver sensitivity. The avalanche excess-noise power was theoretically derived in 1966 by McIntyre [1] and experimentally confirmed in 1966 and 1967 by Baertsch [2, 3] and Melchoir and colleagues [4] among others. However, a knowledge of the noise power is not enough because the APD avalanche gain distribution is highly non-Gaussian. The complete APD characterization came with the theoretical derivation of the APD gain distribution by McIntyre [5] of RCA Laboratories and by Personick [6, 7] and Mazo and colleagues [8] of Bell Laboratories, and with experimental verification by Conradi [9], also of RCA Laboratories.

The receiver-sensitivity analysis of optical receivers is more involved than traditional communication theory, which mainly treats signal detection with additive Gaussian noise. The complication is due to the quantum nature of the photodetection effect creating a signal-dependent and time-variant Poissonian noise process. Thus there exists an upper limit for the receiver sensitivity (referred to as the quantum limit) even in the theoretical case of perfect transmission medium and noiseless receiver. The problem is much further complicated by the non-Gaussian distribution of the avalanche gain as discussed previously.

In his pioneering work Personick [10] formulated the theoretical sensitivity analysis for digital optical receivers which has been used extensively to date. To simplify the analysis the Gaussian approximation for the APD gain distribution was used. Personick's analysis was later extended to multilevel operation by Muoi and Hullet [11]. A simplified form of Personick's theory was also developed by Smith and Garrett [12]. The validity of the Gaussian approximation was checked with various exhaustive and more exact calculations by Personick and coworkers [13].

To maximize received sensitivity Personick [10] proposed a high-impedance optical receiver with an integrating front end. This approach was implemented and verified by Goell [14] and Runge [15]. However, the dynamic range of the high-impedance receiver is limited. The transimpedance receiver design was proposed and implemented by Hullett and Muoi [16, 17] and by Ueno and colleagues [18]. Because of its improved dynamic range and simplicity the transimpedance design has been widely used for practical repeaters.

The amplifier noise is a critical factor in determining the receiver sensitivity. Noise characterization and modeling of amplifiers with FET and bipolar-transistor front ends were investigated by Goell [19] for the high-impedance amplifier design. The analysis was later extended to transimpedance amplifiers by Hullett and Muoi [20]. Amplifier-noise modeling in various configuration was also reported by Witkowicz [21, 22].

The basics of optical receiver design have thus been fairly well understood. A great number of receivers have been implemented in optical-fiber communication systems operating worldwide. The subject of receiver design has also been reviewed in many technical papers and book chapters by Personick [23–25], Smith and Personick [26], and Muoi [27]. Most of these papers, however, discuss mainly the subject of receiver sensitivity and contain but little information on other receiver characteristics.

The purpose of this chapter is twofold. Firstly, the basis of optical receiver design will be presented in a unified theory and rigorously proved for the first time. The noise modeling and sensitivity analysis of high-impedance and transimpedance receiver amplifiers are combined into one single theory. The high-impedance design can thus be considered as a particular case of the transimpedance design. The second purpose of this chapter is to discuss the effects of receiver requirements other than sensitivity (namely wide dynamic range, bit-pattern independency, bit-rate transparency, fast acquisition time) on the receiver design. The trade-offs between these requirements are increasingly important because applications of fiber systems become more and more diversified.

This chapter is organized into twelve sections. In Section 2, important receiver characteristics are defined and their relative importance in typical fiber communication systems is indicated. The properties of photodiodes are reviewed in Section 3. The signal and noise characteristics which will be used later are emphasized. Modeling and signal and noise analysis of optical receivers are then discussed in Section 4. The design of receiver amplifiers is presented in Section 5. Both the high-impedance and transimpedance approach using either field-effect or bipolar transistors are discussed. Section 6 is devoted to optical receiver sensitivity. Both short- and long-wavelength receivers are discussed. The dependence of receiver sensitivity on various system parameters is discussed in detail. Then, in Section 7, dynamic range and its trade-off with receiver sensitivity are discussed. The requirement of bit-rate transparency and its implication in the receiver design are discussed in Section 8. Receiver acquisition time, which is important for burst mode type of communication, is discussed in Section 9. Finally, in Section 10 the analysis and design of optical receivers for analog applications are considered. Section 11 gives some concluding remarks on the chapter. Section 12 lists the references cited in the chapter.

2. Important Receiver Characteristics

A block diagram of a typical digital optical receiver is shown in Fig. 1. The incident optical power can be detected by either a pin or an avalanche photodiode. The photodiode is followed by a low-noise preamplifier (for high detection sensitivity), an automatic gain control (agc) main amplifier (to accommodate different incident optical power levels), and a shaping filter. The amplifier chain including the shaping filter is normally referred to as the *receiver linear channel*. The regenerator then samples the detected signal and regenerates the original data that is being transmitted. The purpose of the shaping filter is to minimize the effects of noise and intersymbol interference at the regenerator input. The clock required for sampling is recovered by the timing extraction circuitry, which can either be a phase-locked-loop, a SAW filter, or an *LC* tank circuit. Automatic gain control is achieved by controlling the amplifier gain and the avalanche gain if an APD is used.

Of most importance to the optical receiver are the photodetector and the following low-noise preamplifier. Together, these two elements dictate many of the receiver characteristics and its performance. In addition, they distinguish the optical receiver against the traditional receiver for coaxial cable and microwave transmission systems. The rest of the circuitry is more or less standard and has been implemented in many operational communication systems.

As the applications of optical-fiber communication systems become more diversified, the optical receivers have to meet many different requirements. In the rest of this section we will discuss important receiver characteristics and their relative importance in typical applications.

Receiver Sensitivity

The receiver sensitivity is a measure of the minimum optical power level required at the receiver input so that it will operate reliably with a bit error rate (BER) less than a desired value. It is often given as the average incident optical power $\bar{P}$ required for a BER of 10^{-9} and is often expressed in the unit of dBm (0 dBm=1 mW). The photodiode has a quantum efficiency η, which is the ratio of the number of primary photoelectrons being generated and the incident photons. The optical power that is directly converted into photoelectrons is thus $\eta\bar{P}$. This optical power level is referred to as the *average detected optical power* and is also used to describe the receiver sensitivity. It is a measure of the incident optical power required if the photodiode quantum efficiency is 100%.

The measurement of receiver sensitivity in terms of the detected optical power $\eta\bar{P}$ is popular because of several reasons. Firstly, it can be measured very accurately by simply monitoring the detected photocurrent (thus no optical measurement is involved). Secondly, this sensitivity measurement excludes the quantum efficiency of the photodiode. Thus it is very useful in comparing the sensitivity performances of various low-noise preamplifier designs. In this chapter the receiver sensitivity is quoted in terms of the average detected optical power $\eta\bar{P}$.

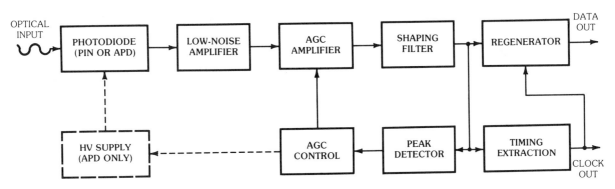

FIG. 1. Block diagram of a typical digital optical receiver.

The receiver sensitivity is a function of the signal and noise performance of both the photodetector and the following preamplifier. It is noted that the receiver sensitivity is only a measure of the operating limit of the optical receiver. In practical systems the optical receiver is rarely, if never at all, operated at its sensitivity limit. This is because some system margin (normally 3 to 6 dB) has to be allowed to guard against system degradation due to temperature variations, manufacturing tolerances, component aging, and the like.

A high receiver sensitivity is essential for achieving the maximum spacing between the two transmission terminals or between repeaters when they are used. For applications such as submarine telecommunication systems it is desirable to minimize the total number of repeaters that have to be placed in deep seawater for obvious reliability and maintainability reasons. Thus the requirement of high receiver sensitivity is very critical. In contrast, the repeater spacings for terrestrial communication systems and data networks may be predetermined by existing regenerator sites (e.g., existing central offices). The high receiver sensitivity requirement is less critical in these applications.

Receiver sensitivity is a fundamental issue of optical receiver design and a great part of this chapter will be devoted to its theoretical analysis as well as practical achievements.

Dynamic Range

In practical systems the optical receiver has to operate not only at the minimum allowable optical power level but also at optical power levels which can be significantly higher. This wide range of received optical power levels is caused by variations in repeater spacings, fiber losses, connector and splice losses, transmitter output changes with temperature, and aging. The minimum allowable received optical power is determined by the receiver sensitivity. The maximum allowable power is dictated by nonlinear distortion and saturation characteristics of the optical receiver. The difference (in decibels) between these two levels is referred to as the *receiver dynamic range*. In this definition the receiver dynamic range is expressed in optical decibels. The electrical dynamic range of the receiver will be twice as large as this.

Suppose the optical receiver is operating at its sensitivity limit with a desired BER of 10^{-9}. As the received optical power increases from this minimum level, the BER decreases because a higher signal-to-noise ratio is obtained. This improved performance continues until saturation or overloading occurs (normally in the front-end preamplifier). At this point the received signal waveform becomes distorted and the BER starts to increase due to intersymbol interference.

The maximum allowable optical level at the receiver input is sometimes defined to be the level where the measured BER starts to be worse than the desired value (normally 10^{-9}). However, as defined, the receiver dynamic range includes some nonlinear operation of the amplifier at the high-power-level end. This nonlinear operation is difficult to predict and may not be repeatable between different amplifiers built from the same design. Thus a more reliable (but more conservative) definition of the maximum allowable received optical power level is when the amplifier output signal waveform starts to become distorted (end of linear operation).

A wide receiver dynamic range is desirable since it allows flexibility and convenience in system configuration. The ability to accommodate a wide range of optical power levels means that the same receiver can be used for both short and long repeater spacings. This requirement is important in terrestrial communication systems because of existing repeater sites as discussed previously. The wide dynamic range requirement is also very critical in local-area network applications, where the transmitting source may be at different distances from the receiver and the transmitting optical power may have to go through different number of optical couplers and splitters before reaching the receiving end.

Bit-Rate Transparency

Bit-rate transparency refers to the ability of the optical receiver to operate over a range of bit rates. In applications such as data bases and local-area networks, the bit rates are normally set by the users according to their specific requirements. The bit-rate transparency capability allows the same receiver to be used for several networks operating at different bit rates. It also allows the users the capability of upgrating the network capacity to accommodate terminal and traffic growth without major hardware changes. In addition, a variable–data-rate capability for local-area networks is also desirable where fast exchanges of data bases between certain high-speed terminals can be carried out over a low-cost network designed for low–data-rate transmission [28]. Thus, for these applications, it is desirable that the optical receiver can operate over a range of bit rates with no or minimum component change and modification.

In submarine and terrestrial telecommunication systems the data rates are well established and each system is designed to operate at a particular bit rate in the digital hierarchy. The flexibility of a bit-rate-transparent optical receiver, however, is still of much interest.

It should be noted that a bit-rate–transparent transmission system can be realized more effectively using fiber-optic systems as compared with coaxial cable systems. This is because the loss of coaxial cables increases rapidly with frequency (as the square root of frequency). Thus for variable- or multiple–bit-rate operation a variable equalizer is required at the receiver to compensate for the increased cable loss with frequency. In addition, the signal-to-noise ratio at the higher bit rates is severely degraded. The optical-fiber cable loss, however, is independent of frequency (up to its bandwidth). Thus as long as the highest data rate of the bit-rate-transparent system does not exceed the fiber cable bandwidth, there is no need for the variable equalizer and the same signal power is obtained at all bit rates.

Bit-Pattern Independency

Bit-pattern independency refers to the capability of the optical receiver to operate with an unrestricted data format. The line code that imposes the most constraint on the optical receiver is probably the nonreturn-to-zero (nrz) code. The average level of the nrz code is dependent on the specific bit pattern being transmitted. The worst case is when a long string of "ones" or "zeros" is transmitted. In this case the dc reference for the average signal value is lost if the receiver is ac coupled. In addition, the absence of transitions in the received data stream can cause problems in the timing extraction circuit.

In synchronous telecommunication systems the above problems are circumvented by using an appropriate line-coding format for transmission or by scrambling and descrambling the transmitted data. In data links and local-area networks, however, the ability of the optical receiver to operate with unrestricted data format is often desirable.

Acquisition Time

In local-area networks using time-division multiple-access (tdma) and burst-asynchronous modes of operation, the information is transmitted in short bursts of data. The optical receiver is thus normally at the idle state of receiving no incident

TABLE 1. Important Receiver Characteristics

	Features	Submarine Communications	Terrestrial Communications	Point-to-Point Data Links	Local-Area Networks
High receiver sensitivity	Maximum repeater spacing	Critical	Moderate-critical	Moderate	Moderate
Wide dynamic range	Flexible convenient system configuration	Moderate	Moderate-critical	Moderate-critical	Critical
Bit-rate transparency	Variable–bit-rate operation	Not required	Not required	Desirable	Desirable
Bit-pattern independency	Flexible	Accommodated by the use of appropriate line codes and scrambling/descrambling		Desirable	Desirable
Fast acquisition time	Short preamble bit sequence	Not required	Not required	Moderate-critical	Moderate-critical

optical power. When the data bursts arrive the optical receiver amplifier has to establish its stable linear operating point quickly to detect the data. In addition, the receiver clock has to synchronize itself to the frequency and phase of the incoming data burst. The time period the receiver takes to achieve the above functions is referred to as the *receiver acquisition time.*

Normally, to ensure that there is no loss of data message during the acquisition period, a short preamble or training sequence of bits is transmitted before the data itself. Thus it is obvious that, in order to maximize the data transmission efficiency, the preamble sequence and hence the receiver acquisition time has to be as short as possible.

The receiver characteristics considered above are summarized in Table 1. Their relative importance in typical applications of optical-fiber communication systems are also indicated. In addition to the listed requirements, there are other constraints for the receiver design such as low power consumption, specified supply voltage, low cost and size, high reliability, etc. These requirements can also be quite important for certain applications, such as submarine telecommunications or tactical military applications.

In the following sections we will consider the basics of receiver design and the trade-offs involved in satisfying the above requirements.

3. Photodetectors

Before proceeding with the receiver analysis and design we review in this section the characteristics of photodetectors for use in optical-fiber systems. The discussion will emphasize the signal and noise characterization and circuit application aspects which will be required later in the receiver analysis and design.

As discussed previously the two most useful photodetectors for optical-fiber systems are the pin and avalanche photodiode. The APD is more sensitive than the simple pin diode because of its internal avalanche gain. The APD is designed such that each primary electron-hole pair generated by the incident optical power creates further electron-hole pairs because of impact ionization phenomenon. The secondary photocurrent can

therefore be several hundred times larger than the primary photocurrent.

Recently there has been much interest in other types of photodetectors for use in optical-fiber communication systems. Among the devices investigated are heterojunction phototransistors (HPTs) [29–32], modulated barrier photodiodes [33], photoconductive detectors [34–36], and FET photodetectors [37–39]. The motivations behind these novel devices are the possibility of complete monolithic optoelectronic integration, and the lack of low-noise APDs in the long-wavelength (1.0 to 1.7 μm) region. It has been predicted that they will offer comparable performance to photodiodes. However, they are not well developed and their performance in actual optical receivers has not been demonstrated. In this chapter we will confine our discussion to pin and APD detectors. The interested reader can refer to the references for other detector types.

The function of the photodetector is to convert the input optical signal into electrical current for further amplification and processing. The essential performance requirements of photodiodes for optical-fiber systems are

high quantum efficiency,
fast response (high modulation bandwidth),
low capacitance,
low dark current, and
low avalanche excess noise (for APDs).

For the short-wavelength region (around 0.85 μm), silicon photodiodes are used almost exclusively. The status of silicon photodiodes is well established and high-quality photodiodes (both pin and APDs) have been fabricated and are readily available [40].

For the long-wavelength region (1.0- to 1.6-μm), photodiodes are fabricated from germanium and several III-V compounds (InGaAsP, GaAlAsSb). In general, long-wavelength photodiodes are still in the developmental stage and have yet to approach the excellent performance of silicon devices at short wavelengths. The dark current and avalanche excess noise of long-wavelength detectors are much higher than for silicon photodiodes.

The state-of-the-art characteristics of photodiodes are listed in Table. 2. Included are silicon devices for the short wavelengths and three repre-

TABLE 2. Typical Characteristics of State-of-the Art Photodetectors

	InGaAs PIN	Germanium APD	InGaAs APD	Silicon APD
Quantum efficiency η	0.8	0.8	0.8	0.8
Response time t_r (ps)	60	100	100	100
Capacitance C_d (pF)	<0.5	<1	<0.5	<1
Dark current I_{du} (nA)	1–5	50–500	1–5	1
Dark current I_{dm} (nA)		50–200	1–5	0.001
Ionization ratio k		0.7–1.0	0.3–0.5	0.01–0.03

sentative detectors for the 1.0- to 1.6-μm region: an InGaAs pin photodiode [41–42], a germanium APD [43–45], and an InGaAs APD [46–48].

Photodiode Equivalent Circuits

In operation the photodiodes are reversed biased. The bias voltage for pin photodiodes is typically 5 to 20 V, while APDs are biased at much higher voltage, typically 100 to 300 V, to generate a strong enough electric field for the impact ionization process.

A small-signal equivalent circuit of the photodiode is shown in Fig. 2. Current source $i_s(t)$ represents the photocurrent generated by the incident optical signal, while current source $i_d(t)$ represents the dark current of the photodiode (including any stray background radiation). Capacitance C_d is the depletion capacitance, typically around 1 pF or less for photodiodes with 100-μm–diameter active area.

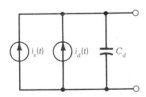

FIG. 2. Small-signal equivalent circuit of the photodiode.

Signal Analysis

For pin photodiodes the signal current $i_s(t)$ generated by the incident optical power $p(t)$ is given by

$$i_s(t) = (\eta e/h\nu)p(t) = R p(t), \qquad (1)$$

where η is the quantum efficiency of the photo-

diode, e is the charge of an electron, and $h\nu$ is the energy of one photon.

The parameter R, which has the dimension of amperes per watt, is call the *responsivity* of the photodiode:

$$R = \eta e/h\nu = \eta e\lambda/hc, \qquad (2)$$

where c is the speed of light and λ is the incident optical wavelength.

For 100% quantum efficiency ($\eta=1$) we have $R=0.684$ A/W at 0.85 μm, $R=1.046$ A/W at 1.3 μm, and $R=1.248$ A/W at 1.55 μm.

For APDs the primary current is multiplied by the avalanche gain process. Since this is a statistically random process the avalanche gain g is a random variable. If we let G be the average value of the avalanche gain ($G=\langle g \rangle$), the signal current output of the APD is given by

$$i_s(t) = (\eta e/h\nu)G p(t) = R G p(t). \qquad (3)$$

The dark current of the APD can be expressed in terms of two components I_{du} and I_{dm} as

$$i_d(t) = I_{du} + G I_{dm}. \qquad (4)$$

Here I_{du} is the dark-current component that is not subject to the avalanche multiplication process. It is mainly the surface leakage current of the APD. The other component I_{dm} undergoes the avalanche multiplication process and is caused by the bulk dark current and stray background radiation. Note that we can apply both of the above equations to pin photodiodes by letting $G=1$ and $I_{dm}=0$.

Noise Analysis

The noise of the photodector consists of two components: (*a*) a signal-dependent noise component which is the shot (or quantum) noise associated with the incident optical power and (*b*) a signal-independent noise component associated with the photodiode dark current.

For the simple pin photodiode the signal-dependent shot noise is caused by the random generation of electron-hole pairs due to the incident optical signal. In any interval of time, say one time slot *T*, the number of electrons generated is a random number *n*. The probability density function of *n* follows the well-known Poisson distribution $P[\{n\}]$:

$$P[n=m]=N^m e^{-N}/m! , \qquad (5)$$

where *N* is the average value of the number of electrons generated in this interval:

$$N=(\eta/h\nu)\int_T p(t) . \qquad (6)$$

The expression $P[n=m]$ in (5) refers to the probability that exactly *m* electrons will be generated in interval *T*.

The randomness of the electron generation sets a fundamental limit on the maximum sensitivity of digital optical receivers. Even in the case of an ideal receiver a finite optical energy must be sent in a pulse if we want to detect it with a desired probability of error. Let us consider a completely noiseless receiver (zero electronic amplifier noise and zero dark current). The transmitter only sends an optical pulse if a "one" bit is being transmitted. The receiver detects the "one" bit only if one or more electrons are generated in that bit interval and it says the bit is "zero" if no electrons are generated. Thus an error will be made if a pulse is transmitted but no electron is generated. The probability that $n=0$ is given from (5) as

$$P[n=0]=e^{-N} . \qquad (7)$$

If an error probability of 10^{-9} is required, we must have from the above equation $N=21$, i.e., on the average 21 photons must be transmitted for each optical pulse (assuming 100% quantum efficiency). The required optical power to achieve this can be evaluated from (6). This minimum limit on the required transmitted optical power is referred to as the *quantum limit*.

For APDs the output current is further complicated by the statistics of the avalanche multiplication process. Since the avalanche gain is a random variable, excess noise is created at the APD output. Further, the probability density function of the output current is modified and is no longer Poissonian.

Because of the random nature of the avalanche multiplication process, the mean-squared value of the avalanche gain is greater than the square of the mean avalanche gain value. This ratio is referred to as the *avalanche excess-noise factor* and is given by [1]

$$F(G)=\langle g^2 \rangle/\langle g \rangle=k\,G+(2-1/G)(1-k) , \qquad (8)$$

where *k* is the ratio of the ionization coefficients of electrons and holes of the APD (*k* is defined such that $k \leq 1$).

The smaller *k* is, the lower $F(G)$ is and the better the photodiode. For short-wavelength silicon photodiodes, *k* is typically 0.02 to 0.03. Long-wavelength photodiodes have higher values of *k* (0.7 to 1 for germanium APDs and 0.3 to 0.5 for InGaAs APDs) as can be observed from Table 2.

The excess-noise factor can also be approximately by a simpler expression [49]:

$$F(G)=G^x , \qquad (9)$$

where the parameter *x* is referred to as the *excess-noise–factor exponent*.

How well the above approximation is can be observed from Fig. 3. In this figure the avalanche excess-noise factor is plotted versus the avalanche gain for typical silicon and InGaAs APDs. Solid lines are accurate expressions (8) and dashed lines are approximate expressions (9). The range of the avalanche gain plotted is limited to the normal operating range of each device. It can be observed from Fig. 3a that for silicon devices the approximation is quite poor. For typical silicon APDs with *k* values from 0.02 to 0.04 the parameter *x* is in the 0.3 to 0.4 range. For InGaAs APDs, with *k* values from 0.3 to 0.5, the approximation is much

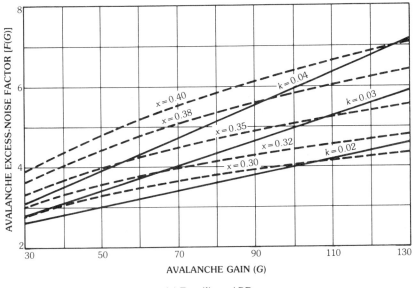

(a) For silicon APDs.

(b) For InGaAs APDs.

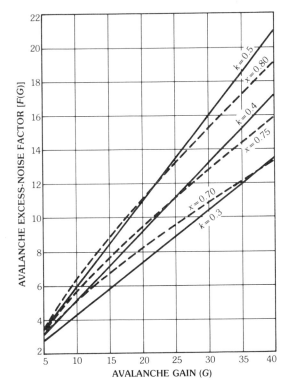

FIG. 3. Avalanche excess-noise factor for typical APDs.

better. Parameter x for InGaAs APDs ranges from 0.7 to 0.8 as seen from Fig. 3b. It is obvious that for germanium APDs, x is equal to unity.

Both parameters k and x are equally used in the literature and in APD component specifications. In this chapter the expression (8), involving

k for the excess noise, will be used wherever possible to make the analysis more accurate. The approximation expression will only be used in cases where the mathematics is more tractable.

The excess-noise power alone, however, does not completely specify the avalanche gain statistics. A knowledge of the probability density function of the avalanche gain is necessary. Such an expression has been derived [5–9]. The secondary current is the product of the primary current and the avalanche gain. Thus the probability density function of the secondary current output can be obtained by a convolution process of the avalanche gain statistics and the Poissonian distribution. The result obtained, however, is exceedingly complex. An approximate function for the output current probability density function has been proposed [40], which although not rigorously proved, has been verified experimentally and by simulation [13] to be fairly accurate. It is given in terms of the average avalanche gain G and the excess-noise factor $F(G)$ as

$$p(i) = (2 \pi \sigma^2)^{-1/2} \, [1 + (i - I_s)/\sigma \lambda]^{-3/2}$$

$$\times \exp \{ -(i - I_s)^2/2 \sigma^2 [1 + (i - I_s)/\sigma \lambda] \} , \quad (10)$$

where

$$\sigma^2 = (e/T)^2 \, n_e \, G^2 \, F(G) , \quad (11)$$

$$\lambda = [n_e F(G)]^{1/2}/[F(G) - 1] , \quad (12)$$

$$I_s = n_e \, e \, G/T . \quad (13)$$

In the above expressions, n_e is the number of primary electrons generated in the bit interval T and I_s is the mean secondary output current of the APD. It is related to the optical power P falling on the APD (which is assumed to be constant during this interval) as [cf. (3)]

$$i_s = (\eta \, e/h \nu) \, G \, P . \quad (14)$$

The above expression for the APD output-current probability density function $p(i)$ is still rather complicated. If is often further approximated by a Gaussian distribution with the same mean and variance as follows:

$$p(i) = (2 \pi \sigma^2)^{-1/2} \exp [-(i - I_s)^2/2 \sigma^2] . \quad (15)$$

The probability density function as given by the simple Gaussian distribution and the more accurate distribution in (10) has been calculated and plotted in Fig. 4. The parameter values taken for the APD are $k = 0.03$ and $G = 100$. Two separate cases are considered for the number of primary electrons: $n_e = 100$, which represents an optical pulse for the "one" bit ("on" state) and $n_e = 5$ for the "zero" bit ("off" state).

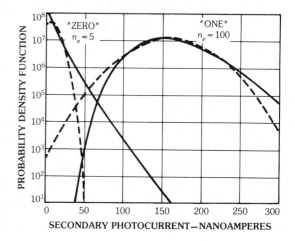

FIG. 4. Probability density function of APD secondary output current.

It can be observed that the actual probability density function is highly asymmetric, particularly for small values of primary electrons. It is close to the Gaussian distribution only for the large values of n_e as can be seen from comparing (10) and (15). However, (10) is not mathematically tractable and the simpler Gaussian distribution is widely used in the optical receiver design and analysis. This point will be expanded on later in the chapter.

4. Receiver Analysis

In the following sections we will consider the signal and noise analysis of optical receivers for use in digital fiber-optic communication systems. The theory presented in this chapter is based on the approach originally proposed by Personick [10]. It is, however, different from Personick's theory

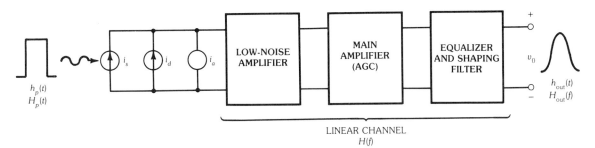

FIG. 5. Model of optical receiver.

in several aspects. The theory presented here is applicable for both high-impedance and transimpedance designs. In addition, the noise representation of receiver amplifiers is also different from Personick's theory.

Receiver Modeling

The optical receiver can be modeled as in Fig. 5. The photodiode is represented by a signal current source i_s and a dark current source i_d. Both of them are secondary currents as given by (3) and (4). The photodiode capacitance is lumped together with the amplifier input capacitance to form a total input capacitance.

The photodiode is followed by the linear channel, which consists of the amplifier chain and the shaping filter. The transfer function of the linear channel (including the photodiode) is represented by $Z(f)$. It should be noted that $Z(f)$ has the dimension of impedance since it converts an input current to an output voltage.

The receiver amplifier noise is normally dominated by the low-noise amplifier at the front end of the linear channel. The receiver amplifier noise is represented by an input equivalent noise-current source i_a which produces the same noise level at the linear channel output. Note that the representation of an equivalent shunt noise-current source and a series noise-voltage source is not used. This single noise-current source representation has the advantage of simplicity and is especially useful for optical receivers in which the input signal is also a current source. The effect of the linear-channel frequency response on the signal and noise current can thus be easily observed. In addition, since the amplifier noise is referred to the input, the noise

performance of different amplifier designs can be compared directly by their input equivalent noise currents, regardless of the amplifier gain or transimpedance.

It should be noted that the model in Fig. 5 is applicable to any optical receiver, regardless of whether the low-noise amplifier design is a high-impedance or transimpedance configuration.

Signal Analysis

We consider a digital system operating at a bit rate B. The time slot (or bit interval) denoted by T, is the reciprocal of the bit rate:

$$T = 1/B. \tag{16}$$

Let the optical power falling on the photodetector be a sequence of pulses of the following form:

$$p(t) = \sum_{k=-\infty}^{\infty} b_k h_p(t - kT), \tag{17}$$

where k is a parameter denoting the kth time slot and $h_p(t)$ represents the pulse shape of an isolated optical pulse at the photodiode input. The function $h_p(t)$ is defined such that

$$(1/T)\int_{-\infty}^{\infty} h_p(t)\, dt = 1. \tag{18}$$

Thus, in (17), b_k is the received optical power in the kth time slot. It can take on two values, b_0 or b_1, depending on whether the data bit in the kth time slot is a "zero" or a "one." Ideally, we would like b_0 to be zero so that no optical power is transmitted for the "zero" bits. However, optical sources such as semiconductor laser diodes require a standing bias current for high-speed modulation. Consequently, there is always a small optical power being transmitted. The ratio of b_0 and b_1 is referred to as the *extinction ratio r* of the optical source:

$$r=b_0/b_1 . \qquad (19)$$

Note that by this definition r is always less than 1.

The average optical power falling on the photodetector is thus given by

$$\bar{P}=(b_0+b_1)/2=(b_1/2)(1+r) , \qquad (20)$$

assuming equal probability of transmitting "ones" and "zeros," which is normally the case.

The linear-channel output voltage $v_0(t)$ has two components: $v_s(t)$ due to the incident optical signal and $v_d(t)$ due to the APD dark current. As far as mean values are concerned, the signal output voltage due to incident optical power $p(t)$ is given by [10]

$$\langle v_s(t)\rangle=(\eta e G/h\nu)\int_{-\infty}^{\infty} z(t-\gamma)p(\gamma)\,d\gamma , \qquad (21)$$

where $z(t)$ is the impulse response of the linear channel and is the inverse Fourier transform of $Z(f)$.

The output signal voltage is thus a sequence of pulses of the form

$$\langle v_s(t)\rangle=\sum_{k=-\infty}^{\infty} s_k h_{\text{out}}(t-kT) , \qquad (22)$$

where s_k takes on two values, s_0 and s_1 (corresponding to b_0 and b_1), and $h_{\text{out}}(t)$ is a function describing the shape of the isolated output pulse. It is chosen such that its maximum value (where the voltage should be sampled) is equal to 1 and occurs at time $t=0$:

$$h_{\text{out}}(0)=1 . \qquad (23)$$

By comparing (21) and (22) we can write the linear-channel transfer function $Z(f)$ in the form

$$Z(f)=(h\nu/\eta e G)(s_k/b_k)H(f) , \qquad (24)$$

where $H(f)$ is defined as

$$H(f)=H_{\text{out}}(f)/H_p(f) . \qquad (25)$$

where $H_{\text{out}}(f)$ and $H_p(f)$ are the Fourier transforms of $h_{\text{out}}(t)$ and $h_p(t)$, respectively. Thus $H(f)$ and its inverse Fourier transform $h(t)$ are functions describing the shape of the linear-channel frequency response and impulse response.

The output voltage due to dark currents is given by

$$\langle v_d(t)\rangle=(I_{du}+GI_{dm})\int_{-\infty}^{\infty} z(\gamma)\,d\gamma$$

$$=(I_{du}+GI_{dm})Z(0) . \qquad (26)$$

This corresponds to a dc shift in the output voltage and is normally removed by ac coupling in the optical receiver.

Noise Analysis

The noise at the linear-channel output consists of three components: a signal-dependent noise term due to the incident optical power, a noise term due to the APD dark current, and a noise term due to the receiver amplifier.

The noise term due to the incident optical power $p(t)$ has been derived and is given by [10]

$$\langle n_s^2(t)\rangle=\langle[v_s(t)-\langle v_s(t)\rangle]^2\rangle=(\eta e^2/h\nu)G^2$$

$$\times F(G)\int_{-\infty}^{\infty} z^2(t-\gamma)p(\gamma)\,d\gamma . \qquad (27)$$

The noise term due to the APD dark current is similarly given by:

$$\langle n_d^2\rangle=\langle[v_d-\langle v_d\rangle]^2\rangle=e[I_{du}+G^2F(G)I_{dm}]$$

$$\times\int_{-\infty}^{\infty} z^2(\gamma)\,d\gamma \qquad (28)$$

or

$$\langle n_d^2\rangle=2e[I_{du}+G^2F(G)I_{dm}]\int_0^{\infty}|Z(f)|^2\,df . \qquad (29)$$

The term outside the integral in the above expression can be recognized as the input-noise spectral density. This is the familiar expression of the shot-noise spectral density. Note that the one-sided frequency representation of spectral density is used throughout this chapter.

The noise term due to the receiver amplifier can be derived very simply. If we let $N(f)$ be the spectral density of the input equivalent noise-current source i_a, the output noise power due to the amplifier is given by

$$\langle n_a^2 \rangle = \int_0^\infty N(f)\,|Z(f)|^2\,df. \qquad (30)$$

The total noise power at the linear-channel output can then be obtained as the sum of the above components:

$$\langle n^2(t) \rangle = \langle n_s^2(t) \rangle + \langle n_d^2 \rangle + \langle n_a^2 \rangle. \qquad (31)$$

However, it is much more convenient to refer all the noise power back at the input of the receiver amplifier as input equivalent-current noise power. Then the noise can be compared directly to the generated signal photocurrent.

The signal-dependent noise term can be evaluated from (27) and referred back to the input. The analysis has been carried out [10, 26] and the results will be given here. This noise term has two components: one due to the noise contribution of the optical pulse in the present time slot and one due to contribution of optical power in all other time slots. This first noise component magnitude is dependent on whether a "one" or "zero" bit is being transmitted (b_k is equal to b_0 or b_1). The second noise component is dependent on the statistics of the incoming data pattern. Since this knowledge is unknown beforehand at the receiver the worst-case condition of maximum level (b_1) for all other optical pulses is assumed. The input equivalent noise power of the signal-dependent noise generated by the incident optical power is therefore given as

$$\sigma_{sk}^2 = 2\,e\,R\,G^2\,F(G)\,B[b_k\,I_1 + b_1(\Sigma_1 - I_1)]. \qquad (32)$$

The term $b_k\,I_1$ is the contribution of the present time slot and the term $b_1(\Sigma_1 - I_1)$ is due to all other time slots as discussed above. In the above equation, R is the photodiode responsivity as defined in (2), and Σ_1 and I_1 are weighting constants originally defined by Personick [10]. They are given by

$$\Sigma_1 = \sum_{k=-\infty}^{\infty} H'_p(k)\,[H'(k)*H'(k)]/2 \qquad (33)$$

and

$$I_1 = \int_0^\infty H'_p(f)\,[H'(f)*H'(f)]\,df, \qquad (34)$$

where $H'_p(f)$ and $H'(f)$ are normalized forms of $H_p(f)$ and $H(f)$. They are defined such that

$$H'_p(f) = (1/T)\,H_p(f/T) \qquad (35)$$

and

$$H'(f) = (1/T)\,H'(f/T). \qquad (36)$$

The normalized functions $H'_p(f)$ and $H'(f)$ are thus defined to be independent of the transmitted bit rate. The constants Σ_1 and I_1 are therefore also independent of bit rate and depend only on the shape of the input optical pulse and the output voltage pulse.

The input equivalent noise power due to the APD dark current can be similarly evaluated from (29) to be

$$\sigma_d^2 = 2\,e\,[I_{du} + G^2\,F(G)\,I_{dm}]\,B\,I_2, \qquad (37)$$

where I_2 is another weighting constant and is defined by

$$I_2 = \int_0^\infty |H'(f)|^2\,df. \qquad (38)$$

The input equivalent noise power of the receiver amplifier can be derived from (30). The input equivalent noise spectral density $N(f)$ can normally be written in a truncated power series of the form

$$N(f) = a_0 + a_1 f + a_2 f^2. \qquad (39)$$

Higher-order terms are neglected but can be included in the analysis if necessary.

The input equivalent noise power of the amplifier can be readily shown to be

$$\sigma_a^2 = a_0\,B\,I_2 + a_1\,B^2\,I_f + a_2\,B^3\,I_3, \qquad (40)$$

where I_f and I_3 are weighting constants given by

$$I_f = \int_0^\infty |H'(f)|^2\,f\,df \qquad (41)$$

and

$$I_3 = \int_0^\infty |H'(f)|^2 f^2 \, df. \qquad (42)$$

All the weighting constants Σ_1, I_1, I_2, I_3 have been calculated and plotted out by Personick [10, 26] for three incident optical pulse shapes: rectangular, Gaussian, and exponential, together with a full raised-cosine spectrum output pulse. In practice, the pulse shape is determined by the line coding being used, the optical source response time, and the dispersion in the transmission fiber. As a reference basis for comparison we can consider the case of negligible fiber dispersion and very fast optical-source response time. This is normally the case in practice since we do not want the source to limit the system response and we like to avoid equalization for fiber dispersion since it is unpredictable. If necessary, optical-source bandwidth limitation and fiber dispersion can be considered separately as degradations. For systems using laser diodes and single-mode fibers the above conditions are almost always satisfied.

In this case of negligible dispersion the values of Σ_1 and I_1 are essentially the same. Thus the noise contribution of the signal-dependent noise term in (32) due to all other time slots can be neglected entirely. We can thus summarize all the expressions for signal current and noise current power, all referred to the receiver amplifier input as follows:

Signal:

$$I_{sk} = R G b_k \qquad (k=0 \text{ or } 1). \qquad (43)$$

Noise:

Signal-dependent noise:

$$\sigma_{sk}^2 = 2 e R G^2 F(G) B I_1 b_k \quad (k=0 \text{ or } 1). \quad (44)$$

Dark-current noise:

$$\sigma_d^2 = 2 e [I_{du} + G^2 F(G) I_{dm}] B I_2. \qquad (45)$$

Receiver amplifier noise:

$$\sigma_a^2 = a_0 B I_2 + a_1 B^2 I_f + a_2 B^3 I_3, \qquad (46)$$

with spectral density

$$N(f) = a_0 + a_1 f + a_2 f^2. \qquad (47)$$

Two Common Practical Cases

We now consider two particular cases for the input and output pulse shapes that are of considerable interest in practice. Two incident optical pulse shapes, which are widely used are the rectangular pulse for nrz code and the 50% duty-cycle rz code as shown in Fig. 6. In the frequency domain they can be expressed as

$$H'_p(f) = (\sin \pi f)/\pi f \quad \text{for nrz code} \qquad (48)$$

and

$$H'_p(f) = 2[\sin (\pi f/2)]/\pi f \quad \text{for rz code.} \qquad (49)$$

The output pulse is normally designed to closely approximate the full raised-cosine spectrum pulse shown in Fig. 7. This pulse has zero crossings at $t = kT$ (sampling times) except at $t=0$, where the pulse height is sampled. thus this pulse has zero intersymbol interference. In the frequency domain it can be expressed as

$$H'_{out}(f) = \begin{cases} (1+\cos \pi f)/2 & \text{for } |f| < 1, \\ 0 & \text{for } |f| > 1. \end{cases} \qquad (50)$$

The shapes of the linear-channel frequency response are therefore given from (24) and (25) to be

$$H'(f) = (\pi f/2) \cot (\pi f/2) \qquad \text{for nrz code} \quad (51)$$

and

$$H'(f) = (\pi f/2) \cot (\pi f/2) \cos (\pi f/2) \quad \text{for rz code.} \qquad (52)$$

TABLE 3. Values of Weighting Constants for Two Common Cases

	NRZ	RZ (50% Duty Cycle)
I_1	0.56	0.50
I_2	0.564	0.403
I_3	0.0868	0.0361
I_f	0.184	0.0984

The linear-channel frequency responses above are plotted in Fig. 8. It can be seen that the receiver 3-dB bandwidth is 0.58 and 0.40 times the bit rate for the nrz code and rz code, respectively.

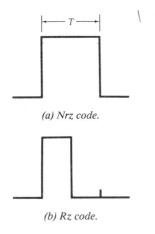

(a) Nrz code.

(b) Rz code.

FIG. 6. Incident optical pulse shape.

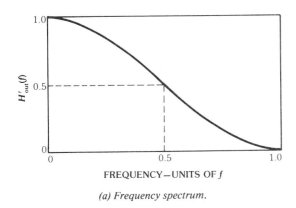

(a) Frequency spectrum.

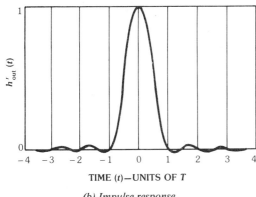

(b) Impulse response.

FIG. 7. Full raised-cosine spectrum output pulse shape.

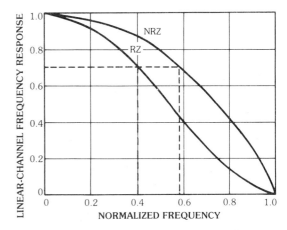

FIG. 8. Linear-channel frequency response.

The weighting constants I_1, I_2, I_3 and I_f for the two cases above have been evaluated and listed in Table 3. Thus the various noise power terms in (44) to (46) can be readily calculated. From the knowledge of the noise power and the signal current from (43), the required optical power for a desired bit-error rate (or signal-to-noise ratio) can be evaluated. The analysis will be carried out in Section 6 on receiver sensitivity.

5. The Receiver Amplifier

Before proceeding with the receiver sensitivity analysis let us consider the low-noise amplifier following the photodetector. Its design and performance are critical in determining the sensitivity and dynamic range of the optical receiver.

The Receiver Amplifier Configuration and Signal Analysis

Depending on their configuration, front-end amplifiers for optical receivers can be classified into two types: high-impedance and transimpedance.

High-Impedance–Amplifier Design—A block diagram of the high-impedance receiver amplifier is shown in Fig. 9. Here R_L is the load, which includes the bias resistance of the photodiode and the front-end transistor, and C is the total input capacitance of the amplifier, including the detector and stray capacitance. For simplicity the gain A can be assumed to be flat over the frequency range of interest.

389

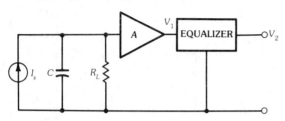

FIG. 9. High-impedance receiver amplifier.

The frequency response of the front-end amplifier before equalization is given by

$$V_1(f)/I_s(f) = A R_L/(1+j2\pi R_L C f). \quad (53)$$

In the high-impedance design, R_L is chosen to be large in order to reduce its noise contribution, as we shall see later. Consequently, the input time constant $R_L C$ is very large compared with the bit interval T. The front-end rolloff frequency $1/2\pi R_L C$ is thus much smaller than the bit rate. The amplifier therefore tends to integrate the detected signal waveform. Because of this fact the high-impedance receiver is also referred to as an *integrating receiver.* An equalizer in the form of a differentiating network is necessary to extend the receiver bandwidth out to the desired value. The frequency response plot of the front-end amplifier and equalizer in the high-impedance design can be seen from Fig. 10. Here K is the gain of the equalizer at high frequencies. The equalizer can be a simple passive RC network as in Fig. 11. The transfer function of this passive equalizer is given by

$$V_2(f)/V_1(f) = [R_2/(R_1+R_2)]\,[1+j\,(f/f_1)]$$

$$/[1+j\,(f/f_2)]. \quad (54)$$

where

$$f_1 = 1/2\pi R_1 C_1, \quad (55)$$

$$f_2 = (R_1+R_2)/2\pi R_1 R_2 C_1. \quad (56)$$

For perfect equalization the equalizer zero f_1 has to be matched with the front-end amplifier pole at $1/2\pi R_L C$. The bandwidth of the amplifier with

equalization is thus extended to the equalizer pole at f_2. The transfer function amplitude, however, has been reduced from $A R_L$ to $A R_L f_1/f_2$ for this passive equalizer.

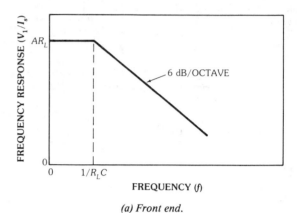

(a) Front end.

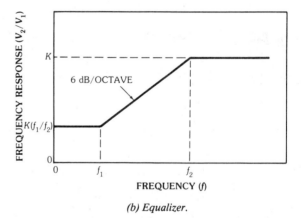

(b) Equalizer.

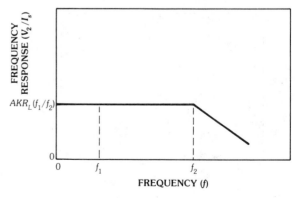

(c) Complete receiver amplifier.

FIG. 10. Frequency responses in the high-impedance receiver amplifier.

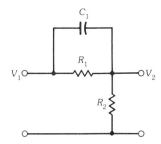

FIG. 11. Passive *RC* equalizer.

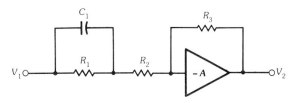

(a) Op-amp network.

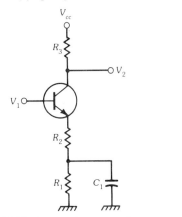

(b) Bipolar-transistor network.

FIG. 12. Two active equalizers.

The ratio f_2/f_1 is often referred to as the *equalization ratio* and can be as high as several decades in practice. For this passive network, K is equal to unity and the attenuation of the equalizer (for low frequencies) is equal to the equalization ratio.

An active equalizer is desirable in certain cases to improve dynamic range and noise performance. Two possible realizations of an active equalizer are shown in Fig. 12. The transfer functions of both networks are given by

$$V_2(f)/V_1(f) = -[R_3/(R_1+R_2)]$$

$$\times [1+j(f/f_1)]/[1+j(f/f_2)], \quad (57)$$

where f_1 and f_2 are given by the same expressions in (55) and (56). In this case we have $K=R_3/R_2$. Thus the gain of the equalizer is determined by R_3, which can be chosen independently of the equalizing frequencies f_1 and f_2.

It can be observed that since the passive equalizer attenuates in-band signals, the amplifier gain A has to be large enough to ensure that noise sources from amplifying stages after the equalizer do not degrade the signal-to-noise ratio. Thus the voltage at the amplifier output for a given detected signal current I_s is large, particularly for low-frequency components (since both R_L and A are large). The maximum voltage swing at the amplifier output is, however, limited by the supply voltage and biasing conditions. Thus the dynamic range (ratio of maximum and minimum allowable input signals) of the receiver amplifier is limited. This problem can be alleviated somewhat by using the active equalizer, which combines gain and equalization in one stage.

It should be noted that in order to simplify the equalization process, a single-pole rolloff (6 dB/

octave) is desirable in the equalizing frequency band (f_1 to f_2). Thus care has to be taken in the amplifier design to ensure that the amplifier gain A does not introduce any extra pole within this frequency band.

Transimpedance Amplifier Design—The transimpedance amplifier is a shunt-feedback amplifier as shown in Fig. 13a. In this figure, R_f is the feedback resistance with stray capacitance C_f, R_b is the photodiode and transistor bias resistance (if present), and, as previously, C includes the photodetector and amplifier capacitance. When the photodiode is dc coupled to the receiver amplifier the feedback resistance can be used to bias the photodiode. Thus the bias resistance R_b is omitted in this case.

The transfer function of the transimpedance amplifier in Fig. 13a can be shown to be

$$V_2(f)/I_s(f) = -R_f/[1+R_f/AR_b+j2\pi fR_f(C_f$$

$$+C/A)]. \quad (58)$$

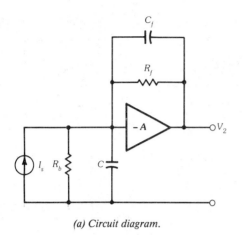

(a) Circuit diagram.

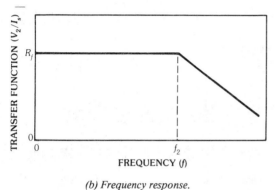

(b) Frequency response.

FIG. 13. Transimpedance receiver amplifier and response.

In practice, R_b is chosen to be larger than R_f and $A \gg 1$. The frequency response of the amplifier is shown in Fig. 13b. The 3-dB bandwidth is given by

$$f_2 = A/2\pi R_f(C + A\,C_f).\qquad (59)$$

If stray capacitance C_f is small enough so that $A\,C_f \ll C$, we have $f = A/2\pi R_f C$. This can be compared with the high-impedance–amplifier roll-off frequency $1/2\pi R_L C$. We see that for the same value of load resistance (R_L or R_f) the bandwidth of the transimpedance amplifier is increased by a factor of A. Another way of explanation is that because of shunt feedback, the input impedance of the receiver amplifier is R_f/A (neglecting C_f). The transimpedance-amplifier input time constant is therefore only $R_f C/A$, a factor of A smaller than the input time constant of the high-impedance amplifier.

The transimpedance amplifier is normally designed so that its bandwidth is high enough to accommodate the operating bit rate. Therefore no equalization is required. For a given feedback resistance the bandwidth can be increased by increasing the open-loop gain A (see (59)). This, however, cannot continue forever, because of two reasons. Firstly, as A increases, the effect of stray capacitance C_f in (59) becomes dominant and the bandwidth is finally limited by $1/2\pi R_f C_f$. Secondly, a high value of A requires a large number of amplifying stages within the shunt-feedback loop (since the bandwidth of the open-loop voltage amplifier has to be high as well, we cannot trade bandwidth for gain in the individual amplifying stage). However, for wide-bandwidth applications this introduces additional propagation delay and phase shift, which degrade the noise and phase margin and can potentially cause instability. Above 100 MHz the number of stages within the feedback loop is normally limited to three or less. Above 1 GHz, feedback is normally applied over only one stage.

We can also observe that the transimpedance receiver amplifier offers a higher dynamic range than the high-impedance design. As discussed previously, the dynamic range of the high-impedance amplifier is limited because of the increased gain of the low-frequency signal components which causes saturation in the amplifier before equalization. For the transimpedance amplifier no equalization is required. Thus the dynamic range of the transimpedance amplifier is better by a factor equal to the equalizing ratio, which is also the same as the ratio of the input time constants of the two amplifiers. Therefore the dynamic-range improvement of the transimpedance amplifier as compared with the high-impedance design is approximately equal to the value of voltage gain A (neglecting C_f).

Noise Analysis of Receiver Amplifiers

As stated previously the receiver amplifier noise can be represented by a single equivalent noise-current source i_a at the amplifier input. We now proceed to derive expressions for the spectral density $N(f)$ of this equivalent noise-current source, which is required for the receiver sensitivity analysis.

For the high-impedance amplifier design the input equivalent noise current can be derived using the approach suggested by Goell [19]. This approach is very useful for determining the equivalent input noise of a cascade of amplifying stages. In this approach the input equivalent noise current of a particular stage can be referred back to the input of the preceding stage by dividing by the short-circuit gain of this preceding stage. Thus by referring successively stage by stage from the output of the cascade back to the input, the input equivalent noise current of the whole cascade can be derived.

The noise analysis of the transimpedance amplifier is complicated because of the feedback loop. This is particularly true when several amplifying stages are inside the feedback loop. However, the analysis can be considerably simplified by using the referred-impedance noise analysis for feedback amplifiers as proposed by Hullett and Muoi [20]. In this approach, as far as the derivation of the input equivalent noise current is concerned, the feedback impedance of the transimpedance amplifier can be replaced by a shunt impedance at the amplifier input of exactly the same value. The feedback loop is thus broken and the analysis is considerably simplified. This referred-impedance approach is illustrated in Fig. 14. Using this approach the noise analysis of the transimpedance amplifier is similar to the high-impedance design except for the shunt impedance Z_f. One single-noise analysis can thus be used for the two different amplifier configurations.

In the design of receiver amplifiers one would

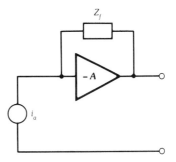

(a) With feedback impedance Z_f.

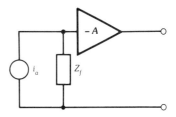

(b) With input shunt impedance Z_f.

FIG. 14. Referred-impedance approach for noise analysis.

like to achieve the lowest noise possible, but not without satisfying other requirements, such as bandwidth and dynamic range. When properly designed the noise contribution of the front-end amplifying stage should dominate the noise contributions of the following stages. We consider two separate cases for the front-end amplifying device: the field-effect transistor (FET) and the bipolar junction transistor (BJT). The common-emitter and common-source configurations for the first stage will be considered since they are optimum for low-noise performance [19].

TABLE 4. Typical Characteristics of FETs

Characteristic	Silicon JFET	Silicon MOSFET	Gallium-Arsenide MESFET
Gate-source capacitance C_{gs} (pF)	3–6	0.5–0.8	0.2–0.5
Gate-drain capacitance C_{gd} (pF)	0.5–1	0.05–0.1	0.05–0.1
Cutoff frequency f_T (GHz)	0.1–0.2	10–20	10–20
Gate leakage current I_g (nA)	Negligible	<0.01	1–3
1/f Noise corner frequency f_c (MHz)	Negligible	5–10	20–50

FET Front-End Amplifiers—Depending on the frequency range of operation, three different types of FET devices can be used for optical receivers: silicon junction FETs [14, 15], gallium-arsenide MESETs [50, 51], and short-channel silicon MOSFETs [52]. Their main characteristics are indicated in Table 4. It can be observed that silicon JFETs are useful in the low–bit-rate range because of their negligible leakage current and low 1/f-noise corner frequency. In contrast, both gallium-arsenide MESEFs and silicon MOSFETs are more suitable for high bit rates because of lower capacitance and higher transconductance. Later in the noise analysis we will see where the crossover point is.

The small-signal and noise equivalent circuit of the FET front-end amplifier is shown in Fig. 15. In this figure C_{ds} is the detector and stray capacitance at the amplifier input, R_b is the bias or load resistance, and R_f is the feedback resistance with stray capacitance C_f (if present as in the case of transimpedance design). Note that the feedback impedance is from the output of the front-end amplifier, which can have more than one amplifying stage within the feedback loop. The parameters C_{gs}, C_{gd}, and g_m are standard hybrid-π parameters of the front-end FET, which are the gate-to-source capacitance, gate-to-drain capacitance, and transconductance, respectively.

The noise generated by the FET can be represented by two noise-current sources with spectral densities S_1 and S_2 as shown in Fig. 15. They are given by [53–56]

$$S_1 = 2\,e\,I_g \qquad (60)$$

and

$$S_2 = 4\,k\,T\Gamma g_m + 4\,k\,T\Gamma g_m f_c / f. \qquad (61)$$

The noise source S_1 at the FET input represents the shot noise of the gate leakage current I_g. The first term ($4\,k\,T\Gamma g_m$) of noise source S_2 at the FET output represents the channel thermal noise (which includes the induced gate noise and its correlation). The second term of S_2 represents the 1/f noise, where the noise corner f_c is the frequency where the 1/f noise is equal to the channel-noise contribution. The parameter Γ is a numerical constant and can be taken to be 0.7 for silicon JFETs, 1.03 for silicon MOSFETs, and 1.75 for gallium-arsenide MESFETs [56]. The value of Γ for gallium-arsenide MESFETs is higher than the usually assumed value of 1.1 due to the contribution of the induced gate noise.

The other noise sources are the thermal noise of bias resistance R_b and feedback resistance R_f with spectral densities given by

$$S_b = 4\,k\,T/R_b \qquad (62)$$

and

$$S_f = 4\,k\,T/R_f, \qquad (63)$$

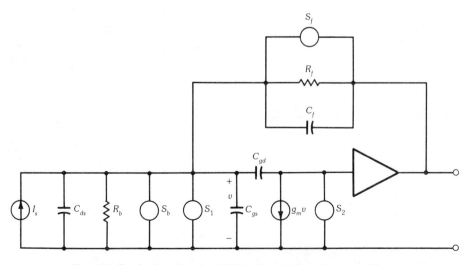

FIG. 15. Equivalent circuit of FET–front-end receiver amplifier.

where kT is the product of Boltzmann's constant and the absolute temperature.

Using the referred-impedance approach outlined previously, the equivalent circuit for noise analysis can be simplified as in Fig. 16. In this figure, R_F and C_T are the equivalent total load resistance and input capacitance given by

$$R_F = R_b R_f / (R_b + R_f) \qquad (64)$$

and

$$C_T = C_{ds} + C_{gs} + C_{gd} + C_f. \qquad (65)$$

It should be noted that C_T is the equivalent total input capacitance for noise analysis and is not subject to the Miller effect as would be the case for signal frequency-response analysis. A similar observation can be made for the equivalent load resistor R_F.

The noise sources S_f and S_b can be combined into one noise current source with spectral density

$$S_F = 4kT / R_F. \qquad (66)$$

For the high-impedance design, R_F is simply equal to R_b and $S_F = S_b$.

The input noise current can now be derived from Fig. 16. The noise source S_2 can be referred back to the input simply by dividing by the short-circuit current gain. We can write out almost by inspection the input equivalent spectral density as

$$N(f) = S_F + S_1 + S_2 (1/g_m^2)(1/R_F^2 + 4\pi^2 f^2 C_T^2) \qquad (67)$$

or by substituting in the values for S_F, S_1 and S_2:

$$N(f) = 4kT/R_F + 2e I_g + (4kT\Gamma/g_m R_F^2)$$

$$\times (1 + f_c/f) + (4kT\Gamma/g_m)(2\pi C_T)^2 f^2 (1 + f_c/f). \qquad (68)$$

In practice, we have $g_m R_F \gg 1$ and thus the third term in the above equation can be neglected. The equivalent noise spectral density is therefore

$$N(f) = 4kT/R_F + 2e I_g + (4kT\Gamma/g_m)$$

$$\times (2\pi C_T)^2 f_c f + (4kT\Gamma/g_m)(2\pi C_T)^2 f^2. \qquad (69)$$

It can be observed that the noise contribution due to the feedback (or load) resistance and gate leakage current is flat with frequency. In addition, the input spectral density of the $1/f$ noise contribution is proportional to frequency, and the channel-noise contribution is proportional to f^2.

The input equivalent noise-current power of the FET–front-end receiver amplifier can therefore be written from (46) as

$$\sigma_a^2 = (4kT/R_F) B I_2 + 2e I_g B I_2 + (4kT\Gamma/g_m)$$

$$\times (2\pi C_T)^2 [1 + (I_f/I_3)(f_c/B)] B^3 I_3. \qquad (70)$$

Several observations can be made from the above equation. For the high-impedance design, R_F is chosen to be large enough so that its noise contribution is negligible. The noise then originates mostly from the FET. In order to obtain the lowest noise performance it is desirable to choose an FET with small gate-leakage current, low capacitance, high transconductance, and low $1/f$ noise corner frequency.

The input equivalent noise power σ_a^2 for high-impedance FET–front-end amplifiers calculated from (70) is plotted in Fig. 17. Two representative curves are shown: one for silicon JFETs and the other for both gallium-arsenide MESFETs and

TABLE 5. Transistor Parameter Values Used in the Receiver Sensitivity Calculation

	Silicon JFET	Gallium-Arsenide MESFET	Silicon BJT
	$g_m = 6$ mS	$g_m = 15$ mS	$\beta = 100$
	$C_{gs} = 4.5$ pF	$C_{gs} = 0.2$ pF	$C_0 = 1.5$ pF
		$I_g = 2$nA	$\alpha = 0.6$ pF/mA
		$f_c = 30$ MHz	$r_{bb'} = 20\ \Omega$

FIG. 16. Simplified equivalent circuit for noise analysis of FET–front-end amplifier.

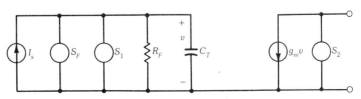

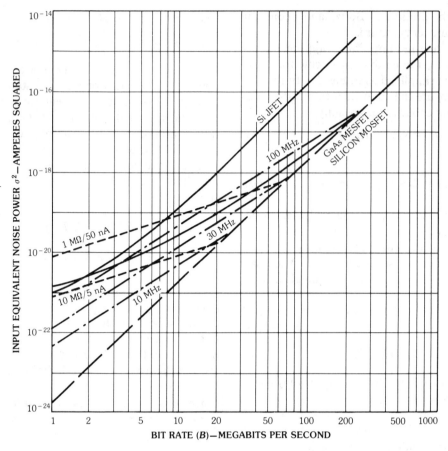

FIG. 17. Typical input equivalent noise powers of the FET–front-end receiver amplifier.

silicon MOSFETs since they offer comparable performance. The FET parameters taken in the calculation are listed in Table 5. The contribution of each noise component is also shown in Fig. 17.

Consider first the gallium-arsenide MESFET/ silicon MOSFET case. The FET channel-noise contribution is proportional to B^3, as indicated by the dashed line of slope equal to 3. It can be seen that the FET noise approaches this line at high bit rates. The contributions of load resistance, thermal noise, and leakage-current shot noise are proportional to the bit rate. They are shown by the two dashed lines with the value 1 MΩ/50 nA and 10 MΩ/5 nA. The 10-MΩ value can be taken to be the maximum value of load resistance that can be implemented in practice. Thus the 10-MΩ/5-nA line limits the amplifier noise level at low bit rates.

If $1/f$ noise contribution is proportional to B^2 and is indicated by the three dot-dash lines with

the values 10 MHz, 30 MHz, and 100 MHz noise corner frequency. It can be observed that the $1/f$ noise can be totally neglected if the noise corner frequency is 10 MHz or lower. For the parameter values used in Table 5 the calculated solid curve is bounded by the 10-MΩ/5-nA line at low bit rates, the 30-MHz line in the midrange, and the channel-noise line at high bit rates.

For silicon JFETs, $1/f$ noise and gate leakage noise are negligible. Thus the silicon JFET noise is limited only by the 10-MΩ/5-nA line at low bit rates and is purely channel noise at high bit rates. It can be observed from Fig. 17 that silicon FETs offer low noise for bit rates up to several megabits per second, while for higher bit rates, gallium-arsenide MESFETs and short-channel silicon MOSFETs are preferable. At present, gallium-arsenide MESFETs are more popular than silicon MOSFETs because of their ready availability. However, silicon MOSFETs offer the

396

opportunity of integrating the whole receiver front-end amplifier on a single monolithic chip. On the other hand, with gallium-arsenide MESFETs we have the potential of integrating the amplifier circuit with the GaAlAs or GaInAs photodetectors. It can be observed from (70) that the noise at high bit rates is proportional to C_T^2. Thus the reduction of capacitance by monolithic or hybrid circuit is critical for low-noise performance.

Bipolar-Transistor–Front-End Receiver Amplifiers—The small-signal and noise equivalent circuit of the common-emitter BJT-front-end receiver amplifier is shown in Fig. 18. The hybrid-π model for the bipolar transistor is used. All other parameters are the same as in Fig. 15. The bipolar-transistor noise consists of three noise-current sources: shot noise of the base bias current I_b, shot noise of the collector bias current I_c, and thermal noise of the base spreading resistor $r_{bb'}$. The spectral densities of these noise

sources are

$$S_1 = 2\,e\,I_b, \tag{71}$$

$$S_2 = 2\,e\,I_c, \tag{72}$$

$$S_3 = 4\,k\,T/r_{bb'}. \tag{73}$$

Using the referred-impedance approach the equivalent circuit for noise analysis can be simplified as in Fig. 19. In this figure, R_F is defined as previously and C_{dsf} is the sum of the detector, stray, and feedback capacitances:

$$C_{dsf} = C_{ds} + C_f. \tag{74}$$

We now need to refer noise sources S_2 and S_3 back to the amplifier input in order to derive the input equivalent noise. Since $r_{bb'}$ (typically 20 Ω for microwave transistors) is small compared with R_F and R_π, its effect on the reference of S_2 back to the input can be neglected. Thus the circuit for referring S_2 can be simplified as in Fig. 20, where

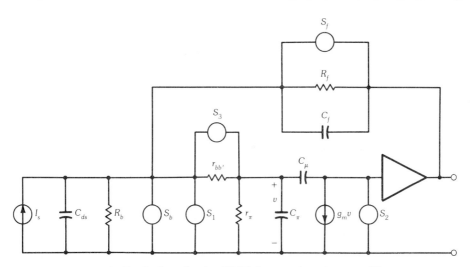

FIG. 18. Equivalent circuit of BJT–front-end receiver amplifier.

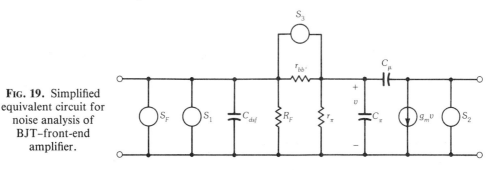

FIG. 19. Simplified equivalent circuit for noise analysis of BJT–front-end amplifier.

C_T is the equivalent total input capacitance:

$$C_T = C_{ds} + C_\pi + C_\mu + C_f. \qquad (75)$$

Fig. 20 is exactly the same as Fig. 16 for the FET–front-end amplifier. Thus we can write out the input equivalent noise spectral density of S_2 as

$$[S_2(f)]_{eq} = (S_2/g_m^2)\,[(1/R_F + 1/r_\pi)^2 + 4\,\pi^2 f^2\, C_T^2]\ . \qquad (76)$$

For the referring of the spreading resistor noise S_3 back to the input, the relevant circuit is shown in Fig. 21. Using the Thevenin equivalent circuit we can write out simply by inspection the input equivalent noise spectral density for S_3 as

$$[S_3(f)]_{eq} = S_3\, r_{bb'}^2\,(1/R_F^2 + 4\,\pi^2 f^2\, C_{dsf}^2). \qquad (77)$$

The input equivalent noise spectral density of the receiver amplifier is thus given by

$$N(f) = 4\,k\,T/R_F + 2\,e\,I_b + (2\,e\,I_c/g_m^2)$$

$$\times [(1/R_F + 1/r_\pi)^2 + 4\,\pi^2 f^2\, C_T^2] + 4\,k\,T r_{bb'}$$

$$\times (1/R_F^2 + 4\,\pi^2 f^2\, C_{dsf}^2). \qquad (78)$$

The above equation can be considerably simplified. The transistor parameters are related by the following equations:

$$I_b = I_c/\beta \qquad (79)$$

and

$$g_m = \beta/r_\pi = I_c/V_T, \qquad (80)$$

where β is the transistor current gain (assumed to be the same for both small signal and dc) and V_T is a parameter given by

$$V_T = k\,T/e \qquad (81)$$

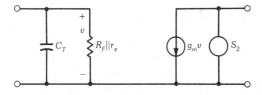

FIG. 20. Simplified circuit for referring collector noise source S_2.

and has the dimension of voltage and is equal to 25.85 mV for $T=300$ K.

In addition, we have $g_m R_F \gg 1$ and $\beta \gg 1$. Therefore (78) for the input equivalent noise spectral density can be simplified to

$$N(f) = 4\,k\,T/R_F + 2\,e\,I_c/\beta + (2\,e/I_c)$$

$$\times (2\,\pi\,V_T\,C_T)^2 f^2 + 4\,k\,T r_{bb'}\,(2\,\pi\,C_{dsf})^2 f^2. \qquad (82)$$

We see that the feedback- (or load-) resistance thermal noise and base-current shot noise are flat with frequency. The collector-current noise and base spreading resistance noise have an f^2 dependence and therefore are more important at high frequencies.

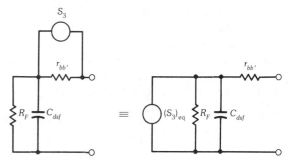

FIG. 21. Equivalent for referring the base spreading resistance noise.

Having derived the input equivalent noise spectral density, the input equivalent noise power of the receiver amplifier can be obtained as

$$\sigma_a^2 = (4\,k\,T/R_F)\,B\,I_2 + (2\,e\,I_c/\beta)\,B\,I_2 + (2\,e/I_c)$$

$$\times (2\,\pi\,V_T\,C_T)^2\,B^3\,I_3 + 4\,k\,T r_{bb'}\,(2\,\pi\,C_{dsf})^2\,B^3\,I_3\,. \qquad (83)$$

The receiver amplifier noise power is a function of the collector bias current I_c. From the above equation there exists an optimum value of I_c that minimizes the receiver noise power. In order to evaluate the optimum bias current it is necessary to consider the variation of C_T with I_c.

The base-collector capacitance C_μ is essentially independent of I_c. However, the base-emitter capacitance C_π is dependent on the bias currents due to the base-charging capacitance component. The variation of C_π with I_c can be approximated by a first-order linear relationship [57]:

$$C_\pi = C_{je} + \alpha I_c, \qquad (84)$$

where C_{je} is the emitter-junction space-charge capacitance and

$$\alpha = \tau_F / V_T, \qquad (85)$$

with τ_F being the forward transit time of the transistor.

The total input capacitance can thus be written in the form

$$C_T = C_0 + \alpha I_c, \qquad (86)$$

where C_0 is the total input capacitance under a zero bias-current condition:

$$C_0 = C_{ds} + C_f + C_{je} + C_\mu. \qquad (87)$$

By substituting the expression for C_T in (86) into (83), the dependence of σ_a^2 on I_c can be written in the form

$$\sigma_a^2 = a + b I_c + c/I_c, \qquad (88)$$

where a, b, and c are constants independent of I_c given as

$$a = (4 k T/R_F) B I_2 + 4 e \alpha C_0 (2 \pi V_T)^2 B^3 I_3$$

$$+ 4 k T r_{bb'} (2 \pi C_{dsf})^2 B^3 I_3, \qquad (89)$$

$$b = (2 e/\beta) B I_2 + 2 e (2 \pi \alpha V_T)^2 B^3 I_3, \qquad (90)$$

$$c = 2 e (2 \pi V_T C_0)^2 B^3 I_3. \qquad (91)$$

The optimum value of I_c for minimizing the receiver noise power σ_a^2 can be found by differentiating (88) to be:

$$(I_c)_{opt} = (c/b)^{1/2} \qquad (92)$$

or

$$(I_c)_{opt} = 2 \pi V_T C_0 B (\beta I_3/I_2)^{1/2} \{1/[1 + (2 \pi \alpha V_T B)^2$$

$$\times \beta I_3/I_2]^{1/2}\}. \qquad (93)$$

The minimum value of σ_a^2 is therefore given by

$$(\sigma_a^2)_{opt} = a + 2(b c)^{1/2} \qquad (94)$$

or

$$(\sigma_a^2)_{opt} = (4 k T/R_F) B I_2 + 8 \pi k T C_0 B^2 (I_2 I_3/\beta)^{1/2}$$

$$\times [1 + (2 \pi \alpha V_T B)^2 \beta I_3/I_2]^{1/2} + 4 e \alpha C_0 (2 \pi V_T)^2$$

$$\times B^3 I_3 + 4 k T r_{bb'} (2 \pi C_{dsf})^2 B^3 I_3. \qquad (95)$$

The optimum collector bias current has been calculated and plotted in Fig. 22 as a function of bit rates. The parameter values used for the calculation are shown in Table 5 and represent state-of-the-art devices. It can be observed that below 1 Gb/s the $[1 + (2 \pi \alpha V_T B)^2 \beta I_3/I_2]^{1/2}$ term in the two equations above can be neglected and the optimum bias current is proportional to bit rate. As the data rate increases above 1 Gb/s the term involving B^2 in the bracketed term becomes more dominant and eventually the optimum bias current becomes independent of bit rate.

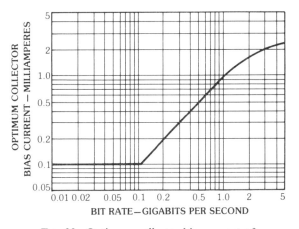

Fig. 22. Optimum collector bias current of bipolar-transistor–front-end receiver amplifiers. *(After Muoi [27], ©1984 IEEE)*

We can also observe that the optimum bias current is quite small at low bit rates. A low bias current, however, is undesirable because of two factors. Firstly, the transistor current gain β and cutoff frequency f_T tend to decrease as the bias current decreases. Thus not only the frequency response is degraded but the noise level may be actually higher at the optimum bias current because of the higher base-current shot noise. The second factor is in relation to dynamic range. When the bias current is too low the dynamic range of the receiver amplifier is limited because the signal current swing has to be less than the bias current. Therefore it is necessary to put a lower limit on the collector bias current. This lower limit is chosen to be 0.1 mA as shown in Fig. 22.

It is therefore of interest to determine the receiver amplifier noise penalty for off-optimum bias operation. Suppose we operate the transistor at a collector bias current which is δ times its optimum value (δ can be less or greater than 1):

$$I_c = \delta\,(I_c)_{\text{opt}}\,. \qquad (96)$$

It can be shown from (88) and (92) that the amplifier noise becomes

$$\sigma_a^2 = a + (b\,c)^{1/2}\,(\delta + 1/\delta)\,. \qquad (97)$$

The amplifier noise increase over its optimized value in (94) can thus be evaluated. For the high-impedance design the constant term a can be neglected (except at very high data rates when the B^3 terms become significant). Thus for this case the amplifier noise simply increases by a factor of $(\delta + 1/\delta)/2$. In the case of the transimpedance design, where the constant σ becomes a significant portion of the amplifier noise, the amplifier noise increase due to off-optimum biasing is considerably less. This amplifier noise increase has been evaluated and plotted in Fig. 23, where the running parameter is defined as $s = a/2(b\,c)^{1/2}$. It can be observed that if we operate at a factor of 2 away from the optimum current ($\delta = 0.5$ or $\delta = 2$), the noise penalty is about 1 dB for the high-impedance design ($s \simeq 0$) while it is only 0.5 dB or less for the transimpedance design ($s > 1$).

Using the parameter values in Table 5, the input equivalent noise power has been calculated and plotted in Fig. 24. The optimum collector bias current lower limit is taken to be 0.1 mA as shown in Fig. 22. Both high-impedance and transimpedance receiver amplifiers are shown in Fig. 24 (BJT–HZ and BJT–TZ curves, respectively). For the transimpedance design the feedback resistance has to be reduced as the bit rate increases in order to accommodate the wider-bandwidth requirement. In Fig. 24 we assume that the feedback resistance is inversely proportional to

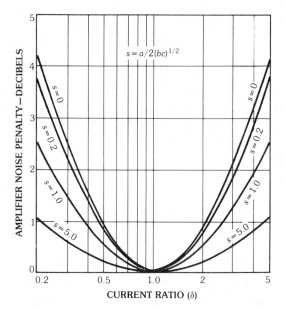

FIG. 23. Amplifier noise penalty for off-optimum bias operation.

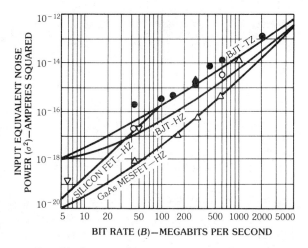

FIG. 24. Input equivalent noise power of typical receiver amplifiers.

the bit rate. The product of feedback resistance and bit rate is thus a constant and taken to be 750 k$\Omega \cdot$Mb/s.

The curves for the FET–front-end high-impedance amplifier design from Fig. 17 are also included in Fig. 24 for comparison. It can be observed that gallium-arsenide MESFETs offer lower noise levels than BJTs up to at least 1 Gb/s. However, FET noise increases with bit rate at a faster rate (as

B^3) than the BJTs (as B^2). Thus, above several gigabits per second, BJTs are more suitable.

Practical Receiver Amplifiers

A great number of optical receiver amplifiers have been implemented for laboratory experiments as well as actual system installations. In Fig. 24 we have included measured data of amplifier noise levels of some of the best results reported in the literature. The circles indicate BJT–front-end receivers, the triangles indicate gallium-arsenide MESFET receivers, and the inverted triangles indicate silicon-JFET design. Open circles (or triangles) indicate high-impedance design while filled circles (or triangles) indicate transimpedance design. Most of the results shown are obtained from either hybrid or monolithic receiver circuits, especially at high data rates. For the transimpedance design most of the receivers reported use bipolar-transistor–front-end amplifiers ranging from data rates of several tens of megabits per second to several gigabits per second [27, 58–63]. Experimental noise levels are somewhat higher than calculated values mainly because practical receivers often have to be designed to satisfy other requirements (such as wide dynamic range). As a result receiver sensitivity is degraded.

For the high-impedance design, silicon FETs have been used up to 50 Mb/s [14, 15] and gallium-arsenide MESFETs used up to 1 Gb/s [50, 64, 65] with excellent results.

In the remainder of this section we will consider typical examples of receiver amplifier design.

Transimpedance-Amplifier Design—The transimpedance receiver amplifier design is normally used when a wide dynamic range is required. The most important parameter of the transimpedance design is the feedback resistance value since it affects the amplifier bandwidth (hence operating data rate), gain (hence dynamic range), and noise (hence receiver sensitivity).

Recall that the bandwidth of the transimpedance amplifier as shown in Fig. 13a is given by (59) as

$$f_2 = A/2\pi R_f(C + A C_f),$$

which has an upper limit of $1/2\pi R_f C_f$ as A increases to large values. The amplifier bandwidth decreases as R_f increases.

The output signal voltage swing of the transimpedance amplifier is directly proportional to R_f. This signal swing is limited due to biasing conditions (bias currents and supply voltages) of the transistors within the transimpedance amplifier. Thus, if a high signal input level is to be tolerated (i.e., wide dynamic range), a low value of R_f is required. On the other hand, a high value of R_f is desirable in order to minimize the receiver amplifier noise as seen from (70) and (83).

Thus the feedback resistance value should be chosen as high as possible design and yet still satisfy the dynamic range and bandwidth requirements. A reasonable design procedure is to establish first the upper limit on R_f based on the dynamic range requirement. If this limit is low enough, then the bandwidth requirement can be readily satisfied with reasonably low values of open-loop voltage gain A. Thus only one or two amplifying stages is necessary and the transimpedance amplifier design can be readily carried out. A more difficult case is when the bandwidth requirement cannot be satisfied at the above limit of R_f, which is due to the dynamic range requirement. In this case the value of R_f is dictated by the bandwidth requirement. In order to make R_f as large as possible, it is desirable that the voltage gain A has to be as high as possible (within limit of stability margin). Thus, depending on the required bandwidth, three or more amplifying stages may be required within the feedback loop.

It has been established that a cascade of common-emitter (or common-source) amplifying stages is the optimum configuration for minimum noise level [19]. However, interaction between coupling stages and parasitic capacitances tends to limit the bandwidth of the open-loop voltage amplifier. Thus, for wideband applications the technique of impedance mismatching between coupling stages to minimize interaction and parasitic capacitances [27] is preferred. The first amplifying stage is a common-emitter or common-source stage with a high output impedance. Therefore, a second stage with a low input impedance is desirable. This second stage can be either a common-base stage or a localized shunt-feedback stage as shown in Fig.

25. It has been shown that the common-emitter (-source)/shunt-feedback (ce/sf or cs/sf) configuration offers slightly better noise performance than the common-emitter/common-base (which is the classical cascode) configuration [17]. Additional amplifying stages, if required, can be provided by the alternate feedback pair of localized series and shunt feedback. The last stage can be an emitter follower to provide a low output impedance before the overall transimpedance feedback is applied.

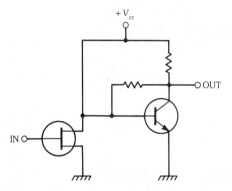

(a) Common-source (-emitter)/shunt-feedback circuit.

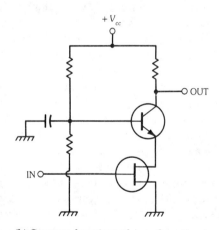

(b) Common-base (cascode) configuration.

FIG. 25. Typical first and second stages.

Two examples of a three-stage transimpedance amplifier design are shown in Fig. 26. The first stage can use either an FET as shown or a bipolar transistor. Note that alternate npn and pnp transistors are used to obtain direct-coupling simplicity. The common-emitter (-source)/shunt-feedback approach is shown in Fig. 26a. In this case a series-feedback third stage provides additional gain and inverts the amplifier's output for negative feedback. This configuration was initially proposed by Hullett and Muoi [17] and also later realized by Hullett and Moustakas [66]. The open-loop voltage gain in this case is given approximately as

$$A = g_m R_{f2} R_{c3}/R_{e3},$$

where g_m is the transconductance of the first-stage active device.

The cascode approach is shown in Fig. 26b. In this case the third stage is an emitter follower. This configuration was initially proposed by Ogawa [51] and is widely used by commercial vendors. The open-loop voltage gain for this case is

$$A = g_m R_{c2}.$$

By comparing the two expressions above, it is evident that the configuration in Fig. 26a can provide a higher open-loop gain than the cascode configuration in Fig. 26b. Thus, a higher value of R_f can be used in Fig. 26a, provided that it still satisfies the dynamic range requirement.

With the cascode configuration in Fig. 26b a bandwidth of 112 MHz can be obtained with a feedback resistance of 5.1 kΩ [51]. By using the configuration in Fig. 26a the same bandwidth can be achieved with a tenfold increase in the feedback resistance value (51 kΩ) [66].

The three-stage configuration in Figs. 26a and 26b can be used for bandwidth up to approximately 500 MHz (corresponding to 1 Gb/s). For wider-bandwidth applications the propagation delay and phase shift through the three stages tend to cause unacceptable gain and phase margin and may even cause instability. Thus the transimpedance feedback is normally applied over only one stage. The resulting front end is thus a local shunt-feedback stage. An example of such a design is shown in Fig. 27, where the second stage is a series-feedback stage as suggested by the interstage impedance-mismatching technique discussed previously. A bandwidth of over 2 GHz is obtained with a feedback resistance of 400 Ω. [27].

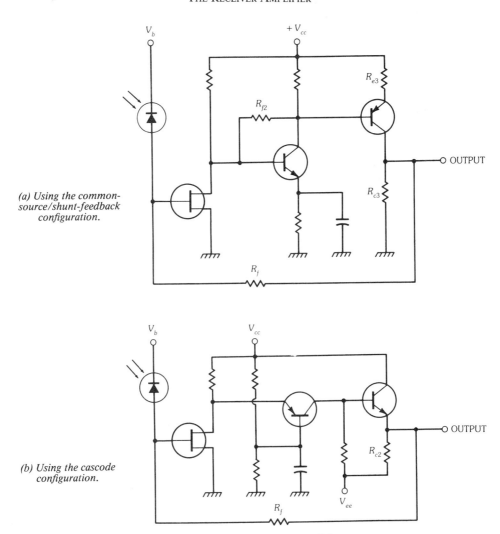

(a) Using the common-source/shunt-feedback configuration.

(b) Using the cascode configuration.

FIG. 26. Transimpedance amplifier.

High-Impedance–Amplifier Design—There are two basic configurations for the high-impedance design. The first configuration is a cascade of amplifying stages where there is no feedback applied to the first stage. Thus the load resistance is only due to the bias resistance of the photodiode and transistor. This load resistance is chosen to be as high as possible so as to minimize its noise contributions. This configuration can be referred to as the true high-impedance approach. The second high-impedance design approach actually uses a transimpedance configuration where the feedback resistance has been increased to the point where its contribution to the amplifier noise is negligible. For this condition to be satisfied the feedback resistance is normally so high that the transimpedance-amplifier bandwidth is severely limited. Thus an equalizer is required to equalize out the excessive low-frequency gain as in the first approach. Therefore this is also classified as a high-impedance (or integrating–front-end) design even though the amplifier configuration is transimpedance.

With the true high-impedance design the optimum configuration for lowest noise is a cascade of common-emitter (common-source) stages. This approach has been used up to 50 Mb/s as shown in Fig. 28a [14, 15]. For wider-bandwidth applications the ce (cs)/sf or cascode configuration as discussed previously is preferable. The cascode

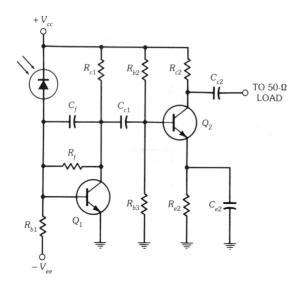

FIG. 27. Transimpedance-amplifier design with 2-GHz bandwidth.

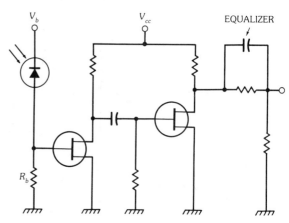

(a) Using cascode of common-source (-emitter) stages.

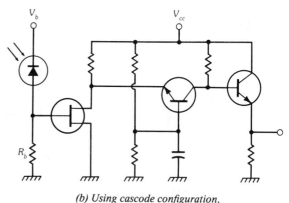

(b) Using cascode configuration.

FIG. 28. High-impedance amplifier.

configuration as shown in Fig. 28b has been used up to 565 Mb/s by British Telecom Research Laboratories [64].

The second high-impedance design approach of using the transimpedance configuration has also been implemented. Using the cs/sf configuration in Fig. 26a the feedback resistance has been increased to 200 kΩ for a 250-Mb/s data rate [67]. With the cascode configuration in Fig. 26b a feedback resistance of 500 kΩ has been used at 45 Mb/s and 90 Mb/s by Bell Laboratories [68].

6. Receiver Sensitivity

One of the primary objectives in optical digital-receiver design is to minimize the incident optical power required to achieve a desired error rate. In this section we evaluate this minimum required power theoretically and see how well it can be approached in practice.

Receiver Sensitivity Analysis

Noise sources at the receiver impose a limit on the receiver sensitivity that can be achieved. They consist of the detector noise and the receiver amplifier noise. The probability density function (pdf) of the APD output current is given previously in (10) and shown in Fig. 4. The receiver amplifier noise can be assumed to be Gaussian distributed with variance σ_a^2. The pdf of the resultant signal can thus be obtained as the convolution of the APD output distribution and the Gaussian-noise distribution.

The pdf's of the resultant signal for the "one" and "zero" bits are shown in Fig. 29. It is well known from communication theory that the optimum detection threshold D_0 for minimum error rate is placed at the intersection of the two pdf curves. The error probability is represented by the shaded area under the two curves. The bit-error rate is given by

$$P[E] = p(0) \int_{D_0}^{\infty} p_0(i)\, di + p(1) \int_{0}^{D_0} p_1(i)\, di, \quad (98)$$

where $p(0)$ and $p(1)$ are the probabilities of occurrence of "zero" and "one" bits in the transmitted data pattern and $p_0(r)$ and $p_1(r)$ are the pdf's for the "zero" and "one" bits.

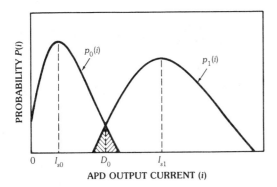

FIG. 29. Probability density function of the detected signal current.

Normally, we can assume that

$$p(0)=p(1)=\tfrac{1}{2},\qquad(99)$$

and therefore

$$P[E]=\tfrac{1}{2}\left[\int_{D_0}^{\infty}p_0(i)\,di+\int_0^{D_0}p_1(i)\,di\right].\qquad(100)$$

Even with the approximate expression for the APD output distribution in (10), the expressions for $p_0(i)$ and $p_1(i)$ are still too mathematically complex and cannot be readily handled. The error-rate calculation has to be carried out by numerical techniques. Among the methods proposed are "exact" calculation [13], Monte Carlo simulation using importance sampling [13, 69], Chernoff bound [13], Gram-Charlier series expansion [70, 71], Gauss quadrature rule [72, 73], and the characteristic function method [74, 75]. These calculations are time-consuming and do not provide much insight to the receiver design process. Consequently, the Gaussian approximation for the APD statistics [10] is widely used. It has been verified that the receiver sensitivity result obtained by the Gaussian approximation approach is in close argument (within 1 dB) with the exact calculation [13]. The inaccuracy of the Gaussian-approximation approach is in its overestimation of the optimum avalanche gain and underestimation of the decision threshold level. These two parameters, however,

can be determined experimentally by optimizing the bit-error rate. The Gaussian approximation is thus most useful in the process of receiver design and trade-off analysis of different system parameters. Once the system has been designed, one of the more complicated numerical techniques can be used to confirm the proper operation of the system.

In the Gaussian approximation approach the APD noise is represented by a Gaussian distribution with variance as given in (44) and (45). The total noise power is thus the sum of the photodetector noise and receiver amplifier noise power.

For the "one" bit the detected signal is obtained from (43) as

$$I_{s1}=R\,G\,b_1,\qquad(101)$$

and the total noise power is

$$\sigma_1^2=2\,e\,R\,G^2\,F(G)\,B\,I_1\,b_1+\sigma_d^2+\sigma_a^2.\qquad(102)$$

Similarly, for the "zero" bit the detected signal and the total noise power are given by the following two relations:

$$I_{s0}=R\,G\,b_1\,r\qquad(103)$$

and

$$\sigma_0^2=2\,e\,R\,G^2\,F(G)\,B\,I_1\,b_1\,r+\sigma_d^2+\sigma_a^2,\qquad(104)$$

with r being the optical-source extinction ratio as defined in (19).

The pdf's of the detected signals for the "one" and "zero" symbol in the case of the Gaussian approximation approach are shown in Fig. 30. In addition, to further simplify the analysis the decision threshold level is assumed to be placed such that the probability of error produced by each of the two symbols is the same, i.e., such that

$$\int_{-\infty}^{D}p_1(i)\,di=\int_{D}^{\infty}p_0(i)\,di.\qquad(105)$$

Since $p_1(i)$ and $p_0(i)$ are now Gaussian distributions with means I_{s1} and I_{s0} and variances σ_1^2 and σ_0^2, the above equation reduces to

$$(D-I_{s0})/\sigma_0=(I_{s1}-D)/\sigma_1.\qquad(106)$$

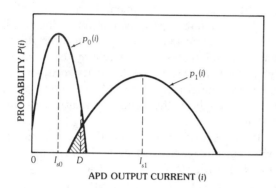

FIG. 30. Gaussian approximation of the detected–signal-current probability density function.

Thus the decision threshold level is given by

$$D = (\sigma_1 I_{s0} + \sigma_0 I_{s1})/(\sigma_1 + \sigma_0) . \tag{107}$$

Recall that the optimum threshold level D_0 for minimum error rate is placed at the intersection of the two pdf curves. It has been shown that the maximum penalty resulting from the use of the nonoptimum threshold level in (107) is only a doubling of the minimum attainable error rate [11]. This nonoptimum threshold level, however, considerably simplifies the analysis.

The probability of error is obtained from (100) as

$$P[E] = \int_D^\infty p_0(i)\, di = \int_{-\infty}^D p_1(i)\, di \tag{108}$$

or

$$P[E] = (1/2\pi)^{1/2} \int_Q^\infty \exp(-x^2/2)\, dx , \tag{109}$$

where Q is a parameter defined as the value of either side of (106), i.e.,

$$Q = (D - I_{s0})/\sigma_0 = (I_{s1} - D)/\sigma_1 . \tag{110}$$

The numerical values of Q for a desired error rate $P[E]$ can be found from mathematic tables. A plot of $P[E]$ versus Q is shown in Fig. 31. For error rate less than 10^{-4}, which is almost always the case of practical interest, the error-rate expression in (109) is bounded by two simple functions as follows:

$$(2\pi)^{-1/2} Q^{-1} (1 - 1/Q^2) \exp(-Q^2/2) < P[E]$$

$$< (2\pi)^{-1/2} Q^{-1} \exp(-Q^2/2) .$$

These two bounds are indistinguishable from the value of $P[E]$ on the plot of Fig. 31.

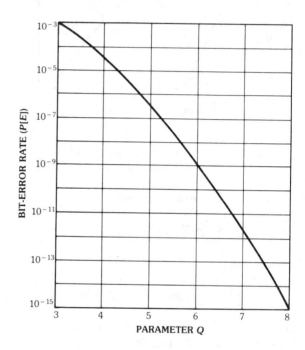

FIG. 31. Plot of error rate versus Q term.

Eliminating D from (110) we obtain

$$Q(\sigma_1 + \sigma_0) = b_1 - b_0 = b_1(1 - r) . \tag{111}$$

When the expressions for σ_1 and σ_0 from (102) and (104) are substituted into (111), we obtain an equation in b_1. Solving this equation for b_1 and using (20) relating b_1 to $\bar{P}$, we finally obtain an expression for the average detected optical power required at the receiver:

$$\eta \bar{P} = Q h \nu e^{-1} (1+r)(1-r)^{-1}$$
$$\times \Big((1+r)(1-r)^{-1} Q e B I_1 F(G)$$
$$+ \Big\{ [Q e B I_1 F(G)/(1-r)]^2 4r$$
$$+ (\sigma_a^2 + \sigma_d^2)/G^2 \Big\}^{1/2} \Big) . \tag{112}$$

The above expression for the optical power required is expressed as a function of the avalanche gain G. Recall that the avalanche excess-noise factor $F(G)$ increases with G as in (8). In addition, the dark-current noise term σ_d^2 also increases with G as in (45).

It can be observed from (112) that as the avalanche gain G increases, the contribution of the receiver amplifier noise term σ_a^2 to the required optical power decreases. However, the contributions of the APD noise (term involving I_1) and the multiplied dark-current noise (I_{dm}) increase. Thus there exists an optimum value for G where the required optical power is minimized.

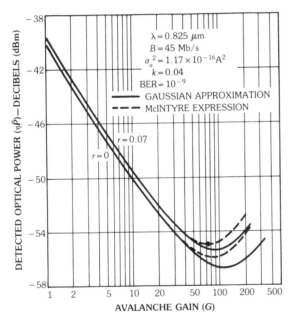

FIG. 32. Detected optical power requirement as a function of avalanche gain.

The optimum avalanche gain value can be found by differentiating (112) and equating the derivative to zero. However, the resulting expression is rather complicated. Thus it can best be found by plotting out the curve of required optical power $\eta \bar{P}$ against the avalanche gain G and looking for the minimum.

An example of such a calculation is shown in Fig. 32 where the two solid lines represent the Gaussian approximation results for two cases of extinction ratio: $r=0$ and $r=0.07$. The parameter

values used for the calculation are obtained from the experimental data reported by Personick and colleagues [13] as shown in Fig. 32. The two dashed curves are calculated from an expression proposed by McIntyre [76] which approximately takes into account the non-Gaussian nature of the avalanche noise statistics. This alternative expression is given as

$$\eta \bar{P} = \tfrac{1}{2} h \nu Q^2 B (1+r)(1-r)^{-1} \Big((1+r)(1-r)^{-1}$$

$$\times F(G) + \{[(1+r)(1-r)^{-1} F(G)]^2 - 2F(G) + 1$$

$$+ (2\sigma_a^2/B e Q G)^2\}^{1/2} \Big). \quad (113)$$

Referring to Fig. 32, the value $G=1$ corresponds to the case of a simple pin photodiode. In this case the detector shot noise is negligible and the receiver amplifier noise dominates. As the avalanche gain increases from unity, initially the receiver amplifier noise is still the dominant noise source. Therefore the required optical power simply decreases proportionally with the avalanche gain increase (e.g., 10-dB reduction in power requirement when the avalanche gain increases by a factor of 10). Then, as the avalanche gain becomes higher, the detector noise becomes significant and the required power starts to depart from the above linear relationship. Finally, as the avalanche gain increases even further, the detector noise becomes dominant and the required optical power increases with the avalanche gain.

The optimum avalanche gain can be obtained from Fig. 32. It can be noted that the minimum of the curves is quite broad. Thus the exact prediction of the optimum avalanche gain is not necessary. Similarly, the optimum avalanche gain need not be accurately controlled in practice.

The effect of a nonzero extinction ratio can be observed from Fig. 32. Compared with the ideal case of zero extinction ratio, the optimum avalanche gain decreases and the required optical power increases. For the plotted extinction ratio value of 0.07 the receiver sensitivity penalty is approximately 1.2 dB for the APD case and 0.6 dB for the case of a pin photodiode ($G=1$). The effect of nonzero extinction will be expanded on in a later section.

407

Also included in Fig. 32 is the experimental data for receiver sensitivity at optimum avalanche-gain operation reported by Personick and coworkers [13]. It can be seen that the Gaussian approximation predicts a slightly higher optimum avalanche gain and slightly better receiver sensitivity (by about 0.3 dB). The agreement between the approximate McIntyre expression and the experimental data in this case is quite good.

The position of the decision threshold level as predicted by the Gaussian approximation approach is shown in Fig. 33. The parameter d plotted is defined as

$$d=(D-I_{s0})/(I_{s1}-I_{s0}) . \tag{114}$$

It can be observed that for low values of avalanche gain (below 10) the threshold level is midway between the two signal levels. This is because in this case the receiver amplifier noise dominates and is the same for both signal levels. As the avalanche gain increases further, the signal-dependent detector noise is more significant. Therefore the noise in the "one" signal level is higher than the noise in the "zero" signal level. Thus the threshold level has to be placed lower than the 50% mark position.

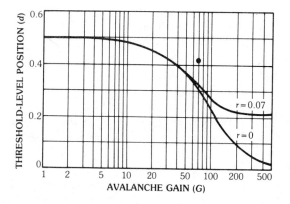

FIG. 33. Decision threshold-level position as a function of avalanche gain.

Also shown in Fig. 33 is the experimental data point for the threshold level in the case $r=0.07$ as reported by Personick [13]. It can be observed that the actual threshold level is higher than the value predicted by the Gaussian approximation.

This is expected since the pdf for the "zero" signal level in Fig. 4 has a long tail which is neglected by the Gaussian approximation.

It can be summarized that the Gaussian approximation provides a reasonably accurate prediction of the receiver sensitivity and can be effectively used, provided we keep in mind its overestimation of the optimum avalanche gain and underestimation of the threshold level. The approximate "McIntyre expression" also predicts the receiver sensitivity very well. In addition, it predicts an optimum avalanche gain which is lower than the Gaussian approximation and is thus closer to the experimental value.

The expression for the receiver sensitivity simplifies considerably for the case of a pin photodiode. In this case $G=1$ and the signal-dependent photodector noise can be neglected (except for analog systems requiring large snr's). The dark current noise is normally lumped together as part of the receiver amplifier noise term σ_a^2. Both (112) and (113) now reduce to

$$\eta \bar{P}=Q h \nu e^{-1}(1+r)(1-r)^{-1}\sigma_a . \tag{115}$$

The Ideal Case of Zero Extinction Ratio

It is convenient to use the receiver sensitivity in the case of zero extinction ratio as a reference baseline. The effect of the extinction ratio can be considered as a separate receiver sensitivity penalty.

For the case $r=0$ the Gaussian approximation expression for the receiver sensitivity from (112) reduces to

$$\eta \bar{P}=Q h \nu e^{-1} \{Q e B I_1 F(G)+[(\sigma_a^2+2 e I_{du} B I_2)$$

$$/G^2+2 e I_{dm} F(G) B I_2]^{1/2} \} , \tag{116}$$

where the expression for the dark-current noise in (45) has been used.

The noise due to unmultiplied dark current component I_{du} can be considered as part of the receiver amplifier noise term σ_a^2 since it is independent of G. The above expression thus becomes

$$\eta \bar{P}=Q h \nu e^{-1} \{Q e B I_1 F(G)+[\sigma_a^2/G^2$$

$$+2 e I_{dm} F(G) B I_2]^{1/2} \} . \tag{117}$$

In order to further simplify the analysis let us consider the case of zero multiplied dark current ($I_{dm} \simeq 0$). This case is true for silicon APDs and possibly InGaAs APDs but certainly not applicable for germanium APDs (see Table 2). When the above condition is applicable, we obtain

$$\eta \bar{P} = (Q h \nu / e)[Q e B I_1 F(G) + \sigma_a / G] . \quad (118)$$

The optimum avalanche gain can be obtained by equating the derivative of the above expression to zero. It can be shown that the optimum avalanche gain is approximately given by

$$G_{opt} = (\sigma_a / k Q e B I_1)^{1/2} . \quad (119)$$

The minimum required optical power at the optimum gain is therefore given by

$$\eta \bar{P} = 2 h \nu Q^2 B I_1 (k G_{opt} + 1 - k). \quad (120)$$

The dependence of the receiver sensitivity on the bit rate can be estimated from the above equation, once the amplifier noise variation with bit rate is known.

bipolar-transistor front-end (TZ–B). The receiver sensitivity is shown for both silicon pin and avalanche photodiodes. It can be observed that when the pin photodiode is used, a low-noise amplifier is necessary to achieve high receiver sensitivity. If an APD is used, the requirement of a low-noise amplifier is much alleviated due to the excellent characteristics of silicon APDs. As a result, most practical optical receivers using silicon APDs use the transimpedance design (normally with bipolar transistors) so that a wide dynamic range can be achieved at the same time. This fact is evident from the experimental data points as reported in the literature [15, 63, 77–78] shown in Fig. 34. The quantum limit of receiver sensitivity is also shown in Fig. 34. It can be seen that practical APD receiver sensitivity values are 12 dB to 15 dB away from the quantum limit.

The sensitivity of long-wavelength optical receivers can be seen from Fig. 35. Similarly as before, for the simple pin photodiode an ultralow-noise receiver amplifier is necessary in order to achieve the best performance. Thus the amplifier design normally follows the high-impedance design with an FET front-end (for example as in Fig. 28b)

TABLE 6. Photodiode Parameter Values Used in the Receiver Sensitivity Calculation

Parameter	InGaAs PIN	Gernamium APD	InGaAs APD	Silicon APD
Detector and stray capacitance C_{ds} (pF)	0.5	0.5	0.5	0.5
Unmultiplied dark current I_{du} (nA)	1	50	1	1
Multiplied dark current I_{dm} (nA)		50	1	0.001
Ionization-coefficients ratio k		1	0.3	0.03

The Sensitivity of Short- and Long-Wavelength Optical Receivers

Using the receiver amplifier noise levels calculated and shown previously in Fig. 24, the sensitivity of optical receivers can be evaluated using (116). The photodiode parameters used in the calculations are listed in Table 6.

The sensitivity of short-wavelength optical receivers is shown in Fig. 34. Three different amplifiers are shown: a high-impedance FET front-end (HZ–F), a high-impedance bipolar-transistor front-end (HZ–B), and a transimpedance

in a hybrid circuit. The resulting receiver is often referred to as a pinFET receiver. This approach is very attractive in the bit-rate range from 10 Mb/s to 500 Mb/s because of its simplicity and relatively low cost. Excellent receiver sensitivity results have been reported up to 565 Mb/s [64] as shown by the inverted triangular data points in Fig. 34. The pinFET receiver has also been used at 1 Gb/s [65]. Hybrid pinFET receiver amplifiers using gallium-arsenide MESFETs are commercially available and have been used in many practical system installations. It is noted here that the pinFET receiver am-

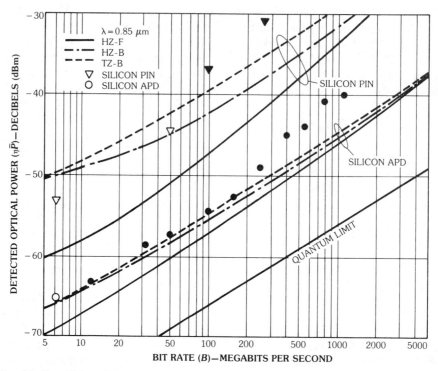

FIG. 34. Sensitivity of short-wavelength optical receivers. (*After Muoi [27], ©1984 IEEE*)

plifier can also be fabricated in monolithic form. This monolithic integration approach has the advantage of lower parasitic capacitance, yielding better receiver sensitivity, higher reliability, and lower cost. Such integrated pinFET receivers have been fabricated using junction FETs [79, 80] and MISFETs [81]. However, their performance is still inferior to hybrid pinFET receivers, probably because they are still in early development stage at this point in time.

Germanium APDs are also often used, particularly in Japan, because of their ready availability. The sensitivity of germanium APD receivers is shown in Fig. 34 for the three amplifier designs considered previously. It can be observed that germanium APDs offer better receiver sensitivity than pinFET receivers at high bit rates. The exact crossover point is dependent on the particular characteristics of photodiode and receiver amplifier being used. It is, however, normally in the range of 100 Mb/s to 600 Mb/s. It has to be remembered that the germanium APD dark current plays a dominant role in determining the germanium-APD receiver sensitivity. Consequently,

the germanium-APD receiver sensitivity is much more sensitive to temperature changes than the pinFET receiver sensitivity. Thus the comparison between the two detector types has to have been carried out at the worst-case operating temperature. Included in Fig. 35 (circular data points) are experimental values of germanium-APD receiver sensitivity [82] using the transimpedance amplifier design. They compare favorably with the predicted performance.

It can be observed from Fig. 35 that by using InGaAs APDs the receiver sensitivity can be improved by 5 to 10 dB. The triangular data points in Fig. 35 show experimental sensitivity of InGaAs APD receiver at 45 Mb/s [48], 420 Mb/s, and 1 Gb/s. These data points are worse than the predicted values due to higher amplifier and detector noise. With continuing improvement in InGaAs APD and amplifier characteristics, experimental performance close to the calculated curve can be expected.

We can also observe that practical receiver sensitivity values obtained so far are about 20 dB away from the quantum limit. This is worse than

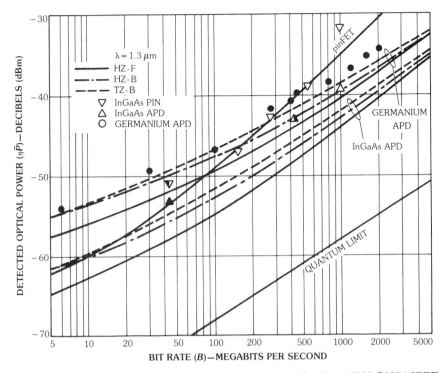

FIG. 35. Sensitivity of long-wavelength optical receivers. (*After Muoi [27], ©1984 IEEE*)

the situation at short wavelengths due to the higher noise of long-wavelength APDs.

The optimum avalanche gain required has also been calculated and is plotted in Fig. 36 for various detectors and amplifier designs. In general, the optimum avalanche gain is such that the detector shot noise is comparable to the receiver amplifier noise. Thus the better the detector (lower dark current and excess noise), the higher the optimum avalanche gain. Similarly, the better the amplifier (lower noise level), the lower the optimum avalanche gain. It can be seen that the optimum gain requirement of long-wavelength photodiodes is quite moderate, ranging from 2 to 15 for germanium APDs and from 10 to 35 for InGaAs APDs as compared with 50 to 100 for short-wavelength silicon APDs. This is because of the high avalanche excess noise and high dark current of these long-wavelength detectors.

The Effect of the Optical-Source Extinction Ratio

As mentioned previously, the effect of the optical-source extinction ratio can be considered as a separate receiver sensitivity penalty. For the case of pin photodiodes this receiver sensitivity penalty can be evaluated from (115) as

$$\eta \, \bar{P}(r)/\eta \, \bar{P}(r=0) = (1+r)/(1-r) . \qquad (121)$$

For the case of APDs the sensitivity penalty can be evaluated from (112) or (113). The calculation, however, is complicated and an explicit expression for the sensitivity penalty in terms of the extinction ratio as above cannot be obtained. Thus we revert back to the approximate (9) for the avalanche excess noise. By using this approximation we can express the sensitivity degradation in closed form.

To further simplify the analysis we assume that the dark current is negligible. Thus (112) now becomes

$$\eta \, \bar{P} = Q \, h \, \nu \, e^{-1} \, (1+r) \, (1-r)^{-1}$$

$$\times \Big((1+r) \, (1-r)^{-1} \, Q \, e \, B \, I_1 \, G^x$$

$$+ \{ [Q \, e \, B \, I_1 \, G^x \, (1-r)^{-1}]^2 \, 4 \, r$$

$$+ \sigma_a^2 \, G^{-2} \}^{1/2} \Big) . \qquad (122)$$

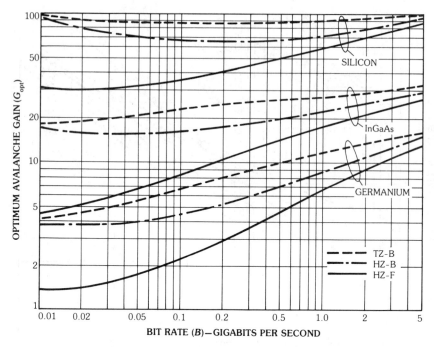

FIG. 36. Optimum avalanche gain of APD receivers. (*After Muoi [27], ©1984 IEEE*)

The optimum avalanche gain can be obtained by equating the derivative of the above equation to zero. Then the minimum required optical power at this optimum gain can be found. After some tedious mathematical manipulation we obtain for the optimum avalanche gain

$$G_{\text{opt}} = [(1-r)\sigma_a(\gamma_1)^{1/2}/Q e B I_1 \gamma_2]^{1/(x+1)} \quad (123)$$

and for the minimum required optical power

$$\eta\bar{P} = h\nu(2e)^{-1}(1+r)(1-r)^{-1}[BI_1 e/(1-r)]^{1/(x+1)}$$

$$\times(Q\gamma_2)^{(x+2)/(x+1)}(\sigma_a^2\gamma_1)^{x/(2x+2)}, \quad (124)$$

where parameters γ_1 and γ_2 are defined as

$$\gamma_1 = 4(1+x)x^{-2}$$

$$\times\{[(1+r)^2+16r(1+x)x^{-2}]^{1/2}+1+r\}^{-1} \quad (125)$$

and

$$\gamma_2 = (2+1/\gamma_1)^{1/2}+(2r+1/\gamma_1)^{1/2}. \quad (126)$$

Note that we have obtained explicit expressions for the optimum avalanche gain and receiver sensitivity in terms of the extinction ratio r. The receiver sensitivity penalty due to a nonzero extinction ratio can thus be obtained from (124) as

$$\eta\bar{P}(r)/\eta\bar{P}(r=0)=(1+r)[2(1+x)]^{-1}(x^2\gamma_1)^{x/(2x+2)}$$

$$\times[\gamma_2/(1-r)]^{(x+2)/(x+1)}. \quad (127)$$

The above receiver sensitivity penalty is plotted as a function of the extinction ratio for various values of x in Fig. 37. For silicon APDs we recall from Section 3 under "Noise Analysis," that x ranges from 0.3 to 0.4. The sensitivity penalty as shown in Fig. 37 can be well approximated by

$$\text{sensitivity penalty (dB)} = 19r \quad \text{(silicon APDs)}. \quad (128)$$

Similarly, we can approximate

$$\text{sensitivity penalty (dB)} = 16r \quad \text{(InGaAs APDs)}. \quad (129)$$

These two approximations are within 5% (for InGaAs devices) and 10% (for silicon devices) of more accurate calculations.

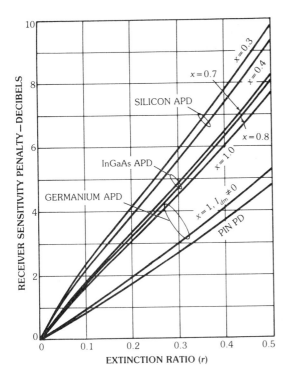

Fig. 37. Receiver sensitivity penalty due to extinction ratio.

When the dark current is negligible the curves for the three APD types are quite close to one another. Results for other values of receiver amplifier noise levels are not much different. For germanium APDs the dark current cannot be neglected, and the reduction in avalanche gain is much less as shown by the upper curve.

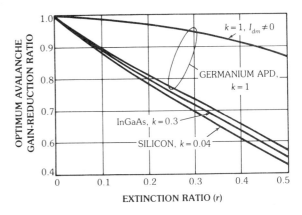

Fig. 38. Reduction in optimum avalanche gain due to extinction ratio.

For germanium APDs the receiver sensitivity penalty due to extinction ratio is highly dependent on dark current (particularly the multiplied component I_{dm}) and the operating data rate. The case of negligible dark current is shown by the line labeled $x=1$ in Fig. 37. The line labeled $x=1$, $I_{dm} \neq 0$ represents the case of $I_{dm}=50$ nA at a data rate of 45 Mb/s. We also note that the effect of APD dark current on receiver sensitivity decreases at high bit rates. Thus the receiver sensitivity penalty of germanium APDs ranges between these two curves for $I_{dm} \leq 50$ nA.

The reduction of the optimum avalanche gain due to the nonzero extinction ratio can be evaluated from (123). However, more accurate determination of the optimum avalanche gain can be obtained from the "McIntyre expression" (113) with the accurate expression (8) for the avalanche excess-noise factor. Using these two expressions the calculation has to be carried out numerically. An example of the results is shown in Fig. 38, where the reduction in optimum avalanche gain is plotted as a function of the extinction ratio.

It can be summarized that as long as dark current can be neglected, the reduction in optimum avalanche gain due to extinction ratio can be well approximated by

$$G_{\text{opt}}(r)/G_{\text{opt}}(r=0)=1-r. \qquad (130)$$

The Effect of Dark Current on Receiver Sensitivity

We have seen that dark currents of long-wavelength APDs are quite significant and can contribute to the total receiver noise, thereby degrading the receiver sensitivity. The effect of the dark current can be quantified by evaluating the maximum allowable dark current for a given receiver sensitivity penalty [27]. This sensitivity penalty can be given in terms of a parameter ϵ such that

$$\eta \bar{P}=(1+\epsilon) \eta \bar{P}_0, \qquad (131)$$

where $\eta \bar{P}_0$ is the receiver sensitivity in the case of zero dark current.

For receiver sensitivity penalty less than 1 dB, ε is approximately proportional to the penalty in decibels. For example, 0.1-dB penalty corresponds to $\varepsilon=0.023$ and 1-dB penalty corresponds to $\varepsilon=0.259$. The minimum allowable dark current is evaluated for the case of zero extinction ratio since this is the case when the dark current effect is most significant.

The effect of dark current on receiver sensitivity can be obtained from (116). For pin photodiodes only the unmultiplied component I_{du} is present and $G=F(G)=1$. The maximum allowable un-multiplied component I_{du} can be shown from this equation to be

$$I_{du}=\varepsilon(2+\varepsilon)(2\,e\,B\,I_2)^{-1}\,\sigma_a^2. \qquad (132)$$

For typical pinFET receivers, σ_a^2 is proportional to B^3 above 100 Mb/s. Thus the maximum allowable dark current increases with bit rate as B^2. In other words, the higher the bit rates, the less importance the effect of the dark current on receiver sensitivity. For typical pinFET receivers a dark current of 10 nA corresponds to a sensitivity penalty of approximately 0.5 dB at 10 Mb/s and 0.1 dB at 100 Mb/s.

In the case of APDs the multiplied component I_{dm} has a dominant effect on receiver sensitivity. The maximum allowable dark current I_{dm} has been shown to be given approximately by [27]

$$I_{dm}=2\,I_2^{-1}\,\varepsilon\,(1+\varepsilon)\,(k\,B\,e\,I_1^3\,Q^3\,\sigma_a)^{1/2}. \qquad (133)$$

Depending on the receiver amplifier being used, σ_a^2 is proportional to B^2 or B^3. Therefore the maximum allowable I_{dm} is proportional to B or $B^{1.25}$. Thus, for APDs the allowable dark current increases less rapidly with bit rate than pin receivers.

It can be observed from (133) that the maximum allowable I_{dm} is proportional to k and to σ_a. Thus, at a given bit rate a reduction in 2 of the maximum allowable I_{dm} requires either a 12-dB improvement in amplifier noise power σ_a^2 or a factor of 4 reduction in the k value of the APD. For example, for 0.5-dB receiver sensitivity penalty at 1Gb/s, the maximum allowable I_{dm} is about 10 nA for germanium APDs and 6 nA for InGaAs APDs.

Dependence of Receiver Sensitivity on Amplifier Noise

For pin photodiodes the receiver sensitivity is proportional to the square root of the receiver amplifier noise power σ_a^2 as can be seen from (115). For APDs the avalanche gain can be increased to offset the increased amplifier noise level. Thus the dependence of receiver sensitivity on amplifier noise can be expected to be less critical. The explicit dependence can be obtained from (124) where $\eta\bar{P}$ is proportional to $(\sigma_a^2)^{x/(2x+2)}$. Thus for typical silicon APDs ($x=0.3$ to 0.4), the receiver sensitivity is proportional to $(\sigma_a^2)^{1/7}$ to $(\sigma_a^2)^{1/8}$. For InGaAs APDs ($x=0.7$ to 0.8) the receiver sensitivity is proportional to $(\sigma_a^2)^{1/4.5}$ to $(\sigma_a^2)^{1/5}$. For germanium APDs, the receiver sensitivity is proportional to $(\sigma_a^2)^{1/4}$ when the dark-current effect can be neglected. When dark current is taken into account, the receiver sensitivity of germanium APDs is approximately proportional to $(\sigma_a^2)^{1/5}$.

For example, a 3-dB improvement in receiver amplifier noise corresponds to a 1.5-dB improvement in receiver sensitivity for pin photodiodes, but only about 0.6 dB for germanium APDs and InGaAs APDs and 0.4 dB for silicon APDs.

7. Dynamic Range

The dynamic range is a measure of the range of input optical power levels that the receiver can operate reliably at without exceeding the specified error rate. It is limited at the low end mainly by noise (receiver sensitivity) and at the high end by saturation and overloading in the receiver amplifier.

In a properly designed receiver the dynamic range is limited by the the low-noise amplifier at the front end since an agc loop can always be incorporated in the following amplifier chain to handle the wide range of signal levels. Since the front-end amplifier has a limited output voltage swing (due to biasing conditions), the achievable dynamic range is a function of the load resistance or feedback resistance (depending on whether the front-end amplifier design is high-impedance or transimpedance). As the load (feedback) resistance decreases, the maximum allowable input optical power increases. Thus the dynamic range is increased. As can be seen from (69) and (83), however, a reduction in the load (feedback) resistance results in an increase in the amplifier noise level. Thus there is a trade-off between receiver sensitivity and dynamic range.

A simple way to trade off receiver sensitivity for dynamic range is reduction of the load resistance. The dynamic-range improvement obtained by reducing the feedback resistance has been calculated as a function of the receiver sensitivity degradation [27]. The result is shown in Fig. 39. The sensitivity degradation increases sharply when pin photodiodes are used. Thus the trade-off is only effective when APDs are used.

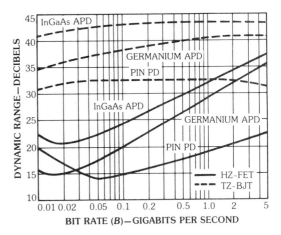

FIG. 40. Typical dynamic range of long-wavelength optical receivers. (*After Muoi [27],* ©*1984 IEEE*)

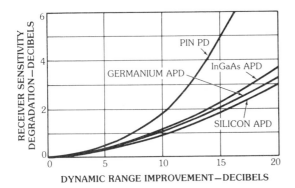

FIG. 39. Trade-off between receiver sensitivity and dynamic range. (*After Muoi [27],* ©*1984 IEEE*)

Typical dynamic range of optical receivers using high-impedance FET and transimpedance bipolar transistors is shown in Fig. 40. We see that high-impedance pinFET receivers typically have a dynamic range of 15 to 20 dB. The dynamic range can be improved by either using an APD or a transimpedance design.

The dynamic range of APD receivers is higher than the corresponding pin receiver because the avalanche gain can be varied by an agc loop. Thus, as the input optical-signal level increases, the avalanche gain can be reduced by lowering its bias voltage to keep the photocurrent constant. For silicon APDs at short wavelengths the bias voltage has to be maintained above a certain value to ensure that the APD response time is not degraded. Thus silicon APDs have a minimum requirement for avalanche gain which is normally about 10. For long-wavelength APDs the avalanche gain can be reduced to 1 without response-time degradation.

Another technique to increase the pinFET receiver dynamic range is by adding more control circuitry. This has the advantage of negligible sensitivity degradation but the receiver is more complex. A popular approach is that by adding a variable shunt resistor at the amplifier input the signal current can be diverted away from its normal path [83, 84]. The gate-source Schottky diode of the front-end gallium-arsenide MESFET itself has also been used to achieve this function [85]. The pinFET receiver dynamic range can be extended to 40 dB using these techniques. The receiver amplifier bandwidth, however, may vary as a function of the variable resistor. Thus adaptive equalization may be necessary, which increases the receiver complexity.

A dual agc loop with resistive capacitive feedback approach to realize a wide dynamic range pinFET receiver has also been reported [86]. A dynamic range of over 50 dB was obtained.

As mentioned previously an agc loop can be incorporated in the amplifier chain following the front-end amplifier in order to handle the wide range of signal levels at its output. An alternative approach to the agc amplifier is the use of limiting amplifier. This has the advantage of simpler circuitry and is very attractive in achieving hybrid or monolithic repeaters (e.g., for submarine applications) [87, 88]. A dynamic range of 60 dB electrical can be achieved by the limiting amplifier approach [89].

8. Bit-Rate Transparency

Normally the optical receiver is designed to operate at a certain data rate. The receiver circuits which need to be optimized at the operating bit rate include the front-end amplifier, the shaping filter, and the timing recovery circuit (see Fig. 1). Thus, for multiple-bit–rate or variable-bit–rate capability the receiver is normally optimized at the highest bit rate. Thus at the lower bit rates some components need to be changed in order not to sacrifice performance. It can be noted that if the shaping filter is unchanged, then there is no advantage in changing the front-end amplifier since its noise is already minimized in the given noise bandwidth. Thus we have the options of changing both the front-end amplifier and shaping filter, or changing the shaping filter, or leaving them both unchanged. The receiver noise-power increase and sensitivity degradation of the last two options (relative to the first option) has been calculated [27]. The result is shown in Fig. 41 for both high-impedance and transimpedance amplifier designs. The bit-rate–range factor K on the horizontal axis is the ratio of the highest and lowest bit rates. The receiver noise-power penalty is shown on the left vertical axis and the receiver sensitivity for various photodiodes is shown on the right vertical axis.

It can be observed from Fig. 41 that if we are willing to change the shaping filter, the sensitivity penalty due to leaving the receiver amplifier unchanged can be kept to a reasonable value over a wide range of bit rates (particularly when an APD is used). For applications involving a range of different fixed bit rates, this can be achieved by different filter modules which can be manually plugged in or automatically switched.

The preceding discussion is based on the assumption that the clock recovery circuitry is optimized for each operating bit rate. The need to provide a timing recovery circuitry that can achieve this function over a wide range of data rates is more difficult to implement. This is because the clock extraction circuitry is basically a narrow-bandpass filter (whether it is a tuned circuit or a phase-locked loop). For applications where only a small number (of the order to 10) of repeaters are used in the system, jitter and noise accumulation due to the clock extraction circuitry is not a main concern. Thus, with some frequency-acquisition aid and sacrifice in jitter and noise, the data range coverage of phase-locked loops can be increased to possibly 2 to 1.

For applications involving a number of different fixed data rates, a wider coverage range can be achieved by switching in different components. For example, a divide-by-n circuit can be included at the vco output so that the operating frequency of the phase-locked loop can be selected (either manually or automatically) by changing n. In addition, different vco's can be selected for dif-

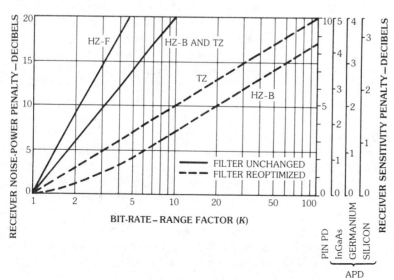

FIG. 41. Receiver noise power and sensitivity penalty for variable-bit-rate operation. (*After Muoi [27], ©1984 IEEE*)

ferent bit-rate ranges. By combining different vco's and dividing ratios *n* as in Fig. 42, a wide range of data rates can be covered.

A simple way to avoid timing recovery problems is not to use timing at all at the receiver. The optical receiver then performs the function of reshaping and regenerating the detected signal (without retiming). This results in the so-called 2R (reshaping and regenerating) receiver instead of the normal 3R (reshaping, retiming, and regenerating) receiver. This approach is attractive not only in local-area networks but also in telecommunication systems where the no-retiming option results in a repeater which is simpler, more reliable, and lower in size, cost, and power consumption. Normally these 2R repeaters are placed in outdoor environments (huts, manholes, etc.) and the full 3R repeaters with the retiming function are located at central offices (or terminal ends). The resulting configuration, shown in Fig. 43, is sometimes referred to as a digital hybrid transmission system. This approach has been investigated

in both coaxial-cable [90–91] and optical-fiber transmission systems [92–98].

The performance of the 2R repeater is degraded because of jitters in the transition edges of the regenerated-data waveform. Jitters can arise from two sources: random jitters and systematic jitters. Random jitters are caused by random perturbations of the threshold-level–crossing time due to receiver noise. Random jitters can also be caused by the laser noise from the preceding repeaters. Systematic jitters arise from several sources: pulse-width distortion due to intersymbol interference (caused by an imperfect shaping filter), data pattern dependence in the preceding laser optical output, and turn-on delay of the preceding laser. By increasing the complexity of the shaping filter and laser driver, systematic jitters can be greatly reduced but cannot be made to be zero.

Jitter accumulation through a chain of 2R repeaters is different for random jitters and systematic jitters. Random jitters tend to accumulate as the square root of the number of 2R repeaters

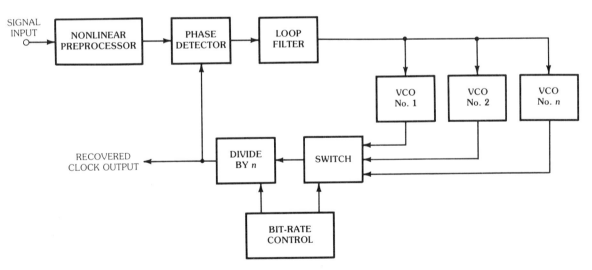

FIG. 42. Phase-locked-loop configuration for variable–bit-rate operation. (*After Muoi [27], ©1984 IEEE*)

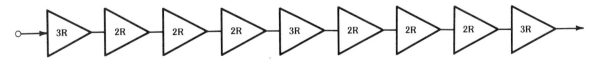

2R: RESHAPING AND REGENERATION
3R: RESHAPING, TIMING, AND REGENERATION

FIG. 43. A typical hybrid digital communication system.

(square-law addition) while systematic jitters tend to add linearly. Thus if θ_{jr} and θ_{js} are the random and systematic jitters in one 2R repeater, respectively, then the total accumulated jitter after n 2R repeaters is given by

$$\theta_j = \theta_{jr}\, n^{1/2} + \theta_{js}\, n \,.$$

The final jitter present at the 3R repeater input determines the error rate of the system. Thus the 2R receiver sensitivity (and hence the 2R repeater spacing) is determined by the amount of jitter introduced by the 2R repeater, the number of 2R repeaters between two 3R repeaters, and the error rate versus jitter characteristic of the 3R repeater. The analysis is thus rather complicated.

Experiments have shown that about ten 2R repeaters can be cascaded in tandem prior to a 3R repeater for a receiver-sensitivity degradation of 3 to 5 dB [98]. Despite this optical power penalty 2R repeaters have been used in actual field trial [96] and installation carrying live traffic [97] of optical-fiber communication systems because of the advantages discussed earlier.

9. Bit-Pattern Independency

The requirement that the optical receiver can operate with an unrestricted data sequence places severe constraints on its design. The amplifier and agc circuitry normally requires some dc reference for measuring the signal level. In addition, the timing extraction circuitry needs some transitions in the incoming-data format to operate properly.

Telecommunication systems normally avoid this problem by using an appropriate line code at the transmitting end. Various line codes have been investigated for optical-fiber communication systems: mBnB code [99–101], mB1C code [102], Manchester code [63] and dicode [103] among others. The aim of these coding schemes is to obtain a signal format which has a negligible dc baseline wander, a large number of transitions, and provisions for error detection. The trade-offs involved here are alleviation of receiver constraints versus implementation complexity of coder/decoder and increase in the transmitted baud rate (hence possible degradation in receiver sensitivity). An alternative technique for obtain-

ing bit-pattern independency is using scrambling/descrambling stages to break up the long strings of "ones" and "zeros" to obtain a dc balanced signal.

The receiver design is more complicated if the input data is not or cannot be randomized as above (for example in asynchronous systems). The timing recovery circuitry has to be carefully designed to handle the longest possible string of "ones" and "zeros." The loss of the dc reference level when a long string of "ones" and "zeros" is encountered also makes the coupling design of the receiver amplifier more difficult. If ac coupling is used in the receiver, the coupling capacitor has to be large enough to ensure that it will not discharge completely during the long period of no transition. Because of the limited capacitor size that can be used, there is a limit as to how long the period without transition can be.

The other alternative of dc coupling is not without problems. Direct-current–coupled receivers are very susceptible to drift and offset and threshold-level variations which result in severe sensitivity degradation. A dc feedback loop can be used to stabilize the dc operating point. However, there still exists the ambiguity of the dc reference level between a long string of "ones" and a long string of "zeros."

Other techniques to obtain an optical communication system which is data-pattern independent are reviewed in [27].

10. Receiver Design for Analog Fiber Systems

Even though digital format is ideally suited for optical-fiber communication systems, analog modulation is also attractive for certain applications. For example, analog optical-fiber systems are often used for transmission and distribution of single-channel and multichannel video and broadband analog signals since they are at present simpler and more cost-effective.

The optical receiver analysis and design discussed previously can be applied to analog systems with slight modifications. Instead of the system bit-error rate the main parameters of interest for analog systems are signal-to-noise ratio (or carrier-to-noise ratio) and signal-to-distortion ratio.

Analog optical-fiber systems can be broadly classified into two types. A simple system is shown in Fig. 44a, where the analog signal directly modulates the intensity of the optical source. The optical receiver therefore consists of a receiver amplifier chain and a filter to recover the transmitted signal. Depending on the original analog signal this filter can be either a low-pass filter (for baseband signals) or a bandpass filter (for rf signals). The signal-to-noise ratio of the detected signal can be readily evaluated from the incoming optical power level and the receiver amplifier noise.

the same as the cnr.

Carrier-to-Noise–Ratio Analysis

Let us consider an optical receiver consisting of an APD followed by an amplifier with input equivalent noise-current spectral density $N(f)$ as calculated from the theory presented earlier. For simplicity of analysis the following filter is assumed to be a block wall filter with lower and upper cutoff frequencies of f_l and f_u. The filter bandwidth (required to pass the modulated subcarrier signal) is therefore

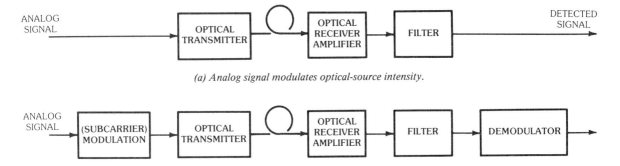

ANALOG SIGNAL → OPTICAL TRANSMITTER → OPTICAL RECEIVER AMPLIFIER → FILTER → DETECTED SIGNAL

(a) Analog signal modulates optical-source intensity.

ANALOG SIGNAL → (SUBCARRIER) MODULATION → OPTICAL TRANSMITTER → OPTICAL RECEIVER AMPLIFIER → FILTER → DEMODULATOR →

(b) Analog signal modulates a subcarrier.

FIG. 44. Analog optical transmission systems.

A more complicated system is shown in Fig. 44b, where the analog signal first modulates a subcarrier and the resultant signal then modulates the optical power output. The aim of the first level of modulation, which can be either frequency modulation, phase modulation, pulse frequency modulation, or pulse-position modulation, is to obtain trade-offs between signal-to-noise and signal-to-distortion ratios with transmission bandwidth. At the repeater a receiver amplifier chain and a filter (usually bandpass) are needed to recover the modulated subcarrier signal. This signal then modulates the optical-source intensity and the optical output is retransmitted. At the receiving terminal an additional demodulator is required to recover the original analog signal being transmitted. To analyze the system's performance we need to calculate the carrier-to-noise ratio (cnr) at the filter output. Then the detected snr can be derived from conventional modulation/demodulation theory. It can be observed that Fig. 44a can be treated as a special case of Fig. 44b where the detected snr is

$$B_R = f_u - f_l. \tag{134}$$

The input equivalent noise power in the signal bandwidth can therefore be calculated from

$$\sigma_a^2 = \int_{f_l}^{f_u} N(f)\, df. \tag{135}$$

The optical power falling on the photodetector can be written in the form

$$P(t) = \bar{P}\,(1 + m \cos \omega t), \tag{136}$$

where ω is the subcarrier frequency and m is the modulation index (m is normally confined to be less than unity).

The detected primary current (ac signal-component only) is

$$I_s = R\,\bar{P}\, m \cos \omega t. \tag{137}$$

The total noise power at the receiver consists of the shot noise due to the incident optical power, the dark-current noise, and the receiver amplifier noise. The sum of the dark current and receiver amplifier noises (referred to as the circuit noise) is given by

$$\sigma_c^2 = \sigma_a^2 + 2\,e\,[I_{du} + I_{dm}\,G^2\,F(G)]\,B_R\,, \quad (138)$$

where the dark-current noise is given by an expression similar to (45).

The rms power carrier-to-noise ratio is therefore given by

$$\text{cnr} = \tfrac{1}{2}(m\,R\,G\,\bar{P})^2/[\sigma_c^2 + 2\,e\,R\,\bar{P}\,G^2\,F(G)\,B_R]\,. \quad (139)$$

The required detected optical power at the receiver in order to achieve a desired cnr can be obtained from the above equation as

$$\eta\,\bar{P} = 2\,h\,\nu\,m^{-2}\,B_R\,F(G)\,(\text{cnr})\left(1 + \{m^2\,\sigma_c^2\right.$$
$$\left. /2\,(\text{cnr})\,[e\,B_R\,G\,F(G)]^2 + 1\}^{1/2}\right). \quad (140)$$

The required optical power can thus be calculated as a function of the avalanche gain. From this calculation the optimum avalanche gain and the minimum required optical power can be determined.

We can distinguish two special cases of interest: shot-noise–limited and circuit-noise–limited operation. The receiver detection is shot-noise limited when the modulation index is low and/or the circuit noise term is negligible. In this case the term under the exponent $\tfrac{1}{2}$ in (140) is approximately unity and the required optical power is given by (for shot-noise–limited detection only a simple pin detector is used and thus $F(G) = 1$)

$$\eta\,\bar{P} = 4\,h\,\nu\,m^{-2}\,B_R\,(\text{cnr})\,. \quad (141)$$

When the circuit noise dominates, the unity terms in the bracket of equation (140) can be neglected. The required optical power in this case is given by

$$\eta\,\bar{P} = h\,\nu\,\sigma_c\,e^{-1}\,[2\,(\text{cnr})]^{1/2}/m\,G\,. \quad (142)$$

It can be observed that when the receiver detection is shot-noise limited, the required optical power is directly proportional to the cnr (i.e., a 10-dB in-crease in optical power input corresponds to a 10-dB increase in the cnr). In addition, it is inversely proportional to the modulation index squared. However, for circuit-noise–limited detection, the required optical power is proportional to the square root of the cnr (i.e., a 10-dB increase in optical power corresponds to only a 5-dB increase in the cnr). Also in this case, the required optical power is inversely proportional to the modulation index.

An Example of Baseband Video Transmission

For the purpose of illustration let us consider an optical-fiber system to transmit a single-channel baseband video transmission at 0.85-μm wavelength. In this simple case the cnr is the same as the video snr (unweighted). The video bandwidth is $B_R = 4.2$ MHz and the receiver amplifier noise is $\sigma_a^2 = 10^{-19}\,\text{A}^2$ (a silicon-FET high-impedance receiver amplifier design is assumed as shown in Fig. 24). The modulation index is chosen to be 50%, a compromise between the snr and distortion. The dark-current term can be neglected for silicon photodiodes.

The detected optical power required at the receiver to obtain a desired video snr (unweighted) has been calculated and shown in Fig. 45. Both

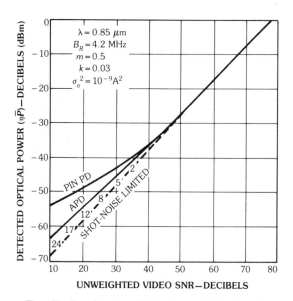

FIG. 45. Receiver sensitivity for a single-channel baseband video transmission.

cases of pin and avalanche photodiodes are plotted with the optimum avalanche gain indicated. It can be observed that the optimum avalanche gain decreases rapidly as the required snr increases. When the snr is higher than 40 dB there is no advantage in using the APD as compared with the simple pin diode. Also, the detection becomes shot-noise limited at high-snr values.

Receiver analysis for multichannel video transmission in a frequency-multiplexed system can be carried out in the same way. However, the effects of intermodulation distortion and cross modulation have to be determined in this case.

11. Concluding Remarks

In this chapter we have considered the basics of receiver design for optical-fiber communication systems. It can be summarized that the theory of receiver design is well established. The current area of interest is in development and implementation of high-speed optical-fiber systems.

As we have seen, receiver sensitivity of optical receivers operating in the long-wavelength region is still 20 dB or more away from the quantum limit due to the lack of low-noise, long-wavelength APDs. A great deal of effort is being spent in searching for such a high-performance APD. Recently, InGaAs/InGaAsP/InP APDs with separated absorption, grading and multiplication layer (referred to as SAGM structure) have been fabricated with very promising performance. Dark current of several nanoamperes (for I_{dm}), coefficient ratios k of 0.35, and avalanche gain-bandwidth product of 60 GHz have been obtained. These detectors have been successfully used in optical receivers operating up to bit rates of 8 Gb/s. Results of receiver sensitivity obtained on recent high-speed fiber transmission systems experiments are summarized in Table 7.

Photoconductive detectors have also been used for optical-fiber systems. The advantages here are low cost and simplicity. They are attractive in the high–bit-rate range (above several hundred megabits per second), where the effect of dark current noise is minimized. Receiver sensitivity levels comparable to pinFET results have been obtained up to 2 Gb/s.

As far as the implementation of the complete

TABLE 7. Recent Receiver Sensitivity Results

Bit Rate (Mb/s)	Wavelength (μm)	Detector	Sensitivity (dBm)
400	1.3	Ge APD	−40.2
420	1.3	InGaAs APD	−43.0
420	1.55	InGaAs APD	−41.5
650	1.55	InGaAs PIN	−38.0
800	1.3	Ge APD	−37.8
1000	1.3	InGaAs APD	−38.0
1000	1.55	InGaAs APD	−37.5
1000	1.3	InGaAs PC*	−34.4
1200	1.55	InGaAs PIN	−36.5
2000	1.55	InGaAs PC*	−28.8
2000	1.3	Ge APD	−33.5
2000	1.55	InGaAs APD	−37.4
4000	1.55	InGaAs APD	−32.6
8000	1.55	InGaAs APD	−22.7

*PC: photoconductive detector.

optical receiver is concerned, the trend is toward monolithic integration. The advantages of integrating all the receiver electronics are high reliability, small size, low cost, and high-performance potential. These advantages are particularly important for submarine applications (where high reliability and small size are critical) and local-area network and subscriber-loop applications (where low cost and small size are critical). A great deal of the receiver electronic circuitry has been integrated using both silicon bipolar and gallium-arsenide FET IC technology.

Another area still in early stage of development but receiving more interest is optoelectronic integration. The aim of this technology is to integrate both the optical device (which is the detector in the receiver case) and the following electronic circuits together in an optoelectronic integrated circuit (often referred to as OEIC). OEICs including the detector and following amplifier have been fabricated although the performance of these OEICs is still very poor. With continued research and development, however, the ultimate goal of realizing the complete receiver, including the detector, in one single OEIC will be obtained eventually.

12. References

[1] R. J. McIntyre, "Multiplication Noise in Uniform Avalanche Diodes," *IEEE Trans. Electron Devices,*" Vol. ED-13, pp. 164–168, January 1966.

[2] R. D. Baertsch, "Noise and Ionization Rate Mea-

surements in Silicon Photodiodes," *IEEE Trans. Electron Devices,* Vol. ED-13, p. 987, December 1966.

[3] R. D. BAERTSCH, "Noise and Multiplication Measurements in InSb Avalanche Photodiodes," *J. Appl. Phys.,* Vol. 38, pp. 4267–4274, October 1967.

[4] H. MELCHIOR and W. T. LYNCH, "Signal and Noise Response of High-Speed Germanium Avalanche Photodiodes," *IEEE Trans. Electron Devices,* Vol. ED-13, pp. 829–838, December 1966.

[5] R. J. MCINTYRE, "The Distribution of Gains in Uniformly Multiplying Avalanche Photodiodes: Theory," *IEEE Trans., Electron Devices,* Vol. ED-19, pp. 703–713, June 1972.

[6] S. D. PERSONICK, "New Results on Avalanche Multiplication Statistics with Applications to Optical Detection," *Bell Syst., Tech. J.,* Vol. 50, pp. 167–189, January 1971.

[7] S. D. PERSONICK, "Statistics of a General Class of Avalanche Detectors with Applications to Optical Communication," *Bell Syst. Tech. J.,* Vol. 50, pp. 3075–3095, December 1971.

[8] J. E. MAZO, and J. SALZ, "On Optical Data Communication via Direct Detection of Light Pulses," *Bell Syst., Tech. J.,* Vol. 55, pp. 347–369; March 1976.

[9] J. CONRADI, "The Distribution of Gains in Uniformly Multiplying Avalanche Photodiodes: Experimental," *IEEE Trans. Electron Devices,* Vol. ED-19, pp. 713–718, June 1972.

[10] S. D. PERSONICK, "Receiver Design for Digital Fiber Optic Communication Systems, Part I and II," *Bell Syst. Tech. J.,* Vol. 52, pp. 843–886, July–August 1973.

[11] T. V. MUOI and J. L. HULLETT, "Receiver for Multilevel Digital Optical-Fiber Systems," *IEEE Trans. Commun.,* Vol. COM-23, pp. 987–994, September 1975.

[12] D. R. SMITH and I. GARRETT, "A Simplified Approach to Digital Optical Receiver Design," *Optical and Quantum Electron.,* Vol. 10, pp. 211–221, 1978.

[13] S. D. PERSONICK et al., "Detailed Comparison of Four Approaches to the Calculation of the Sensitivity of Optical Fiber System Receivers," *IEEE Trans. Commun.,* Vol. COM-25, pp. 541–548, May 1977.

[14] J. E. GOELL, "An Optical Repeater with High-Impedance Input Amplifier," *Bell Syst. Tech. J.,* Vol. 53, pp. 629–643, April 1974.

[15] P. K. RUNGE, "An Experimental 50-Mb/s Fiber-Optic PCM Repeater," *IEEE Trans. Commun.,* Vol. COM-24, pp. 413–418, April 1976.

[16] J. L. HULLETT and T. V. MUOI, "A Modified Receiver for Digital Optical-Fiber Transmission Systems," *IEEE Trans. Commun.,* Vol. COM-23, pp. 1518–1521, December 1975.

[17] J. L. HULLETT and T. V. MUOI, "A Feedback Receive Amplifier for Optical Transmission Systems," *IEEE Trans. Commun.,* Vol. COM-24, pp., 1180–1185, October 1976.

[18] T. UENO et al., "A 40-Mb/s and 400-Mb/s Repeater for Fiber-Optic Communication," *Proc. 1st Eur. Conf. Opt. Commun.* pp. 147–150, September 1975.

[19] J. E. GOELL, "Input Amplifiers for Optical PCM Receivers," *Bell Syst. Tech. J.,* Vol. 53, pp. 1771–1793, November 1974.

[20] J. L. HULLETT and T. V. MUOI, "Referred Impedance Noise Analysis for Feedback Amplifiers," *Electron. Lett.,* Vol. 13, pp. 387–389, June 1977.

[21] T. WITKOWICZ, "Design of Low-Noise Fiber-Optic Receiver Amplifiers Using JFETs," *IEEE J. Solid-State Circ.,* Vol. SC-13, pp. 195–197, February 1978.

[22] T. WITKOWICZ, "Transistor Noise Model for Photodetector Amplifiers," *IEEE J. Solid-State Circ.,* Vol. SC-13, pp. 722–724, October 1978.

[23] S. D. PERSONICK, "Design of Repeater for Fiber Systems," Chapter 6 in *Fundamentals of Optical Fiber Communication,* ed. by M. K. BARNOSKI, New York: Academic Press, 1976.

[24] S. D. PERSONICK, "Receiver Design for Optical-Fiber Systems," *IEEE Proc.,* Vol. 65, pp. 1670–1678, December 1977.

[25] S. D. PERSONICK, "Receiver Design," Chapter 19 in *Optical Fiber Telecommunications,* ed. by S. E. MILLER and A. G. CHYNOWETH, New York: Academic Press, 1979.

[26] R. G. SMITH and S. D. PERSONICK, "Receiver Design for Optical-Fiber Communication Systems," Chapter 4 in *Semiconductor Devices for Optical Communication,* ed. by H. KRESSEL, New York: Springer-Verlag, 1980.

[27] T. V. MUOI, "Receiver Design for High-Speed Optical-Fiber System," *IEEE J. Lightwave Tech.,* Vol. LT-2, pp. 243–267, June 1984.

[28] C. R. HUSBANDS, "The Application of Variable Data Rate Transmission to Local-Area Fiber-Optic Networks," *Proc. IEEE Sixth Conf. Local Computer Network,* pp. 102–107, 1981.

[29] K. T. ALAVI and C. G. FONSTAD, JR., "Performance Comparison of Heterojunction Phototransistors, pin FETs and APD FETs for Optical-Fiber Communication Systems," *IEEE J. Quantum Electron.*, Vol. QE-17, pp. 2259–2261, December 1981.

[30] J. C. CAMPBELL and K. OGAWA, "Heterojunction Phototransistors for Long-Wavelength Optical Receivers," *J. Appl. Phys.*, Vol. 55, pp. 1203–1208. February 1982.

[31] R. A. MILANO et al., "An Analysis of the Performance of Heterojunction Phototransistors for Fiber-Optic Communications," *IEEE Trans. Electron Devices,* Vol. ED-29, pp. 266–273, February 1982.

[32] M. C. BRAIN and D. R. SMITH, "Phototransistors in Digital Optical Communication Systems," *IEEE J. Quantum Electron.* Vol. QE-19, pp. 1139–1148, June 1983.

[33] C. Y. CHEN et al., "Modulated Barrier Photodiode: A New Majority-Carrier Photodetector," *Appl. Phys. Lett.,* Vol. 39, pp. 340–342, August 1981.

[34] J. C. GAMMEL et al., "High-Speed Photoconductive Detectors Using GaInAs," *IEEE J. Quantum Electron.,* Vol. QE-17, pp. 269–272, February 1981.

[35] C. Y. CHEN et al., "Modulation-Doped GaInAs/Al InAs Planar Photoconductive Detectors for 1.0-1.55 μm Applications," *Appl. Phys. Lett.,* Vol. 43, pp. 308–310, August 1983.

[36] Y. M. PANG et al., "1.5-GHz Operation of an AlGaAs/GaAs Modulation-Doped Photoconductive Detector," *Electron Lett.,* Vol. 19, pp. 716–717, September 1983.

[37] C. BAACK et al., "GaAs MESFET: A High-Speed Optical Detector," *Electron. Lett.,* Vol. 13, pp. 193–194, March 1977.

[38] J. P. NOAD et al., "FET Photodetectors: A Combined Study Using Optical and Electron-Beam Stimulation," *IEEE Trans. Electron. Devices.,* Vol. ED-29, pp. 1792–1797, November 1982.

[39] C. Y. CHEN et al., "Ultra High Speed Modulation-Doped Heterostructure Field-Effect Photodetectors," *Appl. Phys. Lett.,* Vol. 42, pp. 1040–1042, June 1983.

[40] P. P. WEBB et al., "Properties of Avalanche Photodiodes," *RCA Rev.,* Vol. 35, pp. 234–278, June 1974.

[41] T. P. LEE et al., "InGaAs/InP PIN Photodiodes for Lightwave Communications at the 0.95-1.65 μm Wavelength," *IEEE J. Quantum Electron.,* Vol. QE-17, pp. 232–238, February 1981.

[42] K. AHMAD et al., "Optimized PIN Photodiodes for Longer-Wavelength Fiber-Optic Systems," *Proc. Sixth Eur. Conf. Opt. Commun.,* pp. 218–221, September 1980.

[43] O. MIKAMI et al., "Improved Germanium Avalanche Photodiodes," *IEEE J. Quantum Electron.,* Vol. QE-16, pp. 1002–1007, September 1980.

[44] T. MIKAWA et al., "A Low-Noise n^{+}np Germanium Avalanche Photodiode," *IEEE J. Quantum Electron.,* Vol. QE-17, pp. 210–216, February 1981.

[45] S. KAGAWA et al., "Germanium Avalanche Photodiode in the 1.3-μm Wavelength Region," *Fujitsu Sci. Tech. J.,* Vol. 18, pp. 397–418, September 1982.

[46] N. SUSA et al., "New InGaAs/InP Avalanche Photodiode Structure for the 1–1.6 μm Wavelength Region," *IEEE J. Quantum Electron.,* Vol. QE-16, pp. 864–869, August 1980.

[47] N. SUSA et al., "Characteristics in InGaAs/Inp Avalanche Photodiodes with Separated Absorption and Multiplication Regions," *IEEE J. Quantum Electron.,* Vol. QE-17, pp. 243–249, February 1981.

[48] S. FORREST et al., "Excess Noise and Receiver Sensitivity Measurements of InGaAs/InP Avalanche Photodiodes," *Electron., Lett.,* Vol. 17, pp. 917–918, November 1981.

[49] H. MELCHIOR et al., "Photodetectors for Optical Communication Systems," *IEEE Proc.,* Vol. 58, pp. 1466–1486, October 1980.

[50] D. R. SMITH et al., "PINFET Hybrid Optical Receiver for 1.1-1.6 μm Optical Communication Systems," *Electron. Lett.,* Vol. 16, pp. 750–751, September 1980.

[51] K. OGAWA and E. L. CHINNOCK, "GaAs FET Transimpedance Frontend Design for a Wideband Optical Receiver," *Electron Lett.,* Vol. 15, pp. 650–652, September 1979.

[52] K. OGAWA et al., "A Long-Wavelength Optical Receiver Using a Short Channel Si-MOSFET," *Bell Syst. Tech. J.,* Vol. 62, pp. 1181–1188, May–June 1983.

[53] A. VAN DER ZIEL, *Noise: Sources, Characterization, Measurement,* Englewood Cliffs: Prentice-Hall, 1970.

[54] M. B. DAS, "FET Noise Sources and Their Effects on Amplifier Performance at Low Frequency," *IEEE Trans. Electron. Devices,* Vol. ED-19, pp. 338–348, March 1972.

[55] W. BAECHTOLD, "Noise Behavior of GaAs Field-Effect Transistor with Short Gate Length," *IEEE Trans. Electron. Devices,* Vol. ED-19, pp. 674–680, May 1972.

[56] K. OGAWA, "Noise Caused by GaAs MESFETs in Optical Receivers," *Bell Syst. Tech. J.,* Vol. 60, pp. 923–928, July–August 1981.

[57] P. E. GRAY and C. L. SEARLE, *Electronic Principles: Physics, Models and Circuits,* New York: John Wiley & Sons., 1969.

[58] R. G. SMITH et al., "Optical Detector Package," *Bell Syst. Tech. J.,* Vol. 57, pp. 1809–1822, July-August 1978.

[59] T. OGAWA et al., "Low-Noise 100 Mbits/s Optical Receiver," *Proc. Second Eur. Conf. Opt. Commun.,* pp. 357–363, September 1976.

[60] M. J. N. SIBLEY and R. T. UNWIN, "Transimpedance Optical Preamplifier Having a Common-Collector Front End," *Electron. Lett.,* Vol. 18, pp. 985–986, November 1982.

[61] R. E. WAGNER et al., "Lightwave Undersea Cable System," *Intl. Conf. Commun.,* pp 7D.6.1–7D.6.5, June 1982.

[62] M. BOENKE et al., "Transmission Experiments Through 101 km and 84 km of Single-Mode Fibre at 274 Mbits/s and 420 Mbits/s," *Electron. Lett.,* Vol. 18, pp. 897–898, October 1982.

[63] T. V. MUOI, "Receiver Design of Digital Fiber-Optic Transmission Systems Using Manchester (Biphase) Coding," *IEEE Trans. Commun.,* Vol. COM-31, pp. 608–619, May 1983.

[64] D. R. SMITH et al., "Experimental Comparison of a Germanium Avalanche Photodiode and InGaAs PINFET Receiver for Longer-Wavelength Optical Communication Systems," *Electron. Lett.,* Vol. 18, pp. 453–454, May 1982.

[65] R. A. LINKE et al., "An 84-km Transmission Experiment at 1 Gb/s Using a 1.55-μm Mode Stabilized Laser," *Technical Digest of Post Deadline Papers, Fourth Integrated Optics and Optical Fiber Commun. (IOOC) Conf.,* pp. 32–33, June 1983.

[66] J. L. HULLETT and S. MOUSTAKAS, "Optimum Transimpedance Broadband Optical Preamplifier Design," *Optical and Quantum Electron.,* Vol. 13, pp. 65–59, 1981.

[67] T. V. MUOI, unpublished.

[68] D. GLOGE et al., "High-Speed Digital Lightwave Communication Using LEDs and PIN Photodiodes at 1.3 μm," *Bell Syst. Tech. J.,* Vol. 59, pp. 1365–1382, October 1980.

[69] P. BALABAN, "Statistical Evaluation of the Error Rate of the Fiberguide Repeater Using Importance Sampling," *Bell Syst. Tech. J.,* Vol. 55, pp. 745–766, July-August 1976.

[70] G. L. CARIOLARO, "Error Probability in Digital Fiber-Optic Communication Systems," *IEEE Trans. Inform. Theory.,* Vol. IT-24, pp. 213–221, March 1978.

[71] M. MASURIPUR et al., "Fiber Optics Receiver Error Rate Prediction Using the Gram-Charlier Series," *IEEE Trans. Commun.,* Vol. COM-28, pp. 402–407, March 1980.

[72] W. HAUK et al., "Calculation of Error Rates for Optical-Fiber Systems," *IEEE Trans. Commun.,* Vol. COM-26, pp. 1119–1126, July 1978.

[73] R. DOGLIOTTI et al., "Error Probability in Optical-Fiber Transmission Systems," *IEEE Trans. Inform. Theory,* Vol. IT-25, pp. 170–178, March 1979.

[74] R. M. RUGEMALIRA, "The Calculation of Average Error Probability in a Digital Fibre Optical Communication System," *Optical and Quantum Electron.,* Vol. 12, pp. 131–141, 1980.

[75] R. M. RUGEMALIRA, "Calculation of Error Probability in an Optical Communication Channel in the Presence of Intersymbol Interference Using a Characteristic Function Method," *Optical and Quantum Electron.,* Vol. 12, pp. 119–129, 1980.

[76] R. J. MCINTYRE, "Avalanche Photodiodes for Optical Communications," presented at the Nordic Fiber Optics Conference, Sweden, August 1976.

[77] R. W. BERRY et al., "Optical-Fiber System Trials at 8 Mbits/s and 140 Mbits/s," *IEEE Trans. Commun.,* Vol. COM-26, pp. 1020–1027, July 1978.

[78] S. M. ABBOTT and W. M. MUSKA, "Low-Noise Optical Detection of a 1.1 Gb/s Optical Data Stream," *Electron. Lett.,* Vol. 15, April 1979.

[79] R. F. LEHENY et al., "Integrated InGaAs pinFET Photoreceiver," *Electron Lett.,* Vol. 16, pp. 353–355, May 1980.

[80] K. INOUE et al., "Monolithically Integrated PIN/JFET Photoreceiver for 1.0–1.7 μm Wavelength," *Tech. Dig. Fourth Integrated Optic and Optical-Fiber Commun. (IOOC) Conf.,* pp. 186–187, June 1983.

[81] K. KASAHARA et al., "Integrated PINFET Optical Receiver with High Frequency InP-MISFET," *Tech. Dig. Fourth Integrated Optics and Optical-*

Fiber Commun. (IOOC) Conf., pp. 188–189, June 1983.

[82] J. Yamada et al., "Characteristics of Gbits/s Optical Receiver Sensitivity and Long-Span Single-Mode Fiber Transmission at 1.3 μm," *IEEE J. Quantum Electron.*, Vol. QE-18, pp. 718–727, April 1982.

[83] D. P. M. Chown, "Dymanic Range Extension for PINFET Optical Receivers," *Proc. Seventh Eur. Conf. Opt. Commun. (ECOC)*, pp. 14.5.1–14.5.3, September 1981.

[84] T. V. Muoi, "Optical Receiver With Improved Dynamic Range," U.S. Patent No. 4, 415, 803, November 15, 1983.

[85] B. Owen, "PIN-GaAsFET Optical Receiver with a Wide Dynamic Range," *Electron Lett.*, Vol. 18, pp. 626–627, July 1982.

[86] G. F. Williams, "Wide Dynamic Range Fiber-Optics Receivers," *Tech. Dig. Intl. Solid-State Circ. Conf. (ISSCC)*, pp. 160–161, February 1982.

[87] D. W. Faulkner et al., "A Single Chip Regenerator for Transmission Systems Operating in the Range 2–320 Mbits/s," *IEEE J. Solid-State Circuits*, Vol. SC-17, pp. 553–557, June 1982.

[88] L. Bickers et al., "140 Mbits/s Thick Film Hybrid Regenerator," *Electron. Lett.*, Vol. 18, pp. 553–555, June 1982.

[89] D. W. Faulkner, "A Wideband Limiting Amplifier for Optical-Fiber Repeaters," *IEEE J. Solid-State Circuits*, Vol. SC-18, pp. 333–340, June 1983.

[90] T. Kitami and K. Otake, "An Ultra High Speed Hybrid Digital Transmission System," *Rev. Electron. Commun. Lab.*, Vol. 21, pp. 57–62, January–February 1973.

[91] H. Marko, "A Digital Hybrid Transmission System for 280 Mbits/s and 560 Mbits/s," *IEEE Trans. Commun.*, Vol. COM-23, pp. 274–281, February 1975.

[92] K. Osafune et al., "Fiber-Optic Hybrid Digital Transmission System Experiment," *IEEE Trans. Commun.*, Vol. COM-27, pp. 789–794, May 1979.

[93] K. Osafune et al., "Hybrid Digital Transmission System Over Optical-Fiber Cable," *Tech. Dig. Intl. Conf. Integrated Optics and Optical-Fiber Commun. (IOOC)*, pp. 553–556, July 1977.

[94] C. R. Patisaul, "Performance Prediction For a High-Speed Digital Optical Cable Video Trunking System," *Proc. Intl. Conf. Commun.*, pp. 21.1.1–21.1.7, June 1978.

[95] F. M. Banks, "An Experimental 45 Mb/s Digital Transmission System Using Optical Fibers," *Proc. Intl. Conf. Commun.*, pp. 17.D.1–17.D.2, June 1974.

[96] D. G. Monteith et al., "The First Canadian Fiber-Optic CATV Trunking System," *IEEE Trans. Cable TV*, Vol. CATV-4, pp. 63–69, April 1979.

[97] J. W. Toy et al., "FT4 Trunking System Field Experience," *Proc. Intl. Conf. Commun.*, pp. 6D.4.1–6D.4.5, June 1982.

[98] J. J. O'Reilly and P. Cochran, "Potential Role of Untimed Repeaters in Optical Submarine Systems," *Proc. IEEE Conf. Subm. Telecommun. Syst.*, pp. 165–169, February 1980.

[99] Y. Takasaki et al., "Optical Pulse Formats for Fiber-Optic Digital Communications," *IEEE Trans. Commun.*, Vol. COM-24, pp. 404–412, April 1976.

[100] R. Petrovic, "New Transmission Code for Digital Optical Communications," *Electron Lett.*, Vol. 14, pp. 541–542, August 1978.

[101] A. X. Midmer and P. A. Franaszek, "Transmission Code for High-Speed Fiber-Optic Data Networks," *Electron. Lett.*, Vol. 19, pp. 202–203, March 1983.

[102] N. Yoshikan et al., "mBIC Code and Its Performance In An Optical Communication System, *IEEE Trans. Commun.*, Vol. COM-32, pp. 163–168, February 1984.

[103] R. C. Hooper, "Analysis of Dicoding In Optical Fibre Communication Systems," *Electron. Lett.*, Vol. 19, pp. 304–306, April 1983.

Communication Theory
for
Fiber-Optic Transmission Systems

J. W. KETCHUM

GTE Laboratories, Inc.

1. Introduction

The technology of optical guided-wave communication is the subject of the intense interest of many in the communications business because of its many desirable features. Perhaps foremost amongst these is the enormous bandwidth potential of optical waveguides. Complete exploitation of this potential requires a thorough understanding of both the physics of the process and the more abstract body of knowledge which generally goes by the name of communication theory, or information theory.

This discussion of modulation methods for fiber-optic communication is intended as a general introduction to some of the fundamental concepts of both analog and digital communication theory. Although many of the problems associated with designing reliable fiber-optic communication systems appear to be quite different from those associated with other communication media, the basic task confronting the communication engineer designing a fiber-optic communication system is the same as that encountered in any other transmission medium. The engineer must design a system which overcomes the fundamentally random nature of the communication process, as well as the many nonideal aspects of any practical medium, in order to achieve efficient and reliable communication.

In the following, many of the approaches to these problems which have been developed over the years are introduced in the context of the constraints imposed by the medium and the technology of optical-waveguide communication. This discussion begins with general models for both signal and noise, and proceeds with first a discussion of methods for analog communication, followed by a discussion of digital communication techniques. This presentation is far from complete; a complete exposition of these topics would occupy a sizable volume at least. It is hoped, however, that the reader who is unfamiliar with these concepts will find this chapter a useful introduction to the topic.

A General Signal Model

For the purposes of the following discussion of analog and digital modulation techniques for fiber-optic communication, it will serve our purposes to define a simple and general signal model. In this model the transmitted signal is given as

$$s(t) = a(t) \cos [2 \pi f_c t + \phi(t)]$$
$$= a(t) \cos \phi(t) \cos 2 \pi f_c t$$
$$- a(t) \sin \phi(t) \sin 2\pi f_c t . \quad (1)$$

We have assumed in writing (1) that the transmitted signal consists of some information-bearing signal which has been imposed on a sinusoidal carrier wave with a constant frequency of f_c Hz. This assumption is not always accurate, particularly in direct-detection optical communication systems where the carrier source is a semiconductor laser or LED with many modes. For the purposes of this introductory material on modulation it is best to ignore these issues, however, in order to avoid obscuring the details of the various modulation methods.

The information is imposed on the carrier sinusoid in (1) through the signals $a(t)$ and $\phi(t)$, which modulate the amplitude and phase, respectively, of the carrier. Alternatively, the transmitted signal may be viewed as the superposition of two sinusoids of the same frequency but with a phase difference of $\pi/2$ radians (90°), multiplied by the information-bearing signals $a(t) \cos \phi(t)$ and $a(t) \sin \phi(t)$, respectively. This approach is illustrated in Fig. 1; $\cos 2\pi f_c t$ and $\sin 2\pi f_c t$ are referred to as the *quadrature carriers,* and $a(t)\cos \phi(t)$ and $a(t)\sin \phi(t)$ are referred to as the *in-phase* and *quadrature signal components,* respectively. The manner in which the transmitted information is imposed on $a(t)$ and $\phi(t)$ distinguishes the various forms of analog and digital modulation.

Another way of representing $s(t)$ which we will use is known as *complex envelope representation.* We begin by noting that (1) may be rewritten as

$$s(t)=\text{Re}\,\{a(t)\,c^t\,e^{\,j\,2\pi f_c\,t}\}=\text{Re}\,\{\alpha(t)\,e^{\,j\,2\pi f_c\,t}\}\,, \quad (2)$$

where the signal $\alpha(t)=a(t)e^{\,j\phi(t)}$ is the complex envelope of $s(t)$. Suppose that $\alpha(t)$ has the Fourier transform $A(f)$, then $\alpha(t)e^{\,j2\pi f_c\,t}$ has the Fourier transform $A(f-f_c)$. Then

$$s(t)=(1/2)[\alpha(t)\,e^{\,j2\pi f_c\,t}+\alpha^{*}(t)\,e^{\,-j2\pi f_c\,t}]$$

has the Fourier transform

$$S(f)=(1/2)\,A(f-f_c)+(1/2)\,A^{*}(-f-f_c)\,. \quad (3)$$

If $s(t)$ is strictly contained in a bandwidth of W Hz, which is less than the carrier frequency f_c,

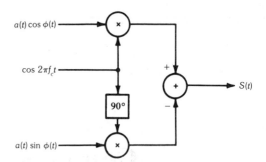

FIG. 1. Quadrature carrier-signal generation.

then $A(f)$ can be viewed simply as the positive frequency portion of $S(f)$ translated by f_c Hz so that its center is at 0 Hz, and scaled by a factor of two. Typically this representation is useful only when $W \ll f_c$, in which case linear systems may be translated in the same way, and analysis becomes independent of carrier frequency.

Noise in Communication Systems

Many kinds of electrical noise are found to interfere with the operation of communication systems. Typically, electrical noise interferes with the communication process by superposition, that is, the noise signal gets added to the desired information-bearing signal at some point, as illustrated in Fig. 2. Some sources of such noise are human-made, some sources are natural, and some sources are a fundamental consequence of the process of communicating. Sources which fall into the first category include electromagnetic energy generated by machines, such as automobiles, telecommunications switching equipment, and electrical power generating and transmission equipment. Also in this category is interference from other communicators who are trying to use a common medium or whose power "leaks" from one medium to another. Thunderstorms and other atmospheric processes are typical sources of natural electrical noise. Noise sources which fall into the third category include thermal-noise sources present in any communication receiver, and the fundamentally random nature of the transmission of electromagnetic waves due to their quantum nature.

Typically, the performance of a given communication system is limited by only one or two of the above types of noise or interference. One communication link might be dominated by thermal noise; another might be dominated by interference from other users. Thus, in analyzing the performance of a given communication technique in a given environment, the operative noise sources must be determined, and appropriate statistical models for these sources found. Unfortunately, in many cases accurate modeling of the noise sources leads to intractable analysis. The most notable exceptions to this are sources which can be modeled as generating noise which has a Gaussian probability distribution function. Such a distribution function is characteristic of thermal noise and other noise sources which are the result of the superposition of many random events.

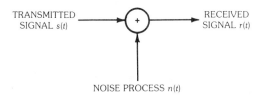

FIG. 2. Signal corruption by noise superposition.

Guided-wave optical communication systems are immune to many of the noise sources mentioned above, since transmission occurs through a dielectric medium and electromagnetic energy from external sources is not easily coupled into this medium. The dominant noise processes in such systems are thermal noise sources in the receivers, and quantum noise effects due to the fundamentally discrete nature of electromagnetic radiation. The quantum noise is most accurately modeled by considering the individual arrivals of photons, the elementary quanta of electromagnetic energy. The appropriate description of the photon arrival process is that of a Poisson random process. In most cases, however, the arrivals are so closely spaced that the effects of the random arrivals can be accurately modeled as an additive Gaussian noise. This will be dealt with in more detail in Section 3. For the purposes of this introductory material on modulation techniques the Gaussian model will be used exclusively.

Representation of White Gaussian Noise

Thermal noise is typically modeled as having a Gaussian probability density function with zero

mean and variance σ^2 at any point in time. This probability density function has the form

$$P_n(x) = (2\pi\sigma^2)^{-1/2} \exp(x^2/2\sigma^2). \qquad (4)$$

In addition, the noise waveform $n(t)$ is assumed to have a very wide bandwidth, with two-sided power spectral density

$$\Phi_{nn}(f) = \begin{cases} N_0/2 & \text{if } |f| < f_{max}, \\ 0 & \text{if } |f| > f_{max}, \end{cases}$$

where f_{max} is very large. In the limit, as $f_{max} \to \infty$, the autocorrelation function of this noise process, the inverse Fourier transform of the power spectral density, is

$$\phi_{nn}(t) = E[n(t)n(t+\tau)] = (N_0/2)\delta(\tau),$$

where $\delta(\tau)$ is the Dirac delta (or impulse) function.

This of course implies that the noise variance σ^2 is infinite. While this may cause philosophical problems, in practice it is not a problem since virtually all receivers include a bandpass or low-pass filter at the input for the purpose of limiting the bandwidth of the noise, and thus its variance.

For the sake of simplicity we assume that the receiver has a filter whose frequency response has unit magnitude over a frequency interval of W Hz and a magnitude of zero everywhere else. Thus the noise waveform $w(t)$ at the output of the bandpass filter has the power spectral density

$$\Phi_{ww}(f) = \begin{cases} N_0/2 & f_c - W/2 \le |f| \le f_c + W/2, \\ 0, & \text{otherwise.} \end{cases} \qquad (5)$$

where f_c is the center frequency of the receiver. The corresponding autocorrelation function is the inverse Fourier transform of this quantity:

$$\phi_{ww}(\tau) = N_0 W[\sin(\pi W\tau)/(\pi W\tau)]\cos(2\pi f_c\tau). \quad (6)$$

The variance of this noise process is $\phi_{ww}(0) = N_0 W$.

Frequently it is convenient to model this type of noise by its low-pass equivalent process. To do this we define the waveform $z(t)$ as the complex envelope of $n(t)$, which satisfies the relationship

$$n(t) = \text{Re} \{z(t) e^{j2\pi f_c t}\},$$

where f_c is usually chosen to be equal to the center frequency of the receiver. The resulting noise process $z(t)$ is a low-pass, complex-valued Gaussian process. In the limit, as the bandwidth of this process goes to infinity (as the bandwidth of $n(t)$ goes to infinity), we can show that the autocorrelation function of $z(t)$ is

$$\phi_{zz}(\tau) = \text{E}[z(t) z^*(t+\tau)] = 2 N_0 \delta(\tau), \qquad (7)$$

and the power spectral density is

$$\Phi_{zz}(f) = 2 N_0. \qquad (8)$$

Although, strictly speaking, the complex envelope of a signal with infinite bandwidth does not exist, we justify (7) and (8) with the statement that the bandwidth of $z(t)$ is very large relative to the receiver bandwidth, but not as large as the carrier frequency. Since we are ultimately only interested in $w(t)$, the noise at the output of the receiver's bandpass filter, or its complex envelope, the description given by (7) and (8) serves as a mathematical convenience which makes it possible for us to deal with the signals of interest in complex envelope form.

It will sometimes be useful to represent $z(t)$ in magnitude/angle form or rectangular form. To this end we define $n_c(t)$, $n_s(t)$, $R_n(t)$, and $\phi_n(t)$ as follows:

$$z(t) = n_c(t) + j n_s(t) = R_n(t) \exp [j \phi_n(t)]. \quad (9)$$

Using these definitions we can represent $n(t)$ as

$$n(t) = n_c(t) \cos 2\pi f_c t - n_s(t) \sin 2\pi f_c t$$

$$= R_n(t) \cos [2\pi f_c t + \phi_n(t)]. \qquad (10)$$

The quadrature noise components $n_c(t)$ and $n_s(t)$ are independent, identically distributed white Gaussian noise processes with autocorrelation function $\phi(\tau) = N_0 \delta(\tau)$. The envelope $R_n(t)$ has a Rayleigh probability density function, and the phase $\phi_n(t)$ is uniformly distributed on $0 \le \phi < 2\pi$.

2. Analog Modulation

The term "analog modulation" refers to communication techniques which operate by varying some parameter of the carrier wave as a continuous function of time so that it is an analog of the information waveform to be transmitted. According to (1) the two parameters available are phase and amplitude, although phase modulation may also be thought of as frequency modulation, since the frequency of a sine wave is just the derivative of the phase. In this section the most common analog modulation methods are described, and their performance in terms of output signal-to-noise ratio analyzed. The descriptions are, of necessity, brief. For a more complete development of these topics see [1, 2, 3].

Intensity Modulation

The most convenient parameter at the disposal of the designer of a fiber-optic communication system is the intensity of the light source; thus most fiber-optic communication systems ultimately rely on the modulation of this parameter to transmit information. This is known as "intensity modulation." Since the intensity of the light source is directly proportional to its output power, intensity modulation involves varying the instantaneous power of the light source in direct proportion to the information signal which is being transmitted. Assuming that the carrier frequency is stable (does not vary with the time), intensity modulation may be characterized in the form of (1), with $\phi(t)$ a constant, and $a(t)$ given by

$$a(t) = K [1 + m x(t)]^{1/2}, \qquad (11)$$

where $x(t)$ is the information signal being transmitted, m is referred to as the *modulation index,* and K is some positive constant which determines the transmitted power.

Since the argument of the square root must be positive for (11) to be a real quantity, the condition $m x(t) \ge -1$ must always be satisfied for (11) to have physical meaning. It is usually assumed that $x(t)$ has an average value of 0, and thus this condition is satisfied by forcing the condition $|x(t)| \le 1$, and using m ($0 < m \le 1$) to control the depth of the modulation.

The receiver for an intensity-modulated signal

simply detects the intensity, or instantaneous power, of the received signal. The output of this receiver is thus given by

$$y(t)=(K^2/2)[1+m\,x(t)]\,,$$

assuming that there is no noise introduced by the receiver or channel. This is obviously not a realistic assumption, since both optical transmission and detection are random processes due to their quantum nature. In addition, some level of additive thermal noise is always added in the receiver electronics. An accurate statistical description of these effects is beyond the scope of this discussion. We will content ourselves with modeling the output of the receiver for intensity-modulated signals as consisting of a component due to the desired signal as well as an additive white Gaussian noise component. In a great many cases this model is sufficiently accurate for engineering purposes. Since the receiver also must remove the dc term $K^2/2$, the output of the intensity-modulation receiver (neglecting the proportionality constant $m\,K^2/2$ for convenience) is

$$y(t)=x(t)+n(t)\,, \qquad (12)$$

where $x(t)$ is the information signal, and $n(t)$ is the white Gaussian noise.

It is important to note that the detection process outlined above does not depend on the stability of the carrier frequency; detection of the intensity of the received optical signal can take place with only a very crude knowledge of the carrier frequency of the received signal. This is an important property of intensity modulation, since, until quite recently, virtually all available optical sources for use in fiber-optic communication systems were multimodal devices with large line widths. Such devices may at any given time emit light at any of a number of frequencies, and change frequencies in a rapid and unpredictable fashion. The carrier in such a case is not strictly a sinusoid, but rather is a randomly varying waveform.

The modulation methods discussed in the sequel all rely, to a greater or lesser extent, on the ability of the receiver to accurately recover and track the carrier frequency of the received signal,

as well as the carrier phase in some cases. Such systems have the great advantage of lower received-power requirements than the direct-detection, intensity-modulation approach described above. Realization of this potential requires light sources with very stable frequency, as well as significantly more complex receivers than those required for intensity-modulated waveforms. Sufficiently stable lasers are a very recent development, and such coherent fiber-optic communication links are the subject of intensive development in a number of laboratories. These techniques may as well be used by imposing a subcarrier on the optical carrier using intensity modulation. This subcarrier may be modulated using any of the techniques discussed below. In this sense we may consider the intensity modulator, optical source and waveguide, and intensity detector as a composite channel which transmits the subcarrier.

Amplitude Modulation

Perhaps the most straightforward form of analog modulation is amplitude modulation (am), where the information is carried entirely by the envelope of the carrier sine wave. In this case $\phi(t)$ in (1) is a constant which we will take to be 0 for convenience, and $a(t)$ is

$$a(t)=K[1+m\,x(t)]\,. \qquad (13)$$

Here, $x(t)$ is the information signal which is being transmitted, and we assume that $|x(t)|\le 1$ and that the average value of $x(t)$ is 0; m is the modulation index. It is usually desirable to make m as large as possible without making $a(t)$ go negative. With the restriction on $x(t)$ given above, m can be no larger than 1, a condition known as *100% modulation*. The transmitted signal $s(t)$ and its complex envelope are then given by the following equations:

$$s(t)=K[1+m\,x(t)]\cos 2\,\pi f_c\,t\,, \qquad (14a)$$

and

$$\alpha(t)=K[1+m\,x(t)]\,. \qquad (14b)$$

The appearance of an amplitude-modulated carrier is shown in Fig. 3.

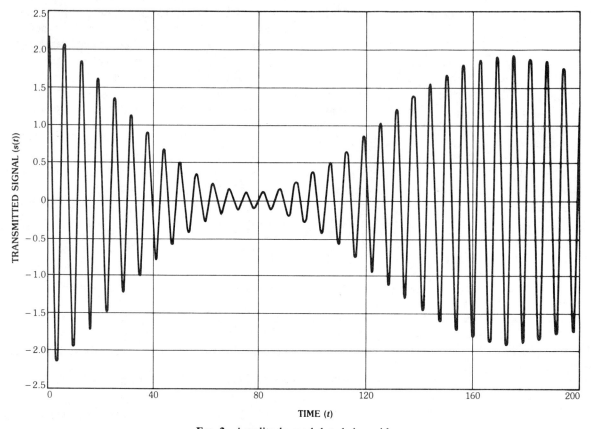

FIG. 3. Amplitude-modulated sinusoid.

The low-pass spectrum for amplitude modulation is

$$A(f) = K\,\delta(f) + m\,K\,X(f),\qquad(15)$$

and the spectrum of the transmitted signal is

$$S(f) = (1/2)\,A(f-f_c) + (1/2)\,A(-f-f_c)$$

$$= (1/2)\,K\,[\delta(f-f_c) + \delta(-f-f_c)$$

$$+ m\,X(f-f_c) + m\,X(-f-f_c)].\quad(16)$$

This spectrum consists of an impulse at $\pm f_c$ Hz, and the spectrum of the information signal, $X(f)$, translated to the carrier frequency, $\pm f_c$, as shown in Fig. 4. Since we are only trying to transmit the information in $x(t)$, the power in the impulses at $\pm f_c$ represents power which is in some sense wasted, since it does not convey any useful infor-

mation. It does, however, make the job of the receiver more straightforward than it might otherwise be.

The simplest receiver for amplitude modulation consists of an envelope detector which finds the envelope of the received signal $r(t)$ or, equivalently, the magnitude of the complex envelope of the received signal. In this case the received signal is $r(t) = s(t) + n(t)$, and its complex envelope is $y(t) = \alpha(t) + z(t)$, where $n(t)$ and $z(t)$ are a white Gaussian noise waveform and its complex envelope, respectively.

An important figure of merit for any analog communication system is the signal-to-noise ratio at the output of the receiver. We will proceed to calculate the output signal-to-noise ratio for amplitude modulation, but first it will be of interest to find the signal-to-noise ratio at the input of the receiver.

The average transmitted power is found by

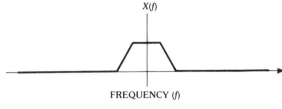

(a) Message signal $x(t)$.

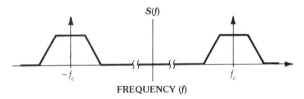

(b) Amplitude-modulated message signal $s(t)$.

FIG. 4. Signal spectra.

averaging over the square of (14), which leads to

$$P_T = (K^2/2)[1 + m^2 P_x] , \qquad (17)$$

where P_x is the power in the information signal $x(t)$. For the sake of simplicity we will assume that the channel has a gain of 1 (i.e., no attenuation); thus the received power is the same as the transmitted power. We further assume that the spectrum of the information signal $x(t)$ is nonzero only over the frequency interval $|f| \leq B/2$, and thus the receiver may use a bandpass filter with a bandwidth of B Hz to limit the noise power, without distorting the desired signal. Given a two-sided noise-power spectral density of $N_0/2$, the noise power at the output of the bandpass filter is given by (6) as

$$\sigma_n^2 = N_0 B . \qquad (18)$$

Thus the input signal-to-noise ratio is

$$P_T/\sigma_n^2 = (K^2/2 N_0 B)[1 + m^2 P_x] . \qquad (19)$$

In order to compute the output signal-to-noise ratio for envelope-detected amplitude modulation, we start by noting that we may write the complex envelope of the received noise in terms of its real and imaginary parts:

$$z(t) = n_c(t) + j n_s(t) .$$

Proceeding to expand the magnitude of the complex envelope of the received signal,

$$d(t) = |y(t)| = |\alpha(t) + z(t)|$$

$$= \{[a(t) + n_c(t)]^2 + n_s^2(t)\}^{1/2} . \qquad (20)$$

The envelope $d(t)$ has the well-known Rician probability density function [4]. At moderate to high signal-to-noise ratios this can be approximated as having a Gaussian probability density function. Thus a useful high-snr approximation for $d(t)$ is

$$d(t) \simeq a(t) + n_c(t) = K[1 + m x(t)] + n_c(t) . \qquad (21)$$

Once the envelope detection has taken place, the receiver must remove the dc component in order to recover the noise-corrupted information signal:

$$s_D(t) = K m x(t) + n_c(t) . \qquad (22)$$

The detected signal power in (22) is $K^2 m^2 P_x$, and the noise power is $N_0 B$, so that the output snr is

$$(\text{snr})_{\text{out}} = K^2 m^2 P_x / N_0 B . \qquad (23)$$

Comparing the input and output snr's we note that the ratio of the output snr to the input snr is $2 m^2 P_x / (1 + m^2 P_x)$.

Double-Sideband/Suppressed-Carrier Modulation

Double-sideband/suppressed-carrier modulation (dsb/sc) is similar to amplitude modulation with the exception that there is no unmodulated carrier transmitted along with the modulated carrier. In this case the transmitted signal is

$$s(t) = K x(t) \cos (2 \pi f_c t + \theta) , \qquad (24)$$

the complex envelope of which is

$$\alpha(t) = K x(t) e^{j\theta} . \qquad (25)$$

Referring to (3) it is apparent that the spectrum of the transmitted signal is simply

$$S(f) = (K/2)[e^{j\theta} X(f - f_c) + e^{-j\theta} X(-f - f_c)] , \qquad (26)$$

as illustrated in Fig. 5. Transmitted power is $P_T = (K^2/2) P_x$, and assuming that the receiver has a bandwidth of B Hz and input noise with a two-sided power spectral density of $N_0/2$, the input signal-to-noise ratio $(\mathrm{snr})_{\mathrm{in}}$ is

$$(\mathrm{snr})_{\mathrm{in}} = (K^2 P_x)/(2 N_0 B) . \qquad (27)$$

Since the unmodulated carrier is not transmitted as in amplitude modulation, the receiver for dsb/scm must employ coherent detection, which requires precise knowledge at the receiver of the carrier frequency f_c and the carrier phase θ. The receiver must derive this knowledge from the received signal; there are various ways of doing this which typically work well as long as the carrier phase is fairly stable. None of these techniques will be covered here; there are many papers and text books on the subject to which the reader may refer [5, 6]. Given this knowledge of carrier frequency and phase, the coherent receiver can recover a noise-corrupted version of the transmitted information

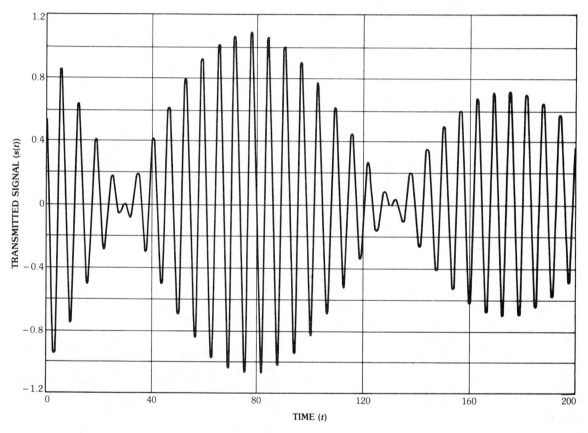

(a) Dsb/sc modulated message signal.

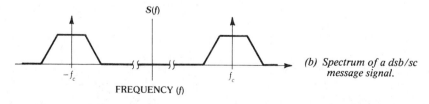

(b) Spectrum of a dsb/sc message signal.

FIG. 5. Double-sideband/suppressed-carrier signals.

signal by multiplying the received signal $r(t)$ $=s(t)+n(t)$ by cos $(2\pi f_c t+\theta)$:

$$d(t)=r(t)\cos(2\pi f_c t+\theta)=Kx(t)$$

$$\times[1/2+(1/2)\cos(4\pi f_c t+2\theta)]$$

$$+n(t)\cos(2\pi f_c t+\theta),\quad (28)$$

where

$$n(t)\cos(2\pi f_c t+\theta)=n_c(t)\ [1/2+(1/2)$$

$$\times\cos(4\pi f_c t+2\theta)]+n_s(t)$$

$$\times[(1/2)\sin(4\pi f_c t+2\theta)].$$

However, since the receiver has a bandpass filter which has a bandwidth roughly equal to that of $s(t)$, we can assume that the double-frequency terms in (28) are removed by the bandpass filter. Thus the detected signal may be represented as

$$s_D(t)=(K/2)x(t)+(1/2)n_c(t).\quad (29)$$

The signal power in $s_D(t)$ is $K^2 P_x/4$, and the noise power is $N_0 B/4$, so the output signal-to-noise ratio is

$$(\text{snr})_{\text{out}}=K^2 P_x/N_0 B=2\ (\text{snr})_{\text{in}}.\quad (30)$$

Thus the output signal-to-noise ratio is twice the input signal-to-noise ratio. This is due to the fact that the coherent receiver effectively rejects the quadrature noise component, which accounts for half the noise power at the receiver input.

Note that a coherent receiver can also be used for am, in which case the detected signal, after subtracting off the component due to the unmodulated carrier, is given by

$$s_D(t)=(Km/2)x(t)+(1/2)n(t),\quad (31)$$

yielding the output signal-to-noise ratio

$$(\text{snr})_{\text{out}}=K^2 m^2 P_x/N_0 B.\quad (32)$$

Referring to (23) we see that performance of coherently detected am is identical to the high-snr

performance of envelope-detected am. Noting, however, that at least half the transmitted power in am must go into the unmodulated carrier, dsb/scm has at least a 3-dB advantage over am.

Single-Sideband Modulation

Single-sideband (ssb) modulation takes advantage of symmetries inherent in narrow-band signals to cut the bandwidth requirements of dsb/scm by a factor of two. To see how this is possible it is first necessary to define the Hilbert transform of a signal, which we will denote by the subscript h, i.e., the Hilbert transform of $x(t)$ is $x_h(t)$. This is defined as

$$x_h(t)=(1/\pi)\int_{-\infty}^{\infty}[x(\tau)/(t-\tau)]\,d\tau.\quad (33)$$

Note that this is just the convolution of $x(t)$ with $1/(\pi t)$. The Fourier transform $X_h(f)$ of $x_h(t)$ is given by

$$X_h(f)=-jX(f)\,\text{sgn}(f),\quad (34)$$

where sgn $(f)=1$ for $f>0$; sgn $(f)=-1$ for $f<0$. A simple interpretation of (34) is that $x_h(t)$ is generated by passing $x(t)$ through a perfect 90° phase shifter. If we define a complex-valued signal whose real part is $x(t)/2$ and whose imaginary part is $\pm x_h(t)/2$, the Fourier transform of this signal is 0 for $f<0$ and $X(f)$ for $f>0$, or $X(f)$ for $f<0$ and 0 for $X(f)>0$. The resulting signal is the complex envelope of an ssb signal:

$$\alpha(t)=(K/2)[x(t)\pm jx_h(t)],\quad (35)$$

whose Fourier transform is

$$A(f)=(K/2)[X(f)\pm X(f)\,\text{sgn}(f)].\quad (36)$$

The corresponding bandpass signal is

$$s(t)=(K/2)[x(t)\cos 2\pi f_c t\mp x_h(t)\sin 2\pi f_c t],\quad (37)$$

whose Fourier transform is

$$S(f)=(K/2)\{X(f-f_c)\,[1\pm\text{sgn}(f-f_c)]$$

$$+X(-f-f_c)\,[1\pm\text{sgn}(-f-f_c)]\}.\quad (38)$$

Comparing (38) with (26) it is apparent that ssb occupies exactly half the bandwidth of dsb/sc, as stated above.

Fig. 6 shows a block diagram for the generation of a single-sideband signal. The Hilbert transform of the baseband signal is generated by applying a 90° phase shift, and this modulates the quadrature carrier, while the original signal modulates the in-phase carrier. The problem with this approach is that the realization of a good approximation to a perfect 90° phase shifter is not easy. The alternative is to apply filtering at the carrier frequency, or at some intermediate frequency, in order to remove the lower or upper sideband from an ssb/sc signal, as shown in Fig. 7. The quality of this approach depends on the quality of the filtering; since the ideal bandpass filter characteristics shown in Fig. 7b are not achievable in practice, energy in the lower sideband in Fig. 7c cannot be made exactly zero. Single-sideband signals generated with real bandpass filters are sometimes referred to as *vestigial sideband* (vsb) signals.

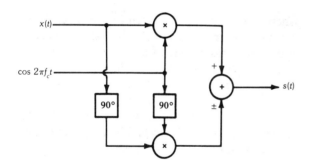

FIG. 6. Single-sideband signal generation by phase shift.

Analysis of the performance of ssb in noise reveals that the output snr of an ssb receiver is given by (30), the same as for dsb/sc. This result is intuitively reasonable if we note that since the signal bandwidth of ssb is half that of dsb/sc, we can make the noise bandwidth of the receiver half what it would be for dsb/sc, and thus the two snr's should be the same. Thus there is no performance penalty incurred for reducing the bandwidth by using ssb; the price paid for this increased bandwidth efficiency is in complexity. Both the transmitter and receiver for ssb must be more complex than comparable units for dsb/sc.

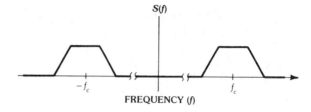

(a) Dsb signal spectrum.

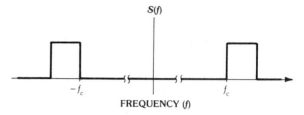

(b) Bandpass-filter frequency response.

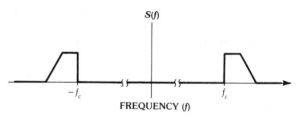

(c) Resulting ssb signal spectrum.

FIG. 7. Single-sideband signal generation using bandpass filters.

Frequency Modulation

Frequency modulation (fm) is a technique for transmitting information as variations in the instantaneous frequency of a sinusoidal carrier. This technique has numerous desirable characteristics, including a constant envelope and the potential for significantly better noise performance than that of the methods discussed above based on varying the amplitude of the sinusoidal carrier. Since the instantaneous frequency is the derivative of the phase it follows that the phase of an fm signal must be proportional to the integral of the information signal $x(t)$. Thus the phase of an fm signal is given by

$$\phi_s(t) = 2\pi h \int_{-\infty}^{t} x(\tau)\, d\tau, \qquad (39)$$

where h, the frequency deviation constant, controls the amount of variation in instantaneous frequency which is obtained for a given variation in $x(t)$. We again adopt the convention that $x(t)$ is normalized so that $|x(t)| \leq 1$. In this case, since the instantaneous frequency is $f_c + h x(t)$ Hz, the deviation constant h is the peak deviation of the instantaneous frequency from the carrier frequency. Thus we will refer to h as the *peak frequency deviation*.

From (39) we see that the complex envelope of $s(t)$ is $\alpha(t) = K \exp [j \phi_s(t)]$, and the bandpass signal is

$$s(t) = K \cos [2 \pi f_c t + \phi_s(t)] . \qquad (40)$$

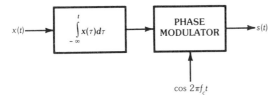

(a) Frequency modulator.

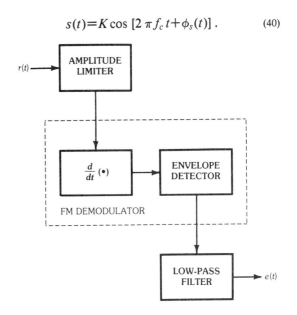

(b) Limiter-discriminator fm receiver.

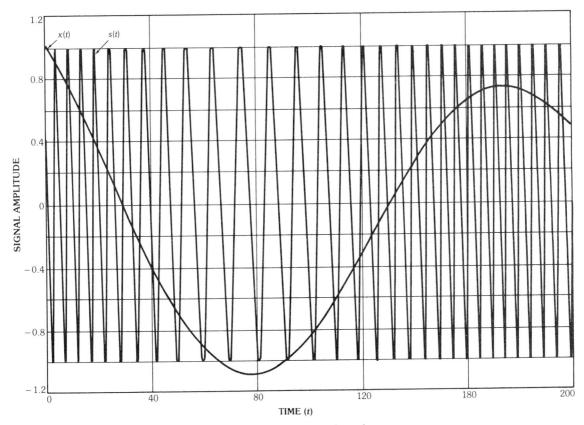

(c) Frequency-modulated signal.

FIG. 8. Frequency-modulation components and signal.

Fig. 8a illustrates one method for generating an fm signal, using a phase modulator whose input is the integral of the baseband signal. A typical fm waveform, along with the modulating waveform, is shown in Fig. 8c.

Calculation of the spectral characteristics of an fm signal is not as straightforward as for modulations discussed previously. It is necessary, however, to have some knowledge of the bandwidth occupancy of an fm signal. It should be clear that as the deviation constant h is increased, the occupied bandwidth will increase. This notion is made concrete in the rule of thumb known as Carson's rule, for transmission bandwidth of an fm signal:

$$B_T \approx 2(h+B), \qquad (41)$$

where h is the peak instantaneous frequency-deviation constant, and $B/2$ is the baseband bandwidth of $x(t)$. It is apparent from (41) that for small values of h, fm bandwidth depends primarily on the signal bandwidth, and for large values of h, fm bandwidth depends primarily on h.

We proceed next to consider the performance of fm in noise. To begin, it is useful to recall the form for expressing bandpass noise given in (9):

$$n(t) = R_n(t) \cos[2\pi f_c t + \phi_n(t)]$$

$$= n_c(t) \cos 2\pi f_c t + n_s(t) \sin 2\pi f_c t, \qquad (42)$$

and the low-pass equivalent form given by (10):

$$z(t) = R_n(t) \exp[j\phi_n(t)]$$

$$= n_c(t) + j n_s(t). \qquad (43)$$

The received signal is then

$$r(t) = K \cos\left[2\pi f_c t + 2\pi h \int_{-\infty}^{t} x(\tau)\,d\tau\right]$$

$$+ R_n(t) \cos[2\pi f_c t + \phi_n(t)]$$

$$= R_r(t) \cos[2\pi f_c t + \phi_r(t)], \qquad (44)$$

and its complex envelope is

$$y(t) = K \exp[j\phi_s(t)] + n_c(t) + j n_s(t)$$

$$= [K + n_c(t) \cos\phi_s(t)$$

$$+ n_s(t) \sin\phi_s(t) + j n_s(t) \cos\phi_s(t)$$

$$- j n_c(t) \sin\phi_s(t)] \exp[j\phi_s(t)]. \qquad (45)$$

The magnitude $R_r(t)$ and phase $\phi_r(t)$ of the right-hand side of (45) are then given by

$$R_r(t) = \{K^2 + R_n^2(t) + 2K R_n(t)$$

$$\times \cos[\phi_s(t) - \phi_n(t)]\}^{1/2} \qquad (46)$$

and

$$\phi_r(t) = \phi_s(t) - \tan^{-1}\{[n_c(t) \sin\phi_s(t)$$

$$- n_s(t) \cos\phi_s(t)]$$

$$\times [K + n_c(t) \cos\phi_s(t) + n_s(t) \sin\phi_s(t)]^{-1}\}. \qquad (47)$$

In order to recover the information signal $x(t)$ from $r(t)$ we must be able to recover $\phi_s(t)$ from $r(t)$ and take the derivative. In general, the relationship between $\phi_s(t)$ and $\phi_r(t)$ and $R_r(t)$ is highly nonlinear. However, it can be shown using simple geometric arguments that in the case of high signal-to-noise ratios, i.e., when with high probability, $R_s(t) \gg R_n(t)$, $\phi_r(t)$ may be approximated by

$$\phi_r(t) \simeq \phi_s(t) + [R_n(t)/K] \sin[\phi_n(t) - \phi_s(t)]. \qquad (48)$$

Since $\phi_n(t)$ is a random variable uniformly distributed on $[0,2\pi)$, $\phi_s(t)$ is a bias term in the second term of (48) which does not effect its statistics, so we can rewrite (48) as

$$\phi_r(t) \simeq \phi_s(t) + [R_n(t)/K] \sin[\phi_n(t)]$$

$$= \phi_s(t) + n_s(t)/K, \qquad (49)$$

where $n_s(t)$ is given by (10).

The desired signal is recovered by computing the derivative of the received signal:

$d(t) = (1/2\pi h)(d/dt)\cos[2\pi f_c t + \phi_r(t)]$

$= [f_c/h + (1/2\pi h)(d/dt)\phi_r(t)]$

$\times \sin[2\pi f_c t + \phi_r(t)]$. (50a)

The envelope of this is the desired signal:

$e(t) = (1/2\pi h)(d/dt)\phi_r(t)$

$\simeq x(t) + (1/2\pi h K)(d/dt)n_s(t)$. (50b)

In the noise-free case (50) reduces to simply the message waveform $x(t)$. Since $n_s(t)$ has the power spectral density N_0, its derivative has the power spectral density $(2\pi f)^2 N_0$, and the noise term in (50) has the power spectral density $(f/h k)^2 N_0$. In order to compute the noise power in (50) we should note that at the receiver input there is a bandpass filter wide enough to pass the transmitted signal with bandwidth B_T, as given by (41). However, at the output of the differentiator of (50) it is desirable to filter $d(t)$ with a filter no wider than the bandwidth B of $x(t)$, in order to limit the output noise variance. Given this filtering, the noise power is

$\sigma_n^2 = [N_0/(hK)^2] \int_{-B/2}^{B/2} f^2 \, df$

$= N_0 B^3/12(hK)^2$. (51)

It is apparent from (51) that the output noise variance is inversely proportional to the square of the deviation constant. Thus, as the bandwidth of the fm signal increases, the signal-to-noise ratio increases. This improvement can be enhanced by a further step known as *preemphasis*. Since the output noise-power spectral density increases as the square of frequency, the output noise will be more disruptive at higher frequencies. This condition may be alleviated by prefiltering the transmitted signal so that its power spectral density more closely matches that of the noise at the output of the demodulator. This prefiltering step is preemphasis, since it accentuates the high-frequency content of the information signal. Subsequent to demodulation at the receiver, the inverse filter to the preemphasis filter is applied in order

to recover the original information signal. This filter is referred to as the *deemphasis* filter. The noise-power spectral density at the output of the deemphasis filter is more nearly white, thus enhancing the quality of the received signal.

A typical fm receiver implementation is illustrated in Fig. 8b. The received signal is first processed by a limiter to remove any amplitude variations which may have been induced by the channel, as well as to limit the amplitude of any impulsive noise at the receiver input. This is followed by the fm demodulator, which is implemented by taking the derivative of the limiter output and detecting the envelope of the resulting waveform, as given in (50a) and (50b). The output of the envelope detector is then lowpass filtered in order to remove out-of-band noise. The receiver in Fig. 8b is frequently referred to as a limiter-discriminator receiver. This is certainly not the only approach to fm receiver design; there are other, superior, approaches based on phase-locked loops, which will not be treated here.

3. Digital Modulation

In the previous section we considered various methods of imposing analog information on a carrier. Although analog-modulation techniques are employed in optical guided-wave transmission, a great deal more interest is focused currently on the use of fiber-optic waveguides for the transmission of digital information. The reasons for this are related to the advantages that digital transmission has on any medium: high reliability and greater control over noise and other channel impairments than is generally afforded by analog methods. The digital revolution that has pervaded much of the electronics industry and the rest of our society is also a driving factor. In particular, the telephone companies, which started to install digital transmission facilities some 25 years ago, are increasing the pace of digital facilities installation. Digital transmission over optical waveguides is a very attractive alternative to other possibilities due to the inherent large bandwidth, high reliability, and low susceptibility to interference of optical waveguides.

Signal Model

The object of digital modulation methods is to

transmit information which is presented to the transmitter in the form of a sequence of symbols chosen from a finite, discrete alphabet. In practice, this alphabet is almost always the binary alphabet, so that the information is presented to the transmitter as a sequence of "ones" and "zeros." The job of the transmitter is to map this input sequence into a waveform which can then be transmitted over the channel. Typically the transmitted waveform consists of a sequence, or superposition, of elementary signaling waveforms, each of which is chosen from the signaling alphabet according to some rule for mapping the input sequence onto the signaling alphabet. Taking the input sequence to be binary, and the signaling alphabet to be M-ary, where $M=2^k$, then the simplest type of modulator is one which simply performs a one-to-one mapping of the input sequence in k-bit blocks onto the signaling alphabet. More complicated modulators map sequences of bits onto sequences of 2^k-ary symbols using any of a variety of coding techniques. Since a development of coding theory is beyond the scope of this discussion, we will restrict our attention to the simpler forms of digital modulation mentioned above.

We will represent the basic signal to be transmitted in any digital modulation scheme as

$$x_m(t)=\text{Re}\ \{u_m(t)\exp{(j\,2\,\pi f_c\,t)}\}\,, \qquad (52)$$

where $x_m(t)$ is the mth basic signal chosen from the signaling alphabet containing M signals, $u_m(t)$ is its complex envelope, and f_c is the carrier frequency in hertz. Typically, M is some power of 2, i.e. $M=2^k$. In magnitude/angle form, $u_m(t)$ may be represented as

$$u_m(t)=a_m(t)\exp{[j\,\phi_m(t)]}\,, \qquad (53)$$

and $x_m(t)$ may be represented as

$$x_m(t)=a_m(t)\cos{[2\,\pi f_c\,t+\phi_m(t)]}\,. \qquad (54)$$

During each symbol interval T_s the transmitter chooses one member of the channel symbol alphabet to transmit over the channel, based on the input information sequence. Thus the trans-

mitted signal $s(t)$ may be represented as the superposition of the symbols transmitted during each symbol interval:

$$s(t)=\sum_i x_i(t-i\,T_s)\,. \qquad (55)$$

The complex envelope of (55) may be expressed by

$$y(t)=\sum_i u_i(t-i\,T_s)\,. \qquad (56)$$

The summations in (55) and (56) are over the duration of the transmitted sequence, which may be considered finite or semi-infinite.

Any practical digital communication system includes filtering or pulse shaping both at the output of the transmitter and at the input of the receiver. In fact, as will be shown later in this section, the receiver which is in some sense optimal includes a filter which is matched to the shape of the transmitted pulse. Thus we should view the pulses $x(t)$ and $u(t)$ of (53) and (54) as pulses which would be observed at the output of the receiver filter under noise-free conditions.

Pulse Shaping for Bandlimited Signaling—In many communication systems the pulses $u_m(t)$, with $1\leq m\leq M$, differ only in some complex weighting factor so that they are of the form $a_m u(t)$. Choice of the pulse $u(t)$ determines the spectral characteristics of the resulting digitally modulated waveform. In particular, many, if not all, communication systems are under some bandwidth constraint which requires that the energy in the modulated waveform must be contained in some finite bandwidth. However, the requirements of bandlimited signaling and distortion-free reception of the transmitted signal are generally in conflict.

The process of correctly detecting the transmitted symbols on the reception of a noise-corrupted version of $y(t)$ as given by (56) involves sampling the received waveform at intervals of T_s seconds. For best performance it is desirable that at the ith sampling instant, the value of $y(t)$ is unaffected by any symbol other than $u_i(t-i\,T_s)$. In other words, it is required that

$$u(i\,T_s)=0, \qquad i\neq 0\,.$$

Violation of this condition results in a type of distortion known as *intersymbol interference* (*isi*). The presence of such interference generally degrades the performance of the communication link. Fig. 9 illustrates pulses with and without isi.

It is well known that forcing any signal, such as a pulse, to be bandlimited, results in increasing the time duration of the pulse. In fact, any signal which is strictly bandlimited must be doubly infinite in duration and thus be noncausal. The minimum amount of bandwidth required to signal at baseband at a rate of $1/T_s$ symbol per second without intersymbol interference was found by

criterion, but which may be constructed to have a more gradual rolloff in the frequency domain, and more rapid decay in the time domain, than the sinc pulse. In order to construct this more general class of signals first define

$$U'(f)=\sum_i U(f+i/T_s),$$

$$-(1/2\,T_s)\le f\le(1/2\,T_s),\quad(59)$$

where $U(f)$ is again the Fourier transform of the signaling pulse $u(t)$. It can be readily shown

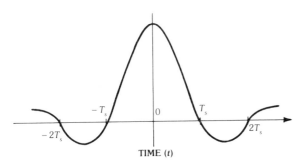

(a) Sinc pulse for bandlimited signaling without intersymbol interference.

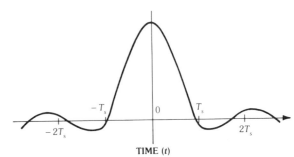

(b) Pulse with nonzero intersymbol interference.

FIG. 9. Pulses with and without intersymbol interference.

Nyquist [7] in 1924 to be $1/2\,T_s$ Hz. A pulse which satisfies this frequency constraint exactly is the *sinc pulse* (shown in Fig. 9):

$$u(t)=\text{sinc }(t/T_s)=[\sin(\pi t/T_s)]/(\pi t/T_s),\quad(57)$$

whose Fourier transform is

$$U(f)=\begin{cases}T_s, & |f|\le1/2\,T_s,\\0, & |f|>1/2\,T_s.\end{cases}\quad(58)$$

Unfortunately, this pulse is a noncausal signal, since it is doubly infinite in extent. Furthermore, the tails of this pulse decay as $1/t$, and as a result it is difficult to achieve a useful approximation to this pulse shape using realizable filters. Equivalently, it is difficult to build a filter which closely approximates the "brick wall" response of (58).

There is, however, a more general class of signals which satisfy the intersymbol interference

that any pulse, $u(t)$, whose Fourier transform, $U(f)$, satisfies the constraint that $U'(f)$ as defined in (59) has the brick-wall characteristic of (58), is free of intersymbol interference. In other words, $U(f)$ may be allowed to roll off gradually, as long as the "folded" spectrum $U'(f)$ is strictly bandlimited to $\pm1/2\,T_s$ Hz.

Obviously, many pulses can be found which satisfy this criterion. One class of such pulses which has been widely used has the raised cosine spectrum:

$$U(f)=\begin{cases}T_s, 0\le|f|\le(1-\beta)/2\,T_s\\(T_s/2)\{1+\cos[(\pi T_s/\beta)f-\pi(1-\beta)/2\beta]\},\\\quad(1-\beta)/2\,T_s\le|f|\le(1+\beta)/2\,T_s.\quad(60)\end{cases}$$

The pulse is the inverse Fourier transform of (60):

$u(t) = \sin(\pi t/T_s)\cos(\beta \pi t/T_s)$

$$\times \{(\pi t/T_s)[1-(2\beta t/T_s)^2]\}^{-1}. \quad (61)$$

Fig. 10 shows the pulse shape and spectrum of a raised cosine pulse for several values of β. The parameter β controls the rolloff rate; for $\beta=0$ this reduces to the sinc pulse; for $\beta=1$ the rolloff takes place over the interval $0\leq|f|\leq 1/T_s$.

Channel Model

As in the previous section we will model the channel as a memoryless, additive, white Gaussian noise channel. The complex envelope of the received signal is then

$$v(t) = \alpha y(t)\exp(j\theta) + z(t), \quad (62)$$

where α is the channel attenuation parameter, and θ is a phase shift induced by the channel. Here $z(t)$ is the complex envelope of the white Gaussian noise process. Thus the received signal is

$$r(t) = \mathrm{Re}\{v(t)\exp(j2\pi f_c t)\} = s_r(t) + n(t),$$

$$s_r(t) = \mathrm{Re}\{\alpha y(t)\exp[j(2\pi f_c t + \theta)]\}. \quad (63)$$

The channel attenuation and phase shift may or may not vary with time; any time dependence is suppressed in (62) and (63). There are many situations in which the behavior of the physical channel departs significantly from this model, and modifications must be made in the model to account for these departures. Some common examples include the intersymbol interference channel, where the

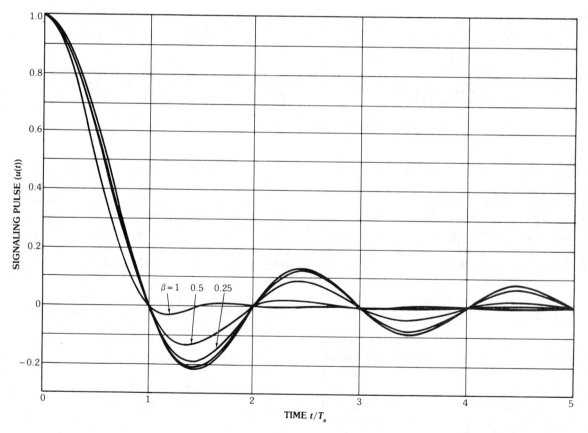

(a) Raised cosine pulses.

FIG. 10. Pulse shape and spectrum

memoryless property no longer holds; channels (or receivers) with signal-dependent noise, such as direct-detection optical receivers; and channels in which the noise is not Gaussian. We will ignore most of these issues and concentrate on the memoryless Gaussian channel, since a full development of these topics is beyond the scope of this exposition. However, a brief discussion of the sources of intersymbol interference in fiber-optic systems follows.

Intersymbol interference in fiber-optic communication systems is caused by a variety of physical effects. For the purposes of this discussion we may classify these effects in two ways. One of these may be characterized as multipath effects. These are due to the fact that radiation from the source may be propagated by a number of modes in the fiber, each of which has a different propagation constant and therefore a different delay through the fiber. This phenomenon occurs only in multimode fiber.

Intersymbol interference can also occur in single-mode fibers. The cause of this is sometimes referred to as *chromatic dispersion*, which is due to nonlinearities in the propagation constant of the single propagation mode. In many single-mode systems this is the dominant distortion mechanism, in which case the fiber may be modeled as a linear time-invariant system whose impulse response is the inverse Fourier transform of the system function:

$$H(\omega) = \exp\left[-j\,\beta(\omega)\,z\right], \qquad (64)$$

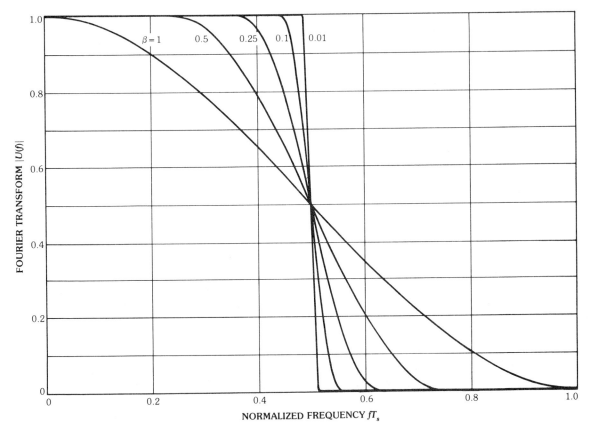

(b) Fourier transforms of raised cosine pulses.

of a raised cosine pulse.

where z is the fiber length, β is the frequency-dependent propagation constant of the fiber, and ω is the frequency in radians per second.

We can also refer to the group delay imposed by the fiber which is given by the derivative of the propagation constant:

$$\tau(\omega) = z \, (d/d\omega) \, \beta(\omega) \,. \tag{65}$$

In terms of the group delay we can refer to the effects due to nonlinearities in the propagation constant as effects due to nonconstant group delay.

In general, $\beta(\omega)$ is a nonlinear function of ω, in which case it may be expanded in a Taylor series around the center frequency ω_c of the pulse:

$$\beta(\omega) = \beta(\omega_c) + \beta'(\omega_c) \, (\omega - \omega_c)$$
$$+ (1/2) \, \beta''(\omega_c) \, (\omega - \omega_c)^2 + \cdots \,. \tag{66}$$

It follows that the group delay may also be expanded in a Taylor series,

$$\tau(\omega) = \tau(\omega_c) + \tau'(\omega_c) \, (\omega - \omega_c) + \cdots \,, \tag{67}$$

where $\tau(\omega_c) = z \, \beta'(\omega_c)$, $\tau'(\omega_c) = z \, \beta''(\omega_c)$, etc.

If $\beta(\omega)$ is a linear function, then the first two terms of (65) are the only nonzero terms, in which case (64) is

$$H(\omega) = \exp \left\{ -j \left[\beta(\omega_c) \, z + \beta'(\omega_c) \, (\omega - \omega_c) z \right] \right\} \,. \tag{68}$$

Evaluating the inverse Fourier transform of (67) yields the impulse response:*

$$h(t) = \delta[t - \beta'(\omega_c) \, z] \,. \tag{69}$$

Thus the effect of a linear propagation constant is simply to delay the transmitted pulse by an amount equal to $\beta'(\omega_c) \, z$. In this case the exact value of $\beta'(\omega_c)$ is of little concern, since the pulse shape is not affected.

More generally β is a nonlinear function of ω or, equivalently, the group delay τ is a nonconstant function of ω. Typically, either the second term in (67) dominates, or, if this term is very

---

*The phase offset $\beta(\omega_c)z - z\omega_c\beta'(\omega_c)$ has been neglected.

small, then the third term dominates. The effects of these two terms are referred to as *linear* and *quadratic group-delay distortion,* respectively.

In order to examine the effects of these two conditions we start by assuming, without loss of generality, that the low-pass equivalent transmitted pulse, $u(t)$, is real valued. We then note that $u(t)$ can be expressed as the inverse transform of $U(f)$:

$$u(t) = \int_{-\infty}^{\infty} U(f) \exp \left(j 2 \pi f t \right) df \,. \tag{70}$$

Furthermore, denoting the output of the fiber as $v(t)$, its Fourier transform is

$$V(f) = U(f) H(f) = U(f) \exp \left[j \Phi(f) \right], \tag{71}$$

where $\Phi(f) = \beta(2 \pi f) z$ is the phase function of the fiber. Note that if $u(t)$ has been designed to satisfy the conditions for freedom from intersymbol interference, from (71), $v(t)$ in general does not satisfy these conditions, and the phase-distortion function $\Phi(f)$ has introduced intersymbol interference. One exception is the case where $\Phi(f)$ is linear.

A well-known property of the Fourier transform is that the Fourier transform of a real function is conjugate symmetric. In this case, since $u(t)$ is real, then $U(f) = U^*(-f)$. As a consequence, if the phase function $\Phi(f)$ possesses odd symmetry, i.e., if $\Phi(-f) = -\Phi(f)$, then $V(f)$ retains the conjugate symmetry of $U(f)$, and $v(t)$ is a real function. However, if $\Phi(f)$ does not have this symmetry, then $V(f)$ is not conjugate symmetric, and as a result $v(t)$ is not real. This means that some of the intersymbol interference caused by the nonlinear phase distortion is contained in the quadrature component of the fiber output. In this case coherent systems have an advantage over noncoherent systems, in that the coherent receiver rejects the quadrature component of the intersymbol interference, while the noncoherent receiver does not. Since linear delay distortion corresponds to quadratic phase distortion, linear delay distortion results in a quadrature intersymbol interference component. Quadratic delay distortion, on the other hand, corresponds to cubic phase distortion, which results in conjugate symmetry in $V(f)$ and no quadrature intersymbol interference component.

Both linear and quadratic delay distortion are typically specified as the product of d, the differential delay across half the transmission bandwidth, and f_{max}, the frequency interval between the center of the band and the band edge, where the bandwidth is taken as twice the bit rate, as in the case for a raised cosine pulse with $\beta=1$. This can be viewed also as the ratio of the differential delay across half the band and the symbol interval, since $R_s=1/T_s$. Sunde [8] has investigated the effects of both linear and quadratic delay distortion as a function of dR_s, in terms of the power penalty caused by a given value of dR_s. Typically the effects of this delay are not significant unless the differential delay is a significant fraction of $1/T_s$. For instance, when binary phase-shift keying (bpsk) is used with conventional coherent detection which makes symbol decisions on a symbol by symbol basis, a power penalty of 1 dB occurs when $dR_s \simeq 0.8$. This power penalty can be reduced if maximum-likelihood detection methods, such as the Viterbi algorithm discussed briefly below, are used.

Yamamoto and Kimura [9] have applied these results of Sunde to the performance of coherent transmission systems using various modulation schemes on monomode fiber systems. They give transmission capacities of typical monomode fibers, which correspond to a power penalty of 1 dB. (This is not channel capacity in the strict information-theoretic sense, but simply the maximum signaling rate which can be obtained without inducing a power penalty of greater than 1 dB relative to the performance of the particular modulation scheme without intersymbol interference.) These results are given graphically in terms of distance-bandwidth product as a function of carrier wavelength.

A method for computing the capacity is given as follows. Linear delay distortion in monomode fibers is usually specified in units of picoseconds per nanometer·kilometer. This parameter gives the amount of differential delay, in picoseconds, across a band with a width of 1 nm on a fiber 1 km long. To make this parameter useful in this context it is necessary to convert it to units of nanoseconds per gigahertz·kilometers. The correct conversion factor is $(\lambda^2/c)\times10^{15}$ nm·ns/GHz·ps, where λ is the operating wavelength in meters, and

c is the speed of light in meters per second. If we denote the material dispersion coefficient in units of picoseconds per nanometer·kilometer by m_1, and the material dispersion coefficient in units of nanoseconds per gigahertz·kilometer by M_1, then d, in nanoseconds, is given by $M_1 L R_s$, where L is the fiber length in kilometers and R_s is the signaling rate in gigabits per second. For a given modulation scheme and power penalty the product dR_s is fixed at some value α_1, as given by the results of Sunde. Thus the fiber transmission capacity is

$$R_s^2 L = (\alpha_1 c/M_1 \lambda^2)\times10^{-15} \qquad (72)$$

in kilometers·(gigabits per second)^2.

At the zero-dispersion wavelength of a monomode fiber, m_1 goes to zero, and the higher-order terms in the expansion (67) must be considered. In particular, the effect of quadratic delay distortion now dominates. The transmission capacity of monomode fibers limited by quadratic delay distortion can be found in a manner similar to the linear delay distortion limits given above. The dispersion due to quadratic delay distortion in monomode fibers is governed by a second dispersion coefficient, which we will denote by m_2, which is typically given in units of picoseconds per (nanometers-squared·kilometers). For our use we must convert this to units of nanoseconds per (gigahertz-squared·kilometers); we will denote the resulting coefficient by M_2. The conversion is

$$M_2 = m_2 (\lambda^4/c^2)\times10^{33}.$$

For a given modulation scheme and power penalty, the product dR_s is again fixed at some value α_2, and the associated transmission capacity of the cable is

$$R_s^3 L = (\alpha_2 c^2/m_2 \lambda^4)\times10^{-33}. \qquad (73)$$

As a numerical example, typical fiber may have a material dispersion coefficient m_1 of 15 ps/nm·km at a wavelength of 1550 nm. Using this value in (72) and taking $\alpha=0.8$ (the value associated with a 1-dB penalty for bpsk signaling) results in a value of $R_s^2 L$ of 6.7×10^3 km·(Gb/s)^2. This translates to a maximum transmission rate of 8.2 Gb/s of a fiber 100 km long. Furthermore, near the zero-disperson wavelength (on a normal fiber this is about 1300 nm) the material dispersion coefficient m_1 typically has a value close to 1 ps/nm·km, in which case the value of $R_s^2 L$ is 1.4×10^5. On a 100-km fiber this would result in a maximum transmission rate of about 38 Gb/s.

At the zero-dispersion wavelength, m_1 goes to zero, and m_2 becomes important. From Sunde [8] the value of α_2 which yields a 1-dB penalty for bpsk is roughly 0.75. Yamamoto and Kimura give a value of $R_s^3 L$ of roughly 4×10^8 km·(Gb/s)^3 at 1300 nm for a 1-dB penalty for bpsk signaling. This implies a maximum rate of about 160 Gb/s on a 100-km fiber.

It should be emphasized that the numbers arrived at above are based on the results of Sunde, who assumed a highly stable carrier and perfect carrier recovery. In terms of optical communication this requires very narrow linewidth lasers, relative to the signaling rate. In this sense the numbers given above represent the potential of coherent optical-fiber transmission which might be achieved with practical, reliable, narrow-line-width lasers and optical recovery schemes.

Maximum-Likelihood Receiver for Digital Signaling

The optimum receiver for a digital signal is the receiver which minimizes the probability of error or, equivalently, the receiver which maximizes the *a posteriori* probability $p(s_n(t)|r(t))$. This is the probability that the signal $s_n(t)$ was transmitted, given that the signal $r(t)$ is observed at the receiver. Under fairly general conditions it can be shown that the receiver which maximizes the *a posteriori* error probability is equivalent to the receiver which maximizes the likelihood function $p(r(t)|s_n(t))$. Such a receiver is referred to as the *maximum-likelihood receiver*.

For the additive Gaussian noise channel the maximum-likelihood receiver maximizes the likelihood function [10, 11]:

$$p(r(t)|s_n(t)) = K \exp \left\{ -(1/N_0) \int_0^T [r(t) - s_n(t)]^2 dt \right\}$$

$$= K \exp \left[-(1/2 N_0) \right.$$

$$\left. \times \int_0^T |v(t) - \alpha y_n(t) \, e^{j\theta}|^2 \, dt \right], \quad (74)$$

assuming that the transmitted message is of duration T seconds. This form of the maximum-likelihood receiver assumes that perfect knowledge of the phase θ and channel attenuation α is available at the receiver. If this is not the case, then an appropriate likelihood function is obtained by averaging over the unknown parameters in (74).

The maximum-likelihood receiver thus computes the likelihood function in (74) for each possible transmitted signal, and selects as the most likely transmitted signal that signal which yields the largest-likelihood function. Equivalently, the log-likelihood function may be used, i.e., the natural logarithm of (74). Thus the maximum-likelihood receiver chooses the transmitted signal which minimizes the log-likelihood function,

$$\int_0^T |v(t) - \alpha y_n(t) \, e^{j\theta}|^2 \, dt, \quad (75)$$

or, equivalently, maximizes the function

$$\Lambda_n = \mathrm{Re} \left\{ e^{-j\theta} \int_0^T v(t) \, y_n^*(t) \, dt \right\}$$

$$- (\alpha/2) \int_0^T |y_n(t)|^2 \, dt. \quad (76)$$

If the message set is an equal-energy set, that is, if all possible messages havbe the same energy, then the second term in (76) is the same for all possible transmitted signals, and (76) simplifies further to

$$\Lambda'_n = \mathrm{Re} \left\{ e^{-j\theta} \int_0^T v(t) \, y_n^*(t) \, dt \right\}. \quad (77)$$

The integral in (77) has the simple interpretation of a correlation operation; the received signal $v(t)$ is correlated with each of the possible transmitted signals. When the received signal consists of independent symbols plus white noise (in other words there is no intersymbol interference, and no other correlation introduced by coding at the transmitter), then the receiver can make decisions on a symbol by symbol basis. In this case $y_n(t)$ is replaced by $u_n(t)$, a single symbol from the symbol alphabet; the receiver computes a decision variable of the form (77) for each symbol in the alphabet during each symbol interval, and decides on the most-likely transmitted symbol for that interval. The receiver, consisting of a bank of correlators of the form of (77), is shown in Fig. 11.

Another way of generating the decision variable Λ'_n is to define the *matched filter,* $\beta_n(t) = u_n*(T_s-t)$, where T_s is the symbol duration. If we further define the output of the nth matched filter,

$$ w_n(t) = \int_{-\infty}^{\infty} v(\tau) \beta_n(t-\tau) \, d\tau \, , $$

then it is easy to verify that $\Lambda'_n = w_n(T_s)$. The bank of matched filters which realizes the receiver of (77) is shown in Fig. 12.

If the received signal consists of correlated symbols, due to the presence of intersymbol interference, or coding, the maximum-likelihood receiver can no longer make decisions on a symbol by symbol basis, but must instead examine many received symbols before making a decision on the most-likely transmitted sequence. The most well-known method for doing this was introduced by A. J. Viterbi as a method for decoding convolutional codes, and is commonly known as the *Viterbi algorithm.* A complete treatment of this topic and other suboptimum methods of dealing with intersymbol interference is beyond the scope of this discussion. For further details, see [10, 11].

Maximum-Likelihood Receiver for Noncoherent Detetection—In some cases it is not possible or not

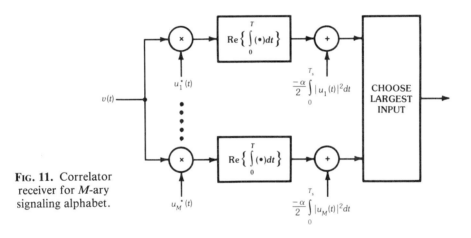

FIG. 11. Correlator receiver for *M*-ary signaling alphabet.

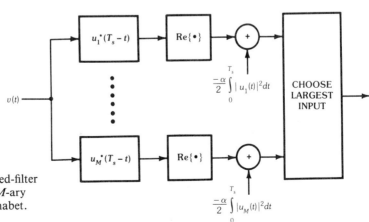

FIG. 12. Matched-filter receiver for *M*-ary signaling alphabet.

desirable to obtain a stable phase reference for coherent detection of the transmitted signal. In order to derive the maximum-likelihood receiver for this case, we treat the phase θ as a random variable uniformly distributed on $[0, 2\pi]$. The likelihood function for the signal with random phase is found by averaging (74) over θ. The resulting likelihood function can be shown to have the form

$$\Lambda_n = C \exp [E] I_0[D] , \qquad (78)$$

where

$$E = -(\alpha^2/2) \int_0^T |y_n(t)|^2 \, dt,$$

and

$$D = \left| \alpha \int_0^T v(t) \, y_n^*(t) \, dt \right| ,$$

and $I_0[x]$ is the modified Bessel function of order zero. When the signal set is an equal-energy set the detector can be implemented using simply D, since the Bessel function is a monotonically increasing function. This is the well-known envelope detector.

Since D does not contain any phase information this detector is clearly inappropriate for any modulation scheme in which the phase carries all or part of the transmitted information. For this reason use of envelope detection is restricted to modulation schemes such as frequency-shift keying, on-off keying, or pulse-position modulation.

Binary Error Probability

As a basis for computing the error probability of various signaling schemes, we will derive a general expression for the binary error probability for the maximum-likelihood receiver. This is the probability that some particular message is chosen by the maximum-likelihood receiver as the most-likely transmitted sequence, given that some other particular message was actually transmitted. More precisely, the binary error probability is $P[q|p]$, the probability that message q is chosen by the receiver, given that message p was transmitted. This error probability is simply the probability that the likelihood function associated with message q is

greater than the likelihood function associated with message p:

$$P[q|p] = P[\Lambda_q > \Lambda_p | p] = P[\Lambda_p - \Lambda_q < 0 | p] . \quad (79)$$

In order to compute this error probability we must first compute the difference between the two log-likelihood functions: $D_{pq} = \Lambda_p - \Lambda_q$. Since this quantity is a Gaussian random variable we need only compute its mean and standard deviation to calculate the desired binary error probability. Thus

$$D_{pq} = \text{Re} \left\{ e^{-j\theta} \int_0^T v(t) \left[y_p^*(t) - y_q^*(t) \right] dt \right\}$$
$$- (\alpha/2) \int_0^T \left[|y_p(t)|^2 - |y_q(t)|^2 \right] dt . \quad (80)$$

Taking $v(t) = \alpha y_p(t) e^{j\theta} + z(t)$, then we have

$$D_{pq} = \alpha \, \text{Re} \left\{ \int_0^T y_p(t) \left[y_p^*(t) - y_q^*(t) \right] dt \right\}$$
$$- (\alpha/2) \int_0^T \left[|y_p(t)|^2 - |y_q(t)|^2 \right] dt$$
$$+ \text{Re} \left\{ e^{-j\theta} \int_0^T z(t) \left[y_p^*(t) - y_q^*(t) \right] dt \right\}$$
$$= (\alpha/2) \int_0^T |y_p(t) - y_q(t)|^2 \, dt$$
$$+ \text{Re} \left\{ e^{-j\theta} \int_0^T z(t) \left[y_p^*(t) - y_q^*(t) \right] dt \right\} . \quad (81)$$

Given this expression for D_{pq}, the binary error probability is then

$$P[q|p] = (2 \pi \sigma_{pq}^2)^{-1/2} \int_{-\infty}^0 \exp \left[-(x - \mu_{pq})^2 / 2\sigma_{pq}^2 \right] dx$$
$$= (1/2) \, \text{erfc} \, (\mu_{pq} / 2^{1/2} \sigma_{pq}) , \quad (82)$$

where

$$\mu_{pq} = \text{E}[D_{pq}] = (\alpha/2) \int_0^T |y_p(t) - y_q(t)|^2 \, dt ,$$

$$\sigma_{pq}^2 = \text{var} \, [D_{pq}] = N_0 \int_0^T |y_p(t) - y_q(t)|^2 \, dt . \quad (83)$$

The error probability expression (82) gives the exact probability that an error will be made in distinguishing between the two messages p and q. In the case of binary signaling with independent bits over a memoryless channel, this expression applies directly to the bit error probability. When p and q are chosen from larger signal alphabets such as an M-ary symbol alphabet, or a sequence of symbols, (82) can be used to form an exact expression or a bound on the bit error probability.

On-Off Keying

Binary on-off keying (ook) is a simple form of binary signaling in which the signal alphabet consists of a pulse and the absence of a pulse. Typically, the pulse is a square pulse, although other shapes may be used due to filtering in the transmitter, and/or a bandlimited channel. Thus $u_0(t)=0$, and $u_1(t)=E^{1/2}p(t)$, where E is the pulse energy and $p(t)$ is a pulse normalized so that $(1/2)\int|p(t)|^2\,dt=1$.

The decision variables λ_0 and λ_1 from (76) are in this case: $\lambda_0=0$, and

$$\lambda_1=E^{1/2}\,\mathrm{Re}\left\{e^{-j\theta}\int_0^{T_s}v(t)p(t)\,dt\right\}$$

$$-(\alpha/2)\,E\int_0^{T_s}|p(t)|^2\,dt$$

$$=E^{1/2}\,\mathrm{Re}\left\{e^{-j\theta}\int_0^{T_s}v(t)p(t)\,dt\right\}-\alpha E.\quad (84)$$

The symbol λ is used for the decision variable instead of Λ here and in the following to emphasize the fact that the decisions are being made on a bit-by-bit basis, as opposed to making a decision on the entire transmitted sequence, as in (76). The term λ_0 is always 0 because $u_0(t)=0$. Because of this the receiver declares a 1 if $\lambda_1>0$, and a 0 if $\lambda_1<0$. Equivalently, a 1 is declared if

$$E^{1/2}\,\mathrm{Re}\left\{e^{-j\theta}\int_0^{T_s}v(t)p(t)\,dt\right\}>\alpha E,$$

and a 0 if

$$E^{1/2}\,\mathrm{Re}\left\{e^{-j\theta}\int_0^{T_s}v(t)p(t)\,dt\right\}<\alpha E.$$

We refer to αE as the *threshold*. A block diagram of the correlation receiver for on-off keying is shown in Fig. 13.

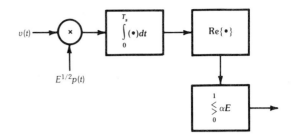

FIG. 13. Correlator receiver for on-off keying.

To compute the bit error probability we first find D_{10} and D_{01}, then, assuming that we have equally likely 1s and 0s, the bit error probability is given by

$$P_b=(1/2)\,P(D_{10}<0|1)+(1/2)\,P(D_{01}>0|0).\quad (85)$$

Because of the symmetry of this signaling scheme the two probabilities on the right-hand side of (85) are equal, so we only need to compute one of them:

$$D_{10}=(\alpha/2)\,\mathrm{E}\int_0^T|p(t)|^2\,dt$$

$$+\mathrm{Re}\left\{e^{-j\theta}\int_0^{T_s}z(t)\left[E^{1/2}\,p^*(t)\right]dt\right\}.\quad (86)$$

The mean and variance of (86) are

$$\mu_{10}=\alpha E,$$

$$\sigma_{10}^2=(1/2)\,E\int_0^T\int_0^T E[z(t)\,z^*(\tau)]\,p^*(t)\,p(\tau)\,dt\,d\tau$$

$$=2\,E\,N_0.\quad (87)$$

From (87), we have

$$\mu_{10}^2/2\,\sigma_{10}^2=\alpha^2\,E/4\,N_0=\alpha^2\,E_b/2\,N_0,\quad (88)$$

and

$$P(1|0)=P(0|1)=P_b=(1/2)\,\mathrm{erfc}\,[(\alpha^2/2)\,E_b/N_0]^{1/2}.\quad (89)$$

This expression is plotted in Fig. 14.

In a direct-detection fiber-optic system using on-off keying the noise is sometimes more accu-

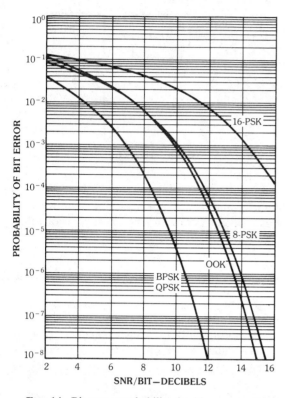

FIG. 14. Bit-error probability for binary on-off keying (ook), and M-ary psk for $M=2, 4, 8,$ and 16.

rately modeled as signal-dependent noise, due to the characteristics of the avalanche-photodiode detection process. In this case the white Gaussian noise model is still acceptable; however, the noise variance when a pulse is transmitted is different from the noise variance when a pulse is not transmitted.

In order to account for the different noise variances we must modify the detection process. Suppose that instead of discriminating between a 1 and a 0 based on the comparison $\lambda_1 <> 0$, as stated above, we assume that instead the comparison $\lambda_1 <> \kappa$ must be made, where κ is a constant which is yet to be determined. This is equivalent [12] to altering the threshold αE to $\alpha E + \kappa$. We can proceed to calculate the bit error probability in this case with κ as a parameter, and then solve for the optimum value of κ. In the following we will denote the noise-power spectral density when a 1 is transmitted as N_{o1}, and the noise-power spectral density when a 0 is transmitted as N_{o0}.

The error probabilities of interest are

$$P(1|0) = P(\lambda_{10} - \kappa > 0) , \qquad (90a)$$

and

$$P(0|1) = P(\lambda_{11} - \kappa < 0) , \qquad (90b)$$

where λ_{10} and λ_{11} are λ_1 given that a 0 was transmitted, and that a 1 was transmitted, respectively. Thus

$$\lambda_{10} = E^{1/2} \operatorname{Re} \left\{ e^{-j\theta} \int_0^T z_0(t)\, p(t)\, dt \right\} - \alpha E , \quad (91a)$$

and

$$\lambda_{11} = E^{1/2} \operatorname{Re} \left\{ e^{-j\theta} \int_0^T z_1(t)\, p(t)\, dt \right\} + \alpha E . \quad (91b)$$

The mean and variance of these two decision variables are

$$\mu_{10} = -\alpha E \quad \text{and} \quad \sigma_{10}^2 = 2 E N_{o0} \qquad (92a)$$

and

$$\mu_{11} = \alpha E \quad \text{and} \quad \sigma_{11}^2 = 2 E N_{o1}. \qquad (92b)$$

Since λ_{10} and λ_{11} are Gaussian random variables it is a straightforward matter to compute $P(1|0)$ and $P(0|1)$ using (92). The resulting error probabilities are

$$P(1|0) = (1/2) \operatorname{erfc} [(\kappa + \alpha E)/2 E^{1/2} N_{o0}] \quad (93a)$$

and

$$P(0|1) = (1/2) \operatorname{erfc} [(\alpha E - \kappa)/2 E^{1/2} N_{o1}] . \quad (93b)$$

Assuming that the probabilities of a 1 and a 0 being transmitted are equal, the bit error probability is

$$P_b = (1/2) P(0|1) + (1/2) P(1|0) . \qquad (94)$$

We wish to minimize this quantity with respect to κ in order to choose the optimum value of κ. This can be done by taking the derivative of (94) with respect to κ, setting this equal to zero and solving for κ. However, the result is a messy equation which yields little insight into the problem. Instead, assume that the minimum occurs when $P(1|0)=P(0|1)$. To enforce this condition it is necessary to simply set the arguments of the error functions in (93) equal and solve for κ. Doing so yields

$$\kappa=\alpha E \left[(N_{o0})^{1/2}-(N_{o1})^{1/2}\right]$$
$$\Big/ \left[(N_{o0})^{1/2}+(N_{o1})^{1/2}\right], \quad (95)$$

and

$$P_b=(1/2)\,\text{erfc}\left\{\alpha(2\,E_b)^{1/2}\right.$$
$$\left.\Big/ \left[(N_{o0})^{1/2}+(N_{o1})^{1/2}\right]\right\}. \quad (96)$$

It should be noted that this detection method is not optimal in the maximum likelihood sense for the problem of detecting on-off keying with different noise variances in the presence and absence of a pulse, since the detector does not incorporate any information about the variance of the received waveform. We will not address the problem of formulating a maximum-likelihood receiver for this case, although the development is a straightforward extension of the material discussed under "Maximum-Likelihood Receiver for Digital Signaling" in this section.

Phase-Shift Keying

Phase-shift keying (psk) is a form of modulation for which the phase of the carrier contains all the transmitted information. Thus the magnitude and phase pulses of (53) and (54), for M-ary psk, are

$$\phi_m(t)=2\,\pi\,(m-1)/M+\pi/M,$$
$$0\leq m\leq M-1\,, \quad (97a)$$

and

$$a_m(t)=(E_s)^{1/2}\,p(t),$$
$$(1/2)\int_0^{T_s}|p(t)|^2\,dt=1\,, \quad (97b)$$

where $1/T_s$ is the transmission rate in symbols per second, and E_s is the transmitted energy per

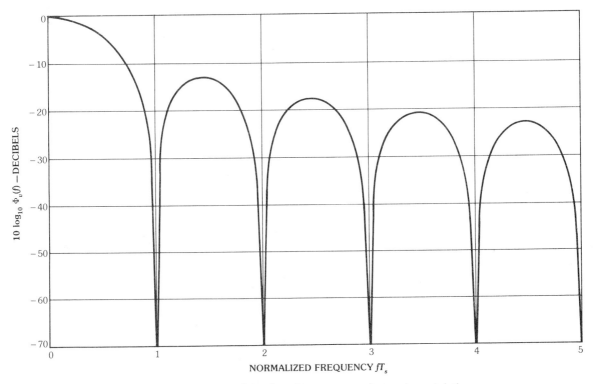

FIG. 15. Power spectral density of constant-envelope psk modulation.

M-ary symbol. For constant-envelope psk the normalized pulse $p(t)$ is constant over the symbol interval, and zero outside this interval: $p(t)= 2/T_s$ for $0 \leq t < T_s$; $p(t)=0$ otherwise. This pulse shape has the disadvantage that its power spectrum is not well contained; it has a main lobe of width $2/T_s$, and falls off with $1/f$ outside this interval. The power spectral density of constant-envelope psk is shown in Fig. 15. In environments where bandwidth is at a premium the pulses must be bandlimited at the expense of the constant-envelope property of the psk. In addition, it is difficult to accomplish the bandlimiting without introducing intersymbol interference (isi), which degrades the bit error rate performance.

The optimum receiver for psk, from (77), consists of a single correlator, or matched filter matched to the pulse $p(t)$, which is sampled at intervals of T_s seconds, as shown in Fig. 16. These samples are complex numbers whose arguments determine the decision which is made for the received symbol. As an example, Fig. 17 shows the phase constellation of quaternary psk (qpsk) and the associated decision regions.

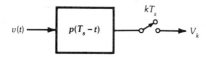

FIG. 16. Matched-filter receiver for psk.

For binary phase-shift keying (bpsk) this yields the low-pass equivalent and bandpass signal sets:

$$u_0(t)=(E_b)^{1/2} p(t) e^{j\pi/2},$$
$$u_1(t)=(E_b)^{1/2} p(t) e^{-j\pi/2}, \tag{98}$$

or

$$x_0(t)=(E_b)^{1/2} p(t) \cos(2\pi f_c t + \pi/2),$$
$$x_1(t)=(E_b)^{1/2} p(t) \cos(2\pi f_c t - \pi/2), \tag{99}$$

where in this case $E_b = E_s$.

To compute the error probability we first find

$$D_{01}=(\alpha/2)\int_0^T |(E_b)^{1/2} p(t)|^2 dt$$
$$+2(E_b)^{1/2} \operatorname{Re}\left\{e^{-j\theta}\int_0^T z(t) p^*(t) dt\right\}. \tag{100}$$

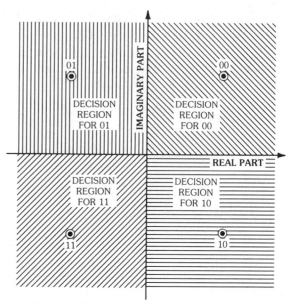

FIG. 17. Signal-space constellation of qpsk.

The term D_{10} has a similar form; the mean and variance of the two decision variables are given by

$$\mu_{10}=\mu_{01}=4\alpha E_b$$

$$\sigma_{01}{}^2=\sigma_{10}{}^2=E_b \operatorname{E}\left\{\left[e^{-j\theta}\int_0^T z(t) p^*(t) dt \right.\right.$$
$$\left.\left. +e^{j\theta}\int_0^T z^*(t) p(t) dt\right]^2\right\}$$

$$=8 E_b N_0. \tag{101}$$

Due to the symmetry of the signal set, $P(1|0) =P(1|0)= P_b$. Using (101) in (82) gives

$$P_b=(1/2)\operatorname{erfc}[(\alpha^2 E_b/N_0)^{1/2}]. \tag{102}$$

This is plotted in Fig. 14; note that bpsk has a 3-dB advantage over binary on-off keying.

Four-phase psk, or qpsk, may be regarded as two independent bpsk signals transmitted on quadrature carriers, and thus the bit error probability performance for qpsk is the same as for bpsk, so (102) applies to qpsk as well as bpsk. For higher-level digital phase-modulation schemes, exact computation of the bit error probability becomes more complex. We can, however, compute an approximate error probability based on the assumption that at moderate to high signal-to-noise ratios the most-likely error will be an error which changes one phase to an adjacent phase. The probability of such a binary error can be computed using (81) with $y_p(t)$ and $y_q(t)$ any two symbols with adjacent phases. Taking these two phases to be $\pm\pi/M$, it is easy to see that

$$\int_0^T |y_p(t)-y_q(t)|^2\,dt=8\,E_s\sin^2(\pi/M). \quad (103)$$

Using this, the mean and variance of D_{pq} are found to be

$$\mu=4\,\alpha\,E_s\sin^2(\pi/M) \quad (104a)$$

and

$$\sigma^2=8\,E_s\,N_0\sin^2(\pi/M). \quad (104b)$$

Thus the binary probability of error is given by

$$P(q|p)=(1/2)\,\text{erfc}\,[\alpha\sin(\pi/M)\,(E_s/N_0)^{1/2}]$$

$$=(1/2)\,\text{erfc}\,[\alpha\sin(\pi/M)\,(K\,E_b/N_0)^{1/2}]. \quad (105)$$

But since for each possible transmitted symbol two such errors are possible, and with Gray coding, approximately one bit error is made for each symbol error,

$$P_b\simeq(1/K)\,\text{erfc}\,[\alpha\sin(\pi/M)\,(K\,E_b/N_0)^{1/2}]. \quad (106)$$

This expression is plotted in Fig. 14 for $M=8$ and $M=16$. It should be apparent from these plots that increasing the symbol set size beyond $M=4$ in psk systems leads to performance penalties in terms of achievable bit-error probability at a given signal-to-noise ratio per bit. This penalty is compensated by increased bandwidth utilization achieved by sending multiple bits per period.

Differential Phase-Shift Keying—Operation of a communication link using psk signaling requires that a very stable phase reference be maintained at the receiver. This phase reference must be derived from the received signal. For a variety of reasons, such as unstable phase in the received signal, or unstable carrier reference at the demodulator, it may not be possible to maintain a sufficiently stable phase reference. In this case psk modulation

may still be used, by encoding the transmitted information differentially. Instead of encoding the symbol to be transmitted during the ith symbol interval as some absolute phase, the symbol is encoded as the phase difference between the symbol transmitted during symbol interval $i-1$ and symbol interval i. In order to demodulate the transmitted information the receiver only needs to be able to estimate phase changes, and does not need to maintain a phase reference with long-term stability.

This simplification comes at the price of a degradation in performance, however, since the accuracy of demodulating the ith symbol depends on two noise observations. The bit error probability for binary differential phase-shift keying (dpsk) can be shown to be

$$P_b=(1/2)\exp(\alpha\,E_b/N_0). \quad (107)$$

The difference between this and the bit error probability of coherent bpsk is less than 1 dB at high signal-to-noise ratios. The derivation of the performance of dpsk with larger symbol sets is rather involved, but can be found in various references [6, 11]. The performance degradation for larger symbol sets is significantly greater than that for binary dpsk.

Orthogonal Symbol Sets

On-off and phase-shift keying are signaling methods which are typically used in environments where bandwidth is at a premium. A properly designed qpsk system, for instance, can achieve a bandwidth efficiency exceeding 1 b/s/Hz. The price paid for this is relatively high signal-to-noise ratio requirements. For example, to achieve a bit error rate of 10^{-9} with qpsk on a white Gaussian noise channel requires an E_b/N_0 of approximately 12.6 dB. Fiber-optic waveguides, on the other hand, have the potential for extremely large bandwidths. It is possible to make use of this large bandwidth to achieve greatly reduced receiver sensitivities, by employing signal sets with very low bandwidth efficiency. In doing this, power efficiency is bought at the expense of bandwidth efficiency.

One class of signals which achieves this trade-off is the class of M-ary orthogonal signal sets.

These signals are characterized by the fact that they are mutually orthogonal:

$$(1/2)\int_0^T u_p(t)\,u_q^*(t)\,dt = E_s\,\delta_{pq},$$

$$0 \le p,\ q \le M-1, \quad (108)$$

where δ_{pq} is the Kronecker delta ($\delta_{pq}=1$ when $p=q$; $\delta_{pq}=0$ otherwise.) This type of signal has the property that as the symbol set size M increases, the bandwidth requirement increases without bound. It can also be shown that as long as the signal-to-noise ratio, $\alpha^2 E_b/N_0$, remains above a lower limit of ln 2, or -1.6 dB, the error probability may in principle be reduced to an arbitrarily small value by increasing M (see, for example [10, 11, 13, 14, 15]).

A bound on the binary error probability may be found from (82) and (83), with $y_p(t)=u_p(t)$, by observing that, due to the orthogonality of the signals,

$$\int_0^T |u_p(t)-u_q(t)|^2\,dt = \int_0^T |u_p(t)|^2\,dt$$

$$+\int_0^T |u_q(t)|^2\,dt = 4\,E_s. \quad (109)$$

Thus the mean and variance, μ_{pq} and σ_{pq}^2, are

$$\mu_{pq}=2\,\alpha\,E_s, \quad (110a)$$

and

$$\sigma_{pq}^2 = 4\,E_s\,N_0, \quad (110b)$$

and the binary error probability is

$$P(q|p) = (1/2)\,\text{erfc}\,[\alpha\,(E_s/2\,N_0)^{1/2}]$$

$$= (1/2)\,\text{erfc}\,[\alpha\,(K\,E_b/2\,N_0)^{1/2}]. \quad (111)$$

For binary orthogonal signaling ($K=1$), (111) gives the bit error probability directly, i.e, in this case $P_b = P(q|b)$. For $K>1$ ($M>2$) the exact analysis of the bit error probability is considerably more complicated. However, we can achieve a loose upper bound on the symbol error probability using the union bound:

$$P_M \le \sum_{q\ne p} P(q|p) = [(M-1)/2]$$

$$\times\ \text{erfc}\,[\alpha(k\,E_b/2\,N_0)^{1/2}]. \quad (112)$$

This can be converted to a bound on the bit error probability by a proportionality constant [11]:

$$P_b \le 2^{k-2}\,\text{erfc}\,[\alpha(k\,E_b/2\,N_0)^{1/2}]. \quad (113)$$

This is a rather loose bound on the bit error probability. Tighter bounds may be found through more involved analysis which will not be dealt with here.

Equation 113 is plotted in Fig. 18 for $M=2, 4, 8$, and 16. Note that contrary to psk systems, as the size of an orthogonal signal set is increased, the performance improves. This is accompanied by a decrease in bandwidth utilization. Thus it would seem that bandwidth and signaling energy or signal-to-noise ratio are parameters which may be traded off against each other. This is a fundamental notion of information theory which is discussed further in the section on channel capacity.

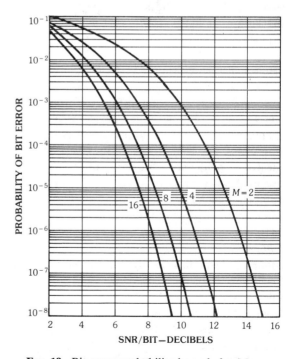

FIG. 18. Bit-error probability bounds for M-ary orthogonal signaling, $M=2, 4, 8,$ and 16.

The two most commonly used orthogonal signal sets are M-ary frequency-shift keying (M-ary fsk), and pulse-position modulation (ppm). For M-ary fsk the bandpass signal set is composed of sinusoids of the form

$$x_m(t)=(2\,E_s)^{1/2}/T_s\ p(t)\cos 2\,\pi f_m t,$$

$$0\le m\le M-1\,,\quad(114)$$

where $p(t)$ is a square pulse of unit amplitude and duration T_s seconds. It is easy to show that if the frequencies f_m are spaced at intervals of $k/2\,T_s$ radians per second, the signals are orthogonal:

$$\int_0^T x_m(t)\,x_n(t)\,dt=0\qquad(115)$$

for any integer k. Thus with this frequency spacing M-ary fsk is an orthogonal signal set (k is usually taken to be 1, to minimize bandwidth occupancy).

The maximum-likelihood receiver for M-ary fsk consists of a bank of M matched filters, each matched to a sinusoid with one of the M frequencies. Alternatively, the receiver can consist of a bank of M correlators, each computing the cross correlation of the received signal with one of the M possible transmitted waveforms. The receiver compares the outputs of the M correlators, or matched filters, every T_s seconds, and chooses as the most-likely transmitted signal that signal which corresponds to the correlator or matched filter with the largest output.

The above comments apply to the case of coherently detected M-ary fsk. Incoherently detected M-ary fsk is also commonly used; in this case orthogonality requires a frequency spacing of k/T_s. The receiver consists of a bank of matched filters followed by envelope detectors. Performance results for noncoherent detection of M-ary fsk are available widely in the literature (e.g., [11]).

The signal set for pulse-position modulation consists of a set of M pulses of duration T_s/M which are identical except for the instant at which they are transmitted. Thus

$$x_m(t)=(M/T_s)\,2^{1/2}\,E_s p(t-m\,T_s/M)$$

$$\times\cos\,[2\,\pi f_c(t-m\,T_s/M)+\theta]\,,\quad(116)$$

where $p(t)$ is typically a square pulse of unit magnitude and duration T_s/M. Since such pulses do not overlap when shifted in increments of T_s/M seconds, the integral (108) is satisfied, and the signal set is orthogonal.

The maximum-likelihood receiver for this signal set may consist of a matched filter or correlator for the elementary signal $p(t)\cos 2\,\pi f_c t$, which is sampled every T_s/M seconds over a period of T_s seconds. The transmitted symbol corresponding to the largest output sample is chosen as the most-likely transmitted signal. This receiver is more complex than the M-ary fsk receiver described above, since the receiver must derive a timing reference in order to synchronize to the symbol frame epoch T_s as well as to the epoch for an individual symbol, T_s/M.

Channel Capacity and Fundamental Limits

So far we have considered a number of specific modulation techniques for digital signaling over a fiber-optic channel, as well as some practical limitations on the capabilities of the optical-waveguide communications channel. It is important, however, to put this discussion in perspective by considering some of the theoretical limitations on fiber-optic communication. Thus we will conclude with a brief discussion of the implications of the results of information theory and quantum mechanics on the ultimate capabilities of fiber-optic communication.

Before proceeding with a discussion of channel capacity it is of interest to discuss noise sources in an optical receiver. It is well known that there are two fundamental and inescapable sources of noise which limit the performance of a linear amplifier. These are thermal noise, whose power spectral density is a function of both the temperature of the source and of frequency, and quantum noise, which is a consequence of the discrete nature of light, and whose power spectral density is a function only of frequency. The power spectral density of the noise of an ideal amplifier is

$$\Phi_{nn}(f)=\frac{hf}{e^{hf/kT}-1}+hf,\quad(117)$$

where h is Planck's constant, T is the temperature in kelvins, and k is Boltzmann's constant. A full development of this expression can be found, for example, in [16].

The first term on the right-hand side of (117) is the thermal-noise term. For $hf \ll kT$, this term dominates and (117) becomes

$$\Phi_{nn}(f) = kT, \qquad (118)$$

and when $hf \gg kT$, the second term, due to quantum noise, dominates, and (117) becomes

$$\Phi_{nn}(f) = hf. \qquad (119)$$

The two terms are equal when $f = kT/h$, which takes on a value of 6.04×10^{10} Hz at room temperature (290 K). For frequencies above this value the quantum term dominates, and for frequencies well below this frequency the thermal term dominates. Thus the thermal term dominates at frequencies up to and beyond microwave frequencies, and is the only term which is of concern in conventional radiocommunication systems. Fiber-optic communication systems operate at wavelengths which are within an order of magnitude of 10^3 nm, which corresponds to a frequency of 3×10^{14} Hz. Thus, at optical frequencies the quantum noise dominates in an ideal amplifier.

The quantum noise whose power spectral density is given in (119) is due to shot noise in the amplifier, caused by the random nature of the amplification process or, equivalently, by the fundamentally discrete nature of the received electromagnetic radiation (see, for example, [16, 17]). These shot-noise, or photon-arrival, effects can be accurately modeled as having Poisson statistics. However, in the limiting case of a high Poisson arrival rate, relative to the receiver bandwidth, the random nature of the received signal may be accurately modeled as the effect of an additive Gaussian noise source. Furthermore, although the power spectral density is a function of frequency, any currently conceivable optical-waveguide communication system will operate at bandwidths that are at least several orders of magnitude less than the optical carrier frequency, and thus (119) will be constant over the system bandwidth, to a very good approximation. Thus we can treat this noise as white Gaussian noise with power spectral density

$$\Phi_{nn}(f) = hc/\lambda_0, \qquad (120)$$

where h is Planck's constant, c the speed of light, and λ_0 is the wavelength of the optical carrier.

With this knowledge we can address the question of the ultimate transmission capacity of an optical-waveguide communication system. In his well-known work of the 1940s Claude Shannon showed that the transmission capacity of a band-limited white Gaussian noise channel is upper bounded by the channel capacity [13, 14, 15]:

$$C = W \log_2 (1 + P/N) \qquad (121)$$

in bits per second, where P is the received signal power, N is the noise variance, and W is the bandwidth of the channel. The significance of this quantity is given by the noisy-channel-coding theorem, a rough statement of which is as follows. Given an information source with a rate of B bits per second, and a channel with capacity C bits per second, there exists an encoding scheme which, when used to encode the source, will allow the source to be transmitted over the channel with an arbitrarily small probability of error, as long as $B < C$. If $B > C$, it is not possible to transmit the source over the channel with an arbitrarily small probability of error. Furthermore, achieving throughputs which approach the channel capacity require highly complex coding schemes. Thus the channel capacity represents an upper limit on the achievable throughput of a band-limited channel.

Further insight into (121) can be found by first noting that when operating at a rate of C b/s, the signal power is given by $P = C E_b$, where E_b is the average signal energy transmitted per bit. Also, from (120) the noise power is $N = hcW/\lambda_0$. Using these in (121) we get

$$(C/W) = \log_2 [1 + (C/W)(E_b \lambda_0/hc)], \qquad (122)$$

or

$$[E_b/(hc/\lambda_0)] = (2^{C/W} - 1)/(C/W). \qquad (123)$$

Since hc/λ_0 is the energy of a single photon of wavelength λ_0, and E_b is the energy required to transmit a bit, the quantity on the left-hand side of (123) can be interpreted as transmission efficiency, in units of photons required per bit, and the right-hand side gives the minimum number of photons per bit required to transmit at the rate C b/s over a channel with a bandwidth of W Hz.

Since optical waveguides offer the potential of extremely high bandwidth, the limiting value of the right-hand side of (123), as C/W goes to zero, is of interest. Taking the limit of the right-hand side of (122) gives $\ln 2 \simeq 0.693$. Increasing the transmission bandwidth with a fixed information throughput leads to an increased power efficiency, which in the limit, as the bandwidth efficiency goes to zero, becomes 0.693 photon per bit. This is the limiting energy efficiency of digital communication with a receiver which employs a linear amplifier, when the carrier frequency is in the region where quantum noise in the receiver is dominant.

It can be shown [16, 18] that a heterodyne receiver can also achieve the performance of a linear amplifier receiver. A schematic diagram of a heterodyne receiver is shown in Fig. 19. In this figure f_0 is the optical carrier frequency, f_ℓ is the local-oscillator frequency, ϕ_ℓ is the phase of the local oscillator, and $a(t)$ and $\phi(t)$ are the information signals imposed on the carrier wave, as described in previous sections. In principle a heterodyne receiver is implemented by multiplying the received signal by a sine wave generated by a local oscillator, and low-pass filtering the resulting product, as shown in Fig. 19. The frequency of the local sinusoid is chosen so that the signal at the output of the low-pass filter is centered around a frequency, called the intermediate frequency, which is lower than the carrier frequency of the received signal. In practice, there are numerous ways of implementing this type of receiver; at optical frequencies the local oscillator is a laser, and the signals are multiplied by combining them using a half-reflecting mirror, and detecting the combination with a photodetector.

Another approach to detection is referred to as *homodyne detection*. This approach is similar to heterodyne detection, except that in this case the frequency of the local oscillator is the same as the optical carrier frequency of the received signal, i.e., $f_\ell = f_0$ in Fig. 19. It can be shown that in this case the noise-power spectral density is effectively half that obtained when a linear amplifier or het-

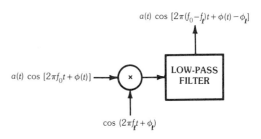

FIG. 19. Heterodyne receiver.

erodyne receiver is used. In this case the channel capacity is

$$(C/W) = \log_2 [1 + (C/W)(2 E_b \lambda_0/h c)] , \quad (124)$$

and the power efficiency is

$$[E_b/(h c/\lambda_0)] = (1/2)(2^{C/W} - 1)/(C/W) . \quad (125)$$

In the limit, as $C/W \to 0$, the potential efficiency of homodyne systems is $(\ln 2)/2 \simeq 0.347$ photon per bit. Comparing (124) and (125) with (122) and (123), it can be seen that homodyne detection has a potential 3-dB advantage over heterodyne.

Fig. 20 is a plot of energy efficiency versus channel capacity for heterodyne and homodyne detection, as given by (123) and (125). Realizable systems will operate above and to the left of the two curves. As stated above, these results are based on the assumption that the noise at the receiver is white Gaussian noise. This is a valid approximation when the bit or symbol decision is based on many photon arrivals. Care must be taken, however, in interpreting the lower left-hand end of the curves, where the efficiencies associated with operation at capacity are less than 1 photon per bit. In many signaling schemes this would imply that a symbol decision is based on very few photon arrivals. In this case the Gaussian assumption is questionable, as is the validity of these expressions for channel capacity.

Of course, there are many practical problems in implementing either heterodyne or homodyne receivers at optical frequencies. Perhaps the most significant such problem is the fact that there are as yet no commercially available devices which generate an optical carrier with the high degree of frequency stability required for coherent trans-

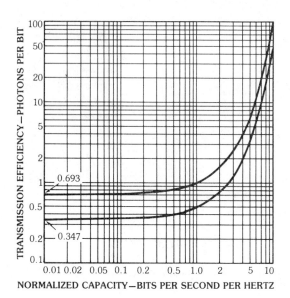

FIG. 20. Transmission efficiency in photons per bit as a function of channel capacity.

mission. State of the art semiconductor lasers typically have line widths of tens of megahertz. However, lasers with greater frequency stability are being developed, as well as methods for tracking the frequency of optical carriers.

The homodyne and heterodyne techniques are classified as coherent detection schemes, since for proper operation the receiver must generate a local sinusoid which tracks closely the phase of the optical carrier. Another approach to optical communication is known as *direct detection,* which involves direct observation of received photons in the signaling bandwidth, and counting the number of photon arrivals. In the absence of background radiation (generally the case in guided-wave communication) the only fundamental source of uncertainty in detecting the received signal is the random nature of photon arrivals.

As stated earlier, the photon arrival process can be described by the Poisson probability distribution function

$$P_k(t) = e^{-rt}(rt)^k/k! , \qquad (126)$$

where $P_k(t)$ is the probability that k photon arrivals occur in t seconds, and r is the average arrival rate, given by

$$r = \eta P_s/(hc/\lambda_0) , \qquad (127)$$

and η is referred to as the *quantum efficiency* of the detector. Thus, in order to ensure that detection of a transmitted pulse occurs with high probability, it will suffice to make r large enough that $P_0(T)$ is very small, where T is the pulse duration. Then a simple pulse-position-modulation encoding scheme will provide arbitrarily large power efficiency. A pulse-position-modulation system encodes information by choosing 1 out of N possible time slots for transmitting a pulse of duration T. In such a scheme an average of rT photons is used to transmit $\log_2 N$ bits, resulting in a power efficiency of $rT/(\log_2 N)$ photons per bit. This number may be made arbitrarily small by simply increasing N.

The problem with this approach is that increasing N without bound requires both unlimited bandwidth and unlimited complexity. In such a case the presence of thermal photons may not be ignored. Pierce [17], and Pierce, Posner, and Rodemich [19], give the resulting efficiency in photons per bit as

$$(\ln 2) kT/(hc/\lambda_0) .$$

At room temperature (290 K) and operating at a wavelength of 10^3 nm, this results in a potential efficiency of 1.4×10^{-2} photon per bit, or about 71 bits per photon. Such an efficiency is achieved through pulse-position modulation only when N is extremely large. Assuming that a pulse is detected by the reception of a single photon, then in order for that photon to represent 71 bits, we must have $N = 2^{71}$, which is clearly too large a number to be practical, due both to decoding complexity, and to the fact that it represents a bandwidth expansion of some 20 orders of magnitude relative to on-off–keyed direct-detection operation. This, however, does demonstrate the potential for power efficiencies which are well beyond those attainable through the use of coherent modulation techniques. Realization of power efficiencies approaching these values depends on the development of codes which are more efficient than pulse-position modulation, as well as the development of detectors which are efficient enough to detect individual photons.

Thus, although direct detection offers the promise of highly power-efficient operation, there are many practical problems associated with its implementation. At present, direct-detection systems do not approach the power efficiency promised by the above arguments. On the other hand, although there are challenging problems associated with implementation of coherent modulation systems for optical-fiber communication, these techniques seem to be more within the reach of current technology. Intensive development of both techniques is proceeding, and the future promises to bring interesting developments in both areas.

Acknowledgments

The author wishes to acknowledge his gratitude to the teaching and guidance of John Proakis, the influence of whose ideas should be strongly evident in a number of places in this exposition. Also, the patience, encouragement, and many helpful suggestions of E. E. Basch are gratefully acknowledged.

4. References

[1] A. B. CARLSON, *Communication Systems,* 2nd ed., New York: McGraw-Hill Book Co., 1974.

[2] H. TAUB and D. L. SCHILLING, *Principles of Communication Systems,* New York: McGraw-Hill Book Co., 1971.

[3] J. B. THOMAS, *An Introduction to Statistical Communication Theory,* New York: John Wiley & Sons, 1969.

[4] A. D. WHALEN, *Detection of Signals in Noise,* New York: Academic Press, 1971.

[5] A. J. VITERBI, *Principles of Coherent Communication,* New York, McGraw-Hill Book Co., 1966.

[6] W. C. LINDSEY and M. K. SIMON, *Telecommunication Systems Engineering,* Englewood Cliffs: Prentice-Hall, 1973.

[7] H. NYQUIST, "Certain Factors Affecting Telegraph Speed," *Bell Syst. Tech. J.,* Vol. 3, pp. 324–346, April 1924.

[8] E. D. SUNDE, "Pulse Transmission by AM, FM and PM in the Presence of Phase Distortion," *Bell Syst. Tech. J.,* Vol. 40, No. 2, pp. 353–422, March 1961.

[9] Y. YAMAMOTO and T. KIMURA, "Coherent Optical-Fiber Transmission Systems," *IEEE J. Quantum Electron.,* Vol. QE-17, No. 6, pp. 919–934, June 1981.

[10] A. J. VITERBI and J. K. OMURA, *Principles of Digital Communication and Coding,* New York: McGraw-Hill Book Co., 1979.

[11] J. G. PROAKIS, *Digital Communications,* New York: McGraw-Hill Book Co., 1983.

[12] T. OKOSHI et al., "Computation of Bit-Error Rate of Various Heterodyne and Coherent-Type Optical Communication Schemes," *J. Opt. Commun.,* Vol. 2, No. 3, pp. 89–96, 1981.

[13] C. E. SHANNON and W. WEAVER, *The Mathematical Theory of Communication,* Urbana: University of Illinois Press, 1949.

[14] J. M. WOZENCRAFT and I. M. JACOBS, *Principles of Communication Engineering,* New York: John Wiley & Sons, 1965.

[15] R. C. GALLAGER, *Information Theory and Reliable Communication,* New York: John Wiley & Sons, 1968.

[16] B. M. OLIVER, "Thermal and Quantum Noise," *Proc. IEEE,* Vol. 53, No. 5, pp. 436–454, May 1965.

[17] J. R. PIERCE, "Optical Channels: Practical Limits with Photon Counting," *IEEE Trans. Commun.,* Vol. COM-26, No. 12, pp. 1819–1821, December 1978.

[18] D. MARCUSE, *Principles of Quantum Electronics,* New York: Academic Press, 1980.

[19] J. R. PIERCE, E. C. POSNER, and E. R. RODEMICH, "The Capacity of the Photon Counting Channel," *IEEE Trans. Inf. Theory,* Vol. IT-27, No. 1, January 1981.

Digital Optical System Design

H. CARNES

Telco Systems Fiber-Optic Corp.

R. KEARNS & E. BASCH

GTE Laboratories, Inc.

1. Introduction

The preceding chapters have reviewed the funda-mental characteristics of the sources, detectors and fibers used in an optical communication system and in some instances clarified trade-offs that need to be made in selecting component characteristics when designing a system. In this chapter we shall examine how individual components interact and how their characteristics determine the design and performance of a complete optical-fiber link. The focus of our attention will be on synchronous digi-tal point-to-point transmission links since this type of a system lends itself to straightforward analysis while providing the basis for examining more complex architectures, such as multiterminal data networks and multichannel wavelength–division–multiplexed (wdm) systems. In order to provide a background for design issues we will briefly review the evolution of optical-fiber communication sys-tems from the past to the foreseeable future through five generations of technology.

Evolution of Optical-Fiber Communication
Since the early 1970s optical-fiber communication has evolved from a laboratory curiosity to com-mercially engineered fiber systems that are being used in many applications, perhaps the best known

and most widespread of which are in the telephone network. The rapid evolution from research and demonstration to practical technology received its start with the attainment of a 20-dB/km doped-silica fiber in 1970 [1]. At the time, a 20-dB/km fiber attenuation was regarded as the threshold of usefulness for telecommunication applications. The development of semiconductor light sources rugged enough for field environment and steady progress in other components needed to make optical-fiber communications practical resulted in the first field trials and commercial traffic installa-tions using standard digital carriers [2, 3, 4, 5].

Five generations of optical-fiber communica-tion technology are now being recognized. Fig. 1 shows the progressively increasing capabilities of each generation in terms of repeater spacing and bit rate. We will note the main features of each generation, along with the reasons for its develop-ment [6]. The short-wavelength (approximately 800–850 nm) first generation exceeded conven-tional copper systems in channel capacity and repeater spacing. A1GaAs lasers and light-emit-ting diodes (LEDs) were used as sources, silicon pin or avalanche photodiodes (APDs) as detec-tors, and multimode step-index or graded-index fiber as the transmission medium. The major lim-itation in these early systems was one of fiber

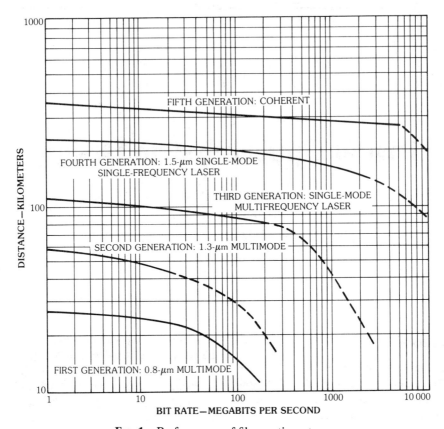

FIG. 1. Performance of fiber-optic systems.

losses, and, if LEDs were used, excessive material dispersion in the fiber.

Although the relative narrow spectral width of semiconductor lasers minimizes the effect of material dispersion, the use of highly coherent (i.e., narrow–line-width) lasers with multimode fiber results in interference between the propagating modes of the fiber and a granular optical intensity distribution across the core of the fiber. This so-called speckle pattern varies randomly with time and when combined with nonuniform attenuation in the fiber will result in excess noise, known as modal noise [7]. This system impairment is described in a later section in more detail. Fig. 2 shows that a shift in the operating wavelength from around 800 nm to around 1300 nm will allow a substantial increase in repeater spacing due to the reduced fiber attenuation at this longer wavelength. Also, material dispersion effects are minimal in this wavelength region. The desire to take

advantage of the improved fiber properties at the longer wavelengths led to the development of sources and detectors based on InGaAsP. The practical implementation of systems operating at 1300 nm marks the beginning of second-generation optical communication technology, [8, 9]. Second-generation technology proved to be very attractive in many ways, but implementation problems soon emerged. Probably the most severe was that the longer repeater spacing put very stringent requirements on fiber bandwidth while production yields of high-bandwidth graded-index fiber were relatively low. This, coupled with the problems of estimating the bandwidth in concatenated fibers [10] set the scene for a changeover to single-mode fiber. The introduction of single-mode fibers, which have both lower dispersion and lower loss than multimode fibers, marked the beginning of third-generation optical communication. Single-mode fiber has lower loss than multimode fiber at any given

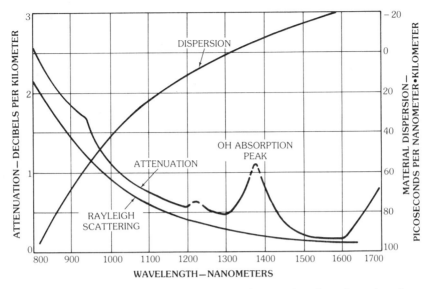

Fig. 2. Optical-fiber attenuation and dispersion as a function of wavelength.

wavelength (about 0.4 dB/km versus about 0.7 dB/km at 1300 nm) due to a reduction of dopants used in the fiber core. The lower doping density significantly reduces the Rayleigh scattering losses. The biggest advantage of the single-mode fiber, however, is the absence of modal dispersion since the single-mode fiber supports only one waveguide mode. This doesn't mean that dispersion effects are entirely nonexistent in single-mode fiber since the group velocity of the propagating mode is in general a function of wavelength, i.e., material and waveguide dispersion cause different wavelengths to have slightly different travel times through a fiber. For standard single-mode fiber, dispersion effects are minimal (less than 4 ps/nm·km) for a band of wavelengths in the region of 1300 nm as shown in Fig. 2.

A typical laser spectrum at 1300 nm consists of several longitudinal modes spaced about 1 nm apart as shown in Fig. 3. Therefore a pulse generated by such a laser will broaden as it travels along the fiber. The pulse dispersion is obtained by multiplying the spectral width of the laser with the dispersion coefficient of the fiber at the center wavelength. Thus the bandwidth of a single-mode fiber is a function of the spectral width of the transmitter source as well as its operating wavelength. A more important spectral problem of

multilongitudinal-mode lasers is that even though the total power output of the laser is constant, the instantaneous fluctuations of the power distribution among the laser longitudinal modes can be quite large. The combination of these partitioning fluctuations and fiber dispersion causes random fluctuations in the received signal and leads to an increase in bit error rate (BER) [11] beyond that expected from the intersymbol interference of the average distribution of power among the modes. Mode-partition noise is a complicated phenomenon and plays a key role in limiting the bit-rate distance product of the third-generation systems; its characteristics and effects on system performance are treated in more detail in Section 4. Using multilongitudinal-mode lasers at or very close to the zero-dispersion wavelength of the fiber (i.e., operating with lower dispersion) would somewhat ameliorate the impairment due to mode-partition noise [12], but imposes very tight manufacturing tolerances and temperature limitations, and adds some stringent new aging criteria as well.

Further examination of Fig. 2 shows that around 1550 nm fiber attenuation is even lower (about 0.2 dB/km) than around 1300 nm. Dispersion, however, has increased to 15 to 20 ps/nm·km. This high dispersion significantly aggravates the mode-partition noise problem caused

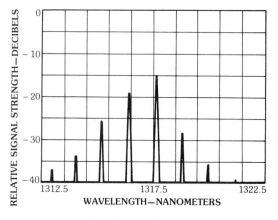

(a) Without modulation.

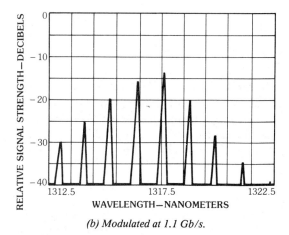

(b) Modulated at 1.1 Gb/s.

FIG. 3. Laser mode structure.

by the use of multilongitudinal-mode sources and limits repeater spacing to far less than that indicated by fiber loss for all but the lowest bit rates, thus negating the advantage of lower fiber attenuation. There are two promising solutions to this problem. The first has been to develop single–longitudinal-mode lasers (also called single-frequency lasers), the second to optimize the fiber design to reduce the dispersion. The latter approach uses a fiber design, referred to as dispersion-shifted fiber, in which material dispersion and waveguide dispersion cancel each other at around 1550 nm. This results in a shift of the zero-dispersion wavelength and the band of wavelengths with low fiber dispersion to the 1550-nm region, yielding a fiber in which attenuation and dispersion are minimized at

the same wavelength [13]. In "dispersion-flattened" fiber this principle is extended still further to design fibers with almost zero dispersion over a wavelength range of several hundred nanometers [14, 15]. The design of these "modified dispersion" fibers is treated in more detail in Chapter 4, Section 4. The primary advantage of this approach is that it allows multilongitudinal-mode laser technology to be used while operating where fiber attenuation is small [16, 17]. The use of dispersion-shifted fiber permits a big leap in performance; however, the bit-rate distance product is still limited by mode-partition noise and therefore it represents a transition between third- and fourth-generation systems. To achieve maximum loss-limited transmission distances, single longitudinal-mode sources are required. The use of such sources characterizes fourth-generation systems. As detailed in Chapter 9, Section 4, in a conventional semiconductor laser the feedback is provided by a Fabry-Perot resonator which doesn't provide any longitudinal-mode discrimination. Hence the only longitudinal-mode discrimination in such a laser is provided by the gain spectrum, which is usually much wider than the longitudinal-mode spacing, and oscillation in several longitudinal modes results.

Single longitudinal-mode operation can be obtained by modifying the basic Fabry-Perot laser structure so that the cavity loss is different for different longitudinal modes. Two laser structures have been found very effective in obtaining single longitudinal-mode operation, and they are known as distributed-feedback (DFB) lasers and coupled-cavity lasers. In a DFB laser [18] frequency-selective feedback is achieved through the use of a diffraction grating etched along the cavity length which provides reflections, i.e., the feedback is distributed throughout the cavity length, rather than located at the mirrors at the cavity ends. This results in unequal cavity losses and, with proper design of the grating, laser operation in a single longitudinal mode [19, 20, 21]. Excellent results have been achieved with DFB lasers in several system experiments [22, 23].

Coupled-cavity lasers use the interaction between two coupled Fabry-Perot cavities to achieve single longitudinal-mode operation. Each Fabry-Perot cavity has its own set of longitudinal modes produced by different cavity lengths. Longitudinal modes of the coupled system are formed by modes for which both cavities are in resonance. Selection of appropriate cavity length results in large separation between pairs of coincident modes and single longitudinal-mode operation. The best known example of a coupled-cavity laser is the cleaved–coupled-cavity or C^3 laser [24]. As with DFB lasers excellent performance has been achieved in system experiments, in particular when external modulation rather than direct modulation of the laser diode itself was used [25, 26].

The improvements reflected in each successive generation were primarily due to advances in optical-source and fiber technology, whereas optical receiver sensitivity has not significantly improved. In coherent optical transmission systems which have been (tentatively) designated fifth-generation systems, a significant improvement in receiver sensitivity is obtained through the use of optical heterodyne or homodyne detection [27, 28]. Unlike direct detection, where the optical signal is converted directly in a demodulated electrical signal, heterodyne detection first adds to the optical carrier signal a locally generated optical signal, the optical local oscillator, and then detects the combination as shown in Fig. 4. The detected signal is a replica of the original signal but shifted down in frequency from the optical domain to a frequency low enough so that conventional electronic signal processing can be used for demodulation. This method offers 15- to 20-dB improvement in receiver sensitivity compared with direct detection using state-of-the-art avalanche photodiodes. Furthermore, there is a significant improvement in

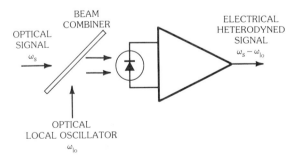

FIG. 4. Optical local oscillator mixes with a received optical signal to produce an electrical frequency equal to the difference between optical frequencies.

wavelength selectivity compared with conventional optical multiplexing technology, and there are other advantages as well.

Coherent optical communication, however, requires extremely stable and narrow–line-width lasers since laser phase noise causes severe degradation of receiver sensitivity. There are many challenging engineering problems, in particular reducing the line width of semiconductor lasers, that remain to be solved before commercial application of coherent systems becomes practical. Coherent optical communication is explored in more detail in Chapter 16.

2. System Design

Although several component options are available in fiber-optic system design, the link design methodology is similar. The primary measure of performance of a digital system usually is the bit error rate (BER), which is the probability that an error will be made in detection of a received bit. The maximum tolerable BER obviously depends on the desired system performance, but in optical communication systems a BER $\leq 10^{-9}$ is commonly specified. In the case of repeatered systems a style of regenerator where both the amplitude is regenerated and the repeater output is retimed is assumed. This provides a degree of independence among individual links. One potential exception to link independence is the effect of jitter on end-to-end system performance. For design completeness one should consider the amount and class of jitter introduced by the regenerator sections. The study of jitter is complex and beyond the scope of this chapter. However, the study of jitter generation and jitter accommodation in the literature is extensive [29, 30, 31, 32].

Design Considerations

To ensure that the desired system performance will be achieved it is usual to perform, first, a power budget or system gain analysis for the selected set components to determine whether or not the system is loss limited, that is, if the received power is sufficient to meet or exceed the BER specification. A power budget analysis, however, is not a sufficient condition to guarantee a minimum BER performance. System performance can also be dispersion or bandwidth

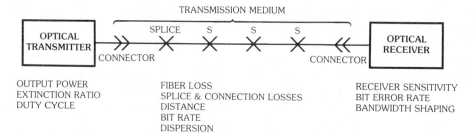

FIG. 5. Optical digital link model.

limited. In such a system the maximum spacing between transmitter and receiver is limited by (excessive) pulse distortion in the fiber which causes an attendant reduction in BER performance. Effectively, the pulse distortion has lowered receiver sensitivity and thus the optical link power budget. Since the design of a link involves many interrelated variables, among the source, detector, and fiber operating characteristics as well as noise phenomena caused by interaction of components, the actual link design and analysis could require several iterations before it is completed satisfactorily. With the above assumptions a simple physical link model can be constructed. The model to be used along with associated parameters is shown in Fig. 5.

We will begin our consideration of the model with the transmitter. The leading light sources for fiber-optic communication at present are the light-emitting diode (LED) and the injection laser diode (ILD). LED devices are incoherent sources that emit spontaneously, while lasers or ILDs are coherent, to a degree, and emit by stimulated emission. Important differences between the two sources are the amount of optical power that can be launched into a fiber, the switching speed, and the optical bandwidth or spectral width of the device. Some of the important parameters are listed in Chart 1.

CHART 1. Transmitter System Parameters

- Average output power
- Extinction ratio
- Operating wavelength
- Optical-source spectral width
- Pulse duty cycle
- Line code

The output signal strength is characterized by the first two parameters: average power and extinction ratio. Power level measurements performed with a photometer read in terms of average power. In order to have complete knowledge of the optical signal more information is needed beyond its average power. The pulse duty cycle is key to interpreting the average power. The extinction ratio is the parameter that completes the information about the signal power. Equations 1a and 1b are the expressions defining average power and extinction ratio:

$$P_{avg} = (P_{max} + P_{min})/M \tag{1a}$$

and

$$r_{ex} = P_{min}/P_{max}, \tag{1b}$$

where

P_{avg} = average power,
P_{max} = maximum or peak power,
P_{min} = minimum power,
r_{ex} = extinction ratio,
M = 2 for nrz (nonreturn-to-zero) signals and 4 for rz (return-to-zero) pulses.

Thus average power, pulse duty cycle, and optical-source extinction ratio are needed so that the receiver power sensitivity can be established. The extinction ratio is sometimes expressed as the reciprocal of (1b) by some manufacturers and authors. The above relationship gives a closed range from 0 to 1.0 for the extinction ratio. Fig. 6 shows a curve of power penalty versus extinction ratio for a pin photodetector receiver. That is, the amount of optical power needed to overcome the effect of not completely extinguishing the source. Penalties will be somewhat larger if an avalanche-type photo-detector is employed because of the excess multiplication noise. Extinction ratios greater than 0.1 will produce power penalties greater than 0.5 dB and are avoided except under special circumstances. (Very high extinction ratios are sometimes used to reduce frequency "chirping" which will be covered in a later section.)

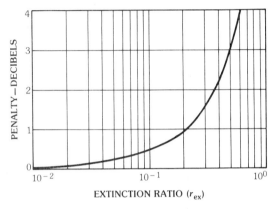

FIG. 6. Extinction-ratio penalty for pin receiver.

The selection of optical wavelength is an example of interrelation between elements of the transmission model. Since fiber attenuation and dispersion are functions of wavelength, one needs not only to be aware of the nominal value but also to take into account the effects of component tolerances. There are device-to-device variations as well as changes due to temperature variations and aging. The spectral width of the optical source is also a parameter requiring attention during link design. The line width of commercially available laser devices is usually less than 2.5 nm rms, while LEDs can be high as 20 to 50 nm rms. Moreover, spectral width may be affected by the modulation process, and ideally should be specified under modulation.

The remaining two parameters, pulse shape and line code, further define the signal generated by the transmitter. The pulse shape relates to the average power (e.g., return to zero and nonreturn to zero). There are advantages to be gained by various pulse shapes that relate to receiver sensitivity, patterning effects, and band-limited operation. Line code is important primarily to avoid base line wander and

to increase the density of level transitions. Both of these parameters are of more significance to ac-coupled amplifiers and clock recovery circuits in the receiver. However, the pulse shape and line code will determine the manufacturer's specified receiver sensitivity value.

The next segment of our model is the transmission medium. For purposes of link calculation the parameters in Chart 2 give the important link design parameters for the fiber waveguide.

CHART 2. Fiber Waveguide System Parameters

- Fiber insertion loss
- Splice/connector insertion loss
- Link distance
- Bit rate
- Fiber dispersion coefficient
- Fiber temperature coefficient (dB/°C)
- Single-mode fiber cutoff wavelength

The above chart gives generic parameters that are common to the different fiber profiles and wavelength bands. Each of the parameters partially determine the longest fiber link capable of maintaining a specified bit error rate. Initially we will consider attenuation-limited operation to demonstrate the system gain concept. Bit rate and fiber dispersion effects on system performance will be considered in a later section.

The receiver is the final model element in Fig. 5. A receiver simply detects the optical signal and reconstructs an estimate of the signal used to drive the transmitter. The important performance criterion here is the amount of optical signal power needed to maintain a particular bit error rate (BER).

The BER is the primary measure of performance of a digital transmission system. Essentially the BER is a ratio of the number of errors to the number of bits in a known time interval. Mathematically, it is proportional to the complementary error function of the received signal-to-noise ratio, if the noise is assumed to be Gaussian:

$$\text{erfc}\,(x) = (2/\pi^{1/2})\int_x^\infty \exp\,(-y^2)\,dy\,, \qquad (2)$$

where $x = \gamma/(Y/2)^{1/2}$ and γ is the normalized decision (or slice) level used by the amplitude comparator in the receiver, and Y is the receiver signal-to-noise ratio.

Instrumentation is readily available that allows routine measurement of the BER and avoids tedious calculations. However, the BER alone is not sufficient by itself to describe the receiver's operation. The additional information needed is the receiver's sensitivity at a stated BER. Sensitivity is specified in terms of average optical power at the receiver's input. When the transmitter was discussed earlier we mentioned extinction ratio and pulse shape. It is at the receiver where these transmitter parameters take on their full meaning. In order for the receiver sensitivity to be truly meaningful one requires knowledge about the transmitter extinction ratio and pulse shape. Other system conditions may influence the receiver sensitivity in an adverse fashion. Examples are fiber dispersion induced bandwidth limiting [33], laser/fiber noise phenomena, temperature effects, and signal patterning effects.

System Gain

Using the link model just described one can begin a rudimentary optical link design. Physically, the designer is interested in how long a distance a link can span while maintaining an acceptable BER. Practically, distance must be transformed into a succession of terms of insertion losses. Rather than specifying a physical span distance in kilometers it is more sensible to specify a system in terms of system gain. Then, knowing the total insertion loss in decibels per kilometer (dB/km) it is simple to compute distance. Equation 3 expresses system gain:

$$G = P_t - P_{rs},\qquad(3)$$

where

G = system gain in decibels
P_t = the average transmitted power in decibels referred to 1 mW (dBm),
P_{rs} = receiver sensitivity, for a 10^{-9} BER, in decibels referred to 1 mW.

System gain represents the maximum power difference that may exist between the transmitted power and receiver sensitivity figure. If the insertion loss exceeds the system gain value, the link

BER will exceed 10^{-9} (or whichever BER is used to define transmission quality).

Fig. 7 illustrates the distribution of absolute power in a typical optical link. The attenuation of a link is the aggregate of all loss elements. The total link loss is expressed as the summation of individual fiber lengths, splice insertion loss, and optical connector loss. Equation 4 represents the total link loss:

$$L = \sum L_s + \sum L_c + \sum \alpha_k d_k + M,\qquad(4)$$

where

L = total link insertion loss in decibels,
L_s = individual splice loss in decibels,
L_c = individual connector loss in decibels,
α_k = attenuation per kilometer of fiber section k in decibels per kilometer,
d_k = length of fiber section k in kilometers,
M = power margin in decibels.

Note that a power margin M has been included into the loss expression. This is essentially a safety factor to allow for unforeseen increases in attenuation and system noise. Examples could include aging, fiber or component temperature variations, and added loss due to cable repairs. The power margin has an additional benefit of allowing the system to operate at a reduced error rate. Usually only the nominal value of the data used in (3) and (4), is known. Sometimes worst-case values or the nominal values with their statistical distributions can be specified. Three approaches are possible: establish a large initial power margin, include statistical variance with all link parameters, or use worst-case values. The first and third are very similar, but using worst-case values at least establishes the maximum link loss. The statistical approach uses mean values with variances to calculate the system gain and insertion loss. Statistical data on the parameters may be difficult to determine or just not available to the system designer. The best approach may be to calculate, measure, or assume worst-case data and use a reasonable system margin. Overstating system margin or using worst-case values may result in unnecessarily stringent component specifications, which increases system costs.

At any rate the inequality of (5) must be maintained in order to preserve the specified end-to-end BER:

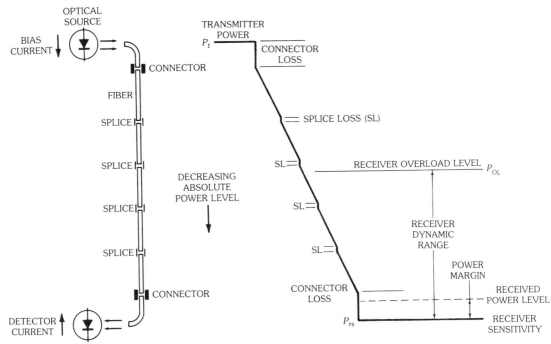

Fig. 7. Link power distribution.

$$P_t - P_{rs} > L \,, \qquad (5)$$

where

P_t = average transmitted power,
P_{rs} = the receiver sensitivity for a 10^{-9} BER,
L = the total link loss including a link power margin of M.

We will, in the following section, review what is important with regard to the receiver sensitivity P_{rs}.

Noise Limitations

An important parameter in system gain calculations is the receiver sensitivity. Detectors and receiver design were treated in detail in Chapters 11 and 12, respectively. Chart 3 lists elements that contribute to the receiver sensitivity.

The last five items are actually laser, fiber, or combined laser and fiber phenomena, but their effect is to reduce the receiver ability to estimate the incoming signal value. As such, they are usually stated as receiver degradation factors. These degradations will be discussed in a following section of this chapter.

A simple estimate of the receiver sensitivity can be obtained using the first five items in the parameter list as follows:

$$P_{avg} = Q\, i_n\, f_b^{1/2} / G\, R + Q^2\, q\, F_e\, f_b / 2\, R \,, \qquad (6a)$$

where Q is defined in the integral

$$\mathrm{BER} = (2\,\pi)^{-1/2} \int_Q^\infty e^{-t^2/2}\, dt \,,$$

with $Q=6$ for a BER of 10^{-9}. Also,

i_n = noise spectral density in picoamperes per hertz$^{1/2}$,
f_b = bit rate in megabits per second,
G = detector gain,
R = detector responsivity,
q = charge on an electron $(1.6 \times 10^{-19}\,\mathrm{C})$,
F_e = detector excess noise
$= k\, G + (2 - 1/G)\, (1-k) \simeq G^x$, where
k = detector excess-noise parameter,
x = alternate excess-noise parameter.

For a bit error rate of 10^{-9} the equation can be simplified to

$$P_{avg} = 6 \times 10^{-9} i_n f_b^{1/2}/GR$$

$$+ 2.88 \times 10^{-15} F_e f_b/R \,. \quad \text{(6b)}$$

The first term in (6a) is the optical power needed to overcome the thermal noise of the receiver, while the second term is the optical power required to overcome the detector quantum noise. Thus this term represents the quantum-limited optical power, assuming a Gaussian noise distribution. For a pin diode, $G=1$ and the first term of the equation generally dominates. When an APD detector is used, an optimum value of gain can be found that minimizes the optical power. The value of this gain can be approximated by

$$G_{opt} = [(10^4 i_n) - (1-k)/k]^{1/2} \,. \quad \text{(7)}$$

The noise spectral density will be specified if a commercial amplifier is used or can be calculated from the equations derived in Chapter 12. For an FET–front-end receiver, from (70) of Chapter 12, the noise spectral density is

$$i_n = \{4kTBI_2/R_F + 2eI_g BI_2 + (4kT\Gamma/g_m)$$

$$\times (2\pi C_T)^2 [1 + (I_f/I_3)(f_c/B)]B^3 I_3\}^{1/2}, \quad \text{(8)}$$

while for a bipolar-transistor–front-end receiver the noise spectral density can be calculated from (83) of Chapter 12.

$$i_n = [4kTBI_2/R_F + 2eI_c BI_2/\beta + (2e/I_c)$$

$$\times (2\pi V_T C_T)^2 B^3 I_3 + 4kT r_{bb'}$$

$$\times (2\pi C_{dsf})^2/B^3 I_3]^{1/2}. \quad \text{(9)}$$

CHART 3. Receiver Sensitivity Considerations

- Detector responsivity
- Pin or APD detector structure
- APD excess noise
- Preamplifier noise
- Receiver noise bandwidth
- Detector leakage current
- Intersymbol interference (ISI)
- Modal noise (multimode systems)
- Mode-partition noise (single-mode systems)
- Laser mode hopping
- Reflections backward into the laser cavity

Chapter 12 has a complete discussion of the noise in receiver amplifiers and therefore we will limit our comments to a few generalizations about system performance using an FET amplifier. At low bit rates the noise spectral density is proportional to the square root of the bit rate. This is the condition where the receiver noise is mainly quantum noise from the detector and thermal noise from the source resistance of the input stage of the receiver. At high bit rates the noise spectral density becomes proportional to three-halves power of the bit rate. Channel noise of the FET is the main contributor in this region. Channel noise is a function of the square of the total junction capacity of the detector FET junction, the FET noise factor Γ, and the g_m of the FET. Significant improvements can be made in receiver performance by reducing the detector FET junction capacity. Improvements have been achieved reducing this capacity using gallium-arsenide FETs with hybrid techniques. Still further gains are anticipated with the implementation of an integrated optical receiver [34].

Fig. 8 is a simplified example of a partial regenerator. The components shown follow the optical detector, preamplifier, and agc postamplifier. Also, a clock recovery circuit and noise filter will directly precede the shown components. After signal amplification and noise filtering the signal is applied to an amplitude comparator, which is sometimes referred to as a slicer. The noisy signal is compared to a noise-free, stable reference level. This reference level is set to a value less than half the amplitude of the maximum signal level. The slicer is a high-gain device which saturates in one of two states, depending on the polarity of the signal and reference difference. In the presence of noise with the signal, the slicer may at any instant make a wrong decision. That is, the slicer saturates in a state other than it would if noise was not present. The greater the amount of noise present on the received signal, the greater the probability of a false decision. Following the slicer the signals are applied to a bistable device, and locked or stored on each successive transition of the recovered clock. Occasionally the clock will occur simultaneously with a noise spike and an erroneous state will be stored. This erroneous state is a bit error. The number of errors in a given time period represents a bit error rate. From the illustration of Fig.

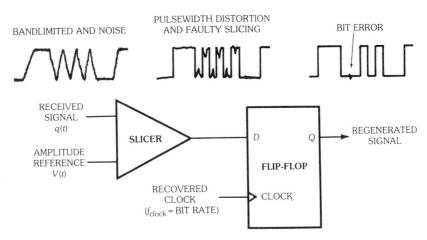

FIG. 8. Effects of noise on signal regeneration.

8 one can visualize that the rate of bit errors is increased by the noise intensity added to the signal. The ratio between signal power and noise power is commonly referred to as the signal-to-noise ratio.

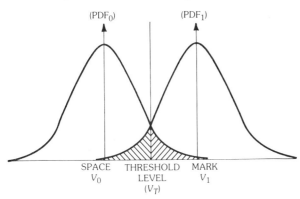

AREA UNDER PDF CURVE = 1.0

(a) Probability density functions for mark and space.

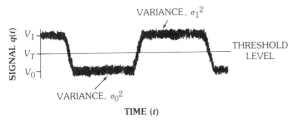

TIME (*t*)

(b) Signal levels.

FIG. 9. Gaussian noise statistics of a binary signal.

Fig. 9 examines the effect of signal-to-noise ratio on bit error rate in more detail.

Each level in the signal $q(t)$ is represented by V_0 and V_1. The binary levels are considered to be the statistical mean of a probability density function (pdf) with variances σ_0^2 (around V_0) and σ_1^2 (around V_1). So in Fig. 9 the mean of the pdf is either the mark or space amplitude, and the variance of the pdf is the noise power. As shown in Fig. 9a each pdf intersects the other at a point midway between the mean values. It is at this intersection that the amplitude reference V_T of the slicer is set for optimum operation. Assuming the noise is Gaussian, the error rate can be calculated in a straightforward manner. Graphically one can normalize the area under the Gaussian curve as equal to unity and the shaded areas on either side of V_T as the probability of error. Strictly speaking, the noise statistics generated at an optical receiver front end are not Gaussian, but the analysis described is a very good approximation, especially when pin-diode detectors are used in the receiver.

The probability of error $P(e)$, as represented by the shaded area under the Gaussian curves, can be plotted in the form of the familiar bit error rate curve of Fig. 10 and is given by

$$P(e) = \tfrac{1}{2}\,\text{erfc}\,(Y/2^3)^{1/2}, \qquad (10)$$

where $P(e)$ is the probability of error and Y is the signal-to-noise ratio of the received signal.

It is an accepted practice to measure the bit error rate $P(e)$ as a function of average received optical power so that the plotted data is identical with the curve of Fig. 10. Note the steepness of the curve in the vicinity of 10^{-9}. This results in large orders of magnitude changes in the BER for fractional decibel (dB) changes in the received power level. Thus the receiver sensitivity is a point on the curve relating probability of error $P(e)$ or BER to received average power.

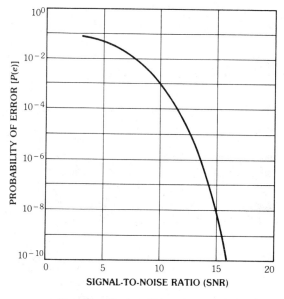

FIG. 10. Curve of bit error rate.

Eye Patterns

Calculating or measuring the BER and receiver sensitivity can be tedious and time-consuming. Communication engineers have a tool useful for quick means of determining received signal quality, known as the eye pattern. It is useful for evaluating noise levels, pattern distortion, intersymbol interference, and receiver group delay. The measurement uses an oscilloscope that is externally synchronized to the signal clock. A display emerges that shows all signal patterns for each bit period. Fig. 11 shows an ideal eye pattern for a noise-free nrz signal. In this case all levels and transition overlay in a fashion that shows an eye that is 100% open. In reality an optical receiver has noise and perhaps other forms of signal distortion. Some distortion in the form of bandwidth

limiting is actually beneficial to the operation of an optical transmission link. Noise power or the variance of the pdf is actually a noise power spectral density (watts per hertz) multiplied by the receiver noise bandwidth. Hence the total noise power in a received signal increases with bandwidth of the receiver. There is a trade-off in performance with receiver bandwidth. An optimum bandwidth exists where the signal-to-noise ratio is maximized. At that point the BER is at its lowest for the level of received optical power. Generally, bandwidth limiting will result in intersymbol interference. Bandwidth-limiting filters which satisfy the Nyquist constraint have zero interference at the nominal sampling times. The tolerance of the filter to jitter should also be taken into account. It can be shown that receiver filter bandwidths of 60% to 70% of the bit rate yield very nearly optimum performance, depending on the type of filter employed [35].

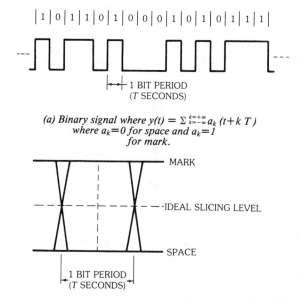

(a) Binary signal where $y(t) = \sum_{k=-\infty}^{k=+\infty} a_k(t+kT)$ where $a_k=0$ for space and $a_k=1$ for mark.

(b) Eye pattern is an overlay, "in one bit space," of all possible states in the sequence.

FIG. 11. Eye pattern concept for noise-free nrz signal.

The time-domain response of such a filter approximates a "raised-cosine" function. Fig. 12 is an example of a properly bandwidth limited signal with moderate noise present. The upper half of Fig. 12 is a pulse showing a noise spike occur-

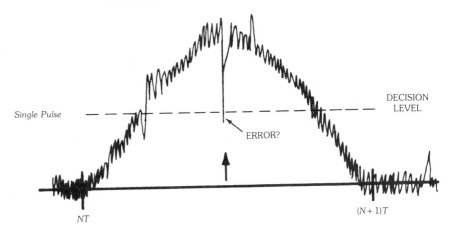

DECISION LEVEL

Single Pulse

ERROR?

NT

$(N+1)T$

FIG. 12. Raised-cosine pulse with noise.

Eye Diagram

ring at the sample time. Note that the spike crosses the amplitude reference of slice level. This example would result in an erroneous decision by the regenerator and become a bit error. The lower half of Fig. 12 shows the eye pattern of the signal. The area in the center of the pattern is referred to as the eye. As the receiver signal-to-noise ratio decreases, the eye becomes increasingly filled with noise, thus making amplitude decisions more difficult, i.e., the likelihood of error increases. The amount of eye opening or closure thus becomes a measure of quality, at least for comparison purposes. It does not result in a firm value such as a BER figure but does give a sense of performance.

3. Bandwidth Effects

In the previous analysis of the power budget it was tacitly assumed that there was no pulse broad-

ening in the optical signal. Pulse broadening would result in intersymbol interference at the receiver with an attendant reduction in receiver sensitivity. Fiber dispersion causes the transmitted pulses to spread and overlap as they propagate. Furthermore, limited transmitter and receiver bandwidth may also significantly limit system speed. Thus it is necessary to determine the end-to-end system bandwidth (or the equivalent pulse dispersion or rise time) to ensure that the desired system performance has been met. Eye patterns provide a quick way to check if intersymbol interference is present. Fig. 13 shows an eye pattern of an nrz signal displaying severe eye closure due to bandwidth limiting. This signal would produce a significant increase in the BER over a signal with a fully opened eye. The eye closure is rated by the ratio of the open area, marked *a*, to the fully opened eye, marked *A*. The percent eye closure is a relative measure of receiver degradation. Equation 11 is a simple expression that relates the receiver degradation to eye closure due to insufficient bandwidth:

$$\Delta P_{bw} = 10 \log_{10} (a/A),\qquad(11)$$

where

ΔP_{bw} = receiver degradation factor due to eye closure (bandwidth limiting), in decibels,

a = amplitude of the eye opening,

A = maximum or 100% eye opening.

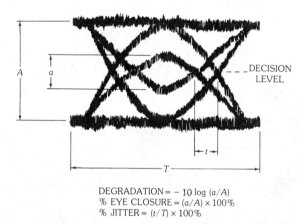

DEGRADATION = $-10 \log (a/A)$
% EYE CLOSURE = $(a/A) \times 100\%$
% JITTER = $(t/T) \times 100\%$

FIG. 13. Highly degraded bandwidth-limited signal.

The factor ΔP_{bw} can be thought of as the amount of power, at the receiver, needed to restore the BER of a 100% opening. The eye closure illustrated in Fig. 13 could be caused by any one of the major building blocks that make up a fiber system.

Dispersion Budget

Analysis of optical receivers has shown that if the product of the bit rate B and the root-mean-square (rms) width of the impulse response of the system, σ_t, i.e., the total system dispersion is less than 0.25, the loss in receiver sensitivity will be less than 1 dB [36, 37]; that is,

$$\sigma_t B \le 0.25\qquad(12)$$

The dispersion σ of a system or one of its elements is defined by the rms width of the output

pulse when the input pulse approaches a delta function. The output pulse is usually referred to as the impulse response $h(t)$.

$$\sigma^2 = \int_{-\infty}^{\infty} h(t)\,(t-\tau)^2\,dt,\qquad(13)$$

where

$$\tau = \int_{-\infty}^{\infty} h(t)\,t\,dt$$

and

$$\int_{-\infty}^{\infty} h(t)\,dt = 1.$$

The total dispersion σ_t is the square root of the sum of the squares of the pulse dispersion σ_i of each independent contributor [38].

$$\sigma_t = \left(\sum_i \sigma_i^2\right)^{1/2}.\qquad(14)$$

It is sometimes more convenient to analyze the rise time in a system than to determine the impulse response. As a rough rule of thumb, for intersymbol interference to cause a small (less than or equal to 1 dB) penalty in receiver sensitivity, the pulse rise time of a digital system should not exceed 70% of the bit period for nonreturn-to-zero (nrz) pulses, or 35% of the bit period for return-to-zero (rz) pulses where the bit period is defined as the reciprocal of the data rate. The total system rise time t_{sys} can be calculated by combining the rise times of each contributor t_i as the square root of the sum of the squares:

$$t_{sys} = \left(\sum_i t_i^2\right)^{1/2}.\qquad(15)$$

The bandwidths of the transmitter and receiver are generally known to the designer. The transmitter bandwidth is determined primarily by the speed of the electronic drive circuit, electrical parasitics, and the light source [39, 40]. It should be noted that not all lasers are capable of modulation at high speed and that there is a correlation between laser structure and laser bandwidth; some designs have modulation bandwidths of only a few hundred magahertz, while others provide several gigahertz [41, 42]. The speed limitation of the receiver results from the response time of the photodetector [43, 44] and the limited bandwidth of the receiver preamplifier. If the 3-dB electrical bandwidth of the receiver (or transmitter) is known, then the receiver (or transmitter) rise time can be calculated from [45]

$$t_r = 350/f, \qquad (16)$$

where t_r is the rise time in nanoseconds and f is the 3-dB electrical bandwidth in gigahertz. The fiber-related rise time can be very roughly estimated by using the full-width half-maximum pulse (FWHM) broadening due to fiber dispersion.

The analysis of fiber dispersion is rather complicated since the pulse broadening it causes is dependent on the type of optical fiber used, the fiber length, the width and shape of the spectrum of the optical source, and the operating wavelength. We shall now examine fiber dispersion in more detail.

Fiber Dispersion

Fiber dispersion can be classified as intermodal or intramodal dispersion. Intermodal dispersion (commonly referred to as modal dispersion) is found in multimode fibers and is due to the differential delay between modes at a single frequency. This type of dispersion is eliminated in single-mode fibers since only one mode can propagate.* Intramodal or chromatic dispersion is due to the variation of group velocity (the speed at which energy in a particular mode travels along the fiber) of a mode with wavelength. Intramodal dispersion and the use of sources with finite spectral width result in pulse spreading that increases with the spectral width of the source. Intramodal dispersion plays a role in both multimode and single-mode fibers. The total fiber dispersion is found by combining the intermodal and intramodal dispersion in the following way:

$$\sigma_f{}^2 = (\sigma_{\text{inter}})^2 + (\sigma_{\text{intra}})^2, \qquad (17)$$

where
σ_f = fiber dispersion,
σ_{inter} = intermodal dispersion,
σ_{intra} = intramodal dispersion.

---

*We are neglecting polarization mode dispersion, which is usually very small.

For a multimode step-index fiber the dominant cause of pulse spreading is likely to be intermodal dispersion, which can vary from 10 ns/km to as high as 50 ns/km. This rather high value of dispersion in multimode step-index fiber spurred the development of graded-index multimode fiber. The refractive index of the core in a graded-index multimode fiber decreases in a near-parabolic fashion radially away from the fiber axis; this causes higher-order modes to propagate faster than the lower-order (axial) modes in order to equalize intermodal delay differences. Typical values for intermodal dispersion in graded-index fibers range from 20 to 500 ps/km [46]. It should be noted that the optimum graded-index profile is a material property and therefore a function of operating wavelength [47, 48]. Thus a profile designed to be optimal at a specific operating wavelength might be far from optimal at another operating wavelength. Consequently, a fiber with a given index profile will exhibit different pulse spreading according to the source wavelength used.

An optical-fiber link seldom consists of one continuous fiber, but of several fiber sections (e.g., 1 km long) that are spliced together, and it is necessary to calculate the end-to-end intermodal dispersion (or bandwidth) of the concatenated fiber sections. Unfortunately, such calculations are not very precise due to mode mixing at the splice points. This mode coupling tends to average out the propagation delays associated with the modes, thereby reducing intermodal dispersion. When mode mixing is minimal, variations from the optimum graded-index profile can offset each other, which would also lead to a less than linear increase in intermodal dispersion. Two limits can be set by assuming complete mode mixing and a second case where no mode mixing occurs. In the former case the intermodal dispersion increases in proportion to the square root of the number of fiber sections, while the latter case results in dispersion increasing linearly with the number of fiber sections.

$$\sigma_0 L \leq \sigma_T \leq \sigma_0 L^{1/2}, \qquad (18)$$

where
σ_T = total intermodal dispersion,
σ_0 = intermodal dispersion of a 1-km section,
L = the number of sections.

This inequality implies that the total intermodal dispersion in concatenated fibers is proportional to the length raised to an exponent between 0.5 and 1.0:

$$\sigma_T(L) = \sigma_0 L^\gamma, \tag{19}$$

where γ is the concatenation exponent.

Usually not all sections have equal bandwidth and (19) is modified as

$$\sigma_T(L) = \left[\sum_i (\sigma_i)^{1/\gamma}\right]^\gamma, \tag{20}$$

where $\sigma_T(L)$ is the total intermodal dispersion in L fiber sections in which the intermodal dispersion in individual sections is given by σ_i.

The concatenation exponent for commonly used fiber typically falls between 0.7 and 0.8. However, since mode mixing is not a controlled effect, the concatenation exponent should be used with caution.

In multimode fiber, intermodal dispersion should be combined with intramodal dispersion to determine the total fiber dispersion. In single-mode fibers intramodal dispersion is the only source of pulse broadening. Intramodal dispersion increases linearly with fiber length. As detailed in Chapter 3, intramodal dispersion can be separated into material dispersion, waveguide dispersion, and profile dispersion [49]. In multimode fibers material dispersion provides a good approximation of the composite chromatic dispersion. In single-mode fibers material, waveguide, and profile dispersion are interrelated in a complicated way and depend on the dispersive effects of both core and cladding material [50]. The distinction between the various components of intramodal dispersion is important to fiber designers (by modifying the individual components they can tailor the composite dispersion), but for our purposes of determining the effects of fiber dispersion on system performance, we can use the total chromatic dispersion coefficient.

Intramodal dispersion is dependent on wavelength as shown in Fig. 14. The dispersion is due to propagation velocity variations with wavelength. Curve A shows the time delay τ (or transit time)

of a mode of wavelength λ.* Note that at about 1310 nm the time delay has a minimum value. If one differentiates time delay with respect to wavelength, the dispersion coefficient D results. The term D is usually stated in picoseconds per nanometer-kilometer. If D is known, pulse broadening due to intramodal dispersion can be computed:

$$\sigma_t = \sigma_\lambda |D(\lambda)| L, \tag{21}$$

where

σ_t = pulse broadening in picoseconds,
σ_λ = root-mean-square (rms) spectral width of the source,
$D(\lambda)$ = first-order fiber-dispersion coefficient in picoseconds per nanometer-kilometer,
L = length of fiber in kilometers.

For pulses with a Gaussian temporal response, a reasonable assumption in most fiber-optic systems, one can derive simple relations between the rms pulse width, full-width half-maximum pulse width and the 3-dB electrical and optical bandwidth [57]:

$$f_{-3dB}(\text{optical}) \simeq 187/\sigma_{rms}, \tag{22}$$

$$f_{-3dB}(\text{electrical}) \simeq 133/\sigma_{rms}, \tag{23}$$

$$\sigma_{rms} \simeq 0.425\,\sigma_{FWHM}, \tag{24}$$

where σ is the pulse broadening in picoseconds and f is in gigahertz.

It has become common to design fiber-optic systems in such a way that when the optical-fiber bandwidth is chosen to be equal to the system bit rate B the receiver penalty is less than 1 dB:

$$B = f_{-3dB}(\text{optical}). \tag{25}$$

For such a system we can estimate from (21), (22), and (25) the bit-rate distance product limitation due to intersymbol interference:

*Measured data, representing curve A, is usually modeled by a mathematical function known as the three-term Sellmeier delay fit which was standardized in Electronics Industry Association (EIA) dispersion test methods and in an International Telegraph & Telephone Consultative Committee (CCITT) recommendation.

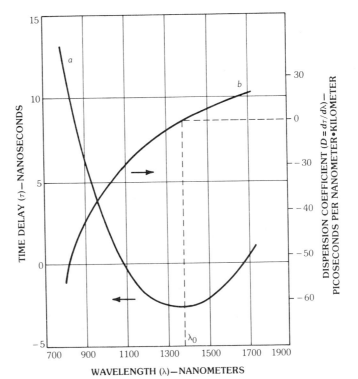

FIG. 14. Single-mode chromatic dispersion.

$$B \times L \leq 187/\sigma_\lambda \, |D(\lambda)| \, . \qquad (26)$$

Further examination of Fig. 14 shows that the chromatic dispersion coefficient $D(\lambda)$ passes through zero at the wavelength of minimum time delay. Thus one could draw the conclusion that at the fiber's "zero-dispersion" wavelength λ_0 there is no pulse broadening and that at this wavelength the fiber would have infinite bandwidth. In the vicinity of λ_0, τ varies slowly with wavelength so pulse-broadening effects are minimal. For a source operating at or near the zero-dispersion wavelength, however, a second-order dispersion effect becomes important [52]. This residual dispersion coefficient is usually expressed in nanoseconds per nanometer-squared kilometer (ns/nm²·km) and pulse broadening due to this effect is proportional to the square of the source spectral width. For the zero-dispersion region, pulse broadening can be calculated from [53]

$$\sigma^2 \simeq \sigma_\lambda^2 D^2 + \sigma_\lambda^4/2 \, (D^1)^2 \, , \qquad (27)$$

where D^1 is the residual dispersion coefficient

(typically $D^1 < 0.1$ ps/nm²·km).

It should be noted that right at λ_0 the propagating pulse experiences oscillatory pulse distortions and that broadening might be minimized by the selection of an optimum output pulse width [54, 55].

With most practical system design cases it is difficult to have complete knowledge about the location of the fiber's zero-dispersion wavelength and the exact operating wavelength of the optical source. Generally the fiber manufacturer will specify a worst-case dispersion coefficient in a given band of wavelengths (e.g., ± 3.5 ps/nm·km in the 1285-nm to 1325-nm band). Sources are also specified within a band bounded by a maximum and minimum wavelength. This allows the designer to calculate the worst-case pulse broadening.

4. Other System Impairments

In our discussion so far we have determined that the maximum repeaterless transmission distance is governed by the maximum available power budget or system gain, fiber attenuation, and link bandwidth. We have ignored some of the other significant system impairments associated with interactions between lasers and fibers. In multimode-fiber–based systems, receiver sensitivity can be seriously degraded by modal noise. In single-mode systems mode-partition noise and chirping are going to play a role. We shall now examine these systems impairments in more detail.

Modal Noise

Modal noise [7, 56, 57] is caused by the combination of mode-dependent losses and changes in phase between modes or fluctuations in the distribution of energy among modes. If light from a coherent laser is launched into a multimode fiber, a speckle pattern results across the core of the fiber from interferences between propagating modes. Any change in phase between the modes or energy of the modes will change the interference and result in a different speckle pattern. Laser chirping (wavelength shifts) or vibration of the fiber causes the speckle pattern to constantly change; therefore any loss that is speckle pattern dependent will cause a reduction in the received power. Hence, intensity noise can be generated at splice points, optical connectors, fiber core discontinuities, or by illuminating a region on the detector larger than the active area. Fig. 15 shows a speckle pattern and an illustration of a modal-noise generator. The example shows a case of two spliced fibers with core offset. As the power passes from the first to the second fiber, mode selection takes place. That is, some of the power is lost in the cladding of the next fiber. Noise is generated when the speckle pattern changes in time so as to vary the power crossing the splice interface. With highly coherent lasers the noise may often greatly exceed the receiver noise.

Since the amount of modal noise is so much dependent on the particular installation, link performance is difficult to predict [58]. The most straightforward approach to modal noise is to take measures to avoid it. As the light propagates through the fiber a reduction in speckle structure will take place. As the speckle pattern becomes less granular, less modal noise is generated by misaligned splices and other sources of mode-dependent optical losses. This indicates that modal noise

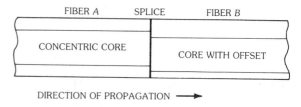

(a) Cross section showing core offset.

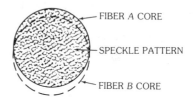

(b) Speckle pattern changes with time and when the fiber is mechanically disturbed.

FIG. 15. Modal noise generation.

can be reduced by using low-coherence lasers [59]. These lasers are characterized by a great number of longitudinal modes, perhaps ten or more. Increasing the number of laser longitudinal modes increases the graininess of the speckle pattern, thus reducing intensity fluctuations at fiber disruptions. Modal noise can be totally avoided by the use of LEDs which are incoherent sources.

Mode-Partition Noise—Single-Mode Fiber

The combination of a multilongitudinal-mode source with a dispersive single-mode fiber in high-speed digital systems can result in degradation due to the random "partitioning" of the optical power between different longitudinal modes of the semiconductor laser. To a first approximation in digital systems employing direct detection, the total power from the laser remains nearly constant, i.e., the noise spectral density for am noise from the laser is usually low, and flat up to several gigahertz. The fluctuations of individual modes, however, can be quite large. In systems with low total dispersion (relative to the bit period) the receiver sees little time delay between the longitudinal modes of the source, and the signal-to-noise ratio on the received optical signal is quite high. In contrast, under high fiber dispersion, when each mode has a unique propagation velocity, noise may occur through the combination of random mode partition and interlongitudinal-mode delay. At any given instant along the receiver pulse the modes no longer add to a constant value. Theoretical predictions of system performance have focused on two limiting cases:

(a) the case of a highly multimode source in which the modes form nearly a continuum in the laser spectrum, and

(b) the case of a "nearly single-mode" laser, with a single side mode below threshold.

The conventional description of a laser spectrum is given by the mean amplitude of each mode such as would be seen on a spectrum analyzer. The static spectral characteristics, while having an important influence on mode-partition noise, do not give sufficient information about the dynamics of the laser to successfully predict the effects of mode-partition noise. The complete mode-partition information requires knowledge of the time-dependent joint probability distribution between the number of modes, $p(a_1, a_2, \ldots, a_N)$, where N represents the number of modes and a_i the amplitude of the ith mode at time t. An additional restriction is placed on the amplitudes a_i, namely, at any given time the total power out of the laser is assumed to be constant, so we may normalize the a_i:

$$\sum a_i = 1 . \qquad (28)$$

This has the effect of reducing the total number of random variables from N to $N-1$. Thus, for the case of a two-mode laser, all the information may be obtained from the probability distribution of a single mode.

Considerable effort has been directed toward identifying a figure of merit for the semiconductor laser spectrum which would be experimentally measurable, yet give an accurate prediction of the laser performance in a dispersive system. The first approach, due to Ogawa [60], applies to the case of a laser with many modes above threshold, while the second applies to the case of a two-mode laser, where the side mode is below the lasing threshold [61].

The mean spectrum $p(\lambda_i)$ is defined as the ensemble average over the ith mode $\langle a_i \rangle$. If the threshold pulse shape is given by $\lambda(t)$, Ogawa has

shown that the mean-squared fluctuations at the receiver may be expressed as

$$\sigma^2 = k^2 \Big(\sum_i [1 - \lambda(t_0 - \tau_i)]^2 p(\lambda_i) - \Big\{ \sum_i [1 - \lambda(t_0 - \tau_i)] p(\lambda_i) \Big\}^2 \Big) \quad (29)$$

where k, referred to as the mode-partition coefficient, is defined by

$$k^2 \equiv \frac{\sum_i \sum_{j>i} [\lambda(t_0 - \tau_i) - \lambda(t_0 - \tau_j)]^2 (\langle a_i \rangle \langle a_j \rangle - \langle a_i a_j \rangle)}{\sum_i \sum_{j>i} [\lambda(t_0 - \tau_i) - \lambda(t_0 - \tau_j)]^2 \langle a_i \rangle \langle a_j \rangle} . \quad (30)$$

Such an unwieldy figure of merit does not lend itself easily to experimental measurement without simplifying assumptions. If all modes are above threshold, it may be assumed that the normalized cross-correlation between any two modes is completely independent of the choice of modes. In statistical terms, the expression $(\langle a_i a_j \rangle - \langle a_i \rangle \langle a_j \rangle)/\langle a_i \rangle \langle a_j \rangle$ is independent of the mode references i and j (provided $i \neq j$) and may be removed from the summation defining k^2, yielding

$$k^2 = (\langle a_i \rangle \langle a_j \rangle - \langle a_i a_j \rangle)/\langle a_i \rangle \langle a_j \rangle . \quad (31)$$

If the normalization condition is imposed, the mode-partition coefficient may be expressed in terms of the mean-squared fluctuations of any individual mode:

$$k^2 = \langle (\Delta a_i)^2 \rangle / (\langle a_i \rangle - \langle a_i \rangle^2) \quad (32)$$

Within the assumptions of uniform cross-correlations, we therefore have a figure of merit which is easily measurable experimentally. Equation 29 may therefore be applied to a calculation of receiver sensitivity under the Gaussian assumption. If the receiver has a raised-cosine response, Ogawa has shown that the signal-to-noise ratio due to mode-partition noise can be expressed as

$$\text{snr} = 0.25 k^2 \pi B \sum_i \Big\{ \big[(\Delta \tau_i)^4 p(\lambda_i) \big] - \big[\sum_i (\Delta \tau_i)^2 p(\lambda_i)^2 \big] \Big\} , \quad (33)$$

where k, $p(\lambda_i)$, and N are already defined and $\Delta\tau_i$ denotes the relative time delay between the ith mode and central mode (τ_0), and B denotes the bit rate. The figure of merit k can vary between 0 and 1. A value of 0 indicates no mode partitioning (stable spectrum), while a value of 1 indicates that only one mode exists at any given time and thus contributes to the maximum amount of delay distortion to the received signal. The value of k can be estimated according to (32) by measuring the average and rms of the output produced by any mode.

For operation away from the zero-dispersion wavelength an estimate of the bit-rate distance product, which will cause a 1.5-dB power penalty at a bit rate of 10^{-9}, can be expressed in gigabits per second-kilometer, as [62]

$$B \times L \le 130/|D(\lambda)|\,\sigma_\lambda\,k^{1/2}. \qquad (34)$$

Fig. 16 shows the BER curves with varying amounts of mode-partion noise, or mpn. Note the error rate floor for the severe mpn case. Unlike most other noise types, mpn cannot be compensated by increasing the receiver's power level.

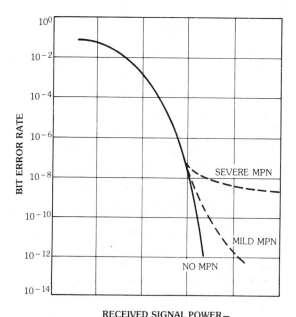

FIG. 16. BER curves showing mode–partition-noise floor.

The desire is, of course, to employ a truly single-mode laser. However, even for a well-designed semiconductor laser there will in all likelihood be one or two satellite modes associated with the main mode. For the two-mode laser, the expression (32) for k is exact, and the main-to-side mode ratio can be expressed as a function of k as follows:

$$\text{mode ratio} = 1/k^2. \qquad (35)$$

And the power I_0 in the main mode is

$$I_0 = k^2/(1-k^2), \qquad (36)$$

while the power J_0 in the satellite mode is

$$J_0 = 1 - I_0. \qquad (37)$$

Most cases of interest concern a satellite mode which is well below the value of the central mode; on the average the side mode is below the lasing threshold and consists of amplified spontaneous emission (ASE). Amplified spontaneous emission is statistically similar to a thermally generated light source, and is described by a negative exponential distribution whose variance is equal to the mean value J_0:

$$p(J) = \exp\,(-J/J_0)/J_0, \qquad J \ge 0. \qquad (38)$$

When the mode-partition noise (21) is used to calculate a bit error rate floor (the error rate in the absence of receiver noise) using these assumptions, the result is as shown in Fig. 17, which displays the bit error rate floor for three values of the mode ratio. For intermodal delays greater than one bit period an asymptotic error probability is realized, in which a mode ratio of 42 corresponds to a bit error rate floor of 10^{-9}.

While this is a very useful piece of data, receiver noise also contributes to the error rate as mentioned earlier. An analysis has been performed [63] that combines the effects mpn BER and the receiver BER due to additive Gaussian noise. Fig. 18 shows curves of total system errors of 10^{-9} and 10^{-12} for different degrees of mpn BER and receiver noise BER. While the error rate mechanism for the two noises are quite different, the curve shows that they can be traded off between each other in order to achieve the system target performance.

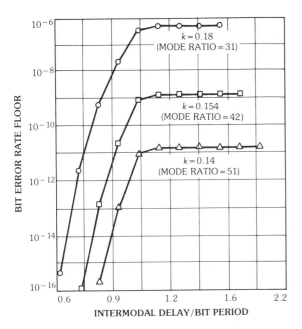

FIG. 17. Dependence of BER on mode ratio for a two-mode laser.

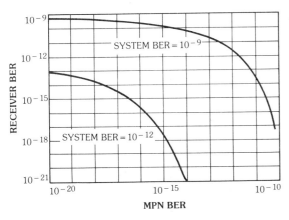

FIG. 18. Dependence of BER on mode-partition noise and thermal receiver noise.

As mentioned in the introduction, dispersion-shifted fiber has advantages of increasing performance of the fiber-optic system. One of the main merits of using this type of fiber is the reduction in mode-partition noise. For other than single-fre-

quency lasers it is necessary to reduce this factor in order to take advantage of the lower optical loss at the longer wavelengths. Fig. 19 shows a comparison of long-wavelength technology using conventional and dispersion-shifted fiber. Note that dispersion-shifted fiber yields far superior performance to standard "high-dispersion" fiber and this is due to the reduction in mode-partition noise. Not surprisingly, the dispersion-shifted fiber at 1500 nm outperformed the standard fiber at 1300 nm due to the lower optical loss at the longer wavelength. The best performance can be achieved with a single-frequency laser; for single-frequency lasers with an external modulator the lower limit to the spectral width is essentially determined by the modulation rate of the optical source [26].

Chirping

As mentioned, the use of single-mode frequency lasers is effective in minimizing degradation due to fiber dispersion and mode-partition noise. It is possible, however, for a laser which oscillates in a single longitudinal mode under cw current injection to experience dynamic line broadening, or chirping under direct modulation of the injection current. This effect may have significant impact in very high speed systems, i.e., systems in which the delay distortion of the frequency modulated laser signal becomes important.

In order to estimate the effects of chirping in digital optical communications systems, it is important to understand the physical mechanism, which finds its root in the strong coupling of the free carrier density and index of refraction in any semiconductor structure. Thus, even a small change in the carrier density, in addition to producing relaxation oscillations in the laser output, will result in a phase shift of the optical field and an associated change in the resonance frequency of either a Fabry-Perot or distributed-feedback laser.

As detailed in Chapter 9 the relaxation due to spontaneous emission is the mechanism for *fundamental* line broadening, and the same model accounts for dynamic frequency shifts of lasers under modulation of the injection current. The cavity relaxation oscillation is a well-understood phenomenon that has been described theoretically and observed experimentally in many types of

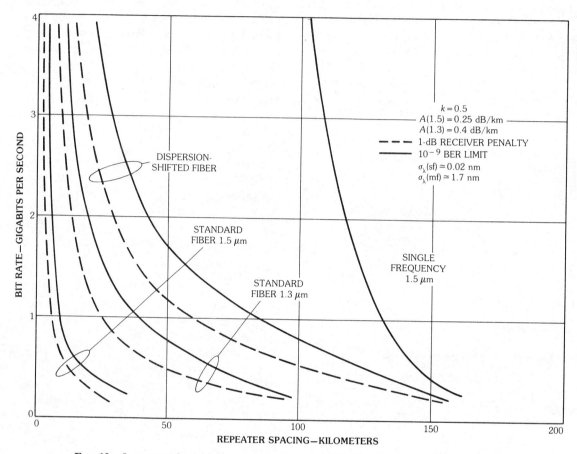

Fig. 19. Long-wavelength performance of conventional and dispersion-shifted fibers.

lasers [64, 65]. The uniqueness of semiconductor lasers lies in the associated relaxation oscillation of the laser emission frequency due to the carrier density-induced phase modulation.

Experiments by Linke [66] and Lin [67] have indicated that the magnitude of the frequency chirp is proportional to the time derivative of the injection current, while the chirp duration t_c is approximately one-half the relaxation oscillation period:

$$t_c = 1/2f_c \simeq \pi[1/\tau\,\tau_p\,(J/J_{th}-1)]^{-1/2}, \quad (39)$$

where f_c is the relaxation oscillation frequency, τ and τ_p are the carrier and photon lifetimes, and J and J_{th} are current density and threshold current density, respectively. It is then possible, following the method of Linke, to estimate the penalty in received power due to the chirping.

Fig. 20 illustrates a simplified picture of the intensity, carrier density, and wavelength response of a semiconductor laser to the application of a square current pulse with rapid rise and fall times. For the region of positive dispersion in the fiber (i.e., for wavelengths shorter than the zero-dispersion point) that portion of the laser pulse which is shifted to the shorter wavelengths (blue shifted) will experience additional delay in propagation, while the red-shifted component will arrive sooner. In the asymptotic limit of high fiber dispersion, all of the chirped energy will be in neighboring time slots, resulting in intersymbol interference and an associated power penalty.

If the chirp magnitude $\Delta\lambda$ of the blue- and red-shifted portions of the pulse are equal, and the fraction of energy in each portion equal to $t_c B$, where B denotes the data rate, the power penalty has been evaluated:

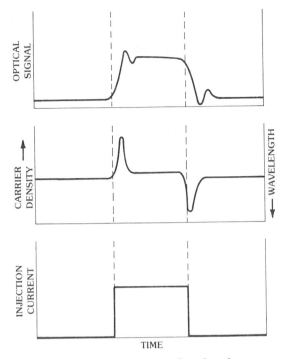

FIG. 20. Typical response of semiconductor laser to square injection current.

nm systems) by the use of dispersion-shifted fiber. It is also possible to adjust the shape of the injected pulse to critically damp the oscillations in carrier density which result from a fast turn-on [68].

Additional solutions lie in the laser structure itself. Structures which raise the relaxation oscillation frequency, or damp the relaxations, such as vapor phase transported distributed-feedback lasers, result in significant reduction in the chirp penalty. For an asymptotic power penalty of 1 dB, (39) and (40) indicate that the relaxation oscillation frequency must be at least 20 times the data rate. Thus the question of whether frequency chirping will become an important source of degradation in a digital optical communication system is highly dependent on the laser transmitter, but small degradation due to chirping can in some cases be present for data rates as low as 100 Mb/s.

$$\text{penalty (dB)} = 10 \log_{10} \left[1/(1 - 4 \, L \, |D(\lambda)| \, B \, \Delta\lambda) \right],$$

$$L \, |D(\lambda)| < t_c/\Delta\lambda, \quad (40)$$

$$\text{penalty (dB)} = 10 \log_{10} \left[1/(1 - 4 \, t_c \, B) \right],$$

$$L \, |D(\lambda)| \geq t_c/\Delta\lambda.$$

where L denotes the fiber length and D the dispersion (typically in picoseconds per nanometer-kilometer). Thus, for dispersion below the asymptotic value $L \, |D(\lambda)| = t_c/\Delta\lambda$, the penalty is dependent only on the magnitude of the chirp and not on its duration. The power penalty above the asymptotic limit, however, is dependent not on the chirp magnitude, but the duration t_c. The power penalties due to chirping have led to the use of external modulation in systems experiments with laser structures that exhibit significant dynamic spectral broadening [25].

Several other approaches may also be considered to minimize this effect. The most obvious approach is to operate very near the zero-dispersion wavelength of the fiber, perhaps (for 1500-

5. References

[1] F. P. KAPRON, D. B. KECK, and R. D. MAURER, "Radiation Losses in Glass Optical Waveguides," *Appl. Phys. Lett.*, Vol. 17, pp. 423–425, November 1970.

[2] E. E. BASCH, R. A. BEAUDETTE, and H. A. CARNES, "Optical Transmission for Interoffice Trunks," *IEEE Trans. Commun.*, Vol. COM-26, No. 7, pp. 1007–1014, July 1978.

[3] J. S. COOK and O. I. SCENTESI, "North American Field Trials and Early Applications in Telephony," *IEEE J. Selected Areas in Commun.*, Vol. SAC-1, No. 3, pp. 393–397, April 1983.

[4] A. MONCALVO and F. TOSCO, "European Field Trials and Early Applications in Telephony," *IEEE J. Selected Areas in Commun.*, Vol. SAC-1, No. 3, pp. 398–403, April 1983.

[5] H. ISHIO, "Japanese Field Trials and Applications in Telephony," *IEEE J. Selected Areas in Commun.*, Vol. SAC-1, No. 3, pp. 404–412, April 1983.

[6] J. E. MIDWINTER, "Current Status of Optical Communication Technology," *IEEE J. Lightwave Tech.*, Vol. LT-3, No. 5, pp. 927–930, October 1985.

[7] K. O. HILL, Y. TREMBLAY, and B. S. KAWASAKI, "Modal Noise in Multimode Fiber Links," *Opt. Lett.*, Vol. 5, No. 6, pp. 270–272, June 1980.

[8] D. GLOGE et al., "High-Speed Digital Lightwave Communication Using LEDs and PIN Photodiodes

at 1.3 μm," *Bell Sys. Tech. J.,* Vol. 59, No. 8, pp. 1365–1382, October 1980.

[9] E. E. BASCH and D. L. HANNA, "GTE Long Wavelength Optical Fiber Communication Systems," *IEEE Intl. Conf. Commun.,* ICC'82, pp. 5D.4.1–5D.4.5.

[10] J. E. MIDWINTER and J. R. STERN, "Propagation Studies of Graded-Index Fiber Installed on Cable in Operational Duct Route," *IEEE Trans. Commun.,* Vol. COM-26, No. 7, pp. 1015–1020, July 1978.

[11] Y. OKANA, K. NAKAGAWA, and T. ITO, "Laser Mode Partition Noise Evaluation for Optical Fiber Transmission," *IEEE Trans. Commun.,* Vol. COM-28, No. 2, pp. 238–243, February 1980.

[12] A. H. GNAUCK, J. E. BOWERS, and J. C. CAMPBELL, "8 Gbit/s Transmission Over 30 km of Optical Fiber," *Electron. Lett.,* Vol. 2, No. 11, pp. 600–602, May 22, 1986.

[13] T. D. CROFT, J. E. RITTER, and V. A. BHAGAVATULA, "Low-Loss Dispersion-Shifted Single-Mode Fiber Manufactured by the OVD Process," *IEEE J. Lightwave Tech.,* Vol. LT-3, No. 5, pp. 931–934, October 1985.

[14] V. A. BHAGAVATULA, M. S. SPOTZ, W. F. LOVE, and D. B. KECK, "Segmented-core Single-mode Fibers With Low Loss and Low Dispersion," *Electron. Lett.,* Vol. 19, No. 9, pp. 317–318, April 28, 1983.

[15] L. G. COHEN, W. L. MAMMEL, and S. J. JANG, "Low-Loss Quadruple-Clad Single-Mode Light Guides With Dispersion Below 2 ps/km·nm Over the 1.28 μm–1.65 μm Wavelength Range," *Electron. Lett.,* Vol. 18, No. 24, pp. 1023–1024, November 25, 1982.

[16] L. C. BLANK, L. BICKERS, and S. D. WALKER, "220 km and 233 km Transmission Experiments Over Low-Loss Dispersion-Shifted Fibre at 140 Mbit/s and 34 Mbit/s," *Post Deadline Tech. Dig.,* OFC'85, PD 7.1–PD 7.4. See also ibid, "Long Span Optical Transmission Experiments at 34 and 140 Mbit/s," *IEEE J. Lightwave Tech.,* Vol. LT-3, No. 5, pp. 1017–1026, October 1985.

[17] R. GOODFELLOW et al., "Practical Demonstration of 1.3 Gbit/s Over 107 km of Dispersion-Shifted Monomode Fiber Using 1.55 μm Multimode laser," *Post Deadline Tech. Dig.,* OFC'85, PD 5.1-PD 5.4.

[18] H. KOGELNIK and C. V. SHANK, "Coupled-Wave Theory of Distributed Feedback Lasers," *J. Appl. Phys.,* Vol. 43, pp. 2327–2335, May 1972.

[19] K. SEKARTEDJO et al., "1.5 μm Phase-Shifted DFB Lasers for Single-Mode Operation," *Electron. Lett.,* Vol. 20, No. 2, pp. 80–81, June 19, 1984.

[20] Y. ITAYA et al., "Longitudinal Mode Behaviors of 1.5 μm Range GaInAs/InP Distributed Feedback Lasers," *IEEE J. Quantum Electron.,* Vol. QE-20, No. 3, pp. 230–235, March 1984.

[21] G. MOTOSUGI, Y YOSHIKUNI, and T. IKEGAMI, "Single Longitudinal Condition for DFB Lasers," *Electron. Lett.,* Vol. 21, No. 8, pp. 352–353, 1985.

[22] A. H. GNAUCK et al., "4 Gbit/s Transmission Over 103 km of Optical Fiber Using a Novel Electronic Multiplexer/Demultiplexer," *IEEE J. Lightwave Tech.,* Vol. LT-3, No. 5, pp. 1032–1035, October 1985.

[23] B. L. KASPER et al., "A 130 km Transmission Experiment at 2 Gbit/s Using Silica-Core Fiber and a Vapor Phase Transported DFB Laser," *Post Deadline Conf. Proc. Tenth Eur. Conf.,* Optical Commun., September 3–6, Stuttgart, W. Germany. Also in *Globecom '84,* pp. 1145–1147, 1984.

[24] W. T. TSANG, "The Cleaved-Coupled-Cavity (C³) Laser," Chapter 5, *Lightwave Communications Technology,* Vol. 22, Part B of Semiconductors and Semimetals, ed. by W. T. TSANG, New York: Academic Press, 1985.

[25] S. K. KOROTKY et al., "4 Gbit/s Transmission Experiment over 117 km of Optical Fiber Using a Ti-LiNbO₃ External Modulator," *IEEE J. Lightwave Tech.,* Vol. LT-3, No. 5, pp. 1027–1031, October 1985.

[26] A. H. GNAUCK et al., "Information-Bandwidth-Limited Transmission at 8 Gbit/s Over 68.3 km of Single-Mode Optical Fiber," *Post Deadline Tech. Dig.,* OFC'86, PD9, pp. 39–42.

[27] Y. YAMAMOTO and T. KIMURA, "Coherent Optical Fiber Transmission Systems," *IEEE J. Quantum Electron.,* Vol. QE-17, No. 6, pp. 919–935, June 1985.

[28] E. E. BASCH, H. CARNES, and R. F. KEARNS, "Heterodyne System Research: Update," *Tech. Dig. Topical Meeting on Optical Fiber Communication,* OFC'83, pp. 56–58.

[29] F. M. GARDNER, *Phaselock Techniques,* 2nd ed., New York: John Wiley & Sons, 1979.

[30] Members of the Technical Staff Bell Telephone Laboratories, *Transmission Systems for Communication,* 5th ed., pp. 728–741, 1982.

[31] D. L. DUTTWEILER, "The Jitter Performance of Phase-Locked Loops Extracting Timing From Base-

band Data Waveforms,'' *Bell Sys. Tech. J.,* Vol. 55, No. 1, pp. 37–58, January 1976.

[32] M. AMEMIYA, M. AIKI, and T. ITO, "Jitter Accumulation for Periodic Pattern Signals," *IEEE Trans. Commun.,* Vol. COM-34, No. 5, pp. 475–480, May 1986.

[33] S. D. PERSONICK, "Baseband Linearization and Equalization in Fiber Optic Digital Communication Systems," *Bell Sys. Tech. J.,* Vol. 52, No. 7, pp. 1175–1194, September 1973.

[34] S. R. FORREST, "Monolithic Optoelectronic Integration: A New Component Technology for Lightwave Communications," *IEEE J. Lightwave Tech.,* Vol. LT-3, No. 6, pp. 1248–1263, December 1985.

[35] K. FEHER, *Digital Communications: Microwave Applications,* Englewood Cliffs: Prentice-Hall, pp. 47–57, 1981.

[36] S. D. PERSONICK, "Receiver Design for Digital Fiber Optic Communication Systems I," *Bell Sys. Tech. J.,* Vol. 52, No. 6, pp. 843–874, July/August 1973.

[37] D. R. SMITH and I. GARRETT, "A Simplified Approach to Digital Optical Receiver Design," *Optical and Quantum Electron.,* Vol. 10, No. 3, pp. 211–221, May 1978.

[38] J. GOWAR, *Optical Communication Systems,* London: Prentice-Hall International, pp. 59–68, 1984.

[39] R. S. TUCKER, "High Speed Modulation of Semiconductor Lasers," *IEEE J. Lightwave Tech.,* Vol. LT-3, No. 6, pp. 1180–1192, December 1985.

[40] A. SUZUKI et al., "1.3 μm Wavelength Surface Emitting LEDs for High-Speed Short Haul Optical Communication System," *IEEE J. Lightwave Tech.,* Vol. LT-3, No. 6, pp. 1217–1222, December 1985.

[41] R. LINKE, "Direct Gigabit Modulation of Injection Lasers—Structure Dependent Speed Limitations," *IEEE J. Lightwave Tech.,* Vol. LT-2, No. 1, pp. 40–43, February 1984.

[42] A. VALSTER et al., "Improved High-Frequency Response of InGaAsP Double Channel Buried-Heterostructure Lasers," *Electron. Lett.,* Vol. 22, No. 1, pp. 16–18, January 2, 1986.

[43] S. R. FORREST, "Gain-Bandwidth-Limited Response in Long-Wavelength Avalanche Photodiodes," *IEEE J. Lightwave Tech.,* Vol. LT-2, No. 1, pp. 34–39, February 1984.

[44] K. MATSUO, M. C. TEICH, and B. E. A. SALEH, "Noise Properties and Time Response of the Staircase Avalanche Photodiode," *IEEE J. Lightwave*

Tech., Vol. LT-3, No. 6, pp. 1223–1231, December 1986.

[45] J. MILLMAN and H. TAUB, *Pulse Digital and Switching Waveforms,''* Chapter 2, New York: McGraw Hill Book Co., 1965.

[46] M. HORIGUCHI, M. NAKAHARA, and N. INAGAKI, "Transmission Characteristic of Ultra-Wide Bandwidth VAD Fibers," *Eighth Eur. Conf. on Optical Commun.,* pp. 75–80, Cannes, September 21–24, 1982.

[47] F. M. E. SLADEN, D. N. PAYNE, and M. J. ADAMS, "Profile Dispersion Measurements for Optical Fibers Over the Wavelength Range 350 nm to 1900 nm," *Eur. Conf. Opt. Commun., (ECOC),* Proc. Fourth, pp. 48–57, 1978.

[48] F. M. E. SLADEN, D. N. PAYNE, and M. J. ADAMS, "Definitive Profile-Dispersion Data for Germania-Doped Silica Fibres Over an Extended Wavelength Range," *Electron. Lett.,* Vol. 15, No. 15, pp. 469–470, 1979.

[49] M. J. ADAMS, *An Introduction to Optical Waveguides,* New York: John Wiley & Sons, pp. 243–250, 1981.

[50] W. A. GAMBLING, H. MATSUMURA and C. M. RAGDALE, "Mode Dispersion, Material Dispersion and Profile Dispersion in Graded-Index Single-Mode Fibers," *Microwaves, Optics and Acoustics,* Vol. 3 (6), pp. 239–246, November 1979.

[51] J. E. MIDWINTER, *Optical Fibers for Transmission,* New York: John Wiley & Sons, pp. 397–399, 1979.

[52] F. P. KAPRON, "Maximum Information Capacity of Fibre-Optic Waveguides," *Electron. Lett.,* Vol. 13, No. 4, pp. 96–97, February 17, 1977.

[53] D. GLOGE, "Effect of Chromatic Dispersion on Pulses of Arbitrary Coherence," *Electron. Lett.,* Vol. 15, No. 21, pp. 686–687, October 11, 1979; errata in *Electron. Lett.,* Vol. 16, p. 240, 1980.

[54] D. MARCUSE, "Pulse Distortion in Single-Mode Fibers," *Appl. Opt.,* Vol. 19, No. 10, pp. 1653–1660, May 15, 1980.

[55] D. MARCUSE and C. LIN, "Low Dispersion Single-Mode Fiber Transmission—The Question of Practical Versus Theoretical Maximum Transmission Bandwidth," *IEEE J. Quantum Electron.,* Vol. QE-17, No. 6, June 1981.

[56] R. E. EPWORTH, "The Phenomenon of Modal Noise in Analog and Digital Optical Fiber Systems," *Proc. Fourth Eur. Conf. on Optical Commun.,* pp. 492–501, Genova, Italy, 1978.

[57] K. PETERMANN, "Nonlinear Distortion and Noise

in Optical Communication Systems due to Fiber Connectors," *IEEE J. Quantum Electron.,* Vol. QE-16, No. 7, pp. 761–770, July 1980.

[58] T. KANADA, "Evaluation of Modal Noise in Multimode Fiber-Optic Systems," *IEEE J. Lightwave Tech.,* Vol. LT-2, No. 1, pp. 11–18, February 1984.

[59] R. E. EPWORTH, "Modal Noise—Causes and Cures," *Laser Focus,* pp. 109–115, September 1981.

[60] K. OGAWA, "Analysis of Mode Partition Noise in Laser Transmission Systems," *IEEE J. Quantum Electron.,* Vol. QE-18, No. 5, pp. 849–855, May 1982.

[61] C. H. HENRY, P. S. HENRY, and M. LAX, "Partition Fluctuations in Nearly Single Longitudinal Mode Lasers," *IEEE J. Lightwave Tech.,* Vol. LT-2, pp. 209–216, June 1984.

[62] K. OGAWA, "Considerations for Single-Mode Fiber Systems, *Bell Sys. Tech. J.,* Vol. 61, No. 8, pp. 1919–1931, October 1982.

[63] E. E. BASCH, R. F. KEARNS, and T. G. BROWN, "The Influence of Mode Partition Fluctuations in Nearly Single-Longitudinal-Mode Lasers on Receiver Sensitivity," *IEEE J. Lightwave Tech.,* Vol. LT-4, No. 5, pp. 516–519, May 1986.

[64] O. SVELTO, *Principles of Lasers,* New York: Plenum Press, pp. 150–164, 1976.

[65] H. KRESSEL and J. K. BUTLER, *Semiconductor Lasers and Heterojunction LEDs,* Chapter 17, New York: Academic Press, 1977.

[66] R. A. LINKE, "Modulation Induced Transient Chirping in Single Frequency Lasers," *IEEE J. Quantum Electron.,* Vol. QE-21, No. 6, pp. 593–597, June 1983.

[67] C. LIN et al., "Picosecond Frequency Chirping and Dynamic Line Broadening of InGaAsP Injection Lasers Under Fast Excitation," *Appl. Phys. Lett.,* Vol. 42, No. 2, pp. 141–143, January 15, 1983.

[68] T. M. SHEN and G. P. AGRAWAL, "Pulse-Shape Effects on Frequency Chirping in Single-Frequency Semiconductor Lasers Under Current Modulation," *IEEE J. Lightwave Tech.,* Vol. LT-4, No. 5, pp. 497–503, May 1986.

15

Design of Multichannel Analog Fiber-Optic Transmission Systems

M. F. MESIYA
American Lightwave Systems, Inc.

1. Introduction

Fiber optics is more naturally suited as a communications technology for the transmission of digital rather than analog signals.* Video signal transmission represents the most common application of analog communication over optical fibers and it is also the most stringent in requirements. It demands a high signal-to-noise ratio (snr), accurately controlled frequency response, and good linearity.

Digital transmission of video signals requires wide bandwidth and/or sophisticated coding and multiplexing equipment. Consequently, the use of digital video transmission is favored in long-haul trunking applications where expensive code cs and muldems can be justified. For medium and short-haul fiber-optic trunks, analog video-transmission techniques are attractive because of their simplicity and cost effectiveness. Multichannel

analog transmission of video signals over optical fibers can employ either am or fm subcarriers. The frequency–division–multiplexed signal (am/fdm or fm/fdm) then intensity modulates the optical carrier. The relative advantages and disadvantages of each modulation method are summarized in Table 1. The block diagram of the system model considered in this chapter is shown in Fig. 1. In many applications all the system elements may not be present. For instance, in satellite video entrance links, fm modulators are not required. No modem equipment may be required in certain analog microwave radio entrance links.

The organization of this chapter is as follows: Section 2 deals with transmission of multiple am carriers where a carrier-to-noise ratio (cnr) analysis and optimization is done. The system optimization usually involves trade-offs between noise and linearity performance. The latter is examined in Section 3. Section 4 derives an expression for snr enhancement obtained using fm. System design examples are considered in Section 5 to achieve familiarity with results derived so far. Section 6 treats the intrinsic noise and distortion characteristics of laser diodes. The role of fiber and connectors in generating the new noise and distortions is examined in Section 7. Section 8 provides system design examples using ILDs in

---

* A digital transmission system requires 20 to 30 dB less cnr performance than an analog system. This, in turn, causes a double penalty for analog transmission over optical fibers. The high-cnr requirement is met by providing higher optical power to the receiver. Consequently, there is, in general, a reduced transmission-loss budget for the system. The latter is further deteriorated by dominance of shot noise, which accompanies the required power levels in analog transmission over optical fibers.

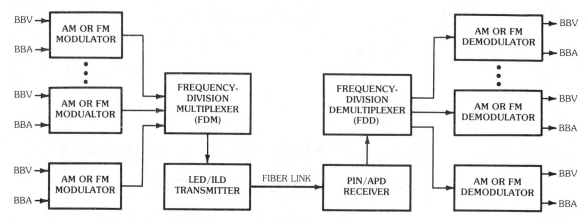

FIG. 1. Block diagram of a multichannel am/fm supertrunk system.

TABLE 1. Advantages and Disadvantages of the Modulation Methods

Modulation/ Multiplexing Scheme	Advantages	Disadvantages	Comment
Am/fdm	Simple & cost effective	High snr requirements limits the transmission distance	The use of injection laser diodes in multimode fiber systems may not be permissible to achieve acceptable performance
	No bandwidth expansion implies more channels can be packed per fiber	Sensitivity to source fiber interaction induced distortions	
Fm/fdm	Snr enhancement	Higher bandwidth occupancy implies less number of channels per fiber	Currently most popular for medium- and short-haul analog fiber-optic supertrunks
	Extra headroom for nonlinearties of the source as well as those produced as a result of source fiber interactions	Additional cost of fm modem equipment	

both multimode and single-mode fiber systems. The conclusions for system design of multichannel analog transmission systems are summarized in Section 9.

2. Multichannel AM Transmission

This section derives an expression for cnr achievable in a fiber-optic am/fdm transmission link. The intrinsic noise of the source as well as the noise produced by source-fiber interactions is neglected in this section. The case of single carrier is considered first. The output current of the am modulator is given by

$$i(t) = A_c [1 + m x(t)] \cos \omega_0 t, \qquad (1)$$

where

A_c = carrier amplitude of am signal,
$x(t)$ = message waveform $-1 \le x(t) \le 1$,
m = am modulation index,
ω_0 = $2\pi f_0$, where f_0 is the carrier frequency.

The source drive current then becomes

$$i_d(t) = I_B [1 + m_I \, i(t)/(1+m) A_c], \qquad (2)$$

where I_B is the light-source bias current and m_I is the optical modulation depth.

The input optical power corresponding to (2) is given by

$$p_{in}(t)=P_B\{1+[m_I/(1+m)]$$

$$\times[1+m\,x(t)]\cos\omega_0 t\},\quad(3)$$

where P_B represents the average power, corresponding to bias current I_B, coupled into the optical fiber. The signal portion of the optical receiver output is given by

$$I_s=R\,\bar{G}\,m_I P_{av}[1+m_I x(t)/(1+m)]$$

$$\times\cos\omega_0 t,\quad(4)$$

where

R = responsivity of the photodiode in amperes per watt,
$\bar{G}$ = mean avalanche gain of the avalanche photodiode,
P_{av} = average received optical power.

The cnr is defined, in accordance with NCTA recommendations, as the ratio of the rms power of the rf signal during the sync pulse over the rms noise power in a 4-MHz bandwidth:

$$cnr=0.5\,(R\,\bar{G}\,m_{eff}P_{av})^2/\Big(\{\overline{i_a^2}+2e[(R\,P_{av}+I_{dm})$$

$$\times\overline{G^2}+I_{du}]\}\,B\Big),\quad(5)$$

where

B = video bandwidth (4 MHz),
m_{eff} = effective modulation depth (see below),
$\overline{G^2}$ = mean-square avalanche gain of the photodiode,
$(\overline{i_a^2})^{1/2}$ = rms value of the amplifier noise current referred to the input in amperes per hertz$^{1/2}$,
e = electron charge,
I_{dm} = primary dark current which undergoes multiplication,
I_{du} = unmultiplied portion of the dark current.

The total attenuation of the fiber link, including splices, is $10\log(P_B/P_{av})$ in decibels. The term

m_{eff} is related to the nominal modulation depth m_I of the optical source by the following expression:

$$m_{eff}(f)=m_I 10^{-0.1\alpha(f)},\quad(6)$$

where

$$\overbrace{\alpha(f)=-3(f/f_{mod})^2 L^{2\zeta}-3(f/f_{mat})^2 L^2}^{\text{rolloff produced by the fiber}}$$

$$-\mathscr{L}(f),\quad 0.5\le\zeta\le1,$$

and

$\mathscr{L}(f)$ = optical-power rolloff characteristics of the LED or ILD,
L = length of the fiber-optic link in kilometers,
f = frequency of operation of the link,
f_{mod} = 3-dB modal-dispersion–limited cutoff frequency of the fiber,
f_{mat} = 3-dB material-dispersion–limited cutoff frequency of the fiber.

The definition of effective modulation depth m_{eff} at the optical receiver takes into account the degradation of the cnr caused by rolloff of the optical-source output power as well as that resulting from the modal and material dispersion in optical-fiber link. It can be seen from (5) that for a given received power level, the cnr increases as square of modulation depth. We shall see in Section 3 that nonlinear distortion beats also increase with m_I. Consequently, the optimum system design specifies the maximum modulation depth that can be advantageously used (to maximize the cnr) without exceeding the specified levels for nonlinear distortion products. In an analog fiber-optic transmission system consisting of N repeater sections the $(cnr)_t$ at the end of the system is related to cnr's ρ_i of the constituent links as follows:

$$1/(cnr)_t=\sum_i 1/\rho_i,$$

where ρ_i is the cnr of the ith link.

If the system design specifies equal cnr's ($\rho_i = \rho$) for the constituent links, then the derating law in decibels becomes

$$(\text{cnr})_t = 10 \log \rho - 10 \log N. \qquad (7)$$

For systems carrying a large number of channels in the fdm mode the optical modulation index m_I is related to per-channel modulation index m_c as

$$m_c = m_I / K^\xi, \qquad (8)$$

where K is the number of fdm carriers and $0.5 \leq \xi \leq 1$. If $\xi = 0.5$, the carriers add on a power basis. This condition is approached as a result of averaging effect produced by a large number of fdm carries with random phases. If $\xi = 1$, the addition is on the voltage basis and represents the worst-case penalty. In an hrc (harmonically related carrier) system, since the carrier frequencies are harmonics derived from the same fundamental frequency, the carriers can add on the voltage basis. However, by suitably selecting the phases of the carriers, the peak signal can be reduced as demonstrated by [1, 2]

$$m_I = K^{0.75} m_c.$$

Estimate of Received Power Level and Optimum Avalanche Gain

In order that performance of a fiber-optic analog trunk meets the specified cnr objective, an estimate of average power level at the optical receiver P_{av} and optimum avalanche gain $\bar{G}$ is required. The required average power P_{av} to achieve a given cnr ρ is obtained from (5) as

$$P_{av} = A \bar{G}^x + (A^2 \bar{G}^{2x} + C/\bar{G}^2)^{1/2}, \qquad (9)$$

where $A = 2 e B \rho / R m_c^2$ and $C = 2 \overline{i_a^2} \rho B / m_c^2 R^2$. The optimum avalanche gain $\bar{G}_{opt}$ using the approximation $\overline{G^2} = \bar{G}^{2+x}$ is given by

$$\bar{G}_{opt} = [C/x (x+2) A^2]^{1/(2 x + 2)}, \qquad (10)$$

where

$$x = \begin{cases} 0.35 & \text{silicon APDs,} \\ 0.95 & \text{germanium APDs.} \end{cases}$$

3. Effect of Nonlinearities in AM Systems

The effect of optical-source nonlinearities on the transmission performance of an am/fdm system for video trunking and distribution is analyzed in this section. The nonlinearity of photodiodes has been measured, and their contribution is negligible [3]. The source ac output can be approximated by the following transfer characteristic of a nonlinear memoryless device:

$$p_{out}(t) = k_1 i_{sd}(t) + k_2 i_{sd}^2(t) + k_3 i_{sd}^3(t), \qquad (11)$$

where $p_{out}(t)$ is the ac optical power output as a function of time, $i_{sd}(t)$ is the signal portion of the driving current, and k_1, k_2, k_3 depend on the device and its operating point. The third-order approximation in (11), although quite popular in analyzing LEDs' and ILDs' nonlinear characteristics, is valid only in a very restricted operating range.

The manufacturers of light sources, LEDs and ILDs usually provide the information about the nonlinear behavior of their devices in terms of the ratios of the second-harmonic and the third-harmonic components to the fundamental:

$$M_2 = 10 \log (P_{2f}/P_f)$$

and

$$M_3 = 10 \log (P_{3f}/P_f)$$

where

P_{2f} = electrical power level of the second harmonic,

P_{3f} = electrical power level of the third harmonic,

P_f = electrical power level of the fundamental.

It is necessary that M_2 and M_3 are specified along with the value of the modulation depth. If the modulation depth m_I changes to m'_I, the new values M'_2 and M'_3 are given by

$$M'_2 = M_2 + 20 \log (m'_I/m_I) \qquad (12)$$

and

$$M'_3 = M_3 + 40 \log (m'_I/m_I). \qquad (13)$$

The nonlinear characteristics of the LEDs and ILDs give rise to intermodulation and cross-modulation distortions in a vsb am/fdm system.

Intermodulation

Intermodulation distortion is due to the generation of harmonics and interfering beats between two or more carriers. In a vsb am/fdm system the picture carriers largely dominate the intermodulation-distortion generation because the aural and the color subcarriers are about 15 dB down. Assuming equal video carrier levels for all channels, the power of the different intermodulation beats, relative to picture carrier level, can be expressed in decibels by

$$(\text{im})_{2P} = M_2 - 20 \log (m_I/m_c) + C_{2P} + 10 \log N, \quad (14)$$

$$(\text{im})_{3P} = M_3 - 40 \log (m_I/m_c) + C_{3P} + 10 \log N, \quad (15)$$

$$(\text{im})_{3V} = M_3 - 40 \log (m_I/m_c) + C_{3V} + 20 \log N, \quad (16)$$

where $(\text{im})_{2P}$, $(\text{im})_{3P}$, and $(\text{im})_{3V}$ are generated by second-order, third-order power-additive, and third-order voltage-additive intermodulation products, respectively. The terms C_{2P}, C_{3P}, and C_{3V} are frequency-dependent coefficients which take into account the type $(f_1 \pm f_2, 2f_1 \pm f_2, f_1 \pm f_2 \pm f_3)$ of beats and the number accumulating at a frequency. The contributions of different beats occurring at the same frequency are added on a power basis. Reference 4 has summarized the worst-case values of C_{2P}, C_{3P}, and C_{3V} for standard 5-, 12- and 21-channel catv systems.

Cross Modulation

Cross-modulation distortion is the transfer of modulation from one rf carrier to another. The degree of cross modulation, according to the NCTA, is defined as the peak-to-peak variation of the test signal, as a result of cross modulation, to its amplitude with the interferring carriers removed. The measurement is usually made by modulating all the rf carriers except the one under test with a 15.75-kHz square wave. An expression for the cross modulation in decibels has been obtained in [5]:

$$xm = M_3 + 15.5 - 40 \log (m_I/m_c)$$

$$+ 20 \log (K - 1), \quad (17)$$

where K is the number of channels. It should be noted that, unlike intermodulation, the cross-modulation performance of a system cannot be optimized by selecting a suitable frequency assignment for the system. The compounding law of cross-modulation is $20 \log N$ for a small number of channels and $15 \log N$ for K large.

4. Multichannel FM Transmission

In an fm/fdm system the frequency of each carrier is varied in proportion to the amplitude of one of the channels being transmitted before the channels are frequency-division multiplied. The resultant fm/fdm signal then intensity modulates the optical source. The signal-to-noise ratio of a video signal that is fm-transmitted is given by [6]

$$\text{snr} = (P_c/N_0) \cdot [12 (\Delta F_s)^2/b_n^3], \qquad (18)$$

where

P_c = carrier power in watts,
N_0 = noise-power spectral density in watts per hertz,
ΔF_s = half of the peak-to-peak deviation, in hertz, produced by that part of the video waveform which is defined to be the signal,
b_n = noise bandwidth of the baseband-filter function (representing the combination of the deemphasis network, measurement bandlimiting filter, and weighting network) with respect to triangular noise.

For our purpose it is convenient to write (18) in the following form:

$$\text{snr} = 12 (P_c/N_0 B) \cdot (\Delta F_s/B)^2 \cdot (B/b_n)^3.$$

Using the CCIR definition of the signal (blanking-reference white) and preemphasis spectrum shaping per CCIR 405, we obtain

$$\text{snr} = \text{cnr} + 20 \log (\Delta F_v/B) + 20.37, \qquad (19)$$

where ΔF_v is half of the peak-to-peak deviation produced by the video waveform (including sync tips).

The relationship between snr and cnr in (18) and (19) is valid only above threshold encountered in an fm system. For a limiter-discriminator type receiver, threshold effect may be avoided in most practical cases if the cnr in the input bandwidth of the fm demodulator is equal to or greater than 13 dB [7]. An estimate of required fm transmission bandwidth B_T is given by

$$B_T = 2(B + \Delta F_v).$$

If the fm transmitter is ac coupled, the peak-to-peak signal excursion equals 1.6 times peak-to-peak video signal. The transmission bandwidth, in turn, becomes

$$B_T = 2(B + 1.6\,\Delta F_v).$$

An expression of the minimum cnr to avoid the fm threshold can now be written as

$$(\mathrm{cnr})_{\min} = 16 + 10 \log (1 + 1.6\,\beta),$$

where $\beta = \Delta F_v / B$. Table 2 summarizes the results.

TABLE 2. FM Performance Characteristics

ΔF_v (MHz)	$(\mathrm{cnr})_{\min}$ (dB)	B_T (MHz)	FM Enhancement
4	20	21.2	19.95
10.74	23	42.7	28.52
12.5	23.6	48.4	29.84

The most common application of video transmission involves one of the audio subcarriers multiplexed as shown in Fig. 2.

SNR Enhancement of FM/FM Subcarriers

Let us define the following parameters of an fm/fm subcarrier:

Δ = deviation of main carrier by subcarrier,
Δ_{sc} = subcarrier peak deviation,
f_a = maximum frequency of the subcarrier baseband.

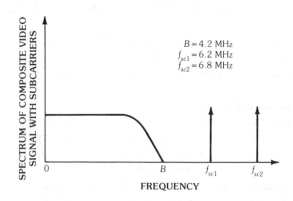

FIG. 2. Spectrum of composite video signals with audio subcarriers.

The cnr of the subcarrier is given by [8]

$$(\mathrm{cnr})_{sc} = (P_c / N_0 B_{sc})(\Delta / f_{sc})^2, \qquad f_{sc} \gg f_a, \quad (20)$$

where $B_{sc} = 2(\Delta_{sc} + f_a)$ is the bandwidth of the subcarrier. The snr of the fm/fm subcarrier is then given by

$$(\mathrm{snr})_{sc} = 1.5\,(\Delta_{sc}/f_a)^2\,(P_c/N_0 f_a)(\Delta/f_{sc})^2$$

$$= 1.5\,(\Delta_{sc}/f_a)^2\,(B_{sc}/f_a)\,(P_c/N_0 B_{sc})(\Delta/f_{sc})^2$$

$$= 3\,\beta_a^2\,(\beta_a + 1)\,(\mathrm{cnr})_{sc}, \quad (21)$$

where β_a is the modulation index of the audio subcarrier. Equation 21 can be put in more convenient form as follows:

$$(\mathrm{snr})_{sc} = (\mathrm{cnr})_{if}$$

$$+ 10 \log [(1.5\,\beta_a^2)(\Delta / f_{sc})^2 (B_{if}/f_a)]. \quad (22)$$

It is clear from (22) that the subcarrier transmission scheme receives the advantage of two consecutive stages of fm improvement. To improve audio snr performance further, preemphasis is used. The snr of audio subcarrier is then given by

$$(\mathrm{snr})_{sc} = (\mathrm{cnr})_{if} + P_{adv}$$

$$+ 10 \log [(1.5\,\beta_a^2)(\Delta / f_{sc})^2 (B_{if}/f_a)],$$

where

(cnr)$_{if}$ = cnr of the main carrier in the if noise bandwidth,

P_{adv} = Preemphasis advantage in decibels (≈ 13.2),

B_{if} = if bandwidth of the main carrier.

Assuming $\beta_a = 5$, $\Delta/f_{sc} = 0.29$, $B_{if} = 32.5$ MHz, and $f_a = 15$ kHz, we get

$$(\text{snr})_{sc} = (\text{cnr})_{if} + 54 \text{ (dB)}.$$

Typically, (cnr)$_{if}$ is greater than or equal to 13 dB, giving (snr)$_{sc} \geq 67$ dB.

5. System Design Using LEDs

The purpose of this section is to utilize the relationships developed in the preceding sections to design a multichannel analog transmission system. In an LED-based fiber-optic system the noise and distortions produced by source-fiber interactions are absent. Consequently, the system analysis presented so far adequately models the trade-offs involved.

We shall consider the design of a standard five- and twelve-channel catv (vsb/fdm) system. The performance objectives of the design are as follows:

$$\text{cnr} > 43 \text{ dB}, \tag{23}$$

$$(\text{im})_{2P} < -56 \text{ dB}, \tag{24}$$

$$(\text{im})_{3P} < -56 \text{ dB}, \tag{25}$$

$$(\text{im})_{3V} < -56 \text{ dB}, \tag{26}$$

$$(\text{xm}) < -50 \text{ dB}. \tag{27}$$

We shall determine the reach of a repeaterless link using realistic values of the state-of-the-art fiber and component parameters. LEDs with $M_2 = -40$ dB and $M_3 = -55$ dB at 50% modulation depth are commercially available. As a first step we want to determine the maximum value of m_c that can be permitted subject to the performance inequalities (24) through (27). The permissible values of m_c for both $K=5$ and $K=1$ are determined by the cross-modulation specification. The values are

$$m_c = \begin{cases} 0.137 & K=5, \\ 0.5 & K=1. \end{cases} \tag{28}$$

Next, we shall calculate the average power level required at the receiver to achieve the cnr specification in (23) subject to maximum m_c as specified in (28). See Table 3. The required P_{av} values were computed using (5) and (6). It is now possible for us to develop the link budgets and estimate the permissible link lengths using various fiber-optic technologies.

The link budgets neglect one very important limitation of the technology: the bandwidth of LEDs. It is very difficult to produce LEDs with 3-dB bandwidth in excess of 100 MHz and obtain enough power out of them. Another point is that we have assumed a link margin of 1 dB. This may be totally inadequate in a large-scale system in the real world.

6. Intrinsic Laser Noise and Nonlinearities

Injection laser diodes exhibit fundamental intensity fluctuations in light output. These fluctua-

TABLE 3. Link Budgets for LED Systems

Parameter	0.8 μm, APD		1.3 μm, PIN	
	$K=5$	$K=1$	$K=5$	$K=1$
Average power coupled into fiber (core dia = 62.5 μm) (dBm)	−10	−10	−15	−15
Average power level required at the receiver, P_{av} (dBm)	−17.42	−22	−19	−22
Connector losses (dB)	2	2	2	2
Link margin (dB)	1	5.5	1.3	3
Fiber attenuation (dB/km)	4.5	4.5	0.7	0.7
Maximum link length (km)				
@0.8 μm	0.98	1.0		
@1.3 μm			2.0	2.79

tions are intrinsic to the statistical nature of the carrier-recombination and the photon-generation process within the gain medium. The theoretical analyses by McCumber [9] and Haug [10] showed that quantum intensity noise peaks at the onset of laser threshold. The noise decreases as the injection-current level increases. The dependence is as follows [11]:

$$\text{rin } \alpha (I_B/I_{th} - 1)^{-3}, \tag{29}$$

where

rin = relative intensity noise = $\overline{\Delta P^2}/\bar{P}^2$,
$\bar{P}$ = average laser light intensity,
$\overline{\Delta P^2}$ = mean-square intensity fluctuation of light output.
I_{th} = laser threshold current.

For injection currents sufficiently above threshold ($I_B/I_{th} > 1.2$) an index-guided laser (CSPBH) exhibits a better intrinsic-noise behavior than the gain-guided laser (V-groove, oxide stripe). The rin of a gain-guided laser lies between -130 and -140 dB. This is about 10 dB inferior compared with an rin of between -140 to -150 dB for an index-guided laser. The intrinsic-noise level is essentially independent of frequency from 1 MHz up to 100 MHz and then increases to a resonance peak at each bias current as shown in Fig. 3. The

resonant frequency corresponds to the relaxation-oscillation frequency of the laser.

The relationship between intrinsic cnr, $(\text{cnr})_Q$, of an ILD and rin is given by

$$(\text{cnr})_Q = -\text{rin} - 63.25 + 20 \log m_I. \tag{30}$$

Assuming $m_I = 0.5$, we get $(\text{cnr})_Q = 65$ dB for a gain-guided laser and 75 dB for an index-guided laser.

Because of the good linearity above threshold of the light/current characteristics of commercially available ILDs, the intrinsic (i.e., without optical fiber) distortion characteristics are quite satisfactory for transmission. Fig. 4 shows [12] the second- (M_2) and third- (M_3) harmonic components for various laser types of different manufacturers. For the sake of comparison the measured distortion products were normalized to a modulated optical power of 0 dBm. The following conclusions can be drawn from Fig. 4:

1. The second- and third-order distortion products are about 50 and 70 dB below the fundamental, respectively

2. There is no significant difference in the linearity properties between index-guided and gain-guided lasers

From the discussion of intrinsic noise and distortion characteristics of laser diodes, one would conclude that they are satisfactory devices for multichannel analog transmission over optical fibers. The noise and distortion characteristics of the laser diodes are significantly altered due to the interaction of the semiconductor laser with the optical fiber.

7. Noise and Distortion Produced by Laser-Fiber Interactions

This section considers the noise and distortions which arise from the interaction of the ILD and the fiber.

Mode-Partition Noise

The instantaneous distribution of power among different lasing modes of an ILD fluctuates randomly, although the total output power remains

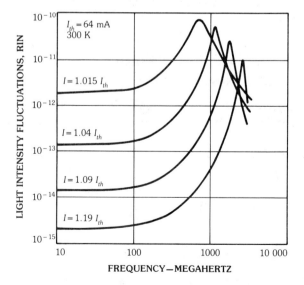

FIG. 3. Intrinsic-noise spectrum of an index-guided laser diode.

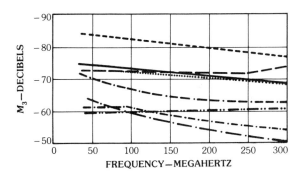

(a) Third-order distortion products.

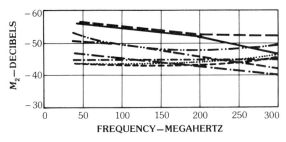

(b) Second-order distortion products.

(c) Legend for (a) and (b).

FIG. 4. Nonlinear second- and third-order distortion products of different laser diodes as a function of frequency. (*After Grosskopf and Kuller [12]*)

constant. The intensity fluctuations of various lasing modes are of opposite phase so that the averaged noise of the whole laser spectrum is less (by 30 to 40 dB) in comparison to the noise power in a single mode. Mode-partition noise is caused if the amplitude and phase relationships of the different lasing modes are distributed by the fiber. This occurs as follows:

1. The material dispersion in a fiber causes the simultaneously emitted longitudinal modes to travel at different speeds, resulting in a wavelength-dependent phase delay in the transmission of each mode. This kind of

mode-partition noise is referred to as *laser mode-partition noise,* which is abbreviated as LMPN

2. The connectors and splices in a multimode-fiber system lead to a wavelength-dependent attenuation of the transmitted laser spectrum. The mode-partition noise caused by the spatial filtering of the laser's longitudinal modes is referred to as *modal mode-partition noise* which is abbreviated as MMPN

The intensity of LMPN strongly depends on the emission spectrum of the laser and the amount of material dispersion produced by the fiber at the wavelength of interest. If the laser would oscillate strictly in a single longitudinal mode, there is obviously no LMPN. Similarly, the LMPN is mostly eliminated in a long-wavelength system operating at close to zero-dispersion wavelength. It can be shown that the spectral density of the LMPN depends on frequency and fiber length and is given by [13]

$$S_{mp}(f) = \sum_i c_{ii} S(f) + \sum_i \sum_{j \neq i} c_{ij}$$
$$\times \cos[(i-j)2\pi f \Delta\tau] S(f), \quad (31)$$

where

$c_{ii} S(f)$ = noise spectrum of the ith mode,

c_{ii} = power variance in the ith laser mode,

c_{ij} = covariance linking the ith and jth modes,

$\Delta\tau$ = $(d\tau/d\lambda)L\delta\lambda$,

$d\tau/d\lambda$ = material dispersion in the fiber in picoseconds per kilometer·nanometer,

L = length of fiber in kilometers,

$\delta\lambda$ = spacing between adjacent modes in nanometers.

It is interesting to consider the behavior of term $c_{i,i+1} \cos(2\pi f \Delta\tau) S(f)$. For $f\Delta\tau \ll 1$

$$c_{i,i+1} \cos(2\pi f \Delta\tau) S(f) = c_{i,i+1} S(f)$$
$$\times [2\cos^2(\pi f \Delta\tau)^2 - 1]$$
$$= c_{i,i+1} S(f) [2(\pi f \Delta\tau)^2 - 1]$$

$$c_{i,i+1} \cos (2 \pi f \Delta \tau) S(f) = 2 \pi^2 c_{i,i+1} S(f)$$

$$\times (f L \, \delta\lambda \, d\tau/d\lambda)^2$$

$$- c_{i,i+1} S(f) \, .$$

From the above it is clear that the terms in the spectral density expression (31) have square-law dependence on the product of fiber length and frequency. The noise enhancement $\gamma(f,L)$ due to the LMPN in a 0.8-μm single-mode fiber link using a gain-guided laser [14] is shown in Fig. 5. The enhancement continues with frequency and fiber length until a maximum of 20 to 22 dB is reached.

Modal Noise

If the laser's coherence time is greater than the time-delay difference of the propagating fiber modes in a multimode fiber, they may interfere with each other, yielding a speckle pattern at the fiber-end face. When a speckle pattern exists at a misaligned splice or connector the transmission through the joint will be dependent on the number of speckles captured by the fiber. Modal noise is caused by the random fluctuation of the speckle pattern, and therefore the coupling efficiency is very sensitive. The speckle pattern is very sensitive with respect to external forces, temperature change, etc., acting on the fiber. It is particularly sensitive to even small changes in laser emission wavelength. The speckle pattern fluctuations are also caused by a change in laser spectral purity. The worst-case cnr in the transmitted signal into the second fiber at a splice or connector is approximately given by [15]

$$\mathrm{cnr} = [N_L \, N \, \bar\eta \, m_I^2/2 \, (1-\bar\eta)B] \exp (\tau_{\mathrm{coh}}/\tau_{\mathrm{rms}}) , \quad (32)$$

where

N = number of modes guided in the fiber,
N_L = number of modes in the laser output spectrum,
$\bar\eta$ = average of coupling efficiency at the splice or connector,
τ_{coh} = coherence time of an individual laser mode,
τ_{rms} = effective rms fiber dispersion.

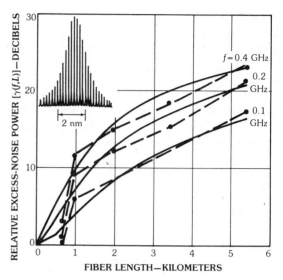

(a) For f = 0.1 to 0.4 GHz.

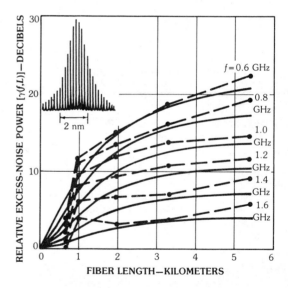

(b) For f = 0.6 to 1.6 GHz.

FIG. 5. Laser mode-partition noise for a gain-guided laser. (*After Grosskopf et al. [14]. Acknowledgement is made of the prior publication of this material by the IEE*)

The modal noise generated mainly due to external fluctuations, which will be slow, shows low-frequency components only. The fiber connector, unfortunately, also affects the noise at high frequencies. The finite width of the laser emission line means that the actual laser emission frequency fluctuates ("phase noise") and these fluctuations of the emission frequency are transformed to fluctuations of the coupling efficiency of the fiber connector. It has been reported [15] that the relative intensity of the phase noise is orders of magnitude worse than the intrinsic laser noise.

Feedback-Induced Intensity Noise

It is well known that laser-diode noise increases significantly when some portion of the light is reflected back into the laser cavity due to discontinuities of the refractive index in the optical transmission path [16–18]. Two kinds of noises are generated in the laser output intensity due to injection of reflected waves. One of them is the low-frequency noise mainly induced by reflected waves at the laser-fiber coupling. This excess noise can be suppressed by various techniques, such as antireflection coating, spherical face formation, and oblique positioning of the fiber's face with respect to the laser beam.

When the laser output light is reflected back into the cavity from a reflection point (such as an optical connector or photodiode) in the transmission path, the laser output signal demonstrates intensity noise whose frequency spectra have periodic peaks. The frequency f_i of the ith harmonic component of the noise spectrum is related to the reciprocal of the round-trip time of the fiber as follows:

$$f_i \simeq i c / 2 n L, \tag{33}$$

where n is the refractive index of the fiber, c is the speed of light, and L is the fiber length between the laser and the reflection point. For $L=2$ m the intensity noise peaks occur at 50-MHz spacing. The spacing between the peaks is reduced to 40 kHz for $L=2.5$ km.

The reflection coefficient R_I of a laser-fiber transmission link is given by [15]

$$R_I = \eta^2 \exp{(-2 \alpha L)} R_f / N, \tag{34}$$

where
  R_f = reflection coefficient of the point of discontinuity,

η = coupling efficiency between laser and fiber,

N = number of propagating modes in the fiber.

α = fiber attenuation per unit length.

Assuming $\eta=30\%$ and 50% for single- and multimode fibers, respectively, $R_f=0.04$, and $\alpha L=1.5$ dB, one obtains a reflection coefficient $R_I \simeq 10^{-3}$ for a single-mode fiber and $R_I \simeq 10^{-5}$ for multimode fiber ($V=31$). Therefore the use of a single-mode fiber is much more critical with respect to reflections than the use of a multimode fiber. The maximum value of R_I which effectively suppresses the reflection-induced laser noise is given by the following [15]:

$$R_I < \begin{cases} -60 \text{ to} -80 \text{ dB} & \text{for single-mode lasers,} \\ -30 \text{ to} -50 \text{ dB} & \text{for multimode lasers.} \end{cases}$$

The range for single-mode lasers covers various types as well as I_B/I_{th} conditions. The range for multimode lasers is determined by the number of modes in the longitudinal-mode spectrum as well as I_B/I_{th} conditions.

The cnr degradation of various single-mode laser types due to feedback-induced intensity noise has been reported in [19]. The noise enhancement of about 20 dB is observed for $R_I=-45$ dB with single-mode lasers. To achieve $R_I < -60$ to -80 dB in a single-mode fiber laser system, it may be necessary to use an optical isolator to reduce the effect of reflection-induced enhancement of laser noise.

Distortions Produced by Laser-Fiber Interactions

The intrinsic distortion characteristics of ILDs were discussed in Section 6 and found adequate. The nonlinear distortion beats, however, are enhanced due to laser-fiber interaction in the form of

(a) modal distortion, and
(b) degradation of laser linearity due to reflections.

Modal Distortion—When a semiconductor laser is modulated by the injection current, not only the optical power but also the emission wavelength is modulated. The simultaneous modulation of the optical power and the emission wavelength yields a modulation of the speckle pattern at the end face of the fiber. Nonlinear distortions therefore occur due to a fiber connector because the coupling efficiency is a nonlinear function of wavelength. The modulation ω_c of the emission frequency may be expressed as

$$\omega_c(t) = \omega_{cf} + \Omega_m \cos(\omega_m t), \qquad (35)$$

where
Ω_m = amplitude modulation of the emission frequency,
ω_m = modulation frequency,
ω_{cf} = center frequency.

Reference 20 has obtained expressions for the second-order harmonic distortion occurring due to fiber connector. It is given in decibels by

$$M_2 = 20 \log [\tau_{rms} \Omega_m \sigma_m / 2 N_L^{1/2}],$$

where τ_{rms} and N_L have the same meaning as in the previous, "Modal Noise" section, and σ_m is the normalized rms value of the derivative of the coupling efficiency with respect to emission frequency for partially coherent sources.

Modal distortion limits the second- and third-order linearity performance of a single longitudinal-mode laser used in a graded-index multimode-fiber system to -38 and -43 dB, respectively. On the other hand, the use of a gain-guided multi-multilongitudinal-mode laser with its low coherence time in a graded-index multimode-fiber system yields second- and third-order harmonic distortions of about -50 to -60 dB. Analog systems, which require linear devices, can in general not provide satisfactory performance using index-guided single-mode lasers with multimode fibers. However, the linearity characteristics of gain-guided lasers in such systems are acceptable for many applications.

Feedback-Induced Distortion—The linearity characteristics of a laser are also influenced by optical feedback. The results of linearity measurements reported in [21] indicate that the reflections from fiber far end can degrade M_2 by about 5 to 10 dB.

8. System Design Using ILDs

Section 5 considered the link design examples using a simple system model which did not include source-fiber interaction generated noise and distortion effects. The system design procedure for links using ILDs must consider the noise and distortion sources discussed in Sections 6 and 7. We make our design effort more manageable by constraining it to following realistic assumptions:

1. The laser diode has a minimum of five modes in the cw mode of operation

2. Modal noise is negligible in single-mode fiber systems. In multimode-fiber systems operating in the frequency range of interest, only phase noise is taken into account by assuming an equivalent rin degradation by 5 dB

3. The enhancement of intensity noise due to fiber reflections is neglected in multimode fiber systems. In single-mode–fiber systems the magnitude of intensity noise peaks due to optical feedback is within 3-dB range subject to the multilongitudinal-mode assumption in (1) above

The trade-offs in multimode and single-mode fiber systems are clearly evident. In multimode fiber systems the assumption of multilongitudinal-mode laser spectrum reduces the modal and feedback-induced intensity noise at the cost of a mode-partition–noise penalty. The assumption of a multimode laser in single-mode fiber systems minimizes the feedback-induced noise peaks at the cost of making the mode-partition noise as one of the prime system performance limiting factors. The cnr of the system, $(\text{cnr})_{syst}$, can now be written in terms of the cnr of the optical receiver, $(\text{cnr})_{rx}$, and transmitter $(\text{cnr})_{tx}$ as

$$1/(\text{cnr})_{syst} = 1/(\text{cnr})_{tx} + 1/(\text{cnr})_{rx}. \qquad (36)$$

An expression for $(\text{cnr})_{rx}$, which takes into account frequency-rolloff characteristics of the fiber and source, is given in (5). An expression of $(\text{cnr})_{tx}$ which takes into account the enhancement of intrinsic noise due to mode partitioning is given by

$$(\text{cnr})_{tx} = 0.5 \, m_c^2 \rho_s / [10^{0.1 \gamma(f, L)} B], \qquad (37)$$

where $\gamma(f,L)$ is the mode-partition excess-noise factor shown in Fig. 5. Substituting (5) and (37) into (36) we get

$$(\text{cnr})_{\text{syst}} = 0.5\,(m_c\,R\,P_{\text{av}}\bar{G})^2 \Big/ B\,\{RP_{\text{av}}\,\bar{G})^2\,\rho_s^{-1}$$

$$\times 10^{0.1[\alpha(f,L)+\gamma(f,L)]} + \overline{i_a^2} + 2\,e$$

$$\times [(R\,P_{\text{av}} + I_{dm})\overline{G^2} + I_{du}]\} , \quad (38)$$

$\rho_s = $ dc snr of the ILD $(= 10^{-0.1\,\text{rin}})$.

In a practical system design one is interested in determining the maximum link length subject to a specified end-to-end performance objective. It is assumed that the choice of laser modulation depth will be made taking into account the linearity requirements of the system and distortion characteristics of the source. The step-by-step procedure for designing the system can, now, be outlined as follows:

1. Assume that N repeaters are required to provide the required snr performance for a system length of L_{syst} kilometers

2. Compute the cnr necessary to achieve the required snr objective. For vsb-am systems, use the following relationship:

$$(\text{snr})_{\text{CCIR}} = (\text{cnr})_{\text{NCTA}} - 0.2 \text{ dB}$$

The corresponding relationship for fm systems is given in (19). The system optimization for fm systems may involve selecting the appropriate value of peak frequency deviation subject to equipment availability constraints

3. Derate the value of cnr computed in step 2 for N repeaters assumed in step 1 using (7)

4. Compute the maximum per channel modulation depth using (8) and subject to linearity requirements

5. Calculate the required average power level P_{av} at the receiver, using (9). The system

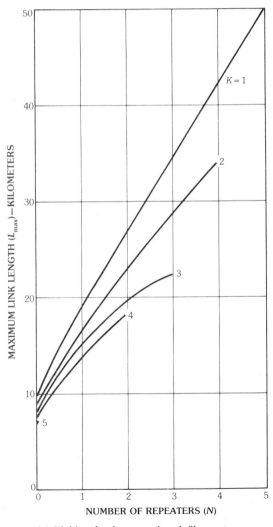

(a) Multimode, short-wavelength fiber system.

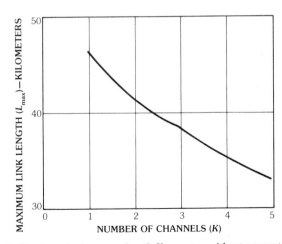

(b) Single-mode, long-wavelength fiber system without a repeater.

FIG. 6. Maximum link length achievable in an fm/fdm tv supertrunk.

optimization involves determining the best receiver front end to achieve the minimum value of P_{av} for a specified cnr objective

6. Estimate the length L_r of repeater span using the expression

$$L_r = (P_B - P_{av} - L_c - L_m)/L_f,$$

where

L_c = connector losses in decibels,
L_m = link margin for temperature, aging, etc., in decibels,
L_f = fiber loss in decibels per kilometer including splice losses

7. Compute $(cnr)_{syst}$ using (37) with $L=L_r$. If $(cnr)_{syst} < (cnr)_{obj}$, let $L=L_r-\Delta L$ where ΔL is a suitable value of decrement in length L. Continue until $(cnr)_{syst} > (cnr)_{obj}$ for some value $L=L_1$. If $L_1 N < L_{syst}$, increase N by 1. If $L_1 N > L_{syst}$, decrease N by 1

8. Continue until N satisfies the inequality $(N-1)L_{N-1} < L_{syst} < (N+1)L_{N+1}$. Here L_{N-1} and L_{N+1} are repeater spacings with $N-1$ and $N+1$ repeaters, respectively

The maximum link lengths achievable using state-of-the-art fiber-optic technologies at short and long wavelengths are shown in Fig. 6. The performance advantage of a 1.3-μm, single-mode fiber system is tremendous—a repeaterless five-channel link with span of approximately 34 km achieves a 55-dB snr performance. The long-wavelength, single-mode technology lends itself as an excellent candidate for an interhub connect link in big metropolitan areas. The absence of any active line equipment in the field is of great value to operating companies from maintenance and operational considerations.

9. Conclusions

The problem of designing a multichannel analog fiber-optic transmission system has been investigated in this chapter. The complexity of the design process is a function of the system model which adequately describes the link. A system using LEDs lends itself to a very predictable design methodology. The use of LEDs, however, is restricted to short-distance links. Multimode fiber systems are the most difficult to analyze because of the presence of modal noise and distortion as well as bandwidth limitation. A long-wavelength, single-mode system does not only offer significant performance advantages, but also lends itself to more predictable system analysis and performance. It is expected that long-wavelength, single-mode technology will dominate fiber-optic fm super-trunk arena. Vestigial-sideband fdm systems may become a reality with maturing of single-frequency, long-wavelength lasers using single-mode fiber-optic technology.

10. References

[1] W. KRICK, "Improvement of CATV Transmission using an Optimum Coherent Carrier System," Symp. Rec. 11th Intl. Television Symp. (Montreux, Switzerland, 1979, Session 1X B-4).

[2] I. SWITZER, "Phase Phiddling," Proc. NCTA '74.

[3] T. OZEKI and E. H. HARA, "Measurement of Nonlinear Distortion in Photodiodes," Electron. Lett., Vol. 12, No. 3, February 5, 1976.

[4] D. CHAN and T. M. YUEN, "System Analysis and Design of a Fiber-Optic VSB FDM System for Video Trunking," IEEE Trans. Commun., Vol. COM-15, pp. 680–686, July 1977.

[5] K. A. SIMONS, "The Decibel Relationships Between Amplifier Distortion Products," Proc. IEEE, Vol. 58, pp. 1071–1086, July 1970.

[6] L. CLAYTON, "FM Television Signal-to-Noise Ratio," IEEE Trans. Cable TV, Vol. CATV, No. 1, pp. 25–30, October 1976.

[7] A. B. CARLSON, Communication Systems, 2nd ed., New York: McGraw-Hill Book Co., 1975.

[8] R. Gagliardi, Introduction to Communications Engineering, New York: Wiley-Interscience, 1978.

[9] D. E. McCumber, "Intensity Fluctuations in the Output of CW Laser Oscillators I," Phys. Rev., Vol. 141, No. 1, pp. 306–322, January 1966.

[10] H. HAUG, "Quantum-Mechanical Rate Equation for Semiconductor Lasers," Phys. Rev., Vol. 184, No. 8, pp. 338–348, August 1969.

[11] H. MELCHIOR, "Noise in Semiconductor Lasers," Topical Meeting Integ. and Guided-Wave Opt., Paper MA2, Nevada, January 1980.

[12] G. GROSSKOPF and L. KULLER, "Measurement of Nonlinear Distortions in Index- and Gain-Guided GaALAs Lasers," *J. Optical Commun.,* Vol. 1, pp. 15–17, September 1980.

[13] A. K. LAUGHTON, and Y. KANABAR, "Mode-Partition Noise in Gain-Guided Lasers at 850 nm and Its Impact on Analog Fiber-Optic Transmission Systems," Paper TU03, *Dig. Tech. Pap.,* Conference on Optical-Fiber Communication, San Diego, February 11–13, 1985.

[14] G. GROSSKOPF et al., "Laser Mode-Partition Noise in Optical Wideband Transmission Links," *Electron. Lett.,* Vol. 18, No. 12, pp. 493–494, June 1982.

[15] K. PETERMANN, and G. ARNOLD, "Noise and Distortion Characteristics of Semiconductor Lasers in Optical-Fiber Communication Systems," *IEEE J. Quantum Electron.,* Vol QE-18, No. 4, pp. 543–555, April 1982.

[16] I. IKUSHIMA, and M. MAEDA, "Self-Coupled Phenomena of Semiconductor Lasers caused by an Optical Fiber," *IEEE J. Quantum Electron.,* Vol. QE-14, pp. 331–332, May 1978.

[17] O. HIROTA and Y. SUEMATSU, "Noise Properties of Injection Lasers Due to Reflected Waves," *IEEE J. Quantum Electron.,* Vol. QE-15, pp. 142–149, March 1979.

[18] O. HIROTA, Y. SUEMATSU, and K. S. KWOK, "Properties of Intensity Noises of Lasers Due to Reflected Waves from Single-Mode Optical Fibers and Its Reduction," *IEEE J. Quantum Electron.,* Vol. QE-17, pp. 1014–1020, June 1981.

[19] K. SATO, "Intensity Noise of Semiconductor Laser Diodes in Fiber-Optic Analog Video Transmission," *IEEE J. Quantum Electron.,* Vol. QE-19, No. 9, pp. 1380–1391, September 1983.

[20] K. PETERMANN, "Nonlinear Distortions and Noise in Optical Communication Systems Due to Fiber Connectors," *IEEE J. Quantum Electron.,* Vol. QE-16, No. 7, pp. 761–770, July 1980.

[21] G. WENKE, and G. ELZE, "Investigation of Optical Feedback Effects on Laser Diodes in Broadband Optical Transmission Systems," *J. Optical Communications,* Vol. 2, pp. 128–133, November 1981.

16

Introduction to Coherent Fiber-Optic Communication

E. Basch

GTE Laboratories, Inc.

T. Brown

Institute of Optics, University of Rochester

1. Introduction

The ultimate goal in the development of any communications medium is to realize its maximum possible information capacity. This generally includes schemes such as error-correcting codes and special signal design, as discussed in Chapter 13, "Communication Theory for Fiber-Optic Transmission Systems," [1]. These approaches seek to optimize the performance of a system in which the fundamental physical limitations have been established. For an optical-fiber system they include fiber attenuation (limited by Rayleigh scattering), maximum power density (limited by stimulated Raman and Brillouin scattering), as well as background radiation. A fourth physical limitation is that imposed by the quantum nature of light. The imposition of the uncertainty principle requires that any detection be statistical in nature, as governed by the quantum fluctuations of the optical fields. Conventional direct-detection systems have been unable to approach these fundamental limits due to detector and preamplifier noise limitations.

A step closer to exploiting the capacity of optical systems can be accomplished through heterodyne or homodyne detection of the optical signal. Heterodyne-type systems have been extensively studied for free-space communications [2], and

their feasibility in conjunction with optical fiber is a topic of active research. Optical communication systems which use heterodyne or homodyne detection are commonly referred to in the literature of fiber optics as coherent optical communication systems.*

Heterodyne techniques have been common in broadcasting, satellite, and microwave radio transmission for many years. It was only in 1955, however, that the first beat signal from the mixing of two light sources on a photocathode was observed [3] and some of the present-day concepts of optical heterodyne detection were born. In the early 1980s, improvements in spectral purity and stability of semiconductor lasers led to experiments with coherent optical transmission through optical fibers. Heterodyning or homodyning techniques, shown in principle in Fig. 1, can provide significant improvement in receiver sensitivity, leading to substantial increases in repeater spacing over conventional optical communication systems [4].

In addition, heterodyning would allow frequency-division multiplexing of several hundred or

---

*The term "coherent" refers to the requirement for highly coherent lasers. Both coherent and noncoherent detection schemes can be used. Theoretically the coherent detection schemes have better receiver sensitivity than the noncoherent schemes but require increased temporal coherence of the source.

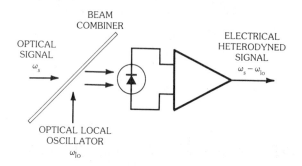

FIG. 1. Optical heterodyne receiver in which an optical local oscillator mixes with a received optical signal to produce an electrical frequency equal to the difference between the optical frequencies. In homodyning, $\omega_s = \omega_{lo}$.

more optical carriers with very narrow separation, in contrast to the multiplexing of a few channels with broad spacing in conventional wavelength—division multiplexing. Exploitation of these techniques could be used to increase channel capacity and result in a more efficient use of the optical fiber as a transmission medium. For many applications, the local telephone loop in particular, such frequency selectivity may be even more important than improved receiver sensitivity. Using such frequency-division multiplexing techniques it would appear that optical fibers have unlimited channel capacity. This is not the case, however, since the transmitted power into optical waveguides is limited by nonlinear interactions [5], as discussed in Chapter 7. The available transmitter power must be divided among the multiplexed channels, and this serves as one limit to the channel capacity.

The following sections summarize the basic principles of coherent optical-fiber communication systems, making connection wherever possible with standard techniques in rf and microwave communications. Also an attempt is made to acquaint the reader with certain concepts in quantum optics which are currently being researched in connection with coherent optical communication.

2. Modulation Formats

The modulation formats for coherent optical-fiber communications are essentially the same as for standard rf and microwave communications. Such modulation formats were discussed in a previous chapter in a slightly different context, i.e., the generation of subcarriers by intensity modulation for frequency-division multiplexing. These subcarriers require only electrical demodulation, and direct detection of the optical signal is used at the receiver. Such systems carry no significant improvements in signal-to-noise ratio over standard digital carriers, since the same receiver noise sources are dominant regardless of the modulation format.

When a narrow–line-width laser source is used in an optical-fiber system it is possible to directly amplitude, phase, or frequency modulate the optical carrier. The techniques for accomplishing such modulation are discussed in the following sections.

Amplitude-Shift Keying

Several techniques are possible for amplitude modulating an optical signal. The intensity modulation used in direct-detection systems is a crude sort of amplitude-shift keying in which the received signal is simply detected using the photodetector as a square-law device. It is clear, therefore, that the simplest approach to amplitude modulation is direct modulation of the laser drive current. The difficulty with such an approach lies in the inability of the laser to maintain a stable frequency with changing drive current. This frequency deviation can be as high as 200 MHz/mA [4] and broadens the spectrum of a modulated laser enough to render it useless in coherent optical communication. This property of the laser can be useful in frequency modulation and stabilization of semiconductor lasers, but external modulation is a much more promising approach for amplitude and phase modulation.

Prior to the rapid rise in optical fiber applications, external modulators were primarily of bulk geometry, utilizing the electro-optic or Kerr effects. In electro-optic materials the application of an electric field **E** produces a refractive-index change δn of [6]

$$\delta n = -n^3 r E/2 + \cdots, \tag{1}$$

where r is the coefficient of the electro-optic tensor connecting the direction of refractive-index change with that of the electric field, and n is the refractive index. The applied voltage can be used to phase shift the signal to produce destructive or constructive interference in an interferometer, or to induce a phase shift between the x and y polarizations of the optical signal (see Fig. 2). By rotating the state of polarization and passing the optical signal through a linear polarizer, it is possible to produce an amplitude-shift keyed signal with good extinction in the "off" state.

direction of the electric field. For LiNbO$_3$, r_{33} has a value of 30×10^{-12} mV^{-1}, yielding a refractive-index change of 1.6×10^{-3} with an applied field of 10 V/μm, making modulation possible in a waveguide geometry [6].

Gallium arsenide (GaAs) also exhibits an electro-optic coefficient sufficient for waveguide modulation and a similar approach can be used. While good-quality planar and channel waveguide structures are relatively easy to fabricate in

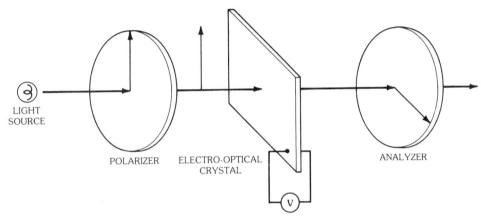

FIG. 2. Bulk electro-optic modulator.

Polarization modulation, while practical in bulk crystals, is somewhat cumbersome in waveguide geometries when compared with other approaches [6, 7]. A 2×2 directional waveguide coupler can be used as a modulator, yielding complementary on-off keyed signals in each output port. Y-channel waveguide couplers can form the integrated optical equivalent of a Mach-Zehnder interferometer. It is also possible to modulate the waveguide cutoff condition by changing the refractive index electrically, achieving amplitude modulation at the output.

Comparatively few crystals have electro-optic coefficients large enough to be useful as modulators. Lithium niobate (LiNbO$_3$) has received a great deal of research and development effort due to its electro-optic properties and the ability to fabricate planar and channel waveguide devices for integrated optic applications. Fig. 3 shows a typical LiNbO$_3$ waveguide modulator using channel waveguides in a Mach-Zehnder interferometric design. The crystal is usually cut to maximize the electro-optic coefficient in the

LiNbO$_3$ by titanium in-diffusion or other techniques, low-loss waveguides in gallium arsenide are much more difficult to fabricate, requiring sophisticated epitaxial growth techniques. However, the true goal of integrated optics—the fabrication of lasers, waveguides, modulators and other elements on a single substrate—requires a semiconductor such as gallium arsenide in order to be realized.

A traveling-wave modulator can use the electro-optic effect to achieve very high speed modulation by closely matching the propagation of a microwave signal on a stripline electrode with the propagation constant of the guided modes in a channel waveguide. Such modulators have been operated at speeds in excess of 15 GHz [7] and will be necessary to fully utilize the transmission bandwidth available in an optical fiber.

Phase-Shift Keying

As will be shown, phase-shift keying (psk) offers the best theoretical receiver sensitivities for fiber-optic systems. Phase-shift keying also offers the

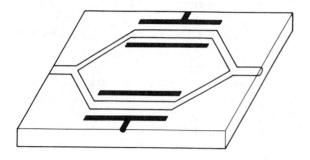

(a) Physical construction.

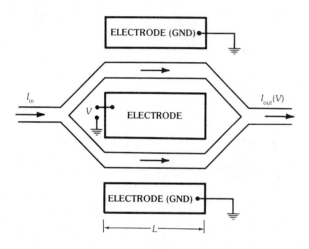

(b) Electrical design.

FIG. 3. Typical LiNbO₃ waveguide modulator employing a Mach-Zehnder interferometer design.

simplest modulation configuration. Any crystal exhibiting the electro-optic effect, if oriented correctly, will have a change in refractive index δn with applied voltage. This results in a phase shift Δ of the transmitted optical signal

$$\Delta = 2\pi\,\delta n\,\ell/\lambda \qquad (2)$$

after traversing a distance ℓ in the material. Thus a phase shift of π radians can easily be accomplished over relatively short interaction lengths given a modest index change δn.

Multilevel as well as binary phase-shift keying can be performed on an optical carrier. The considerations for choosing a particular multilevel

scheme for optical communications can be quite different from those in rf and microwave communications. The usual reasons for adding the complexity of higher-level psk schemes are requirements for increased bandwidth efficiency—such requirements are usually not present in optical systems. On the other hand, coherent optical systems suffer from carrier phase noise, which is often a minor consideration in radio and microwave communication.

Frequency-Shift Keying

As with amplitude-shift keying, frequency-shift keying (fsk) of the optical carrier can be accomplished either by *direct* or *external* techniques. The source central frequency is influenced primarily by the gain spectrum of the active medium and the round-trip propagation time in the laser cavity. Semiconductor lasers typically have a broad gain spectrum and a wide frequency interval between the longitudinal modes of the cavity. Changes in the current density or temperature may produce a shift in the gain profile, resulting in a sudden frequency shift to a neighboring cavity mode. It is the goal of frequency stabilization schemes to avoid this by means of feedback controls to the temperature and drive current.

A slight modulation of the drive current, however, can (through the corresponding change in carrier density in the semiconductor) produce small changes in the index of refraction of the active medium. Mode resonance frequencies (those frequencies for which standing waves can exist in the cavity without interfering destructively) can be shifted by a very slight change in the index of refraction. Thus a controlled frequency shift is possible even in ordinary laser structures. The use of external cavities, a common approach for frequency stabilization, can also be used for frequency-shift keying simply by modulating the optical path length of the external cavity.

External techniques include both acousto-optic and electro-optic approaches. In bulk optics, frequency-shift keying is usually accomplished via a Bragg cell which employs traveling acoustic waves in a crystal to simultaneously diffract and frequency shift the optical signal. This approach is somewhat cumbersome, but the equivalent effect can be achieved by using surface acoustic waves on a slab waveguide. The frequency shift is due to the inelastic scattering of light from acoustic waves in the crystal, and the resultant output is similar to diffraction from a grating. In this case, however, each diffracted order has a different optical frequency. The obvious difficulty in this approach is that any shift in frequency has an associated shift in diffraction angle (spatial modulation), making alignment problems difficult for large frequency shifts.

Electro-optic approaches rely on the relationship of optical frequency ω and phase ϕ,

$$\omega = d\phi/dt \tag{3}$$

and the ease of accomplishing a phase shift using the electro-optic effect. The introduction of any

sawtooth waveform to shift the frequency [8], as shown in Fig. 4a. This device promises low loss; however, an extremely accurate signal is required to accomplish the precise 2π radians phase shift at the transition. An approach which avoids the difficulties of the sawtooth drive signal is the double Mach-Zehnder single-sideband modulator shown schematically in Fig. 4b. Theoretical conversion efficiencies are lower for this device, the loss higher, and the fabrication problems much more difficult. However, the simple sinusoidal drive signal permits this device to be operated well in excess of 1 GHz.

In principle, multilevel fsk yields the best receiver performance in the limit of large channel spacings. As noted in Chapter 13, M-ary fsk approaches the Shannon limit of information

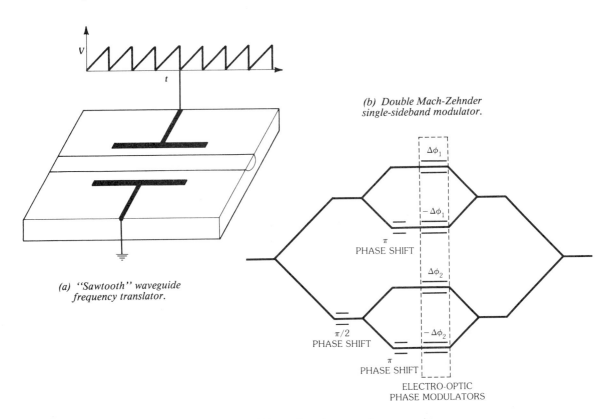

(b) Double Mach-Zehnder single-sideband modulator.

(a) "Sawtooth" waveguide frequency translator.

FIG. 4. Phase shifting using electro-optic approaches.

phase modulation linear in time will produce a constant frequency shift. Since a phase shift of 2π radians leaves the optical signal unchanged, it is possible to use an ordinary phase modulator with a

capacity as the number of levels increases [1]. In contrast to multilevel psk and ask schemes, M-ary fsk provides *bandwidth expansion,* reducing the sensitivity to both am and fm noise. The complexity of modulator design for an optical M-ary fsk system increases considerably with the number of levels, and the advantages of such a scheme over ordinary frequency-division multiplexing in optical frequency ranges are not yet clear.

3. Detection/Demodulation Schemes

The detection and demodulation of an information-bearing optical signal is ordinarily accomplished via *direct detection* in a square-law device, usually a photodetector whose output is proportional to the integrated light intensity incident on the detector surface over some finite resolving time. In this case the signal amplitude at the output of the detector/preamplifier is proportional to the power of the incident optical signal. Thus, if a classical analytic signal (slowly varying complex envelope) representation $e_s(t)$ and its complex conjugate $e_s{}^*(t)$ are used for the electric field of the optical signal,

$$E_s(t)=e_s(t)\,e^{-i\omega t}+e_s{}^*(t)\,e^{i\omega t}, \tag{4}$$

then the optical power incident on the detector is

$$P_s(t)=e_s(t)\,e_s{}^*(t)=|e_s(t)|^2. \tag{5}$$

The mean number $N(t)$ of electron-hole pairs created per second in a photodetector is related to the mean optical power $P(t)$, and to the quantum efficiency η of the detector:

$$N(t)=\eta\,P(t)\,/h\,\nu, \tag{6}$$

where

    $h\nu$ = energy of a photon,
    h = Planck's constant,
    ν = optical frequency.

Clearly, all information relating to the optical frequency and phase of the original signal is lost, the resultant signal being proportional to the mean-squared electric-field amplitude of the optical signal.

For the case of coherent optical communications, detection is accomplished via the mixing of signal and local oscillator in the square-law detector. The incident optical field is the superposition of the weak signal and strong local-oscillator fields:

$$E(t)=E_s(t)+E_{lo}(t). \tag{7}$$

The resultant optical power incident on the detector is

$$P(t)=P_s(t)+P_{lo}(t)$$
$$+\,e_s(t)\,e_{lo}{}^*(t)\exp\{i\,[\theta+(\omega_{lo}-\omega_s)\,t]\,\}$$
$$+\,e_s{}^*(t)\,e_{lo}(t)\exp\{-i\,[\theta+(\omega_{lo}-\omega_s)\,t]\,\}. \tag{8}$$

The angular-frequency difference $\omega_{lo}-\omega_s$ can immediately be recognized as an intermediate frequency, while the phase angle θ reflects the randomly fluctuating phase difference between the signal and local-oscillator signals.

The usual case is one in which the signal is very weak compared with the strong local-oscillator fields, and $P(t)$ is then given by

$$P(t)=P_{lo}(t)+e_s(t)\,e_{lo}{}^*(t)\exp\,[i\,(\theta+\omega_{if}t)]$$
$$+\,e_s{}^*(t)\,e_{lo}(t)\exp\,[-i\,(\theta+\omega_{if}\,t)]. \tag{9}$$

Since the detector output is proportional to $P(t)$, detection using the local oscillator retains the phase and frequency information of the optical signal *relative to that of the local oscillator.*

We shall now discuss specific aspects of the detection and demodulation of coherent optical signals.

Homodyne Detection

The special case of equal signal and local-oscillator frequencies is referred to as homodyne detection. From (9), if $\omega_{if}=0$, then

$$P(t)=P_{lo}(t)+e_s(t)\;e_{lo}{}^*(t)\,e^{i\theta}$$
$$+\,e_s{}^*(t)\,e_{lo}(t)\,e^{-i\theta}. \tag{10}$$

If the information-bearing portion of the signal is contained on the in-phase component

$$e_s(t)=e_s{}^*(t)=[P_s(t)]^{1/2}$$

and

$$e_{lo}(t)=e_{lo}*(t)=[P_{lo}(t)]^{1/2} ,$$

then

$$P(t)=P_{lo}+2\,[P_{lo}\,P_s(t)]^{1/2}\cos\theta. \qquad (11)$$

We see, then, that homodyne detection brings the signal directly to baseband and no further electrical demodulation is necessary.

The requirement that the signal and local oscillators have equal frequencies generally leads to very stringent criteria for the optical sources, since the combination of high frequency stability, extremely narrow line width and tunability is difficult to achieve. There are three possible approaches to optical homodyning:

Injection Locking—Any laser may be considered a multipass amplifier whose input consists of a single spontaneous-emission photon in each mode. With such an input, control of the oscillation frequency depends on maintaining a high cavity stability. If a sufficiently strong signal is injected as input, however, the frequency and phase of the output will follow those of the injected signal. This allows for direct control of the local oscillator by the signal. Unfortunately, injection locking requires excessive power at the input, far exceeding the target sensitivity for most homodyne systems [10].

Optical Phase-Lock Loop—For an ordinary laser oscillator a high cavity stability must be maintained in order to keep the frequency drift between the signal and local oscillator much less than the signal bandwidth, and to provide phase estimation for coherent demodulation. It is possible, in a homodyne receiver, to electrically detect the phase error between the signal and local oscillator, and use the error signal to correct the local-oscillator frequency and phase [10, 11].

Selective Amplification Of The Carrier—In this approach the carrier component of the received signal is selectively amplified (i.e., the modulation sidebands are not amplified) prior to photodetection and the amplified carrier functions as the local oscillator. Such selective amplification is possible by making use of backward-wave stimulated Brillouin scattering within the transmission

fiber, as described in Chapter 7. This process, however, also increases the carrier phase noise, which will degrade system performance, and the utility of the scheme will ultimately depend on the noise processes associated with the Brillouin gain mechanisms [12].

Heterodyne Detection

The case of heterodyne detection, where ω_{if} is nonzero is of primary interest considering both source requirements and the potential of added channel capacity. Assuming once again that the information-bearing portion of the signal is on the in-phase component, we have the direct analogy to (11) for heterodyne detection:

$$P(t)=P_{lo}+2\,[P_{lo}\,P_s(t)]^{1/2}\cos\,(\omega_{if}t+\theta) . \quad (12)$$

The mean signal power can be compared for the homodyne and heterodyne cases, respectively:

$$S_{hom}=4\,P_{lo}\,P_s \qquad (13a)$$

$$S_{het}=2\,P_{lo}\,P_s \qquad (13b)$$

compared to

$$S_{dd}=P_s^2 \qquad (13c)$$

for direct-detection systems.

Through the coherent detection process the signal undergoes preamplification by the local oscillator. Fig. 5 illustrates the change in signal-to-noise ratio with increasing local-oscillator power. Under the "quantum-limited" condition the quantum fluctuations of the field exceed the receiver noise power, yielding an improvement in receiver sensitivity.

Heterodyne Coherent Demodulation—Since for heterodyne detection the signal exists on an intermediate frequency at the preamplifier output, one has the usual choice of demodulation methods to bring the signal to baseband; the choice will generally depend on the modulation format.

Coherent demodulation implies an estimation of the phase of the if signal in bringing it to base-

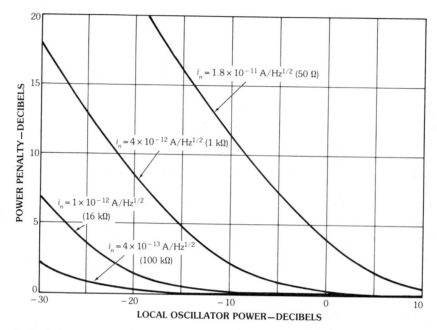

FIG. 5. Penalty in signal-to-noise ratio as a function of local-oscillator power for a family of receivers with thermal noise present at the input of the preamplifier, where i_n is the noise spectral density of the receiver.

band. Such an approach requires phase-locking techniques to track phase fluctuations in the carrier and local-oscillator signals. Ordinary psk modulation requires estimation of the phase of the signal and therefore requires coherent demodulation; however, a differential phase-shift keyed (dpsk) signal can be demodulated by a simple delay loop to compare the phase of each bit with that of the previous bit [13].

All homodyne techniques, of course, require coherent demodulation,* and possible approaches to optical phase estimation were discussed in the previous section. For heterodyne detection, since the signal will be processed on an intermediate-frequency carrier, electrical phase estimation can be used. Any phase-locked loops (pll's) appropriate for rf and microwave communication may be considered, and they are described in many textbooks [14]. Most analyses rely on a linearized pll model, in which the phase error is related to that

of the input by a loop transfer function; the order of the loop and its bandwidth have a strong influence on the performance of the system in the presence of carrier phase noise.

An examination of the spectrum of a psk signal reveals that it is a *suppressed-carrier* type of modulation—no signal energy is present at the carrier frequency if the phase shift from "mark" to "space" states is a full 180°. The introduction of a nonlinear element is therefore necessary for proper carrier recovery. A simple method for binary psk systems which is especially applicable to phase-noise–sensitive optical systems is a "squaring loop," shown in Fig. 6a. Squaring of the psk signal results in a carrier at twice the original frequency which can be filtered and used for phase estimation. It has been shown [15] that this is statistically equivalent to the Costas loop, shown in Fig. 6b.

Another approach to carrier recovery is to reduce the depth of phase modulation slightly. In so doing, a small component of the transmitted energy lies at the carrier frequency, which can be amplified and recovered as a phase reference. The weak carrier would require a much smaller loop bandwidth (i.e., longer integration time) in order

---

*A possible exception is the *phase-diversity receiver,* which results directly in a baseband signal without the use of phase-locked loops. A brief discussion of the phase-diversity receiver will be given in the following section.

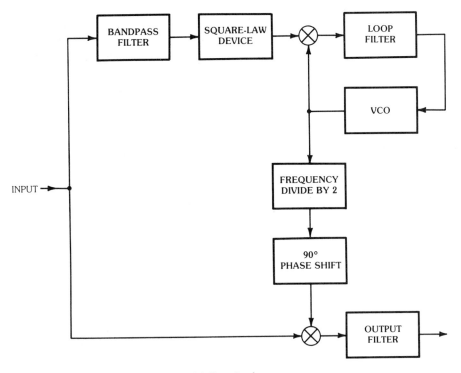

(a) Squaring loop.

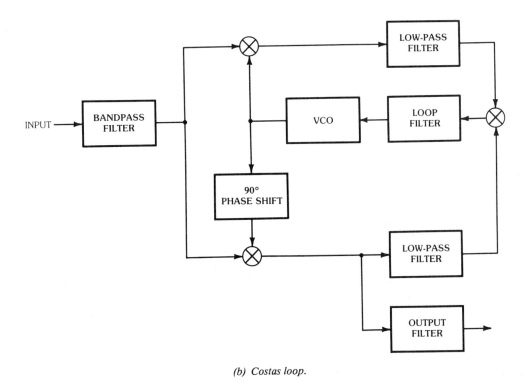

(b) Costas loop.

FIG. 6. Loops for carrier recovery in psk systems.

to adequately recover the carrier. Long integration times, while providing a stable reference, increase the sensitivity to carrier phase noise [16]. A significant portion of the signal energy might have to be sacrificed in order to use this technique to adequately detect the psk signal.

It should be pointed out that a measurement of the phase of a fully modulated psk signal implies the existence of a reference—since no absolute reference exists for the phase of such a signal, any error measurement must be referenced against the phase of the average transmitted carrier in a given time interval. The purpose of a phase-locked loop is to provide that reference; in general the averaging time is defined by the bandwidth of the loop. More specific examples of the effects of phase noise on receiver sensitivity will be given in later sections.

Similar coherent demodulation techniques can also be used for ask and fsk systems, but other techniques are often more reliable in the presence of phase noise. Such techniques require no phase estimation and are referred to as *incoherent* demodulation.

Incoherent Demodulation Techniques—The design and fabrication of sources suitable for coherent demodulation schemes may be impractical in some applications. It is possible, however, to reduce the sensitivity to phase noise by the use of ask envelope, fsk dual filter, or psk phase diversity demodulation [17].

The envelope detection of an ask signal is achieved through the use of an intermediate-frequency bandpass filter followed by a peak detector to recover the signal at baseband. When only receiver noise is considered, envelope detection incurs a small penalty (0.5 dB for a bit error rate of 10^{-9}) in receiver sensitivity due to the nonlinear filtering of Gaussian noise in the peak detection process. The effect of phase noise, although reduced, is not negligible even for envelope detection. Its major effect is a broadening of the signal spectrum in such a way that a significant portion of the energy may lie outside the intermediate-frequency (if) band. The output signal power is then reduced relative to the noise, incurring a penalty in receiver sensitivity. Alternatively, the if filter bandwidth may be increased to maintain the signal strength,

resulting in increased receiver noise. The optimum bandwidth for receiver performance balances the penalties due to phase noise and receiver noise. For combined source and local-oscillator line widths of 10% of the transmitted data rate, this penalty amounts to 3 dB [18].

It is possible to incoherently demodulate a frequency-shift–keyed signal by *dual filter detection*. This technique can best be understood by expressing the fsk signal as complementary ask signals on different carrier frequencies:

$$s(t) = a(t) \cos \omega_1 t + [1 - a(t)] \cos \omega_2 t, \quad (14)$$

where

$$a(t) = \begin{cases} 0 & \text{``space,''} \\ 1 & \text{``mark.''} \end{cases}$$

By employing parallel filters with channels centered around ω_1 and ω_2 it is possible to use envelope detection on each channel and obtain a differential ask signal. The penalty for incoherent detection is therefore similar to the ask case, as is the system performance in the presence of phase noise. Garrett [18] has calculated a 2-dB penalty for source and local-oscillator line widths of 10% of the transmitted data rate. The subtle differences between ask and fsk in this regard are due to the complementary nature of dual filter detection; for large channel spacing it is possible to have a significant spectral broadening of the signal yet still have insufficient energy in the complementary channel to register an error.

An fsk dual-filter detection scheme allows the application of distributed-feedback (DFB) lasers using direct frequency modulation. Such lasers typically have insufficient stability for coherent demodulation, but could be used in noncoherent demodulation at data rates of several hundred megabits per second with little penalty.

An additional demodulation scheme used in microwave systems and applicable to coherent optical communications is the *phase-diversity* receiver [17]. Operable with two or more ports, this technique utilizes a fixed phase relationship between the ports of a multiport mixer to allow direct demodulation to baseband without the need for an optical phase-locked loop. The sensitivity of this approach is poorer than heterodyne detection (1.8 dB for a three-port system) [17] but has the advantage of requiring only baseband filtering. This is particularly important for very high speed applications in which heterodyne detection would require excessively high intermediate frequencies and broadband microwave filters.

A ''six-port'' detection scheme for phase diversity reception is illustrated in Fig. 7. The receiver is constructed using four directional couplers—ports 1 and 2 receive the signal and local-oscillator fields, while ports 5 through 8 lead to matched detectors. Subtraction of the photocurrent from the following ports leads to simultaneous detection of the in-phase and quadrature components of the modulated signal:

$$P_5 - P_6 = |e_s e_{lo}| \cos \theta \qquad (15a)$$

and

$$P_7 - P_8 = |e_s e_{lo}| \sin \theta , \qquad (15b)$$

where P_i denotes the received power from the ith port. Summing the squares of the two outputs results in a baseband signal achieved without prior knowledge of the phase. Such schemes may ease

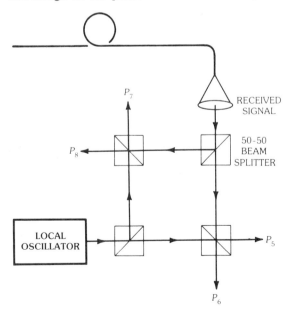

FIG. 7. Six-port phase-diversity receiver. (*After Walker and Carroll [17]. Acknowledgement is made of the prior publication of this material by the IEE*)

the stringent requirements necessarily placed on the phase and frequency stability of the signal and local oscillator. In a later section we shall also see another advantage to be gained from multiport detection—the ability to eliminate excess noise from the local oscillator through coherent subtraction in a two- (or more) port system.

4. System Components

In this section we shall discuss the basic elements of coherent optical transmission systems: optical source and local oscillator, optical fiber, and the detector-preamplifier.

The Optical Source and Local Oscillator

It was pointed out in Section 3 that a distinctive feature of the coherent detection process is that the phase and frequency information of the transmitted signal are retained in the mixing process. We shall also see in a later section that retention of this frequency and phase information allows receiver sensitivities better than the fundamental intensity (shot) noise limits in direct-detection systems. Such improvements are not achieved without a trade-off, and a coherent detection system will always have some penalty due to source frequency and phase noise reflected in a finite spectral line-width.

Perhaps the most critical system parameters in the evaluation of coherent optical communications systems are the line widths of the source and local oscillator. A number of techniques exist for line-width narrowing, most of which consist of modifying the laser cavity in some way. To understand these approaches it is helpful first to understand the mechanism of the laser line broadening, the important material and cavity parameters, and the subtle differences between different lasers. We will follow a simplified derivation due to Henry [19] which was applied to DFB lasers by Kojima and colleagues [20]. The starting point for such a discussion will be the cavity rate equations for the intensity and phase of the laser field, which are presented without proof, with the reader referred to [19].

$$d\phi/dt = \alpha(G-\gamma)/2 \qquad (16)$$

and

$$dI/dt=(G-\gamma)\,I. \qquad (17)$$

Here, $I(t)$ and $\phi(t)$ represent the intensity and phase of the laser field, respectively, G is the net rate of stimulated emission, and γ is the cavity loss due to end facets and waveguide losses. The term α represents the relative change of the real (dispersive) part of the refractive index $\Delta n'$ as influenced by a change in the imaginary (absorptive) part of the index $\Delta n''$, and is primarily a material constant (approximately 5 for InGaAs lasers). This parameter influences the extent to which a single spontaneous-emission photon can perturb the phase of the laser field through changes in the carrier density of the semiconductor. We wish to evaluate the root-mean-square phase error and laser line width due to this coupling.

Fig. 8 illustrates how a single spontaneous-emission event may instantaneously perturb the laser intensity and phase. The laser field has a normalized complex amplitude $\beta(t)$, so $I=\beta\beta^*$ represents the mean number of photons in the cavity, and

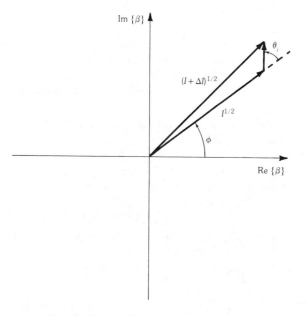

$$\beta=I^{1/2}e^{i\phi}. \qquad (18)$$

A single spontaneous-emission event will have unit amplitude (one photon) and random phase θ_i, yielding an instantaneous phase change $\Delta\phi_i$ of

$$\Delta\phi_i=I^{-1/2}\sin\theta_i \qquad (19)$$

and an instantaneous change in intensity of

$$\Delta I_i=1+I^{1/2}\cos\theta_i. \qquad (20)$$

We wish to evaluate the spontaneous-emission–induced phase shift, which must include not only the "instantaneous" phase shift but the associated "relaxational" phase shift due to the coupled rate equations (16–17) of the amplitude and phase.

The time-dependent factor $G-\gamma$ can be eliminated from (16) and (17) by combining the two equations:

$$d\phi/dI=\alpha/2\,I. \qquad (21)$$

The intensity in the denominator may be approximated as a constant (valid for lasers well above threshold) and the relaxational part of the phase shift $\Delta\phi_r$ may be evaluated in terms of the instantaneous shift in intensity ΔI_i due to the spontaneous emission, which relaxes to zero in the measurement interval t. Thus (21) may be integrated, yielding

$$\Delta\phi_r=-\alpha\,\Delta I_i/2\,I$$

$$=-\alpha(1+2\,I^{1/2}\cos\phi_i)/2\,I. \qquad (22)$$

The total phase change due to a single spontaneous-emission event is then

$$\Delta\phi=\Delta\phi_r+\Delta\phi_i$$

$$=-\alpha/2\,I+I^{-1/2}(\sin\theta_i+\alpha\cos\theta_i). \qquad (23)$$

This yields an average phase change for each spontaneous emission event of

$$\langle\Delta\phi\rangle=-\alpha/2\,I. \qquad (24)$$

In calculating the rms phase error σ_ϕ^2 for each

FIG. 8. Phasor diagram illustrating how individual spontaneous-emission events perturb the amplitude and phase of the laser field.

event, the cross terms $\langle \sin \theta_i \cdot \cos \theta_i \rangle$ vanish, yielding a root-mean-square phase fluctuation for each event of

$$\sigma_\phi{}^2 = (1+\alpha^2)/2\,I. \qquad (25)$$

The total phase drift of the laser field in a time interval t will be the sum of many independent spontaneous-emission events

$$\phi_t = \sum_{i=1}^{n} \; [-\alpha/2I + I^{-1/2}(\sin \phi_i + \alpha \cos \phi_i)] \qquad (26)$$

where n, the total number of spontaneous-emission events in the time interval t, will itself be a Poisson-distributed random variable. The randomness of n can be neglected for lasers well above threshold, and the rms phase deviation over the interval t becomes

$$\sigma_\phi{}^2\Big|_t = (1+\alpha^2)\,R\,t/2\,I, \qquad (27)$$

where R is the average rate of spontaneous-emission events. The phase statistics therefore have the characteristics of Brownian motion, or the one-dimensional random walk, whose probability density is a Gaussian distribution [21]

$$P(\phi;t) = (2\,\pi\,\sigma_\phi{}^2)^{-1/2} \; \exp{(-\phi^2/2\,\sigma_\phi{}^2)}. \qquad (28)$$

The power spectrum of such a signal is Lorentzian, with full width at half maximum of

$$\Delta\nu = R\,(1+\alpha^2)/4\,\pi\,I. \qquad (29)$$

The treatment so far has been made independent of any particular laser structure, but the important parameters in the laser line width are already clear. From experimental investigations, α appears to be primarily a material parameter and is not influenced a great deal by laser structure or cavity design. Therefore, to minimize the laser line width one must either

(a) maximize I, the total photon number in the cavity, or

(b) minimize R, the spontaneous-emission rate, or

(c) externally stabilize the carrier density to minimize amplitude-phase coupling.

The first can be accomplished by changing the cavity geometry so as to increase the total volume, increasing the energy stored per unit volume (such as increasing facet reflectivities), or increasing the output power. The second can be accomplished via the injection of an already stabilized laser source. Thirdly, the carrier density can be stabilized directly via feedback control on the drive current, effectively decreasing the coupling of the amplitude and phase of the laser field.

Henry [19] used (29) to successfully explain experimental data by Fleming and Mooradian [22] indicating that injection diode lasers exhibit line widths up to a factor of 50 above that predicted by conventional laser theory. The significant difference for the semiconductor laser lies primarily in the coupling coefficient α, which is negligible for most gas and nonsemiconductor solid-state lasers. The formula can be applied to conventional Fabry-Perot type laser structures by expressing the line-width in terms of specific cavity and material parameters:

$$\Delta\nu = h\,\nu\,v_g{}^2\,n_{sp}\;\gamma_m(\gamma_L+\gamma_m)\,(1+\alpha^2)/8\,\pi\,P_{out}, \qquad (30)$$

where

$$\gamma_m = L^{-1}\ln{(R_m{}^{-1})}.$$

Material parameters:

v_g = group velocity,

n_{sp} = spontaneous emission factor (approximately 2.7 for InGaAs lasers),

$h\nu$ = photon energy,

α = line-width enhancement factor.

Cavity parameters:

γ_m = distributed mirror loss,

R_m = mirror reflectivity,

γ_L = waveguide scattering losses,

L = cavity length,

P_{out} = output power per facet.

Conventional single-mode semiconductor lasers can typically achieve a line width of several hundred megahertz, which is unsuitable for most coherent systems. Some specific approaches to line-width narrowing are as follows.

Distributed-feedback (DFB) lasers have the attraction of providing high power (about 50 mW) with a comparatively stable, narrow spectral width (5 to 10 MHz typically) which is adequate for noncoherent ask and fsk detection schemes. As illustrated in Fig. 9, a DFB laser provides a cavity resonance via Bragg reflections from a waveguide grating. Kojima and colleagues [20] have shown that the line width for the DFB structure will theoretically follow (30) in the absence of excessive waveguide scattering losses, if the distributed facet reflectivity γ_m *is replaced by* $2\gamma_{th}$, where γ_{th} denotes the ideal threshold gain for the structure. The term γ_{th} is a function of the cavity length, waveguide coupling coefficient and detuning from the Bragg resonance frequency. In a well fabricated DFB laser, γ_{th} can be quite small, resulting in significant line-width reduction over the ordinary Fabry-Perot semiconductor laser. An increase in the length of the active region may provide some improvement, but waveguide scattering losses often become high enough to raise the threshold gain and negate any improvement in line width. For further line-width reduction it is therefore necessary to couple the laser to an external cavity.

function is to extend the photon round-trip time in the cavity using a fiber coupled to an anti-reflective-coated facet, or an integrated structure using planar or channel waveguides [23]. The new cavity line-width is that of the original structure modified by the ratio of the square of the photon round-trip times in each cavity. Thus the reduced spectral width is given by

$$\Delta\nu = \Delta\nu_0[(L_\alpha/v_g)/(L_\alpha/v_g + L_{ext}/v_{ext})]^2 ,$$

where L_α and v_g denote the length of the active region and group velocity in the laser structure, and L_{ext} and v_{ext} those of the external region. A 20-cm single-mode fiber acting as an external cavity for a buried-heterostructure laser can reduce the line width by 10^4.

Passive external cavities can also provide optical feedback to the shorter cavity of the semiconductor laser. In the optical feedback process the laser is injection locked at time t to its own instantaneous frequency at some earlier time, $t - \tau_c$, where τ_c is the round-trip time of the external cavity. This inhibits rapid frequency fluctuations in the laser and serves to narrow the line width. The simplest approach to optical feedback is usually a simple single-mode fiber with feedback provided from the reflection of the endface. Only moderate

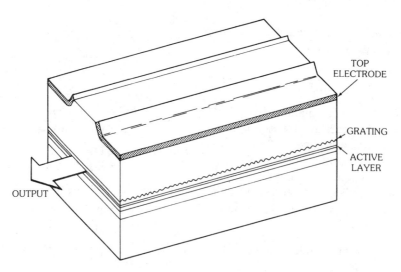

FIG. 9. Typical distributed-feedback laser.

Passive external cavities have been extensively studied for frequency stabilization. The essential

feedback levels can be used for simple constructions such as this, since the injection may reinforce small amplitude fluctuations in the laser and produce self-pulsations. It is possible to increase the optical feedback levels by introducing additional frequency selectivity into the external cavity, narrowing the bandwidth over which feedback is provided. A popular approach has been to use a grating to provide high levels of optical feedback in combination with a tuning etalon. Careful design can result in line-width narrowing to low as 10 kHz [24, 25].

Active external cavities, such as the cleaved coupled cavity laser [23] have also been used to provide a combination of spectral narrowing (several megahertz) and tunability. In practice, the laser line width is quite sensitive to the separation of the two cavities, and stable operation at optimum line widths is difficult to achieve. Long-term stability is the key practical consideration for source line-width narrowing and stabilization. Of all the external cavity approaches, integrated structures achieve the best immunity to mechanical disturbances and drift.

As was shown in (16) and (17) a major mechanism for line broadening is the coupling of amplitude and phase through relaxation oscillations in the carrier density of the semiconductor. Active electrical feedback can damp these oscillations if the drive current is adjusted in response to a phase-error signal. This approach is entirely equivalent to the optical phase-locked loop already discussed, and thus is limited to reduction of the noise inside the loop bandwidth. It is therefore often combined with other techniques, such as external feedback, which are more effective in eliminating high-frequency phase fluctuations.

Phase noise in the signal and local oscillator will degrade the performance of any coherent optical communication system by combining with receiver noise to increase the error rate of the received signal. For psk systems in particular, (28) leads to a minimum error probability (i.e., a bit error rate floor) possible given a particular spectral width and receiver configuration. For coherent demodulation, both fsk and ask systems also have such an error rate floor. Fig. 10 shows a plot of the error probability floor versus spectral width for a dpsk system (in which the time delay for the phase measurement is equal to one pulse period). For a bit error rate floor of 10^{-12} the combined spectral width of carrier and local oscillator should be approximately 0.8% of the transmitted bit rate.

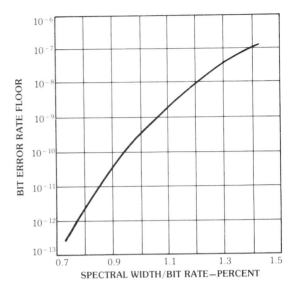

FIG. 10. Error probability floor versus spectral width to bit-rate ratio for a differential phase-shift keyed system.

Optical-Fiber Requirements

For a coherent system receiver to work properly the polarization states of the transmission signal and the local oscillator have to be nearly identical. Ordinary single-mode fibers actually propagate two orthogonally polarized modes which are degenerate under ideal conditions. This degeneracy can easily be broken by stresses or imperfections which break the cylindrical symmetry of the fiber. Any introduction of rotational asymmetry in the fiber will result in two orthogonally polarized modes with slightly different propagation constants. Such splitting is usually referred to as *birefringence.* These two modes need not be linearly polarized; in general, they are two elliptical polarizations. After a short distance the two modes differ by a small phase shift, resulting in yet another polarization state of the signal. If the relative delay between the two modes is less than the source coherence time, the modes add coherently and the signal output is *fully polarized.* In the opposite limit, in which the relative delay is longer than the source coherence time, the signal becomes *partially polarized* or *unpolarized,* the orthogonal modes having little or no statistical correlation.

The intentional introduction of birefringence has been used with some success to provide polarization stability—a large delay between the modes and increased loss for one polarization results in an output with a stable state of polarization. There are two major fabrication approaches for birefringence between linear polarizations; uniaxial internal stress can be introduced in the drawing process [26], or an asymmetric index profile can be introduced in the preform [27]. The latter approach typically suffers from high loss and inadequate polarization stability.

If the two modes are such that each electric-field vector traces a circular path around the fiber axis, the polarizations are referred to as *circular,* and may be decoupled using twisted single-mode fiber [27]. Such fiber would ease the polarization-matching requirements in fiber splicing and mixing with the local oscillator, but little experimental information is currently available on the loss and dispersion characteristics of such fibers.

The fabrication problems may be solved in the future; however, it is clear that a polarization matching method compatible with standard single-mode fibers would have significant advantages since it would allow the use of coherent optical systems in upgrading existing single-mode systems to higher bit rates or longer repeater spacings. In a polarization tracking receiver the polarization state of the incoming signal or the local oscillator is actively modified to maintain a match. Measurements of the polarization state on long cable links have shown that the polarization state drifts slowly [28] and that polarization-control devices should provide correction with adequate response time. Active polarization control can be accomplished using mechanical, electro-optical, or magneto-optical techniques. In all cases a polarization error signal must be generated and fed back to the polarization control device. The state of polarization (SOP) of an optical signal is described in general by the amplitude ratio of the x and y components of the electric-field vector, and their relative phase difference. If it were possible to examine the time evolution of the electric-field vector at a fixed point along the fiber axis, it would trace an ellipse in the plane transverse to the direction of propagation. There are then two parameters which vary randomly at the receiver: the ellipticity of the SOP and its orientation. The error signal must correct for both ellipticity and orientation if the signal and local oscillator are to have identical states of polarization.

The first (and simplest) polarization controllers used simple electromagnetic fiber squeezers to induce sufficient birefringence to correct the SOP. These early devices corrected only for the orientation of the SOP, and penalties were still incurred due to the ellipticity. The introduction of two error signals made possible the simultaneous correction of ellipticity and orientation. Other early mechanical approaches included rotatable phase plates between the fiber and beamsplitter to accomplish, in principle, the same effect as stress birefringence.

Integrated electro-optic polarization-control devices [29] can result in fast response time and high mechanical stability, although fabrication of such devices can be quite involved. It is also possible to use the Faraday effect to rotate the state of polarization via the introduction of a magnetic field in a glass fiber containing magnetically active impurities [30].

Active polarization correction may encounter difficulties in cases of extremely large signal bandwidths such as may be the case with optical frequency-division–multiplexed signals. Significant spectral variation in the state of polarization has been observed over relatively large bandwidths (approximately 100 nm) for short interaction lengths (60 cm) in polarizing couplers [31]. The spectral characteristics of random polarization conversion over lengths of several hundred kilometers have not yet been established, but could present a problem for fdm systems with active polarization correction.

An alternate approach would be to construct a polarization-insensitive receiver [32] as has been used in the past for microwave communication. Such an approach normally employs separate heterodyne detection for each polarization and recombines them in the receiver electronics. A penalty of up to 3 dB in receiver sensitivity is incurred, but the scheme avoids possible problems associated with active polarization correction.

The Detector-Preamplifier

A great deal of effort has been expended in recent years toward the development of high-speed detector-preamplifier combinations using both quaternary pin devices and avalanche photodiodes (APDs) in the 1.3- and 1.5-μm wavelength regions. The merits of each type of detector have been discussed in Chapters 11 and 12, but the emphasis in a coherent optical communication system may be somewhat different. The heterodyne optical receiver is insensitive to receiver noise in the limit of large local-oscillator power, so the avalanche photodiode does not offer the dramatic improvement over pin devices found with direct-detection systems. With a pin detector the dominant receiver noise source is found in the input stage of the preamplifier. The power penalty for a coherent receiver (in decibels) associated with receiver noise spectral density i_n^2 (in square amperes per hertz) is as follows:

$$\text{power penalty (dB)} = 10 \log_{10}(1 + i_n^2/2qRP_{\text{lo}}), \quad (31)$$

where q denotes the electron charge and R the detector responsivity (amperes per watt). This penalty is shown in Fig. 5 for a family of receivers with thermal noise present from the input impedance of the preamplifier. A local oscillator with sufficient power should be chosen to render the receiver noise negligible if the system is to be quantum limited. With high local-oscillator power, emphasis may be placed on high-speed circuitry, as has been developed for conventional microwave systems. Such high speed is necessary since the detection bandwidth of heterodyne receivers must typically be several times the baseband requirements of direct-detection or homodyne systems. If the local-oscillator power is limited, the use of an APD could result in improved receiver sensitivity, depending on the excess noise of the APD and available local-oscillator power.

An additional feature of coherent receivers lies in the fact that the detection process is linear with respect to the electric fields of the optical signal. This has a dual impact: First, the dynamic range of coherent receivers is automatically twice that of direct-detection counterparts. Second, the receiver may now include electrical equalization to compensate for the effects of fiber dispersion on the signal spectrum, which may be necessary as the signal bandwidths increase beyond several gigabits per second.

5. Receiver Sensitivity

Discussions of the receiver sensitivity of coherent optical communication systems can range from the simple classical model to the fully quantum optical picture; each has its important utilities. The semiclassical picture evaluates the receiver sensitivity essentially from three noise sources: the shot noise due to the detector photocurrent, the detector and preamplifier noise (thermal noise, detector leakage, and preamplifier channel noise), and the phase noise of the optical source and local oscillator. This picture neglects the sometimes crucial role of the statistical nature of the quantum light fields. Experimental and theoretical work [33] has indicated that the randomness in photodetection lies not primarily with the statistical generation of charge carriers in response to a given incident field, but from the statistics of the photon numbers themselves. Given that limitation, it is possible to extend the semiclassical treatment to allow for fluctuations in the intensities of the signal and local-oscillator signals. Using that approach we shall compare two possible detector configurations in a heterodyne detection scheme.

Finally, it is instructive to examine the fundamental limits to a coherent optical communication system imposed by the quantum nature of the electromagnetic field. It will be seen that the fundamental limits depend crucially on the quantum state of the field.

Receiver Error Probabilities: Amplitude and Phase Noise

Prior to a detailed discussion of the noise sources, we may compare the sensitivities of several modulation formats in the presence of additive, white, Gaussian-amplitude noise (of arbitrary source) and carrier phase noise as described in (28). For this purpose we write the homodyne output signal $x(t)$ as

$$x(t) = a(t) \cos \theta(t) + n(t) . \qquad (32)$$

Here, $n(t)$ is a zero-mean Gaussian random variable, and $a(t)$ is denoted as follows, according to the modulation format:

$$a(t) = \begin{cases} ask & psk \\ 1 & 1 \text{ ``mark''} \\ 0 & -1 \text{ ``space''} \end{cases}$$

The error criteria on $x(t)$ are as follows:

	ask	psk	
	$x < \gamma$	$x < 0$	"mark"
	$x > \gamma$	$x > 0$	"space"

The threshold γ for the ask case is set to provide equal error probabilities in the "mark" and "space" states. For psk the threshold is zero due to the symmetry of the noise sources. The probability for x to exceed the threshold in the signal-off case (ask) is

$$P(x \geq \gamma) = (2 \pi \sigma_n^2)^{-1/2} \int_\gamma^\infty \exp(-y^2 / 2 \sigma_n^2) dy$$

$$= 0.5 \text{ erfc } [\gamma / 2^{1/2} \sigma_n] . \qquad (33)$$

For the "mark" case the probability for $x(t)$ to go below the decision threshold is obtained from the joint probability of $\theta(t)$ and $n(t)$:

$$P(x \leq \gamma) = \int_{-\infty}^\infty p(\theta) P(n \leq \gamma - \cos \theta) d\theta$$

$$= (2 \pi \sigma_n^2)^{-1/2} \int_{-\infty}^\infty p(\theta) d\theta$$

$$\times \int_{-\infty}^{\gamma - \cos \theta} \exp(-y^2 / 2 \sigma_n^2) dy$$

$$= 2^{-1} (2 \pi \sigma_\theta^2)^{-1/2} \int_{-\infty}^\infty \exp(-\theta^2 / 2 \sigma_\theta^2)$$

$$\times \text{ erfc } \left[(\cos \theta - \gamma) / 2^{1/2} \sigma_n \right] d\theta , \quad (34)$$

where

$$\text{erfc } (X) = (2 / \pi)^{1/2} \int_X^\infty \exp(-y^2) dy .$$

Equations 33 and 34 form coupled equations for the error probability and decision threshold and, as such, must be solved numerically. In the absence of phase noise, $\gamma = 0.5$ and we obtain an expression for the error probability P_e of an ask homodyne receiver in terms of the single-to-noise ratio (snr$= 1 / \sigma_n^2$ for the normalized signal):

$$P_e = 0.5 \text{ erfc } [(0.5 \text{ snr})^{1/2} / 2] . \qquad (35)$$

For psk homodyne we can consider the error probability in the "mark" case by setting $\gamma = 0$ in (34):

$$P(x \leq 0) = 2^{-1} (2 \pi \sigma_\theta^2)^{-1/2} \int_{-\infty}^\infty \exp(-\theta^2 / 2 \sigma_\theta^2)$$

$$\times \text{ erfc } [\cos \theta / 2^{1/2} \sigma_n] d\theta .$$

For the case of zero phase noise this becomes

$$P_e = 0.5 \text{ erfc } [(0.5 \text{ snr})^{1/2}] . \qquad (36)$$

Similar calculations can be performed for the heterodyne cases—the error rates in the absence of phase noise (for ideal binary receivers) are given in Table 1. It is evident from this brief analysis that, within the semiclassical approximation, the relative sensitivities of different modulation schemes are exactly the same as with microwave and rf communications.

We can also consider the influence of phase noise alone in each case. For psk, in the limit $\sigma_n \to 0$, equation 36 becomes

$$P_e = 2 (2 \pi \sigma_\theta^2)^{-1/2} \int_{\pi/2}^\infty \exp(-\theta^2 / 2 \sigma_\theta^2) d\theta$$

$$= \text{erfc } [\pi / 2^{3/2} \sigma_\theta] . \qquad (37)$$

This represents a "floor" to the bit error rate for psk in the presence of phase noise. A quick examination of (33) and (34) reveals that the bit error rate floor for ask approaches that of psk, provided the threshold γ is permitted to become vanishingly small, an awkward requirement for receiver design. If, as is often the case, the decision threshold is fixed at some finite value, the bit error rate floor for ask becomes

$$P_e = 2 (2 \pi \sigma_\theta^2)^{-1/2} \int_{\cos^{-1} \gamma}^\infty \exp(-\theta^2 / 2 \sigma_\theta^2) d\theta$$

$$= \text{erfc } \{\cos^{-1} [\gamma / 2^{1/2} \sigma_\theta] \} . \qquad (38)$$

TABLE 1. Sensitivity of Optical Receivers*

Modulation/Detection Type	P_e	Number of Photons for P_e of 10^{-9}, $\eta=1$
Ask heterodyne	$0.5 \, \mathrm{erfc} \, (\eta \, P_s/4 \, h \, \nu \, B)^{1/2}$	72
Ask homodyne	$0.5 \, \mathrm{erfc} \, (\eta \, P_s/2 \, h \, \nu \, B)^{1/2}$	36
Fsk heterodyne	$0.5 \, \mathrm{erfc} \, (\eta \, P_s/2 \, h \, \nu \, B)^{1/2}$	36
Psk heterodyne	$0.5 \, \mathrm{erfc} \, (\eta \, P_s/h \, \nu \, B)^{1/2}$	18
Psk homodyne	$0.5 \, \mathrm{erfc} \, (2 \, \eta \, P_s/h \, \nu \, B)^{1/2}$	9
Direct detection quantum limit	$0.5 \, \exp \, (-\eta \, P_s/h \, \nu \, B)$	21
practical receiver		400–4000

*The minimum detectable (peak) power level for a 10^{-9} error rate is obtained by multiplying the number of photons with $h \nu B/\eta$, where B is the bit rate and η is the detector quantum efficiency and $h \nu$ is the energy of a photon. P_s is the peak signal power of the received signal.

As noted in Section 4, under "Optical Source and Local Oscillator," the phase noise σ_θ^2 is proportional to the combined spectral width of the signal and local oscillator, and to the time delay associated with the phase estimation/recovery circuit. For first-order phase-locked loops the phase error variance can be expressed in terms of the ratio of combined spectral width to loop bandwidth B_L [10]:

$$\sigma_\theta^2 = \Delta\nu/2 \, B_L \, .$$

It therefore represents a performance limit imposed by the source spectral width on the bit error rate of an optical system.

The expression (36) for the error probability of the psk system is not the most general. The rms phase error was assumed to depend only on the combined spectral width of the signal and local-oscillator sources $\Delta\nu$, and on the bandwidth B_L of the phase-locked loop. In this idealization, B_L should be made as large as possible to adequately track high-frequency phase fluctuations. On the other hand, for minimum fluctuation in the phase of the recovered carrier due to shot noise, the loop bandwidth should be as small as possible. If the system has only receiver noise (i.e., the uncertainty in the recovered phase is only caused by the shot noise and not by the finite line width of the transmitter and local-oscillator lasers), calculations have shown [34] that a 10% ratio of loop to system bandwidth is sufficient for near-ideal loop performance. Proper loop design will optimize the bandwidth to achieve a trade-off between the influences of carrier phase noise and shot noise in the system [14].

The phase statistics, described first by (28) are assumed to retain their Gaussian character through the carrier recovery process. This is consistent with the linearized treatment often used to predict phase-locked loop performance, but may often be unjustified with the highly nonlinear recovery systems which have been developed for psk demodulation. We can write the psk error probability (35) in terms of a more general probability density function (pdf) for the phase error $p(\theta)$:

$$P_e = 0.5 \int_{-\infty}^{\infty} p(\theta) \, \text{erfc} \, [\cos \theta / 2^{1/2} \sigma_n] \, d\theta . \quad (39)$$

The exact form of the pdf will depend on the loop type and the signal-to-shot-noise ratio [15]. A distribution which was shown to be valid for many first- and second-order loops (including data-aided and Costas loops with zero detuning) is the so-called truncated Gaussian:

$$p(\theta) = \exp \, (S_\ell \cos \theta) / 2 \, \pi \, I_0(S_\ell) ,$$

where S_ℓ denotes the signal-to-noise ratio in the loop bandwidth B_L and $I_0(S_\ell)$ is the Bessel function with imaginary argument.

Fig. 11 shows the bit error rate (BER) or receiver error probability P_e as a function of amplitude and phase noise for the two different

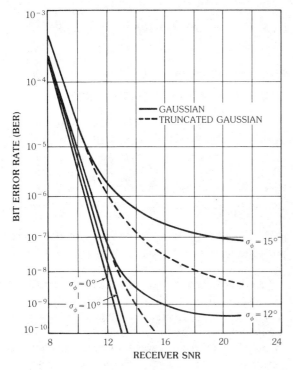

FIG. 11. Receiver error probability as a function of receiver signal-to-amplitude noise ratio and rms phase error for the Gaussian and truncated Gaussian distributions. (*After Prabhu [35]*, ©1976 IEEE)

types of distributions [35]. In the case of small phase noise ($\sigma_\theta \leq 10°$) the truncated Gaussian approximates a Gaussian distribution; this is the region in which linearized treatments are valid. In this case the penalty to the receiver sensitivity due to phase noise is small (less than or equal to 1 dB). It should also be noted that in general for a given degradation in system performance the criterion on the optical source line width varies as a function of the specified bit error rate.

A number of theoretical treatments have dealt with demodulation in the presence of phase noise both at microwave and optical frequencies. As mentioned, microwave design criteria require rms phase deviations σ_θ at less than 10° in order to minimize the penalty due to shot and phase noise in the carrier recovery process. The direct calculation of the penalties associated with phase noise in coherent optical receivers has proved to be an elusive task, and theoretical predictions currently vary significantly in maximum spectral width permissible for various modulation/detection schemes. We will look at a few of the approaches taken. The results of Garrett [18] for ask and fsk systems employing noncoherent detection indicated negligible penalties (less than 1 dB) for these systems at spectral widths up to 5% of the transmitted data rate, and found reasonable agreement with experiment.

Calculations of spectral-width requirements for systems employing coherent demodulation have also been performed. The treatment by Kikuchi and colleagues [10] is conceptually the simplest, assuming Gaussian statistics for the phase error after tracking, and a linearized phase-locked loop in assessing the rms phase error and bit error rate degradation for ask, fsk, and psk schemes. This approach neglects the effects of shot noise on the carrier recovery process due to finite loop bandwidth, and therefore does not attempt to optimize the loop bandwidth but leaves it open as a system parameter. We can therefore assume, for comparison, the loop bandwidth B_L to be 10% of the data rate B for electrical phase-locked loops. For optical phase-locked loops the loop bandwidth may be somewhat narrower due to the delay in electrical feedback frequency-correction schemes. Practical loop bandwidths for optical homodyne schemes may typically be 1% of the data rate or less. For these cases we can let $\sigma_\theta^2 = \Delta\nu/2B_L$, and the results of [10] then indicate that

$$\Delta\nu/B \leq 0.35\%$$
(ask coherent with electrical pll),

$$\Delta\nu/B \leq 0.035\%$$
(ask coherent with optical pll),

$$\Delta\nu/B \leq 0.5\%$$
(psk coherent with electrical pll),

$$\Delta\nu/B \leq 0.05\%$$
(psk coherent with optical pll),

The different results for ask and psk schemes in this model are due to different threshold requirements in each case; just as coherent ask requires a higher signal-to-shot noise ratio for a given BER, it also requires a higher signal-to-phase-noise ratio as expressed in the ratio of data rate to spectral width. For this reason, ask systems only offer advantages when used with envelope detection.

Kazovsky [36] performed linearized analyses of second-order data-aided loops, in which an optimization of the loop bandwidth was performed in order to achieve the most reliable phase lock. The following conditions predicted a 1-dB power penalty:

$$\Delta\nu/B \leq 0.031\%$$
(optical homodyne ask and psk)

$$\Delta\nu/B \leq 0.23\%$$
(psk heterodyne with electrical pll)

For the latter the optimum bandwidth was found to be $B_L = 0.35 B$, in contrast to the assumed $0.1 B$ of the earlier cases. The validity of such linearized pll treatments in systems employing wide loop bandwidths remains an open question. Thus it appears that the superficial agreement shown in the previous approaches hides at least one underlying inconsistency which has not yet been resolved.

Nicholson [37] extended the model of [10] to include the effects of receiver filtering and inter-symbol interference yielding for a 1-dB penalty in a psk heterodyne system

$$\Delta\nu/B \leq 0.065\%$$
(psk heterodyne)

Salz [38] applied solutions of the Fokker-Planck equation describing the nonlinear dynamics of a data-aided loop to the problem of psk system performance. An upper bound on the error rate yielded the following criteria for a received power penalty of less than 1 dB:

$$\Delta\nu/B \leq 0.67\%$$
(psk heterodyne)

$$\Delta\nu/B \leq 0.1\%$$
(psk homodyne)

It is not yet clear why the discrepancy between [10] and [37] should be so large, considering the similarity of the approaches, and it appears at the time of this writing that there is not universal agreement on the theoretical performance of receivers employing coherent demodulation techniques. It is evident, however, that psk systems will require spectral widths somewhat less than 1% of the data rate for reliable system performance.

Semiclassical Evaluation of Receiver Sensitivity

It is helpful in many circumstances to use a measure of receiver sensitivity that is independent of data rate, and to specify the number of photo-generated electrons required at the amplifier input to achieve a specified error probability. We shall first consider the semiclassical regime, where we consider the optical fields to be continuous and the charge flow in the detector to be discrete and random, statistically following the Poisson distribution [39]:

$$p(n) = N^n e^{-N}/n! \,, \qquad (40)$$

where N is the mean number of electrons generated per bit interval.

The semiclassical error probability for ideal direct detection is then simply the probability of detecting no electrons at the input to the pre-amplifier:

$$P_e = e^{-N}.$$

An average of 21 photogenerated electrons are required per bit to ensure an error probability of less than 10^{-9}. Optical energy is not continuously absorbed but arrives at the detector in discrete quanta, known as photons. One can then describe the sensitivity of an ideal receiver in terms of the number of photons per bit incident on a unit–quantum-efficiency photodetector required to achieve a specified bit error rate. An ideal direct-detection system then requires an average of 21 photons per bit to achieve a 10^{-9} bit error rate. In any practical receiver the amplification process following photodetection adds thermal and other circuit noise that requires an increase in the received power level from 13 to 20 dB above this quantum limit.

We shall now calculate the semiclassical receiver sensitivity for homodyne and heterodyne detection, and make comparisons between coherent and direct-detection schemes for the case of an ideal detector. The photocurrent due to an incident signal of optical power $P(t)$ can be expressed in terms of the detector quantum efficiency η and the photon energy $h\nu$:

$$i = q\eta P / h\nu, \qquad (41)$$

where q is the electron charge.

In the semiclassical picture we assume ideal classical fields excite quantum-mechanical charge carriers in such a way that the number of charge carriers present in a short time interval is random, and can be described by the Poisson distribution (40). Such a process is known as a shot-noise process—it is δ-correlated (white noise) with spectral density $2q\langle i_s \rangle$. The output i_s from a realistic receiver can be described by adding a randomly fluctuating noise current i_d which may represent thermal noise, amplifier noise, or detector leakage current:

$$i_s = q\eta P / h\nu + i_d. \qquad (42)$$

For the case of direct detection the optical power P incident on the detector is simply P_s, the signal power arriving at the receiver. For coherent detection the signal and local-oscillator fields interfere at the surface of the detector and produce beat signals at the difference frequency ω_{if}:

$$P = P_{lo} + P_s + 2(P_{lo}P_s)^{1/2}\cos\omega_{if}t. \qquad (43)$$

To calculate the semiclassical signal-to-noise ratio we must compute the signal power in the intermediate-frequency band (or baseband, in the case of homodyne detection), and the mean-squared fluctuations of the output current i_s. We will use brackets to denote expectation values; the term σ_s^2 will denote the mean-squared fluctuations in output current, and σ_d^2 will denote the detector noise. For homodyne detection, $\omega_{if} = 0$ and

$$\langle i_s \rangle = 2q\eta (P_{lo}P_s)^{1/2}/h\nu. \qquad (44)$$

For the shot-noise contribution we assume that the noise bandwidth is one-half of the data rate B, and consider the case of high local-oscillator power such that the dominant shot-noise contribution arises from the first term in (43).

Hence

$$\sigma_s^2 = q^2\eta P_{lo}B/h\nu + \sigma_d^2. \qquad (45)$$

The receiver noise σ_d^2 can be expressed in a more meaningful way as the product of a noise spectral density N_0 and the noise bandwidth $B/2$. The signal-to-noise ratio becomes

$$(\mathrm{snr})_{\mathrm{hom}} = 4\eta P_s / [h\nu B$$
$$\times (1 + N_0 h\nu/2\eta q^2 P_{lo})]. \qquad (46)$$

We are generally interested in the limit of high local-oscillator powers, in which case the homodyne signal-to-noise ratio becomes

$$(\mathrm{snr})_{\mathrm{hom}} = 4\eta P_s / h\nu B. \qquad (47)$$

For heterodyne detection the shot-noise power in the receiver band remains the same, but the signal power is reduced by 3 dB. If we denote the time average over one cycle of the intermediate frequency with an overbar, then

$$\langle \overline{i_s} \rangle^2 = [4\eta^2 P_{lo}P_s/(h\nu)^2]\,\overline{\cos^2 \omega_{if}t}$$
$$= 2\eta^2 P_{lo}P_s/(h\nu)^2. \qquad (48)$$

The signal-to-noise ratio for heterodyne detection becomes

$$(\mathrm{snr})_{\mathrm{het}} = 2\eta P_s / [h\nu B(1 + N_0 h\nu/2\eta q^2 P_{lo})]. \qquad (49)$$

And in the limit of large local oscillator power:

$$(\text{snr})_{\text{het}} = 2 \eta P_s / h \nu B . \qquad (50)$$

We can now consider the case of unit quantum efficiency—one photoelectron for each unit of optical energy ("photon") arriving at the receiver. The average number of photons per pulse in a digital system is given by

$$N_s = P_s / h \nu B ,$$

and the signal-to-noise ratio limit for a unit-quantum-efficiency detector becomes

$$(\text{snr})_{\text{hom}} = 4 N_s \qquad (51)$$

and

$$(\text{snr})_{\text{het}} = 2 N_s . \qquad (52)$$

If we combine these results with the error probabilities for ask, psk, and fsk calculated earlier, we can compare the receiver sensitivity for different modulation/detection schemes. Table 1 shows such a comparison, and represents the fundamental limits for receiver sensitivity using coherent communications or direct detection within the semiclassical framework.

In the discussion regarding the sensitivity of direct-detection receivers it was shown that for an average of 21 photons per bit duration, one measurement out of 10^9 would yield no photons. As shown in Table 1, psk homodyne is the most sensitive of the binary coherent transmission schemes and requires an average of nine photons per bit interval to yield an error rate of 10^{-9}. This represents something of a paradox, since a direct, ideal measurement of the number of photons per bit interval in such a signal would yield no photons in one interval out of 10^4. Experimental studies, however, have verified that extremely weak optical fields can form interference fringes and produce beat signals even when there is a very low probability of photons from both sources coincidentally striking the detector [40]

Such a result immediately shows the inability of the semiclassical picture to adequately describe the coherent detection process. We have, in fact, not yet truly defined what is meant by a "photon"—it was merely described as the amount of optical energy required to produce a single charge carrier (or carrier pair) at the detector. The effect of *interference* on photon statistics, which is of significance in coherent detection, can only be described by a model that treats not only the detector but the optical field itself quantum mechanically.

An additional inadequacy of the semiclassical picture lies in its interpretation of the shot-noise process. As mentioned earlier, experimental and theoretical investigations indicate that the fundamental quantum fluctuations in photodetection

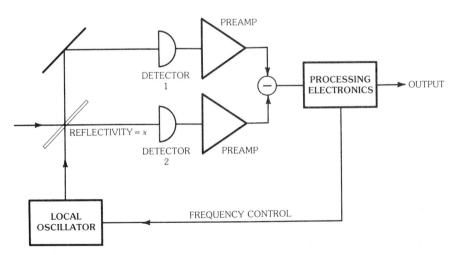

Fig. 12. Two-port balanced-mixer detection scheme. For noise cancellation the detector quantum efficiencies η_1 and η_2 and coupling coefficient κ must be chosen so that $\eta_1(1-\kappa) = \eta_2\kappa$.

are dominated by intrinsic fluctuations in the optical fields. We therefore proceed to a model in which the detectors are ideal but the optical fields fluctuate. In the following derivation, in order to maintain simplicity, we shall consider only the case of unit quantum efficiency. However, as far as is reasonable in the development, we shall keep the character of the fluctuations quite general, treating the case of Poisson statistics in the signal and local oscillator as a special case.

Any form of coherent optical detection requires phase matching between the spatial modes of the signal and local-oscillator fields; the most common configurations use either a beamsplitter or single-mode waveguide coupler. Two output ports are available, the intensities of which are determined by the coupling coefficient κ, and which differ by a π-radian phase shift in the interference term. Most experimental configurations have used only one port for the detection of the heterodyned or homodyned signal, as illustrated in Fig. 1.

The balanced mixer has often been used in microwave communications to suppress local-oscillator fluctuations, and it can also be applied to coherent optical detection schemes as illustrated in Fig. 12. Using one detector at each port of the beamsplitter/coupler and coherently subtracting the two signals, it is possible to eliminate the large dc contribution from the local oscillator and greatly reduce, if not eliminate, the associated fluctuations [41, 42, 43].

Equation 43 can be modified to include the effects of the beamsplitter/coupler as follows:

$$P = \kappa P_{lo} + (1-\kappa)P_s$$

$$+ 2[\kappa(1-\kappa)P_{lo}P_s]^{1/2} \cos \omega_{if} t. \quad (53)$$

We also introduce lowercase p_s and p_{lo} to indicate the fact that the signal and local-oscillator intensities are random variables representing the fluctuating intensities of the signal and local oscillator. The uppercase P_{lo} and P_s will denote the mean intensities. Once again we must compute the signal power and the mean-squared fluctuations of the output current i_s, and since we are dealing with statistical fields, σ_p^2 will denote the variance of the total optical power incident on the detector:

$$\sigma_s^2 = (q/h\nu)^2 \sigma_p^2 + \sigma_d^2 \quad (54)$$

and

$$\langle i_s \rangle = q P/h\nu + \langle i_d \rangle. \quad (55)$$

It is difficult to calculate the expectation values rigorously for the square roots of the local-oscillator power and the signal power (these are, in fact, the expectation values for the envelope of the electric field). It is possible, however, to use an expansion of the square-root valid for high signal-to-noise ratios. If P denotes the expectation value for p, and the fluctuations of p about its mean are small, then

$$p^{1/2} = P^{1/2}[1 + (p-P)/P]^{1/2}$$

$$\simeq P^{1/2}[1 + (p-P)/2P + (p-P)^2/8P^2].$$

We can use this expansion for both signal and local-oscillator powers. While it is sufficient to retain only the second term in expanding p_{lo}, the third term will be retained for the signal, which has a lower signal-to-noise ratio. The mean and variance of the power incident on the detector can then be written

$$\langle p \rangle = \kappa P_{lo} + (1-\kappa)P_s$$

$$+ 2[\kappa(1-\kappa)P_{lo}P_s]^{1/2} \cos \omega_{if} t \quad (56)$$

$$\sigma_p^2 = \kappa^2 \langle(\Delta p_{lo})^2\rangle + (1-\kappa)^2 \langle(\Delta p_s)^2\rangle$$

$$+ 2[\kappa(1-\kappa)P_{lo}P_s]^{1/2} \cos \omega_{if} t$$

$$+ 2[\kappa(1-\kappa)P_{lo}P_s]^{1/2}[(1-\kappa)\langle(\Delta p_s)^2\rangle/P_s$$

$$+ \kappa \langle(\Delta p_{lo})^2\rangle/P_{lo}] \cos \omega_{if} t$$

$$+ \kappa(1-\kappa)[P_{lo}\langle(\Delta p_s)^2\rangle/P_s] \cos^2 \omega_{if} t, \quad (57)$$

where $\langle(\Delta p_s)^2\rangle$ and $\langle(\Delta p_{lo})^2\rangle$ denote the mean-squared fluctuations in the signal and local oscillator, respectively.

Since we are primarily interested in the effect of fluctuations in local-oscillator power, we can divide the local-oscillator fluctuations into two contributions. The first represents the intrinsic fluctuations of any ideal laser far above threshold (a result of the Poisson statistics of the field described earlier). The presence of spontaneously emitted photons usually results in fluctuations exceeding those predicted by the Poisson distribution, and this contribution can be referred to as excess noise N_{exc}:

$$\langle(\Delta p_{lo})^2\rangle = P_{lo}\,h\,\nu\,B + N_{exc}\,,$$

where B is the data rate. If we now consider the case of high local-oscillator power, $P_{lo} \gg P_s$, the mean-squared fluctuations in signal current reduce to

$$\sigma_s^2 = (q/h\nu)^2\,\sigma_p^2\,, \tag{58}$$

where

$$\sigma_p^2 = \kappa^2\,(P_{lo}\,h\,\nu\,B + N_{exc})$$
$$+ \kappa\,(1-\kappa)\,P_{lo}\langle(\Delta p_s)^2\rangle/P_s\,. \tag{59}$$

When the local-oscillator power is large enough to achieve the "shot-noise limited" condition, $\sigma_d^2 \ll (Q/h\nu)^2\sigma_p^2$ and can be neglected. The above expression assumes the power spectrum of signal fluctuations to be broadband compared with the intermediate frequency. This yields a signal power-to-noise power ratio for one-port heterodyne detection of

$$(snr)_{\text{one port}} = 2\,P_s/[\,\langle(\Delta p_s)^2\rangle/P_s + \kappa/(1-\kappa)$$
$$\times\,(h\nu B + N_{exc}/P_{lo})]\,. \tag{60}$$

Two observations can be made:

1. The relative contributions of signal and local-oscillator fluctuations depend critically on the coupling coefficient

2. For the case of a Possonian input signal ($\langle(\Delta p_s)^2\rangle = P_s\,h\,\nu\,B$), and no local-oscillator excess noise, we obtain the simplest result

$$(snr)_{\text{one port}} = 2\,(1-\kappa)\,P_s/h\,\nu\,B\,. \tag{61}$$

For the balanced-mixer configuration the analysis is similar. The photocurrent from each detector

can be subtracted as shown in Fig. 12. For coherent subtraction of the local-oscillator shot-noise fluctuations, the relative time delay between the two arms must be much less than the coherence time of the fluctuations after filtering in the if channel. The resultant signal is

$$i_s = (2\,q/h\,\nu)\,2\,[\kappa\,(1-\kappa)\,p_{lo}\,p_s]^{1/2}\cos\omega_{if}\,t\,.$$

The mean and variance can be calculated as before, yielding a signal-to-noise ratio

$$(snr)_{\text{two port}} = 2\,P_s^2/\langle(\Delta p_s)^2\rangle\,. \tag{62}$$

For the special case of Poisson statistics on the input signal

$$(snr)_{\text{two port}} = 2\,P_s/h\,\nu\,B\,. \tag{63}$$

There are some important observations to be made:

1. The receiver performance is no longer dependent on the coupling coefficient, as was the case in the single-port configuration. The performance of the latter improves with small κ. However, the signal-to-noise ratio for the single-port configuration is valid only in the limit of large local-oscillator powers,

$$\kappa\,P_{lo} \gg (1-\kappa)\,P_s\,.$$

This may put severe demands upon the amount of local-oscillator power required to achieve quantum-limited performance

2. Excess noise in the local oscillator no longer degrades the receiver performance, further relaxing the requirements on the local oscillator. It is therefore clear that the balanced-mixer configuration can most easily achieve quantum-limited performance

In addition, we must note the interesting result that, for the special case of Poisson statistics for the signal and local oscillator, the results for the two-detector configuration are identical with the semiclassical shot-noise limit (when the presence of the beamsplitter is ignored, and unit quantum efficiency is assumed). The two results differ, however, if the input signal has non-Poisson statistics, either in the case of "excess noise" on the signal

$$\langle (\Delta p_s)^2 \rangle > h \nu B P_s , \qquad (64)$$

or, for so-called sub-Poisson light [44]

$$\langle (\Delta p_s)^2 \rangle < h \nu B P_s . \qquad (65)$$

The latter does not exist classically, and therefore the conditions under which a light field will be sub-Poisson can be described only through the tools of quantum optics.

Quantum Statistics

The following section introduces some concepts in quantum optics which are relevant to the study of coherent optical communications. Most treatments of this type are developed in terms of an operator algebra unfamiliar to many engineers and scientists accustomed to dealing with the formalism of classical electrodynamics. Here these concepts are expressed in somewhat more familiar language.

We have already found that the limits to the sensitivity of a coherent system are dependent on the modulation method and detection configuration. We have also calculated these sensitivity limits using semiclassical assumptions, and found signal-to-noise ratios for two different receiver configurations. The validity of the semiclassical picture depends primarily on the photodetection statistics of the quantum field—the signal and local oscillator are assumed to exhibit Poisson statistics and the resultant intensity of the two interfering signals is assumed to be Gaussian. In quantum optics the photon statistics are determined primarily by the "quantum state" of the field.

States of the Quantum Field—An electromagnetic disturbance in any medium can be resolved as a superposition of orthogonal field modes, each characterized by a propagation constant k and a polarization index s. For the case of an electromagnetic wave in free space these modes form a continuum; for a waveguide there generally exists a finite set of guided modes and an infinite set of unbound, or radiative, modes. Each mode is characterized by a dispersion relation $\omega(k)$ relating the angular frequency (or energy) to the propagation constant. For waves in a homogeneous, isotropic medium the dispersion relation is defined by the bulk index of refraction $n(\omega)$ as follows:

$$\omega(k) = c k / n .$$

The dispersion relation for a guided mode must be obtained through a solution of Maxwell's equations.

The simple harmonic time dependence of the electric-field amplitude in each mode is analogous to a harmonic oscillator of angular frequency ω, and can be quantized as such; the allowed energy states, as with the quantum-mechanical harmonic oscillator, are separated by $h\nu$, and these quanta of energy are referred to as photons. For electromagnetic waves at radio and lower microwave frequencies the photon energy is far too small to be detected,* and quantum effects are generally unimportant in those systems; for optical fields, however, the photon energy is on the order of 1 eV and quantum effects are considerable.

If the optical field existed in a pure (deterministic) energy state, a measurement of the energy, or photon flux, would always yield the same result. This is never the case, however. Even a well-stabilized laser source yields fluctuations in the measured field energy. Indeed, all ordinary optical fields exist in a state which is a weighted mixture of energy states.

A state of the field most closely resembling a monochromatic classical field is referred to as a coherent state. Such states are described by a complex amplitude $\nu(t)$, which is equivalent to the classical analytic signal (complex envelope) representation. A measurement of the photon number when the field is in a coherent state is statistically described by the Poisson distribution as follows:

$$p(n) = e^{-N} N^n / n! ,$$

---

* This is due to the fact that the background energy in each mode, kT is much greater than the photon energy; large numbers of photons per mode must be present for a detectable signal. In the limit of large photon numbers, quantum fields go to their classical limit.

where $N=|\nu|^2$ is the average number of photons in the mode. A more detailed description of the coherent state is given in [45]. An ideal single-mode laser approaches a coherent state when the phase fluctuations due to spontaneous emission are extremely small—this occurs when the laser is far above threshold. Actual sources, and semiconductor lasers in particular, may deviate considerably from this description, and a more general description may be appropriate.

For states of the field other than pure coherent states it is often possible to form a superposition of coherent states weighted over a function of the complex amplitudes $\nu(t)$; the function $\Phi(\nu,\nu^*)$ is alternately referred to as a P-representation or a phase-space density [46]. A detailed treatment of the phase-space density is beyond the scope of this chapter, but certain properties relevant to coherent optical transmission can be described. The phase-space density exhibits the behavior of a generalized probability distribution of the complex envelope $\nu(t)$.

According to classical statistics a probability distribution $p(\nu)$ must be nonnegative, normalized, and must exhibit nothing more singular than a δ-function. In many cases the phase-space density is equivalent to a classical probability distribution for the complex amplitude—most fields having a classical analog can be described in this way. In some instances, however, it may become singular or negative—the fields it then describes are true quantum fields and are not physically describable using classical electrodynamics. For example, the sub-Poisson statistics mentioned earlier can arise only from a highly singular phase-space density.

The phase-space density for a pure coherent state of complex envelope $\nu_0(t)$ can be written

$$\phi(\nu;t)=\delta(\text{Re }\{\nu\}-\text{Re }\{\nu_0\})\,\delta(\text{Im }\{\nu\}-\text{Im}\{\nu_0\})$$

$$\equiv\delta^2(\nu-\nu_0)\,.$$

Phase-space densities for two other types of fields are listed in Table 2.

For coherent detection systems we are concerned with the superposition of two statistically independent optical sources on a single spatial mode of the field. The phase-space density for the resultant field can then be written as the convolu-

TABLE 2. Phase-Space Densities

Radiation	Phase-Space Density
Thermal light	$\phi(\nu)=\exp\left(-|\nu/\nu_0|^2\right)/\pi|\nu_0|^2$
Randomly phased single-mode laser	$\phi(\nu)=\delta(|\nu|-|\nu_0|)/2\,\pi\,|\nu_0|$

tion of the signal and local-oscillator phase-space densities:

$$\phi(\nu;t)=\int\!\!\int\phi_{\text{sig}}(\nu'-\nu;t)\,\Phi_{\text{lo}}(\nu'e^{i\omega_{\text{if}}t})\,d^2\nu',\quad(66)$$

where ω_{if} denotes the angular-frequency difference between the two modes. It is now possible to examine the detection statistics in terms of the phase-space densities of the signal and local oscillator.

There is an important class of states which cannot be described using an ordinary phase-space density. These are the so-called squeezed states [47], in which the uncertainty of one quadrature component of the electric field is reduced at the expense of the other. Such a state can be thought of as having a definite (deterministic) complex amplitude, but whose quadrature components fluctuate according to the minimum uncertainty condition. If the field has quadrature components p and q,

$$\text{Re }\{\nu\}=p\sin\omega t+q\cos\omega t\,,$$

where $\langle p^2\rangle+\langle q^2\rangle=N,$ the average photon number, minimum uncertainty requires that [48]

$$\langle(\Delta p)^2\rangle\langle(\Delta q)^2\rangle=1/16.\quad(67)$$

If $\langle(\Delta p)^2\rangle=\langle(\Delta q)^2\rangle$, the field is said to be in an ordinary coherent state. If the equality does not hold, however, the state is said to be squeezed. Communication using an optical field in such a state offers the theoretical possibility of quantum-noise reduction if the low uncertainty quadrature component is modulated and homodyne detection is employed [49]. The generation of light in such quantum states requires two-photon processes [50], such as a combination of parametric down-conversion and second-harmonic generation [51], and so presents significant experimental difficulties. Such states would, however, have quantum limits significantly different from coherent states.

Quantum Detection Statistics—We now wish to describe the detection statistics of two interfering light beams. The rate of generation $W(\nu;t)$ of photoelectrons in a detector with resolving time τ can be written

$$W_\tau(\nu;t) = (\eta/\tau) \int_t^{t+\tau} |\nu(t')|^2 \, dt' . \tag{68}$$

For signals changing slowly compared to the resolving time τ, we have

$$W_\tau(\nu;t) = \eta \, |\nu(t)|^2 = \eta \, N(t) .$$

The phase-space density defines a quasi-probability density over all possible coherent states; hence we can express the photodetection statistics for the field in an arbitrary state by performing a weighted average over the Poisson distribution:

$$p_\tau(n;t) = \int [\, W_\tau{}^n(\nu;t)/n! \,]$$

$$\times \exp [-W_\tau(\nu;t)] \, \Phi(\nu) \, d^2\nu . \tag{69}$$

For coherent detection we are interested in the case of large total power incident on the detector. In the limit of large W_τ, n can be approximated as a continuous random variable $x(t)$, yielding

$$p_\tau(x;t) = (2 \pi x)^{-1/2} \int \exp \{ -[x - W_\tau(\nu;t)]^2/2x \}$$

$$\times \Phi(\nu) \, d^2\nu . \tag{70}$$

For the special case of coherent-state communication we can reduce the output statistics to a Gaussian distribution by assuming $x \simeq W_\tau$ in each denominator:

$$p_\tau(x;t) = (2 \pi W_\tau)^{-1/2}$$

$$\times \exp \{ -[x - W_\tau(\nu;t)]^2/2 W_\tau \} . \tag{71}$$

This expression, however, is not necessarily valid in the extreme tails of the distribution. If the condition of sufficiently large local-oscillator power is not met, the error rate estimate must be based on (70). A large local-oscillator power is generally required for the quantum noise to dominate the thermal and active device noise of the receiver

electronics; thus, under normal circumstances the Gaussian approximation can be used.

In summary, it is evident that the quantum statistics reduce to the semiclassical approach in the conditions of *high local-oscillator power,* and *coherent-state communication* (laser well above threshold with narrow line width). These are not new restrictions, however. With the exception of signaling using novel quantum states, semiclassical theories should give satisfactory agreement with experiment, and remain the standard approach for coherent optical system design.

6. References

[1] A. M. Michelson and H. A. Levesque, *Error Control Techniques for Digital Communications,* New York: John Wiley & Sons, pp. 37–42, 1985.

[2] O. E. Delange, "Optical Heterodyne Detection," *IEEE Spectrum,* pp. 77–85, October 1968.

[3] A. T. Forrester, R. A. Gudmundsen, and P. O. Johnson "Photoelectric Mixing of Incoherent Light," *Phys. Rev.,* Vol. 99, p. 1961, 1955.

[4] Y. Yamamoto and T. Kimura, "Coherent Optical Transmission Systems," *IEEE J. Quantum Electron.,* Vol. QE-17, No. 6, pp. 919–934, June 1981.

[5] R. H. Stolen, "Nonlinear Properties of Optical Fibers," Chapter 5, in *Optical Fiber Telecommunications,* ed. by S. E. Miller and A. G. Chynoweth, New York: Academic Press, 1979.

[6] I. P. Kaminow and T. Li, "Modulation Techniques," Chapter 17, in *Optical Fiber Telecommunications,* ed. by S. E. Miller and A. G. Chynoweth, New York: Academic Press, p. 569, 1979. See also W. A. Stallard et al., "LiNbO$_3$ Electro-Optic Waveguide Devices for Coherent Optical Fiber Systems," *Colloquium on Advances in Coherent Optic Development and Technology,* IEE, London, Digest 1985/30.

[7] C. M. Gee and G. D. Thurmond, "Mach-Zehnder 17-Ghz Bandwidth Modulator," *OFC '84 Tech. Dig.,* Paper WG5, New Orleans. See also R. C. Alferness, "Guided Wave Devices for Optical Communications," *IEEE J. Quantum Electron.,* Vol. QE-17, No. 6, pp. 946–958, June 1981.

[8] W. Johnstone et al., "Integrated Optical Frequency Translators," in *Colloquium on Advances in Coherent Optic Development and Technology,* IEE, London, Digest 1985/30.

[9] L. L. Jeromin and V. W. S. Chan, "Performance

Estimates for a Coherent Optical Communication System," *OFC '84 Tech. Dig.,* Paper TuK2.

[10] K. KIKUCHI et al., "Degradation of Bit Error Rate in Coherent Optical Communications Due to Spectral Spread of the Transmitter and Local Oscillator," *IEEE J. Lightwave Tech.,* Vol. LT-2, No. 6, pp. 1024–1033, 1984.

[11] W. R. LEEB et al., "Frequency Synchronization and Phase Locking of CO_2 Lasers," *Appl. Phys. Lett.,* Vol. 41, pp. 592–594, October 1, 1982.

[12] D. W. SMITH et al., "Application of Brillouin Amplification in Coherent Optical Transmission," *OFC '86 Tech. Dig.,* pp. 88–89.

[13] G. NICHOLSON, "Probability of Error for Optical Heterodyne DPSK System with Quantum Phase Noise," *Electron. Lett.,* Vol. 20, No. 24, November 22, 1984.

[14] F. M. GARDNER, *Phaselock Techniques,* New York: John Wiley & Sons, 1979.

[15] W. C. LINDSEY and M. K. SIMON, "Data-Aided Carrier Tracking Loops," *IEEE Trans. Commun. Tech.,* Vol. COM-19, No. 2, pp. 157–168, April 1971.

[16] T. G. HODGKINSON, "Phase-Locked-Loop Analysis for Pilot Carrier Coherent Optical Receivers," *Electron. Lett.,* Vol. 21, No. 25/26, pp. 1202–1203, December 1985. See also T. G. HODGKINSON, "Costas Loop Analysis for Coherent Optical Receivers," *Electron. Lett.,* Vol. 22, No. 7, pp. 394–396, March 27, 1986.

[17] N. G. WALKER and J. E. CARROLL, "Simultaneous Phase and Amplitude Measurements on Optical Signals Using a Multiport Junction," *Electron. Lett.* Vol. 20, No. 23, pp. 981–983, November 8, 1984. See also A. W. DAVIS and S. WRIGHT, "A Phase Insensitive Homodyne Optical Receiver," in *Colloquium on Advances in Coherent Optic Development and Technology,* IEE, London, Digest 1985/30.

[18] I. GARRETT and G. JACOBSEN, "Theoretical Analysis of Heterodyne Optical Receivers for Transmission Systems Using (Semiconductor) Lasers with Non-negligible Linewidth," *IEEE J. Lightwave Tech.,* Vol. LT-4, No. 3, pp. 323–334, March 1986.

[19] C. H. HENRY, "Theory of the Linewidth of Semiconductor Lasers," *IEEE J. Quantum Electron.,* Vol. QE-18, No. 2, pp. 259–264, February 1982. See also C. H. HENRY, "Phase Noise in Semiconductor Lasers," *IEEE J. Lightwave Tech.,* Vol. LT-4, No. 5, pp. 298–311, March 1986.

[20] K. KOJIMA et al., "Analysis of the Spectral Linewidth of Distributed Feedback Laser Diodes," *IEEE J. Lightwave Tech.,* Vol. LT-3, No. 5, pp. 1048–1055, October 1985.

[21] S. STENHOLM, "Introduction to Stochastic Processes," in *Quantum Optics, Experimental Gravitation, and Measurement Theory,* ed. by P. MEYSTRE and M. O. SCULLY, New York: Plenum Press, p. 169, 1981.

[22] M. W. FLEMING and A. MOORADIAN, "Fundamental Line Broadening of Single Mode GaAs Lasers," *Appl. Phys. Lett.,* Vol. 38, p. 511, 1981.

[23] T. P. LEE, "Linewidth of Single-Frequency Semiconductor Lasers," *IOOC-ECOC '85 Proc.,* p. 189–196.

[24] R. WYATT and W. J. DEVLIN, "10-kHz Linewidth 1.5-μm InGaAsP External Cavity Laser With 55-nm tuning range," *Electron. Lett.,* Vol. 19, No. 3, pp. 110–112, February 3, 1983.

[25] C. Y. KUO and J. VAN DER ZIEL, "Linewidth Reduction of 1.5-μm Grating-Loaded External Cavity and Semiconductor Lasers," *OFC '86 Tech. Dig.* pp. 16–17.

[26] S. C. RASHLEIGH and M. J. MARRONE, "Polarization Holding in a High-Birefringence Fiber," *Electron. Lett.,* Vol. 18, No. 8, pp. 326–327, April, 1982. See also T. KATSUYAMA et al., "Low Loss Single Polarization Fibers," *Electron. Lett.,* Vol. 17, No. 13, pp. 473–474, June 25, 1981.

[27] T. OKOSHI and K. KIKUCHI, "Heterodyne Type Optical Fiber Communications," *J. Optical Commun.,* Vol. 2, No. 3, pp. 82–88, September 1981.

[28] T. G. HODGKINSON et al., "Studies of 1.5-μm Coherent Transmission Systems Operating over Installed Cable Links," *Globecom '83 Proc.,* pp. 725–729. See also T. MYOGADANI et al., "Polarization Fluctuation in Singlemode Fiber Cables," *IOOC-ECOC '85,* Vol. 1, pp. 151–154, and R. E. WAGNER et al., "Polarization Measurements on a 147-km Lightwave Undersea Cable," *OFC '86,* Postdeadline Digest, pp. 31–34.

[29] R. C. ALFERNESS, "Electro-Optic Guided-Wave Device for General Polarization Transformations," *IEEE J. Quantum Electron.,* Vol. QE-17, No. 6, pp. 965–969, June 1981.

[30] T. OKOSHI, K. KIKUCHI, and Y. CHENG, "A New Polarization-Control Scheme for Optical Heterodyne Receiver," *IOOC-ECOC '85 Tech. Dig.,* Vol. 1, pp. 405–408. See also T. OKOSHI, "Polarization-State Control Schemes for Heterodyne or Homodyne Optical Fiber Communications,"

IEEE J. Lightwave Tech., Vol. LT-3, No. 6, pp. 1232–1237, December 1985.

[31] M. S. YATAKI et al., "All Fiber Polarizing Beamsplitter and Spectral Filter," in *Colloquium on Advances in Coherent Optic Development and Technology,* IEE, London, Digest 1985/30.

[32] T. OKOSHI, S. RYU, and K. KIKUCHI, "Polarization Diversity Receiver for Heterodyne/Coherent Optical-Fiber Communications," *IOOC '83 Tech. Dig.,* pp. 386–387.

[33] For a compilation of relevant experimental and theoretical work, see P. L. KNIGHT and L. ALLEN, *Concepts of Quantum Optics,* New York: Pergamon Press, 1983.

[34] V. K. PRABHU and J. SALZ, "On the Performance of Phase-Shift-Keying Systems," *Bell Syst. Tech. J.,* Vol. 60, No. 10, pp. 2307–2343, December 1981.

[35] V. K. PRABHU, "PSK Performance With Imperfect Carrier Recovery," *IEEE Trans. Aerospace and Electronic Systems,* Vol. AES-12, No. 2, pp. 275–285, March 1976.

[36] L. G. KAZOVSKY, "Decision Driven Phase-Locked Loop for Optical Homodyne Receivers: Performance Analysis and Laser Linewidth Requirements," *IEEE J. Lightwave Tech.,* Vol. LT-3, No. 6, pp. 1238–1247, December 1985. See also L. G. KAZOVSKY, "Performance Analysis and Laser Linewidth Requirements for Optical PSK Heterodyne Communications Systems," *IEEE J. Lightwave Tech.,* Vol. LT-4, No. 4, April 1986.

[37] G. NICHOLSON, "Optical Source Linewidth Criteria for Heterodyne Communication Systems With PSK Modulation," *Optical and Quantum Electron.,* Vol. 17, pp. 399–410, 1985.

[38] J. SALZ, "Coherent Lightwave Communications," *AT&T Tech. J.,* Vol. 64, No. 10, pp. 2153–2209, December 1985.

[39] A. PAPOULIS, *Probability, Random Variables and Stochastic Processes,* New York: McGraw-Hill Book Co., p. 101, 1965.

[40] R. L. PFLEEGOR and L. MANDEL, "Further Experiments on the Interference of Independent Photon Beams at Low Light Levels," *J. Opt. Soc. Am.,* Vol. 58, No. 7, pp. 946–950, July 1968.

[41] H. P. YUEN and V. W. S. CHAN, "Noise in Homodyne and Heterodyne Detection," *Opt. Lett.,* Vol. 8, pp. 177–179, March 1983.

[42] J. H. SHAPIRO, "Quantum Noise and Excess Noise in Optical Homodyne and Heterodyne Receivers," *IEEE J. Quantum Electron.,* Vol. QE-21, No. 3, pp. 237–250, March 1985.

[43] G. L. ABBAS, V. W. CHAN, and T. K. YEE, "Dual-Detector Optical Heterodyne Receiver for Local Oscillator Noise Suppression," *IEEE J. Lightwave Tech.,* Vol. LT-3, No. 5, pp. 1110–1122, October 1985.

[44] M. C. TEICH, B. E. A. SALEH, and J. PERINA, "Role of Primary Excitation Statistics in the Generation of Antibunched and Sub-Poisson Light," *J. Opt. Soc. Am. B,* Vol. 1, No. 3, pp. 366–389, June 1984.

[45] R. J. GLAUBER, "The Quantum Theory of Optical Coherence," *Phys. Rev.,* Vol. 130, No. 6, p. 2529, 1963.

[46] R. J. GLAUBER, "Coherent and Incoherent States of the Radiation Field," *Phys. Rev.,* Vol. 131, p. 2766, 1963.

[47] D. F. WALLS, "Squeezed States of Light," *Nature,* Vol. 306, pp. 141–146, 1983.

[48] D. MARCUSE, *Principles of Quantum Electronics,* New York: Harcourt Brace Jovanovitch, 1980.

[49] S. MACHIDA and Y. YAMAMOTO, "Quantum Limited Operation of Balanced Mixer Homodyne and Heterodyne Receivers," *IEEE J. Quantum Electron.,* Vol. QE-22, No. 5, pp. 617–624, May 1986.

[50] H. P. YUEN and J. H. SHAPIRO, "Optical Communication with Two-Photon Coherent States— Part 1, Quantum State Propagation and Quantum Noise Reduction," *IEEE Trans. Inform. Theory,* Vol. IT-24, pp. 657–658, November 1978.

[51] L. MANDEL, "Proposal for Almost Noise-Free Optical Communication Under Conditions of High Background," *J. Opt. Soc. Am. B,* Vol. 1, p. 108, 1984.

INDEX